Bioclimatology and Natural Hazards

Katarína Střelcová · Csaba Mátyás · Axel Kleidon · Milan Lapin · František Matejka · Miroslav Blaženec · Jaroslav Škvarenina · Ján Holécy (Eds.)

Bioclimatology and Natural Hazards

Springer

Editors
Ass. Prof. Dr. Katarína Střelcová
Faculty of Forestry
Technical University Zvolen
T.G. Masaryka 24
960 53 Zvolen
Slovakia
strelcov@vsld.tuzvo.sk

Dr. Axel Kleidon
Max-Planck-Institute for Biogeochemistry
Hans-Knoell-Str. 10
07745 Jena
Germany
akleidon@bgc-jena.mpg.de

Dr. František Matejka
Geophysical Institute
Slovak Academy of Sciences
Dúbravská cesta 9
845 28 Bratislava
Slovakia
geofmate@savba.sk

Prof. Dr. Jaroslav Škvarenina
Faculty of Forestry
Technical University Zvolen
T.G. Masaryka 24
960 53 Zvolen
Slovakia
jarosk@vsld.tuzvo.sk

Prof. Dr. Csaba Mátyás
Institute of Environmental and Earth Sciences
Faculty of Forestry
West Hungarian University
Ady Endre Str. 5
Sopron 9400
Hungary
cm@emk.nyme.hu

Prof. Dr. Milan Lapin
Faculty of Mathematics, Physics & Informatics
Comenius University in Bratislava
Mlynská dolina - F1
842 48 Bratislava
Slovakia
lapin@fmph.uniba.sk

Dr. Miroslav Blaženec
Institute of Forest Ecology
Slovak Academy of Sciences
Štúrova 2
960 53 Zvolen
Slovakia
blazenec@savzv.sk

Prof. Dr. Ján Holécy
Faculty of Forestry
Technical University Zvolen
T.G. Masaryka 24
960 53 Zvolen
Slovakia
holecy@vsld.tuzvo.sk

ISBN: 978-1-4020-8875-9 e-ISBN: 978-1-4020-8876-6

DOI 10.1007/978-1-4020-8876-6

Library of Congress Control Number: 2008936211

Printed on acid-free paper

9 8 7 6 5 4 3 2 1

springer.com

Foreword

Man-made changes in the climate system, consisting of atmosphere, hydrosphere, cryosphere, lithosphere and biosphere represent a most serious challenge not only to the planet's ecosystems and their natural environment but to human civilizations. While the Earth system will undoubtedly adapt, human infrastructures and societal organization may be questioned if no action is taken in time to buffer unavoidable consequences related to climate change. As a reaction, scientific disciplines such as bioclimatology, genetics, hydrology, bio-hydrology and eco-physiology are now considered an important part of forestry, agriculture, water management, environmental protection, and natural hazards control (e.g., droughts, floods, windstorms, weather extremes, and wild fires). Bioclimatology provides an integrated, interdisciplinary framework for dealing with contemporary challenges of natural hazards. Bioclimatology has also the potential to assess and predict extreme weather events in a very complex way.

Bioclimatology will help in better understanding the causes and impacts of natural hazards and ways how to prevent them. Improved knowledge of natural hazards is a vital prerequisite for the implementation of integrated resource management. It provides a useful framework for combating current climate variability and for adapting to ongoing climate change.

Climate change explains the occurrence of extreme weather in Central and Eastern Europe (CEE). Today, the increases in precipitation and soil moisture variability and increased temperature are the most important single issue that needs to be addressed. The assessment of impacts caused by extreme weather situations such as heat waves, droughts, floods, windstorms, etc., is even more complicated. Atmospheric General Circulation Models (GCMs) are currently used to predict such situations. These models need to be adjusted to provide downscaled outputs using localized scenarios of selected extreme events. Some of the designed GCM scenarios of extreme weather situations need to be modified according to analogues.

The shift of vegetation zones is the most investigated and obvious response of ecosystems to climate change. Forecasting the shifts of vegetation zones in response to weather extremes and ongoing climate change is based on climatically determined actual distribution models, or so-called "bioclimatic envelopes". Bioclimatic modelling is based on the concept that distributional patterns depend on the physiological tolerance of populations to climatic effects besides ecological and life history factors. These limits are genetically determined and thus more or less fixed. Genetically regulated plasticity enables the adaptation of individuals and populations to changing environments without any change in the inherited genetic resources. Natural selection eliminates the genotypes of low fitness from a population, thus adjusting its gene pool towards better adaptation. It is important to realise that the adaptive response of ecosystems to environmental stress is ultimately regulated by genetics and that bioclimatic modelling has to consider genetically set adaptation mechanisms in plants as important parts of ecosystems.

This book presents a carefully edited and reviewed selection of papers from the International Scientific Conference on Bioclimatology and Natural Hazards held in Slovakia at the Polana Biosphere Reserve on September 17–20, 2007. There 250 participants from the 14 different countries of Europe discussed recent research on the interactions between meteorological, climatological, hydrological and biological processes in the atmosphere and terrestrial environment. All contributing authors come from renowned scientific research institutions and universities in Europe and specialise in issues of climate change, soil-plant-atmosphere interactions, hydrologic cycle, ecosystems, biosphere, and natural hazards. From the total of 215 conference contributions, the 25 most important issues have been selected for this book to highlight a spectrum of topics associated with climate change and weather extremes and their impact on different sectors of the national economy.

Most of the presented papers point out that the damage caused by the occurrence of extreme climate events and its impact on ecosystems seems to have substantially increased over the past decades. Some of these climate extremes can induce disastrous effects. For instance, drought and windstorms can act as promoters of wind throws and can result in increased population sizes of different kinds of insects. This in turn can have effects on landscape wild fire occurrence and enhance the vulnerability of ecosystems and their resilience. The vulnerability and the impacts of disaster on ecosystems and society are influenced by many factors. The combination of methods and knowledge from various academic disciplines provide efficient set of tools and procedures to reduce the vulnerability of ecosystems by strengthening their resilience. The contributions reflect the diversity and the interdisciplinary character of the research concerning the occurrence of natural hazards. Some contributions report results of research in the fields of severe storms, heavy precipitation and floods, soil erosion and degradation resulting from the destruction of forest by wild fire as well as results of modeling the impacts of natural hazards on tree growth.

The editors gratefully acknowledge the enthusiastic support and constructive suggestions made by many colleagues and friends. We express our sincere thanks to all reviewers of the manuscript.

Katarína Střelcová
Csaba Mátyás
Axel Kleidon
Milan Lapin
František Matejka
Miroslav Blaženec
Jaroslav Škvarenina
Ján Holécy

Contents

Contributors

José Alexandre Andrade Geosciences Department, University of Évora, Portugal, zalex@uevora.pt

Peter Baláž Forest Research Institute, National Forest Centre, T.G. Masaryka 22, 960 92 Zvolen, Slovakia, balaz@nlcsk.org

Vilam Bárek Department of Landscape Engineering, Faculty of Horticulture and Landscape Engineering, Slovak University of Agriculture, Hospodárska 7, 949 76 Nitra, Slovakia, viliam.barek@uniag.sk

Ilona Bärlund University of Kassel, Kassel, Germany, baerlund@usf.uni-kassel.de

Judith Bartholy Department of Meteorology, Eötvös Loránd University, Pazmany st. 1/a, H-1117 Budapest, Hungary, bari@ludens.elte.hu

Juraj Bebej Technical University in Zvolen, Faculty of Forestry, Zvolen, Slovakia, bebej@vsld.tuzvo.sk

Sylvi Bianchin Technische Universität Bergakademie Freiberg, Interdisciplinary Environmental Research Centre, Freiberg, Germany, sylvin.bianchin@ioez.tu-freiberg.de

Svetlana Bičárová Geophysical Institute, Slovak Academy of Sciences, Meteorological Observatory Stará Lesná, 059 60 Tatranská Lomnica, Slovakia, bicarova@ta3.sk

Miroslav Blaženec Institute of Forest Ecology, Slovak Academy of Sciences, Zvolen, Slovakia, blazenec@savzv.sk

Ingrid Damborská Faculty of Mathematics, Physics and Informatics, Comenius University, Bratislava, Slovakia, damborska@fmph.uniba.sk

L'ubica Ditmarová Institute of Forest Ecology, Slovak Academy of Sciences, Zvolen, Slovakia, ditmarova@savzv.sk

Marek Fabrika Technical University in Zvolen, Faculty of Forestry, T.G. Masaryka 24, 960 53 Zvolen, Slovakia, fabrika@vsld.tuzvo.sk

Pavol Faško Slovak Hydrometeorological Institute, Bratislava, Slovakia, Pavol.Fasko@shmu.sk

Peter Fleischer Research Station of the Tatra National Park, State Forest of TANAP, 059 60 Tatranská Lomnica, Slovakia, fleischer@post.sk

Marcelo Fragoso Centre of Geographical Studies, University of Lisbon, mfragoso@fl.ul.pt

Ladislav Gaál Faculty of Civil Engineering, Slovak University of Technology, Bratislava, Slovakia

Györgyi Gelybó Department of Meteorology, Eötvös Loránd University, Budapest, Hungary, gyoresz@elte.hu

Dušan Gömöry Technical University in Zvolen, Faculty of Forestry, Zvolen, Slovakia, gomory@vsld.tuzvo.sk

Erika Gömöryová Technical University in Zvolen, Faculty of Forestry, T.G. Masaryka 24, 960 53 Zvolen, Slovakia, egomory@vsld.tuzvo.sk

Peter Halaj Department of Landscape Engineering, Faculty of Horticulture and Landscape Engineering, Slovak University of Agriculture, Nitra, Slovakia, halaj@afnet.uniag.sk

Stephanie Hänsel Technical University, Bergakademie Freiberg, Interdisciplinary Environmental Research Centre, Brennhausgasse, 14, D-09599 Freiberg, Germany, stephanie.haensel@email.de

Hermann Heilmeier Technische Universität Bergakademie Freiberg, Interdisciplinary Environmental Research Centre, Freiberg, Germany, hermann.heilmeier@ioez.tu-freiberg.de

Tomáš Hlásny Forest Research Institute, National Forest Centre, T.G. Masaryka 22, 960 92 Zvolen, Slovakia; Czech University of Life Sciences, Faculty of Forestry and Wood Sciences, Department of Forest Protection and Game Management, Kamýcká 1176, Prague 6 – Suchdol 165 21, Czech Republic

Kamila Hlavčová Department of Land and Water Resources Management, Faculty of Civil Engineering, Slovak University of Technology Bratislava, Bratislava, Slovakia, kamila.hlavcova@stuba.sk

Ján Holécy Technical University in Zvolen, Faculty of Forestry, T.G. Masaryka 24, 960 53 Zvolen, Slovakia, holecy@vsld.tuzvo.sk

Ján Horák Department of Biometeorology and Hydrology, Slovak Agricultural University, Hospodárska 7, 949 01 Nitra, Slovakia, jan.horak@uniag.sk

Matúš Hríbik Technical University in Zvolen, Faculty of Ecology and Environmental Sciences, T.G. Masaryka 24, 960 53 Zvolen, Slovakia, vrchar@gmail.com

Ján Hrvol′ Faculty of Mathematics, Physics and Informatics, Comenius University, Bratislava, Slovakia, hrvol@fmph.uniba.sk

Tat′jana Hurtalová Geophysical Institute, Slovak Academy of Sciences, Bratislava, Slovakia, geoftahu@savba.sk

Dušan Igaz Department of Biometeorology and Hydrology, Faculty of Horticulture and Landscape Engineering, Slovak University of Agriculture, Nitra, Slovakia, dusan.igaz@uniag.sk

Matthias Kändler International Graduate School Zittau, Markt 23, D-02763 Zittau, Germany, kaendler@ihi-zittau.de

Anikó Kern Department of Meteorology, Eötvös Loránd University, Budapest, Hungary, anikoc@nimbus.elte.hu

Axel Kleidon Biospheric Theory and Modelling Group, Max-Planck-Institut für Biogeochemie, Jena, Germany, akleidon@bgc-jena.mpg.de

Zdeňka Klimánková Laboratory of Plants Ecological Physiology, Institute of Systems Biology and Ecology, Academy of Sciences of the Czech Republic, Brno, Czech Republic

Silvia Kohnová Department of Land and Water Resources Management, Faculty of Civil Engineering, Slovak University of Technology, Radlinského 11, 813 68 Bratislava, Slovakia, silvia.kohnova@stuba.sk

Milan Koreň Research Station of the Tatra Nationl Park, State Forest of TANAP, Tatranská Lomnica, Slovakia

Vladimír Kunca Technical University in Zvolen, Faculty of Ecology and Environmental Sciences, Zvolen, Slovakia, vkunca@vsld.tuzvo.sk

Daniela Kyselová Slovak Hydrometeorological Institute, Regional Center Banská Bystrica, Slovakia, Daniela.Kyselova@shmu.sk

Milan Lapin Faculty of Mathematics, Physics and Informatics, Comenius University, Mlynská dolina - F1, 842 15 Bratislava, Slovakia, lapin@fmph.uniba.sk

António Lopes Centre of Geographical Studies, University of Lisbon, 1600-214 Lisbon, Portugal, antlopes@fl.ul.pt

Andrea Majlingová Technical University in Zvolen, Faculty of Forestry, Zvolen, Slovakia, amajling@vsld.tuzvo.sk

František Matejka Geophysical Institute, Slovak Academy of Sciences, Dúbravská cesta 9, 845 28 Bratislava, Slovakia, geofmate@savba.sk

Jörg Matschullat Technische Universität Bergakademie Freiberg, Interdisciplinary Environmental Research Centre, Freiberg, Germany, joerg.matschullat@ioez.tu-freiberg.de

Csaba Mátyás Institute of Environmental Sciences, Faculty of Forestry, West Hungarian University, Ady Endre Str. 5, Sopron 9400, Hungary, cm@emk.nyme.hu

Marián Melo Faculty of Mathematics, Physics and Informatics, Comenius University, Bratislava, Slovakia, melo@fmph.uniba.sk

Mariusz Merta International Graduate School Zittau, Zittau, Germany, merta@ihi-zittau.de

László Nagy Forest Research Institute Experimental Station, Sárvár, Hungary, lnagy@ertisarvar.hu

Viliam Novák Institute of Hydrology, Slovak Academy of Sciences, Račianska 75, 831 02 Bratislava, Slovakia, novak@uh.savba.sk

Sandra Oliveira Centre of Geographical Studies, University of Lisbon, sisoliveira@hotmail.com

Juraj Parajka Institute of Hydraulics, Hydrology and Water Resources Management, Vienna University of Technology Vienna, Austria, parajka@hydro.tuwien.ac.at

Pedro Pedro Lisbon City Council, RSBL, pedro.pedro@cm-lisboa.pt

Sylvia Petzold Technische Universität Bergakademie Freiberg, Interdisciplinary Environmental Research Centre, Freiberg, Germany, sylvia.petzold@gmx.de

Radek Pokorný Laboratory of Plants Ecological Physiology, Institute of Systems Biology and Ecology, Academy of Sciences of the Czech Republic, Brno, Czech Republic, eradek@brno.cas.cz

Rita Pongrácz Department of Meteorology, Eötvös Loránd University, Pazmany st. 1/a, H-1117 Budapest, Hungary, prita@nimbus.elte.hu

Markku Puustinen Finnish Environment Institute, Helsinki, Finland, markku.puustinen@ymparisto.fi

Elke Richert Technische Universität Bergakademie Freiberg, Interdisciplinary Environmental Research Centre, Freiberg, Germany, richert@ioez.tu-freiberg.de

Stan Schymanski Biospheric Theory and Modelling Group, Max-Planck-Institut für Biogeochemie, Jena, Germany, sschym@bgc-jena.mpg.de

Christina Seidler International Graduate School Zittau, Zittau, Germany, seidler@ihi-zittau.de

Bernard Šiška Department of Biometeorology and Hydrology, Slovak Agricultural University, Nitra, Slovakia, bernard.siska@uniag.sk

Jaroslav Škvarenina Technical University in Zvolen, Faculty of Forestry, T.G. Masaryka 24, 960 53 Zvolen, Slovakia, jarosk@vsld.tuzvo.sk

Jana Škvareninová Technical University in Zvolen, Faculty of Ecology and Environmental Sciences, Zvolen, Slovakia, janask@vsld.tuzvo.sk

Marc Stieglitz Department of Civil and Environmental Engineering, Georgia Institute of Technology, Atlanta, Georgia, USA

Katarína Střelcová Technical University in Zvolen, Faculty of Forestry, Zvolen, Slovakia, strelcov@vsld.tuzvo.sk

Péter Szabó Department of Meteorology, Eötvös Loránd University, Budapest, Hungary, szabpet83@gmail.com

Ján Szolgay Department of Land and Water Resources Management, Faculty of Civil Engineering, Slovak University of Technology Bratislava, Bratislava, Slovakia, jan.szolgay@stuba.sk

Jozef Takáč Soil Science and Conservation Research Institute, Gagarinova 10, 827 13 Bratislava, Slovakia, takac@vupu.sk

Ján Tomlain Faculty of Mathematics, Physics and Informatics, Comenius University, Bratislava, Slovakia, tomlain@fmph.uniba.sk

Ján Tuček Technical University in Zvolen, Faculty of Forestry, T.G. Masaryka 24, 960 53 Zvolen, Slovakia, tucek@vsld.tuzvo.sk

Marek Turčáni Czech University of Life Sciences, Faculty of Forestry and Wood Sciences, Department of Forest Protection and Game Management, Prague 6 – Suchdol, Czech Republic, turcani@fld.czu.cz

Éva Ujvári-Jármay Forest Research Institute Experimental Station, Mátrafüred, Hungary, eva.ujvari@axelero.hu

Tomáš Vaculčiak Technical University in Zvolen, Faculty of Forestry, Zvolen, Slovakia, vaculciak@vsld.tuzvo.sk

Reviewers

Jan Bednář Department of Meteorology and Environmental Protection, Faculty of Mathematics and Physics of Charles University, Prague, Czech Republic

Istvan Bogardi University of Nebraska-Lincoln, USA

Pavel Cudlín Institute of Systems Biology and Ecology, Academy of Sciences of the Czech Republic

Wilfried Endlicher Humboldt-Universität zu Berlin, Institute of Geography, Geography of Climates and Environmental Climatology, Germany

Jürgen Friedel University of Natural Resources and Applied Life Sciences, Vienna, Austria

Igantavičius Gytautas Vilnius University, Environmental Studies Centre, Latvia

Tove Heidman Danish Institute of Agricultural Sciences, Department of Agroecology, Research Centre Foulum, Denmark

Jaroslav Holuša The Forestry and Game Management Research Institute, Prague, Czech Republic

Dušan Húska Slovak Agricultural University, Nitra, Slovakia

Paul Jarvis School of GeoSciences, University of Edinburgh, Scotland, UK

Jan Kouba Czech University of Life Sciences Prague, Faculty of Forestry and Wood Sciences, Department of Forest Management, Czech Republic

Jan Kyselý Institute of Atmospheric Physics, Academy of Sciences of the Czech Republic

Viliam Novák Institute of Hydrology, Slovak Academy of Sciences, Slovakia

Ladislav Paule Technical University in Zvolen, Faculty of Forestry, Slovakia

Edward Pierzgalski Warsaw Agricultural University, Poland

Kálmán Rajkai Research Institute for Soil Science and Agricultural Chemistry of the Hungarian Academy of Sciences, Hungary

Christian-D. Schönwiese Institut für Atmosphäre und Umwelt (IAU) der Johann-Wolfgang Goethe Universität Frankfurt am Main, Germany

Miloslav Šír Institute of Hydrodynamics, Academy of Sciences of the Czech Republic

Miroslav Tesař Institute of Hydrodynamics, Academy of Sciences of the Czech Republic

Józef Zwoliński Department of Industrial Region Forest Management, Forest Research Institute, Poland

Part I
EXTREME EVENTS, RISKS AND CLIMATE VARIABILITY

What Climate Can We Expect in Central/Eastern Europe by 2071–2100?

J. Bartholy, R. Pongrácz, Gy. Gelybó and A. Kern

Keywords Regional climate change · Temperature · Precipitation · Central/Eastern Europe · Regional climate model

Introduction

According to the Fourth Assessment Report of the Intergovernmental Panel on Climate Change (IPCC) Working Group I, published on 2 February 2007 (IPCC 2007), the key processes influencing the European climate can be summarized as follows: (i) water vapour transport from low to high latitudes has increased; (ii) variation of atmospheric circulation has changed on interannual as well as longer time scales; (iii) snow cover during winter has reduced in the northeastern part of the continent; (iv) the soil has dried in summer in the Mediterranean and Central European regions. For instance, the heat wave in summer 2003 in Europe can be considered as a consequence of a long period of anticyclonic weather (Fink et al. 2004), which coincided with a severe drought in the region (Black et al. 2004). For Europe, it is likely that the increase of annual mean temperature will exceed the global warming rate in the twenty-first century. The largest increase is expected in winter in northern Europe (Benestad 2005) and in summer in the Mediterranean area. Minimum temperatures in winter are very likely to increase more than the mean winter temperature in northern Europe (Hanssen-Bauer et al. 2005), while maximum temperature values in summer are likely to increase more than the mean summer temperature in southern and Central Europe (Tebaldi et al. 2006). For precipitation, the annual sum is very likely to increase in northern Europe (Hanssen-Bauer et al. 2005) and decrease in the Mediterranean area. Central Europe is located at the boundary of these large regions. Different seasonal trends can therefore be expected, namely, precipitation is likely to increase in winter, while it decreases in summer. In case of summer drought events, the risk is likely to increase both in Central Europe and in the Mediterranean area due to a decrease in the mean summer precipitation and an increase in spring evaporation (Pal et al. 2004; Christensen and Christensen, 2004). As a consequence of the European warming, the length of the snow season and the accumulated snow depth are very likely to decrease over the entire continent (IPCC 2007).

Global climate models (GCMs) are inappropriate to describe regional climate processes due to their coarse spatial resolution. GCM outputs may therefore be misleading to compose regional climate change scenarios for the twenty-first century (Mearns et al. 2001). In order to determine better estimations for regional climate parameters, fine resolution regional climate models (RCMs) can be used. RCMs are limited-area models nested in GCMs, so the initial and the boundary conditions of RCMs are provided by the GCM outputs (Giorgi 1990). Due to computational constrains the domain of an RCM does not cover the entire globe, sometimes not even a continent. On the other hand, their horizontal resolution can be as fine as 5–10 km.

The PRUDENCE (Prediction of Regional scenarios and Uncertainties for Defining EuropeaN Climate change risks and Effects) project involved 21 European research institutes and universities and was completed

J. Bartholy (✉)
Department of Meteorology, Eötvös Loránd University, Pazmany st. 1/a, H-1117 Budapest, Hungary
e-mail: bari@ludens.elte.hu

K. Strelcová et al. (eds.), *Bioclimatology and Natural Hazards*,
DOI 10.1007/978-1-4020-8876-6_1,

in 2004 in the context of the European Union 5th framework programme. The primary objectives of PRUDENCE were to provide high-resolution climate change scenarios of 50 km × 50 km for Europe for 2071–2100 using dynamical downscaling methods with RCMs with the reference period of 1961–1990 and to explore the uncertainty in these projections (Christensen 2005). Results of the PRUDENCE project are disseminated widely via Internet (http://prudence.dmi.dk) and several other media, and thus, they support socioeconomic and policy-related research studies and decisions.

In the frame of PRUDENCE project, the following sources of climate uncertainty were studied (Christensen 2005):

1. Sampling uncertainty: Simulated climate is considered as an average over 30 years (2071–2100, with a reference period of 1961–1990).
2. Regional model uncertainty: The RCMs use different techniques to discretize the differential equations and to represent physical processes on sub-grid scales.
3. Emission uncertainty: The RCM runs used two IPCC-SRES emission scenarios, namely, the A2 and B2. Twenty-two experiments from the PRUDENCE simulations considered the A2 scenario, while only eleven of them used the B2 scenario.
4. Boundary uncertainty: The RCMs were run with boundary conditions from different GCMs. Most of the PRUDENCE simulations used the HadAM3H model as the driving GCM. Only a few of them used the ECHAM4 model or the ARPEGE model (Déqué et al. 2005).

Detailed intercomparison and analysis of the results of the PRUDENCE project is published in a special issue of the journal *Climatic Change* edited by Christensen et al. (May 2007). In this chapter, we focused on our region, and therefore the regional climate change projections are summarized for Central/Eastern Europe using the outputs of all available PRUDENCE simulations. The results of the expected temperature change by the end of the twenty-first century are discussed, as well as the expected change of precipitation. More detailed results are presented for Hungary.

Data

The adaptation of RCMs with 10–25 km horizontal resolution is currently proceeding in the frame of EU-funded projects: CECILIA (Central and Eastern Europe Climate Change Impact and Vulnerability Assessment) and CLAVIER (Climate Change and Variability: Impact on Central and Eastern Europe). In Hungary, the Department of Meteorology at the Eötvös Loránd University (Bartholy et al. 2006a, 2006b) and the Hungarian Meteorological Service (Horányi 2006) play an active and important role in regional climate modelling. Since these EU projects will end in June 2009, results of the RCM experiments are expected within 1–3 years. However, impact studies and end-users need and would like to have access to climate change scenario data much earlier. In order to fulfil this instant demand with preliminary information, outputs of the PRUDENCE simulations are evaluated and can be offered for Central/Eastern Europe. For the A2 scenario outputs of 16 RCM experiments, while for the B2 scenario, only outputs of 8 RCM simulations are available. Since the project PRUDENCE used only these two emission scenarios, no other scenario is discussed in this chapter. The A2 scenario projects a very heterogeneous world with an emphasis on family values and local traditions, while the B2 scenario considers a world with an emphasis on local solutions to economic and environmental sustainability (IPCC 2007). The projected CO_2 concentrations may reach 850 ppm (in the A2 scenario) and 600 ppm (in the B2 scenario) by the end of the twenty-first century (IPCC 2007), which are more than double of the pre-industrial concentration level (280 ppm).

Table 1 lists the name of the contributing institutes, the RCMs, the driving GCMs and the available scenarios we used in the composite maps. Composite maps of expected temperature and precipitation change cover Central/Eastern Europe (44.75°–55.25°N, 9.75°–27.25°E). The climate projections of PRUDENCE are available for the end of the twenty-first century (2071–2100) using the reference period of 1961–1990.

Temperature Projections for Central/Eastern Europe

Composite maps of the mean expected seasonal temperature change are shown for both A2 and B2 scenarios in Fig. 1. In order to represent the uncertainty of these composites, standard deviation values of the

Table 1 List of RCMs with their driving GCMs used in the composite analysis

Institute	RCM	Driving GCM	Scenario
Danish Meteorological Institute	HIRHAM	HadAM3H	A2, B2
	HIRHAM	ECHAM5	A2
	HIRHAM high resolution	HadAM3H	A2
	HIRHAM extra high resolution	HadAM3H	A2
Hadley Centre of the UK Met Office	HadRM3P (ensemble/1)	HadAM3P	A2, B2
	HadRM3P (ensemble/2)	HadAM3P	A2
ETH (Eidgenössische Technische Hochschule)	CHRM	HadAM3H	A2
GKSS (Gesellschaft für Kernenergieverwertung in Schiffbau und Schiffahrt)	CLM	HadAM3H	A2
	CLM improved	HadAM3H	A2
Max Planck Institute	REMO	HadAM3H	A2
Swedish Meteorological and Hydrological Institute	RCAO	HadAM3H	A2, B2
	RCAO	ECHAM4/OPYC	B2
UCM (Universidad Complutense Madrid)	PROMES	HadAM3H	A2, B2
International Centre for Theoretical Physics	RegCM	HadAM3H	A2, B2
Norwegian Meteorological Inst.	HIRHAM	HadAM3H	A2
KNMI (Koninklijk Nederlands Meteorologisch Inst.)	RACMO	HadAM3H	A2
Météo–France	ARPEGE	HadCM3	A2, B2
	ARPEGE	ARPEGE/OPA	B2

RCM model results are also determined and mapped for all seasons (Fig. 2). Similar to the global and the European climate change results, larger warming can be expected for the A2 scenario in Central/Eastern Europe than for the B2 scenario. The largest temperature increase is expected in summer, while the smallest increase in spring. The expected summer warming ranges in Hungary are 4.5–5.1°C and 3.7–4.2°C for the A2 and B2 scenarios, respectively. In case of spring, the expected temperature increase in Hungary is 2.9–3.2°C (for A2 scenario) and 2.4–2.7°C (for B2 scenario). On the basis of seasonal standard deviation fields (Fig. 2), the largest uncertainty of the expected temperature change occurs in summer for both emission scenarios.

Figure 3 summarizes the expected mean seasonal warming for Hungary in case of A2 and B2 scenarios. In general, the expected warming by 2071–2100 is more than 2.5°C and less than 4.8°C for all seasons and for both scenarios. The expected temperature changes for the A2 scenario are larger than that for the B2 scenario. The smallest difference is expected in spring (0.6°C), while the largest difference is expected in winter (1°C). The largest temperature increase is expected in summer, 4.8°C (A2) and 4.0°C (B2). The smallest temperature increase is expected in spring (3.1°C and 2.5°C in case of A2 and B2 scenarios, respectively).

In order to evaluate the model performance, the temperature bias is determined for each RCM output field using the GCM-driven simulations for the reference period (1961–1990) and the CRU (Climate Research Unit of the University of East Anglia) database (New et al. 1999). In general, the RCM simulations slightly overestimate the temperature in most of the Central/Eastern European regions; however, small underestimation can be seen in the southwestern part of the selected domain, at the mountainous areas of the Alps. The temperature bias does not exceed 1.5°C.

Similar to the mean temperature, expected seasonal warming of daily maximum and minimum temperatures is summarized for Hungary in Fig. 4 The largest warming is expected in summer for both scenarios: in case of maximum temperature the interval of the expected increase is 4.9–5.3°C (A2) and 4.0–4.4°C (B2), while in case of minimum temperature these intervals are 4.2–4.8°C (A2) and 3.5–4.0°C (B2). The expected

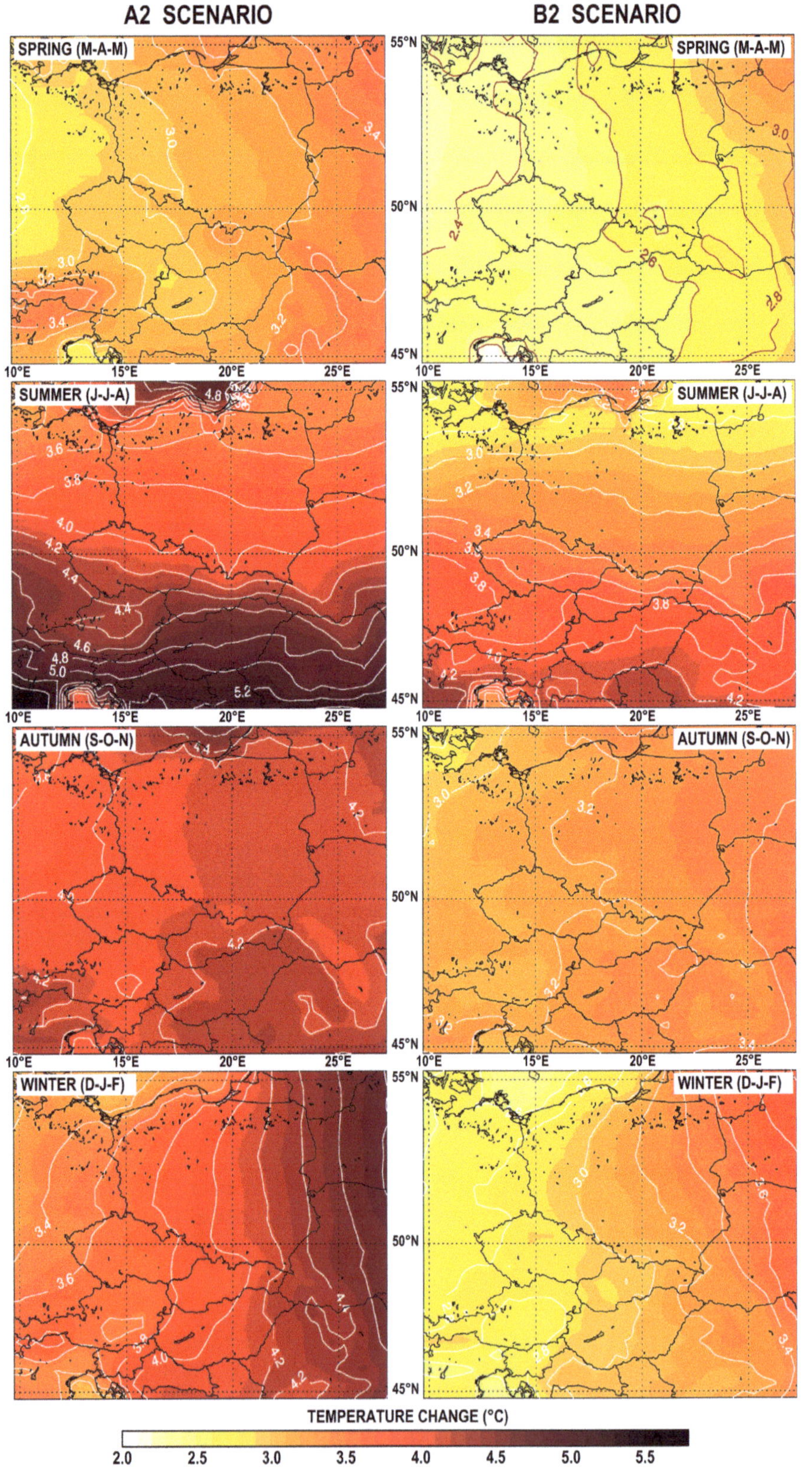

Fig. 1 Seasonal temperature change (°C) expected by 2071–2100 for Central/Eastern Europe using the outputs of 16 and 8 RCM simulations, A2 (*left panel*) and B2 (*right panel*) scenarios

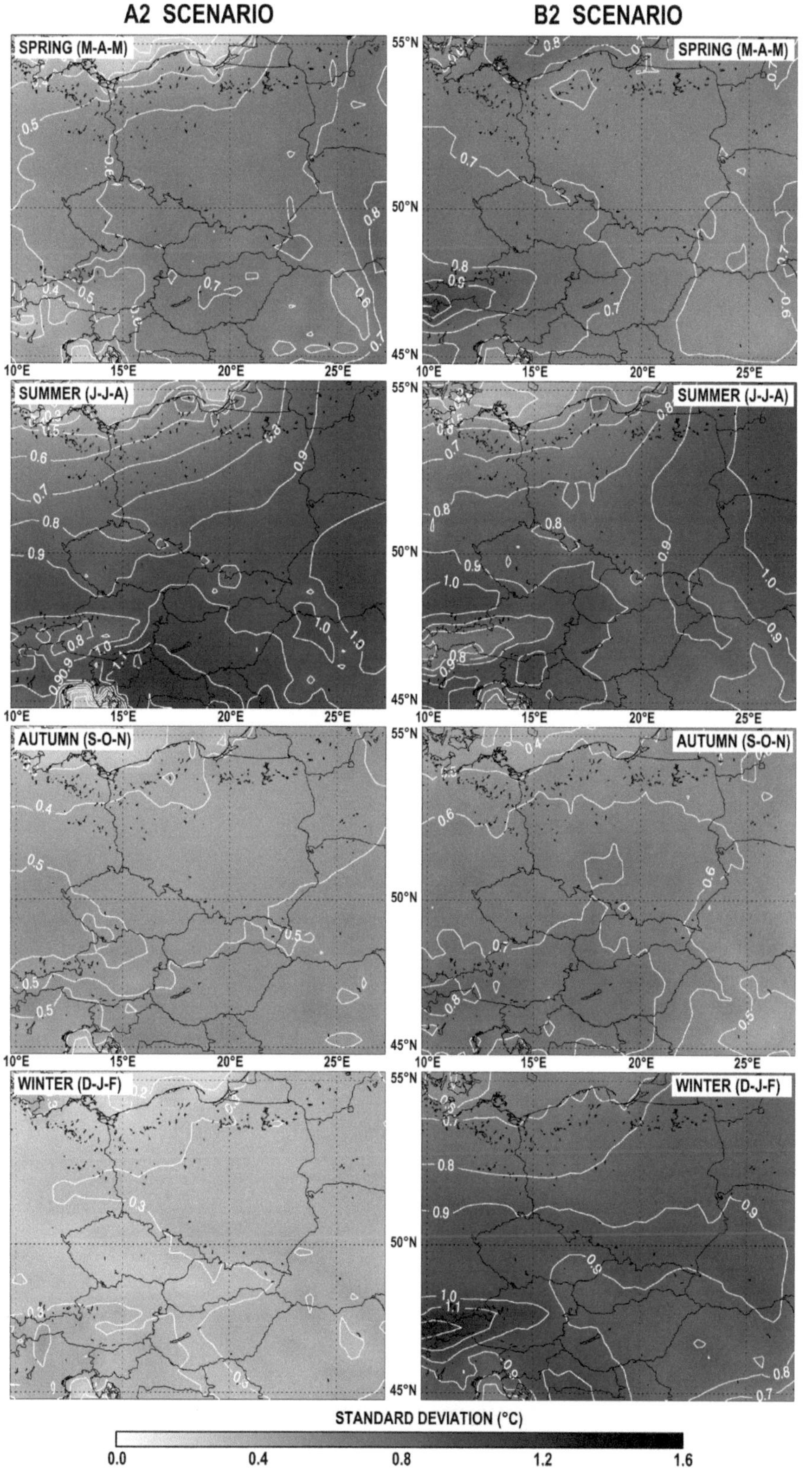

Fig. 2 Standard deviation of seasonal temperature change (°C) expected by 2071–2100 for Central/Eastern Europe using the outputs of 16 and 8 RCM simulations, A2 (*left panel*) and B2 (*right panel*) scenarios

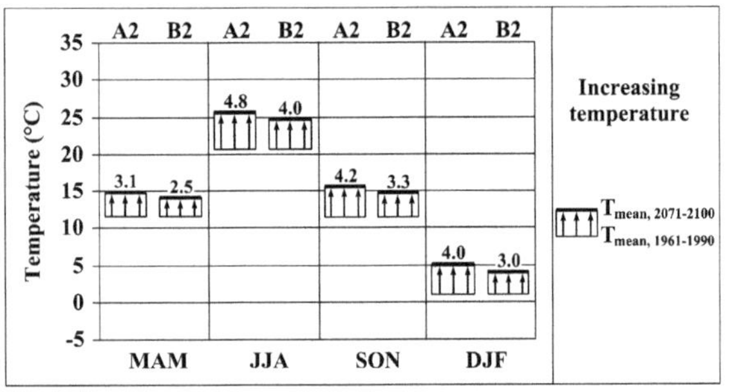

Fig. 3 Expected seasonal increase of mean temperature (°C) for Hungary (temperature values of the reference period, 1961–1990, represent the seasonal mean temperature in Budapest)

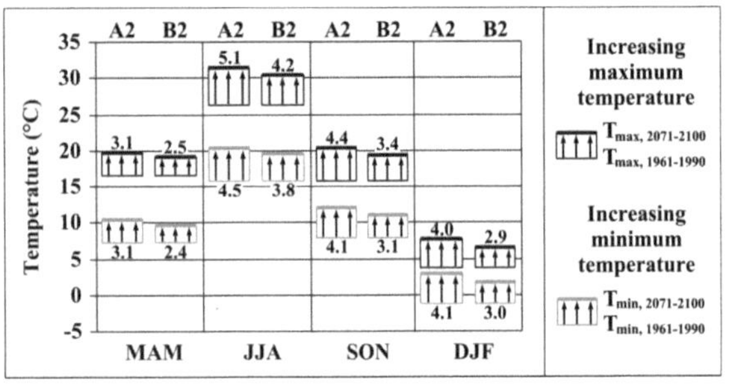

Fig. 4 Expected seasonal increase of daily minimum and maximum temperatures (°C) for Hungary (temperature values of the reference period, 1961–1990, represent the seasonal mean temperature in Budapest)

increase of mean temperature in summer is between the expected warming of the maximum temperature and that of the minimum temperature. Summarizing the expected mean seasonal increase of daily extreme temperature for Hungary, the entire interval of the expected warming includes values from 2.4 to 5.1°C, which is 0.4°C larger than in case of the mean temperature. The largest temperature increases are expected in summer for both scenarios, which is not surprising when the above results are considered. The expected increase of maximum temperature is generally not smaller than the expected increase of minimum temperature, winter being the only exception.

In order to provide a better overview on the spatial structure of expected temperature changes (both mean and extremes) for Central/Eastern Europe by the end of the twenty-first century, Table 2 summarizes the spatial differences of warming for summer and winter. In summer, zonal structure of warming (i.e. increasing values from north to south) can be detected for all parameters. The zonal temperature change difference values for the entire region exceed 1.4°C, and in general, they are larger for the A2 scenario than for the B2 scenario. The largest south–north difference is expected for maximum temperature, the A2 scenario (1.9°C). In winter, generally a meridional structure of warming is expected (i.e. increasing values from west to east). Similar to the summer values, the winter temperature change difference values are larger for the A2 than for the B2 scenario. The meridional difference values exceed 1.0°C (except maximum temperature, in case of the B2 scenario). The largest difference value is expected for the minimum temperature, the A2 scenario (1.8°C). In spring and autumn, the meridional temperature change difference values are smaller and do not exceed 0.8°C and 0.4°C, respectively.

Precipitation Projections for Central/Eastern Europe

Similar to the temperature projections, composite maps of mean expected seasonal precipitation change (Fig. 5) and standard deviations (Fig. 6) are mapped for both A2 and B2 scenarios for the 2071–2100 period. The annual precipitation sum is not expected to change significantly in this region (Bartholy et al. 2003), but this does not hold for seasonal precipitation. According to the results shown in Fig. 5, summer precipitation is very likely to decrease (also, slight decrease of autumn precipitation is expected), while winter precipitation is likely to increase considerably (somewhat less increase is expected in spring).

In summer, the projected precipitation decrease exceeds 25% (A2) and 10% (B2). The largest decrease (exceeding 35% and 10% in case of A2 and B2 scenarios, respectively) is expected at the southern part of the selected region. In winter, the expected precipitation increase exceeds 20% (A2) and 15% (B2). The largest precipitation increase is expected in the Transdanubian subregion located in Hungary (more than 35% and 20% in case of A2 and B2 scenarios, respectively). Furthermore, large increase can be expected in the north-eastern part of the selected region. As the composite maps of Fig. 5 suggest, the precipitation is expected to increase in spring in most of the selected region, except the southernmost subregions. This increase is somewhat less than the expected increase for winter. Note that in the eastern part of Central/Eastern Europe, the expected seasonal change is larger in case of the B2 than in the A2 scenario (about 10–18% and 5–8%, respectively). The expected precipitation change

Table 2 Spatial differences of expected summer and winter temperature change for Central/Eastern Europe for 2071–2100 (zonal difference is positive in case of increasing change from north to south, while the meridional difference is positive in case of increasing change from west to east)

	Scenario	Summer (J-J-A)	Winter (D-J-F)
Mean temperature	A2	Zonal: +1.6°C	Meridional: +1.4°C
	B2	Zonal: +1.4°C	Meridional: +1.0°C
Maximum temperature	A2	Zonal: +1.9°C	Meridional: +1.4°C
	B2	Zonal: +1.6°C	Meridional: +0.8°C
Minimum temperature	A2	Zonal: +1.4°C	Meridional: +1.8°C
	B2	Zonal: +1.4°C	Meridional: +1.4°C

in autumn is between –5% and +5%, which is not significant for the B2 scenario. For the A2 scenario, the expected precipitation decrease exceeds 10% in the eastern part of the selected region. The expected decrease is smaller in autumn than in summer. Based on the seasonal standard deviation values, the largest uncertainty of precipitation change is expected in summer, especially in case of the A2 scenario when the standard deviation of the RCM results exceeds 20% in the Carpathian basin.

Figure 7 summarizes the expected seasonal change of precipitation for Hungary for the A2 and the B2 scenarios. Black and grey arrows indicate increase and decrease of precipitation, respectively. According to the reference period (1961–1990) the wettest season was summer, then, less precipitation was observed in spring, even less in autumn and the driest season was winter. If the projections are realized then the annual distribution of precipitation will be totally restructured, namely, the wettest seasons will be winter and spring (in this order) for both scenarios. The driest season will be the summer in case of the A2 scenario and autumn in case of the B2 scenario. On the basis of the projections, the annual difference between seasonal precipitation amounts is expected to decrease significantly (by half) for the B2 scenario (which implies more similar seasonal amounts), while it is not expected to change for the A2 scenario (nevertheless, the wettest and the driest seasons are completely changed).

In order to evaluate the model performance, the precipitation bias is determined for all the RCM output fields using the GCM-driven simulations for the reference period (1961–1990) and the CRU database (New et al. 1999). The RCM simulations overestimate the precipitation in most of the Central/Eastern European region; however, slight underestimation can be seen in the southwestern part of the region. In Hungary, the bias does not exceed 15% in absolute values (Bartholy et al. 2007).

Discussion of Country-Based Temperature and Precipitation Projections Relative to 1°C Global Warming

The target period of the PRUDENCE simulations covers the end of the twenty-first century (2071–2100). Thus, the above results presented for Central/Eastern Europe provide climate projections for this period. On the other hand, impact studies would require regional climate change scenarios for earlier periods, preferably for the next few decades. The only information source currently available with fine (i.e. 50 km) horizontal resolution for all the European countries is a special comprehensive assessment based on the PRUDENCE simulations (Christensen 2005). This country-based analysis is conducted for both the mean temperature values and the precipitation amounts. In order to avoid the specific characteristics of the A2 or B2 scenario, a pattern-scaling technique has been applied, in which the changes are expressed relative to a 1°C global warming. Uncertainties in the estimates of projected changes are due to the use of different GCMs and RCMs as well as natural variability. As a result, mean and standard deviation of 25 estimates of temperature and precipitation change are provided for each country. Furthermore, these main statistical parameters are used to fit a normal probability distribution function for the projected change. Table 3 summarizes the mean, the standard deviation, the 5th and the 95th percentiles of the annual, the winter and the summer projected temperature changes for the Central/Eastern European countries. Table 4 summarizes the same in case of precipitation.

For annual and seasonal temperature changes (Table 3) the expected increase in the Central/Eastern European countries is larger than the global 1°C

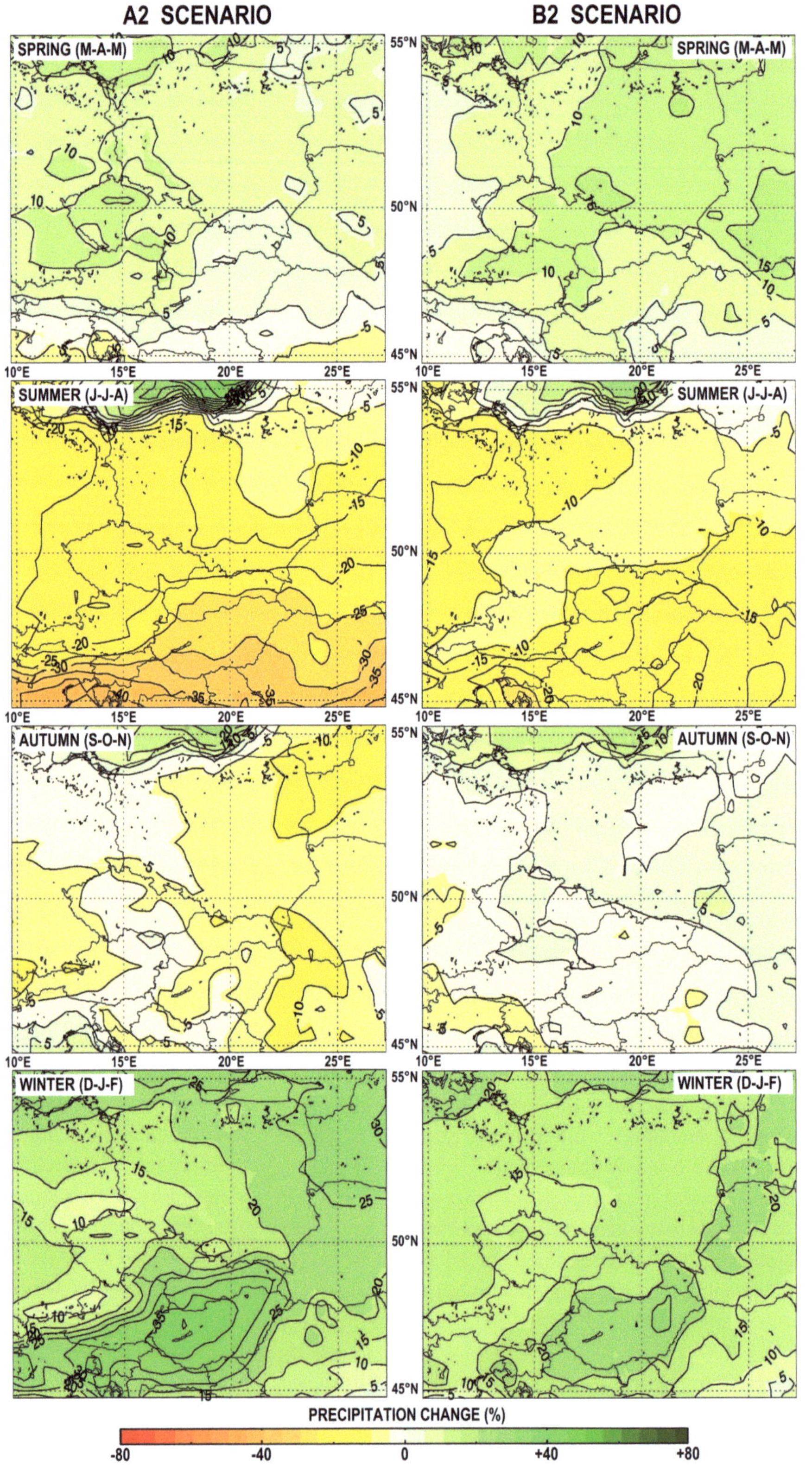

Fig. 5 Seasonal precipitation change (%) expected by 2071–2100 for Central/Eastern Europe using the outputs of 16 and 8 RCM simulations, A2 and B2 scenarios

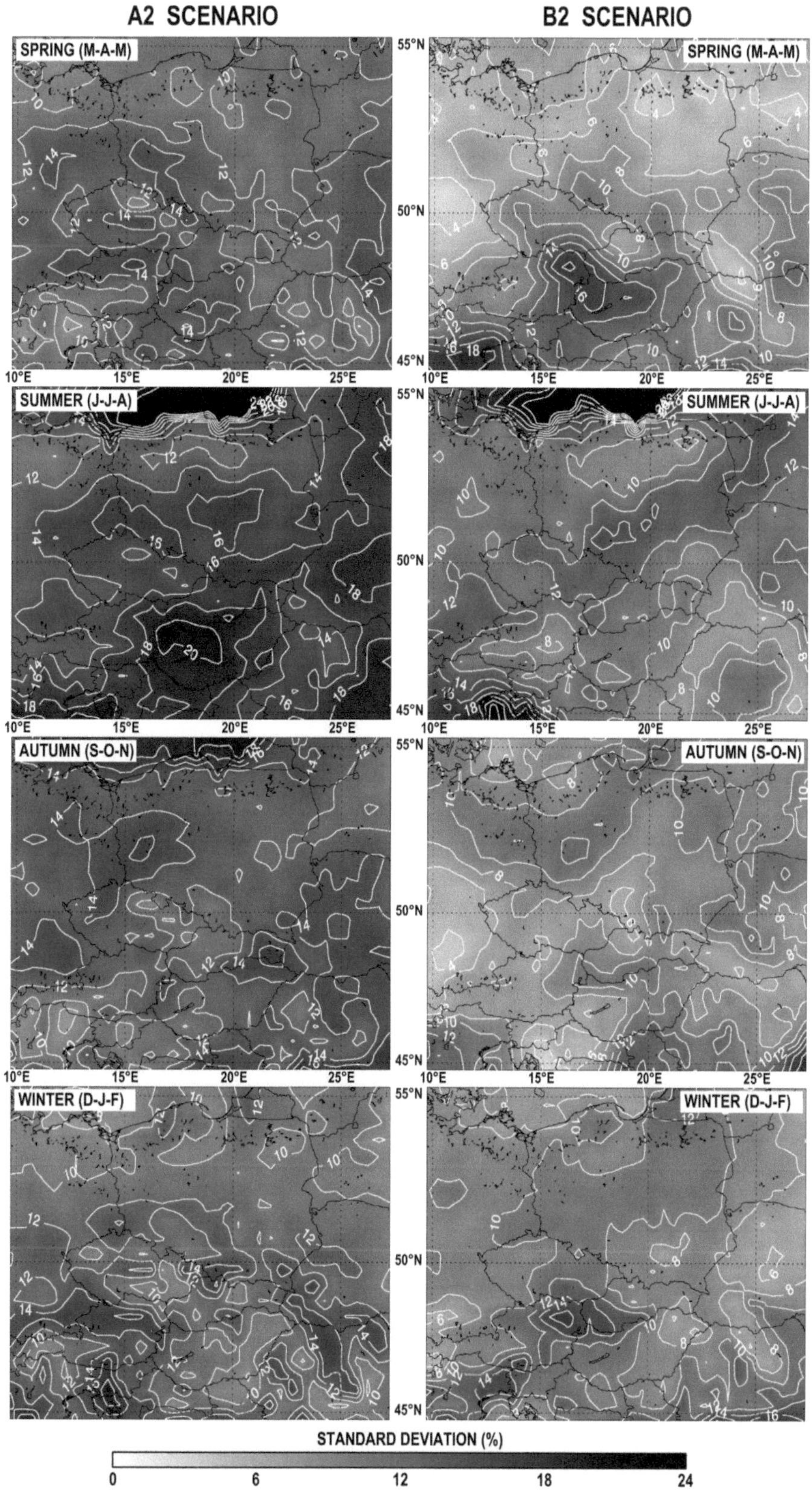

Fig. 6 Standard deviation of seasonal precipitation change (%) expected by 2071–2100 for Central/Eastern Europe using the outputs of 16 and 8 RCM simulations, A2 and B2 scenarios

Table 3 Statistical characteristics of expected increase of temperature (°C) for Central/Eastern European countries relative to 1°C global warming using 25 RCM simulations (Christensen 2005)

	Year		Winter (D-J-F)		Summer (J-J-A)	
Country	Mean ± SD	5th and 95th percentiles	Mean ± SD	5th and 95th percentiles	Mean ± SD	5th and 95th percentiles
Czech Republic	1.3 ± 0.4	0.6–2.0	1.3 ± 0.4	0.6–1.9	1.5 ± 0.6	0.5–2.4
Hungary	1.4 ± 0.3	0.9–1.9	1.3 ± 0.3	0.8–1.9	1.7 ± 0.4	1.0–2.4
Poland	1.3 ± 0.3	0.8–1.8	1.3 ± 0.3	0.8–1.8	1.3 ± 0.4	0.6–2.0
Slovakia	1.4 ± 0.4	0.7–2.0	1.3 ± 0.4	0.7–1.9	1.6 ± 0.6	0.6–2.5
South Germany	1.3 ± 0.4	0.6–2.0	1.1 ± 0.4	0.5–1.7	1.6 ± 0.6	0.6–2.5
North Germany	1.2 ± 0.4	0.6–1.8	1.1 ± 0.4	0.5–1.8	1.3 ± 0.5	0.6–2.0

warming, which implies that this region is quite sensitive to the global environmental change. In general, the projected mean summer regional warming (1.3–1.7°C) is larger than the mean annual temperature increase (1.2–1.4°C), while the expected mean winter (1.1–1.3°C) warming is smaller than that. In case of Poland, the expected temperature increases are equal for summer and for winter, which are also equal to the projected annual warming (1.3°C). In the Czech Republic, the expected summer warming (1.5°C) is larger than the projected temperature increase in winter (1.3°C) and in the whole year (1.3°C). The largest standard deviation values are expected in summer, while the winter and the annual values are equal in case of all Central/Eastern European countries. The 5th percentiles of the expected regional warming are between 0.5 and 1.0°C. The 95th percentiles of the projected annual temperature increase are between 1.8 and 2.0°C, and for winter and summer 1.7–1.9°C and 2.0–2.5°C, respectively.

According to the results presented in Table 4, the annual amount of precipitation in the Central/Eastern European countries is not expected to change significantly. The mean values are between −0.6 and +0.7%, and the standard deviation values are less than 2.4%. On the other hand, considerable precipitation decrease

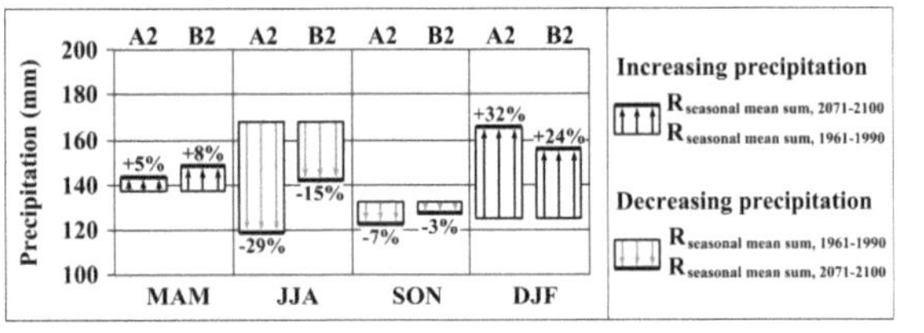

Fig. 7 Expected seasonal change of mean precipitation (mm) for Hungary (increasing or decreasing precipitation is also indicated in percent). Precipitation values of the reference period, 1961–1990, represent the seasonal mean precipitation amount in Budapest

by 4.0–8.2% and increase by 4.5–9.0% are projected for the summer and winter seasons, respectively. These results confirm the conclusions drawn from the precipitation maps in the previous section, which implies that the expected shift in the annual distribution of precipitation starts quite early, that is when the global warming reaches 1°C.

Conclusions

On the basis of our research results shown in this chapter, the following conclusions can be drawn:

Table 4 Statistical characteristics of expected increase of precipitation (%) for Central/Eastern European countries relative to 1°C global warming using 25 RCM simulations (Christensen 2005)

	Year		Winter (D-J-F)		Summer (J-J-A)	
Country	Mean± SD	5th and 95th percentiles	Mean ± SD	5th and 95th percentiles	Mean ± SD	5th and 95th percentiles
Czech Republic	+0.1 ± 2.0	−3.5–3.5	+4.5 ± 3.3	−0.9–10.0	−4.9 ± 5.2	−13.4–3.6
Hungary	−0.3 ± 2.2	−3.9–3.4	+9.0 ± 3.7	3.0–15.0	−8.2 ± 5.3	−16.9–0.5
Poland	+0.7 ± 2.1	−2.6–4.1	+6.0 ± 3.5	0.2–11.8	−4.0 ± 4.9	−12.0–4.1
Slovakia	−0.6 ± 2.4	−4.5–3.3	+7.5 ± 4.4	0.3–14.8	−7.3 ± 5.8	−16.8–2.2
South Germany	−0.3 ± 1.6	−2.8–2.3	+5.1 ± 2.9	0.3– 9.9	−6.4 ± 4.7	−14.0–1.3
North Germany	+0.3 ± 1.7	−2.7–2.9	+6.0 ± 4.1	−0.7–12.6	−7.3 ± 4.9	−15.4–0.9

- Expected seasonal temperature increase for Central/Eastern Europe for the A2 scenario is larger than for the B2 scenario, which is in good agreement with the expected global and European climate change results (IPCC 2007). The largest and the smallest warming are expected in summer and in spring, respectively.
- In summer, a zonal structure of projected warming with increasing values from north to south can be expected for all temperature parameters, while in winter, a meridional structure of warming is expected with increasing values from west to east. In spring and autumn, the spatial difference values of projected temperature change are much smaller; they do not exceed 0.8°C and 0.4°C, respectively.
- In the reference period of 1961–1990, the RCM simulations slightly overestimate the temperature in most of the Central/Eastern European region. A small region of underestimation can be seen in the southwestern part of the selected domain, at the mountainous areas of the Alps. The temperature bias does not exceed 1.5°C.
- In Hungary, for all the four seasons and for both scenarios, the expected warming by 2071–2100 is between 2.5 and 4.8°C. The largest temperature increase is projected for summer, 4.8°C (A2) and 4.0°C (B2), and the smallest seasonal warming is expected in spring, 3.1°C (A2) and 2.5°C (B2). The smallest difference between the A2 and B2 scenarios is projected for spring (0.6°C), while the largest is projected for winter (1°C).
- The largest increase of maximum and minimum temperatures in Hungary is expected in summer for both scenarios. For maximum temperature, the interval of the expected warming is 4.9–5.3°C (A2) and 4.0–4.4°C (B2). For minimum temperature, these intervals are 4.2–4.8°C (A2) and 3.5–4.0°C (B2). In general, the expected increase of maximum temperature is not smaller than the expected increase of minimum temperature with the exception of the winter season.
- The annual precipitation sum is not expected to change significantly in the Central/Eastern European region, but this does not hold for seasonal precipitation sums. Summer precipitation is very likely to decrease; furthermore, slight decrease of autumn precipitation is expected. On the other hand, winter precipitation is likely to increase considerably, and slight increase in spring is also expected.
- The projected summer precipitation decrease in Hungary is 24–33% (A2) and 10–20% (B2), while the expected winter precipitation increase is 23–37% (A2) and 20–27% (B2).
- In case of Hungary, the wettest season was summer, while the driest season was winter in the reference period (1961–1990). If the projections are realized then the annual distribution of precipitation will be totally restructured: namely, the wettest season is found in winter in both scenarios; the driest season is found in summer in the A2 scenario and in autumn in the B2 scenario.

Acknowledgments Climate change data have been provided through the PRUDENCE data archive, funded by the EU through contract EVK2-CT2001-00132. Research leading to this chapter has been supported by the following sources: the Hungarian Academy of Sciences under the program 2006/TKI/246 titled *Adaptation to Climate Change*, the Hungarian National Research Development Program under grants NKFP-3A/082/2004 and NKFP-6/079/2005, the Hungarian National Science Research Foundation under grants T-049824, K-67626 and K-69164, the Hungarian Academy of Science and the Hungarian Prime Minister's Office under grant 10.025-MeH-IV/3.1/2006, the Hungarian Ministry of Environment and Water under the National Climate Strategy Development project and the CECILIA project of the European Union Number 6 programme (contract no. GOCE-037005).

References

Bartholy J, Pongrácz R, Matyasovszky I, Schlanger V (2003) Expected regional variations and changes of mean and extreme climatology of Eastern/Central Europe. In: Combined Preprints CD-ROM of the 83rd AMS Annual Meeting. Paper 4.7, 10pp

Bartholy J, Pongrácz R, Torma Cs, Hunyady A (2006a) Regional climate model PRECIS and its adaptation at the Department of Meteorology, Eötvös Loránd University. In: 31. Meteorological Scientific Days – Dynamical climatological research on objective estimation of regional climate change (ed.: Weidinger, T.) Hungarian Meteorological Service, Budapest. 99–114 (in Hungarian)

Bartholy J, Pongrácz R, Torma Cs, Hunyady A (2006b) Regional climate change scenarios for the Carpathian Basin. In: Bioclimatology and water in the land (eds.: Lapin, M. and Matejka, F.) CD-ROM. FMFI Comenius University, Slovakia. 9pp

Bartholy J, Pongrácz R, Gelybó Gy (2007) Regional climate change expected in Hungary for 2071–2100. Applied Ecology and Environmental Research, 5, 1–17

Benestad RE (2005) Climate change scenarios for northern Europe from multi-model IPCC AR4 climate simulations. Geophysical Research Letters, 32, L17704, doi:10.1029/2005GL023401

Black E, Blackburn M, Harrison G, Hoskins BJ, Methven J (2004) Factors contributing to the summer 2003 European heatwave. Weather, 59, 217–223

Christensen JH (2005) Prediction of Regional scenarios and Uncertainties for Defining European Climate change risks and Effects – Final Report. DMI. 269p

Christensen OB, Christensen JH (2004) Intensification of extreme European summer precipitation in a warmer climate. Global and Planetary Change, 44, 107–117

Christensen JH, Carter TR, Rummukainen M, Amanatidis G (guest editors) (2007) Prediction of regional scenarios and uncertainties for defining European climate change risks and effects: The PRUDENCE project. Climatic Change, 81, Supplement 1, 371p

Déqué M, Jones RG, Wild M, Giorgi F, Christensen JH, Hassell DC, Vidale PL, Rockel B, Jacob D, Kjellström E, de Castro M, Kucharski F, van den Hurk B (2005) Global high resolution versus Limited Area Model climate change scenarios over Europe: results from the PRUDENCE project. Climate Dynamics, 25, 653–670. doi:10.1007/s00382-005-0052-1

Fink AH, Brücker T, Krüger A, Leckebusch GC, Pinto JG, Ulbrich U (2004) The 2003 European summer heatwaves and drought – synoptic diagnostics and impacts. Weather, 59, 209–216

Giorgi F (1990) Simulation of regional climate using a limited area model nested in a general circulation model. Journal of Climate, 3, 941–963

Hanssen-Bauer I, Achberger C, Benestad RE, Chen D, Førland EJ (2005) Statistical downscaling of climate scenarios over Scandinavia: A review. Climate Research, 29, 255–268

Horányi A (2006) Dynamical climatological research on regional scales: International and Hungarian review. In: 31. Meteorological Scientific Days – Dynamical climatological research on objective estimation of regional climate change (ed.: Weidinger, T.) Hungarian Meteorological Service, Budapest. 62–70. (in Hungarian)

IPCC (2007) Climate Change 2007: The Physical Science Basis. Contribution of Working Group I to the Fourth Assessment Report of the Intergovernmental Panel on Climate Change. (eds.: Solomon, S., Qin, D., Manning, M., Chen, Z., Marquis, M., Averyt, K.B., Tignor, M., Miller, H.L.), Cambridge University Press, Cambridge, United Kingdom and New York, NY, USA, 996pp

Mearns LO, Hulme M, Carter TR, Leemans R, Lal M, Whetton PH (2001) Climate scenario development. In: Climate Change 2001: The Scientific Basis. (eds.: Houghton, J. et al.) Intergovernmental Panel on Climate Change, Cambridge University Press, New York. 739–768

New M, Hulme M, Jones P (1999) Representing twentieth-century space-time climate variability. Part I: Development of a 1961–90 mean monthly terrestrial climatology. Journal of Climate, 12, 829–856

Pal JS, Giorgi F, Bi X (2004) Consistency of recent European summer precipitation trends and extremes with future regional climate projections. Geophysical Research Letters, 31, L13202, doi:10.1029/2004GL019836

Tebaldi C, Hayhoe K, Arblaster JM, Meehl GE (2006) Going to the extremes: an intercomparison of model-simulated historical and future changes in extreme events. Climatic Change, 79, 185–211

Detected and Expected Trends of Extreme Climate Indices for the Carpathian Basin

R. Pongrácz, J. Bartholy, Gy. Gelybó and P. Szabó

Keywords Climate extreme index · Daily precipitation · Daily temperature · Carpathian basin · Trend analysis

Introduction

Regional climatological effects of global warming may be recognized not only in shifts of mean temperature and precipitation but also in the frequency and intensity changes of different climate extremes. A joint WMO-CCl (World Meteorological Organization Commission for Climatology)/CLIVAR (a project of the World Climate Research Programme addressing Climate Variability and Predictability) working group formed in 1998 on climate change detection (Karl et al. 1999); one of its task groups aimed to identify climate extreme indices (Peterson et al. 2002) and completed a climate extreme analysis on all parts of the world where appropriate data were available (Frich et al. 2002). The main results of this working group appeared in the *IPCC Assessment Reports* (2001, 2007). Klein Tank and Können (2003) analyzed extreme climate indices on continental scale for Europe (86 and 151 stations were used in case of temperature and precipitation time series, respectively) for the second half of the twentieth century. Their results give a general overview on the European scale but they are not detailed enough for the Carpathian basin. In this chapter, trend analysis of extreme temperature and precipitation indices is discussed for the Carpathian basin for both the past few decades and for the last decades of the twenty-first century.

The next section of this chapter presents the database and the definition of the extreme climate indices. Then, the Detected Trend of Extreme Climate Indices for the Carpathian Basin section discusses the trend analysis of extreme temperature and precipitation indices detected in the Carpathian basin in the second half of the twentieth century. The Future Trends of Extreme Climate Indices for the Carpathian Basin section compares the extreme climate indices for the periods 1961–1990 and 2071–2100 using simulated daily temperature and precipitation datasets. Finally, Conclusions section discusses the main findings of this chapter.

Data and Methodology

In order to compile a global climate database suitable for extreme analysis, the WMO-CCl/CLIVAR task group on extreme indices contacted the national meteorological services and collected daily precipitation, minimum, maximum, and mean temperature time series for the period 1946–1999. In addition to the datasets from the national meteorological services, further data sources, namely, the National Oceanic and Atmospheric Administration (NOAA) and National Climatic Data Center (NCDC) datasets (Peterson and Vose 1997), the European Climate Assessment Dataset (ECAD) (Klein Tank et al. 2002b), and daily meteorological time series for Australia (Trewin 1999), have been included. All these datasets have been quality controlled and adjusted for inhomogeneities. Then, in order to accept a given observation station, the following general criteria have been used: (i) from

R. Pongrácz (✉)
Department of Meteorology, Eötvös Loránd University, Pazmany st. 1/a, H-1117 Budapest, Hungary
e-mail: prita@nimbus.elte.hu

K. Strelcová et al. (eds.), *Bioclimatology and Natural Hazards*,
DOI 10.1007/978-1-4020-8876-6_2, © Springer Science+Business Media B.V. 2009

the entire 1946–1999 period data must be available for at least 40 years, (ii) missing data cannot be more than 10%, (iii) missing data from each year cannot exceed 20%, and (iv) in each year, more than 3 months consecutive missing values are not allowed.

The WMO-CCl/CLIVAR task group decided to map station data instead of gridded database since extreme events (e.g., local floods and droughts, heat waves, local cold spells) often occur on local scale, but on the other hand, they are all important part of global climate patterns, which could disappear in case of a spatial data interpolation. Therefore, maps showing the detected trends and presented in this chapter use similar technique applying station-based analysis. Since simulations of regional climate models (RCMs) use and result in gridded datasets, in case of analyzing the expected future tendencies, we used grid-based maps.

For the evaluation of recent tendency of climate, extreme indices in the Carpathian basin 32 meteorological stations have been used (Fig. 1). Datasets for 21 Hungarian stations were provided by the Hungarian Meteorological Service, while datasets for 11 stations located in the neighboring countries are freely available via Internet from the ECAD (Klein Tank 2003). Two basic constraints are taken into account during the selection of the stations: (i) covering the area of the Carpathian basin with the best spatial homogeneity and the best representation of the main climatic subregions, (ii) minimal number of missing data.

Our datasets are compiled for 1901–2001. However, our previous analyses (Pongrácz and Bartholy 2000) suggest that precipitation and temperature tendencies of the last quarter of the twentieth century and the second half of the century are significantly different. Because of this and due to the limited temporal extent of the time series, the analysis presented in this chapter for the Carpathian basin has been accomplished for 1946–2001.

Table 1 lists the main extreme precipitation and temperature indices (12 and 14, respectively) that the WMO-CCl/CLIVAR task group identified and suggested for global and regional climate extreme analysis. Indices listed in Table 1 include a few precipitation-related parameters, which do not indicate extreme conditions. They belong to the index type annual number of precipitation days exceeding a given threshold (i.e., 5, 1, 0.1 mm). We accomplished the trend analysis for these latter indices, because they add important characteristics to regional precipitation conditions.

Detected Trend of Extreme Climate Indices for the Carpathian Basin

First, extreme climate indices related to daily temperature values are analyzed. On the basis of our previous study of time series of mean and extreme temperature parameters for the Carpathian basin, a strong warming tendency was detected from the middle of the 1970s (Pongrácz and Bartholy 2000). Therefore, the entire 1961–2001 period has been separated into two subperiods, namely, 1961–1975 and 1976–2001. The trend analysis has been accomplished for these subperiods (Bartholy and Pongrácz 2006).

Summary of the trend analysis of the extreme temperature indices is presented in Fig. 2. The distribution of the trend coefficients, determined for each station, can be seen. The whisker plot diagrams provide statistical characteristics of the decadal trends for each extreme temperature index, for the two subperiods (1961–1975 and 1976–2001). Opposite sign of trend coefficients may indicate warming and cooling tendencies. For instance, negative coefficients of the number of cold days (T×10) and positive coefficients of the number of hot days (T×30GE) both indicate warming climate. Therefore, warming and cooling tendencies are indicated by different background colors (gray and white, respectively). In case of the intra-annual extreme temperature range (ETR) index, hatched background is used since the sign of the trend coefficient is not directly connected to a warming or cooling tendency.

According to the results, warming trends are detected in the first subperiod in case of extreme indices indicating negative temperature extremes (i.e., T×10, Tn10, FD, T×0LT, Tn−10LT), while cooling trends are detected in case of positive temperature extremes (i.e., HWDI, T×90, SU, T×30GE, T×35GE, Tn20GT). In case of Tn90, the regional mean trend coefficient is not significant, both warming and cooling trends have been detected. In the last quarter of the twentieth century, warming tendencies are dominant. Warming trend is detected in case of nine indices, namely, HWDI, T×90, Tn10, Tn90, SU, T×30GE, T×35GE, Tn20GT, Tn−10LT, while the regional mean trend coefficient is not significant in case of T×10, FD, and T×0LT.

In this chapter, detailed analysis is presented for the last quarter of the twentieth century when the largest

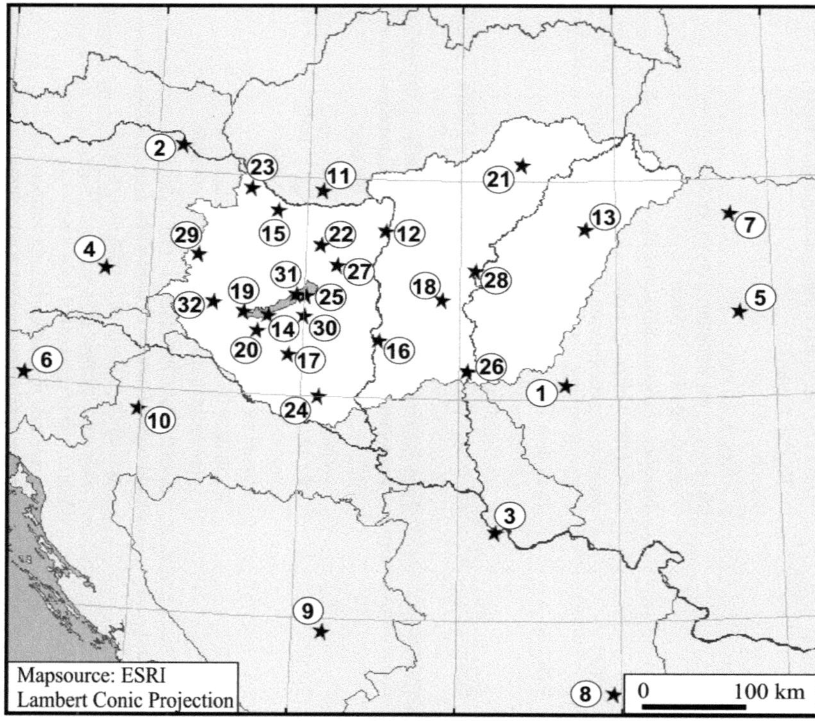

	Station	N.lat.	E.long.	m
	Stations in neighbouring countries			
1	Arad	46°08'	21°21'	117
2	Vienna	48°14'	16°21'	199
3	Beograd	44°48'	20°28'	132
4	Graz	47°05'	15°27'	366
5	Cluj Napoca	46°47'	23°34'	410
6	Ljubljana	46°04'	14°31'	299
7	Baia Mare	47°40'	23°30'	216
8	Nis	43°20'	21°54'	202
9	Sarajevo	42°30'	22°30'	211
10	Zagreb	45°49'	15°59'	157
11	Hurbanovo	47°53'	18°12'	115
	Hungarian stations			
12	Budapest	47°31'	19°02'	118
13	Debrecen	47°33'	21°37'	123
14	Fonyód	46°44'	17°33'	166
15	Győr	47°41'	17°38'	119
16	Kalocsa	46°31'	18°59'	92
17	Kaposvár	46°22'	17°50'	152
18	Kecskemét	46°54'	19°46'	114
19	Keszthely	46°45'	17°15'	115
20	Marcali	46°35'	17°25'	128
21	Miskolc	48°08'	20°48'	118
22	Mór	47°23'	18°12'	203
23	Mosonmagyaróvár	47°53'	17°16'	121
24	Pécs	46°00'	18°14'	202
25	Siófok	46°55'	18°02'	108
26	Szeged	46°15'	20°06'	82
27	Székesfehérvár	47°12'	18°25'	111
28	Szolnok	47°10'	20°12'	90
29	Szombathely	47°15'	16°38'	218
30	Tab	46°44'	18°01'	177
31	Tihany	46°55'	17°54'	150
32	Zalaegerszeg	46°50'	16°51'	152

Fig. 1 Geographical locations of meteorological stations in the Carpatihian basin

Table 1 Definition and idicator of extreme climate indices

Number	Indicator (ECAD)	Definition of the extreme index	Unit
1	CDD	Maximum number of consecutive dry days (when $R_{day} < 1$ mm)	Day
2	R×1	Highest 1-day precipitation amount	mm
3	R×5	The greatest 5-day rainfall total	mm
4	SDII	Simple daily intensity index (total precipitation sum/total number of days when $R_{day} \geq 1$ mm)	mm/day
5	R95T	Fraction of annual total rainfall due to events above the 95th percentile of the daily precipitation in the base period 1961–1990 ($\sum R_{day}/R_{total}$, where $\sum R_{day}$ indicates the sum of daily precipitation exceeding $R_{95\%}$)	%
6	R95	Number of very wet days ($R_{day} > R_{95\%}$, where $R_{95\%}$ indicates the 95th percentile of the daily precipitation in the base period 1961–1990)	Day
7	R75	Number of moderate wet days ($R_{day} > R_{75\%}$, where $R_{75\%}$ indicates the upper quartile of the daily precipitation in the base period 1961–1990)	Day
8	RR20	Number of very heavy precipitation days (when $R_{day} \geq 20$ mm)	Day
9	RR10	Number of heavy precipitation days ($R_{day} \geq 10$ mm)	Day
10	RR5	Number of precipitation days exceeding a given threshold ($R_{day} \geq 5$ mm)	Day
11	RR1	Number of precipitation days exceeding a given threshold ($R_{day} \geq 1$ mm)	Day
12	RR0.1	Number of precipitation days exceeding a given threshold ($R_{day} \geq 0.1$ mm)	Day
13	ETR	Intra-annual extreme temperature range (difference between the observed maximum and minimum temperatures, $T_{max} - T_{min}$)	°C
14	GSL	Growing season length (start: when for > 5 days $T > 5°C$, end: when for > 5 days $T < 5°C$)	Day
15	HWDI	Heat wave duration index (for minimum 5 consecutive days $T_{max} = T_{max}^N + 5°C$, where T_{max}^N indicates the mean T_{max} for the base period 1961–1990)	Day
16	T×10	Cold days (percent of time when $T_{max} < 10$th percentile of daily maximum temperature based on the base period 1961–1990)	Day
17	T×90	Warm days (percent of time when $T_{max} > 90$th percentile of daily maximum temperature based on the base period 1961–1990)	Day
18	Tn10	Cold nights (percent of time when $T_{min} < 10$th percentile of daily minimum temperature based on the base period 1961–1990)	Day
19	Tn90	Warm nights (percent of time when $T_{min} > 90$th percentile of daily minimum temperature based on the base period 1961–1990)	Day
20	FD	Number of frost days ($T_{min} < 0°C$)	Day
21	SU	Number of summer days ($T_{max} > 25°C$)	Day
22	T×30GE	Number of hot days ($T_{max} \geq 30°C$)	Day
23	T×35GE	Number of extremely hot days ($T_{max} \geq 35°C$)	Day
24	Tn20GT	Number of hot nights ($T_{min} > 20°C$)	Day
25	T×0LT	Number of winter days ($T_{max} < 0°C$)	Day
26	Tn−10LT	Number of severe cold days ($T_{min} < -10°C$)	Day

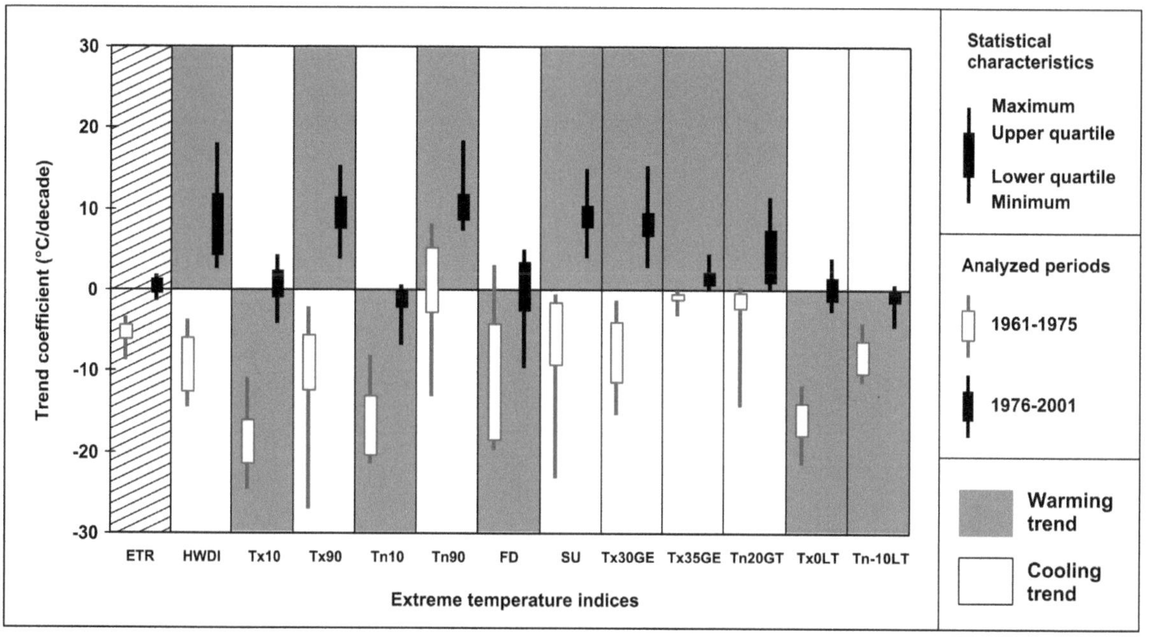

Fig. 2 Distribution of decadal trend coefficients of extreme temperature indices for the Carpathian basin. For each index warming and cooling tendencies are indicated by gray and white background, respectively. In case of the ETR, hatched background indicates that the sign of the trend coefficient is not directly indicative of a warming or cooling tendency

changes occurred. The daily maximum temperature of summer may be indicated by three extreme indices (i) number of summer days (SU, $T_{max} > 25°C$), (ii) number of hot days (T×30GE, $T_{max} \geq 30°C$), and (iii) number of extremely hot days (T×35GE, $T_{max} \geq 35°C$). Increasing trend coefficients of these indices are detected during the entire 1961–2001 period, and the 1976–2001 subperiod, while they are decreasing in the 1961–1975 subperiod. Figure 3 presents the maps containing the trend coefficients of extreme indices SU and T×30GE in the Carpathian basin in the last 26 years. Circles represent decadal trend coefficients of the meteorological stations (using the base period 1961–1990). Black and gray circles indicate increasing and decreasing tendencies, respectively, while circle size depends on the intensity of these positive or negative trends. According to the results, no decreasing tendency can be identified in either map shown in Fig. 3. Large positive trend coefficients dominate both the numbers of summer and hot days, with more than 6 days per decade, in general. These positive trend coefficients are significant at 95% level. Tendency analysis map of the extreme index T×35GE is not presented in this chapter, since the frequency of this event is quite small, however, the trend coefficients are similar to those shown in case of the SU and T×30GE.

Based on the above results, analysis of the extreme temperature indices suggests that similar to the global (Frich et al. 2002) and European trends (Klein Tank et al. 2002a), the regional climate of the Carpathian basin tended to be warmer in the last four decades.

In the second part of this section, extreme precipitation indices are analyzed for the second half of the twentieth century (1946–2001), and for the last quarter of the twentieth century (1976–2001). Results of the trend analysis are summarized in Table 2 for these two subperiods. When similar changes are detected for all stations located in the Carpathian basin, only one "+" or "−" sign indicates the tendency. In case of different tendencies of the western and eastern parts of the selected region, "− +" or "+ −" signs can be found in the table. The detected regional mean trend coefficients for the 1976–2001 subperiod are significant at 95% level, except the indices SDII and RR5. In case of the entire 1946–2001 period, only four precipitation indices exhibit significant regional mean tendency, namely, the consecutive dry days (CDD), R×5, RR1, and RR0.1. In separate columns the numbers in-

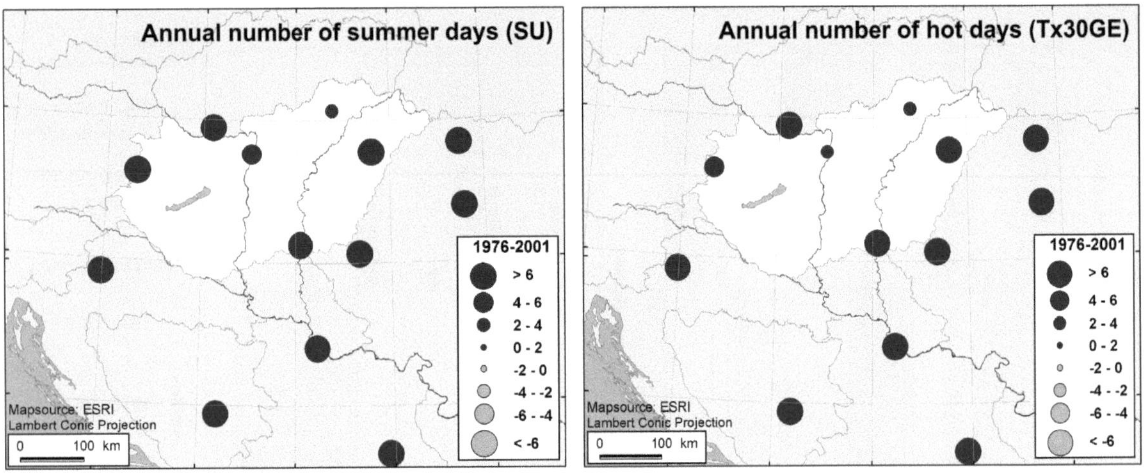

Fig. 3 Increasing tendency of the number of summer days (SU, $T_{max} > 25°C$) and hot days (T×30GE, $T_{max} \geq 30°C$) in the Carpathian basin during the last quarter of the twentieth century. Trend coefficients greater than 0.4 in absolute value are significant at 95% level

dicate the stations where the decadal trend coefficients are significant at 95% level. Note that the numbers of stations used in the trend analysis are different in case of periods 1946–2001 and 1976–2001 (26 and 31, respectively). On the basis of the analysis of tendency maps (Bartholy and Pongrácz 2005a), only some of the extreme indices can be characterized by homogeneous positive or negative trends for both periods. Most of the extreme precipitation indices increased considerably in the Carpathian basin by the end of the twentieth century. Positive trends were detected mostly in the last 26 years. The strongest increasing tendencies appear in case of extreme indices indicating very intense or large precipitation (i.e., R95T, RR20, R75, R95).

Summarizing the results of trend analysis, it can be concluded (Bartholy and Pongrácz 2007) that although in general, precipitation occurred more rarely in the Carpathian basin (RR1, RR0.1) but the ratio of heavy or extreme precipitation days (RR20, RR10, R95, R75, R95T) increased considerably by the end of the twentieth century.

As an example, maps of decadal trend coefficients of annual number of very wet days (R95) are presented in Fig. 4 for 1946–2001 and 1976–2001. Based on the tendency analysis of the entire European continent (Klein Tank et al. 2002a), frequency of very wet days increased during the last quarter of the twentieth century in the northern stations, while they became less

Table 2 Summary of trend analysis of extreme precipitation indices for the Carpathian basin for the periods 1946–2001 and 1976–2001 (based on 26 and 31 stations, respectively). Signs in parentheses indicate regional mean trend coefficients being not significant at 95% level. Numbers indicate the stations with detected trend coefficients significant at 95% level

	1946–2001		1976–2001	
Extreme index	Sign of regional trend coefficient	Stations with significant trend coefficient	Sign of regional trend coefficient	Stations with significant trend coefficient
CDD	+	22	−	28
R×1	(−+)	22	−	27
R×5	−	21	+ −	28
SDII	(+)	2	(+)	8
R95T	(+)	21	+	28
R95	(− +)	4	+	20
R75	(−)	15	+	25
RR20	(− +)	9	+	26
RR10	(− +)	19	+	23
RR5	(−)	16	(−)	23
RR1	−	25	−	29
RR0.1	−	22	−	30

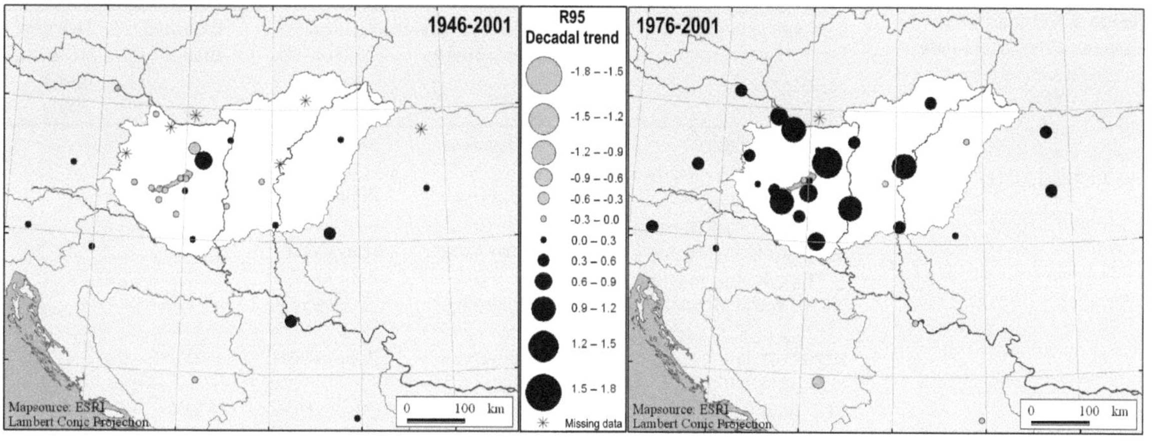

Fig. 4 Tendency of annual number of very wet days (R95) in the Carpathian basin during the last quarter of the twentieth century. Trend coefficients greater than 0.3 in absolute value are significant at 95% level of confidence

frequent in the Mediterranean region. The Carpathian basin is located in the border of the northern and the Mediterranean regions; however, our detailed regional analysis suggests that the trend coefficients are positive similar to the northern European stations. The entire Carpathian basin can be characterized by a strong positive trend in the last 26 years (right panel of Fig. 4), while in the second half of the twentieth century in most of the stations the trend coefficients are not significant (left panel of Fig. 4).

After summarizing the past extreme climate trends for the Carpathian basin, they will be compared to the expected future trends based on climate simulation for the 2071–2100 period in the next section.

Future Trends of Extreme Climate Indices for the Carpathian Basin

In order to compare the past and future trends of the extreme climate indices for the Carpathian basin, simulated daily values of meteorological variables are obtained from the RCM outputs of the Swiss Federal Institute of Technology Zurich (Eidgenössische Technische Hochschule Zürich, ETHZ) in the frame of the completed EU project Prediction of Regional scenarios and Uncertainties for Defining EuropeaN Climate change risks and Effects (PRUDENCE). Results of the project PRUDENCE (Christensen 2005) are disseminated widely via Internet and several other media. The primary objectives of PRUDENCE were to provide high-resolution (50 km × 50 km) climate change scenarios for Europe for 2071–2100 (Christensen and Christensen 2007) using dynamical downscaling methods with RCMs (using the reference period 1961–1990). ETHZ used the climate high-resolution model (CHRM) RCM (Vidale et al. 2003) with 50 km horizontal resolution, for which the boundary conditions were provided by the HadAM3H/HadCM3 (Rowell 2005) global climate model of the UK Met Office. The simulations were accomplished for present-day conditions using the reference period 1961–1990 (the model performance of CHRM is analyzed by Jacob et al. 2007) and for the future conditions in 2071–2100 using scenario A2. According to this scenario, fertility patterns across regions converge very slowly resulting in continuously increasing world population. Economic development is primarily regionally oriented, per capita economic growth and technological changes are fragmented and slow. The projected CO_2 concentration may reach 850 ppm by the end of the twenty-first century (IPCC 2007), which is about three times the pre-industrial concentration level (280 ppm).

Extreme temperature indices are compared in Table 3 for the reference period and the last three decades of the twenty-first century. The annual values are calculated as an average of 10 representative grid points located in Hungary (Bartholy et al. 2007b). It can be seen that negative extremes are expected to decrease, while positive extremes tend to increase significantly. Both imply regional warming in the

Table 3 Comparison of extreme temperature indices in the reference period (1961–1990) and in case of the A2 scenario (2071–2100) based on the daily outputs of the RCM of ETHZ

Extreme temperature index	Reference period (1961–1990)	A2 scenario (2071–2100)	Expected change	Detected trend (1961–2001)
T×10: Cold days ($T_{max} < T_{max,10\%,1961-1990}$)	36 days/year	10 days/year	−72%	−
T×90: Warm days ($T_{max} > T_{max,90\%,1961-1990}$)	36 days/year	78 days/year	+116%	+
Tn10: Cold nights ($T_{min} < T_{min,10\%,1961-1990}$)	36 days/year	9 days/year	−76%	−
Tn90: Warm nights ($T_{min} > T_{min,90\%,1961-1990}$)	36 days/year	79 days/year	+120%	+
FD: Frost days ($T_{min} < 0°C$)	74 days/year	26 days/year	−65%	−
SU: Summer days ($T_{max} > 25°C$)	98 days/year	136 days/year	+39%	+
T×30GE: Hot days ($T_{max} \geq 30°C$)	47 days/year	90 days/year	+91%	+
T×35GE: Extremely hot days ($T_{max} \geq 35°C$)	13 days/year	45 days/year	+250%	+
Tn20GT: Hot nights ($T_{min} > 20°C$)	5 days/year	36 days/year	+625%	+
T×0LT: Winter days ($T_{max} < 0°C$)	24 days/year	6 days/year	−75%	−
Tn−10LT: Severe cold days ($T_{min} < -10°C$)	8 days/year	1 days/year	−83%	−

Carpathian basin. The largest increase due to this warming trend can be expected in case of hot days (T×30GE), warm days (T×90), warm nights (Tn90), extremely hot days (T×35GE), and hot nights (Tn20GT) by +91%, +116%, +120%, +250%, and +625%, respectively. Expected changes of all the temperature indices are completely consistent with the detected trend in the 1961–2001 period (Bartholy and Pongrácz 2005b).

Among the extreme temperature indices, the number of frost days (FD) and the number of summer days (SU) are selected to illustrate the spatial distribution of the index values both in the reference period (upper map) and in the last three decades of the twenty-first century (center map) in Figs. 5 and 6, respectively. In both cases the orography is the main factor determining the spatial patterns of the extreme indices. In the higher elevated mountainous areas (in the range of the Carpathians and in the eastern foothills of the Alps), the annual number of FD exceeds 120 in the reference period, while it is expected to decrease to only about 80 days per year. In the lower elevated areas of Hungary, FD was about 60–80 days in each year on average, which is likely to decrease to 10–40 days by the end of the twenty-first century resulting in a decrease by 60–70% in the country (lower map of Fig. 5). According to the ETHZ model simulations, the decrease of FD is expected to be smaller in the northern part of Hungary (about 60%) than in the southern subregions (about 70%).

In case of the annual number of SU similar patterns can be identified. In 1961–1990, SU was about 80–105 days per year on average in Hungary, and it is expected to increase to 120–145 days by 2071–2100 in case of the A2 scenario. The smallest increase (35%) is projected in the southeastern part of the country, while in the northwestern border the increase may exceed 60% (lower map of Fig. 6). In the higher elevated mountainous subregions, the expected increase of the SU is larger than 150%.

Similar analysis is presented in Table 4 for the extreme precipitation indices. Expected changes of annual indices are generally consistent with the detected trends in the last quarter of the twentieth century. However, the expected regional increase or decrease is usually small (not exceeding 15% in absolute value). Much larger positive and negative changes are projected in January and in July, respectively, on the basis of the RCM simulations in case of the A2 scenario. The only exception is the

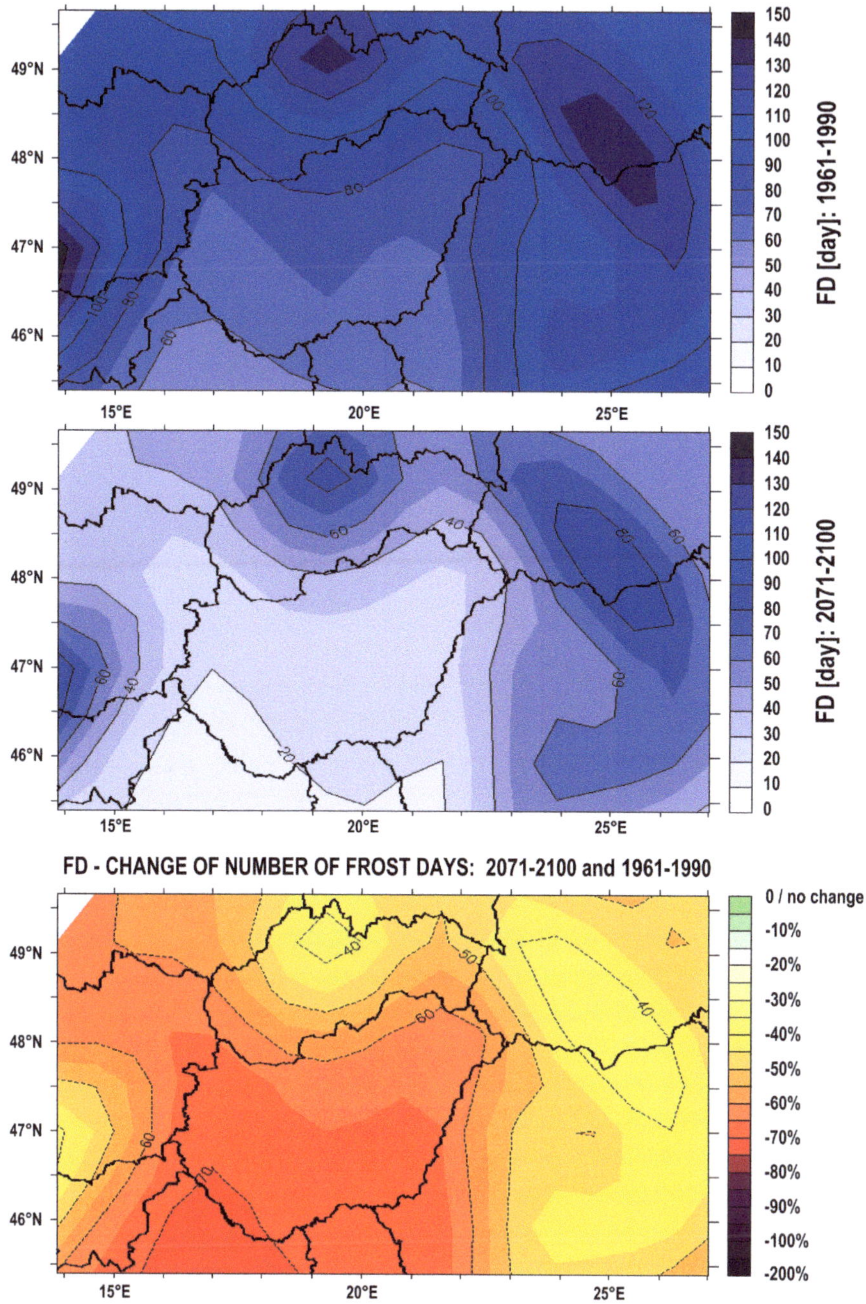

Fig. 5 Expected change of annual number of frost days (FD) in case of the A2 scenario (2071–2100) compared to the reference period (1961–1990), maps are determined using daily minimum temperature values of the RCM of ETHZ

CDD but this is the only precipitation index, which is related to drought instead of rainfall. Therefore, the opposite signs of the projected CDD changes are consistent with the expected regional changes of the other precipitation indices. These results suggest that the climate tends to be wetter in January and drier in July in the Carpathian basin (Bartholy et al. 2007a). Since the projected increases of the RR20, RR10, and R95 (these indices describe very extreme precipitation events) exceed 50% in January, and the expected increases of RR0.1 or RR1 (these indices are not related to extreme precipitation) do not reach 15%, the extreme precipitation events are expected to become more intense and more frequent in January. Furthermore, drought is projected to become more intense in July by the end of the twenty-first

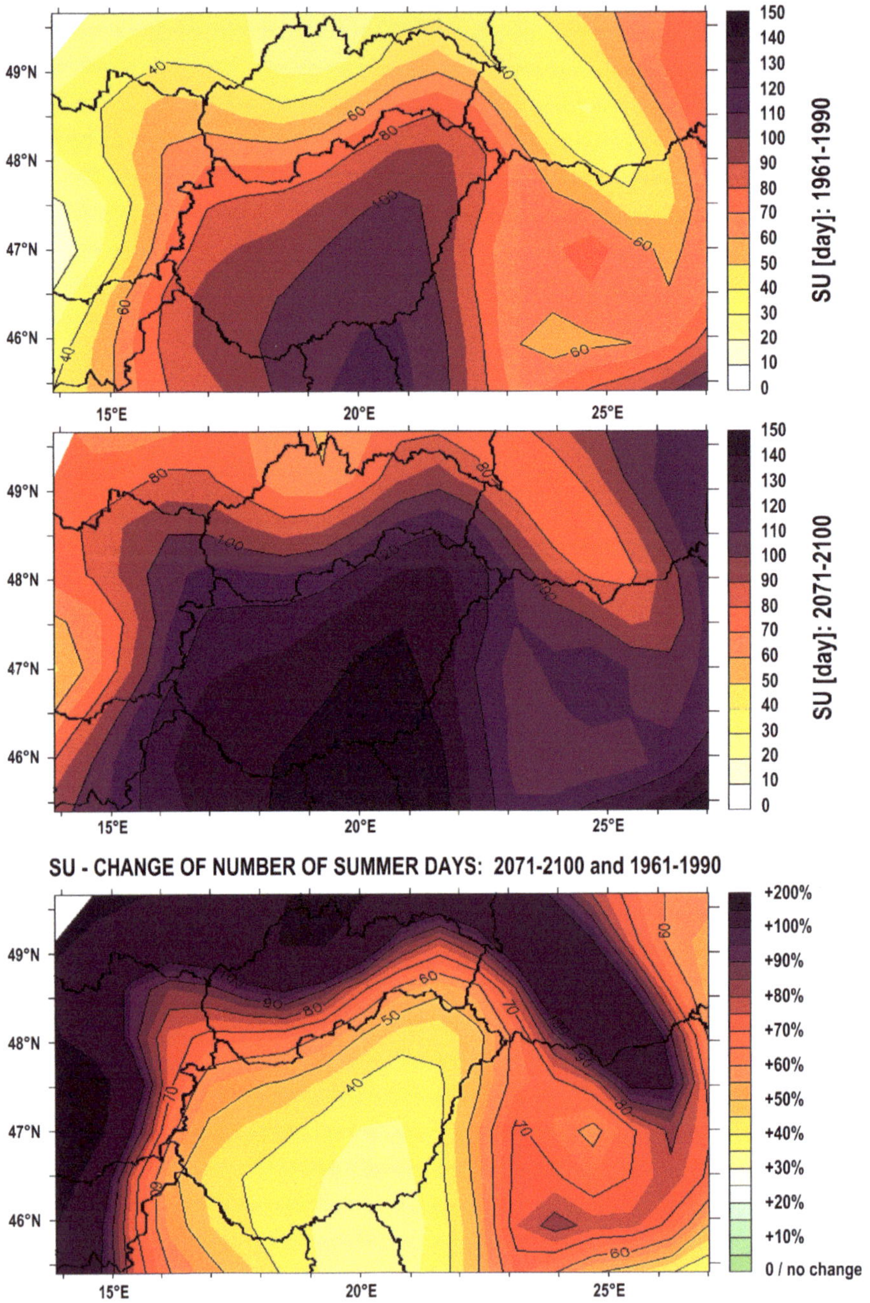

Fig. 6 Expected change of annual number of SU in case of the A2 scenario (2071–2100) compared to the reference period (1961–1990), maps are determined using daily maximum temperature values of the RCM of ETHZ

century. This can be derived from the robust decrease of precipitation indices and the increase of CDD. The largest decrease rates (exceeding 35%) in July are expected in case of the RR0.1, RR1, R75, R95, and RR5.

Detailed comparison of the decadal monthly numbers of precipitation days exceeding 5 mm (RR5) in the periods 1961–1990 and 2071–2100 is shown in Fig. 7 for January and for July using the index values determined for the 10 selected grid points (each of them is assigned to one of the meteorological stations located in the vicinity of the grid point). Regional average changes of the RR5 are −2% (in case of the entire year), +38% (in January), and −39% (in July).

Table 4 Comparison of extreme precipitation indices in the reference period (1961–1990) and in case of the A2 scenario (2071–2100) based on the daily outputs of the RCM of ETHZ. In case of the detected trends, signs in parentheses indicate regional mean coefficients being not significant at 95% level

Extreme precipitation index	Reference period (1961–1990)	A2 scenario (2071–2100)	Expected change			Detected trend (1976–2001)
			Annual	January	July	
CDD (R_{day} < 1 mm)	159 days/year	176 days/year	+10%	−27%	+26%	−
R×1(R_{max})	147 mm	156 mm	+6%	+27%	−17%	−
R×5 ($R_{max,5days}$)	251 mm	252 mm	+0.3%	+18%	−24%	+
SDII (R_{year}/RR1)	68 mm	72 mm	+7%	+15%	−5%	(+)
R95 ($R_{day} \geq R_{95\%,1961-1990}$)	18 days/year	19 days/year	+6%	+55%	−39%	+
R75 ($R_{day} \geq R_{75\%,1961-1990}$)	90 days/year	77 days/year	−14%	+13%	−46%	+
RR20 (R_{day} ≥ 20 mm)	2 days/year	3 days/year	+37%	+308%	−5%	+
RR10 (R_{day} ≥ 10 mm)	11 days/year	12 days/year	+13%	+89%	−28%	+
RR5 (R_{day} ≥ 5 mm)	29 days/year	28 days/year	−2%	+38%	−39%	(−)
RR1 (R_{day} ≥ 1 mm)	77 days/year	67 days/year	−13%	+13%	−45%	−
RR0.1 (R_{day} ≥ 0.1 mm)	133 days/year	114 days/year	−15%	+9%	−47%	−

Although the annual number is not expected to change significantly in any grid point, but the seasonal distribution of RR5 may change very much. The RR5 in January is projected to become about 3–4 times of the RR5 in July.

The difference between the expected changes of RR1 and RR10 is illustrated in Fig. 8 using annual and monthly (January and July) grid point values of the indices. Blue circles in the maps indicate expected decrease, while yellow and red circles imply expected increase. The size of the circles corresponds to the magnitude of the expected changes. In case of the annual change, the expected decreasing rate in the country is about 10–20% in RR1, while the increasing rate of RR10 between 2071–2100 and 1961–1990 is expected to vary between 0 and 40% in Hungary. The largest changes of RR1 are projected in July, namely, 40–60% increase by the end of the twenty-first cen-

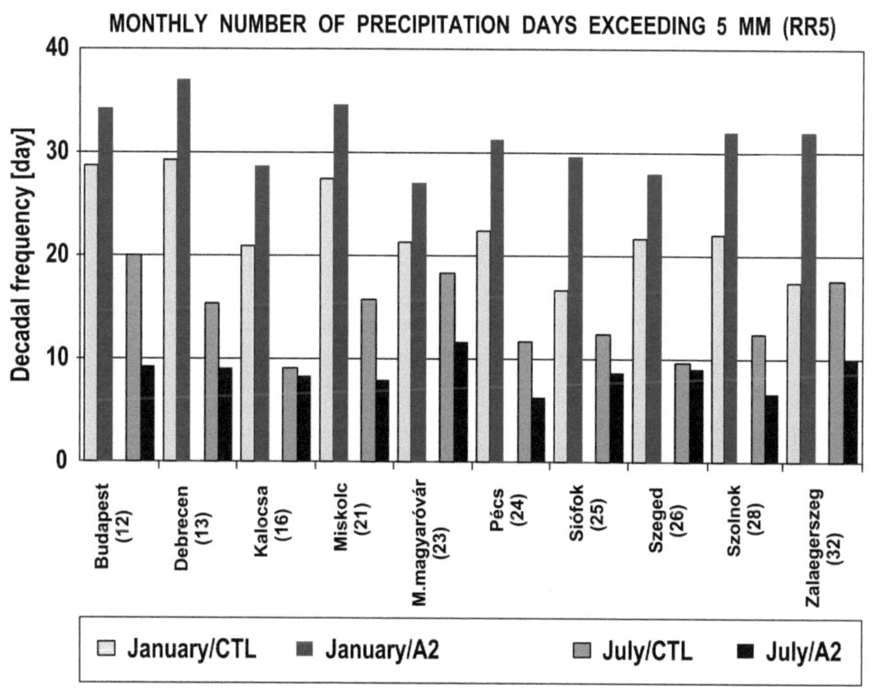

Fig. 7 Comparison of monthly number of precipitation days exceeding 5 mm (RR5) in January and in July, in the reference period (1961–1990, CTL) and in case of the A2 scenario (2071–2100) based on the daily precipitation outputs of the RCM of ETHZ

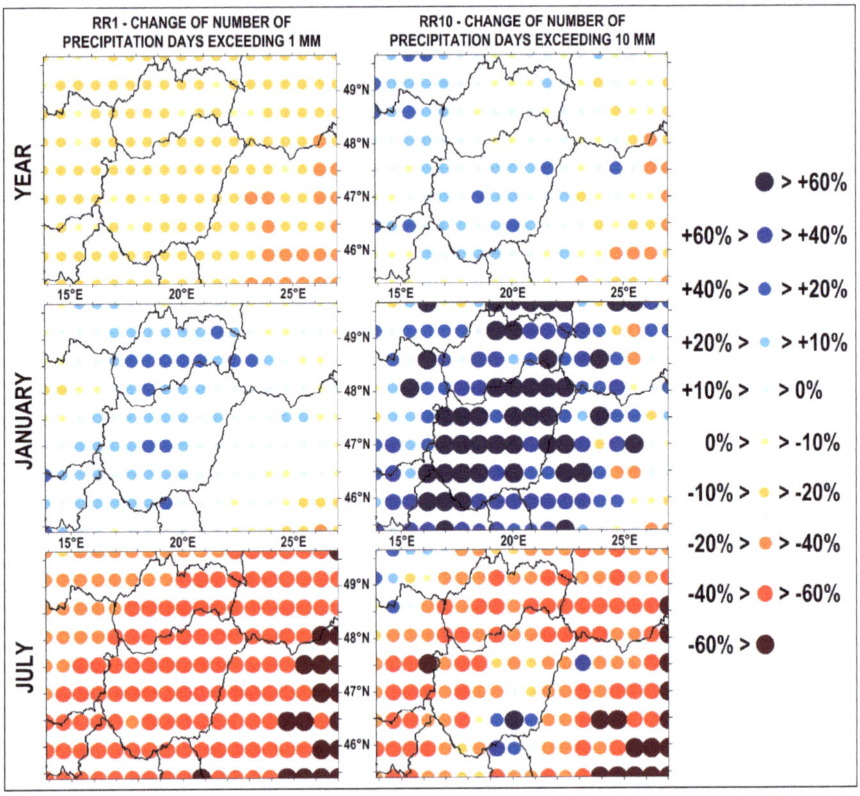

Fig. 8 Expected change of annual and monthly number of precipitation days exceeding 1 and 10 mm (RR1 and RR10, respectively) in case of the A2 scenario (2071–2100) compared to the reference period (1961–1990). Maps are determined using simulated daily precipitation amounts of the RCM of ETHZ

tury. Opposite changes (0–40% increase) can be expected in January. The largest changes of RR10 are expected in January, the projected decrease exceeds 60% in most of the grid points. In July, in general, about 10–60% decrease is expected, except the south-eastern part of the country where the expected increase exceeds 40%.

Conclusions

Analysis of extreme temperature and precipitation indices (according to the suggestions of the WMO-CCl/CLIVAR working group) is presented in this chapter for the second half of the twentieth century and for the last three decades of the twenty-first century (using the A2 scenario). Based on the results, the following conclusions can be drawn for the Carpathian basin:

1. Warming trends are detected in 1961–1975 in case of extreme indices indicating negative temperature extremes (i.e., T×10, Tn10, FD, T×0LT, Tn–10LT), while cooling trends are detected in case of positive temperature extremes (i.e., HWDI, T×90, SU, T×30GE, T×35GE, Tn20GT).
2. Detected trends of the extreme temperature indices imply significant regional warming in 1976–2001.
3. Analysis of simulated daily temperature datasets suggests that the detected regional warming is expected to continue in the Carpathian basin by 2071–2100. The negative temperature extremes are projected to decrease, while positive extremes tend to increase significantly. The largest increase due to the regional warming trend can be expected in case of the T×30GE, T×90, Tn90, T×35GE, and Tn20GT by +91%, +116%, +120%, +250%, and +625%, respectively.
4. In case of the entire 1946–2001 period, only four precipitation indices (i.e., CDD, R×5, RR1, RR0.1) exhibit significant regional mean trend, which imply increasing aridity in the region.
5. In 1976–2001, precipitation occurred more rarely in the Carpathian basin (which is indicated by the decreasing trends of the RR1 and RR0.1), but the

ratio of heavy or extreme precipitation days (RR20, RR10, R95, R75, R95T) increased considerably.

6. Expected changes (for 2071–2100) of annual precipitation indices are small, but generally consistent with the detected trends in 1976–2001.
7. Large positive and negative changes of monthly precipitation indices are projected in January and in July, respectively. Projected increase of very extreme precipitation events exceeds 50% in January, while the expected increases of not extreme precipitation indices do not reach 15%, these results imply that the extreme precipitation events are expected to become more intense and more frequent in January. Furthermore, drought is projected to become more intense in July.

Acknowledgments Research leading to this chapter has been supported by the following sources: the Hungarian Academy of Sciences under the program 2006/TKI/246 titled *Adaptation to Climate Change*, the Hungarian National Research Development Program under grants NKFP-3A/082/2004 and NKFP-6/079/2005, the Hungarian National Science Research Foundation under grants T-049824, K-67626, and K-69164, the Hungarian Academy of Science and the Hungarian Prime Minister's Office under grant 10.025-MeH-IV/3.1/2006, the Hungarian Ministry of Environment and Water under the National Climate Strategy Development project, and the CECILIA project of the European Union Number. 6 program (contract no. GOCE-037005). ESRI software has been used to create maps. Climate change data have been provided through the PRUDENCE data archive, funded by the EU through contract EVK2-CT2001-00132.

References

Bartholy J, Pongrácz R (2005a) Tendencies of extreme climate indices based on daily precipitation in the Carpathian Basin for the 20th century. Időjárás 109, 1–20

Bartholy J, Pongrácz R (2005b) Global and regional tendencies of extreme indices calculated on the base of daily temperature and precipitation for the 20th century. (in Hungarian) Agro-21 40, 70–93

Bartholy J, Pongrácz R (2006) Comparing tendencies of some temperature related extreme indices on global and regional scales. Időjárás 110, 35–48

Bartholy J, Pongrácz R (2007) Regional analysis of extreme temperature and precipitation indices for the Carpathian Basin from 1946 to 2001. Global and Planetary Change 57, 83–95, doi:10.1016/j.gloplacha.2006.11.002

Bartholy J, Pongrácz R, Gelybó Gy (2007a) Regional climate change expected in Hungary for 2071–2100. Applied Ecology and Environmental Research 5, 1–17

Bartholy J, Pongrácz R, Gelybó Gy, Szabó P (2007b) Expected change of temperature extremes in the Carpathian basin by the end of the 21st century. (in Hungarian) Klíma-21 51, 3–17

Christensen JH (2005) Prediction of Regional scenarios and Uncertainties for Defining European Climate change risks and Effects. Final Report. DMI, Copenhagen.

Christensen JH, Christensen OB (2007) A summary of the PRUDENCE model projections of changes in European climate by the end of this century. Climatic Change 81, 7–30

Frich P, Alexander LV, Della-Marta P, Gleason B, Haylock M, Klein Tank, AMG, Peterson T (2002) Observed coherent changes in climatic extremes during the second half of the twentieth century. Climate Research 19, 193–212

IPCC (2001) Climate Change 2001: The Scientific Basis. Contribution of Working Group I to the Third Assessment Report of the Intergovernmental Panel on Climate Change. (eds.: Houghton JT, Ding Y, Griggs DJ, Noguer M, van der Linden PJ, Dai X, Maskell K, Johnson CA), Cambridge University Press, Cambridge, United Kingdom and New York, NY, USA, 881pp

IPCC (2007) Climate Change 2007: The Physical Science Basis. Contribution of Working Group I to the Fourth Assessment Report of the Intergovernmental Panel on Climate Change. (eds.: Solomon S, Qin D, Manning M, Chen Z, Marquis M, Averyt KB, Tignor M, Miller HL), Cambridge University Press, Cambridge, United Kingdom and New York, NY, USA, 996pp

Jacob D, Bärring L, Christensen OB, Christensen JH, de Castro M, Déqué M, Giorgi F, Hagemann S, Hirschi M, Jones R, Kjellström E, Lenderink G, Rockel B, Sánchez E, Schär Ch, Seneviratne SI, Somot S, van Ulden A, van den Hurk B (2007) An inter-comparison of regional climate models for Europe: Model performance in Present-Day Climate. Climatic Change 81, 21–53, doi:10.1007/s10584-006-9213-4

Karl TR, Nicholls N, Ghazi A (1999) Clivar/GCOS/WMO Workshop on Indices and Indicators for Climate Extremes Workshop Summary. Climatic Change 42, 3–7

Klein Tank AMG (2003) The European Climate Assessment and Dataset project. http://www.knmi.nl/samenw/eca/index.html

Klein Tank AMG, Können GP (2003) Trends in Indices of Daily Temperature and Precipitation Extremes in Europe, 1946–99. Journal of Climate 16, 3665–3680

Klein Tank AMG and coauthors (2002a) Daily dataset of 20th-century surface air temperature and precipitation series for the European Climate Assessment. International Journal Of Climatology 22, 1441–1453

Klein Tank AMG, Wijngaard JB, van Engelen A (2002b) Climate of Europe; Assessment of observed daily temperature and precipitation extremes. KNMI, De Bilt, the Netherlands, 36pp

Peterson, T, Folland, CK, Gruza, G, Hogg, W, Mokssit, A, Plummer, N, (2002) Report on the Activities of the Working Group on Climate Change Detection and Related Rapporteurs, 1998–2001. World Meteorological Organisation Rep. WCDMP-47. WMO-TD 1071. Geneva, Switzerland. 143pp

Peterson TC, Vose RS, (1997) An overview of the global historical climatology network database. Bulletin of the American Meteorological Society 78, 2837–2849

Pongrácz R, Bartholy J (2000) Changing trends in climatic extremes in Hungary. (in Hungarian) In: Third Conference on

Forest and Climate. (ed.: Kircsi, A) Kossuth University Press, Debrecen, 38–44

Rowell DP (2005) A scenario of European climate change for the late 21st century: seasonal means and interannual variability. Climate Dynamics 25, 837–849

Trewin BC (1999) The development of a high-quality daily temperature datasets for Australia and implications for the observed frequency of extreme temperatures. In: Meteorology and Oceanography at the Millennium: AMOS'99 Proceedings of the 6th National Australian Meteorological and Oceanographic Society Congress, Canberra, 1999, pp. 87

Vidale PL, Lüthi D, Frei C, Seneviratne SI, Schar C (2003) Predictability and uncertainty in a regional climate model. J. Geophys. Res. 108 (D18), 4586, doi:10.1029/2002JD002810

Precipitation Trend Analysis for Central Eastern Germany 1851–2006

S. Hänsel, S. Petzold and J. Matschullat

Keywords Climate change · Trend analysis · Central Europe

Introduction

Rising global temperatures are accompanied by precipitation changes. This relates to both total precipitation amounts, where an increase has to be expected, and to changing precipitation patterns. The latter being more spatially and seasonally variable than temperature change (IPCC 2007). Northern Europe is one of the regions that have become significantly wetter during the last century (IPCC 2007). In Germany, annual precipitation totals increased by just under 10% from 1901 to 2000 (Schönwiese and Janoschitz 2005). These increases are normally lower in Eastern than in Western Germany. Especially in Saxony, the annual precipitation trends are close to zero; probably related to the increasing continentality of the climate.

Due to altered circulation patterns, a redistribution of precipitation within the year became apparent in different parts of Germany (Bernhofer et al. 2001, 2002; Freydank 2001; Hänsel et al. 2005; Schönwiese and Rapp 1997) with drier summer and wetter winter half years. There is also evidence for changes in the precipitation variance (Jonas et al. 2005; Trömel 2005).

This study focuses on changes in precipitation characteristics in the region of central Eastern Germany. The precipitation shift from the summer to the winter months was quantified. Next to general studies regarding changes in the annual precipitation regime, an analysis concerning the precipitation classes with major changes was performed. A trend comparison of different time intervals allowed qualifying the temporal representativeness of the trends.

Materials and Methods

The analysed monthly precipitation records were taken from the Saxon climate database, developed within the project CLISAX (Statistical Assessment of Regional Climate Trends in Saxony) for the Saxon State Agency for Environment and Geology (Franke et al. 2004). Comprehensive data verification was performed within the CLISAX project, and selected records were already analysed with respect to their homogeneity.

For the time frame 1951–2006, more than 200 stations with monthly precipitation records were analysed for their general precipitation characteristics as well as variations and changes in these characteristics. Some analysis used regional data. For this purpose, the stations were grouped into nine regions with similar natural landscapes and precipitation characteristics (Table 1). Station grouping was supported by correlation and cluster analysis.

These nine sub-regions represent low-lying areas from about 100 m a.s.l. to mountainous areas with average elevations above 500 m a.s.l., and average annual precipitation ranging from below 600 mm to more than 900 mm.

The total number of stations in each sub-region is representative. When looking at the availability and homogeneity of the data, however, it becomes obvious

S. Hänsel (✉)
Technical University, Bergakademie Freiberg, Interdisciplinary Environmental Research Centre, Brennhausgasse 14, D-09599 Freiberg, Germany
e-mail: stephanie.haensel@email.de

K. Strelcová et al. (eds.), *Bioclimatology and Natural Hazards*,
DOI 10.1007/978-1-4020-8876-6_3,

Table 1 Characterisation of the nine regions with average altitude and annual precipitation

Region (number/name)	Number of rain gauge stations	Average altitude (m)	Average annual precipitation for 1951–2000 (mm)
1/Thuringian-Franconian Mountains	13	559	805
2/Vogtland and Thuringian Basin	28	422	697
3/Western Erzgebirge	41	573	916
4/Eastern Erzgebirge	27	446	827
5/Erzgebirge Foreland	14	305	728
6/Western Saxon Hilly Country and Central German Black Earth Area	23	153	572
7/Eastern Saxon Hilly Country	23	187	643
8/Elbe-Mulde Lowlands	24	103	571
9/Lausitz and Spreewald	45	208	687

Table 2 Number of stations with at least 90% data availability for individual periods

Region	1901–2000	1901–2006	1931–2000	1931–2006	1951–2000	1951–2006
1	0	0	5	5	11	11
2	1	1	4	4	27	24
3	2	2	7	7	20	17
4	2	1	4	3	14	9
5	2	2	2	2	39	34
6	2	1	10	7	27	24
7	2	2	3	3	23	19
8	1	1	5	5	21	19
9	3	3	11	7	41	38
Total	15	13	51	43	223	195

that a statistically relevant number of stations are available only as of the early 1950s. Before, only individual stations deliver nearly undisturbed data series from 1901 onwards (Table 2).

Additionally two stations with monthly precipitation records were analysed for the time frame 1851–2006. Those stations are Görlitz in Eastern Saxony and Jena Sternwarte in Thuringia. Jena belongs to the subregion Thuringian Basin and Görlitz to the Lausitz region. Both stations are quite similar in their latitude, altitude and average annual precipitation, but Jena is situated in the west and Görlitz in the east of the study area (Table 3). The data were taken partly (1851–1951) from Anonymus (1961) and were completed by data from the CLISAX database.

Linear regression was used for the trend analysis. Because of missing normal distribution of the data, an assumption for linear regression, the non-parametric Mann–Kendall trend test was performed in addition. This test also delivers information about the significance of the trend (Table 4).

The monthly precipitation distributions of different time intervals as well as their statistical parameters were compared with several tests. The *t*-test was used to compare the means of the distributions. This test has been developed to determine whether the difference between the two means equals 0.0, versus the alternative hypothesis that the difference does not equal 0.0. The *t*-test depends on a normal distribution of the samples. Furthermore, the test assumes that the variances of the

Table 3 Characterisation of the two long-term study sites

Station	Latitude (°min)	Longitude (°min)	Altitude (m)	Average precipitation in 1951–2000 (mm)
Jena Sternwarte	50 56	11 35	155	600
Görlitz	51 10	14 57	211	665

Table 4 Error probability α and confidence limits C of the Mann–Kendall trend test value Q (Rapp 2000)

Q	C	α	Level of significance
>1.282	>80%	<0.2	Low
>1.645	>90%	<0.1	
>1.960	> 95%	<0.05	Medium
>2.576	>99%	<0.01	
>3.290	>99.9%	<0.001	High

two samples are equal. Because of missing normal distribution of the monthly precipitation data, the medians were additionally compared using the Mann–Whitney U-test. This test is constructed by combining the two samples, sorting the data from smallest to largest and comparing the average ranks of the two samples in the combined data.

The distributions themselves were compared with the Kolmogorov–Smirnov test. This test is performed by computing the maximum distance between the cumulative distributions of the two samples.

Results and Discussion

Long-term precipitation trends from 1851 to 2006 were analysed for the two stations Jena Sternwarte and Görlitz. For this period, Jena shows a significant annual precipitation increase, whereas the annual trends for the shorter intervals are smaller and non-significant (Fig. 1a; Table 5). At the station Görlitz, all periods show negative trends, but only the one for 1901–2006 is significant according to the Mann–Kendall trend test (Fig. 1b; Table 5).

When comparing the Mann–Kendall trends for different time periods it becomes visible that both stations behave very similar within the last time slice of 1951–2006 (Table 5). Here, they show distinct precipitation increases during the winter and decreases during the summer half year. The direction of the season trends is the same, but Jena is characterised by a slight significant ($\alpha = 0.2$) precipitation increase in autumn and Görlitz by a significant negative spring trend ($\alpha = 0.1$). The precipitation decrease is most pronounced at both stations during the first vegetation period (April–June) and a progression of this trend into the future might enhance the water stress in natural ecosystems and irrigation problems in agriculture. The availability and quality of drinking water might become another problem. When looking at monthly trends, May (for Görlitz) and June (for Jena) show the most pronounced precipitation decreases and November the highest precipitation increases for the period 1951–2006.

Furthermore, it becomes apparent that the station Görlitz, situated farthest to the east, shows more negative precipitation trends than Jena in all time intervals and seasons (Table 5). It seems that the precipitation deficit increases with rising continentality (increasing longitude). Especially in the period 1901–2006, precipitation has decreased at Görlitz in all seasons and months, except for March.

The precipitation record of Jena for the period 1851–2006 shows positive trends in almost all seasons, except for slight precipitation decreases in June, July, October and the summer season. Most pronounced are the precipitation increases in the winter months January and December and accordingly in the winter season and half year.

Changes in the annual precipitation patterns were analysed not only for these two stations with long-term records but also for more than 200 stations in the 9 different sub-regions in central Eastern Germany. Analysis was done for the half years, the seasons, as well as on a monthly basis. For the time slice 1951–2006, the results of the regional analysis are quite similar to the station trends of the two long-term study sites Görlitz and Jena, suggesting a fine match.

Regarding the half years, the winter shows a positive precipitation trend in all regions, whereas precipitation decreased in the summer half year for the

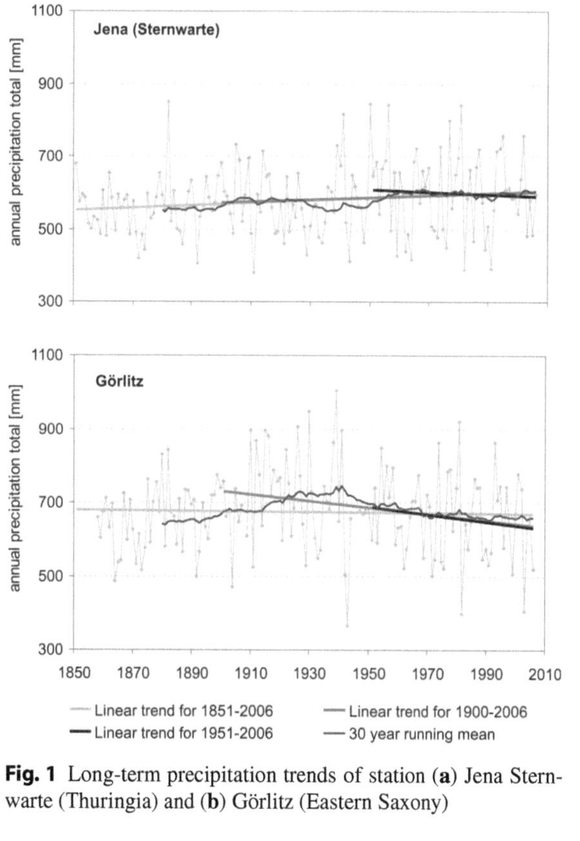

Fig. 1 Long-term precipitation trends of station (**a**) Jena Sternwarte (Thuringia) and (**b**) Görlitz (Eastern Saxony)

Table 5 Mann–Kendall trends for different time periods for the stations Jena and Görlitz

	Jena			Görlitz		
	1851—2006	1901—2006	1951—2006	1851—2006	1901—2006	1951—2006
Year	1.88	1.02	-0.25	0.23	-1.90	-0.66
SHY	0.57	-0.26	-1.30	-0.90	-1.60	-2.04
WHY	3.02	2.52	0.98	1.45	-0.95	1.41
Spring	1.61	1.52	-0.10	-0.66	-1.75	-1.79
Summer	-0.64	-0.31	-0.60	-1.26	-1.28	-1.22
Autumn	0.59	0.23	1.29	0.54	-1.09	0.30
Winter	3.22	1.84	0.23	1.50	-1.21	0.48
January	2.42	-0.45	-1.18	1.53	-1.40	0.00
February	1.57	1.16	1.13	0.22	-0.73	0.38
March	1.28	1.88	1.34	-0.29	0.31	0.91
April	1.48	0.13	-1.10	-0.40	-1.36	-0.83
May	0.27	0.55	-0.56	-0.31	-1.25	-2.08
June	-0.27	0.47	-2.66	-0.97	-0.99	-0.59
July	-0.70	-1.20	1.16	-1.37	-1.03	-1.07
August	0.33	0.36	-0.03	-0.59	-1.14	0.69
September	0.46	-0.82	0.24	-0.18	-0.47	0.57
October	-0.48	-0.44	-0.14	-0.59	-1.29	-0.21
November	1.53	2.16	2.78	1.37	-0.27	1.27
December	1.79	1.56	0.41	0.42	-0.63	-0.24

significance level:	positive trends:	negative trends:
< 0.2		
0.2		
0.1		
0.05		
0.01		

time frames 1951–2006 and 1931–2006 (Fig. 2). The summer trends of the two different time intervals are quite similar within the regions. The winter trends of the shorter period 1951–2006 are generally higher than those of the longer interval 1931–2006. The 1930s have been a quite wet period and thus suppressed the positive precipitation trend.

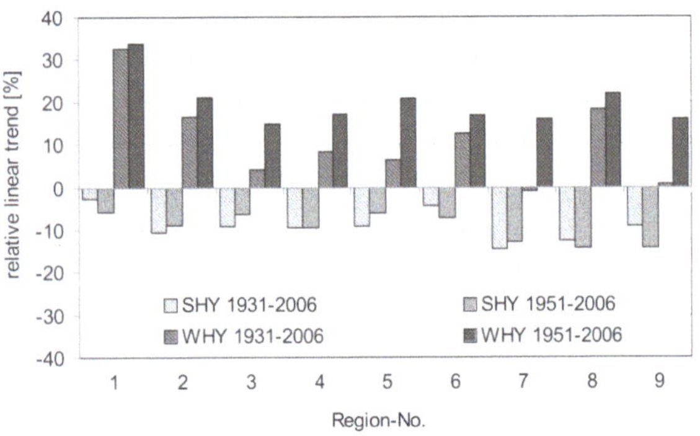

Fig. 2 Average half year precipitation trends of nine sub-regions for the periods 1931–2006 and 1951–2006 (SHY, summer half year; WHY, winter half year)

A distinct annual precipitation increase could be monitored for the Thuringian-Franconian Mountains only (region 1) – for both periods (Fig. 2; Table 6). The annual trends of the other regions are quite indifferent as are the spring trends of all regions. Those indifferent spring trends are a result of the different trend directions in the spring months with very high precipitation increases in March and strong negative trends in April and May. In summer, all regions show an average precipitation decrease that is most pronounced in the agriculture-dominated regions – the Elbe-Mulde Lowlands in the north (8) – and the Lausitz and Spreewald (9) in the east of the study area (Table 6). Autumn and winter became wetter in all regions with the highest increases in the Thuringian-Franconian Mountains (1).

When looking at monthly trends, March and November stand out by very high precipitation

Table 6 Monthly, seasonal and annual precipitation trends of nine different regions for the period 1951–2006

	Region-No.								
	1	2	3	4	5	6	7	8	9
Year	13	4	3	2	7	3	0	2	-1
Summer half year	-6	-9	-6	-9	-6	-7	-13	-14	-14
Winter half year	34	21	15	17	21	17	16	22	16
Spring	3	-2	-5	-3	2	5	-3	7	-1
Summer	-7	-9	-2	-6	-4	-10	-11	-18	-15
Autumn	30	20	15	3	13	8	5	2	4
Winter	30	13	8	19	18	17	16	25	16
January	33	-2	-8	6	0	-3	5	14	9
February	33	18	15	30	28	21	22	27	24
March	46	39	33	33	38	37	34	48	41
April	-21	-28	-30	-37	-29	-6	-30	-13	-29
May	-14	-13	-16	-4	-2	-9	-11	-10	-11
June	-16	-27	-12	-15	-19	-33	-20	-35	-18
July	0	-7	-27	-38	-17	6	-31	-16	-38
August	-5	7	41	39	26	-2	22	-3	12
September	21	12	8	-7	4	4	-8	-4	-2
October	30	-8	-18	-29	-17	-32	-25	-25	-20
November	38	57	56	46	57	53	50	33	32
December	20	22	19	19	23	30	19	25	12

linear Trend: 0 to 5%; 5% to 15%; 15% to 25%; > 25%; 0 to -5%; -5% to -15%; -15% to -25%; < -25%

increases between 1951 and 2006 (Table 6). In all regions precipitation also increased in February and December. This becomes also visible on a station level. Almost all stations show positive precipitation trends above 5% in March and November and in February and December; only few stations show negative or just small trends (Fig. 3).

During the first vegetation period (April–June), and in most regions also in July and October, the precipitation decreases are most pronounced (Table 6). This is also true on a station basis. Figure 3 shows the highest percentage of stations with negative precipitation trends in the months April–July and October. The Thuringian-Franconian Mountains (region 1) in the southwest of the study area is the only region with

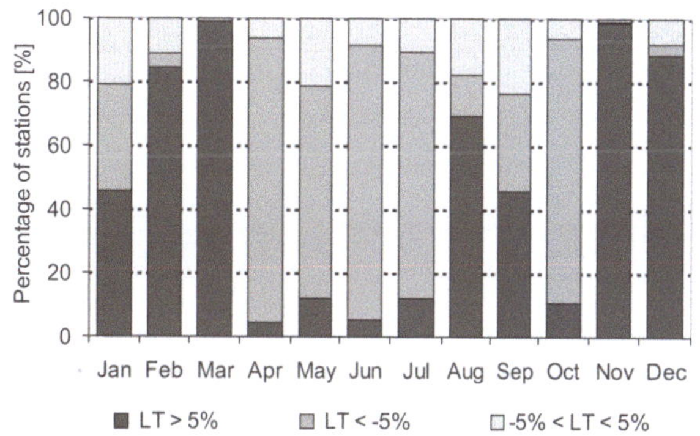

Fig. 3 Spatial homogeneity of monthly precipitation trends (1951–2006)

a strong positive October trend (Table 6). This region generally behaves different from the other regions.

An overview on the temporal and spatial representativeness of the precipitation trends is given in Fig. 4. It shows the trend direction for each 30-year period from 1901–1930 to 1977–2006 for all stations within the 9 regions. Three months have been chosen to display different trend signatures. These are May, as an example for decreasing precipitation trends in 1951–2006 in all regions; August, as an example for indifferent regional trends; and November with a strong precipitation increase.

In May, most stations show negative 30-year trends as of 1955 – single stations already since 1951. Only in the Thuringian-Franconian Mountains (region 1) and Vogtland and the Thuringian Basin (region 2), the negative trends start in the 1960s. Since about 1940, precipitation increases have been observed for all regions.

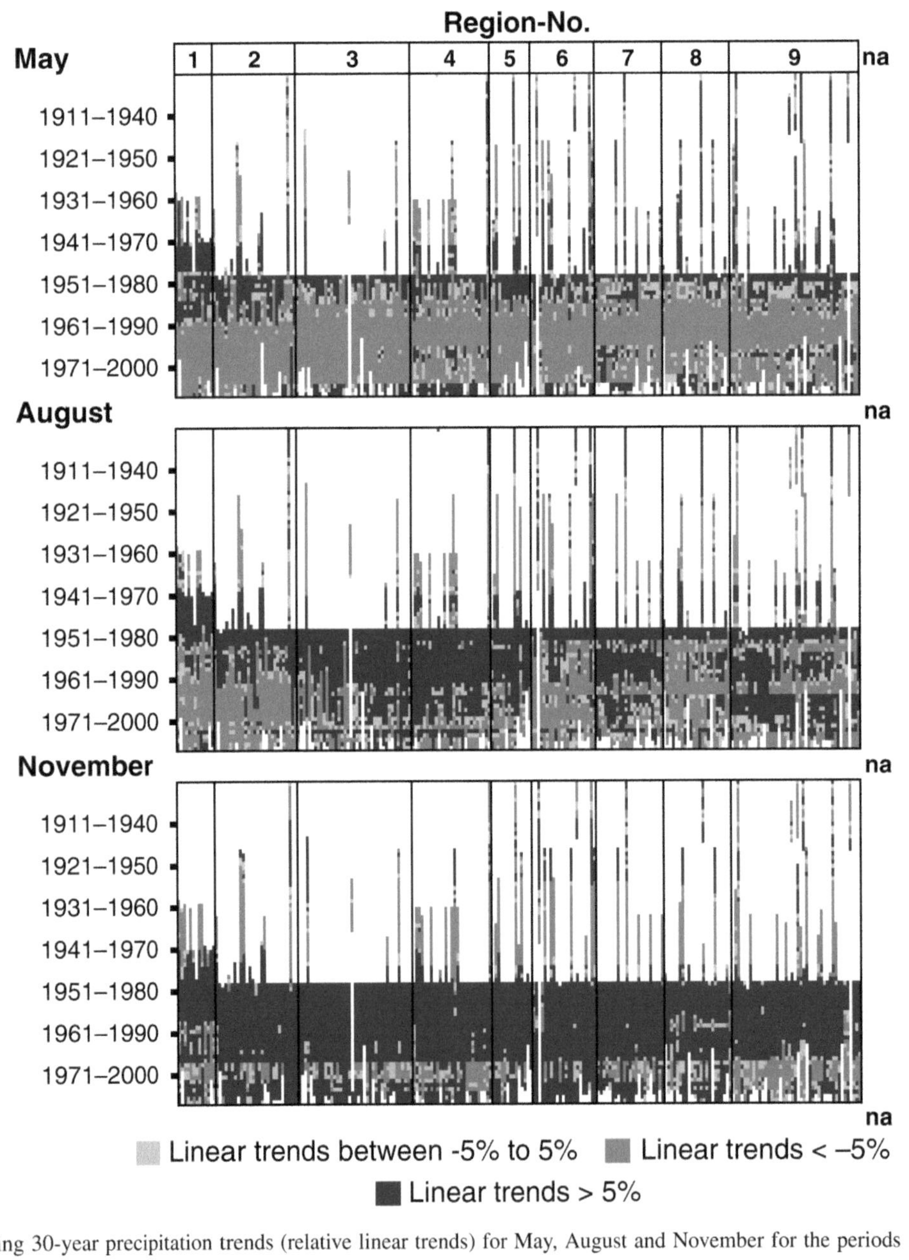

Fig. 4 Running 30-year precipitation trends (relative linear trends) for May, August and November for the periods 1901–1930 to 1977–2006

Before that time, the trend behaviour of the different regions does not give a clear picture. At the end of the studied time frame, the trends in almost all regions switch to positive signs.

August shows predominantly precipitation decreases since 1951 in the regions 1 (Thuringian-Franconian Mountains), 2 (Vogtland and Thuringian Basin), 6 (Western Saxon Hilly Country) and 8 (Elbe-Mulde Lowlands), but the trends are not as temporally stable as in May (Fig. 4). Those four regions extend from the southwest to the north of the study area. The other regions from the Erzgebirge (regions 3 and 4) in the south to region 9 (Lausitz and Spreewald) in the northeast of study area show predominantly precipitation increases since the 1940s. Since about 1910 until the 1940, negative 30-year trends dominate the picture.

For November, the 30-year trends are mostly homogeneous within all regions in these three exemplary months (Fig. 4). For the time intervals 1921–1950 to the 30-year trends, starting in the 1940s, precipitation decreases have been observed in all regions. Positive trends occur predominately since that time and until the end of the study time frame – interrupted by a narrow band of small or negative trends at about the interval 1971–2000.

All stations show positive as well as negative trends for all months. This reflects the natural variability of precipitation and restricts the significance of short-term precipitation trends. In many intervals, especially in May and November, those trend directions are identical for most stations. The spatial representativeness of the trends seems to be quite high. These patterns suggest that there exist some larger scale precipitation trigger(s).

According to the monthly trend patterns, the regions 1 (Thuringian-Franconian Mountains) and 2 (Vogtland and Thuringian Basin) behave similarly. The trend patterns of the Eastern Erzgebirge (region 4) are in some months, like May, more similar to the Eastern Saxon Hilly Country (region 7) than to the Western Erzgebirge (region 3) which itself shows similarities to the Erzgebirge Foreland (region 5) and the Western Saxon Hilly Country (region 6). Generally, these trend patterns support the chosen classification into sub-regions.

The precipitation cycles of the nine different regions have to be characterised before looking at changes in the annual precipitation cycles. The regions show quite similar annual precipitation cycles at different precipitation levels, with a precipitation maximum in summer.

The highest monthly precipitation averages occur in most regions in July, showing a distinct peak (Fig. 5). A second smaller precipitation maximum occurred in December or January. Corresponding to the two maxima, the annual precipitation cycle shows two minima – one in February and the other in October or November.

Figure 5 shows changes within the annual precipitation cycle, using region 1 (the Thuringian-Franconian Mountains) and region 7 (the Eastern Saxon Hilly Country) as examples. Displayed are the average precipitation sums of the period 1951–1975 compared to the period 1976–2000. The summer maximum of the annual precipitation cycle seems to have shifted to later times in the year at region 7 (Fig. 5b), due to decreasing precipitation averages in May, June and July and increasing ones in August. It stretches over two months with almost equal rainfall sums

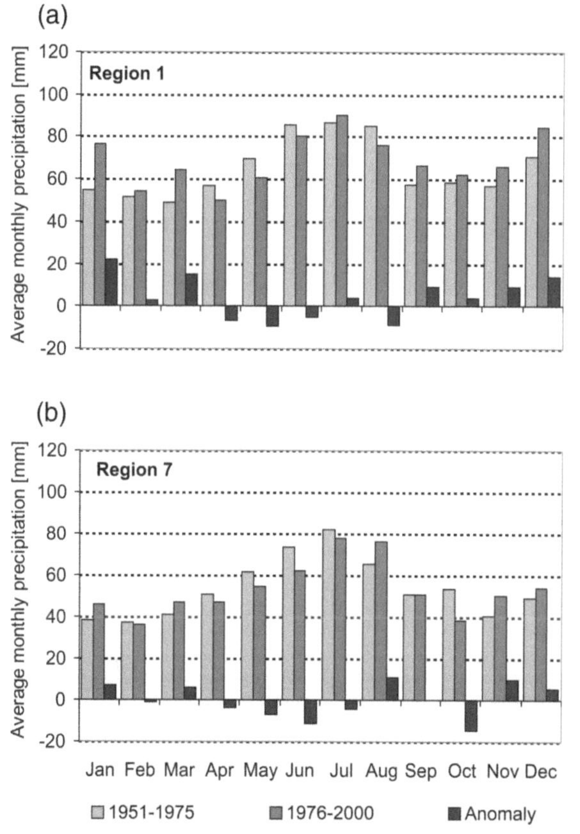

Fig. 5 Changes in the annual precipitation cycle for region 1 (Thuringian-Franconian Mountains) and region 7 (Eastern Saxon Hilly Country) for 1951–1975 compared to 1976–2000

(1976–2000), whereas July clearly showed the highest precipitation amounts in 1951–1975. The autumn minimum shifted from November to October. The winter maximum in December increased, but is still considerably smaller than the summer maximum, while the strong winter precipitation increases in region 1 led to a winter maximum of almost the same value as the summer maximum (Fig. 5a). Primarily region 1 is discriminated from the other regions by a higher summer maximum in July as well as a higher autumn minimum in the second period compared to 1976–2000. A third small precipitation maximum in March and a minimum in April became visible in some regions like region 1. In other regions like region 7, the precipitation totals of March and April are almost equal. The shifted increase to the summer maximum from early to late summer and the higher winter maximum in period 1976–2000 compared to 1951–1975 are common trend characteristics of all regions.

Changes in the frequency distribution of monthly rainfall totals became visible (Fig. 6) next to the changes in the annual precipitation cycle (Fig. 5). A general difference between the distribution of the summer and winter half year months can be seen. The distribution of the precipitation totals of the summer months is more expanded than the one of the winter months (Fig. 6a). The highest frequency of monthly precipitation totals is reached in class 40–60 mm for the summer half year and class 20–40 mm for the winter half year.

The frequency distribution of the summer half year shifted to smaller classes; small precipitation classes became more frequent and those near or beyond normal conditions became less frequent. The winter precipitation distribution has shifted in the inverse direction with a decreasing frequency in small precipitation classes and an increase in the frequency of high-rainfall classes.

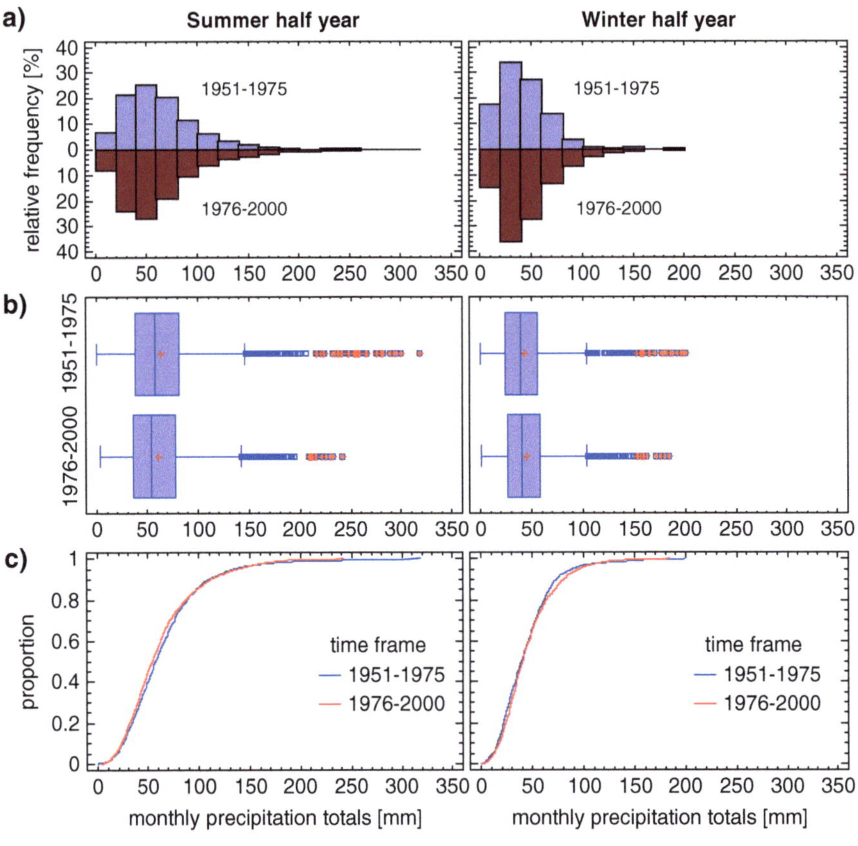

Fig. 6 Changes in (**a**) the frequency distribution of the monthly precipitation totals, (**b**) the box and whisker plots, and (**c**) the quantile plots of the summer and the winter half years for region 7 (Eastern Saxon Hilly Country) for 1951–1975 compared to 1976–2000

The box and whisker plots (Fig. 6b) also show those distribution shifts. The mean, the median and the 50th percentile of monthly precipitation totals for 1951–1975 within region 7 is higher for the summer half year and lower for the winter half year as compared to 1976–2000. The first period shows noticeably more and larger outliners than 1976–2000, true for both half years. The described changes in mean, median and the distributions themselves are significant for region 7 (Table 7).

The other regions behave similar to region 7 with most of them showing different distributions in the two time intervals 1951–1975 and 1976–2000; especially in the winter half year with high significance. For the summer half year just the changes in mean, median and distribution of the regions 1 (Thuringian-Franconian Mountains) and 5 (Erzgebirge Foreland) are non-significant, whereas the Erzgebirge regions 3–5 show non-significant trends during the winter half year with equal medians in regions 4 and 5 and a non-significant increase in mean in region 3.

The results of the *t*-test for comparing the means have to be interpreted with care since the condition of normal distribution is not met by the data. Since the test results for the significance of the medians are similar to those of the means one can assume that there are significant changes in the mean of the distributions in most regions.

Conclusions

The changes in precipitation characteristics and patterns of nine sub-regions in central Eastern Germany were analysed, based on a comprehensive database of more than 200 official rain gauge stations. A precipitation trend analysis with high spatial resolution was possible for the time frame 1951–2006 only. Prior to 1951, only single stations deliver nearly undisturbed data. Additionally, two long-term precipitation records were analysed for the period 1851–2006. Those records show different trend characteristics, especially the long-term trends for 1851–2006 and 1901–2006 differ from each other. The monthly and seasonal trends for the period 1951–2006 are quite similar to each other as well as to those of the other analysed stations. The differences could be a result of the different maritime and continental influences. The station Jena Sternwarte, situated in the west of the study area, shows predominantly precipitation increases. At the more continentally influenced station Görlitz, at the

Table 7 Test statistics for the comparison of the monthly rainfall frequency distributions of the half year for the time intervals I (1951–1975) and II (1976–2000)

Region number	Mean (mm)		*p*-value	Median (mm)		*p*-value	ND	*p*-value
Summer half year	I	II	(*t*-test)	I	II	(*U*-test)		(KS-test)
1	73.5	70.6	0.02	65.8	63.4	0.17	No	0.111
2	70.0	66.9	<0.001	63.0	61.9	0.025	No	0.055
3	88.4	84.2	<0.001	80.0	77.0	0.004	No	<0.001
4	80.0	77.4	0.012	71.0	68.9	0.04	No	0.008
5	71.4	70.2	0.314	64.0	65.0	0.757	No	0.12
6	58.4	55.0	<0.001	52.7	50.2	<0.001	No	<0.001
7	64.1	61.5	0.005	58.0	54.0	<0.001	No	<0.001
8	57.4	52.1	<0.001	51.6	48.0	<0.001	No	<0.001
9	67.0	63.9	<0.001	59.0	56.0	<0.001	No	<0.001
Winter half year								
1	56.8	67.9	<0.001	48.4	58.7	<0.001	No	<0.001
2	45.6	49.7	<0.001	40.8	45.0	<0.001	No	<0.001
3	66.1	67.0	0.194	59.0	58.0	0.099	No	<0.001
4	57.8	60.3	0.002	52.0	52.0		No	<0.001
5	49.3	51.8	0.007	45.0	45.0		No	<0.001
6	37.5	40.6	<0.001	34.0	36.6	<0.001	No	<0.001
7	43.3	45.3	0.001	39.0	40.0	<0.001	No	<0.001
8	39.3	42.2	<0.001	36.0	38.5	<0.001	No	<0.001
9	47.3	49.8	<0.001	42.0	43.2	<0.001	No	<0.001

ND, normal distribution; *KS*, Kolmogorov–Smirnov

border of Poland, precipitation decreased in many months and seasons. This fits the observations of Schönwiese and Rapp (1997), who found strong annual precipitation increases in Western Germany and just slight changes in Eastern Germany.

When looking at the annual precipitation cycles – that are quite similar for the different regions – a distinct redistribution of precipitation becomes visible. Regarding the half years, winter shows positive precipitation trends in all regions, whereas precipitation decreased in the summer half year (time frames: 1951–2006 and 1931–2006). This summer precipitation decrease is especially high in the agriculture-dominated lowlands in the north of the study area. A continuation of this negative summer precipitation trend would not only increase the irrigation challenge in agriculture but also lead to changes in the natural vegetation and may enhance problems in the availability and quality of drinking water. Since the precipitation decreases are highest during the first vegetation period (April–June), water stress and productivity are major problems for natural as well as agricultural ecosystems.

Due to opposite monthly trends, spring and autumn show indifferent trend behaviour. While March and November stand out by very high precipitation increases in 1951–2006, April, May and October are characterised by pronounced precipitation decreases. These reversed monthly trends also appear in shifts in the annual precipitation cycle. The summer maximum has shifted from early to late summer and the autumn minimum from November to October. Furthermore, the negative summer and positive winter trends lead in some regions to an equalisation of the summer and winter maximum of the annual precipitation cycle that has been in the past characterised by a summer rain season.

Highly significant changes in the frequency distributions could be monitored in most regions. The frequency in high precipitation classes increased for the months of the winter half year, while the distribution shifted to smaller classes for the summer months.

The natural variability of precipitation is reflected by the moving 30-year trends that show variable trend reversals in all months and seasons. These point out the limited validity of short-term trends. A simple extrapolation of linear trends in the future is quite possible but not reasonable. Since the trend reversals occur at similar times in almost all regions there must exist some larger scale triggers of precipitation. Furthermore, it shows a high spatial representativeness of the trends.

References

Anonymus (1961) Klimatologische Normalwerte für das Gebiet der Deutschen Demokratischen Republik (1901–1950)/2. Lieferung, Meteorologischer Dienst der DDR (in German)

Bernhofer C, Queck R, Schneider F (2001) CLISAX - Statistische Untersuchungen regionaler Klimatrends in Sachsen. Zwischenbericht 1 zum Forschungs- und Entwicklungsvorhaben des Sächsischen Landesamtes für Umwelt und Geologie; TU Dresden in Tharandt

Bernhofer C, Goldberg V. Frank J (2002) CLISAX II - Assimilation von standardisierten und abgeleiteten Klimadaten für die Region Sachsen und Ausbau der Sächsischen Klimadatenbank. Vorläufiger Abschlussbericht zum Forschungs- und Entwicklungsvorhaben des Sächsischen Landesamtes für Umwelt und Geologie, AZ 13-3-9902.2521/50; TU Dresden in Tharandt

Franke J, Goldberg V, Eichelmann U, Freydank E, Bernhofer C (2004) Statistical analysis of regional climate trends in Saxony, Germany. Clim Res 47: 145–150

Freydank E (2001) Statistische Untersuchungen Regionaler Klimatrends in Sachsen. DWD, Regionales Gutachterbüro Dresden-Radebeul (in German)

Hänsel S, Küchler W, Matschullat J (2005) Regionaler Klimawandel Sachsen. Extreme Niederschlagsereignisse und Trockenperioden 1934–2000. UWSF – Z Umweltchem Ökotox 17/3: 159–165 (in German)

IPCC (2007) Climate Change 2007: The Physical Science Basis. Contribution of Working Group I to the Fourth Assessment Report of the Intergovernmental Panel on Climate Change [Solomon, S., D. Qin, M. Manning, Z. Chen, M. Marquis, K.B. Averyt, M. Tignor and H.L. Miller (eds.)]. Cambridge University Press, Cambridge, United Kingdom and New York, NY, USA, 996 pp.

Jonas M, Staeger T, Schönwiese C-D (2005) Berechnung der Wahrscheinlichkeiten für das Eintreten von Extremereignissen durch Klimaänderungen – Schwerpunkt Deutschland. Bericht zum UBA-Forschungsvorhaben 201 41 254; Bericht Nr. 1, Inst. Atmosphäre Umwelt, Univ. Frankfurt/Main (in German)

Rapp J (2000) Konzeption, Problematik und Ergebnisse klimatologischer Trendanalysen für Europa und Deutschland. Berichte des DWD 212 Selbstverlag des DWD Offenbach am Main (in German)

Schönwiese C-D, Janoschitz R (2005) Klima-Trendatlas Deutschland 1901–2000. Bericht Nr. 4, Inst. Atmosphäre Umwelt, Univ. Frankfurt/Main (in German)

Schönwiese C-D, Rapp J (1997) Climate Trend Atlas of Europe Based on Observations 1891–1990. Kluwer Academic Publ., Dordrecht

Trömel S (2005) Statistische Modellierung von Klimazeitreihen. Dissertation; Bericht Nr. 2, Inst. Atmosphäre Umwelt, Univ. Frankfurt/Main (in German)

Some Facts on Extreme Weather Events Analysis in Slovakia

M. Lapin, I. Damborská, P. Faško, L. Gaál and M. Melo

Keywords Extremes · Extreme events · Statistical elaboration · Climate change scenarios

Introduction

The issue of climate change has been presented mostly as a change in long-term means, mainly in terms of air temperature and precipitation totals. In the IPCC (2001) report there was for the first time a discussion on a possible change in the shape of distribution curves of individual climatic phenomena, specifically the change in probability densities in the time series from different periods. The basic alternatives of shift of distribution curves are presented in Fig. 1. It is obvious that besides the shift of mean, central values, the change of values at the left and right margins of distribution curve (extreme design values) can be considered as a very important aspect.

The presented simplified schemes clearly show that there is a chance for increase of population variability, that is, increase of extremes at both margins of distribution curve. Most of the human socioeconomic activities as well as the processes in the natural ecosystems have been adjusted to the natural climatic variability that occurred in several past decades. It means that most of them may cope with the occurrence of some extreme values without any significant difficulties. Direct measurements and observations confirmed that the return period of 50 years is probably the limit for a natural adaptation for the majority of biological species. Also, human activities fulfill the limit of 50-year return period. On the other hand, this time roughly corresponds to the mean active period of individual human's existence. Therefore, people should be prepared to master extreme problems naturally occurring during one's active life. The complex of natural ecosystems is probably adapting to the weather and climate variability in a more complicated way. The genetic memory of species depends on many factors: some of them can adapt more quickly and others require a very long time. It is anticipated that the return period of 50 years suits well also in this case (IPCC 1998, 2001, 2007).

The statistical analysis of time series based on meteorological as well as hydrological observations enables us to design and to estimate certain values in the distribution curves (e.g., probability of occurrence $p = 0.02$ and $p = 0.98$) only when some conditions and limits are met. We consider as important the following four preconditions:

1. The time series of observations have to fulfill the precondition of temporal homogeneity from the aspect of observing methods and representativeness. Observations and measurements need to be taken at the same place, no significant change in the surroundings of the observing site should have occurred, the methods of observations should not have changed, and measuring instruments should be comparable. In this case, all the detected inhomogeneities may be explained by climatologic causes, such as temporarily changed atmospheric circulation, temporal trends of some climatologic variable – mainly temperature, etc. (Nosek 1972; Guide WMO 100 1983; Peterson et al. 1998; Lapin and Tomlain 2001).

M. Lapin (✉)
Faculty of Mathematics, Physics and Informatics, Comenius University, Mlynská dolina - F1, 842 15 Bratislava, Slovakia
e-mail: lapin@fmph.uniba.sk

K. Strelcová et al. (eds.), *Bioclimatology and Natural Hazards*,
DOI 10.1007/978-1-4020-8876-6_4,

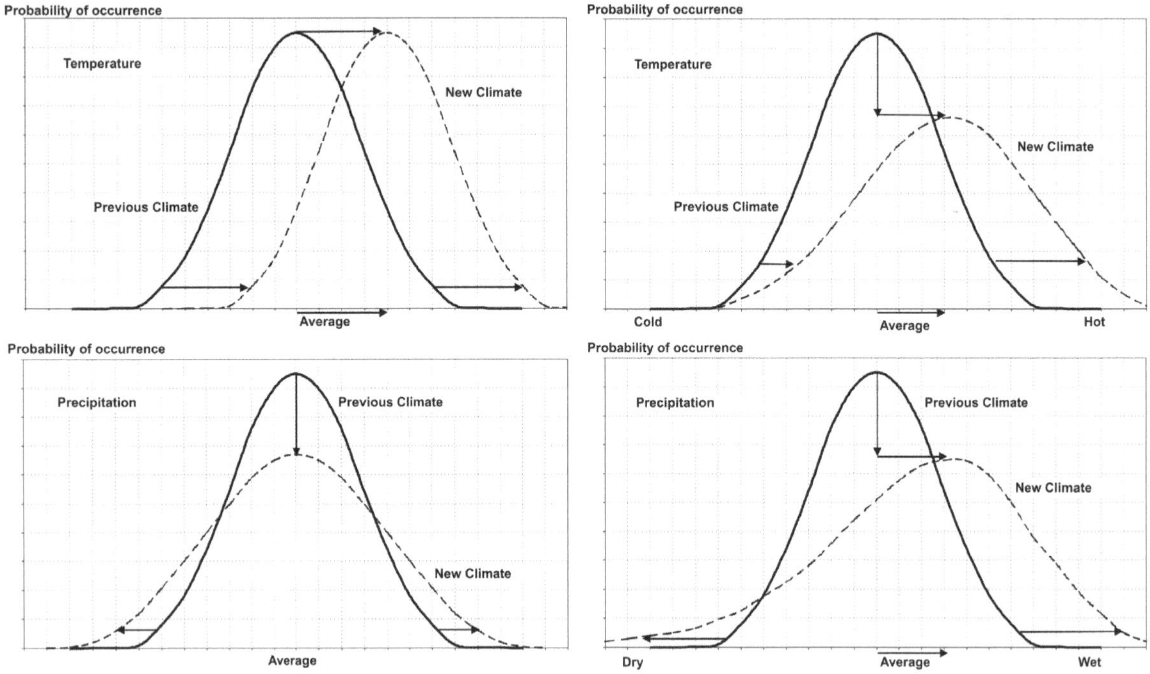

Fig. 1 Possible changes in means and variability of air temperature (*top*) and precipitation totals (*bottom*). Distribution curves represent deviations from the past long-term averages – modified after IPCC (2001)

2. The time series have to meet further preconditions: independent random choice and sufficiently great population. Otherwise, the selected time series should have statistical properties comparable, if not identical, with much longer ones from the same place, and its duration should be at least 30 years (Nosek 1972; Guide WMO 100 1983). Sometimes the series are selected according to some intention (standard period 1961–1990); however, in this case, the selection should not be motivated by additional reasons, for example, an examination of unusual weather, because it would remove other degrees of freedom from the statistical analysis (in the case of climatologic series, $n - 2$ degrees of freedom is the best option, where n is the number of observations).
3. The independent choice has also another dimension. Sometimes it is necessary to examine time series created under not fully correct conditions. For instance, individual data may not be taken at the same time of the year, or under different synoptic situations, or from the events with different radiation and atmospheric stability conditions. This is connected mainly with extreme precipitation totals, heat waves, drought spells, etc. Therefore, the reliability testing of time series and the testing of the independence of all the items in the sample must be done very carefully, otherwise the final statistical characteristics can be misleading. Some details on this aspect are presented further below.
4. The temporal stationarity of time series may be mentioned as the last important condition. Examples like that may be found frequently; in Fig. 2 the annual mean temperatures from the Hurbanovo Observatory are shown. It is clear that within the whole 1871–2006 period only limited number of stationary 30-year series can be selected. The 1951–1980 period is nearly stationary, but the 1971–2000 period is surely not. The issue of the stationarity will also be addressed in the following parts of the chapter.

Methods

The statistical elaboration of design values with low probability of occurrence is aimed at estimating climatic characteristics for practice in socioeconomic sphere, research, engineering, and state administration.

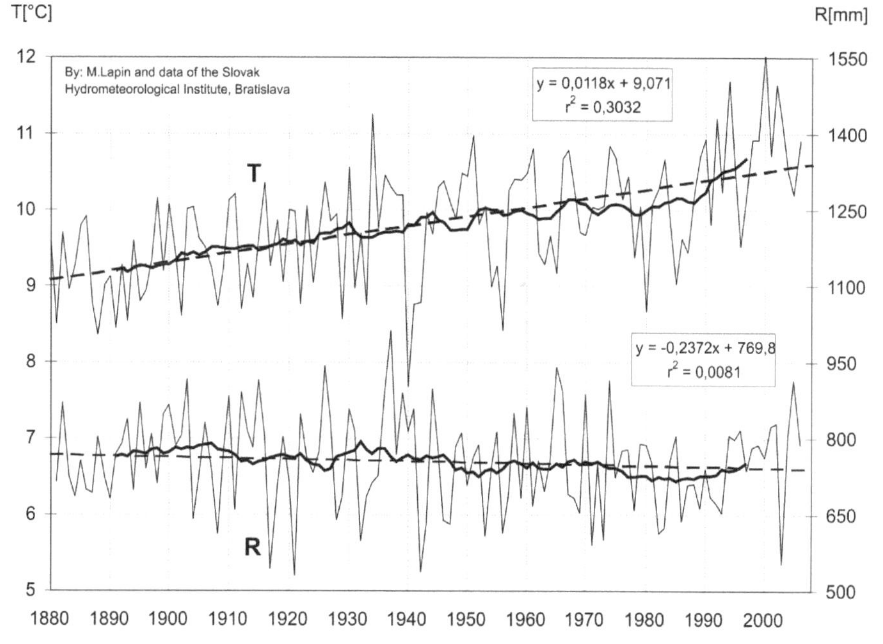

Fig. 2 Annual air temperature means (T) at Hurbanovo and annual precipitation totals (R) in Slovakia for the period of 1881–2006 (based on double-weighted averages of spatial means from 203 stations); changes in T and R are shown by 20-year moving averages and linear trend lines of T and R

In the subsequent parts of the chapter, we would like to point out to the most essential principles of this analysis (Nosek 1972; Guide WMO 100 1983; Lapin and Tomlain 2001).

Input Data

The input data represent some series of observations regularly or irregularly distributed in time and space – one should use data and series from regular station network of meteorological and hydrometeorological services. These institutions provide stable long-term measurements from the point of view of methods and management.

In the case of "extreme value statistics," a question concerning the definition of extremes emerges. Strictly in terms of mathematical statistics, the extreme is the lowest or the highest value within the selected time interval – the local extreme. The basic time interval may be a day or a month; hence there is a single maximum and a single minimum in each day or month, respectively. Nevertheless, some problems arise if all the values within the selected basic interval are the same, for example, in a month without any precipitation, all the daily totals are zero, or during a completely overcast week, all the daily values of cloudiness are 100%, etc. The duration of the basic interval can also be the same as the definition length of the selected variable. In this case, each value is a maximum and minimum as well (e.g., daily precipitation totals on July 1st, during the period of 1871–2000 at Hurbanovo Observatory; in Fig. 3, there are only absolute daily maxima for each day in the annual course).

A further important precondition of a correct climatological analysis is that for statistical elaboration, one needs at least 30 local extremes from entirely comparable basic intervals (days, months, years…). The independence of the extremes may also be a matter of discussion if we suppose that various conditions of the occurrence of extremes that may reduce the degrees of freedom exist in different basic intervals (different conditions at the beginning and the end of a basic interval, different synoptic situations, etc.). For instance, considerably different conditions for observing high daily precipitation totals may be at the beginning and the end of April (regarding the convection, circulation, specific humidity…). Even more complex problems arise if an entire season is selected as the basic interval, for example, the growing period (April 1st–September 30th). Consequently, different conditions significantly influence variables such as runoff, soil moisture, relative humidity, etc. The entirely subjective application of "extreme weather events" is not suitable for every statistical elaboration.

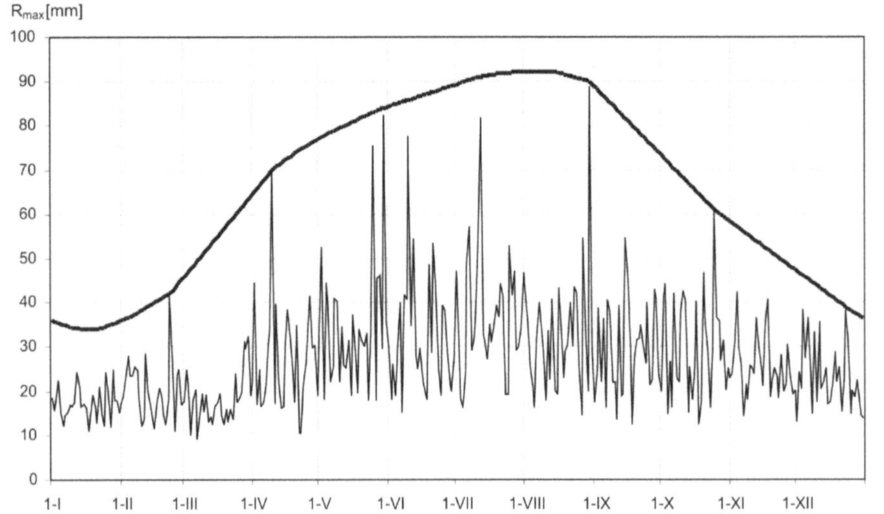

Fig. 3 Annual course of absolute maxima in daily precipitation totals at Hurbanovo for the period of 1871–2000 and their approximate envelope curve. In the 2001–2006 period only 10 corrections of maximum daily totals occurred with no effect on the depicted envelope curve

The absolute maxima in daily precipitation totals at Hurbanovo measured during the period of 1871–2000 are presented in Fig. 3 as an example of the difficulties in the analysis of precipitation extremes. There is a clear annual course in maximum precipitation totals on one side, and irregular occurrences of absolute extremes to the utmost on the other. In the 2001–2006 period, only 10 corrections of maximum daily totals occurred with no effect on the depicted envelope curve. The daily air temperature means show slightly better patterns (Fig. 4), where the considered interval is identical with the definition interval of the variable. The values have been normalized in order to eliminate the annual course of the variable (in this case, each value represents a deviation from the 1951–2000 average). In the case of precipitation totals there are 130 values available for each day, while for air temperature only 50 values are available. For temperature it is sufficient for statistical elaboration (Nosek 1972); however, the annual course of design values with low probability of occurrence is surely a serious problem also for temperature.

It needs to be stressed that nearly ideal time series of measured data have been presented here; generally, however, much shorter series with disputed data selection are statistically processed (individual rain spells for rain intensities, extraordinary weather events, *k*-day precipitation totals, etc.). The regional assessment of weather extremes can be considered as a specific problem, for example, maxima in daily precipitation totals within some climatologically homogeneous region (Gaál 2006). Figure 5 (annual maxima in precipitation totals from 557 stations in Slovakia in 1951–2000) serves as an illustration of the problems mentioned before.

Even though Fig. 5 was constructed using the best data available in Slovakia for the statistical evaluation of characteristics of 1-day precipitation totals (Faško et al. 2002), it does not provide all information on absolute precipitation totals (Table 1) or on their temporal and spatial distribution. A drawback of such an approach is that the annual maxima are selected without any consideration of different physical conditions during the year and different synoptic situations. That is the reason why some events from the cold half-year connected with long-lasting precipitation in central cyclones (C) or troughs (B) over central Europe are included in the last column of Table 1. Precipitation maxima in the warm half-year may occur either due to localized convective situations (one or several thunderstorms) or due to long-lasting rain in central cyclones and troughs occurring occasionally during this time of year (Table 1).

A selection of values above some threshold ($R \geq 100$ mm) is a significantly different method of basic extreme data selection (Fig. 6, Faško et al. 2002, completed in 2005). Such an approach enables us to get a relatively good glance on high daily precipitation totals in the individual years (based on observations from about 700 stations each year). On the other hand, such a method of selection is far from correct data preparation for statistical elaboration, mainly for the following reasons: (a) there is a varying number of stations each year in operation; (b) some

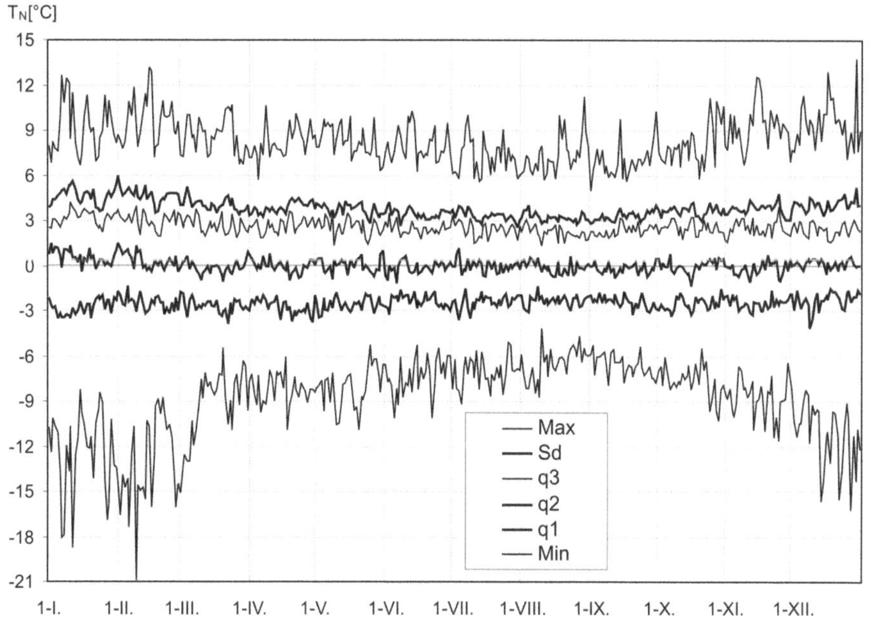

Fig. 4 Annual course of statistics for normalized daily mean temperatures at Hurbanovo (deviations from long-term averages) for the period of 1951–2000; Max (Min) – absolute maximum (minimum) in daily means; q3 (q1) – upper (lower) quartiles; q2 – median, and Sd – standard deviation of daily means

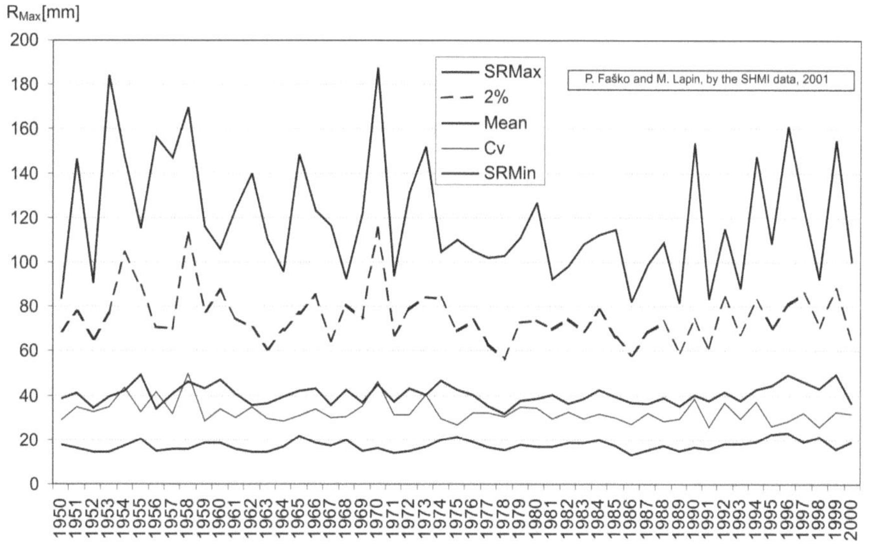

Fig. 5 Annual maxima in daily precipitation totals at 557 stations in Slovakia in 1950–2000; SRMax (SRMin) – absolute maxima (minima) in Slovakia; 2% – 98th percentile; Mean – mean value; and Cv – coefficient of variation of the individual annual maxima throughout the country (Faško et al. 2000, 2002)

stations have only short observations, and (c) it is impossible to assess whether this selection is independent. A similar analysis has been carried out for southern Poland by Cebulak et al. (2000). The extreme on June 29, 1958 represents an outstanding event, when there were observed daily precipitation totals $R \geq 100$ mm at 36 stations in Slovakia; no other extreme events exceeding this threshold occurred during 1958 (in south Poland more than 100 events of $R \geq 100$ mm were registered on the same day).

Obviously, in the case of complex data related to extraordinary weather events where the extremes are considered as a result of interaction of several meteorological variables, the selection is even more problematic. In an analysis of complex extreme events such as droughts, heat waves, flood risk, wild fire risk, avalanche risk, etc., one needs to take into account several meteorological variables simultaneously, including soil moisture, vegetation, preceding weather, land use, etc. Extreme events of duration ranging from one to several days might be a subject of any statistical

Table 1 List of stations with daily precipitation totals $R \geq 150$ mm in Slovakia for the period of 1951–2000 (based on all stations with precipitation measurements, about 700 each year; Faško et al. 2002); φ – latitude, λ – longitude, H – altitude, Date – date of occurrence, Sit CS – Czecho-Slovak classification of synoptic situations (C – central cyclone, B – trough of low air pressure, VFZ – air pressure saddle, Wal – westerly anticyclonal of summer type, SWc3 – southwesterly cyclonal of 3rd type, Ec – easterly cyclonal)

#	Station	φ(N)		λ(E)		H	R	Date	Sit
		[°]	[']	[°]	[']	[m]	[mm]	(dd-mm-yy)	CS
1	Trnava	48	22	17	38	155	162.8	03-VI-51	C
2	Sokolovce	48	31	17	50	165	156.4	19-VII-56	VFZ
3	Salka	47	53	18	45	111	231.9	12-VII-57	B
4	Chata Zbojnícka	49	11	20	10	1958	169.0	29-VI-58	C
5	Hrebienok	49	09	20	13	1285	165.0	29-VI-58	C
6	Skalnaté pleso	49	11	20	14	1778	170.0	29-VI-58	C
7	Železnô	48	57	19	23	990	153.7	18-X-61	B
8	Veľké Pole	48	33	18	34	556	164.0	19-VIII-66	C
9	Oravská Lesná	49	22	19	10	780	163.2	18-VII-70	B
10	Novoť	49	25	19	15	770	187.6	18-VII-70	B
11	Oravská Polhora – Roveň	49	33	19	26	704	182.1	18-VII-70	B
12	Oravice	49	16	19	45	853	154.3	18-VII-70	B
13	Zverovka	49	14	19	42	1027	154.5	18-VII-70	B
14	Podspády, VT	49	16	20	10	910	152.3	30-VI-73	Wal
15	Luková	48	57	19	35	1661	153.0	29-X-90	SWc3
16	Jasná	48	58	19	35	1196	154.0	29-X-90	SWc3
17	Cífer	48	19	17	29	147	161.5	12-VIII-96	B
18	Limbach	48	17	17	13	181	155.0	10-VII-99	Ec
19	Pezinok, Grinava	48	05	17	05	159	151.5	10-VII-99	Ec

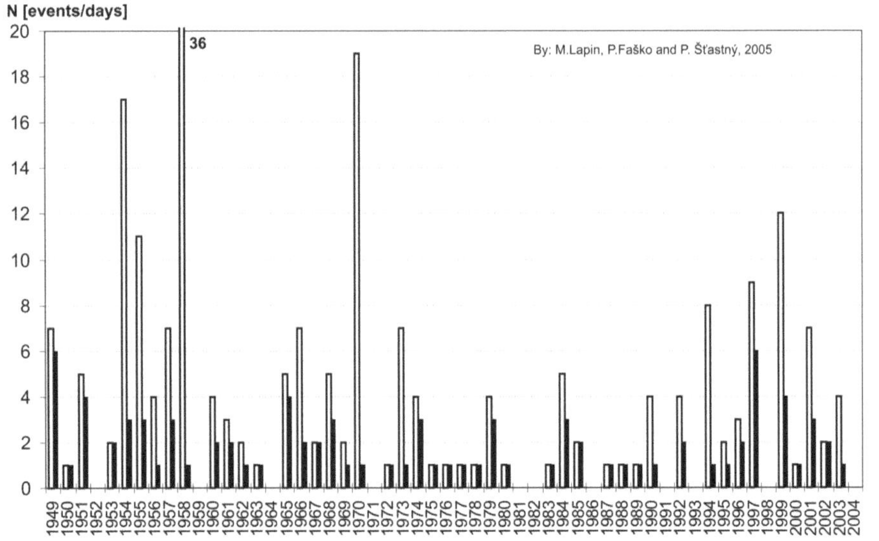

Fig. 6 Number of events (each station is considered as one event each day) and days (at least one event in Slovakia) with daily precipitation totals $R \geq 100$ mm in Slovakia for the period of 1949–2004 (Lapin et al. 2005)

analysis only following a thorough reliability testing of all variables. In such cases, the series of selected items representing extreme weather events need to be tested also according to the four basic preconditions mentioned in Introduction section. Otherwise, the statistical elaboration can be misleading.

Reliability Testing

Generally, the climatological data comprises a number of potential shortages. Therefore, the reliability testing and the necessary modification of data should consist of several steps. In this section, we concentrate mainly

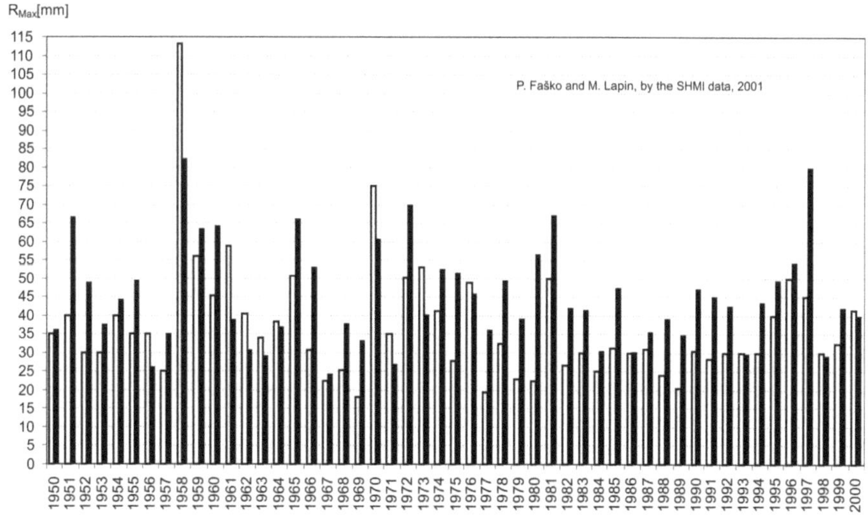

Fig. 7 Time series of annual maxima in daily precipitation totals at two close stations (6 km distance) in the Liptov region (the first one, Bobrovček, 667 m, has 16 filled-in data; the second one, Huty, 795 m, (black columns) is complete)

on the completeness and the temporal homogeneity of the time series. Further details concerning the testing of statistical characteristics, such as central values, marginal percentiles and extremes are addressed in the relevant scientific literature (e.g., Nosek 1972; Storch and Zwiers 2000; Lapin and Tomlain 2001; Wilks 2005).

A key problem of any analysis of extreme data or extreme weather events is usually the incompleteness of the data series. In general, the longer the series are the more reliable statistical results can be obtained. A time period spanning at least 30 years is deemed to be satisfactory for a correct analysis. Our experiences confirmed that a rapid decrease of the number of usable stations occurs if the desired length of series exceeds 50 years. Sometimes only 10% of stations fulfill all the requirements. Relatively long series are necessary also from the point of view of the outliers. For instance, a maximum value $R \geq 100$ mm of daily precipitation totals appears nearly each year at some of 700 stations in Slovakia. However, these totals may be considered as outliers at least in half of such cases due to the fact that the length of the time series does not reach 30 years. Daily minimum of air temperature shows similar potential biases, mainly in the winter months. In statistical elaboration, mainly in the case of an at-site analysis, the outliers should be taken into account very carefully. Regional approaches that analyze time series of several stations simultaneously, however, are able to cope with the problem of estimation of the design values with low probabilities more reliably even when outliers are present (Gaál 2006).

Completion of missing data in time series is shown through the example of annual extremes in daily precipitation totals. In Fig. 5, there are data from 557 stations in Slovakia for the period of 1950–2000. During the process of elaboration, we have found several stations with gaps (missing values) in individual months that did not allow us to derive annual maxima in the particular year. However, the missing monthly totals can easily be estimated using isopluvial maps of the monthly precipitation totals, and the neighboring station's data. This is also quite easy to do for daily totals in the case of cyclonic weather (central cyclone C, or trough of low pressure B) with equally distributed high daily totals. On the other hand, the selection and filling of missing maxima is not so straightforward at unequally distributed high daily totals connected with severe summer thunderstorms. We have accomplished the filling of missing daily maxima in two steps following Cebulak et al. (2000): first, we have identified all days with potential extreme daily totals and then, second, the precipitation map of some area around the selected station has been constructed using all measuring stations (in some cases, synoptic maps have been applied as well). As an example of difficulties, we show two stations in Fig. 7. Both time series are from the Liptov region in northern Slovakia. The first station (Huty, filled columns) is complete and the second one (Bobrovček, unfilled columns) has gaps for 16 years at the beginning and the end of the observation period. The distance between the stations is about 6 km.

It is obvious that such a method may also be applied at other climatic variables, even though such a

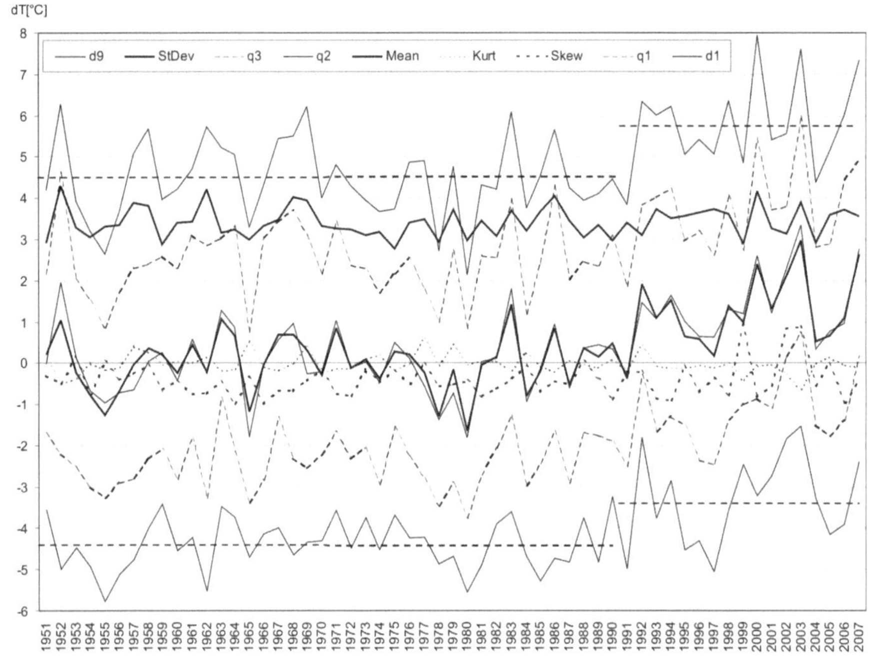

Fig. 8 Temporal course of basic statistical characteristics of normalized daily mean temperatures dT (deviations from normal) at Hurbanovo for the season April 1st–August 31st for the period of 1951–2007 (from the top: d9 – upper decile, StDev – standard deviation, q3 – upper quartile, q2 – median, mean, Kurt – kurtosis, Skew – skewness, q1 – lower quartile, and d1 – lower decile); no significant change in StDev, Kurt, and Skew, increases in other values since 1991 by about 1–1.5°C (dashed horizontal lines) indicate climate change category Number 1 from Fig. 1 (the year 2007 seems comparable with the extreme years 2000 and 2003)

high number of filled-in data (31%) is unusual. The Bobrovček station is an extraordinary one with its 16 missing data among the 557 stations. Completing the missing station's data was possible mainly due to the dense gauge network, and well-defined upwind effects, respectively. Otherwise, it is sometimes very difficult to complete even a single missing value because of special synoptic situations (e.g., isolated thunderstorms).

After completing the time series of annual maxima of daily precipitation totals at 557 stations in Slovakia, we considered the testing of temporal homogeneity of individual time series (Craddock test, double-mass analysis, etc., Lapin and Tomlain 2001). Nevertheless, none of the tests has brought usable result. A plenty of fake inhomogeneities have been detected, similarly to Gaál (2006), since none of them has been confirmed by metadata. Therefore, we came to the conclusion that if there are homogeneous series of annual precipitation totals, the annual maxima of daily totals are probably homogeneous as well. Note that in the case of time series of daily air temperature maxima and daily specific air humidity maxima, there are much better possibilities of temporal homogeneity testing using only the Cradock test (Lapin and Tomlain 2001).

Stationarity of time series is a further serious problem in the analysis of climatic extremes (Guide WMO 100 1983). As it can be seen from Fig. 1, there are reasons for unreliable assessments of design values in the case of climatic change. To some extent, it is also evident among the series of normalized daily temperature averages for the modified growing period April 1st–August 31st in the period of 1951–2007 (Figs. 8 and 9). After the year 1990, there are significantly different conditions, not only in the trend of mean values of daily temperature, but also at the margins of the distribution spectrum (upper and lower deciles, etc.). Evaluation of design values with high return periods (low probability of occurrence) based on the statistical evaluation of the series from 1951 to 1990 has, therefore, new dimensions recently. It seems that the upper and the lower deciles have shifted by about 1.3°C to higher values and this probably holds also true for the extremes. In this specific example, the mean of upper deciles is only 4.41°C in the period 1951–1990 but 5.80°C in 1991–2007 (dashed line in Fig. 8). For the mean values, the shift is 0.00 and 1.27°C, for the lower decile, it is −4.40 and −3.31°C, respectively. Figure 8 also indicates that the standard deviation (StDev), skewness (Skew), and the kurtosis (Kurt) have not changed significantly after 1990. Figure 9 shows the shift in distribution curves of new climate (1991–2007) compared to the past (1951–1990). It follows from the foregoing discussion that the first option of climate change (Fig. 1) might be the most probable

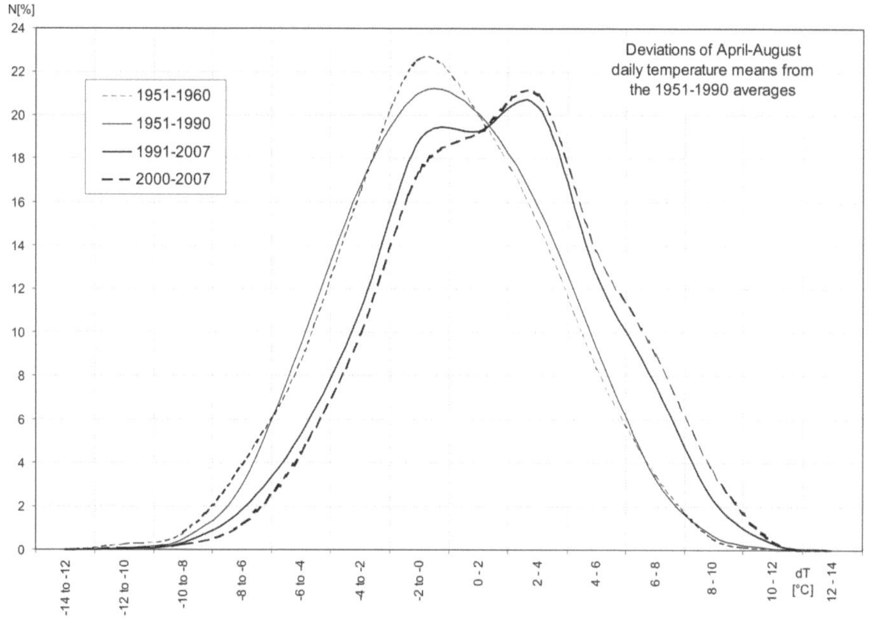

Fig. 9 Shift in the daily air temperature distribution curves between the periods of 1951–1990 and 1991–2007 in the April–August season (the dashed lines represent shorter periods 1951–1960 (*left*) and 2000–2007 (*right*)). Each daily value in this graph has been calculated as deviation from the 1951–1990 average for every day in annual course (the lines have been slightly smoothed)

one since the distribution curve has shifted nearly linearly without any significant change of variability.

Selecting the Theoretical Distribution Function for the Estimation of Design Values

The shortness and the temporal nonstationarity of time series as well as the complex patterns and temporal course of basic statistical characteristics of climatic variables result in the fact that the theoretical statistical models of probability distribution have several limitations in climatology. Only seldom is it possible to apply the Gaussian (normal) distribution with a clear interpretation of all the statistical characteristics. This is important mainly for extreme value analysis. Climatic extremes occur very rarely, and their magnitude is only partly influenced by random processes. Exact forecast of such events is basically not possible, but there are several ways of how to assess them by means of mathematical tools of statistical climatology (Nosek 1972; Storch and Zwiers 2000; Wilks 2005).

The process of estimation of the magnitude of extremes (the so-called design values) expected to occur in the future with a given probability of occurrence p (or with a given return period t, where $t=1/p$) is called frequency analysis. A common assumption of the frequency analysis of hydrological and meteorological extremes is the independence of the observed phenomena. This implies that natural climatic variability does not affect the distribution of extreme events (Bradley 1998). Besides, the stationarity of the climate is supposed, that is, regardless of the prevailing climate forming mechanisms, the probability of occurrence of an extraordinary phenomenon is constant from year to year (Franks 2002). The hypothesis about the identical distribution, independence, and stationarity of the climatic extremes, however, seems to be violated in the light of the global climate change (Fig. 1; IPCC 2001, 2007). Consequently, the design values estimated in the past may not be valid recently, and their applicability in possible future engineering constructions have to be seriously reconsidered.

For a long time, according to the guidelines of Nosek (1972), the two-parameter Gumbel and the three-parameter Pearson distributions have been used for the estimation of design values of climatological/hydrological extremes in the former Czechoslovakia. The recent studies in frequency analysis worldwide, and the statistical handbooks, however, offer a much broader palette of possible theoretical distribution functions for these purposes. A general consensus among scientists exists that the two-parameter distributions should be used very carefully since they have limited flexibility: using these distributions may lead to enhanced bias of the estimated quantiles at the right tails of the frequency curves. The application of distribution functions with three parameters (e.g., generalized extreme value (GEV), generalized logistic, generalized normal, Pearson III, Weibull, or Pareto) or more (kappa with four or Wakeby with five parameters) is generally recommended. The advantage of the four-parameter or five-parameter distributions lies in their ability to mimic a wide range of plausible distributions if their parameters are properly set (Hosking and Wallis 1997). Especially, the GEV distribution function (Coles 2001) has been proved to be a suitable one for modeling extremes of precipitation (e.g., Kohnová et al. 2004; Gaál 2006; Kyselý and Picek 2006; Alila 1999; Smithers and Schulze 2001), floods (Kohnová and Szolgay 2003, Kohnová et al. 2006; Castellarin et al. 2001; Kumar and Chatterjee 2005), or temperature (Kharin and Zwiers 2000; Kyselý 2002). Note that the Gumbel distribution that has been popular in the hydrologic community is a special case of the GEV distribution when the shape parameter is equal to zero.

The aim of the frequency analysis is to assess reliable estimates of design values corresponding to relevant return periods. Given a n-year series of observations, the design values with return period t can be reliably calculated by means of traditional (at-site) analysis only in case $t \leq n$ (Hosking and Wallis 1997). A rule of thumb is to confine the design value estimation to the return period t not exceeding three times the value of n; the design values cannot be assessed sufficiently reliably for longer return periods (Malitz 1999). Nevertheless, having relatively short series of observations (30, 40, and sometimes up to 100 years), one runs up against problems, usually in engineering applications, when trying to estimate design values with return periods of $t \geq 100$ years. In order to overcome the difficulties with the insufficient length of series of observations, the so-called regional frequency analysis was developed. The regional approach successfully replaces the temporal dimension of the problem by the spatial one since it uses data simultaneously from several sites within a given region (multisite analysis). Such a spatial extension of the data sample requires to group sites together with similar statistical properties of their on-site probability distributions. Having a much wider dataset to analyze from a given region, one is able to obtain a more accurate and reliable estimation of the design values of the underlying variable, mostly at the tail of the selected distribution function (Hosking and Wallis 1997). Even though the regional frequency analysis has originally been developed for flood data (hence the popular term flood frequency analysis), Hosking and Wallis (1997) emphasize that the methodology is also applicable for any other environmental applications.

The so-called 5t-rule (Jakob et al. 1999) is the basic principle recommended for the regional frequency analysis. It specifies that "the pooled stations should collectively supply five times as many years of record as the target return period t." For example, for a reliable estimation of design values with $t = 200$ years, at least 1000 station-years of data are needed. Supposing an average record length of 40 years (a realistic case for the precipitation data in Slovakia), each group of sites should consist of at least 25 stations.

In frequency analysis, there are basically two ways of selecting extreme values out of long climatological or hydrological series. The first one, the annual maximum series (AMS) considers the largest events each year; the second one, the partial duration series (PDS) or peaks-over-threshold (POT) approach includes all values above a selected threshold level. The AMS approach is the most frequently used method of constructing data samples even though it usually leads to loss of some information (a secondary maximum from a given year may exceed the maximum value from another one). The POT approach avoids these drawbacks by considering all the independent events above a certain level; however, it is applied less due to its complexity and difficulties in definition of the threshold level (Madsen et al. 1997). The requirement of independent selection is easily obtained for the AMS method, whereas it is not a straightforward task for the PDS approach where multiple peaks corresponding to the same hydrologic event may occur. In such case, additional options for the definition of independent events have to be introduced.

Further details related to the issue of frequency analysis, a historical overview, a comparison of the methods, and their application on the precipitation data in the area of Slovakia are presented in Gaál (2006) and Kohnová et al. (2005b).

Examples of Elaboration

In this section, several examples based on the recent examination of precipitation extremes are shown. The

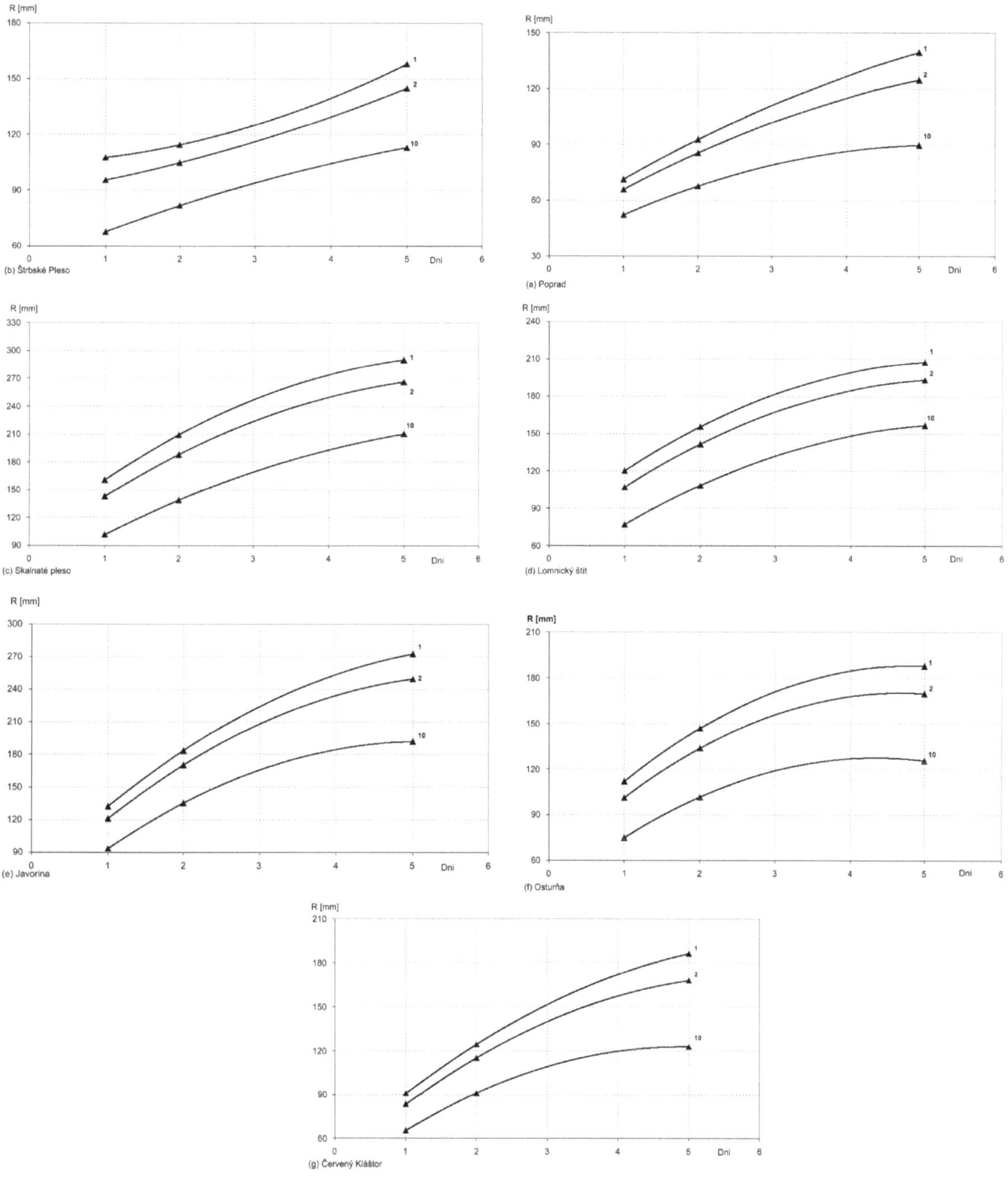

Fig. 10 Design values daily precipitation total R (mm) corresponding to the probability of occurrence of 1%, 2%, and 10% at the stations in the High Tatras region: (**a**) Poprad, (**b**) Štrbské Pleso, (**c**) Skalnaté Pleso, (**d**) Lomnický štít, (**e**) Javorina, (**f**) Osturňa, and (**g**) Červený Kláštor based on data observed in the 1951–2005 period, and the Pearson III theoretical distribution function (Faško et al. 2006). The trend lines only roughly illustrate the increase of R with increasing number of days

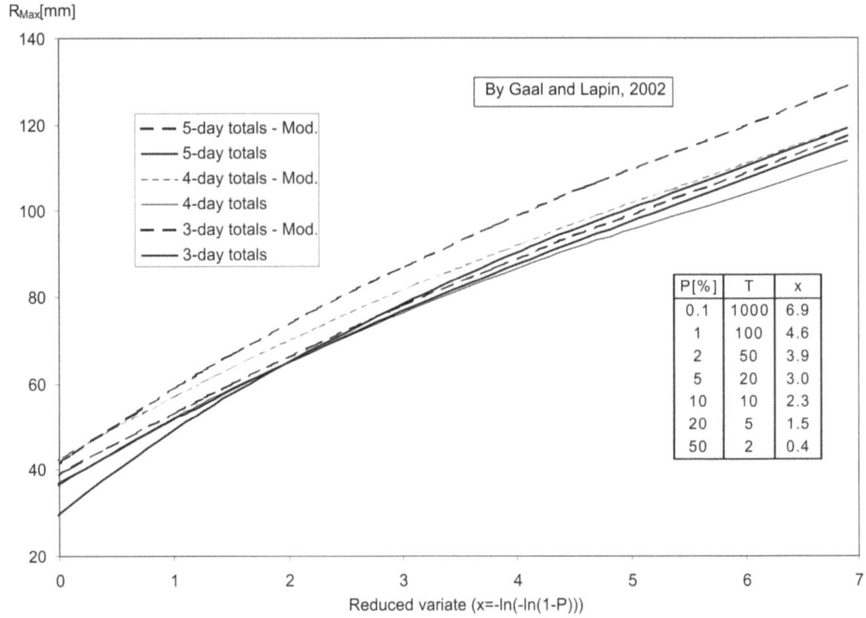

Fig. 11 Basic and modified DDF curves (Mod. – with 1-day gap in precipitation totals) of maximum 3–5-day precipitation totals (R_{Max}) for the warm season (April–September) at the Hurbanovo Observatory for the period of 1901–2000 (Gaál and Lapin 2002)

illustrations clearly show the problems related to a correct assessment of usable design values.

In Fig. 10, the 1-, 2-, and 5-day precipitation totals are presented, based on 55-year long series (1951–2005), according to Faško et al. (2006). In the case of 5-day precipitation totals, only events with daily precipitation values of at least 0.1 mm per day have been selected. Although the area of the High Tatras mountains is relatively small (about 500 km^2), the design values show considerable variability for the same return periods. The altitude does play an important role here; on the other hand, much more significant are the upwind effects, mainly at stations such as Javorina (northern slope foot) and Skalnaté Pleso (southeastern slope). Concerning the 5-day totals, the design values are nearly two times higher than those of other stations in the comparable altitudes (Poprad and Štrbské Pleso). The fact that the deviations between the design values of 1- and 2-day totals are not so relevant may be related to different physical conditions of short-term (thunderstorms) and long-term (cyclonic) precipitation events. The trend lines in Fig. 10 represent roughly the increase of extreme precipitation totals with rising number of precipitation days. In the case of 30–180 min precipitation intensities, one would probably observe similarly insignificant differences based on altitude (Šamaj and Valovič 1973). Further details related to this issue may be found, for example, in Dzubák (1969), Šamaj et al. (1985), Kyselý and Picek (2006), Kyselý et al. (2007), Lapin et al. (2003), and Parajka et al. (2004).

The second example presents similar evaluations for the Hurbanovo Observatory (Gaál and Lapin 2002; Gaál et al. 2004). The lowland localities in southern Slovakia have a different precipitation regime compared to the northern mountains. That is why a slightly modified methodology was introduced here. Events with continuous precipitation of a duration of 3–5 days rarely occur at Hurbanovo. Therefore, a modified method of selection of several-day periods with precipitation has been proposed, where a possible 1-day gap is allowed. Great differences between the basic and modified method of selection may be recognized mainly in the warm season (April–September, Fig. 11). The modified method has been applied also by Kohnová et al. (2005a) for 10-day precipitation events, where the evapotranspiration effects during 1- or 2-day gaps in significant precipitation events has been considered. It can be stressed that 5–10-day precipitation events play a very important role in the regional floods occurrence in Slovakia.

Conclusions

The presented chapter demonstrates a series of possible difficulties related to the reliable selection and processing of extreme weather events. The Slovak Hydrometeorological Institute Observatory at Hurbanovo (SHMI, 115 m a.s.l., since 1871) was the main source of observed data series for this chapter. Obviously, it was impossible to mention and analyze all potential possibilities and risks in the reliability of the data series. However, the most essential principles may be summarized as follows: (1) Extreme weather events must be correctly defined, selected, and tested, otherwise the result of the evaluation is not reliable; (2) Several theoretical distribution functions may be applied in design values estimation; nevertheless, the distribution function with at least three parameters is advised to be used; (3) The return period of design values t should be lower than three times the number of observing years n; the best way is to confine the estimation to the values $t \leq n$; (4) Any complex event of extraordinary weather (droughts, wild fire risk, etc.) should be selected very carefully, otherwise the evaluated statistical characteristics will be impossible to interpret and apply correctly; (5) Special attention should be devoted to the stationarity of long-term time series.

The chapter also outlined the importance of climate change impacts on the frequency distribution and return periods of extreme events. Climate change may cause the design values calculated from the historical observations to be invalid even at present and likely in the next decades. Therefore, the methods of design value estimation should be modified. From this point of view of more detailed reliability testing of input data, it is recommended to reprocess also the design values of extremes published in the past decades. It is clear that the authors of the aforementioned studies have sometimes not kept the basic principles of selection, testing, and statistical elaboration. The scientific interpretation and practical applications of these results could therefore be problematic.

Other climatic and hydrologic elements, from the aspect of extreme events, have been studied by further authors (e.g., Faško et al. 2003a, 2003b; Szolgay et al. 2003; Majerčáková et al. 2004; Lapin et al. 2005, 2007, etc.). In some other papers, the difficulties in scenarios of extreme events design under climate change impacts have been emphasized (Lapin et al. 2001, 2004; Lapin 2003; Lapin and Hlavčová 2003; Lapin and Melo 2002, 2003).

Acknowledgments Some parts of this chapter are based on the results of projects VEGA No. 1/1042/04, APVT-51-017804, and the observed SHMI data; the authors thank for the availability of these data series.

References

Alila Y (1999) A hierarchical approach for the regionalization of precipitation annual maxima in Canada. J. Geophys. Res., 104(D24), 31, 645–31, 656

Bradley AA (1998) Regional frequency analysis methods for evaluating changes in hydrologic extremes. Water Resour. Res., 34(4), 741–750

Castellarin A, Burn DH, Brath A (2001) Assessing the effectiveness of hydrological similarity measures for flood frequency analysis. J. Hydrol., 241, 270–287

Cebulak E, Faško P, Lapin M, Šťastný P (2000) Extreme precipitation events in the Western Carpathians. In: Prace geograficzne – Zeszyt 108, Images of Weather and Climate, Cracow, Inst. of Geography, Jagelonian Univer., Cracow, Poland, 117–124

Coles S (2001) An introduction to statistical modeling of extreme values. Springer-Verlag, London, 222 pp

Dzubák M (1969) Estimation of maximum daily precipitation totals principle using global probability curves of exceeding. Vodohospodársky časopis SAV, Bratislava, Slovakia, 17(3), 209–225 (in Slovak)

Faško P, Lapin M, Šťastný P, Vívoda J (2000) Maximum daily sums of precipitation in Slovakia in the second half of 20th century. In: Prace geograficzne – Zeszyt 108, Images of Weather and Climate, Cracow, Inst. of Geography, Jagelonian Univer., Cracow, Poland, 131–138

Faško P, Lapin M, Šťastný P (2002) Evaluation of high daily precipitation totals in Slovakia. In: Proc. of the Conference "100th anniversary of academician Dub's birth", Slovak Hydrometeorological Institute, Bratislava, Slovakia, Oct. 10, 2002, 37–38. (in Slovak)

Faško P, Lapin M, Sekáčová Z, Šťastný P (2003a) Extraordinary climatic anomaly in 2003. Slovak Meteorological Journal, 6(3), 3–7

Faško P, Šťastný P, Vívoda J (2003b) The orography influence on the maximum daily sums of precipitation in Slovakia. In: Proc. of the Int. Conf. on Alpine Meteorol. and MAP2003 Meeting, Brig, Switzerland, May 19–23, 2003. Publ. of MeteoSwiss, No. 66

Faško P, Gaál L, Lapin M, Pecho J, Šťastný P (2006) Contribution to the issues of the estimation of design values of daily precipitation totals. In: Proc. of the Seminary "80th anniversary of Milan Dzubák's birth", Slovak Hydrometeorological Institute, Bratislava, Slovakia, 19–26 (in Slovak)

Franks SW (2002) Identification of a change in climate state using regional flood data. Hydrol. Earth Syst. Sc., 6(1), 11–16

Gaál L, Lapin M (2002) Extreme several day precipitation totals at the Hurbanovo observatory (Slovakia) during the 20th century. Contributions to Geophysics and Geodesy, 32(3), 197–213

Gaál L, Lapin M, Faško P (2004) Maximum daily precipitation totals in Slovakia. In: Rožnovský J, Litschmann T (eds): Proc. of the Seminary "Extremes of Weather and Climate", Brno, Czech Republic, March 11, 2004. Czech Bioclimatological Society, 15pp. [CD ROM]. (in Slovak)

Gaál L (2006) Methods for estimation of statistical characteristics of design *k*-day precipitation totals in Slovakia. Ph.D. Thesis, Comenius University Bratislava, Slovakia, 227pp. (in Slovak)

Guide WMO, No. 100 (1983) Guide to climatological practices. 2nd ed. WMO, Geneva, Switzerland

Hosking JRM, Wallis JR (1997) Regional frequency analysis: An approach based on L-moments. Cambridge University Press, Cambridge, 224pp

IPCC (1998) The regional impacts of climate change. An assessment of vulnerability. Watson RT, Zinyowera MC, Moss RH, (eds.): A special report of IPCC WG II. Cambridge University Press, 518 pp

IPCC (2001) Climate change 2001: The scientific basis. Contribution of Working Group 1 to the Third Assessment Report of the IPCC. Cambridge University Press, 944pp

IPCC (2007) IPCC Fourth Assessment Report. Working Group I Report "The Physical Science Basis". Oct. 2007. 996pp. Available at: http://www.ipcc.ch/ipccreports/ar4-wg1.htm

Jakob D, Reed DW, Robson AJ (1999) Selecting a pooling-group. In: Flood Estimation Handbook, Vol. 3. Institute of Hydrology, Wallingford, UK

Kharin VV, Zwiers FW (2000) Changes in the extremes in an ensemble of transient climate simulations with a coupled atmosphere-ocean GCM. J. Clim., 13, 3760–3788

Kohnová S, Szolgay J (2003) Regional estimation of the index flood and the standard deviation of the summer floods in the Tatry mountains. J. Hydrol. Hydromech., 51(4), 241–255

Kohnová S, Szolgay J, Gaál L (2004) Estimation of design 5-day maximum precipitation totals in the upper Hron region. Slovak Meteorological Journal, 7(2), 79–83

Kohnová S, Lapin M, Szolgay J, Gaál L (2005a) Methodology for the selection of 10-day maximum precipitation totals and their statistical analysis in the upper Hron region. Contributions to Geophysics and Geodesy, 35(3), 299–318

Kohnová S, Gaál L, Szolgay J, Hlavčová K (2005b) Analysis of heavy precipitation amounts in the Upper Hron region. Faculty of Civil Engineering, Slovak University of Technology, Bratislava, Slovakia, 120pp (in Slovak)

Kohnová S, Szolgay J, Solín Ľ, Hlavčová K (2006) Regional methods for prediction in ungauged basins. Key Publishing, Ostrava, Czech Republic, 113pp

Kumar R, Chatterjee Ch (2005) Regional flood frequency analysis using L-moments for North Brahmaputra region of India. J. Hydrologic Engrg., 10(1), 1–7

Kyselý J (2002) Comparison of extremes in GCM-simulated, downscaled and observed central-European temperature series. Clim. Res., 20, 211–222

Kyselý J, Picek J (2006) Regional growth curves and improved design value estimates of extreme precipitation events in the Czech Republic. Clim. Res., 33, 243–255

Kyselý J, Picek J, Huth R (2007) Formation of homogeneous regions for regional frequency analysis of extreme precipitation events in the Czech Republic. Studia Geophysica et Geodaetica, 51(2), 327–344

Lapin M, Tomlain J (2001) General and Regional Climatology. Comenius University, Bratislava, Slovakia, 184pp (in Slovak)

Lapin M, Damborská I, Melo M (2001) Scenarios of several physically plausible climatic elements. Slovak National Climate Program, VI, No. 11, Slovak Hydrometeorological Institute and Ministry of the Environment of the Slovak Republic, Bratislava, Slovakia, 5–30. (in Slovak with English summary)

Lapin M, Melo M (2002) Scenarios of time series for 10 climatic elemets and the 2001–2090 period based on the GCMs CCCM2000 and GISS98. In: Rožnovský J, Litschmann T (eds): Proc. of the XIVth Czecho-Slovak Bioclimatological Conference "Bioclimate-Environment-Economy", Lednice, Czech Republic, Sept. 2–4, 2002. Czech Bioclimatological Society, 254–266. [CD ROM]. (in Slovak)

Lapin M (2003) Changes in meteorological conditions of flood risks during changing climate. Životné prostredie, 37(4), 184–190 (in Slovak)

Lapin M, Hlavčová K (2003) Changes in summer type of flash floods in the Slovak Carpathians due to changing climate. In: Proc. of the Int. Conf. on Alpine Meteorol. and MAP2003 Meeting, Brig, Switzerland, May 19–23, 2003. Publ. of MeteoSwiss, No. 66, 105–108

Lapin M, Melo M (2003) Possibilities of climatic scenarios design with daily values and extremes, based on the newest GCMs. In: Proc. of the Int. Conf. "The function of the energy and water balance in bioclimatological systems", Račková dolina, Slovakia, Sept. 2–4, 2003. [CD ROM]. (in Slovak)

Lapin M, Hlavčová K, Petrovič P (2003) Climate change impacts on hydrological processes. In: Proc. of the Int. Conf. "Hydrology at the beginning of the 21st century", Smolenice, Slovakia, May 5–7, 2003. [CD ROM]. (in Slovak)

Lapin M, Melo M, Damborská I, Gera M (2004) Scenarios of precipitation totals during extreme weather events in Slovakia. In: Rožnovský J, Litschmann T (eds): Proc. of the Seminary "Extremes of Weather and Climate" , Brno, Czech Republic, March 11, 2004. 18pp. [CD ROM] (in Slovak)

Lapin M, Šťastný P, Chmelík M (2005) Detection of climate change in the Slovak mountains. Croatian Meteorological Journal, 40, 101–104

Lapin M, Faško P, Pecho J (2007) Snow cover variability and trends in the Tatra mountains in 1921–2006. In: Proc. of the XXIXth ICAM Conf., June 4–8, 2007, Chambery, France, 683–686

Madsen H, Rasmussen PF, Rosbjerg D (1997) Comparison of annual maximum series and partial duration series methods for modeling extreme hydrologic events: 1. At-site modeling. Water Resour. Res., 33(4), 747–757

Majerčáková O, Šťastný P, Faško P (2004) Review of extraordinary hydrological and meteorological events in Slovakia in recent years. Vodohospodársky spravodajca, 47(2–3), 10–11 (in Slovak)

Malitz G (1999) Starkniederschlag in Deutschland – Messergebnisse, statistische Auswertungen, Schätzungen. In: Klimastatusbericht 1999, Deutscher Wetterdienst, 35–41. Available at: http://www.dwd.de/de/FundE/Klima/KLIS/prod/KSB/ksb99/malitz.pdf

Nosek M (1972) Methods in Climatology. Praha, Academia, 431 pp. (in Czech)

Parajka J, Kohnová S, Szolgay J (2004) Mapping the maximum daily precipitation totals in the upper Hron region using stochastic interpolation methods. Acta Hydrologica Slovaca, 5(1), 78–87 (in Slovak)

Peterson TC et al. (1998) Homogeneity adjustments of in situ atmospheric climate data: A review. Int. J. Climatol., 18(13), 1493–1517

Šamaj F, Valovič Š (1973) Intensities of short-term rains in Slovakia. In: Proceedings of Works of the Hydrometeorological Institute Bratislava. SPN, Bratislava, Slovakia, 93pp (in Slovak)

Šamaj F, Valovič Š, Brázdil R (1985) Daily precipitation totals with extraordinary substantiality in Czechoslovakia in 1901–1980. In: Proceedings of Works of the Slovak Hydrometeorological Institute, Vol. 24. Alfa, Bratislava, Slovakia, 9–113 (in Slovak)

Smithers JC, Schulze RE (2001) A methodology for the estimation of short duration design storms in South Africa using a regional approach based on L-moments. J. Hydrol., 241, 42–52

Storch H von, Zwiers FW (2000) Statistical Analysis in Climate Research. Cambridge University Press, 484pp

Szolgay J, Hlavčová K, Lapin M, Danihlík R (2003) Impact of climate change on mean monthly of runoff in Slovakia. Slovak Meteorological Journal, 6(3), 9–21

Wilks D (2005) Statistical methods in the atmospheric sciences. 2nd ed. Academic Press, San Diego, 648pp

Wind Risk Assessment in Urban Environments: The Case of Falling Trees During Windstorm Events in Lisbon

A. Lopes, S. Oliveira, M. Fragoso, J.A. Andrade and P. Pedro

Keywords Strong winds · Trees in urban streets · Urban environment · Lisbon

Introduction

Planting trees brings many benefits to the urban environment (Fabião 1996; Jim and Liu 1997; Nilsson et al. 2000; Saebe et al. 2003; Soares and Castel-Branco 2007). Trees have a moderating effect on the urban microclimate and improve the physical, biological and chemical aspects of the urban environment, namely by reducing the urban 'heat island' effect (Alcoforado 1992), by acting as a barrier against strong wind channelling (Lopes 2003), by protecting urban surfaces from direct sunlight (McPherson and Muchnick 2005), by capturing air pollutants and dust in urban areas (Freer-Smith et al. 2004) and by increasing the biodiversity, providing habitat for birds and small mammals (Clergeau 1996). Trees also protect urban surfaces by reducing the impact of rainwater, while their roots remove nutrients that can be harmful to the water in urban soils, hence improving urban hydrology and controlling erosion. In addition, trees in urban settings sequester carbon, providing a helpful hand to combat climate change, reduce the energetic demand of the city, influence thermal and mechanic comfort and foster citizens' wellbeing (e.g. affecting physical and mental health, aesthetic and socio-economic values, common heritage, recreation benefits, etc.). A 1999 World Health Organization report revealed that in three European countries (Austria, France and Switzerland), more people died prematurely due to the effects of pollution from vehicle emissions than due to car accidents. Since a large proportion of the population is expected to continue living in urban areas, where current EU standards for PM and long-term average nitrogen dioxide are exceeded, owing mainly to road-traffic emissions (Krzyzanowski et al. 2005), the plantation of trees in these areas must be strongly encouraged. However, excessive plantation of trees and obstructive 'green masses' in ventilation paths should be avoided, because it can reduce mean wind speed and deplete dispersion conditions (Lopes 2003; Alcoforado et al. 2005).

On the other hand, trees' benefits for the urban environment are frequently accompanied by hazards caused to the human population and infrastructure, particularly during strong wind events, due to total or partial tree falls in public and private open spaces. This is one of the major causes of human injuries (and, sometimes, death) during extreme events. For example, the extreme windstorms of 26–28 December 1999 were directly responsible for killing 95 people, injuring 11 and affecting approximately 3,400,000 people in France; killing 15 in Germany; 11 in Switzerland; 11 in the United Kingdom and 5 in Spain. The total cost of the disaster was estimated at more than USD 10 billion (EM-DAT: the OFDA/CRED International Disaster Database). During this event, wind speed exceeded 160 km/h in many places along the French coast and along the 49°N parallel. Near Paris, in Versailles, more than 4000 trees were windthrown as a result of the storm and the French *Office National des Forêts* reported that an estimated 1,300,000 trees were blown away

A. Lopes (✉)
Centre of Geographical Studies, University of Lisbon, 1600-214 Lisbon, Portugal
e-mail: antlopes@fl.ul.pt

K. Strelcová et al. (eds.), *Bioclimatology and Natural Hazards*,
DOI 10.1007/978-1-4020-8876-6_5,

throughout the country (http://www.onf.fr/reg/Ile-de-France/versailles/tempet.htm).

The Constraints of Trees in the Urban Environment

Windstorm damages to trees greatly depend on the physical conditions imposed by the urban environment. In fact, the environment in urban areas is generally unfavourable for most arboreous species because climate, sunlight, soil conditions and air quality are more restrictive than in rural areas. The severity of urban conditions is not uniform, since it depends on the location of urban trees (Fabião 1996; Nilsson et al. 2000): trees planted in paved areas (*street trees*) are exposed to a relatively higher level of stress than those planted in urban parks (*park trees*) or in small urban forests (*urban woodlands*). Therefore, the average life span of *street trees* is generally shorter than those of *parks trees* or trees planted in *urban woodlands*. Furthermore, the number of species used in urban planning in paved areas is fewer than those used in the other urban environments. For example, about 55% of all trees planted in Spanish paved areas (Garcia-Martin and Garcia-Valdecantos 2001) or in Lisbon streets (Soares and Castel-Branco 2007) are represented by only five genera.

Trees planted in the urban environment, namely street trees, must be adapted to higher temperatures (*heat island effect*) and higher shadow levels than those observed outside urban areas. Building construction and paving are responsible for some important changes in soil conditions, such as the lowering of the phreatic water table, the removal of the soil surface layer and the soil compaction. This reduces the soil volume available for root expansion, the available nutrient content and soil permeability in addition to restricting the water and air uptake by the roots. Street trees are also damaged by the effects of air pollution, mainly by gas emissions from industries and cars, poisoned by salts, gas, oil and various piped chemical products from industries. Furthermore, they are damaged by severe and inaccurate pruning practises or by occasional events like windstorms, hail fall and physical impacts caused by cars and building materials. Finally, street trees can also be damaged by inadequate procedures during planting, particularly those concerning the planting depth, the dimension of the planting holes and transplant shocks caused by changes in urban dynamics regarding their proximity to new paving or the opening of ditches, and also by insects and diseases (Fabião 1996; Saebe et al. 2003).

Thus, the survival and failure of urban trees are a consequence of the simultaneous action of all of the abovementioned abiotic and biotic factors. As such, the ability of trees to resist these environmental factors must be taken into account both by selecting wind-resistant species and evaluating the damage to trees caused by windstorms. A simple relationship between the observed damage and the measured wind speed is not enough to evaluate wind damage to trees and other parameters must also be considered.

Wind Damage to Urban Trees

The effects of windstorms on urban trees also depend on the inherent characteristics of each of the species (e.g. age, size, height, bearing, density of the foliage, width of the fustic, density of the wood, etc.), as well as its cultural and health conditions. After two hurricane events in Florida in the United States (Hurricane Erin, 2 August 1995, with 37 m/s sustained wind and peak gusts of 45 m/s; and Hurricane Opal, 4 October 1995 with 56–65 m/s, respectively), Duryea (1997) noted that 'species that grow into large trees were more likely to cause property damage than small trees'. Sand pine (*Pinus clausa*), slash pine (*Pinus elliottii*), longleaf pine (*Pinus palustris*) and laurel oaks (*Quercus laurifolia*) were more likely to cause damage if they fell than the smaller Southern redcedar (*Juniperus silicicola*), Carolina laurelcherry (*Prunus caroliniana*) and Chinese tallow (*Sapium sebiferum*). Besides the size there is a predisposition to Florida native trees, like the dogwood (*Cornus florida*), sand live oak (*Quercus geminate*), live oak (*Quercus virginiana*), sabal palm (*Sabal palmetto*) and Southern magnolia (*Magnolia grandiflora*), to tolerate hurricane-force winds extremely well. Some foreign species have the predisposition to be affected by insects and disease after experiencing strong winds and are less resistant (Duryea 1997).

Trees in natural environments are optimised structures which are subjected to static and dynamic loads (Mattheck and Breloer 1994). According to James (2003), the strength of each structural member depends on the size of the cross-section (the base of the trunk is the strongest part where static loads are the greatest), the shape of the cross-section (a circular section is best adapted to torsional forces) and the strength of the material (young wood is more flexible than old wood). Trees also show some physical features of adaptive growth against wind loads (Niklas 2002). For example, the growth rings are thicker on the side where the dynamic (wind) load is greatest and thinner on the opposite side, while the base of a trunk or branch is massive due to the addition of wood and, thus, represents a localised reinforcement to withstand the high (static) loads. In the urban environment, trees experience a wind dynamic load that produces a complex sway motion in trees: more lateral movements are caused by wind than in forests, where trees are protected by their neighbours. Furthermore, the energy is transferred from the wind to the main structure of the trunk (damping effect) and, consequently, dissipated by the leaf drag, wood and root/soil system, and partly by branch sway (James 2003).

Urban Climate and Environment of Lisbon

Lisbon is the capital and largest city of Portugal and is located at 38°43′N and 9°09′W, about 30 km to the east of the Atlantic shore, on the bank of the Tagus estuary. The climate is Mediterranean (Csa according to Köppen's classification), characterised by mild winters (61–90 climatological normal temperature, in January, from minimum 8.2°C to maximum 14.5°C) and dry summers (in August, from 17.7°C to 27.9°C). The annual precipitation is 751 mm .

The summer wind regime in the city of Lisbon is dominated by a relatively strong northerly wind that occurs near the western coast of the Iberian Peninsula, the *Nortada* (Fig. 1).

This regional circulation occurs when a strong pressure gradient is maintained by a persistent thermal depression in the warm continent and the Azores anticyclone is above the cooled ocean. The upwelling of the cooler ocean waters near the shore reinforces the pressure gradient, increasing wind speeds near the ground. In Lisbon, the *Nortada* is very important in what concerns air quality and comfort, since it promotes the dispersion of pollutants and, at the same time, reduces both the natural and anthropogenic heat loads. Alcoforado (1987) found that N and NW are the most frequent wind directions for 45% of summer days. From June to September (27%) the *Nortada* was considered strong (hourly mean wind speed ≥ 15 km/h), though wind velocities can reach 50 km/h in the early afternoon, and maximum gusts of 70 km/h (19.4 m/s) in the late afternoon. Consequently, the mechanical effect of strong winds on people can be unpleasant, especially in the leisure areas: with a wind-effective speed greater than 9 m/s, the performance of a pedestrian in a street is

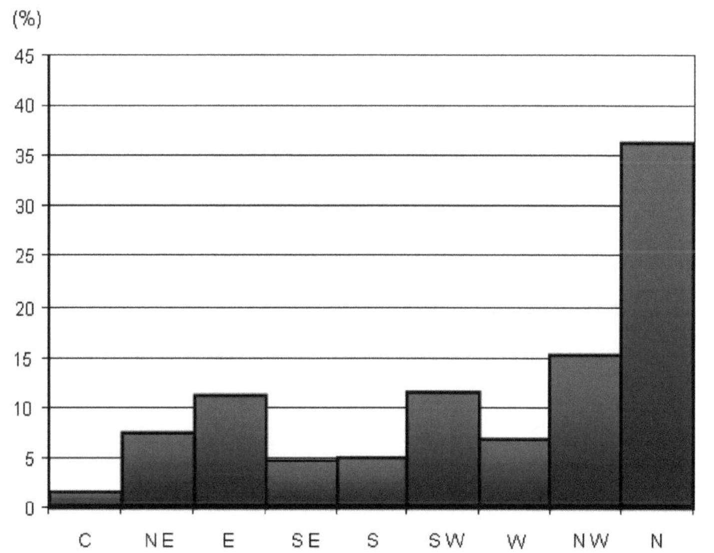

Fig. 1 Wind direction in summer (June–August) at Lisbon Airport (1971–2000), at 12.00 p.m
Source: Alcoforado and Lopes (2003)

significantly affected, while an effective wind greater than 20 m/s is hazardous for weaker persons (Saraiva et al. 1997). Experiments conducted in a wind tunnel with a physical model of the eastern part of Lisbon showed many critical points (from the security point of view) in the streets. Findings show that more than 43 h per year could be critical, with effective wind velocities greater than 20 m/s. Sudden gusts in these areas could be dangerous for weak or unprepared persons and trees should be planted to prevent damages (Saraiva et al. 1997).

In winter, winds blow mostly from SW, W, NW and N. However, from March onwards, there is a great increase in the frequency of northerly winds (Alcoforado 1992). Northeast is the most frequent wind direction in December (30%), but the north direction is still a strong component (13%). In January and February, the west direction is persistent and at the end of the winter season the north direction becomes more important (Fig. 2). Although the wind blowing from south and southwest directions are not the most frequent directions throughout the year, they can cause serious damages to urban trees (Lopes 2003).

In Lisbon, the windy weather conditions occur more frequently during the winter, when the westerly flow is more intense, bringing more tracks of frontal disturbances to the Portuguese latitudes. These winter cyclonic circulations are related to three main large-scale patterns, two of them with a clear connection to low-pressure systems located just to the west of the British Isles: (i) westerly or north-westerly winds could blow over Portugal in connection with travelling frontal systems that extend as far as southern Portugal; (ii) in other cases, windy weather conditions are related with westerly and south-westerly winds from the North Atlantic that occur over Portugal along the southern flank of cyclone systems, which have a high moisture content, particularly over northern and central Portugal' (Santos et al. 2005); (iii) a third type of large-scale circulation is related with southerly winds associated with cyclonic circulations which are, in turn, connected with the more southerly path of the westerlies in the mid-tropospheric flow. These cyclonic circulations at the lower levels are mainly controlled by low-pressure centres located to the west or southwest of Portugal.

In the last 30 years, some windstorms causing damage to trees have been reported in Lisbon, with maximum gusts of 22–33 m/s (80–120 km/h). This corresponds to a statistical return period from 2 to 8 years (Borges 1971). The OFDA/CRED International Disaster Database indicates the occurrence of a small number of deaths and injuries during windstorms in Portugal, most of them caused by fallen trees.

Green Areas and Trees in Lisbon

Since the fifteenth century, the city of Lisbon was the point of departure for the discovery of new lands. During many centuries, Portuguese ships travelled around the world in search of new trade routes and, since then, exotic tree species were introduced in Europe and many gardens of foreign species were built in the city. In the nineteenth century, parks and public gardens were integrated in new urban strategies. These green spaces today contain a wide variety of species, some being Mediterranean (or well adapted to the Lisbon climate), and also many exotics, from Southeast Asia, Central America, Australia, the Pacific Islands and even from cold climates (Moreira 1998).

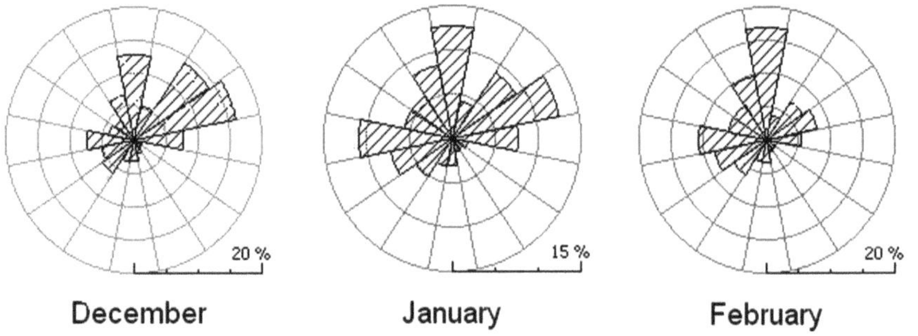

Fig. 2 Winter windroses at the Lisbon Airport (1971–1980)
Source: Lopes (2003)

At present, the *Green Index* of Lisbon (which corresponds to the mean value of green areas, mainly covered by trees) has an average value of 26.8 m^2/inhabitant (including *Parque de Monsanto*, the largest continuous green area in the urban perimeter of Lisbon); excluding the *Parque de Monsanto*, this value decreases to 9.1 m^2/inhabitant (Soares and Castel-Branco 2007) (Fig. 3).

Since the early twentieth century, several inventories and studies of the trees in the city of Lisbon were made. The first list dates back to 1929 and the last one available is from 2003 (Fig. 4). The most representative species found in Lisbon's streets are the European nettle tree (*Celtis australis* L.), the plane tree (*Platanus hybrida* Brot.), different species of limes (*Tilia* spp.), blue jacaranda (*Jacaranda mimosifolia* D. Don), box elder maple (*Acer negundo* L.) and different species of poplars (*Populus alba* L., *Populus nigra* L. and *Populus x canescens*) (Soares and Castel-Branco 2007). It was verified that the type and proportion of species varied along this period: for example, while *Celtis australis* exists in the city since 1929, *Acer negundo* was only found in 2003.

Today, more than 70 monumental trees (or groups of trees) are a part of the Portuguese heritage worth preserving: it includes a variety of species, some exotic (e.g. *Phoenix dactylifera* L., *Ficus macrophylla* Desf. Ex Pers., *Tipuana tipu* (Benth.) Kuntze, *Brachychiton* spp., *Cupressus macrocarpa* Hartweg ex Gordon, etc.) and others representative of our local flora (e.g. *Q. suber L., Q. rotundifolia Lam., Q. robur L., Q. faginea Lam., Q.coccifera L., P. pinea L., Arbutus unedo L., Phillyrea latifolia L., Pistacia lentiscus L., Viburnum tinus L., Rhamnus alaternus L., Juniperus turbinata Guss, Olea europaea L.,* etc.), most of them

Fig. 3 Lisbon's green structure
Source: Lisbon City Council

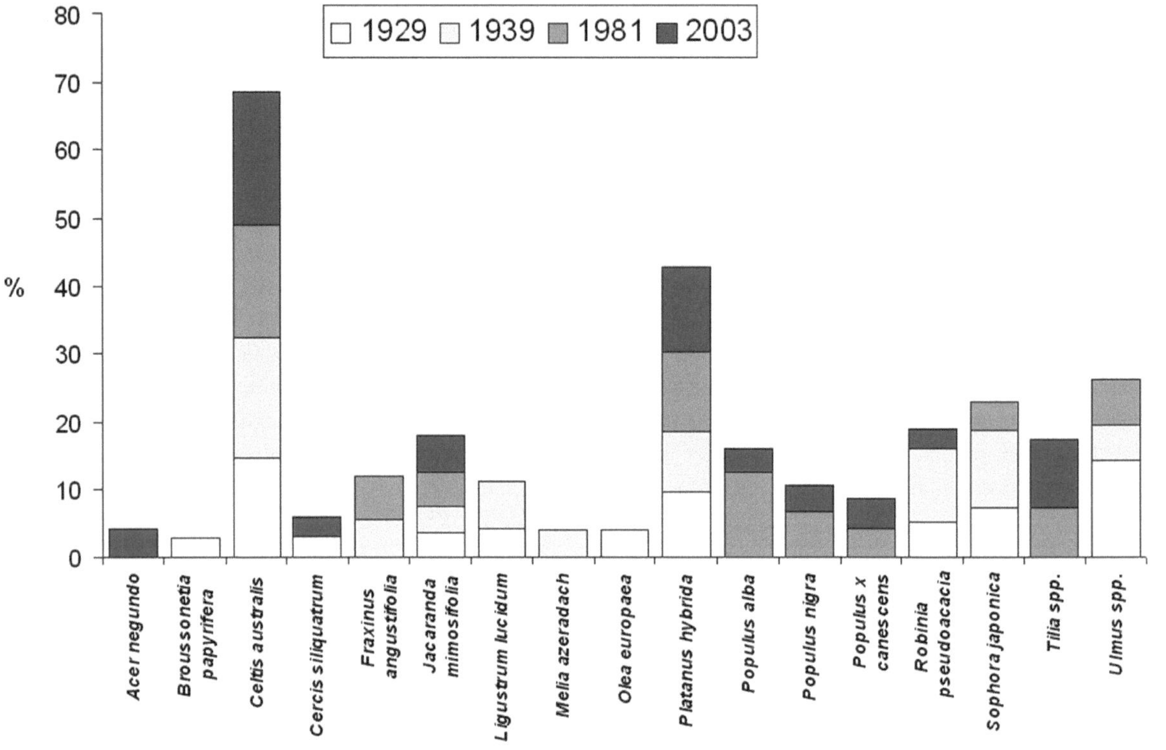

Fig. 4 The dominant tree species found in the streets of Lisbon in each inventory
Source: Soares and Castel-Branco 2007

found in parks and public gardens (11 centenary specimens) (DGRF, n.d.). Many species are adapted to the Mediterranean climate, although the urban heat island can stimulate biological activity by advancing phenological phases, for example.

Data and Methods

For this study, a series of 17 years was used, from 1990 until 2006, corresponding to the latest available data of fallen trees. These data were obtained from the information compiled by the Lisbon Fire Brigade and Rescue Services (*Regimento de Sapadores Bombeiros de Lisboa – RSBL*), which handles emergency calls and is also in charge of cleaning up the areas and notifying the Lisbon City Council. The time of the fallen trees occurrences is registered in the RSBL database as the time of the call to the call centre. Information about the place where it occurs includes the description of the street name and the nearest door number, crossroad, light pole number or a building easy to identify (e.g. school, church, bank, etc.). In few cases, it is possible to find specific details about the species, the dimension of the tree, the probable cause of the occurrence and the damages caused by the event.

However, it is important to point out that the information collected by the RSBL has certain limitations:

- The time registered by the fire brigade corresponds to the time of the emergency call and not to the tree fall; there may be a significant time difference between the occurrence and the call, which remains unknown;
- The occurrences recorded refer only to the incidents that have caused disturbances or damages to people, which excludes the falls that occur where no one is affected;
- Information on the species, dimension, age and health conditions of the trees, which is essential to understand the relations between the meteorological conditions and the fall of trees, is seldom included;
- The methodology applied by the RSBL to register the occurrences changed during the study period. The record books changed in 1994 and 1997

and, consequently, so did the type of information included. In addition, the method of registry and the terminology used among the RSBL technicians are not well defined.

Until 1998, the occurrences were recorded in a book and, after that, on a digital format. The data recorded in paper had to be collected from the archives of the RSBL and inserted into a digital database created specifically for this work. Considering that this was a time-consuming task and that the main purpose of this study was to assess the adequacy of the proposed methodology, it was not possible to make a visual check of all available data. A selection of the days between 1990 and 1998 with a higher probability of occurrences was carried out, based on the analysis of meteorological data (only when strong wind events occurred). The threshold for assuming the occurrence of these events was based upon basic statistics and is explained later in this section.

The meteorological data used were hourly wind speed and direction at the Lisbon/Airport meteorological station. Due to the impossibility of verifying how much time had elapsed between the incident and the emergency call, each occurrence in the database was matched with the wind data (speed and direction) recorded at the start of the hour. In addition, each occurrence also corresponded to the mean and maximum wind speed values recorded in 3, 6, 9 and 12 h prior to the call put in to the fire brigade. To ensure that only occurrences caused by meteorological conditions were included, only days with three or more incidents were selected. At the first stage, this analysis was applied to the set of data available initially for the period of 1999–2003. It was found that the majority of tree falls happened when certain wind speed limits were surpassed, specifically above 7 m/s for the time period that included the 6 h prior to the call. The same methodology was then applied to the data for the period of 1990–1998 and the days with higher probability to the occurrence of tree falls due to meteorological conditions were selected, based on the threshold found. The collection of data from the archives of RSBL was carried out afterwards. It was verified that the methodology applied to select the days was rather precise and efficient; even so, 2 days before and after each occurrence were also visually checked.

Finally, all the information obtained was introduced into a GIS in order to detect the relationships between the tree falls and the characteristics of the places where they occur, such as the H/W ratio, aerodynamic roughness, distance to high-density areas, relation between the wind direction and street orientation, etc. Some technical problems were encountered, namely the precise location of the trees. In such cases, the central coordinate of the street, or the nearest feature found was attributed as the location of the occurrence. The problems encountered will be addressed with the fire brigade in order to facilitate the future collection of data. Some suggestions are put forth in this chapter's 'Conclusions and Further Research' section.

The authors collected newspapers and online news articles regarding the occurrence of extreme climatic events for the last several years. Such data has proven to be an important source of information, with a necessary filter attitude. Some examples are presented later in this chapter.

Although the authors obtained monthly data, here a general view by 'season' has been adopted in order to simplify the results. The months are not aggregated by means of astronomical or thermal criteria. Instead, we used the methodology of Lopes (2003), to obtain the general wind regime at the Lisbon Airport meteorological station. The author used a series of 10 min wind speed and direction data, from the period of 1971–1980. A tree diagram and a principal component analysis of both *A* and *K* of the Weibull function revealed that September has a clearly different wind regime from other summer months (June, July and August), and was included in the autumn months (September–November). Winter comprises the period from December to February. March, April and May were considered to be the months of spring.

Preliminary Results

Tree Falls in Lisbon from 1990 to 2006: A General Overview

Between 1990 and 2006, 1241 occurrences of trees, boughs and branches falls were recorded during strong wind events. As can be seen in Fig. 5, there was a large variation in the number of occurrences by year: the highest percentage of incidents occurred in the years of 2006, 2000, 2001 and 1997. Over 70% of the incidents occurred in the 2000–2006 period, with a maximum of 23.8% in 2006 and 14.3% in 2000. In June of the latter year, strong winds (wind gusts of 85 km/h $\approx$ 24 m/s) were reported in the Lisbon region (where a person who was waiting for the bus was badly injured); in the beginning of November, a wind storm swept throughout the country (with floods and high-intensity precipitation that surpassed 90 mm in a period of 24 h in some places in northern Portugal). On 6–7 December, a very active cold front passed over the country and the Lisbon region was very affected (gusty winds of 100 km/h). The civil protection service alerted the population about the event, and confusion set in the city, with thousands of people working in the city that live in the suburbs trying to escape the storm. This situation led the city into chaos (as noted by Portuguese newspapers), which is a good example of how civil authorities should deal with the divulgation of information. 'The repeated warnings about bad weather, the deterioration of the situation throughout the day and the expectation that it might worsen at nightfall, all left the population in a state of comprehensible anxiety' (*Jornal de Notícias online edition*, 7 December 2000). In the end, the city sustained only light damages, though it led to great fright due to rumours.

When the 17-year period was analysed according to the wind direction, it was found that over 58% of the tree falls occurred with southwest and south winds (31.2% and 27.4%, respectively), 13.4% were caused by west wind and 12.5% from winds from the northwest. Only 1.3% of the occurrences were caused by east winds (Fig. 6a).

When the data was analysed by season, it was found that more than 76% of tree and branch falls in the city occurred during the autumn and winter. In the summer, the N and NW winds are dominant, while in the other seasons, winds from the S and SW cause the majority of falls (Fig. 6b). The first situation is due to the prevailing *Nortada*, as explained above. In the 'coldest' time of the year the 'extreme' synoptic situations are due to a relatively high frequency of south depressions and storms associated to the cold front.

The wind speed also varied according to the direction. At the time of the occurrences, winds recorded from south, southwest and north reached higher speeds, over 7 m/s. The averages of the maximum speeds recorded for the 3, 6 and 12 h prior to the falls were calculated and compared (Fig. 7). It was found that the maximum speed increases with time before the emergency call, meaning that the maximum wind speed for the period of 12 h prior to the occurrence is higher than the maximum speed for the previous 3 and 6 h periods (Fig. 8). These results can be due to the time lag existing between the moment of the occurrence and the moment it is recorded in the database. Also, many situations may occur during the night and only in the morning, after people wake up, the call is made. The west direction is a curious situation, as the value from 6–12 h prior to the occurrence is greater than in all other directions.

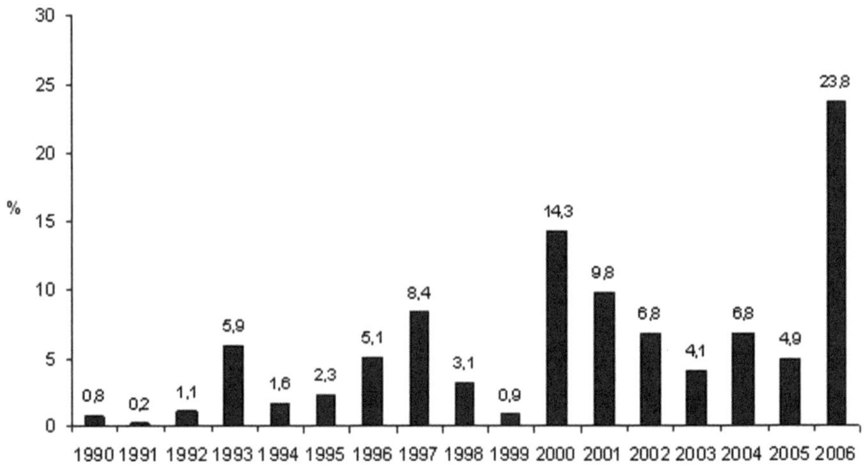

Fig. 5 Percentage of occurrences of trees, boughs and branch falls registered in the database for the period of 1990–2006

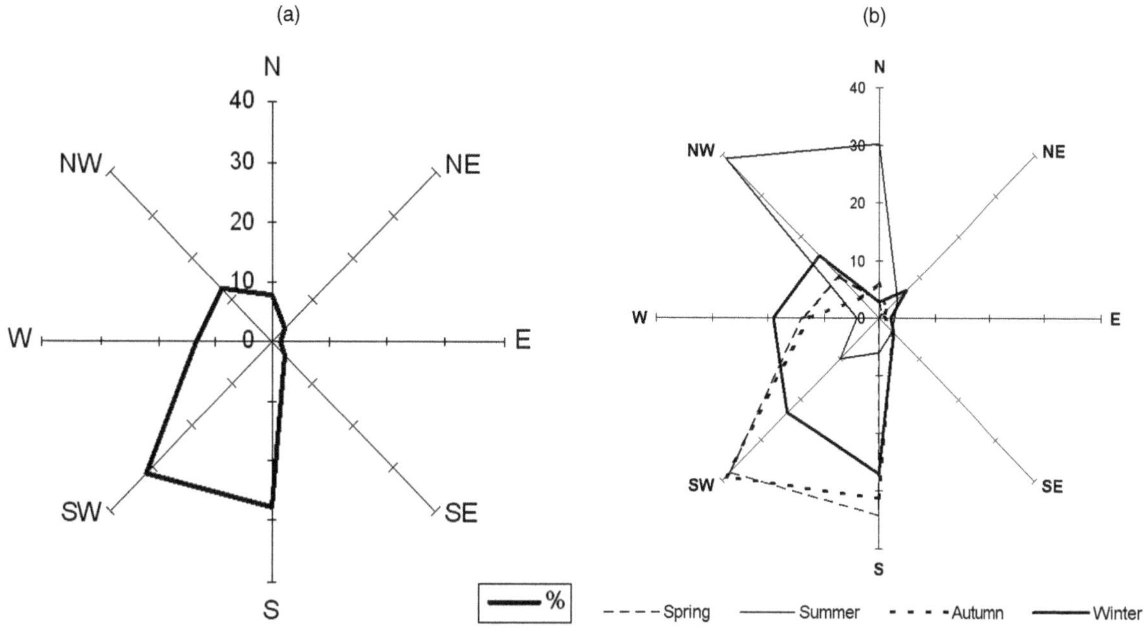

Fig. 6 Tree falls according to the wind direction (**a**) and season (**b**)

The falls occurred relatively disperse in the city. Even so, it is possible to distinguish specific patterns of location in relation to the total number of street trees in the different places.

The location of the falls was analysed by *freguesia* (the smallest administrative division in Portugal and the basis for local actions). Lisbon has 53 *freguesias*, with different characteristics, including dimension, population and number of street trees. It was verified that, in general, a higher number of falls occur in the *freguesias* with a higher number of street trees, with few exceptions, the most important being Marvila (it has a high number of street trees and few occurrences) and Olivais (it has a medium number of street trees, smaller than Marvila, but a higher number of occurrences (Fig. 9).

These exceptions may be explained by the roughness, the topographic features of the areas and

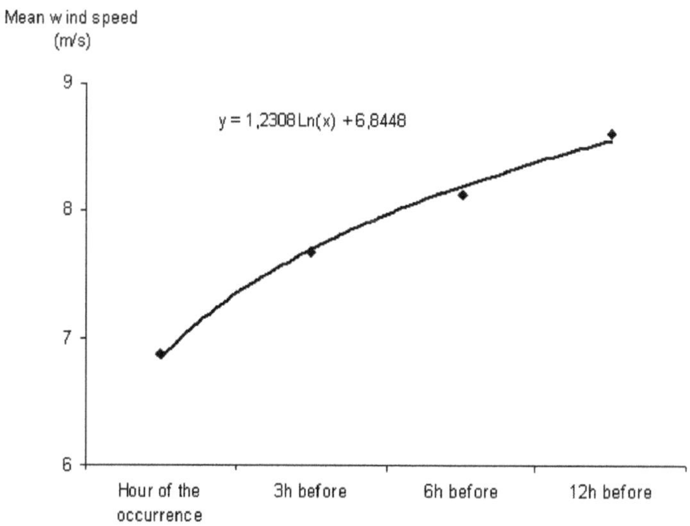

Fig. 7 Mean wind speed at the hour of the falls and in the previous hours

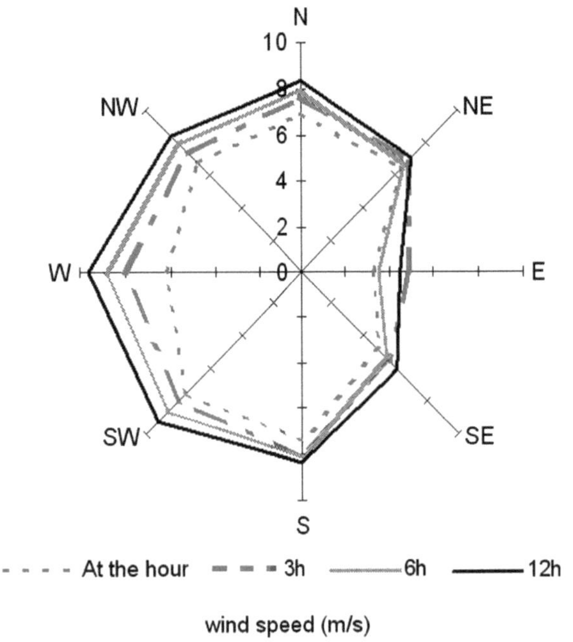

Fig. 8 Maximum wind speed at the hour of the falls and in the previous hours

the type of species found. The inventory of 2003 shows that the most representative species in Olivais are *Platanus* spp. (15.4%), *Tilia* spp. (9.4%), *Celtis australis* (8.7%), *Jacaranda mimosifolia* (8.3%), *Cercis siliquastrum* (7.1%) and *Populus Canadensis* (6,9%), and in Marvila are *Platanus* spp. (19.1%), *C. australis* (16.8%), *Gleditsia triacanthus* (6.4%), *Grevillea robusta* (6.1%), *Cercis siliquastrum* (5.8%) and *Robinia pseudoacacia* (5.7%) (Soares 2006). These differences can be a result of the fyto-sanitary conditions of the trees and of the urban characteristics such as geometry and street orientation (Sashua-Bar et al. 2003), which must be investigated further.

The Atmospheric Environment of the Windstorms: Lisbon's Soundings and Stability Indices (2000–2005)

This part of the study aims to analyse the incidence of thermodynamic instable conditions in the troposphere among the falling trees events in Lisbon. Soundings data are usually applied to assess the potential conditions for the occurrence of severe weather phenomena, like heavy showers, thunderstorms or wind gusts. A systematic analysis of sounding data from Lisbon Airport station seems a useful approach to identify some of the thermodynamic factors of the windstorms. By the moment, this specific research is just in the beginning and it was not yet possible to analyse all the periods studied in this chapter (1990–2006). In this topic, preliminary results obtained from the analysis of a subset of tree falls events are shown, and this secondary sample was constructed by selecting the occurrences registered in the last years of the initial dataset (since 2000 until 2005). The availability of sounding data from the Lisbon Airport station has made possible the collection of 65 soundings related with tree falls events in the described 6-year period. Using RAOB[1] software the sounding data were analysed and different indices and parameters were computed taking into account two objectives: (i) to ascertain the frequency of unstable thermodynamic conditions associated with the tree fall events in Lisbon; (ii) to verify the importance of wind shear conditions to trigger the windy weather conditions in Lisbon area. Here, the results concerning three stability indices will be presented and discussed:

(a) the '*convective available potential energy*' (CAPE), a measure of positive buoyancy:

$$CAPE = g \int_{LFC}^{EL} \frac{\theta - \theta_e}{\theta_e} \qquad (1)$$

where

θ is the potential temperature (°C) of the lifted parcel

θ_e is the potential temperature (°C) of the environment

g is the acceleration due to gravity

(b) '*severe weather threat*' (SWEAT, nondimensional), an index that combines the role of humidity at the lower levels, the 850–500 hPa thermal lapse and the 850–500 hPa wind shear:

$$\begin{aligned} \text{SWEAT} = {} & 12.\text{Td}_{850} + 20.(\text{TT} - 49) \\ & + 2.f_{850} + f_{500} + 125.(S + 0.2) \end{aligned} \qquad (2)$$

where

[1] Version 5.7, Environmental Research Services, Matamoras, USA.

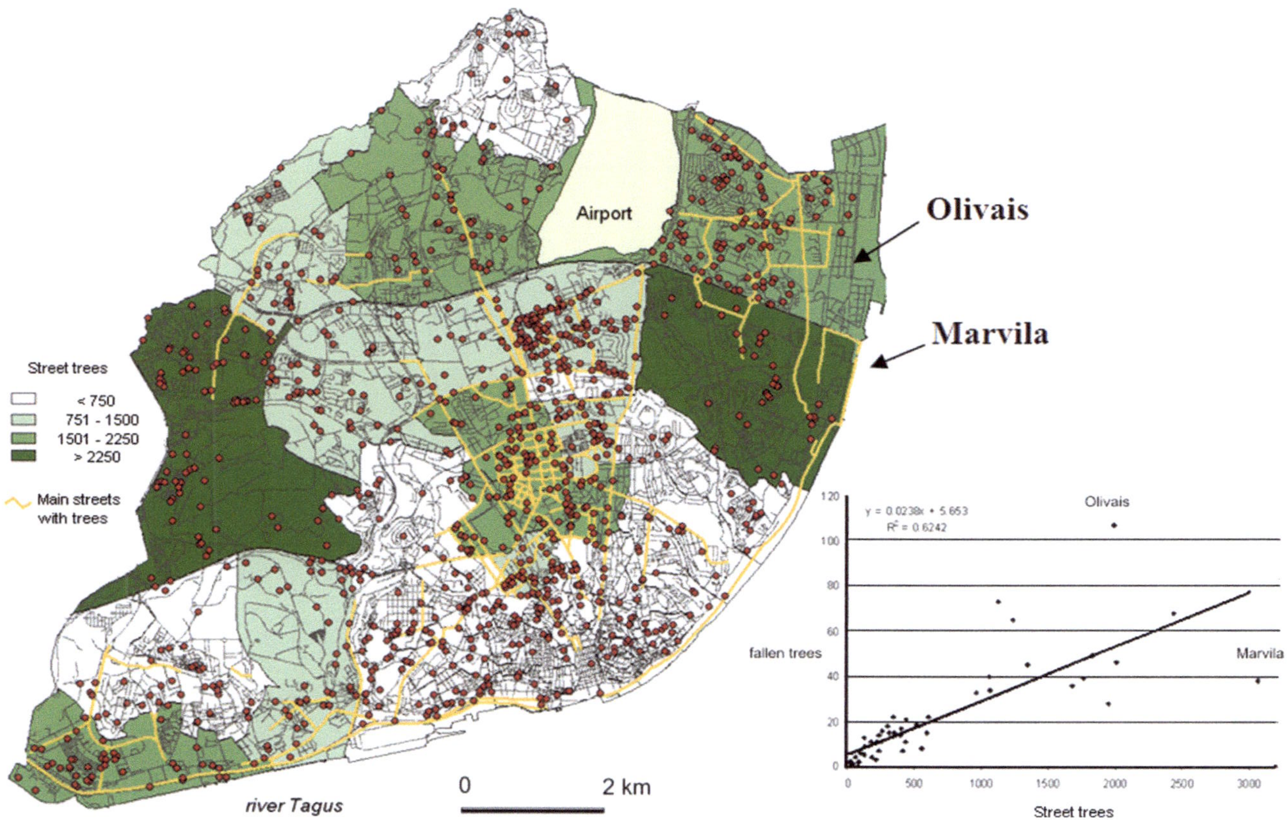

Fig. 9 Location of the falls occurred between 1990 and 2006. The *freguesias* are classified by the total number of street trees. The graph shows the regression results and the peculiar cases of Olivais and Marvila

TT is the Total Totals índex (and TT-49 is set to zero if value is negative),

f_{850} is the speed of 850 hPa wind in knots,

f_{500} is the speed of 500 hPa wind in knots and

S is the sine of (500 hPa–850 hPa wind directions)

(c) Wind shear (m/s), defined by the magnitude of the vector difference between the 0–6 km mean wind and the 0–500 m mean wind.

The results of the computed indices for the 65 soundings related with tree fall events need to be compared with correspondent values associated with 'non-windy' weather conditions in Lisbon Airport, in order to ascertain the significance of this analysis. Therefore, a random sample of Lisbon soundings related with weak wind conditions in Lisbon Airport station was also created. This random set of soundings includes only soundings at 12.00 p.m. from days with wind below 7 m/s between 10.00 a.m. and 3.00 p.m., and it has the same amount of cases (65 soundings) observed in the previously mentioned 6-year period (2000–2005). Thus, two samples of soundings are compared in this study: one comprises soundings related with falling trees days (FTD sample) and the other one includes soundings associated with weak wind conditions (WWD sample). It is expected that this comparison between the indices values from the two opposite samples could contribute to ease the detection of significant thresholds related with windstorms triggering.

The events with positive buoyancy (positive CAPE) are quite more frequent in the falling trees days sample than in the weak wind days sample. In Table 1, the frequency (absolute and relative) of soundings with positive CAPE in the two samples is indicated and it must be underlined that the number of cases related with falling trees is much higher (46% against 23%). However, it should also be emphasised that both kinds of events could happen with positive buoyancy conditions. The CAPE value has exceeded 1000 J/kg in four cases in the FTD sample. No case with CAPE $\geq$ 1000J/kg has occurred in the WWD sample.

The distributions of CAPE values could be compared in a box-and-whiskers plot (Fig. 10), representing the maximum, the third quartile, the median, the first quartile and the minimum.

The comparison of SWEAT index values reveals a much stronger contrast between the two distributions (Fig. 11). Almost 80% of the soundings from the FTD sample did occur with SWEAT $\geq$ 100 (moderate conditions for convection) and about the same percentage was observed for SWEAT values below the same threshold in the WWD soundings set. This result suggests that the combination of humidity, strong thermal lapse rate and wind shear bellow 500 hPa could be important to induce surface windy conditions in Lisbon. In fact, the clear distinction between the two distributions may indicate that this index could act like an important predictor of windy conditions in Lisbon area, a result that must be confirmed by further research. Furthermore, it can be noticed that, on three occasions (in a 65 days sample), SWEAT values above 300 were observed, suggesting that very severe conditions (potential) for deep convection are not so rare in the Lisbon region, as it could be supposed.

The comparison of the wind shear values distributions (Fig. 12) exhibit, once again, another clear distinction between the two samples, although not so strong as in the previous index. More than half (54%) of the total number of FTD occurred under moderate ($\geq$ 7 m/s) or severe wind shear conditions. Seventeen percent of the FTD occurred with strong wind shear ($\geq$ 10 m/s). By contrast, 80% of the WWD are related with lower wind shear values (7 < m/s). The threshold of 7 m/s could be taken, empirically, as a discriminative value that could roughly distinguish the most part of the cases of the two sets of days, suggesting that it

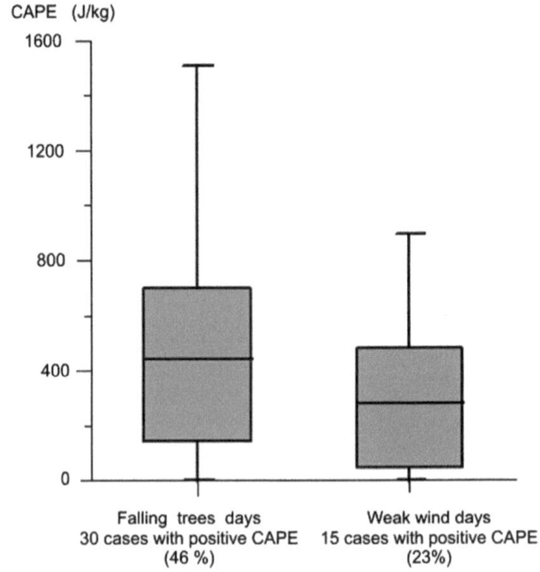

Fig. 10 Box-and-whiskers plot of CAPE values

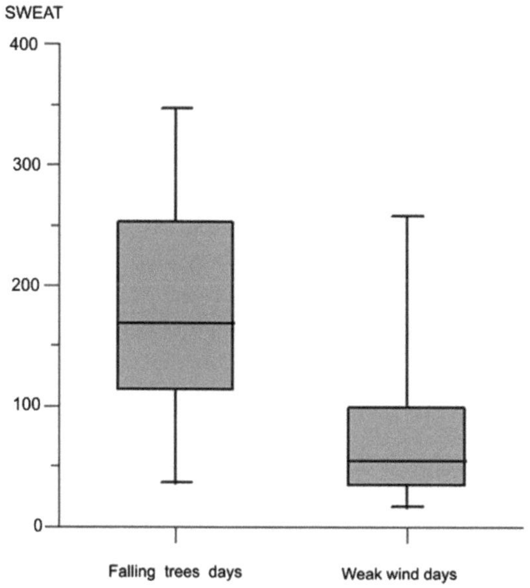

Fig. 11 Box-and-whiskers plot of SWEAT values

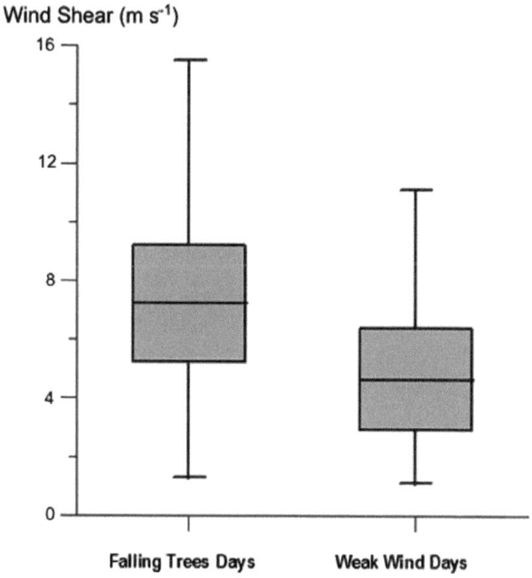

Fig. 12 Box-and-whiskers plot of wind shear values

might be a critical value useful for windstorms prediction in Lisbon Airport. Therefore, it will be necessary to verify this assumption in a larger sample of soundings, in order to achieve more concluding and statistically significant results.

The results described pointed out that the conjugation of positive buoyancy, humidity at the lower troposphere and strong wind shear seems to be an important factor to increase the probability of windy weather conditions in Lisbon. The prior observations allowed us to make a summary of the main thermodynamic and wind shear conditions associated with falling trees events (Table 1). In this table, the main conclusions and remarks, extracted from the frequency analysis based on some critical values, are indicated.

The Example of the Windstorms of 8th October 2004

On 8 October 2004, the Iberian Peninsula was under the influence of a low-pressure centre located in the Atlantic Ocean. Consequently, the flux was from south and southwest over Portugal.

Table 1 Comparison (empirical thresholds) of the stability conditions between the soundings samples of FTD and WWD

Magnitude of instability conditions (thresholds)	FTD (65 soundings in 2000–2005 period) absolute/relative frequency	WWD (65 soundings in 2000–2005 period) absolute/relative frequency
Positive CAPE	30 (46%)	15 (23%)
SWEAT $\geq$ 100	51 (79%)	14 (21%)
$\geq$ 200	25 (39%)	2 (3%)
Wind Shear $\geq$ 7m/s	37 (57%)	14 (21%)
$\geq$ 10m/s	12 (19%)	1 (1.5%)
(Combined conditions)		
SWEAT $\geq$ 100 + Wind shear $\geq$ 7 m/s	35 (54%)	7 (11%)
Positive CAPE + SWEAT $\geq$ 100 + Wind shear $\geq$ 7 m/s	15 (23%)	1 (1.5%)

A relatively high number of trees, boughs and branches (35 occurrences, recorded between 9.30 a.m. and 6.40 p.m.) have fallen in the city. The wind speed began to increase significantly from 9.00 a.m., reaching a maximum of 15 m/s at 2.00 p.m., with its speed considerably decreasing from that time onwards (Fig. 13).

As the wind speed increased, its negative impact on the city became clearer, and the tree falls were directly related to it. Between 12 p.m. and 4 p.m., the wind speed varied between 13 and 15 m/s and 80% of the tree falls occurred (28 of the 35). This peak in occurrences coincided with the period of maximum gust (27 m/s).

Most of the occurrences took place in the south and southwest parts of the city, in relation to the abovementioned synoptic situation. In the morning (prior to 1.00 p.m.), the falls were caused by flows from the south; in the afternoon, a wind shift towards the southwest led to the remaining occurrences (Fig. 14).

The distribution of falls was analysed in relation to the different land uses and ventilation classes defined for the city of Lisbon (Alcoforado et al. 2005). It was found that 41% of the trees fell in densely built-up areas (where at least 50% of the area is covered with high-density buildings) in the southern part of the city and 20% in green areas. Although these districts have greater roughness length ($z_0 \approx 1$ m) and therefore diminish regional wind speed by about 30% when compared to non-occupied areas (Lopes 2003), local acceleration through the narrow streets can occur during extreme wind events. Therefore, an index capable of translating the narrowing of the streets should be included in the analysis. This index could be given by the relation between the height of the buildings and the width of the streets (height/width) and should be applied in future research.

In relation to the species affected by these events, it was possible to identify 24 fallen trees, which were distributed among 8 species (Table 2). It was found that 48% of the trees only suffered damages to their upper parts (e.g. branches and boughs) and over half (52%) were uprooted. Of all the cited species, the black poplar was the most affected by this strong event, corresponding to one-quarter of all tree falls. In this case, the cause of the falls is not very well understood and other constraints, especially the fyto-sanitary conditions, may have a great influence on the resistance of trees. In future research, all the conditions should be included in the database and should be analysed in depth.

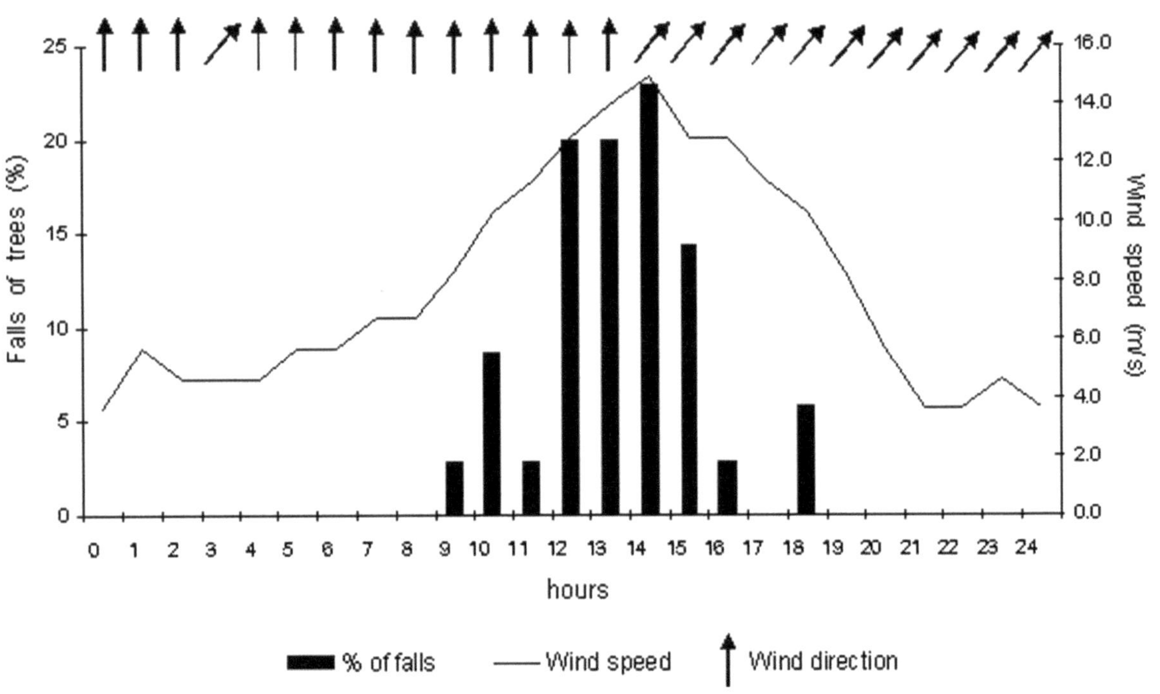

Fig. 13 Relation between tree falls and wind (speed and direction) on the 8 October 2004

Fig. 14 Pattern of fallen trees according to wind direction during the windstorm on the 8 October 2004

Table 2 The characteristics of the trees identified for the episode of 8th October 2004

Species	Percentage of trees	Tree foliage	Average height
Populus nigra L. (black poplar)	25	Deciduous	17
Sophora japonica L. (Japanese pagode tree)	21	Deciduous	12
Acer negundo L. (box elder)	18	Deciduous	15
Jacaranda mimosifolia D. Don. (jacaranda)	12	Semi deciduous	11
Tilia tomentosa Moench (lime)	8	Deciduous	19
Ulmus minor Miller (elm tree)	4	Deciduous	20
Cercis siliquastrum L. (Judas tree)	4	Deciduous	7
Ficus carica L. (fig tree)	4	Deciduous	4

Atmospheric Analysis of Two Windstorm Events in Lisbon: 18th July and 24th November 2006

This section presents a diagnosis of the atmospheric environment associated with two severe windstorms triggered in quite different synoptic situations and thermodynamic conditions.

The first windstorm (18 July 2006) was a local storm, with strong gusts probably induced by convective motions and instability in the troposphere above the Lisbon area. The second windstorm (24 November 2006) was related with a clear synoptic forcing

Table 3 Sounding-derived parameters

Indices (Units)	Lisbon Airport 1200 UTC
CAPE (J/kg)	247
DCAPE (J/kg)	994
WINDEX (knots)	40
T^2Gust (knots)	37
SWEAT (non-dimensional)	227.6

and high pressure gradients affecting the surface atmospheric circulation over the Portuguese area.

The 18 July windstorm occurred under moderately warm weather conditions. In the previous days, the synoptic situation was influenced by the Azores subtropical high. After 17 July, the surface circulation (Fig. 15a) over the western part of the Iberian Peninsula became conditioned by shallow (thermal) low pressure, with southerly (SW to SE) winds into the Lisbon area in the low levels of the troposphere. Nocturnal temperatures remained above 20°C and the maximum temperature reached 32°C at the Lisbon/Airport meteorological station. In this atmospheric environment, local thunderstorms were formed in the southwestern part of Portugal on the 18 July 2007 and the associated convection was responsible for a windstorm affecting Lisbon by the end of the morning, with the most intense wind records observed at the Lisbon/Airport meteorological station. The hourly wind speed at this location increased irregularly during that morning. The maximum hourly wind speed was recorded from SE, with a mean speed of 10.8 m/s (with a maximum gust of 18 m/s) and occurred at 11.00 a.m.

The analysis of the 1200 UTC Lisbon sounding (Fig. 15b) allowed us to understand the dynamic causes of this windstorm, taking into account that the maximum gusts occurred around that moment. Combining different sounding-derived parameters, some features of the thermodynamic factors of the windstorm could be found. Table 3 shows several indices and parameters of instability.

The thermodynamic structure represented in the sounding (Fig. 15b) exhibits different important features favourable to the activation of strong downdrafts, probably triggering gusts in the lower troposphere. The thunderstorm activity has started earlier, but the CAPE value (247 J/kg) suggests its possible continuation to the afternoon period. A jet stream at the upper levels was also present, underlying a troposphere marked by strong wind shear. But it is most interesting to verify the effects of a thick dry air layer below 700 hPa, inducing strong negative buoyancy, with the DCAPE value reaching 994 J/kg (Table 3). This highly instable environment has resulted in downdrafts that, very probably, were at the origin of the strong gusts experienced at the surface in the Lisbon area. This event was responsible for 11 fallen trees, boughs and branches.

The second windstorm studied occurred on 24 November 2006 and its dynamic causes were quite different from those of the previous event. The synoptic circulation over the Iberian Peninsula (Fig. 16a) was controlled by the influence of an intense frontal cyclone (low minimum of 969 hPa at 0000 UTC), its centre being located west of the British Isles. On 24 November, Portugal was affected by strong surface pressure gradients, ensuring the occurrence of persistent windy weather conditions. The daily mean hourly wind speed observations at the Lisbon/Airport meteorological station was 9.5 m/s, from southwest or south in almost all the time records. By 9.00 p.m., the wind gust was at 23 m/s.

By the end of the morning, the Lisbon area was crossed by a cold front moving from west to east, and the most intense winds occurred just after this passage. Wind speed observations between 11 and 13.5 m/s were registered from 12.00 to 2.00 p.m. The sounding (Fig. 16b) reveals strong south-westerly fluxes from the upper troposphere (where a jet stream is visible) until the surface levels. The most interesting feature in this profile is a low-level jet, with 29 m/s, blowing just 520 m above the ground level.

This thermodynamic profile (Fig. 16b) shows a moderate highly instable atmospheric environment, favourable not just for the occurrence of surface gusts but also to thunderstorm activity and thick cloud formation, taking into account the availability of the moisture (precipitable water: 36.1 mm) and convective energy to freely activate the lifting of the air from the lower levels. This event was responsible for the fall of 112 trees, boughs and branches.

Conclusions and Further Research

The fall of trees, boughs or branches occurs quite frequently in the city of Lisbon. To identify the causes of the fall of trees during windstorm events, a database was created and it includes so far 1241 occurrences

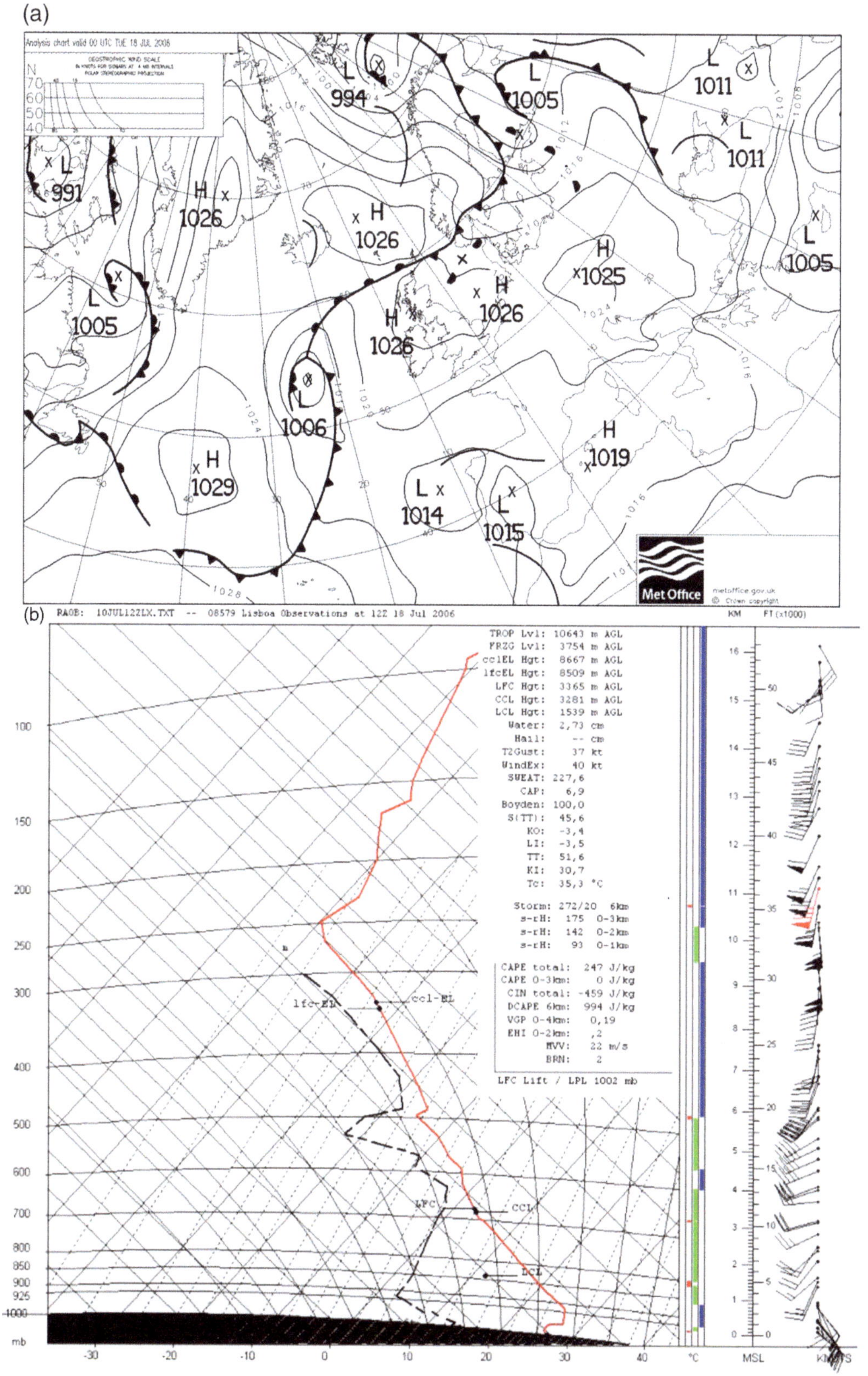

Fig. 15 Windstorm on 18 July 2006: (**a**) Synoptic chart; (**b**) sounding over Lisbon Airport (1200 UTC)
Source: MetOffice (UK) and University of Wyoming, Department of Atmospheric Science

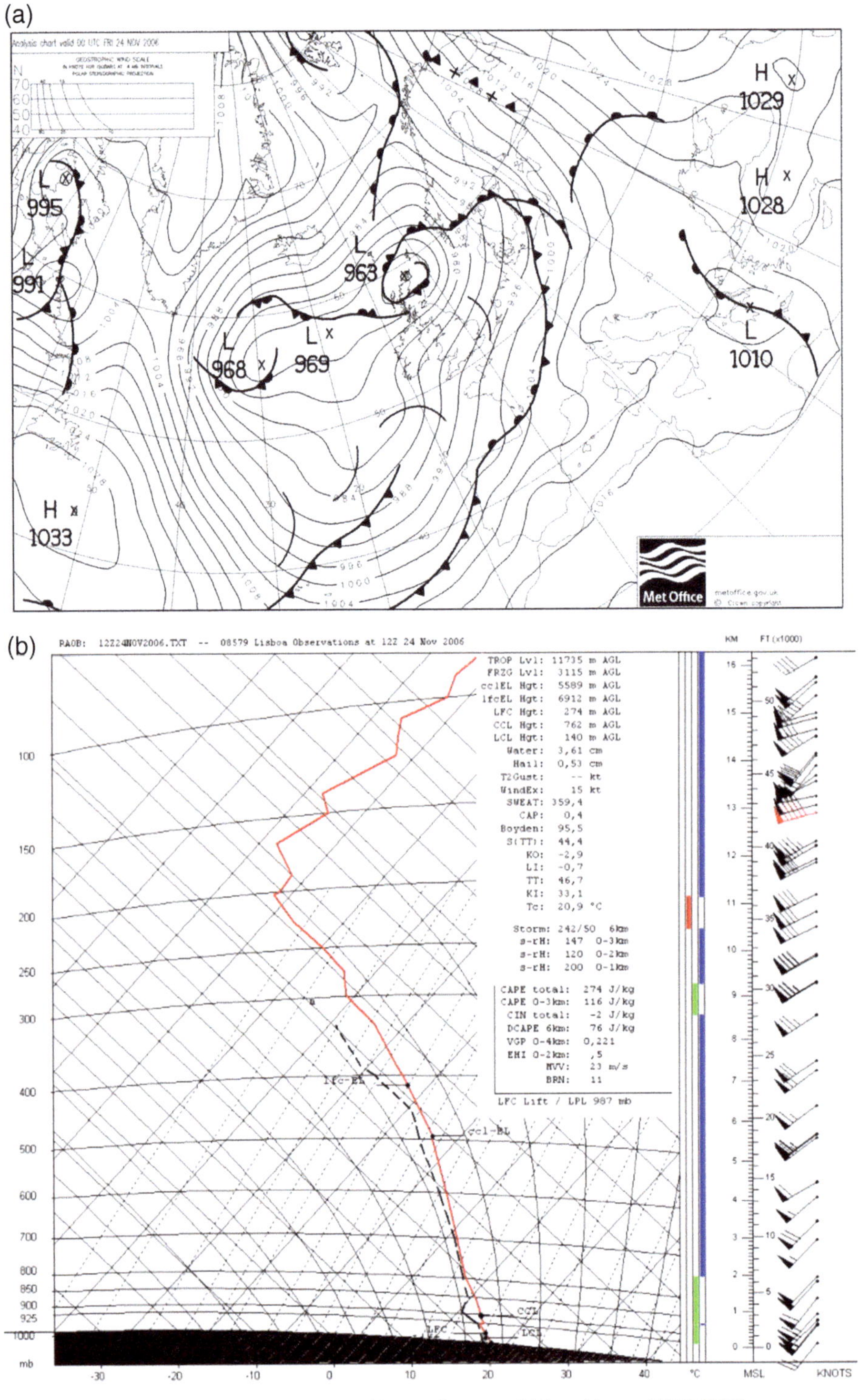

Fig. 16 Windstorm on 24 November. (**a**) Synoptic chart; (**b**) sounding over Lisbon Airport (1200 UTC) Source: MetOffice (UK) and University of Wyoming, Department of Atmospheric Science

of fallen trees (from 1990 to 2006), meteorological data (wind speed and direction) and some of other required information, such as urban parameters (roughness length, H/W ratio, street orientation, street pollution levels, etc.), trees characteristics (diameter of the trunk, crown, height, age) and other conditions like fyto-sanitary state, soil, etc. Wind data and the occurrences were correlated with other data and the preliminary and general results were presented.

The main directions that cause tree falls were identified: in summer, *Nortada* winds are responsible for 74% of the falls and most of the cases are related to the falls of boughs and branches. From autumn to spring, perturbed weather types from west, southwest and south (cyclonic circulations related to low pressure in connection with frontal systems or low-pressure centres located to the west or to the southwest of Portugal), are responsible for the most part of the occurrences (79%), most of them affecting the whole tree, but also some boughs and branches. Infrequently, strong winds from downdrafts may also occur, such as those on 18 July 2006.

The majority of the falls (70.5%) happened in the last six years, with the maximum of 24% in 2006. Only 29.5% of the total occurrences took place in the 1990s (with a maximum of 8.4% in 1997). It is possible that during this later period the trees had suffered higher stress due to urban factors, such as the increase in traffic and in the level of pollutants, which influence the fyto-sanitary conditions, in addition to the influence of age (since older trees are more vulnerable). It was also possible to identify specific areas where tree falls are more frequent, probably due to the combination of the roughness and topography of the areas, the number of street trees, the species and the fyto-sanitary conditions, among other factors, which will be studied in more detail in the near future.

It was also found that the majority of falls happened when specific wind speed limits were surpassed, specifically above 7 m/s for the time period that included the 6 hours prior to the occurrences.

For a successful implementation of an 'alert plan' for the city during strong wind events, a few recommendations must be made:

i. A common language must be defined among the RSBL technicians, potentially to be shared with researchers. This language should include a glossary of terms used in the description of the trees and the type of occurrences, besides the exact definition of the size of trees, boughs and branches and other important parameters;
ii. The precise location of the place of the occurrence is necessary to understand many of the factors that influence the fall of trees. A GPS device should be used for this purpose;
iii. A form, based on a checklist of the factors that may influence the falls, should be created and filled in by RSBL personnel with observations or measurements when the service is undertaken. The implementation of this method would allow for the creation of a reasonably complete database to be used in future analysis;
iv. The development of a training programme for the RSBL personnel, in order to clarify the methodology described and to ensure the correct application of the devices, as well as strengthening the working relations between researchers and RSBL personnel.

The methodology applied has proved suitable for this study, although some limitations were encountered. It is necessary to deepen our research and collect more data, with the purpose of understanding all the probable causes of tree falls in the city and assist in the development of contingency planning necessary to face these events.

Acknowledgments The authors would like to thank to Prof. Maria João Alcoforado and Prof. Maria Eugénia Moreira for their review and discussion on this subject and for their advice concerning the methodology carried out in this research. We also have to thank to the *Regimento de Sapadores Bombeiros de Lisboa,* for their support and for the data supplied. Without their cooperation, it would not have been possible to develop this study.

This research was undertaken within the scope of the project 'Climate and urban sustainability. Perception of comfort and climatic risks (URBKLIM)', funded by FCT (POCI/GEO/61148/2004).

References

Alcoforado M-J (1987) Brisas estivais do Tejo e do Oceano na região de Lisboa. Finisterra. XXII (43): 71–112

Alcoforado M-J (1992) O clima da região de Lisboa. Contrastes e ritmos térmicos. Memórias do Centro de Estudos Geográficos, vol. 15, CEG, Lisboa: 347 pp

Alcoforado M-J, Lopes A (2003) Windfields and temperature patterns in Lisbon (Portugal) and their modification due to city growth. 5th International Conference on Urban Climate (ICUC5), Lodz, Poland: 383–386

Alcoforado M-J, Lopes A, Andrade H, Vasconcelos J (2005) Orientações Climáticas para o Ordenamento em Lisboa, CEG/Geoecology, University of Lisbon, Report 4: 81pp

Borges J (1971) Structural Safety, Course 101, 2° ed., LNEC, Lisbon, 326pp

Clergeau P (1996) Urban biodiversity: is there such a thing? Le courrier du CNRS, n° 82:102–104; and Cybergeo, (1996) – Les Bonnes Feuilles du PIRVILLES mis en ligne le 12 avril 1996, modifié le 31 mai 2007, (http://www.cybergeo.eu/ index.html), Accessed on 11 June 2007

DGRF – Ministry of Agriculture, Rural Development and Fishery/Forestry Resources (n.d.) (http://www.dgrf.min-agricultura.pt) Accessed on June 2008

Doswell CA, Brooks HE, Maddox RA (1996) Flash flood forecasting: an ingredients-based methodology. Weather and Forecasting, 11:560–581

Duryea M (1997) Wind and Trees: Surveys of Tree Damage in Florida Panhandle after Hurricanes Erin and Opal; Institute of Food and Agricultural Sciences; Cooperative Extension Service; University of Florida: 2–7

Fabião AMD (1996) A agressividade do meio urbano e algumas medidas de mitigação do stress em árvores das cidades. Cirurgia das Árvores, Workshop, Lisbon

Freer-Smith PH, El-Khatib A, Taylor G (2004) Capture of Particulate Pollution by Trees: A Comparison of Species Typical of Semi-Arid Areas (Ficus Nitida and Eucalyptus Globulus) with European and North American Species, Water, Air, & Soil Pollution, 155(1): 173–187

Garcia-Martin G, Garcia-Valdecantos JL (2001) El arbolado urbano en las ciudades españolas. Actas del III Congreso Florestal Español: 467–474

James K (2003) Dynamic loading of trees. Journal of Arboriculture 29(3): 165–171

Jim C, Liu H (1997) Storm damage on urban trees in Guangzhou, China. Landscape and Urban Planning 38: 45–59

Jornal de Notícias online edition (2000), 7 December

Kaňák J, Benko M, Simon A, Sokol A (2007) Case study of the 9th May 2003 windstorm in southwestern Slovakia, Atmospheric Research, 83: 162–175

Krzyzanowski M, Kuna-Dibbert B, Schneider J (ed.) (2005) Health effects of transport-related air pollution, WHO

Lopes A (2003) Modificações no clima de Lisboa como consequência do crescimento urbano. Vento, ilha de calor de superfície e balanço energético, PhD thesis, University of Lisbon: 375pp

Mattheck C, Breloer H (1994) The body language of trees. HMSO. London, UK

McPherson E, Muchnick J (2005) Effects of street tree shade on asphalt concrete pavement performance, Journal of Arboriculture 31(6): 303–310

Moreira ME (1998) Estudo Fitogeográfico do Jardim Braancamp Freire (Lisboa), Finisterra, XXXIII (66): 7–24

Niklas K (2002) Wind, size, and tree safety, Journal of Arboriculture, 28: 84–93

Nilsson K, Randrup TB, Wandall BM (2000) Trees in the urban environment. The forest handbook (Ed. Evan J), Blackwell Science, Oxford, Vol 1: 347–361

OFDA/CRED International Disaster Database (n.d.) – (http://www.em-dat.net/) Accessed in June 2007

Peppler R, Lamb P (1989) Tropospheric static stability indices and Central North America growing season rainfall, Monthly Weather Review, 117: 1156–1180

Saebe A, Benedikz T, Randrup TB (2003) Selection of trees for urban forestry in the Nordic countries. Urban Forestry & Urban Greening, 2: 101–114

Santos JA, Corte-Real J, Leite SM (2005) Winter regimes and their connection to the winter rainfall in Portugal, International Journal of Climatology, 25: 33–50

Saraiva JG, Marques da Silva F, Silva FG (1997) O vento, a cidade e o conforto. IV National Meeting on Comfort in Built Environments, Bahia, Brazil

Shashua-Bar L, Hoffman M (2003) – Geometry and orientation aspects in passive cooling of canyon streets with trees. Energy and Buildings, 35: 61–68

Soares AL (2006) – O valor das árvores. Árvores e Floresta Urbana de Lisboa. PhD thesis, Institute of Agronomy, Technical University of Lisbon

Soares AL, Castel-Branco C (2007) As árvores da cidade de Lisboa. Floresta e Sociedade. Uma história em comum. Ed. Público/FLAD: 289–333

World Health Organisation (WHO) (1999) Health Costs Due to Road Traffic-Related Air Pollution, WHO, Copenhagen

Ozone Air Pollution in Extreme Weather Situation – Environmental Risk in Mountain Ecosystems

S. Bičárová and P. Fleischer

Keywords Ground-level ozone · Episode · Temporal and spatial variation · Complex topography · Emissions

Introduction

Extreme weather events including heat waves have assumed significant changes in intensity and frequency in the context of global warming. Unpreceeded 15-day long heat wave with record temperatures and unusually persistent high-ozone concentrations was observed in Europe during August 2003 (Vautard et al. 2005). Special meteorological situation favored the progressive accumulation of ozone. Marked atmospheric stability with very weak winds, which limited the dispersion of pollutants, the high temperatures caused by the advection of a warm African air mass, and strong sun radiation with clear skies led to high ozone concentrations (Rodríguez et al. 2004).

Extraordinary high level of ozone pollution occurred also in Slovakia in August 2003. Monitoring stations recorded maximal O_3 concentrations from 127 to 301 $\mu g\ m^{-3}$. The ambient air quality standards were exceeded frequently at southwest lowland urban stations in Bratislava and north rural mountain station Lomnický Štít: alert threshold $IH_{1h} = 240\ \mu g\ m^{-3}$ 20 times only at Bratislava stations, information threshold $IH_{1h} = 180\ \mu g\ m^{-3}$ total 69 times. Increase of O_3 concentration with altitude documents measurement in the High Tatras vertical profile. Mean monthly values raised from 82 $\mu g\ m^{-3}$ at Stará Lesná (814 m a.s.l.) to 124 $\mu g\ m^{-3}$ at Lomnický štít (2635 m a.s.l.) (Bičárová et al. 2005).

Air pollution is a serious problem in the High Tatra Mountain. Ozone AOT40 index for protection of natural forest vegetation (10,000 ppbh) is commonly exceeded in the middle of vegetation period. A synergic effect of air pollution, extreme weather conditions, and biotic agents related to global climate change has caused serious deterioration of the forest condition in the Tatra National Park since early 1990s (Fleischer et al. 2005).

Recent O_3 concentration in Slovakia is substantially more affected by long-range transport and by climatic changes than by national strategy application for reduction of O_3 precursor emissions (Hrouzková et al. 2004). Results of LOTOS–EUROS model application, including calculation of sequential hourly O_3 concentrations for years 1999 and 2003, show minor effect of Slovak emissions on the mean annual O_3 concentration in all territories in Slovakia (Kremler 2006).

In this chapter, the summer O_3 episode observed during August 12–14, 2003 in the High Tatra Mountain region is investigated using Meteorology and Photochemistry Model (MetPhoMod) (Perego 1999). The goal of MetPhoMod model implementation is to study time and spatial distribution of O_3 pollution and to analyze the contribution of emission sources to maximal O_3 concentrations.

Methods

MetPhoMod Model Description

MetPhoMod is an eulerian, three-dimensional, mesoscale model for simulation of summer smog over

S. Bičárová (✉)
Geophysical Institute, Slovak Academy of Sciences, Meteorological Observatory Stará Lesná, 059 60 Tatranská Lomnica, Slovakia
e-mail: bicarova@ta3.sk

K. Strelcová et al. (eds.), *Bioclimatology and Natural Hazards*,
DOI 10.1007/978-1-4020-8876-6_6, © Springer Science+Business Media B.V. 2009

very complex terrain under fair weather conditions (Perego 1999). It is a single program that includes modules for air motion, turbulence, radiation, ground–atmosphere interactions, gas phase chemistry, and deposition. The model uses a Cartesian grid. Complex topography is considered, by dividing grid points into the two categories: normal and underground. Transport is calculated using the Piecewise Parabolic Methods (PPM) based transport scheme. The nonhydrostatic pressure is evaluated by solving an elliptic equation derived from the mass continuity equation. The turbulence module implements the k–ε turbulence closure scheme with an implicit solver. The program includes a chemical interpreter and predefined input file of chemical equations for the regional atmospheric chemistry mechanism (RACM) (Stockwell et al. 1997). MetPhoMod does not include modules for clouds, aerosols, and heterogeneous chemistry. A grid of rectangular cubes represents the modeling domain. All values are stored in the center of the cube. The dynamics then are solved with the method of Rhie and Chow (1983), in the form proposed by Clappier (1998). The solution procedure strictly keeps mass consistency. The chemical equation system is calculated by a technique based on separation of fast and slow species (Gong and Cho 1993). The fast species are solved with an implicit, the slow ones with an explicit integration step. The MetPhoMod software consists of a single UNIX executable file (http://www.giub.unibe.ch/klimet/metphomod/) and is handled with common UNIX commands.The network common data format (netCDF), a binary data format, defined by UNIDATA/UCAR (http://www.unidata.ucar.edu/) is used for input and output files formulation.

Ozone Episode in August 2003

Unusual high O_3 concentrations were recorded in Europe and Slovakia (301 $\mu g\ m^{-3}$ in Bratislava and 195 $\mu g\ m^{-3}$ at Lomnický štít) during the August 2003 heat wave (Bičárová et al. 2005). Simulation of ozone by model CHIMERE (Vautard et al. 2005) shows the first exceeding of the European standards (90 ppb, or 180 $\mu g\ m^{-3}$) in the Ruhr area, northeast France, Paris, and near Marseille on 2nd and 3rd August. Then in an anticyclonic move the high-ozone region rotated clockwise from southern Germany (4 August) to north-central France (6–7 August) and then to western France (8–9 August). After 12th August, the whole polluted air mass was pushed away eastward by a cold front coming from the Atlantic area (Fig. 1) and contributed to significant increase of O_3 concentrations observed at monitoring stations in Slovakia.

Transfer of ozone pollution documents also the forecast of maximal O_3 concentration (Fig. 1) prepared by PREV′AIR according the results of model CHIMERE. Analysis of the weather situation presented by Slovak Hydrometeorological Institute (SHMI) shows the occurrence of anticyclone occasions (A, Wal) in central Europe for the period 12–14 August. Anticyclones generally bring fair weather (rise of air temperature, decrease relative humidity) and clear skies (increase solar radiation and sunshine duration). The dynamics of an anticyclone lead to downward vertical movement, which suppresses convective activity. Moving of colder air from northwest influenced decrease of O_3 concentration in region of 48–54° N on 14 August.

Model Domain

Selected area (20 km × 18 km) of the High Tatra Mountain region extends from Svit on the west to Kežmarok on the east and from Lomnický štít on the northwest to Poprad basin on the southeast (Fig. 2). Model domain includes fixed ground-level stations in vertical profile:

- Poprad-Gánovce (H = 706 m a.s.l., φ = 49° 02′ N, λ = 20°19′ E);
- Stará Lesná (H = 810 m a.s.l., φ = 49°09′ N, λ = 20° 17′ E);
- T. Lomnica-Štart (H = 1200 a.s.l., φ = 49° 10′ N, λ = 20° 15′ E);
- Skalnaté Pleso (H = 1778 m a.s.l., φ = 49° 11′ N, λ = 20° 14′ E);
- Lomnický štít (H = 2634 m a.s.l., φ = 49° 12′ N, λ = 20° 13′ E).

At these stations, SHMI provides a meteorological and air pollution measurement in cooperation with the Geophysical Institute of the Slovak Academy of Sciences (GPI SAS) and the Research Centre of the Tatra National Park (RC TANAP). Stationary (semiurban, industrial) and mobile sources around the towns Poprad and Kežmarok produce more anthropogenic

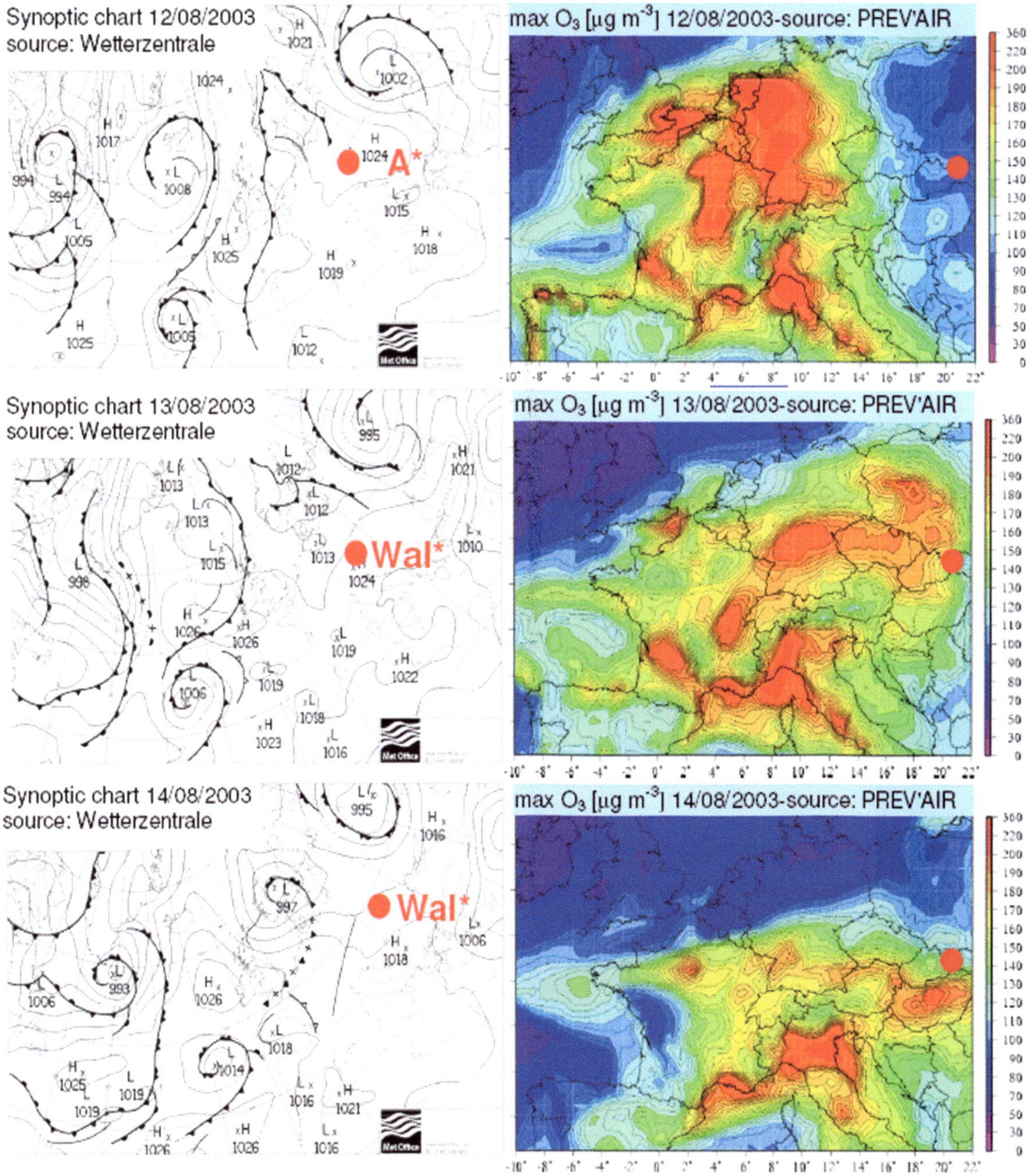

Fig. 1 Synoptic charts (source: Wetterzentrale http://www.wetterzentrale.de/topkarten/tkfaxnwsar.htm), *type of meteorological situations: A – anticyclone over Central Europe, Wal – west anticyclone occasion of summer type (source: SHMI http://www.shmu.sk) and forecast of maximal O_3 concentrations (source: PREV'AIR http://www.prevair.org/en/prevision_o3.php) for the period August 12–14, 2003

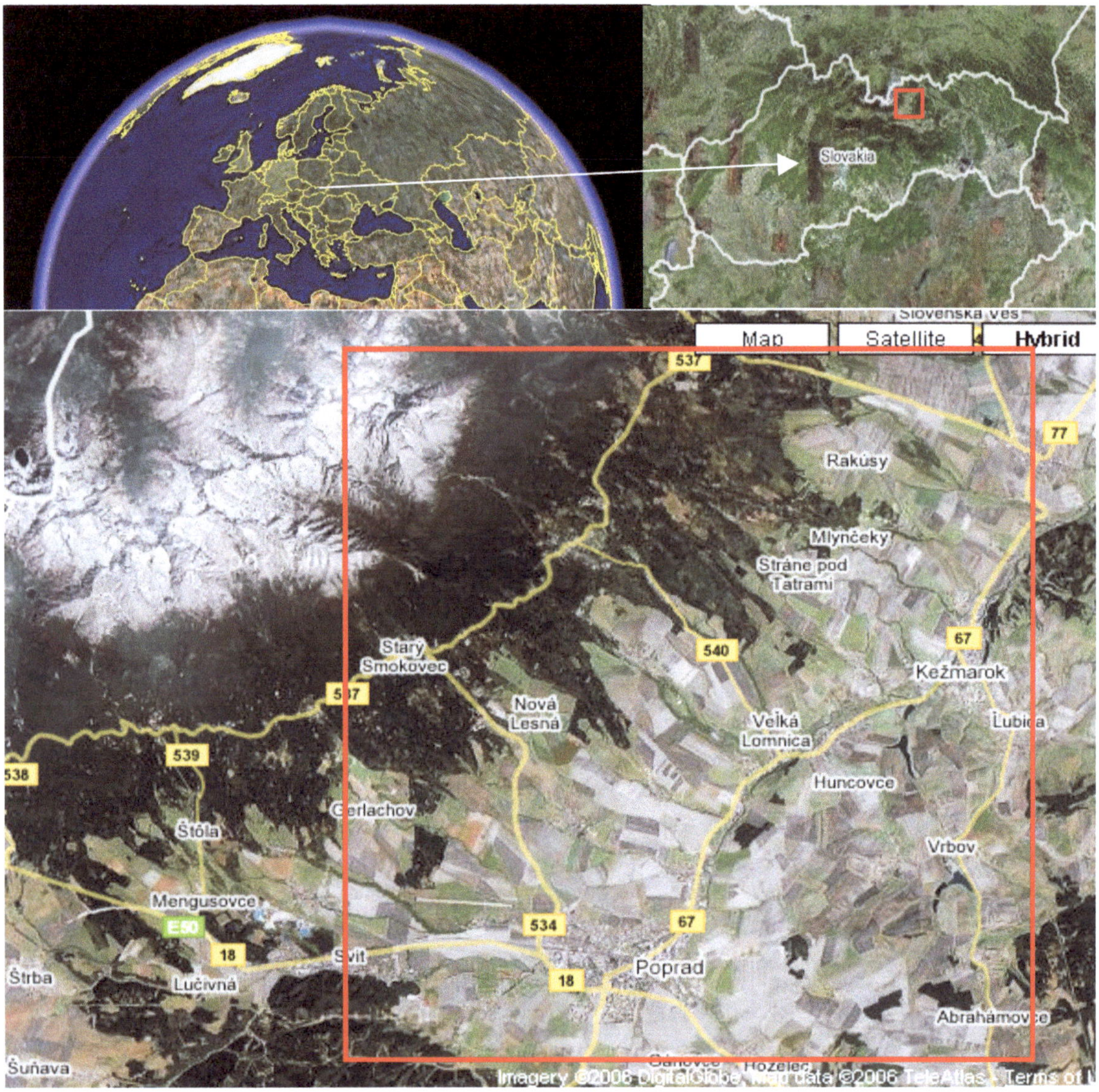

Fig. 2 Satellite images and borders of the High Tatra Mountain model domain (http://geology.com/europe-satellite-images.shtml)

emission in comparison with sparsely populated high-altitude mountain localities. On the other hand, forest vegetation at foothills releases important quantity of biogenic volatile organic compounds (BVOC) emissions. Model grid consists of 21 × 19 horizontal cells of resolution 1 × 1 km^2 with 22 vertical levels in an elevated interval of altitude 600–2700 m a.s.l. Digital topographic data and representative altitude for every cell were obtained by digitalization of the High Tatras paper map (ratio 1:50,000) using the software Didger and Surfer (http://www.goldensoftware. com). Digital Elevation Model (DEM) of investigated domain 20 km × 18 km × 3 km is shown in Fig. 3.

Input Data – Specification and Validation

Static Parameters

Static parameters include relatively stable ground characteristics: the ground roughness, the surface albedo, the ground heat capacity, the ground diffusivity, the

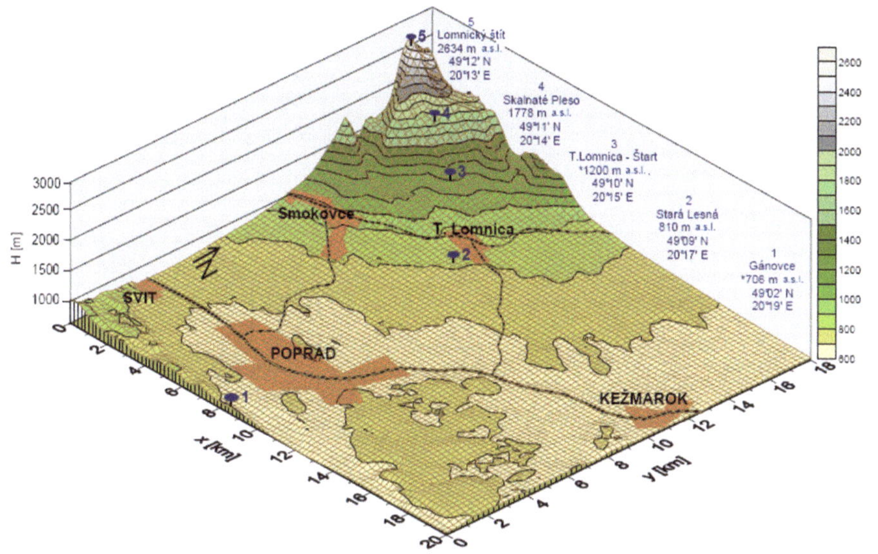

Fig. 3 DEM of the High Tatra Mountain model domain and location of ground meteorological and O_3 stations: 1 – Poprad-Gánovce, 2 – Stará Lesná, 3 – Štart, 4 – Skalnaté Pleso, 5 – Lomnický štít

relative air humidity in ground pores, the evaporation resistance of soil, and the shielding factor of plants. Parameters were defined in accordance with the classification System of U.S.G.S (U.S. Geological Survey Land Use/Land Cover System). Land of the High Tatra Mountain model domain covers mainly vegetation of grass, shrubs, forest, and mixed barren land of alpine rockies. There are also several semi-urban and rural settlements.

Dynamical Parameters

Dynamical parameters involve the basic meteorological data: the air temperature, the air moisture, the wind direction, the wind speed, and the O_3 concentration. These data come from measurements at stations of model domain. The air temperature (Kroneis NTC sensor, YSI 44212 type) and relative air humidity (Vaisala MHP35D type) were measured on thermometer screen at 2 m level above the surface. Wind speed and wind direction were registered by R. Fuess type of anemometer. Continuously recorded data were averaged and stored as hourly values. The global radiation was obtained by thermopile-based pyranometer (Ostrožlík 2004). UV absorption ozone analysators measure O_3 concentration in accordance with the secondary national ozone calibration standard of SMHI. Intercomparisons with the Czech primary ozone standard are regularly organized. Data are collected and validated in central SHMI database.

Hourly data were used to strengthen the model behavior around the edges of vertical layers and on the top of domain. Figure 4 illustrates the modest differences between modeled and measured hourly values for the air temperature (T), the relative air humidity (RH), and the wind speed (Ws) at individual stations. Correlation coefficients show good agreement between the experimental data and values processed by model ($r_{\mathrm{T}} = 0.985, r_{\mathrm{RH}} = 0.951, r_{\mathrm{Ws}} = 0.896$) with an exception of wind direction ($r_{\mathrm{Wd}} = 0.366$) that is very variable in the complex mountain area. Model calculates global radiation according to Paltridge and Platt (1976). The tight relationship between modeled and measured data at Stará Lesná ($r = 0.936$) indicates sufficient accuracy of solar radiation module (Fig. 5).

Emissions

Emissions data of EMEP expert emissions database WebDab were applied (Vestreng et al. 2005). Temporal and spatial disagregation of annual (2003) emissions for EMEP grid square including model domain of the High Tatra Mountain region were performed for gaseous species CO, NO_x, SO_x, and non-methane hydrocarbons (NMHC). Composition and fractions of NMHC groups were assigned according to Stockwell

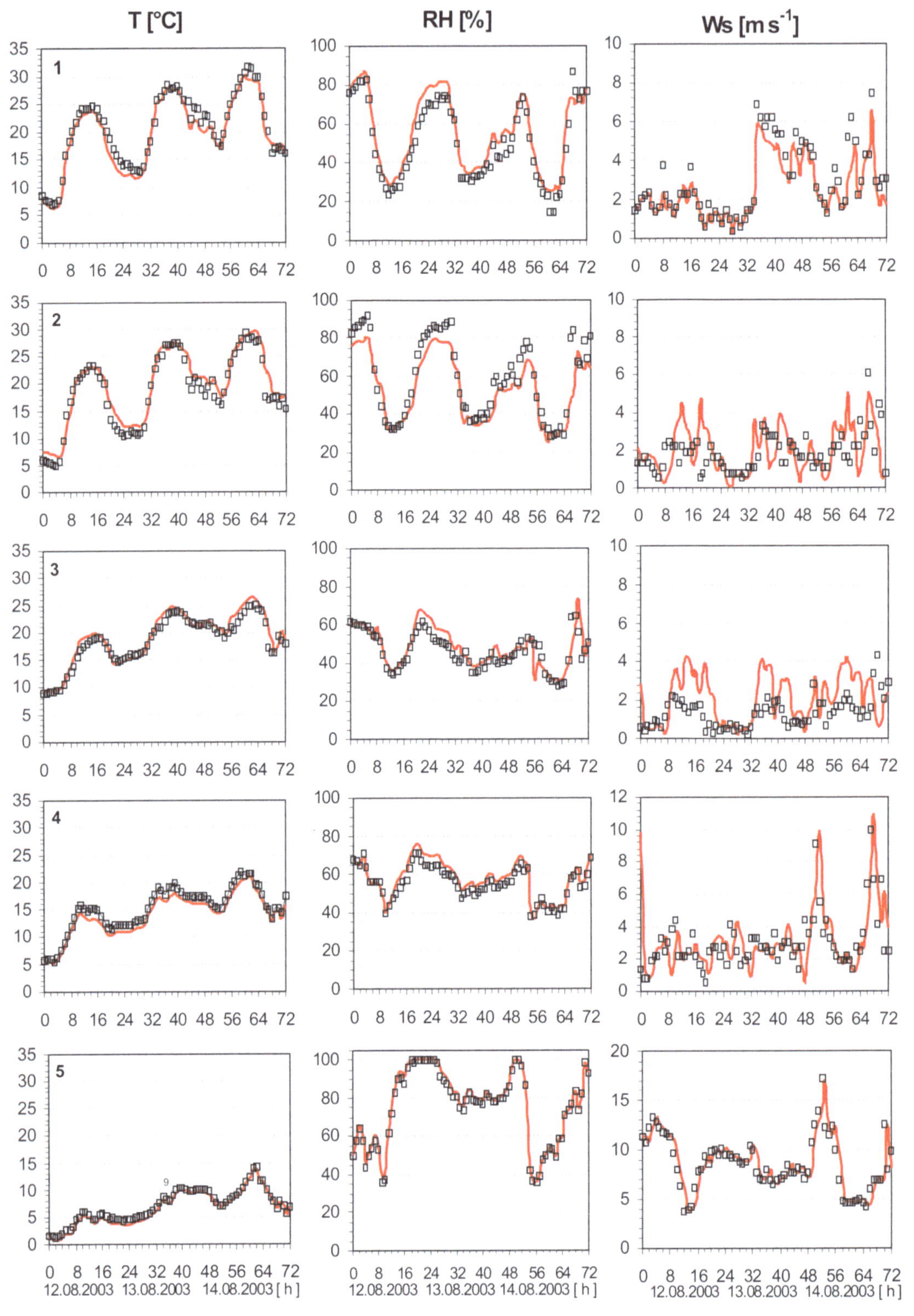

Fig. 4 Comparison of modeled (*red line*) and measured (*black square*) values for air temperature *t* (°C), relative air humidity RH (%), wind speed Ws (m · s^{-1}) at stations: 1 – Poprad-Gánovce, 2 – Stará Lesná, 3 – Štart, 4 – Skalnaté Pleso, 5 – Lomnický štít during the period August 12–14, 2003

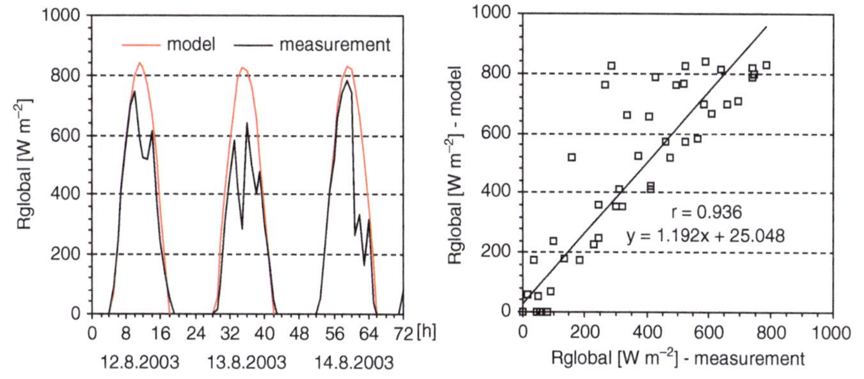

Fig. 5 Global radiation (Rglobal, Wm^{-2}): modeled (*red line*) and measured values (*black line*) at Stará Lesná for the period August 12–14, 2003

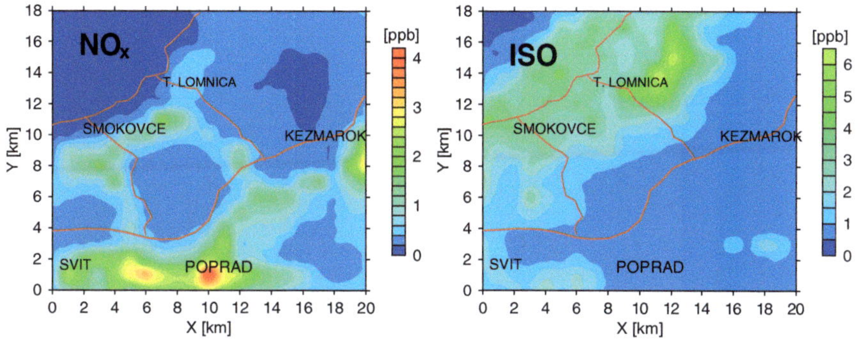

Fig. 6 Spatial distribution of mean daily concentration of NO_x (ppb) and ISO (ppb) processed by MetPhoMod for the High Tatra Mountain domain during the period August 12–14, 2003

et al. (1990). BVOC emissions of model domain (Bičárová and Fleischer 2006) were estimated by GLOBEIS-BEIS2 model (Guenther et al. 1993). Figure 6 shows the spatial distribution of mean daily concentration NO_x and isoprene (ISO) obtained by the MetPhoMod. Maximal NO_x concentration corresponds with the distribution of stationary and mobile emission sources as well as ISO concentration with forested area of model domain.

Results

MetPhoMod was used for study of O_3 episode in the High Tatra Mountain during the period August 12–14, 2003 in two applications. In both cases, identical meteorological emissions and ground parameters and also initial O_3 concentrations were specified as input data. The first application, appointed as interpolation, considered O_3 data from measurement involved into the boundary conditions. The second application simulated O_3 concentration from the local emission sources without measured O_3 data included in the border section. In this case, comparison of modeled and measured O_3 concentration can specify contribution of emissions from local independent sources. Model results for all grid cells just above the ground (ground level) are presented.

Model Interpolation

Model interpolates O_3 data involved as individual data file into the boundary conditions that contains O_3 concentrations measured at the stations: Poprad-Gánovce, Stará Lesná, Štart, Skalnaté Pleso, and Lomnický štít, adjusted for each vertical layer. Initial O_3 concentrations were identical with that measured, and other initial background gaseous species concentrations were assigned as zero. Comparison between modeled and measured O_3 concentrations and also correlation coefficients in range from 0.867 to 0.999 confirm an acceptable accuracy of the model interpolation (Fig. 7). Sufficient agreement between experimental and modeled O_3 concentration (correlation coefficient $r = 0.833$) was detected also at station Starý Smokovec. In this case O_3 measurement was excluded from O_3 data file.

Hourly O_3 concentrations obtained by model interpolation were used for illustration of O_3 vertical

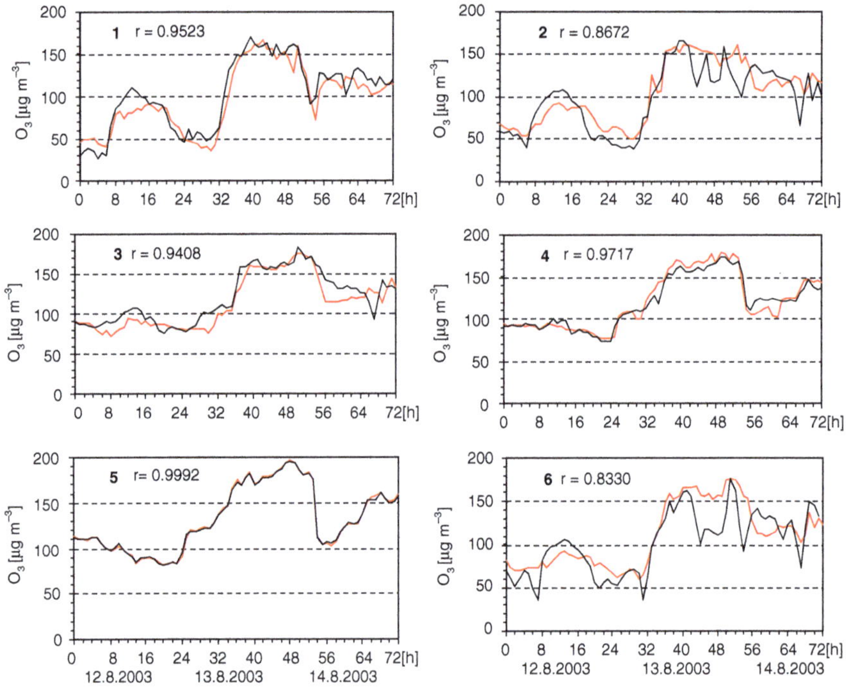

Fig. 7 Hourly O_3 concentrations ($\mu g\ m^{-3}$) of model interpolation (*red line*) and measured data (*black line*): 1 – Poprad-Gánovce, 2 – Stará Lesná, 3 – Štart, 4 – Skalnaté Pleso, 5 – Lomnický štít, 6 – Starý Smokovec for the period August 12–14, 2003

profile (Fig. 8) that shows ozone transport from high troposphere to surface layer of atmosphere in the High Tatra Mountain. The highest O_3 concentration ($\sim$190 $\mu g\ m^{-3}$) occurred in the night from August 13 to 14,2003, on the top layer of model domain cannot be caused by photochemical production from local sources. Results of model interpolation correspond with the ozone forecast provided by the French information system PREV′AIR and also with the evolution of O_3 episode described by Vautard et al. (2005) that documents transport of O_3 polluted air masses from Western Europe to the east by a cold front after August 12, 2003. Analysis of interpolation results indicates that a long-range transport through the high troposphere played more significant role in O_3 concentration increase than its local formation during the extremely high O_3 events in the High Tatra Mountain region.

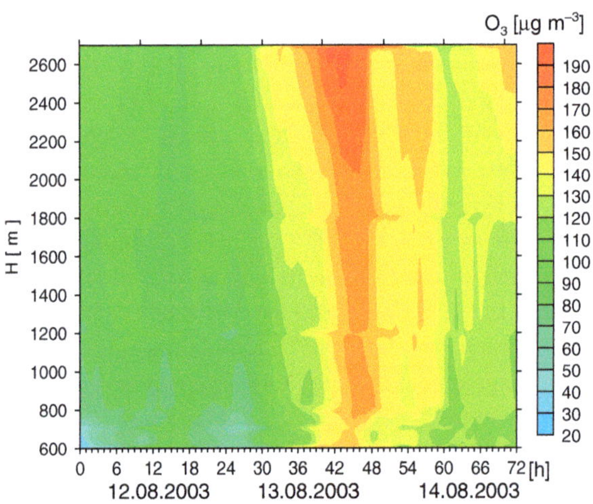

Fig. 8 Vertical profile of hourly O_3 concentrations ($\mu g\ m^{-3}$) obtained by model interpolation for the High Tatra Mountain model domain during August 12–14, 2003

Model Simulation

Model simulation assumed O_3 concentrations that can be caused mainly by local anthropogenic and biogenic emission sources at model domain. However, initial O_3 concentrations were defined by measured values, the effect of O_3 measurement on output O_3 values sharply decreased after initialization because O_3 data file was excluded from boundary conditions. Another initial gaseous species concentrations were derived from output data file of model interpolation. It is considered that calculating process was not controlled by measured O_3 data and simulated O_3 concentrations represent local

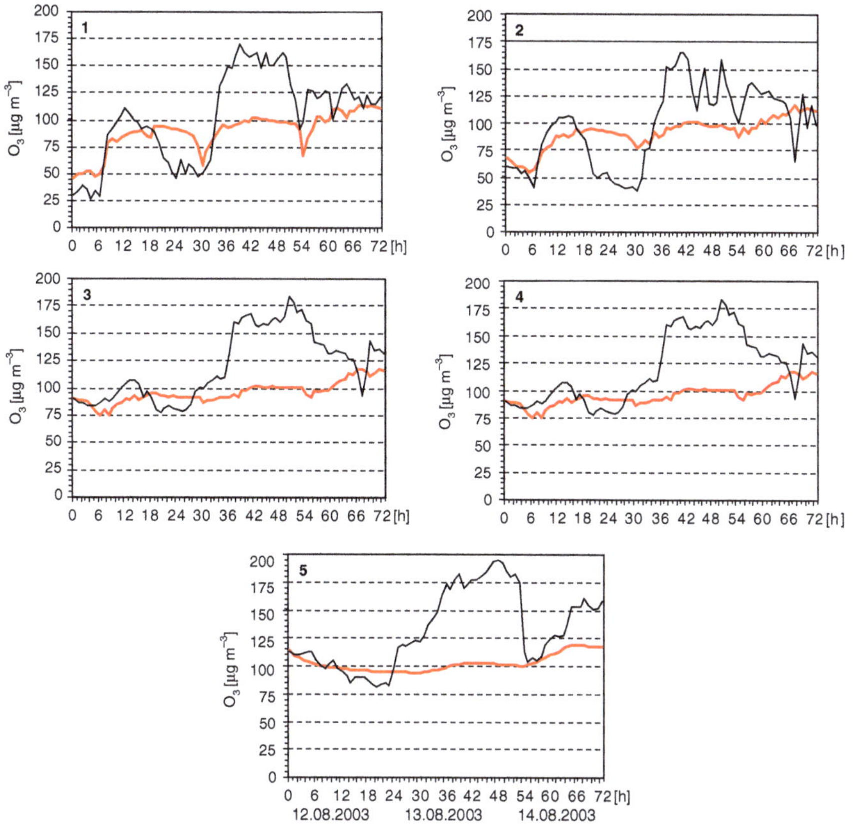

Fig. 9 Hourly O_3 concentrations ($\mu g\ m^{-3}$) measured (*black line*) and calculated by model simulation (*red line*): 1 – Poprad-Gánovce, 2 – Stará Lesná, 3 – Štart, 4 – Skalnaté Pleso, 5 – Lomnický štít during August 12–14, 2003

potential of ozone formation. Daily courses of simulated and measured hourly O_3 concentrations (Fig. 9) show the highest contrast in the night time from August 13 to 14, 2003 (36–54 h of ozone episode) at all stations of vertical profile.

Relationship between measured maximal (*M*) and simulated (*S*) O_3 concentrations presents ratio *M/S* (Table 1a). On the first day (August 12) *M/S* was evidently lower (7–30%) than on the following days (61–92%) of O_3 episode when apparently polluted air mass has moved over the High Tatras area. The measured O_3 concentrations were substantially higher than simulated about: 53% at Poprad-Gánovce, 39% at Stará Lesná, 65% at Štart, 60% at Skalnaté Pleso, and 74% at Lomnický štít during peak phase of O_3 episode from 12 h (UTC) August 12 to 06 h (UTC) August

Table 1a Comparison between measured maximal (*M*) and simulated (*S*) O_3 concentrations ($\mu g\ m^{-3}$) for individual days of O_3 episode in the High Tatra Mountain region during August 2003

Maximal hourly O_3 concentrations ($\mu g\ m^{-3}$)												
Day h UTC	12. 08. 2003				13. 08. 2003				14. 08. 2003			
Station	h	*M*	*S*	*M/S*	h	*M*	*S*	*M/S*	h	*M*	*S*	*M/S*
Poprad-Gánovce	12	111	85	1.30	15	170	96	1.76	01	161	98	1.65
Stará Lesná	14	107	87	1.23	16	165	98	1.69	02	157	98	1.61
Štart	13	107	89	1.21	17	168	100	1.68	02	182	101	1.81
Skalnaté Pleso	11	100	93	1.08	23	165	102	1.62	00	165	101	1.64
Lomnický štít	10	106	99	1.07	23	193	102	1.90	00	194	102	1.92

Table 1b Mean hourly O_3 concentrations ($\mu g\ m^{-3}$) during peak phase of O_3 episode (time period from 12 h August 13 to 06 h August 14, 2003)

	M	*S*	*M/S*	*(M–S)/M* (%)	*S/M* (%)
Poprad-Gánovce	147.1	96.1	1.53	34.7	65.3
Stará Lesná	135.1	97.0	1.39	28.2	71.8
Štart	162.9	99.0	1.65	39.2	60.8
Skalnaté Pleso	161.0	100.5	1.60	37.6	62.4
Lomnický štít	177.0	101.5	1.74	42.7	57.3
Average	156.6	98.8	1.58	36.5	63.5

13, 2003 (Table 1b). Observed O_3 concentrations (*M*) were considered as sum of O_3 contributions from local dependent (*S*) and local independent (*M–S*) emission sources. According to these results, approximate contributions of 63.5% local dependent and 36.5% local independent emission sources to observed O_3 concentration were specified (Table 1b) in the High Tatra Mountain region during peak phase of ozone episode in August 2003.

Discussion

A coupled meteorological and photochemical model, MetPhoMod, simulates sensitive O_3 balance affected by many factors, mainly the quantity of emissions, the composition of precursors, the environmental conditions, and the meteorological situation in troposphere. Input parameters inaccuracy for each factor influences results of the model simulation. Although size of the model domain limits the accuracy determination due to effect of long-range air pollution transport, on the other hand, it enables to analyze contribution of local independent emission sources for the High Tatra Mountain region.

Usability of MetPhoMod in complex terrain shows good agreement between model results and O_3 data obtained by aircraft measurement during intensive field campaign in the Grenoble region in July 1999. Interestingly, the maximum O_3 concentration was not produced in Grenoble city but at higher altitudes, up to 1500–2000 m a.s.l. over the rural area. Above the residual layer between 1300 and 2300 m a.s.l. O_3 concentration decreased and measured values at 3200 m a.s.l. were considered as an ozone background reference level (Coauch et al. 2003). This is contrary to the comparison of the vertical distribution of O_3 concentration in the High Tatra Mountain region in August 2003. Maximal ozone pollution was detected on the top of the model domain at the altitude above 2600 m a.s.l. and toward lower altitudes O_3 concentrations slightly decreased during peak phase of O_3 episode.

Assuming relevant contribution of long-range transport, the national reduction of emissions is not an effective tool to achieve decrease of O_3 concentrations in Slovakia. The first results of model LOTOS–EUROS simulation obtained on base of Dutch-Slovak cooperation also show an insignificant impact of Slovak emissions reduction to O_3 concentrations (Kremler 2006). Furthermore, increase of O_3 concentrations is expected due to effects of variables associated with future changes in climate and ozone precursor emissions. Climatic changes assumed for temperature, atmospheric water vapor, and the biogenic VOC, each individually causes a 1–5% increase in the daily peak ozone in central California (Steiner et al. 2006). Considering unchanged anthropogenic precursor emissions, 10% higher daily maximum values of O_3 concentrations were simulated by coupled climate-chemistry model for the region of the southern Germany (Forkel 2006).

Current O_3 concentrations have adverse effects on vegetation. Risk assessment based on relationships between external concentration and plant response is inadequate for evaluating the consequences of changing global patterns of exposure to ozone. New models linking stomatal flux, and detoxification and repair processes to carbon assimilation and allocation provide a more mechanistic basis for future risk assessments. There is an urgent need to develop more holistic approaches linking the effects of ozone, climate, and nutrient and water availability, on individual plants, species interactions, and ecosystem function (Ashmore 2005).

Conclusion

Air pollution, especially ground-level ozone, affects negatively the sensitive ecosystems of the Tatra National Park (UNESCO Biosphere reserve). Application of MetPhoMod is a tool for analysis of the temporal and the spatial distribution of O_3 air pollution in the

High Tatra Mountain region. The vertical profile of O_3 concentration obtained by model interpolation documents the ozone transport from ozone-enriched high troposphere to the surface layer of the atmosphere during O_3 episode in August 2003. Achieved results show that long-range and descent transfer of air pollution supported by high-pressure system apparently played more significant role in O_3 concentration increase than its local formation during investigated period in the High Tatras Mountain region. A further evaluation of other O_3 episodes is required for generalization of presented results.

Acknowledgments This work was supported by the Slovak Research and Development Agency under the contract no. APVV-51-030205 and partially by the Slovak Grant Agency VEGA (grant no. 2/0036/08). The authors are grateful to ILTER-NGO and SHMI for providing data.

References

Ashmore MR (2005) Assessing the future global impacts of ozone to vegetation, Plant. Cell. Env., 28, 949–964

Bičárová S, Fleischer P (2006) Windstorm effect on forest sources of biogenic volatile organic compound emissions in the High Tatras. Contr. Geophys. Geod., 36, 3, 269–282

Bičárová S, Sojáková M, Burda C, Fleischer P (2005) Summer ground level ozone maximum in Slovakia in 2003. Contr. Geophys. Geod., 35, 3, 265–279

Clappier A (1998) A correction method for use in multidimensional time-splitting advection algorithms: application to two- and three-dimensional transport. Monthly Weather Review 126, 232–242

Couach O, Balin I, Jiménez R, Ristori P, Peregeo S, Kirchner F, Simeonov V, Calpini B, van den Bergh H (2003) An investigation of ozone and planetary boundary layer dynamics over the complex topography of Grenoble combining measurements and modeling. Atmos. Chem. Phys., 3, 549–562

Fleischer P, Godzik B, Bičárová S, Bytnerowicz A (2005) Effects of air pollution and climate change on forests of the Tatra Mountains. In: The 6th International Symposium on Plant Responses to Air Pollution and Global Changes (eds: K. Omasa, I. Nouchi, L. J. De Kok). Tsukuba-Ibaraki, JAPAN, 21.–22. 10. 2004. Springer-Verlag Tokyo, 111–121

Forkel R, Knoche R (2006) Regional climate change and its impact on photooxidant concentrations in southern Germany: Simulations with a coupled regional climate-chemistry model. J. Geophys. Res., 111, D12302, doi:10.1029/2005JD006748

Gong W, Cho HR (1993) A numerical scheme for the integration of the gas phase chemical rate equations in three-dimensional atmospheric models′. Atmospheric Environment 27A(14), 2147–2160

Guenther A, Zimmerman P, Harley P, Monson R, Fall R (1993) Isoprene and monoterpene emission rate variability: Model evaluation and sensitivity analysis. J. Geophys. Res., 98, 12609–12617

Hrouzková E, Kremler M, Sojáková M, Závodský D (2004) Ground level ozone in Slovakia in 2003. Meteorol. čas., 7, 17–24

Kremler M (2006) Modelovanie výmeny látok medzi zložkami prírodného prostredia: Prízemný ozón. PhD thesis. FMFI UK Bratislava, 170 s (in Slovak)

Ostrožlík M (2004) Results of meteorological measurements at the observatories of the Geophysical Institute of the Slovak Academy of Sciences. Year-book 2003, Bratislava, 33pp

Paltridge G, Platt C (1976) Radiative Processes in Meteorology and Climatology. Number 5 in Developments in Atmospheric Science. Elsevier. Amsterdam

Perego S (1999) A numerical mesoscale model for simulation of regional photosmog in complex terrain: model description and application during POLLUMET 1993 (Switzerland). Meteor. Atmos. Phys., 70, 43–69

Rhie C, Chow W (1983) Numerical study of the turbulent flow past an airfoil with trailing edge separation', AIAA Journal 21(11), 1525–1532

Rodríguez P, Caballero S, Galindo N, Torres JG, Orza JAG, Yubero E, Nicolás J, Crespo J (2004) Characterization of an episode of high tropospheric ozone levels in the Iberian peninsula in August 2003. In: Proceedings of the XX Quadrennial Ozone Symposium (Ed. Zeferos). Kos, 906

Steiner AL, Tonse S, Cohen RC, Goldstein AH, Harley RA (2006) Influence of future climate and emissions on regional air quality in California. J. Geophys. Res., 111, D18303, doi:10.1029/2005JD006935

Stockwell WR, Kirchner F, Kuhn M, Seefeld S (1997) A new mechanism for regional atmospheric chemistry modeling. J. Geophys. Res., 102 (D22), 25847–25879

Stockwell WR, Middleton P, Chang JS, Tang X (1990) The second generation Regional Acid Deposition Model chemical mechanism for regional air quality modelling. A new mechanism for regional atmospheric chemistry modeling. J. Geophys. Res., 95 (D10), 16343–16367

Vautard R, Honore C, Beekmann M, Rouil L (2005) Simulation of ozone during the August 2003 heat wave and emission control scenarios. Atmospheric Environment 39, 2957–2967

Vestreng V, Breivik K, Adams M, Wagner A, Goodwin J, Rozovskaya O, Pacyna JM (2005) "Inventory Review 2005. Emission Data report to LRTAP Convention and NEC Directive. Initial review for HMs and POPs". EMEP Technical Report MSC-W 1/2005, 114pp. ISSN 0804-2446

EMEP	http://www.emep.int/
GPI SAS	http://gpi.savba.sk/
GEOLOGY	http://geology.com/europe-satellite-images.shtml
GLOBEIS	http://www.globeis.com/
CHIMERE	http://euler.lmd.polytechnique.fr/chimere/
METPHOMOD	http://www.giub.unibe.ch/klimet/metphomod/
PREV'AIR	http://www.prevair.org/en/index.php
SHMÚ	http://www.shmu.sk/
SURFER	http://www.goldensoftware.com
UNIDATA	http://www.unidata.ucar.edu/
WETTERZENTRALE	http://www.wetterzentrale.de/topkarten/tkfaxnwsar.htm

Part II
DROUGHT, FLOODS AND ECOSYSTEM RESPONSES

Physiological Drought – How to Quantify it?

V. Novák

Keywords Physiological drought, Soil water content, Transpiration, Biomass production, Mathematical modelling, Maize

Introduction

Drought as a phenomenon means deficiency of water – in general. It is used frequently, but its definition is usually qualitative and sometimes contradictive. Definition of drought from Encyclopedia Wikipedia says 'A drought is an extended period of months or years when a region notes a deficiency in its water supply'. According to Multilingual Technical Dictionary (ICID 1996), drought is 'a sustained period of time, with insufficient precipitation'. So, drought is mentioned here as a 'period' of time.

There are different drought definitions; drought is treated as a 'state' or as a time interval in which deficiency of water is noted.

Meteorological drought (Meteorological Vocabulary 1993) is 'frequently used term characterized by deficiency of water in soil, plant and atmosphere'.

Hydrological drought is defined as a deficiency of water in streams and rivers, usually according to number of days with water level or discharge below some defined value. The same criteria can be applied for groundwater and for springs (Burger 2005).

Agronomical drought (Meteorological Vocabulary 1993) is a result of meteorological drought. This term seems to be not appropriate, better term for the state of water in soil below some level should be the expression 'soil drought'. According to Šútor and others (Šútor et al. 2005, Šútor 2006), soil drought occurs if an average soil water content is below the soil water content (SWC) characterised by permanent wilting point. A better term is given by Wikipedia: characterised deficiency of water that negatively affects crop production as 'agricultural drought'.

Physiological drought (Meteorological Vocabulary 1993) is expressed as a 'deficiency of water, not covering plant needs'. It means the state of soil (and water in plants, respectively) limiting plant growth and plant production. Its relation to different types of drought is not unambiguous; even if there is a meteorological drought, it does not mean necessarily physiological or hydrological drought. Accordingly, the stage of physiological drought depends on plant type, especially on the ontogenesis stage of particular plant. Biomass production means usually production of the shoot parts of plants. But crop yield (known as agricultural output) means usually biomass production of final product (like grain, or roots of sugar beet). But, exact and quantitative expression of drought is difficult to formulate. It can mean quite different situation, depending on the aspect we are looking for.

Probably the most important aspect of agricultural drought is deficiency of soil water, limiting biomass production. In this chapter it will be discussed what can be understood under 'drought' from the point of view of plant production.

This chapter presents method of physiological drought evaluation, depending on the relationship between transpiration (evapotranspiration) of particular plants and plant production (Havrila and Novák 2006).

V. Novák (✉)
Institute of Hydrology, Slovak Academy of Sciences, Račianska 75, 831 02 Bratislava, Slovakia
e-mail: novak@uh.savba.sk

K. Strelcová et al. (eds.), *Bioclimatology and Natural Hazards*,
DOI 10.1007/978-1-4020-8876-6_7,

Theory

Rate of photosynthesis, expressed by the carbon dioxide consumption by plant, can be appreciatively expressed by the equation (Bierhuizen and Slayter 1964):

$$P = \frac{\Delta c_{ou}}{r_{ac} + r_{sc} + r_m} \tag{1}$$

Transpiration rate can be expressed by van Honert's (1948) type of equation

$$E_t = \frac{\Delta c_v}{r_a + r_s} \tag{2}$$

P – photosynthesis rate (kg m^{-2} s^{-1})
E_t – transpiration rate (kg m^{-2} s^{-1})
r_{ac}, r_{sc}, r_m – resistance of boundary layer of atmosphere at the leaf surface, the resistance of the stomata and mesophyl resistance to carbon dioxide transport from atmosphere to plant (s m^{-1})
r_a, r_s – resistance of boundary layer of atmosphere at the leaf surface for water vapour transport, the stomata resistance to water vapour from substomatal cavity to the atmosphere (sm^{-1})
Δc_{ou} – the difference of the mass concentration of carbon dioxide between leaf (after carboxylation) and the atmosphere (kg m^{-3})
Δc_v – the difference of the mass concentration of water vapour between the leaf and atmosphere (kgm^{-3})

By combination of Eqs. (1) and (2), we get

$$\frac{P}{E_t} = \frac{r_a + r_s}{r_{ac} + r_{sc} + r_m} \frac{\Delta c_{ou}}{\Delta c_v} \tag{3}$$

Resistances to CO_2 and water vapour transport are complex functions of environmental characteristics and they are changing with time. For particular plant, actual environmental properties and stage of plant ontogenesis it seems to be reasonable to assume the constant value of the resistances ratio at the right side of Eq. (3) – as an approximation – and to express it as B. Then, the photosynthesis rate P can be written as proportional to the transpiration rate E_t.

$$P = B.E_t \tag{4}$$

Equation (4) demonstrates the proportionality between photosynthesis and transpiration rate based on simplified assumptions. So, validity of Eq. (4) is limited by the approximation of the right side of Eq. (3) by constant B, which contains all the environmental properties (plant, soil, agrotechnics, fertilisation). Parameters of atmosphere and partially of the plants are involved in the procedure of transpiration rate calculation. Of course, this procedure is an approximation. There are a few empirical relationships of this type, because of difficulties with transpiration fluxes separation (Vidovič and Novák 1987, Feddes and Raats 2004). There are many empirical relationships between grain yield Y and seasonal evapotranspiration totals E during the vegetation period (Hillel and Guron 1973, Hanks and Hill 1980), some of them for maize canopy are illustrated in Fig. 1. For dense canopy, transpiration and evapotranspiration rates are close; therefore rela-

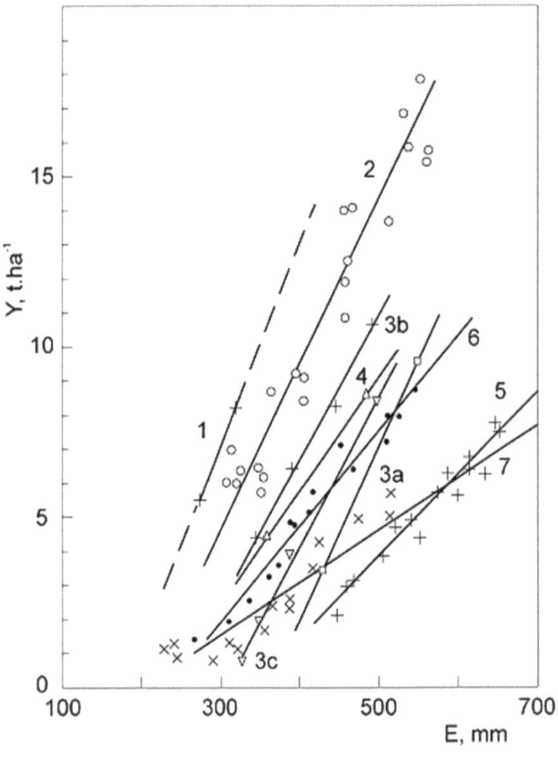

Fig. 1 Empirical relationship of dry maize grains yield Y and seasonal maize evapotranspiration totals E during the vegetation period. 1 – Trnava (1981–1982) – the only case where transpiration totals were presented (Vidovič and Novák, 1987); 2 – Logan, USA (1975); 3a, 3b, 3c – Gilat, Izrael (1968, 1969, 1970); 4 – Cherson, Ukraina (1974–1978); 5 – Greenville, USA (1978); 6 – Farmington, USA (1978), 7 – Evans, USA (1978). Items 2–7 are compiled from Hanks and Hill (1980)

tionships $Y = f(E)$ are close to linear too, as it can be seen in Fig. 1.

There exists a family of 'production' models, which could be used to model production process, based on the photosynthesis modelling. They are relatively complicated, canopy oriented and their weak point is necessity of many parameters estimation, needed as input data (Vanclooster et al. 2004). Therefore, even simplified, appreciative models with minimum input data which can be applied for particular plant with acceptable accuracy are valuable for application. This is the case of presented approach.

Transpiration is frequently used as an indicator of the soil water resources. de Wit (1958) found that dry matter production is proportional to transpiration. Relative transpiration as an index of the soil water resources state was used by Budagovskij and Grigorieva (1991) as the ratio of transpiration E_t and the potential transpiration E_{tp}:

$$\eta_p = E_t / E_{tp} \tag{5}$$

Equation (5) is characterizing the availability of soil water within the range (0;1) and the expression (6) can be noted as drought index in the range (0;1); $\eta_d = 1$ means 'absolute' drought, $\eta_d = 0$ means full, unlimited availability of soil water to plant.

$$\eta_d = 1 - \eta_p \tag{6}$$

The Critical Soil Water Content of Limited Water Availability Concept

The 'critical SWC of limited water availability' (θ_{la}) is characterizing an average SWC of the soil rooting layer at which the transpiration rate starts to decrease followed by biomass production decrease (Novák and Havrila 2006). Usually, upper 1 m layer is considered in which majority of roots are located. The core of the estimation method is based on an analysis of the known relationship between relative transpiration rate and average SWC of the root zone, as it is presented in Fig. 2 (Feddes and Raats 2004).

The principle of θ_{la} evaluation is briefly characterized using Fig. 2. It follows from previous analysis that maximum plant production rate can be reached if transpiration total per vegetation period of particular canopy is maximum, that is potential is equal to 1. From it follows that any transpiration rate below its potential value is limiting plant growth. A decrease in SWC in the soil root zone below θ_{la} has to result in a decrease of biomass production too. Therefore, the water content of the soil root zone below this value can be declared as corresponding to the state characterized as 'physiological drought'.

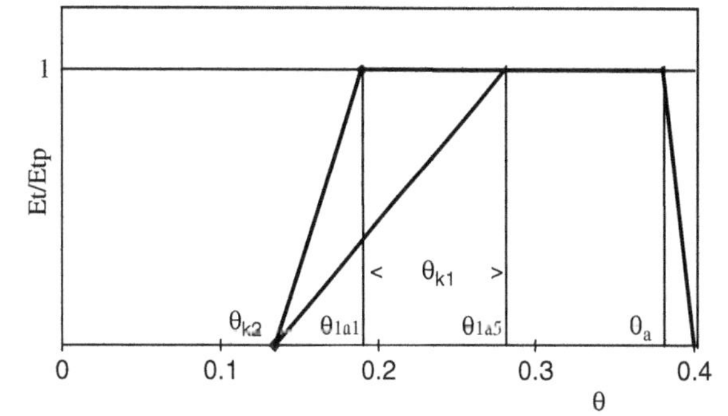

Fig. 2 Schematic relative transpiration E_t/E_{tp} and the SWC θ of the upper 1 m soil layer, where θ_{la5}–θ_{la1} is the range of the 'critical SWCs of limited availability' for plants, for the range of daily transpiration totals $1 \le E_t \le 5$ mm/day, E_p, E_{tp} is potential transpiration and transpiration, respectively. Site Most pri Bratislave (South Slovakia), θ_{k1}, θ_{k2} are 'critical' SWCs, indicating the beginning and the end of the transpiration decrease rate range

Method of the 'critical SWC of limited water availability' (θ_{la}) estimation was described earlier (Novák and Havrila 2006). It can be expressed by the empirical Eqs. (7), (8) and (9) (Novák et al. 1989):

$$\theta_{la} = \theta_{k1} = \frac{1}{\alpha} + \theta_{k2} \tag{7}$$

$$\theta_{k2} = 0.67 \,.\, \theta_v \tag{8}$$

$$\alpha = -2.27 E_p + 17.5 \tag{9}$$

where θ_{k1}, θ_{k2} are the so-called critical SWCs, indicating the beginning and the end of the transpiration decrease rate range and θ_v is SWC of the permanent wilting point (Kutílek and Nielsen 1994). Expected average daily potential transpiration rate E_p allows to calculate coefficient α (Eq. 9); θ_{k2} estimation needs SWC corresponding to wilting point θ_v (Eq. 8). Finally θ_{la} is calculated, using Eq. (7).

The critical SWC is less than θ_v ($\theta_{k2} < \theta_v$); because plant transpires at slow rate even at the SWC corresponding to the 'wilting point'. Coefficient α depends on the potential evapotranspiration rate E_p. It follows that SWC corresponding to the critical SWC of limited availability for plants does not depend on the soil properties only, but it is also a function of the Soil–Plant–Atmosphere Continuum (SPAC) properties. Potential transpiration rate strongly influences critical SWC θ_{k1}, which is a function of meteorological properties. Sensitivity analysis of transpiration process, as it is quantitatively described by Penman–Monteith equation (Monteith 1965) and was performed by Novák et al. (1997), documented the primary importance of net radiation on potential transpiration rate, followed by the air temperature.

Figure 3 presents SWC corresponding to the critical SWC of limited availability to plants of three soils as they depend on the transpiration rate – $\theta_{la} = f(E_t)$. From the analysis it follows that there is a strong dependence of critical SWC of the limited availability θ_{la} on the transpiration rate and it increases with the transpiration rate. It is a fact that decrease in daily transpiration rate during days with maximum energy input (hot days) limited by lack of the soil water is limiting biomass production much more significantly than during cold days.

In Fig. 3 (denoted by circles), SWCs corresponding to 'point of limited availability' (θ_{pla}) calculated according to the frequently used empirical equation (Kutílek 1978) could be seen

$$\theta_{pla} = \theta_v + 0.6(\theta_{fc} - \theta_v) \tag{10}$$

θ_{fc} is SWC corresponding to the field capacity, characterizing the amount of water in the soil after thorough wetting followed by natural draining of water to subsoil (Kutílek and Nielsen 1994). It can be seen that SWC corresponding to the 'point of limited availability' (θ_{pla}) differ from the critical SWC of the limited availability θ_{la}, but it is in the range of SWC estimated by the proposed method (Fig. 3). But values θ_{pla} are corresponding to the high transpiration rates, which are rare. Realistic transpiration rates under conditions of South Slovakia are usually 2–3 mm/day. The range of SWC with available water in soil is estimated by the difference between the water content at field capacity and the water content at the wilting point (Peters 1965). Later, the concept of different water availability in this SWC range was presented and is accepted widely, but in literature the information on how to quantify 'different water availability' to be based physically and physiologically was not found. An example how this problem was treated is expressed by Eq. (10); the 'critical' SWC is assumed to be 0.6 times of the range between field capacity and permanent wilting point.

Proposed estimation method of the critical SWC of limited availability for plants θ_{la} is physically and physiologically clearly interpreted and can be easily calculated using Eqs. (7), (8) and (9). It is not a constant value for particular soil and plant, but it changes, depending on the transpiration rate. For practical purposes it is possible to use the critical SWC (θ_{la}) corresponding to the daily average transpiration rates.

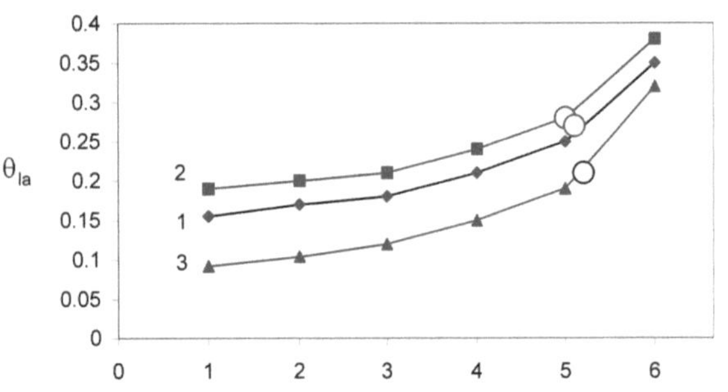

Fig. 3 Critical SWC of limited availability for maize θ_{la}, corresponding to the transpiration rate E_t of the three soils. Circles are denoting values of θ_{pla} estimated according to the Eq. (10). Trnava (1) – chernozem on loess, Láb (3) – sandy soil, Most pri Bratislave (2) – loamy soil

Results – An Illustrative Example

The above described method of 'physiological drought' estimation will be illustrated on results of measurement and modelling at the experimental site of Most pri Bratislave (South Slovakia), with loamy soil and maize canopy. As a tool, simulation model HYDRUS –ET (Šimůnek et al. 1997) was used, with incorporated modified Penman–Monteith method to calculate evapotranspiration and its components (Majerčák and Novák 1992). Calculations were made for 31 seasons and maize canopy. Basic soil characteristics can be found in papers by Havrila and Novák (2006) and Novák and Havrila (2006).

Maximum (V_{max}) and minimum (V_{min}) daily values of SWC in 0–50 cm upper layer of soil with maize canopy – expressed in water layer thickness – calculated by mathematical model HYDRUS–ET, during 31 seasons for Most pri Bratislave site, as well as SWC corresponding to the basic soil water hydrolimits (wilting point, critical SWC of limited availability and field capacity) are shown in Fig. 4. It can be noted that during this period, there was no state when the SWC of the upper 50 cm soil layer was below permanent wilting point. There were found periods, when SWC was below critical SWC of limited availability (V_{la}), which corresponds here to the potential (highest) rates of transpiration. V_{la} is the thickness of soil water layer, which depends on transpiration rate, which changes during the seasons. To evaluate the long time courses of SWCs related to water content of limited availability, V_{la} was evaluated for transpiration rates close to maximum ones; $E_t = 5\,\mathrm{mm\,d^{-}1}$ (see Fig. 2). That means, keeping SWC above the such evaluated $V_{la} = 12.8\,\mathrm{cm}$ in upper 50 cm soil layer, decrease of photosynthesis rate due to water deficiency could be rare.

Cumulative frequency curves of maximum (V_{max}) and minimum (V_{min}) daily values of SWC in 0–50 cm upper layer of soil with maize canopy are shown in Fig. 5. In 60% of seasons analysed, the V_{la} was estimated to drop below the critical level.

Figure 6 presents the time interval expressed in days during the vegetation period of maize with SWC of the upper 50 cm soil layer less than SWC corresponding to the ‘critical SWC of limited availability’ to plants V_{la}, ($V_{la} = 12.8\,\mathrm{cm}$ of water layer) for 31 vegetation periods. In other words, there is the time interval, during which photosynthesis rate is limited by soil water and is less than optimal one. The ‘driest’ growing season for maize canopy was found in 2003, with permanent deficiency of water needed for maximum biomass production.

Cumulative frequency curves of number of days n during the vegetation period of maize with SWC of the upper 50 cm soil layer less than SWC corresponding to the ‘critical SWC of limited availability’ of soil water to plants V_{la}, ($V_{la} = 12.8\,\mathrm{cm}$), calculated by mathematical model HYDRUS–ET, during 31 seasons is

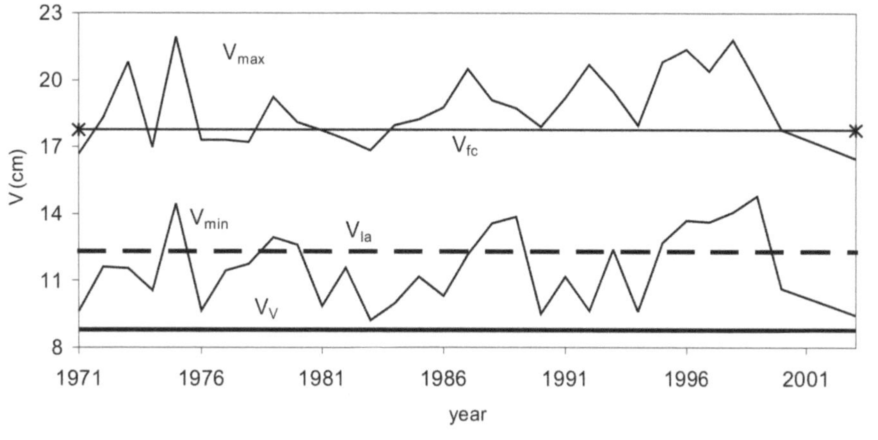

Fig. 4 Maximum (V_{max}) and minimum (V_{min}) daily values of SWC of the 0–50 cm upper layer of soil with maize canopy, calculated by mathematical model HYDRUS–ET, during 31 seasons, expressed in water layer thickness. SWCs corresponding to the basic soil water characteristics (V_v – wilting point, V_{la} – SWC of limited availability and V_{fc} – field capacity) are presented too (Most pri Bratislave)

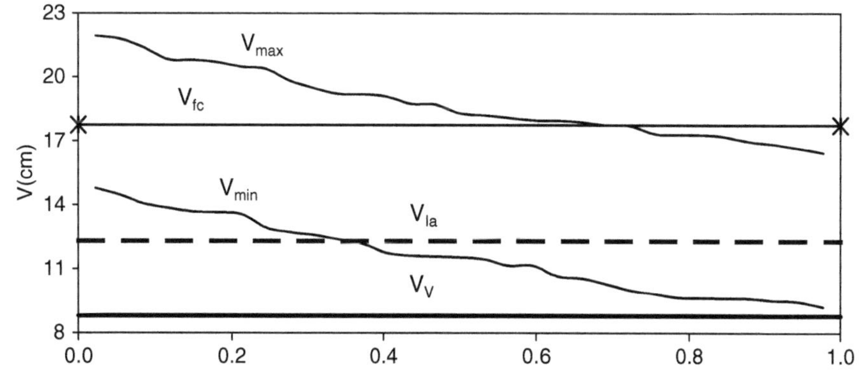

Fig. 5 Cumulative frequency curves of maximum (V_{max}) and minimum (V_{min}) daily values of SWC in 0–50 cm upper layer of soil with maize canopy, calculated by mathematical model HYDRUS–ET, during 31 seasons, corresponding to the basic soil water characteristics (V_v – wilting point, V_{la} – SWC of limited availability and V_{fc} – field capacity), Most pri Bratislave, p – probability

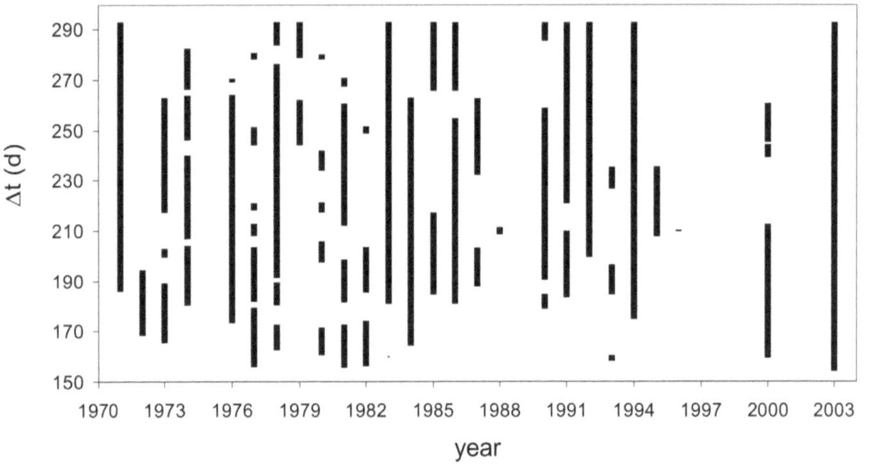

Fig. 6 Duration of time interval Δt in days during the vegetation period of maize with SWC of the upper 50 cm soil layer smaller than SWC corresponding to the 'critical SWC of limited availability' to plants V_{la}, ($V_{la} = 12.8$ cm of water layer) calculated by mathematical model HYDRUS–ET, during 31 seasons (Most pri Bratislave site)

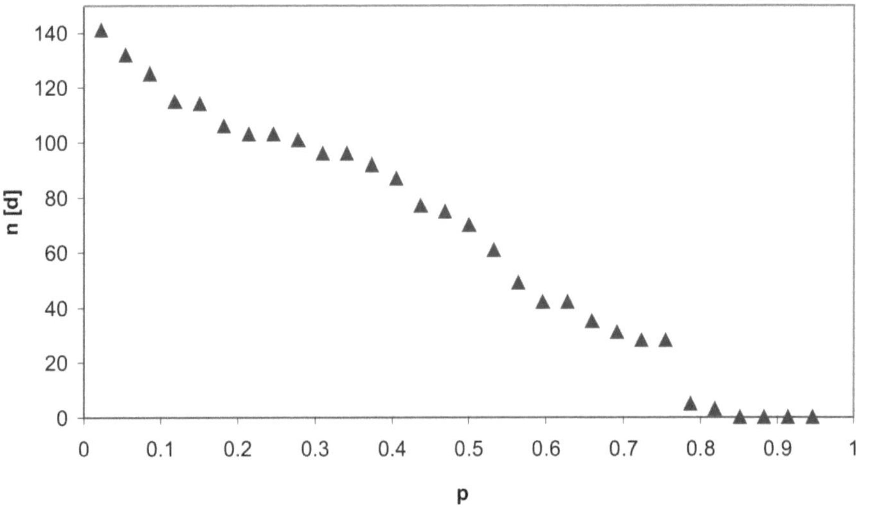

Fig. 7 Cumulative frequency curves for number of days *n* during the vegetation period of maize with SWC of the upper 50 cm soil layer less than SWC corresponding to the 'SWC of limited availability' of soil water to plants V_{la}, ($V_{la} = 12.8$ cm of water layer), calculated by mathematical model HYDRUS–ET, during 31 seasons (Most pri Bratislave site), *p* – probability

shown in Fig. 7. Data from Fig. 6 were used. Only four vegetation periods of maize were wet enough to ensure optimum SWC for maximum biomass production.

Conclusions

A proposal is presented how to define 'soil drought' from biomass production point of view. The presented approach is based on the theoretical considerations and empirical relationship between biomass production of particular plants and transpiration (evapotranspiration) total of this plant during its vegetation period. This relationship is linear and valid for particular plant and environmental conditions (soil, nutrition, agrotechnics).

Optimal plant production can be reached for maximum transpiration total during vegetation period of a particular canopy, therefore potential transpiration total corresponds to maximum possible yield in given conditions. Transpiration rate, less than potential one leads to the biomass production decrease.

This phenomenon was used to define the so-called physiological drought, as the state of soil water expressed by the SWC of soil root zone below so-called critical SWC of limited availability for plants. Then, transpiration rate is below potential one and biomass production is below its potential rate.

Method of calculation for 'critical SWC of limited availability' for plants, presented in this chapter is a function of soil properties, and transpiration rate. It means, the lower is the transpiration rate, the longer are preserved optimal conditions for plant production and the state of physiological drought is postponed.

Another consequence of this analysis is a recognition that the state noted as a 'physiological' drought interpreted through SWC is not characterized by some value of SWC only, but it depends at the particular site, on soil, plant and meteorological characteristics.

Acknowledgments Author is grateful to the Slovak Grant Agency VEGA (Grant No. 2/7091/27) and to the Slovak Grant Agency APVV (Grant No. 51- 030205) for partial support of this work.

References

Bierhuizen JF, Slayter B (1964) An apparatus for the continuous and simultaneous measurement of photosynthesis and transpiration under controlled environmental conditions. CSIRO Australian Division of Land Research Technical Paper,, v. 24

Budagovskij AI, Grigorieva NI (1991) The ways of soil water resources efficiency increasing. Vodnyje Resursy, v. 1, 131–142 (in Russian)

Burger F (2005) Concept of hydrological drought and its identification through groundwater deficit. Acta Hydrologica Slovaca, v. 6, no. 6, 3–10 (in Slovak)

de Wit CT (1958) Transpiration and crop yields. Versl. Landbov. Onderz. No. 646, p. 88

Feddes RA, Raats PAC (2004) Parameterising the soil – water – plant root system. In: Unsaturated Zone Modelling – Progress, Challenges and Applications (RA Feddes, GH De Rooij and JC van Dam Eds.) Kluver Publ., Doortrecht–Boston–London, 95–144

Hanks RJ, Hill RW (1980) Modelling crop responses to irrigation in relation to soils, climate and salinity. Inter. Irrig. Inform. Center, Publ. No. 6, Bet Dagan, Israel, p. 57

Havrila J, Novák V (2006) Method of soil water regime evaluation in relation to biomass production. Journal of Hydrology and Hydromechanics, v. 54, no. 1, 15–25 (in Slovak)

Hillel D, Guron Y (1973) Relation between evapotranspiration rate and maize yield. Water Resources Research, v. 9, 743–748

ICID (1996) Multilingual technical dictionary on irrigation and drainage. 2nd revised edition. ICID–CIID, New Delhi, India

Kutílek M (1978) Hydropedology. Praha, SNTL – ALFA Publishers, (in Czech)

Kutílek M, Nielsen DR (1994) Soil hydrology. Catena Verlag, Cremlingen-Destedt, Germany, 370pp

Majerčák J, Novák V (1992) Simulation of the soil – water dynamics in the root zone during the vegetation period: I. Simulation model. Journal of Hydrology and Hydromechanics,, v. 40, 299–315

Meteorological vocabulary (1993) Academia, MZCR, Praha, p. 594.

Monteith JL (1965) Evaporation and Environment. Symposia of the Society for Experimental Biology, v. 29, 205–234

Novák V (1989) Estimation of critical soil water contents to evapotranspiration. Počvovedenie [Soviet Soil Science], no. 2, 137–141 (in Russian)

Novák V, Hurtalová T, Matejka F (1997) Sensitivity analysis of the Penman type equation for calculation of potential evapotranspiration. Journal of Hydrology and Hydromechanics,, v. 45, 173–186

Novák V, Havrila J (2006) Method to estimate the critical soil water content of limited availability for plants. Biologia, (Bratislava), v. 61, Suppl. 19, 289–293.

Peters, D.B. 1965. Water availability. In: Methods of Soil Analysis, Part 1. (CA Black ed.). Number 9 in Series Agronomy, ASA, Madison, Wisconsin, 279–285

Šimůnek J, Huang K., Šejna M, Van Genuchten ThM, Majerčák J, Novák V, Šútor J (1997) The HYDRUS -ET software package for simulating the one – dimensional movement of water, heat and multiple solutes in variably – Saturated media. Version 1.1. Institute of Hydrology, Slovak Academy of Sciences, Bratislava

Šútor J, Gomboš M, Mati R (2005) Soil drought quantification and its interpretation pôdneho sucha. Acta Hydrologica Slovaca, v. 6, 299–306 (in Slovak).

Šútor J (2006) Soil drought prognosis. Acta Hydrologica Slovaca, v. 7, 176–182 (in Slovak)

Vanclooster M, Boesten J, Tiktak A, Jarvis N, Kroes JG, Munoz-Carpena R, Clothier BE, Green SR (2004) On the use of unsaturated flow and transport models in nutrient and pesticide management. In: Unsaturated Zone Modeling-Progress, Challenges and Applications (RA Feddes, GH De Rooij and JC van Dam Eds.) Kluver Publ., Doortrecht–Boston–London, 331-361

van Honert TH (1948) Water transport in plants as a catenary proces. Discussions of the Faraday Society, 3, 146–153

Vidovič J, Novák V (1987) Yield and maize canopy evapotranspiration. Rostlinní výroba, v. 33, no. 6, 663–670 (in Slovak)

Occurrence of Dry and Wet Periods in Altitudinal Vegetation Stages of West Carpathians in Slovakia: Time-Series Analysis 1951–2005

J. Škvarenina, J. Tomlain, J. Hrvol' and J. Škvareninová

Keywords Energy balance components · Relative evapotranspiration · Bowen ratio · Drought index

Introduction

The meteorological and bioclimatologic vocabulary defines the term drought (dryness) from several points of view. In principle, however, it is defined as the state of water deficit concerning soil, plants and atmosphere (Krevčmer 1980). The grounds underlying lack of moisture are complex and intricate. In countries with generally sufficient water supply, the dryness is considered a rare event or even a natural calamity. In agriculture and forestry, it is understood as an important meteorological factor, distressing landscape ecosystems. We can say that perhaps drought is the most important, because besides the direct damage represented by spells of drought on plants, such event entails a whole range of other consequences. These are synergism following consequences, such as excessive transpiration, sunburnt bark, decline of seedlings and regeneration, premature fall of plant assimilatory organs, premature beginning of autumn phenophases, deficient development of next-year vegetative and generative buds, damage and injury to fine root systems, blocked mycorrhizas, the lowered microbial activity in soil connected with a soil acidification, water eutrophication, the outbreak of biotic pests, forest fires (Ditmarova et al. 2006; Hayes et al. 1999; Heim 2002; Holécy 2004; Paganová 2000; Střelcová et al. 2004).

Drought is a normal, recurrent feature of temperate climate. It occurs almost everywhere, although its features vary from region to region. Drought is therefore difficult to define; its definition depends on differences in regions, needs and disciplinary perspectives (Hayes et al. 2005). Drought is a temporary aberration; it differs from aridity, which is restricted to low-rainfall regions and is a permanent feature of climate.

Drought is an insidious hazard of nature. Drought expresses a relatively short-term negative deviation of water balance from the normal and has significant impacts on various activities (Heim 2002; Dubrovsky et al. 2005; Šiška and Samuhel 2007; Šiška et al. 2005). Using various points of reference, drought can be considered in meteorological, agricultural, hydrological and socio-economic terms (Fig. 1). Without exception, the primary cause of drought under the conditions prevalent in Slovakia is a deficit in precipitation over a certain period, such as during all periods of the growing season. Other climatic factors, such as high temperature, strong wind and low relative humidity, can significantly aggravate its severity (Hayes et al. 1999; Heim 2002; Kharel and Bruins 2004).

The altitude and topography are strong climate-differentiating factors. Consequently, in conditions of the considerably broken topography in Slovakia droughts play an extra-important role. The increase in altitude also causes changes in solar radiation, thermal and water balance of the land. The primary importance of the climate from the viewpoint of the natural vegetation has already been pointed out by Zlatník (1976). The author defines the vegetation stages as basic units characterising the altitudinal climate conditions (vertical differentiation) through vegetation (biocenoses). The diversity of vegetation stages result from the climatic differences due to the

J. Škvarenina (✉)
Technical University in Zvolen, Faculty of Forestry,
T.G. Masaryka 24, 960 53 Zvolen, Slovakia
e-mail: jarosk@vsld.tuzvo.sk

K. Strelcová et al. (eds.), *Bioclimatology and Natural Hazards*,
DOI 10.1007/978-1-4020-8876-6_8,

Fig. 1 Definition understanding drought for different sector activities and time sequence of drought (according to Hayes et al. 1999; Heim 2002)

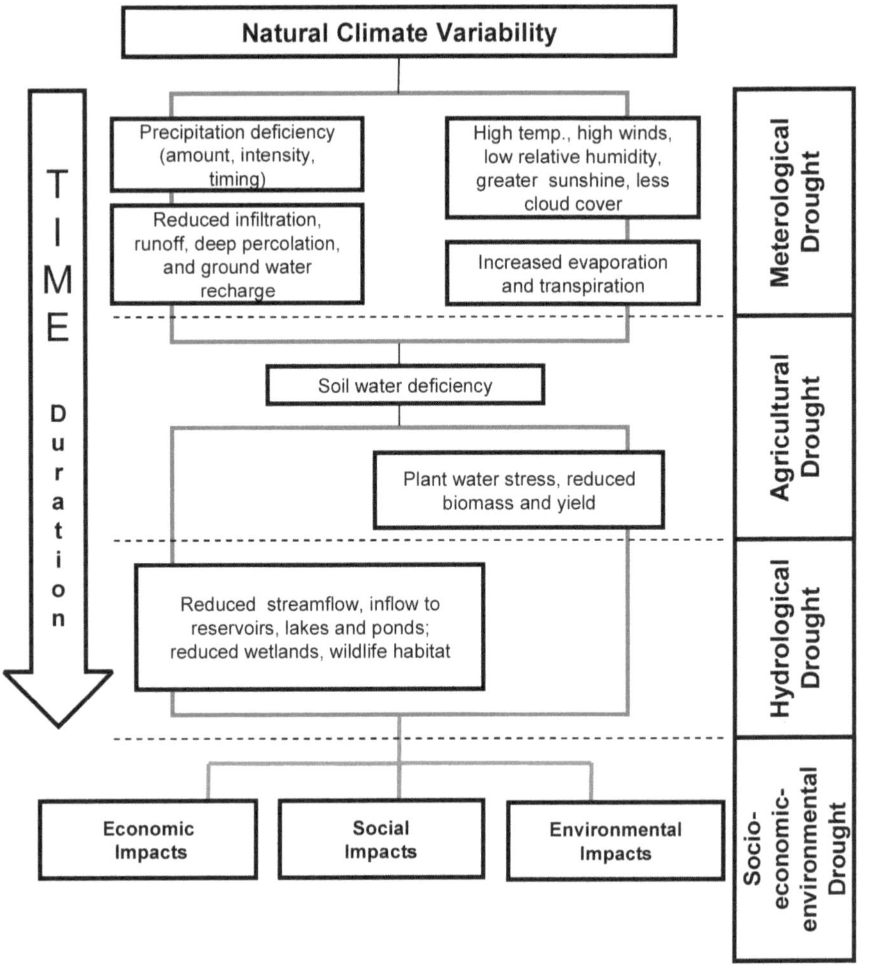

altitude, exposure and topography. The biogeocenoses resulting from variability of these three factors can be classified as belonging to nine vegetation stages. The Slovak territory has been, according to Zlatník (1976), divided into the following vegetation stages named after the significant tree or bush indicator species dominating in the area (what does not need to be in accordance with their occurrence optimum). Altitudinal vegetation stages are characterized by their dominant climax tree species as follows: 1st Oak– *Quercus* (Q), 2nd Beech-Oak –*Fagus-Quercus* (FQ), 3rd Oak-beech– *Quercus-Fagus* (QF), 4th Beech –*Fagus* (F), 5th Fir-beech –*Abies-Fagus* (AF), 6th Spruce-fir-beech –*Picea-Abies-Fagus* (PAF), 7th Spruce– *Picea* (P), 8th Mountain pine*Mughetum* (M), 9th Alpine (non-forest high mountain pastures) (the acronyms in parentheses have been derived from the Latin names).

The vegetation stages of lower elevations, that is the 1st oak vegetation stage, the 2nd oak stage with admixture of beech, and the 3rd beech stage with admixture of oak, are rather arid during the vegetation period (March–September). The precipitation deficit reaches 100–300 mm during the vegetation season. The 4th beech stage is characterized by an equitable climatic water balance. In the higher vegetation stages (the 5th beech stage with fir, and the 6th fir stage with beech and spruce), the climate humidity increases. The humidity of the climatic regime belongs to the fundamental properties of montane forests. The water balance reaches the highest values in the 8th vegetation stage of mountain dwarf pine and the 9th alpine stage, where the amount of precipitations considerably exceeds the evaporation requirements of the atmosphere. Within the annual balance, the surplus of precipitation water is approximately 1000 mm (Škvarenina et al. 2004).

The experiments confirmed that under optimum conditions of plant growth, the actual evapotranspiration (E) is proximate to the potential one (to the maximum possible evapotranspiration under the given climatic conditions from sufficient soil moisture $-E_o$). That is why the ratio E/E_o (relative evapotranspiration) and drought index (E_o/P), P – precipitation total – enable quantification of deficit of water in soil root zone for optimum plant growth (Budyko and Zubenok 1961).

The indices were proposed to characterise general environmental conditions and processes on the Earth's surface. It approximates the ability of precipitation to provide the water required by native vegetation for an undisturbed evapotranspiration process and is often used in ecological studies (Novák 1995; Tomlain 1996; Špánik and Šiška 2007; Arora 2002; Gongalsky et al. 2004; Pražák et al. 1994; 1996).

In this chapter, for the firsttime, we present detailed relative evapotranspiration (E/E_o), and drought index (E_o/P), results of altitudinal vegetation stages in Slovakia, covering the period 1951–2005.

Methodology

Relative evapotranspiration and drought index express functional dependences among all energy and water balance equation components of the locality (net radiation, air temperature and humidity, turbulent state of atmosphere, difference of saturation water vapour pressure at the temperature of evaporating surface and water vapour pressure in the air, precipitation, change of critical soil moisture during the year and heat flux in the soil). The model following from common solution of the energy and water balance equations was performed and discussed for eight localities in the territory of Slovakia (Fig. 2 and Table 1). The input model data are air temperature and humidity, cloudiness, precipitation and number of days with snow cover.

According to Budyko and Zubenok (Budyko 1980) the potential evapotranspiration was computed by the equation of water vapour diffusion in the atmosphere, and the actual evapotranspiration is supposed to be proportional to the potential evapotranspiration:

$$E = E_o \cdot W/W_o \quad (1)$$

The storage W is specified as the moisture stored in the upper soil layer of 1 m depth and W_o as the critical value above which E equals E_o. W_o usually amounts to a layer of 100–200 mm water with seasonal and regional variations. The average soil moisture $W = (W_1 + W_2)/2$ is determined from the water balance equation by the method of step-by-step approximation (W_1 is the moisture stored in the soil layer at the beginning of the month, and W_2 at its end) (Hrvol' et al. 2001, 2003; Tomlain 1996, 2004).

At present, we can only use the data from a limited number of meteorological stations for which the potential evapotranspiration has been calculated. For this reason we decided to use, representing the climate-hydric conditions in the particular vegetation stages, the method of representative climatic station for the given vegetation stage. The climatic stations were classified into the appropriate vegetation stages 1st–7th

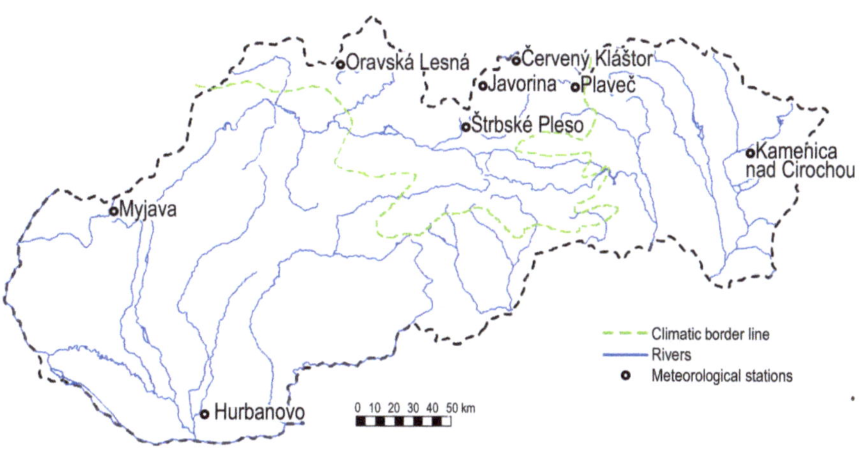

Fig. 2 The map of Slovakia with particular meteorological stations that represent conditions in the vegetations stages. The map displays the climatic border line (*dashed line*) and rivers (*solid line*)

Table 1 Average annual values of precipitation (P), potential (E_0), actual evapotranspiration (E), drought index (E_0/P) and relative evapotranspiration (E/E_0) at chosen localities of Slovakia for the period 1951–2005

Meteorological Station	H (m a.s.l.)	P (mm)	E_0 (mm)	E (mm)	E/E_0 (%)	E_0/P (–)	Vegetation stage
Hurbanovo	115	537	748	432	58	1.44	1st Oak
Myjava	375	671	627	450	72	0.96	2nd Beech-Oak
Kamenica n. C.	178	722	644	501	78	0.92	3rd Oak-beech
Plaveč	488	693	522	451	87	0.78	4th Beech
Červený Kláštor	474	755	507	457	90	0.69	5th Fir-beech
Oravská Lesná	780	1114	456	432	95	0.42	6th Spruce-fir-beech
Tatranská Javorina	1020	1253	422	412	97	0.35	7th Spruce
Štrbské Pleso	1360	995	435	407	94	0.45	7th Spruce and 8th Mountain pine

(8th). For the 9th stage there were no data based on the map of vegetation stages (Raušer and Zlatník 1966) and the typological maps of the forest type groups (scale 1:2,00,000).

Result and Discussion

Average Annual Values of Relative Evapotranspiration and Drought Index

Table 1 shows the average annual values of drought index, potential evapotranspiration, precipitation, actual evapotranspiration and relative evapotranspiration for the period 1951–2005 at chosen stations in Slovakia. On an average, the smallest annual values of E/E_0 for the analysed period 1951–2005 were recorded in the Danube Lowland (1st Oak vegetation stage), that is in region with relative high totals of potential evapotranspiration (above 700 mm) and with annual precipitation totals below 550 mm. In mountain areas of Slovakia, where excess moisture is observed during the year, the actual evapotranspiration shows only small differences relative to the potential evapotranspiration (less than 4%). With the increase of precipitation totals, E grows according to the elevation approaching a definite boundary (to the 3rd Oak-beech vegetation stage). Then, as a result of the net radiation decrease (the growth of cloudiness, relative humidity of the air, number of days with snow cover, the decrease of the air and soil surface temperature) the actual evapotranspiration decreases with height. The average annual values of potential evapotranspiration ranges from 750 mm in the southern part of Danube Lowland to 350 mm and less in the highest mountain positions (7th Spruce, 8th Mountain pine and 9th Alpine stage).

The relative evapotranspiration E/E_0 is an excellent measure of water sufficiency of vegetation. It approaches its lowest values of about 60% in the lowest areas. Towards higher vegetation stages, the E/E_0 increases approaching more than 90% in the 4th beech vegetation stage. However, on the mountainous sites this measure partly loses its accuracy. By the 5th vegetation stages its resolution approaches only about 1–5%. Figure 3 presents the dependency between values of relative evapotranspiration E/E_0 and the terrain altitude.

The drought index E_0/P informs about the relationship between solar energy and precipitation inputs within particular vegetation stages. This index has been already used for the evaluation of bioclimatologic regions at the planetary scale (Budyko 1955, 1971; Larcher 1984; Baumgartner and Liebscher 1990; McCabe and Wolock 2002; Sankarasubramanian and Vogel 2002). Beyond any expectations, the E_0/P index proved to be very sensitive to the altitudinal changes and was able to detect differences in bioclimatological

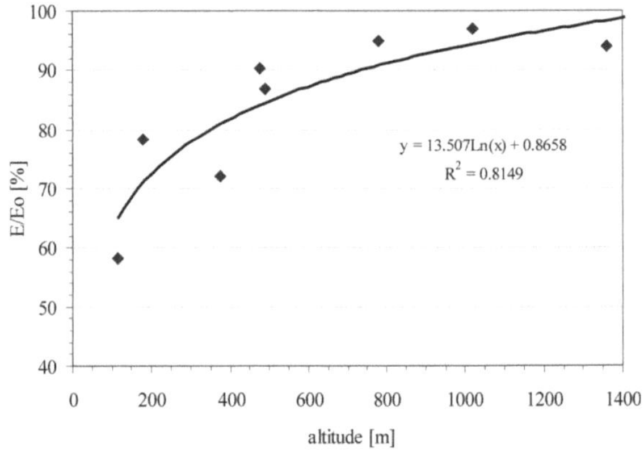

Fig. 3 Logarithmic regression dependence of relative evapotranspiration annual values (E/E_0) on altitude (m a.s.l.) on the territory of Slovakia for the period 1951–2005

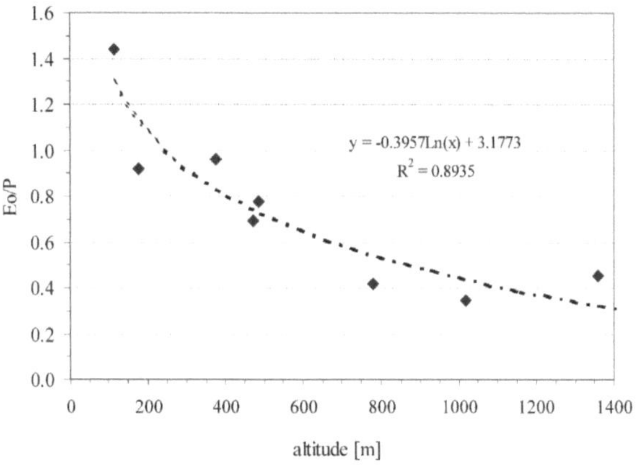

Fig. 4 Logarithmic regression dependence of drought index annual values (E_0/P) on altitude (m a.s.l.) on the territory of Slovakia for the period 1951–2005

conditions also in the comparatively small territory of Slovakia (Fig. 4). As presented by Budyko (1980), the values of $E_0/P > 1$ indicate the territory of the dry (arid) climate (steppe, forest-steppe). The values of $0.3 < E_0/P < 1$ specify forest bioms and values of $E_0/P < 0.3$ reveal the climate ecosystems of tundra, or mountain forest of a temperate zone as well.

The presented distribution of E_0/P values also corresponds to the climate condition in Slovakia. Warm forest-steppe formations dominated by oak, provides E_0/P values about 1. This represents the 1st vegetation stage (in the sense of Zlatnik classification) or a part of the 2nd vegetation stage dominated by Beech-Oak formations. Vegetation stages up to the E_0/P value 0.3 represent the predominant area of Slovak forest; the values of index E_0/P decrease relatively proportionally when related to the both increasing altitude and precipitation amount. The vegetation stages with $E_0/P < 0.3$ are of mountainous (boreal) climate, characterized by its low temperature and high precipitation amount, with Norway Spruce and Dwarf Pine being the predominant tree species growing here.

Bochenek et al. (2005) observed the drought index during vegetation period (April–September). It ranged from 1.6 in the area heavily prone to drought (central part) to 1.0 in areas where sum of summer precipitation exceeds potential evapotranspiration. Such areas exist in mountainous regions in southern Poland and are characterized by high rainfall in summer months.

Long-Term Course of Relative Evapotranspiration and Drought Index

The observed annual values of relative evapotranspiration in the period 1951–2005 varied from 38% (1990, P = 422 mm) to 86% (1965, P = 827 mm) at Hurbanovo, from 54% (2003, P = 445 mm) to 92% (1965, P = 785 mm) at Myjava, from 58% (1961, P = 400 mm) to 95% (1985, P = 911 mm) at Kamenica nad Cirochou, from 70% (1964, P = 627 mm) to 98% (1997, P = 780 mm) at Plaveč, from 71% (1964, P = 530 mm) to 99% (1975, P = 850 mm) at Červený Kláštor, from 74% (1992, P = 980 mm) to 99% (1958, P = 1389 mm; 1965, P = 1320 mm; 1968, P = 1133 mm; 1979, P = 1051 mm; 1991, P = 986 mm; 2004, P = 1256 mm; 2005, P=1342 mm) at Oravská Lesná, from 84% (2000, P=1389 mm) to 100%, (1997, P=1516 mm) at Ždiar-Javorina, from 84% (2003, P = 794 mm) to 100% (1958, P = 1160 mm) at Štrbské Pleso.

Annual values of drought index for the given period varied from 0.81 (1965, P = 827 mm) to 2.65 (2003, P = 333 mm) at Hurbanovo, from 0.69 (1965, P = 785 mm) to 1.71 (2003, P = 445 mm) at Myjava, from 0.59 (1974, P = 1010 mm) to 1.80 (1961, P = 400 mm) at Kamenica nad Cirochou, from 0.50 (1985, P = 930 mm; 2005, P = 883 mm) to 1.28 (1961, P = 477 mm) at Plaveč, from 0.46 (1980, P = 908 mm) to 1.03 (1971, P = 548 mm) at Červený Kláštor, from 0.28 (1974, P = 1463 mm) to 0.55 (1959, P = 860 mm) at Oravská Lesná, from 0.20 (1980, P = 1630 mm) to 0.51 (1951, P = 903 mm; 1961, P = 972 mm) at Ždiar-Javorina, from 0.28 (2004, P = 1299 mm) to 0.73 (1986, P = 690 mm) at Štrbské Pleso.

The differences in the observed extreme annual values of both the E/E_0 and E_0/P were caused by time and space variabilities concerning the energetic possibilities of evapotranspiration and precipitation fields in the territory of Slovakia. Figures 5, 6, 7, 8, 9, 10, 11, 12, 13, 14, 15, 16, 17, 18, 19 and 20 show long-term courses of relative evapotranspiration and drought index at the selected stations in Slovakia during the period of 1951–2005. These figures show the decreasing tendency of annual E/E_0 values in the 1st Oak, 2nd Beech-Oak vegetation stages situated in the Danube and Zahorská Lowlands, while E_0/P values gradually increase. The values of E/E_0 observed in basins and northern parts of East Slovakia lowland gradually increase. The sufficient of precipitation in mountain areas

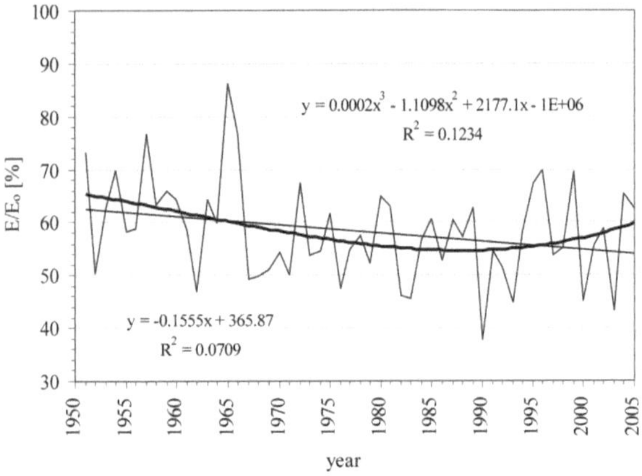

Fig. 5 Long-term course of relative evapotranspiration annual values (E/E_0) at Hurbanovo for the period 1951–2005 with the linear trend and polynomial of third order

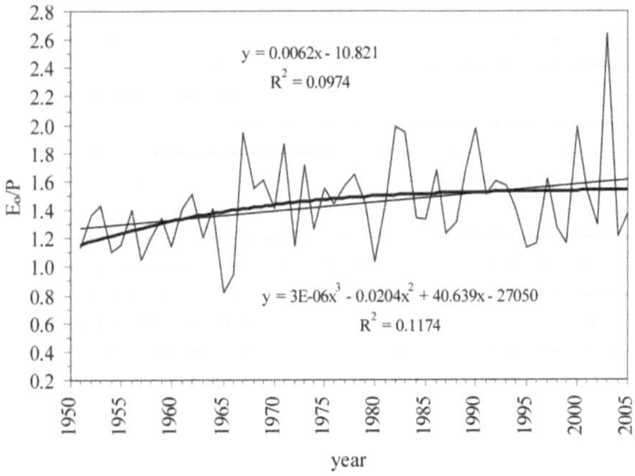

Fig. 6 Long-term course of drought index annual values (E_0/P) at Hurbanovo for the period 1951–2005 with the linear trend and polynomial of third order

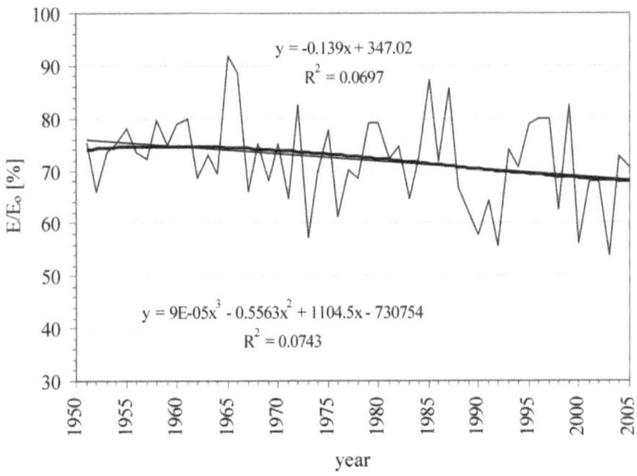

Fig. 7 Long-term course of relative evapotranspiration annual values (E/E_0) at Myjava for the period 1951–2005 with the linear trend and polynomial of third order

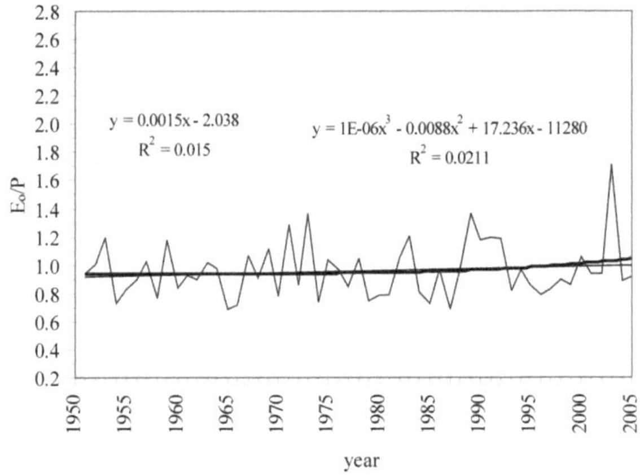

Fig. 8 Long-term course of drought index annual values (E_0/P) at Myjava for the period 1951–2005 with the linear trend and polynomial of third order

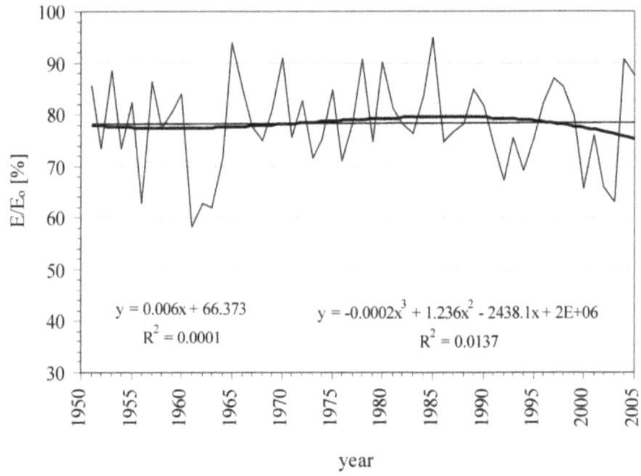

Fig. 9 Long-term course of relative evapotranspiration annual values (E/E_0) at Kamenica nad Cirochou for the period 1951–2005 with the linear trend and polynomial of third order

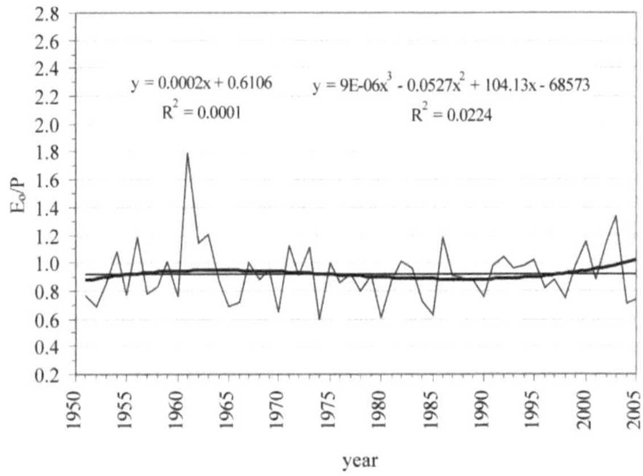

Fig. 10 Long-term course of drought index annual values (E_0/P) at Kamenica nad Cirochou for the period 1951–2005 with the linear trend and polynomial of third order

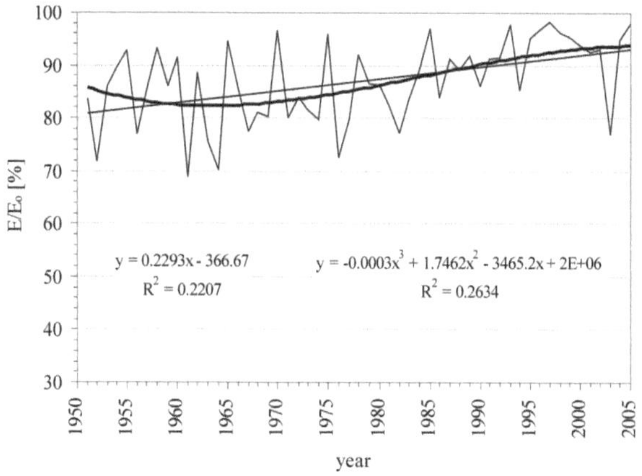

Fig. 11 Long-term course of relative evapotranspiration annual values (E/E_0) at Plaveč for the period 1951–2005 with the linear trend and polynomial of third order

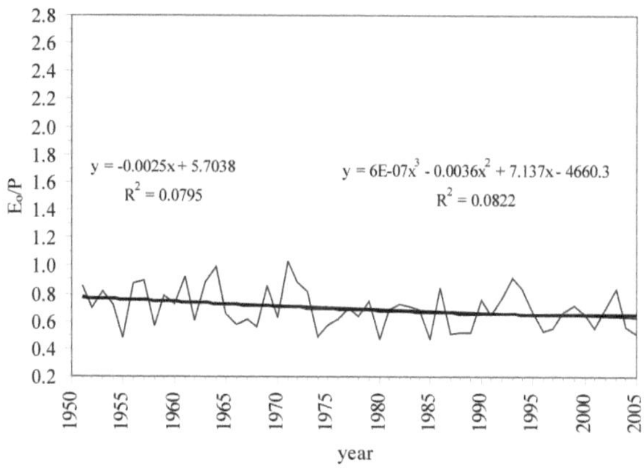

Fig. 14 Long-term course of drought index annual values (E_0/P) at Červený Kláštor for the period 1951–2005 with the linear trend and polynomial of third order

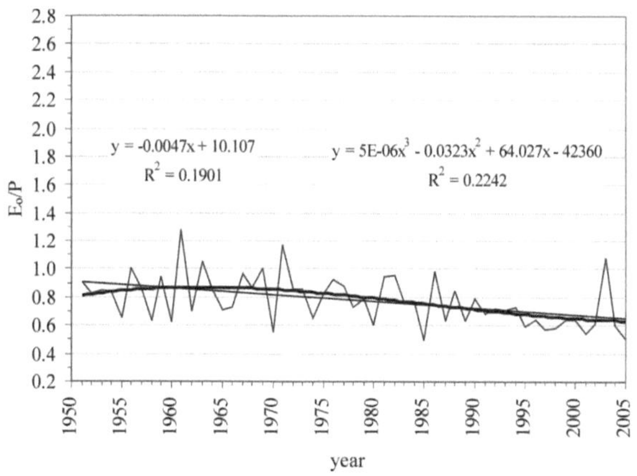

Fig. 12 Long-term course of drought index annual values (E_0/P) at Plaveč for the period 1951–2005 with the linear trend and polynomial of third order

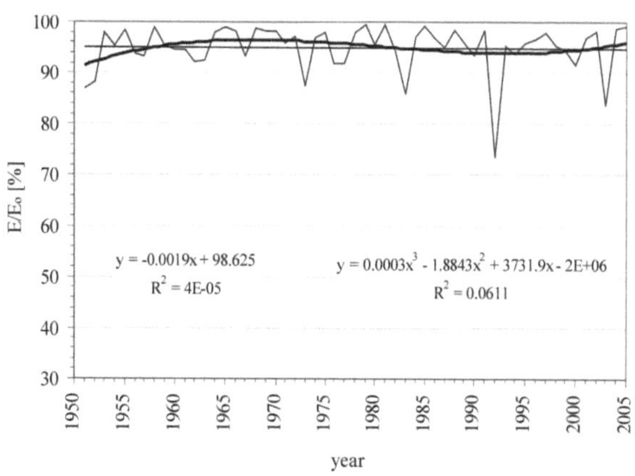

Fig. 15 Long-term course of relative evapotranspiration annual values (E/E_0) at Oravská Lesná for the period 1951–2005 with the linear trend and polynomial of third order

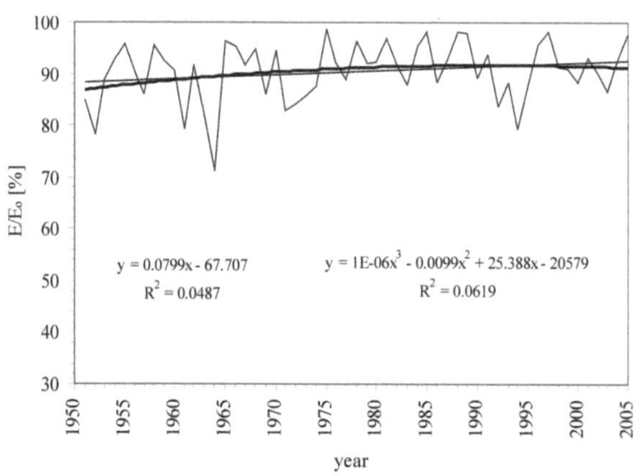

Fig. 13 Long-term course of relative evapotranspiration annual values (E/E_0) at Červený Kláštor for the period 1951–2005 with the linear trend and polynomial of third order

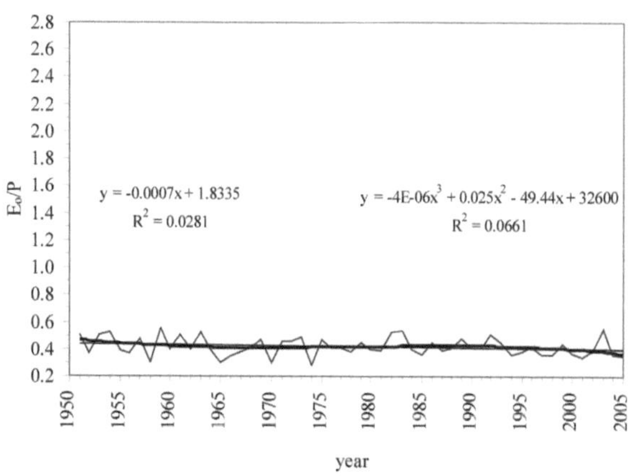

Fig. 16 Long-term course of drought index annual values (E_0/P) at Oravská Lesná for the period 1951–2005 with the linear trend and polynomial of third order

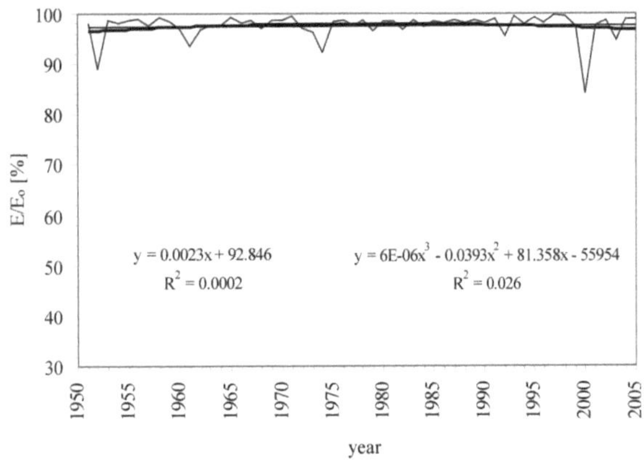

Fig. 17 Long-term course of relative evapotranspiration annual values (E/E_0) at Ždiar-Javorina for the period 1951–2005 with the linear trend and polynomial of third order

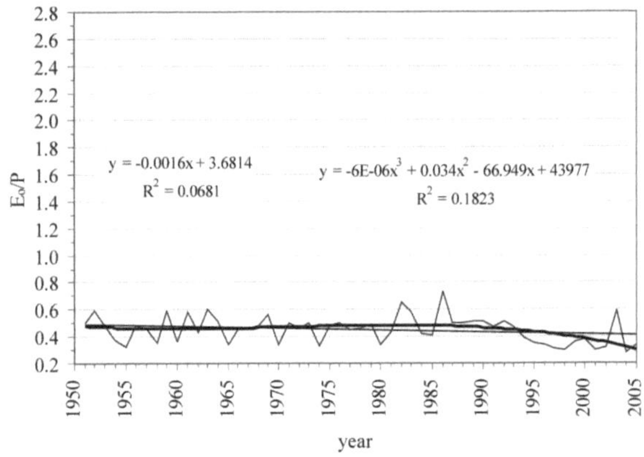

Fig. 20 Long-term course of drought index annual values (E_0/P) at Štrbské Pleso for the period 1951–2005 with the linear trend and polynomial of third order

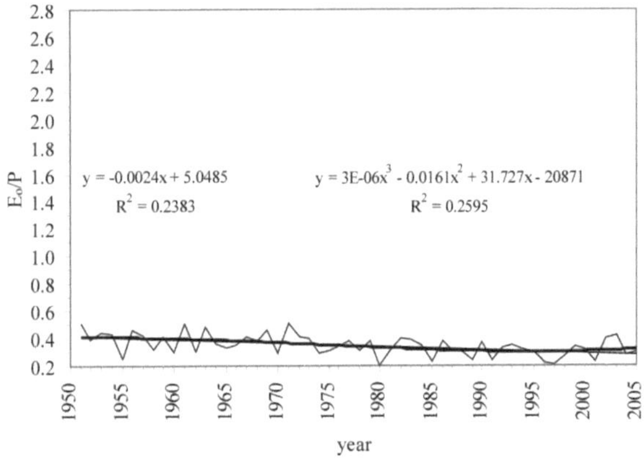

Fig. 18 Long-term course of drought index annual values (E_0/P) at Ždiar–Javorina for the period 1951–2005 with the linear trend and polynomial of third order

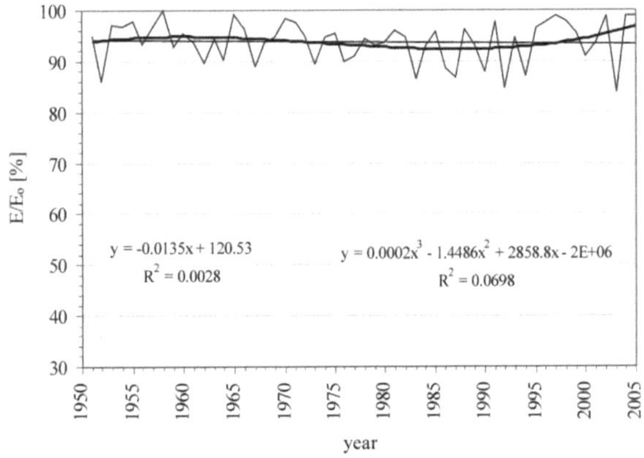

Fig. 19 Long-term course of relative evapotranspiration annual values (E/E_0) at Štrbské Pleso for the period 1951–2005 with the linear trend and polynomial of third order

of Slovakia is observed all year round and so the mean values of E/E_0 vary only in a small extent.

Table 2 presents the results of the Student's *t*-test of significance concerning the correlation coefficients for E/E_0(relative evapotranspiration) and drought index E_0/P and the trend of linear regression as well. The significant trend to aridity was detected only in the southern parts of Slovakia at station in Hurbanovo and partly in Myjava. Highly significant trend to humidity was revealed at station in Plaveč (the north-east part of Slovakia) concerning the both evaluated indices. The highly significant growth of humidity for the period lasting 55 years also shows the northern part of Tatra Mountains at the station in Javorina. The decline of drought index and the growth of humidity here are conditioned mainly by the growth of rainfall. The stations

Table 2 Results of the trend analysis performed on yearly data (1950–2005) for selected meteorological stations. E/E_0 (relative evapotranspiration), drought index (E_0/P) probability level $^{*}p < 0.1$, $^{**}p < 0.05$, $^{***}p < 0.01$, $^{****}p < 0.001$; n.s., not significant

Meteorological Station	$E/E_0 = f$ (years)	Trend	$E_0/P = f$ (years)	Trend
Hurbanovo	**	Aridity	***	Aridity
Myjava	**	Aridity	n.s.	–
Kamenica nad Cirochou	n.s.	–	n.s.	–
Plaveč	****	humidity	****	Humidity
Červený Kláštor	*	humidity	**	Humidity
Oravská Lesná	n.s.	–	n.s.	–
Tatranská Javorina	n.s.	–	****	Humidity
Strbske Pleso	n.s.	–	*	Humidity

of Kamenica nad Cirochou and Oravska Lesna (both situated in the northern part of Slovakia) do not show any significant trend.

The territory of Slovakia is divided to the area of Panonia (Panonia lowland), influenced by the Mediterranean climate, and the area of Inner Carpathians, influenced by subocean mountainous climate and by the climate of both northern and Baltic Seas. The dividing line between these areas has been determined by Zlatník (1959) by so-called main climatic line, where the Carpathian bow separates two important European climatic areas (Fig. 2). The area to the north of this line is remarkably wetter and colder than the southern, which is drier and warmer. The northern part is favourable, for example, for the growth of spruce, unlike the southern part, where spruce is growing only in the highest spruce vegetation zone. The results presented in Table 2 and Figs. 5, 6, 7, 8, 9, 10, 11, 12, 13, 14, 15, 16, 17, 18, 19 and 20 point out the overwhelming arid trend at the stations in Hurbanovo and Myjava both situated in the southern part of Slovakia. The station situated to the north of the climatic line shows an overwhelming humid trend during the same period (1951–2005). Similar results are also presented by Lapin et al. (1997) and Šútor et al. (2002).

Trnka et al. (2007) have evaluated the occurrence of drought (Palmer Drought Severity Index, PDSI) in Morava and Silesia during the period of 1961–2000 and a statistically significant drought is evident for long-term PDSI. With an exception of the Beskydy Mountains, the tendency towards a lower (arid) PDSI is typical for most of the territory of Moravia and Silesia.

Drought can be seen as a structural problem occurring every 10 or 12 years, but which is on its increase (Rožnovský and Janouš 2001; Litschmann et al. 2002; Rebetez et al. 2006; Paltineanu et al. 2007; Vicente-Serrano and Cuadrat-Prats 2007). It is assumed that a climate change will lead to reduced rainfall, higher temperature and more frequent heat waves and drought, which means that forest fires will become more severe and more extensive. Furthermore, draught is not just a 'problem for farmers'. Another consequence of drought concerns drinking water supplies. Hydroelectric production is heavily hit in drought years. This induces serious environmental impacts in terms of increased carbon dioxide emissions.

Conclusion

This work analyses the trends in occurrence of dry and wet periods in altitudinal vegetation stages in Slovakia between 1951 and 2005. We have used the relative evapotranspiration (E/E_0) and drought index (E_0/P). We have used meteorological data from eight meteorological stations representing the predominant vegetation stages of Slovakia.

The relative evapotranspiration expresses functional dependences among particular energy and water balance equation components, as well as drought index, which denotes relation between energetic possibilities of evapotranspiration and precipitation totals in chosen locality. These are used very often for characterizing humid conditions. These ratios enable to characterize humid condition and thus, they can be used in mapping and modelling of biological processes.

The drought index ranges from 1.44 for the area heavily prone to drought (southern part) to 0.35 for the mountainous areas (northern part) where sum of precipitation exceeds potential evapotranspiration. The relative evapotranspiration shows values as high as 97% in the northern mountainous regions, to 58% in the Danubian lowland.

A significant increase in the severity of drought was identified from 1951 to 2000 only in the Danube Lowland (1st Oak vegetation stage) and in Zahorská Lowland (2nd Beech-Oak vegetation stage). We determined a significant trend of increase in humidity in mountains and in the northern part of East Slovakia

Acknowledgments The study was supported by research grants VEGA No. 1/0515/08, 1/4393/07, 1/3528/06, 1/2357/05 from the Slovak Grant Agency for Science and from Slovak Research and Development Agency No. APVV-0022-07.

References

Arora VK (2002) The use of aridity index to assess climate change effect on annual runoff. J Hydrol 265:164–177

Baumgartner A, Liebscher HJ (1990) Allgemeine Hydrologie. Bd 1 (in German). Gebrüder Bornträger, Berlin

Bochenek Z, Dkabrowska-Zielińska K, Ciołkosz A, Drupka S, Boken VK (2005) Monitoring agricultural drought in Poland. In: Boken, VK, Cracknell, AP, and Heathcote, RL (eds) Monitoring and predicting agricultural drought: A Global Study, published by Oxford University Press, New York

Budyko MI (1955) Atlas of the heat balance (in Russian). Gridometeoizdat, Leningrad

Budyko MI (1971) Climate and life (in Russian). Gridometeoizdat, Leningrad

Budyko MI (1980) Climate in the past and in the future (in Russian). Gidrometeoizdat, Leningrad

Budyko MI, Zubenok LI (1961) The determination of evaporation from the land surface (in Russian). Izv Ak Nauk SSR Se Geog 6:3–17

Ditmarova Ľ, Kmeť J, Střelcová K, Gőmőry D (2006) Effects of drought on selected physiological parameters of young beech trees under stress conditions. Ekológia (Bratislava) 25:1–11

Dubrovsky M, Trnka M, Svoboda M, Hayes M, Wilhite D, Zalud Z, Semeradova D (2005) Drought conditions in the Czech Republic in present and changed climate. In: Proceedings of the European Geophysical Union, Vienna, pp. 25–29

Gongalsky KB, Pokarzhevskii AD, Filimonova ZV, Savin FA (2004) Stratification and dynamics of bait-lamina perforation in three forest soils along a north-south gradient in Russia. Appl Soil Ekol 25:111–122

Hayes MJ, Sovoboda MD, Wilhite DA, Vanyarko OV (1999) Monitoring the 1996 Drought using the Standardized Precipitation Index. Amer Meteor Soc 80:429–438

Hayes M, Svoboda M, Le Comte D, Redmond K, Pasteris P (2005) Drought monitoring: New tools for the 21st century. In: Wilhite D.A. (ed.) Drought and Water Crises: Science, Technology, and Management Issues, CRC Press. Boca Raton, FL

Heim RR (2002) A review of Twentieth-Century drought indices used in the United States. Bull Amer Meteor Soc 83:1149–1165

Holécy J (2004) Mathematical model of fire insurance for forests in Slovakia (in Slovak). Technical University, Zvolen

Hrvol' J, Lapin M, Tomlain J (2001) Changes and variability in solar Radiation and evapotranspiration in Slovakia in 1951–2000. Acta Meteorologica Universitatis Comenianae 30:31–58

Hrvol' J, Tomlain J, Horecká V (2003) Total radiation balance of active surface on the territory of Slovakia for period 1951–2000. Acta Meteorologica Universitatis Comenianae 32:1–15

Kharel H, Bruins HJ (2004) Drought and desertification hazards in Izrael: Time-series analysis 1970–2002. In: Malzahn D, Plapp T (eds) Disasters and Society – From Hazard Assessment to Risk Reduction: Universität Karlsruhe (TH): 91–98

Krecmer V (1980) Bioclimatological dictionary – Terminological and explicative (in Czech). Academia, Praha

Lapin M, Závodský D, Majerčáková O, Minďáš J, Špánik F (1997) Vulnerability and adaptation assessment for Slovakia. Final report of the Slovak Republic's Country Study, Element 2, U.S. Country Studies Program, Slovak Ministry of the Environment, Slovak Hydrometeorological Institute, Bratislava

Larcher W (1984) Ökologie der Pflanzen auf physiologischer Grundlage (in German). Ulmer, Stuttgart (Hohenheim)

Litschmann T, Klementová E, Rožnovský J (2002) Palmer's index and its utilization in conditions of Czech Republic (in Czech). In: Xth International Poster Day, Transport of Water, Chemicals and Energy in the system Soil – Crop Canopy – Atmosphere, CD, UH SAS, Bratislava

McCabe GJ, Wolock DM (2002) Trends and temperature sensitivity of moisture conditions in the conterminous United States. Clim Res 20:19–29

Novák V (1995) Evapotranspiration and its estimation (in Slovak). Veda, Bratislava

Paganová V (2000) Water stress impact on growth of Wild pear progenies. Folia Oecologica 27:35–48

Paltineanu C, Mihailescu IF, Seceleanu I, Dragota C, Vasenciuc F (2007) Using aridity indices to describe some climate and soil features in Eastern Europe: a Romanian case study. Theor Appl Climatol 90:263–274

Pražák J, Šír M, Tesař, M (1994) Estimation of plant transpiration from meteorological data under conditions of sufficient soil moisture. J Hydrol 162:409–427

Pražak J, Šír M, Tesař, M (1996) Parameters determining plant transpiration under conditions of sufficient soil moisture. J Hydrol 183:425–431

Raušer J, Zlatník A (1966) Biogeography (map No. 21), (in Czech). Atlas ČSSR, Kartografický a reprodukční ústav, Praha

Rebetez M, Mayer H, Dupont O, Schindler D, Gartner K, Kropp JP, Menzel A (2006) Heat and drought 2003 in Europe: a climate synthesis. Ann For Sci 63:569–577

Rožnovský J, Janouš D (2001) Drought, valuation and prediction (in Czech and Slovak). SHMI, Brno

Sankarasubramanian A, Vogel RM (2002) Annual hydroclimatology of the United States. Water Resour Res 37:701–708

Střelcová K, Matejka F, Kučera J (2004) Beech stand transpiration assessment – two methodical approaches. Ekológia (Bratislava) 23, (Suppl. 2/2004):147–162

Šiška B, Takáč J, Igaz D (2005) Climate change impacts on winter whet yield on Danubian lowland in Slovak republic. Contemp Agric (Savremena poljopriveda):324–328

Šiška B, Samuhel P (2007) Modeling Climate Change Impact on Maize (Zea Mays L.) Yields in Conditions of the Danubian Lowland. Meteor J 10:81–84

Škvarenina J, Križová E, Tomlain, J (2004) Impact of the climate change on the water balance of altitudinal vegetation stages in Slovakia. Ekológia 23 (Suppl. 2/2004):13–19

Španik F, Šiška B (2007) Biometeorology (in Slovak). SPU Nitra

Šútor J, Štekauerová V, Majerčák J (2002) Water balance in agriculture ecosystems. In: Proceedings of 19th European Regional Conference of ICID "Sustainable Use of Land and Water", Brno-Prague, 8pp., CD

Tomlain J (1996) Model computation of the climatic change impacts on potential and actual evapotranspiration changes on the territory of Slovakia (in Slovak). National Climate Programme of Slovak Republic, No. 4, Slovak Ministry of Environment, SHMI, Bratislava, pp. 45–74

Tomlain J (2004) Contribution to humid conditions of Slovakia. Acta Meteorologica Universitatis Comenianae 33:21–30

Trnka M, Dubrovský M, Semerádová D, Žalud Z (2007) Drought, precipitation index, Palmer drought indices in Moravia and Silesia (in Czech). In: Brázdil R, Kirchner K (eds) Selected natural extremes and their impacts in Moravia and Silesia. Masarykova universita, Český hydrometeorologický ústav, Ústav geoniky Akademie věd ČR Brno, Praha, Ostrava

Vicente-Serrano SM, Cuadrat-Prats JM (2007) Trends in drought intensity and variability in the middle Ebro valley (NE of the Iberian peninsula) during the second half of the twentieth century. Theoretical and Applied Climatology 88(3-4): 247–258

Zlatník A (1959) Waldtypengruppen der Slowakei (in German). VŠZ, Brno

Zlatník A (1976) Forest phytocoenology (in Czech). SZN, Praha

Thermodynamics, Irreversibility, and Optimality in Land Surface Hydrology

A. Kleidon, S. Schymanski and M. Stieglitz

Keywords Hydrology · Thermodynamics · Entropy production · Irreversibility · Memory · Vegetation · Biotic effects

Introduction

The water exchange at the land surface is driven by the input of water by precipitation and the loss by runoff generation and evapotranspiration into the atmosphere. It is strongly linked to the surface energy balance by the flux of latent heat associated with evapotranspiration, but also to the dynamics of the atmosphere and the terrestrial biosphere.

From a more general perspective, the exchange fluxes of water at the land surface are embedded in the global hydrologic cycle, which is driven by thermodynamic processes to a state far from thermodynamic equilibrium. This state far from thermodynamic equilibrium is maintained by a range of processes that continuously perform work, dissipate energy, and thereby produce entropy. Quantifying rates of entropy production allows us to measure the dissipative nature of the hydrologic cycle, its irreversibility, and can provide a basis to identify some general functional characteristics. This thermodynamic perspective then allows us to investigate the applicability of the principle of maximum entropy production (MEP, e.g., Ozawa et al. 2003; Kleidon and Lorenz 2005; Martyushev and Seleznev 2006) to surface and soil hydrology, and to test any general direction in fundamental, thermodynamic terms by which the biota alters these thermodynamic processes, such as increasing the rates of entropy production (e.g., Ulanowicz and Hannon 1987; Schneider and Kay 1994; Kleidon 2004; Tesar et al. 2007).

More specifically, evaporation, as one of the dominant processes of the land surface water balance, produces entropy because evaporation of water into unsaturated air is irreversible. Only at saturation is the phase transition of liquid to gas and back to liquid reversible since it then takes place at the same temperature. This can be seen directly in the expression of the chemical potential μ_a of moist air with a temperature T_a and a relative humidity RH (e.g., Campbell and Norman 1998):

$$\mu_a = R_v \cdot T_a \cdot \ln(\mathrm{RH}) \qquad (1)$$

where R_v is the gas constant of water vapor ($R_v = 461.5\,\mathrm{J\,kg^{-1}K^{-1}}$) and the chemical potential is in units $\mathrm{J\,kg^{-1}}$. For a value of RH = 100%, the chemical potential is at its maximum value of 0, that is, at this point the moisture content of the air is in thermodynamic equilibrium with an open water surface of pure water. As relative humidity decreases, the chemical potential becomes increasingly negative, and the air (with respect to its moisture content) is further away from thermodynamic equilibrium.

This state far from thermodynamic equilibrium with respect to atmospheric water vapor (i.e., the occurrence of dry air) is the result of atmospheric motion, specifically convection. The atmospheric circulation acts as both, a heat engine, which generates motion, and as a dehumidifier, which produces dry air (Pauluis and Held 2002; Pauluis 2005). A stronger

A. Kleidon (✉)
Biospheric Theory and Modelling Group, Max-Planck-Institut für Biogeochemie, Jena, Germany
e-mail:akleidon@bgc-jena.mpg.de

K. Strelcová et al. (eds.), *Bioclimatology and Natural Hazards*,
DOI 10.1007/978-1-4020-8876-6_9, © Springer Science+Business Media B.V. 2009

atmospheric circulation will result in a state further away from thermodynamic equilibrium with respect to the average relative humidity of the air. This nonequilibrium state is maintained by the removal of atmospheric moisture by precipitation, which in turn is caused mainly by rising air masses. The consequence of this removal of atmospheric moisture is that the descending air masses are unsaturated, resulting in overall lower relative humidity and the ability of the surface to evaporate more water. The tendency of the atmospheric circulation to maintain a state of MEP (Paltridge 1975; Lorenz et al. 2001; Ozawa et al. 2003; Kleidon et al. 2003, 2006) implies that it circulates as strongly as possible. This in turn would imply that it is associated with strongest convective motion, removing moisture as much as possible. From this perspective, MEP seems to imply that the atmospheric hydrologic cycle is furthest away from thermodynamic equilibrium, with on average the lowest possible relative humidity. However, this coupling with atmospheric motion is not further explored here but would need further investigations.

Before we can apply MEP to hydrology and explore its implications, we first need to establish a basis and identify the dissipative processes at the land surface and how we can quantify their rates of entropy production. The goal of this chapter is to provide this background as well as simple estimates of entropy production rates associated with surface water exchange. Before we go into the thermodynamics of land surface hydrology, we will first briefly review the thermodynamic background of nonequilibrium systems. The global entropy budget is discussed along with estimates of its components in order to demonstrate the basic methodology and to provide a means for comparison of entropy production rates associated with the water-exchange fluxes over land. Many of the aspects described here are not necessarily new, and neither are some of the thermodynamic aspects presented here. What is new is the integrated view of seeing the surface water budget as a thermodynamic system that is maintained far from thermodynamic equilibrium and the derivation of first estimates of the associated entropy production rates. In the Discussion section we then identify how these first estimates can be improved and discuss the role of vegetation on land surface hydrology from this perspective as well as potential implications. We close with a brief summary and conclusion.

Thermodynamic Background

In general, we can distinguish three different, thermodynamic systems (e.g., Kondepudi and Prigogine 1998): (a) isolated systems with no exchange fluxes of energy and mass across the system boundary, (b) closed systems, where only energy is exchanged across the boundary, and (c) open systems with both, energy and mass exchange across the boundary.

A typical example for an isolated system is the ideal gas and its associated relationships (Boltzmann distribution, ideal gas law, etc.). The Earth is largely a closed system, since it exchanges primarily radiation of different wavelength with space, while the mass exchange with its surroundings can be neglected. When we deal with the land surface, we need to account for energy fluxes and mass fluxes, since for example, water supplied by precipitation (as a liquid) enters the system in a different phase than when it leaves by evapotranspiration (as a gas). Hence the land surface system is an open thermodynamic system.

Figure 1 shows the definition of the land surface system in terms of its system boundary and its major fluxes across the boundary. We draw the upper system boundary very near to the interface between the surface and the lower atmosphere (which would be just above the canopy to include vegetation and its effects in the system). The lower boundary in Fig. 1 is chosen such that there is no heat flux through the bottom of the soil column (zero flux boundary condition), and for the water fluxes we choose the rooting zone depth at a lower boundary condition since the change of water content below the rooting zone is negligible, at least for surface processes.

In non-isolated systems, a steady state away from thermodynamic equilibrium can only be maintained through a net export of entropy from the system to the surroundings. This can be written in terms of the entropy balance as (e.g., Kondepudi and Prigogine 1998):

$$dS/dt = \sigma - \text{div } F_e/T - \Sigma \text{ div } (F_{m,i}\mu_i)/T \quad (2)$$

where S is the entropy of the system, σ, the rate of entropy production within the system, F_e and $F_{m,i}$, the net heat and mass fluxes across the system boundary, respectively, with the latter at a chemical potential μ_i and the sum running over all substances. The sign

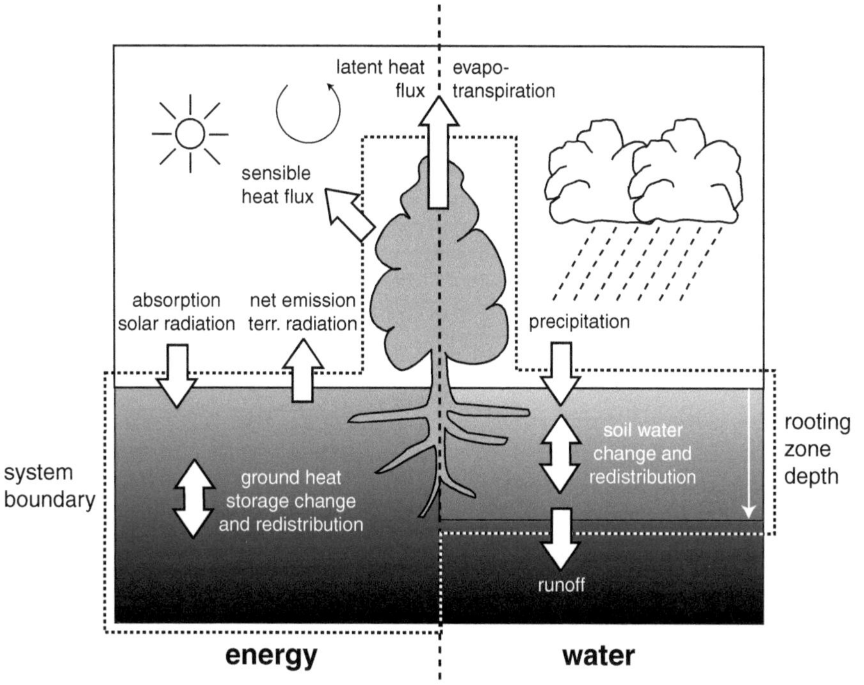

Fig. 1 Definition of the land surface system with respect to energy- and water fluxes. The exchange fluxes of this land surface system are described by the surface energy and water balance (and other mass balances, such as carbon, nitrogen, etc.). The carbon balance of the land surface is omitted here since in the climatological mean steady state the same amount of carbon dioxide is taken up by photosynthesis as is released by respiration (and in the same phase as gas), so that there is no net flux of entropy across the system boundary with respect to carbon dioxide

convention of the fluxes is such that a positive flux adds heat, material, and entropy to the system, while a negative flux exports the same to the surroundings.

The evolution towards thermodynamic equilibrium can be understood in terms of Eq.(2). When the system becomes isolated, the net heat and mass exchange with the surroundings diminishes. Then $\sigma > 0$, but decreases with time and local entropy S converges to its maximum possible value.

The steady state of the land surface system is reached when the long-term averages of the state variables (soil heat- and water content as well as their associated entropies) do not change in time. The steady-state assumption allows us to estimate mean entropy production rates from the mean divergence of energy fluxes and mass fluxes, since S on average no longer changes in time (i.e., $\mathrm{d}S/\mathrm{d}t = 0$ in Eq.(2)).

In most cases, entropy production results from dissipative processes. In steady state, the rate of dissipation D is equal to the rate at which work is being performed ($\mathrm{d}W/\mathrm{d}t$)

$$\mathrm{d}W/\mathrm{d}t = D \tag{3}$$

This then is directly linked to the rate of entropy production σ by

$$\sigma = D/T \tag{4}$$

given that the temperature T at which the dissipation occurs does not change with time.

We next review the global entropy budget for two reasons: (i) to demonstrate how rates of entropy production are estimated; and (ii) to provide a basis for comparison of the entropy production rates estimated for surface water fluxes below.

The steady-state assumption and Eq.(2) allow us to derive an estimate of the global entropy budget (Table 1). As stated before, the divergence of mass fluxes at the global scale can be set to zero, so that the entropy production can be calculated directly from the divergence of energy fluxes at the system boundaries. The entropy production rates associated with the global mass cycles are then subsumed in the energy fluxes. This then requires only knowledge about the heat fluxes Q as well as the temperatures at which heat is transformed. The rate of entropy production is then simply given by

$$\sigma = Q \cdot (1/T_{\mathrm{c}} - 1/T_{\mathrm{w}}) \tag{5}$$

with T_{w} and T_{c} being the temperatures at which the heat was converted. The estimated fluxes Q and

Table 1 Estimate of the planetary entropy budget and its components. Entropy production σ is calculated from the conversion of the energy flux Q from higher temperature T_w to the lower temperature T_c (see Eq.(5)). After Peixoto and Oort (1992) and Kleidon (2008)

Process	Q(Wm^{-2})	T_c(K)	T_w(K)	σ(mWm^{-2}K^{-1})
Atmospheric absorption of solar radiation	68	252	5760	258
Surface absorption of solar radiation	170	288	5760	561
Atmospheric absorption of terrestrial radiation	28	252	288	14
Moist convection (evaporation + precipitation)	79	266	288	23
Dry convection (sensible heat)	24	280	288	2
Large-scale circulation	10	255	300	6
Biotic activity	8	288	5760	5
Planetary	235	255	5760	881

temperatures T_w, T_c for different irreversible processes in the Earth system are given in Table 1.

We can see from Table 1 that the dominant process that produces entropy is radiative transfer, in particular the absorption of solar radiation. The irreversibility associated with the absorption of solar radiation reflects the fact that solar radiation was emitted at the surface temperature of the Sun (of about 5760 K), and once absorbed on Earth cannot be reemitted from Earth to space at this temperature, but inevitably at a much lower temperature and longer wavelengths. The associated entropy production is shown in Table 1 in terms of three components: absorption of solar radiation in the stratosphere at a temperature of about $T = 252$ K, absorption at the surface of about $T_s = 288$K, and radiative exchange in the long wave between the surface and the atmosphere, following Peixoto and Oort (1992).

The next two terms of the budget in Table 1 (moist and dry convection) are related to the turbulent fluxes of sensible and latent heat. The sensible heat flux is assumed here to transport heat from the surface to the boundary layer (with $T_{bl} \approx 280$ K). For the latent heat flux, it is assumed that water evaporates at a surface temperature T_s and condenses higher in the atmosphere at a temperature of about $T_c \approx 266$ K. Entropy production associated with the large-scale transport of heat from the equator to the poles is estimated from the average value of heat transport of about 10 W m^{-2} and mean tropical and polar temperatures of 300 K and 255 K, respectively. For these three fluxes, the irreversible nature is related to the conversion of density differences from differential radiative heating into kinetic energy and subsequent frictional and thermal dissipation.

Also shown in Table 1 is an estimate of entropy production by biotic activity, the combined process of generating chemical free energy by photosynthesis and its dissipation by respiration. Here it is assumed that photosynthesis converts 8 W m^{-2} of sunlight into carbohydrates that are subsequently converted into heat by respiration at roughly the surface temperature T_s. This number is obtained using a global photosynthetic activity of about 200 GtC year^{-1} and an average quantum yield efficiency of about 10 photons per molecule of CO_2 fixed. Note that the estimate of entropy production by photosynthesis is less than $Q(1/T_c - 1/T_w) =$ 29 mW m^{-2} K^{-1} because other factors need to be taken into account as well, for example, not all solar radiation is available for photosynthesis. This contribution, however, is already subsumed in the entropy production rates associated with the absorption of solar radiation at the surface.

The planetary rate of entropy production is calculated from the overall mean absorption of solar radiation ($Q = 235$ W m^{-2}) that is emitted to space as long-wave radiation at Earth's radiative temperature $T_r \approx 255$ K.

For later comparison, note that the typical rate of entropy production by non-radiative processes is of the order of 1–10 mW m^{-2} K^{-1}.

Entropy Production of Surface Water Exchange

In order to understand and estimate the dissipative nature of soil water-exchange processes at the surface, we first describe the simplified surface water balance and introduce the description of the soil water distribution in terms of its binding and potential energy. This description is analogous to the energetics of the atmospheric circulation that was developed by

Lorenz (1955), but applied to the movement of water within the soil column. This energetic consideration then forms the basis for simple estimates of entropy production by the most important processes of the surface water balance. We then discuss entropy production of the various components of the simplified surface water balance in separation and provide some rough estimates to evaluate their significance in the entropy budget of the surface water balance.

Surface Water Balance

The water balance on land can be characterized by the total amount of soil water w_s within the rooting zone D. Note that in the following we use the notation of "w_s" for the amount of soil water, not for performed work, for which we use a capital W. The change of w_s with time is given by the simple budget equation

$$\mathrm{d}w_s/\mathrm{d}t = P - \mathrm{ET} - R \tag{6}$$

where P is precipitation, ET, the evapotranspiration, and R, the runoff.

Less common is the description of the water content of the rooting zone in terms of the energy that is associated with the binding energy of moisture to the soil matrix and with the vertical location (but see, e.g., Hillel 1998; Roderick 2001). We derive a description of this binding energy from the general expression for the change in total internal energy of the soil column $\mathrm{d}U$ (Kondepudi and Prigogine 1998):

$$\mathrm{d}U = T\,\mathrm{d}S - p\,\mathrm{d}V + \mu\,\mathrm{d}N \tag{7}$$

In the following, we neglect changes in total energy due to changes in heat storage (T dS) and volume changes (p dV) and focus on the last term (μ dN) in Eq. (7). This term represents the change in total energy due to the addition of an infinitesimal amount of water dN to the soil column with water at a chemical potential μ. We can express the addition of water dN by the change in water content dΘ and the density of water ($\rho_{\mathrm{water}} \approx 1000\,\mathrm{kg\ m^{-3}}$) as $\mathrm{d}N = \rho_{\mathrm{water}} \cdot \mathrm{d}\Theta$. The chemical potential of soil water is directly related to the matric potential Ψ_m by

$$\Psi_m = \mu_s - \mu_{s0} \tag{8}$$

where μ_{s0} is the chemical potential of pure, unbound water. We assume $\mu_{s0} = 0$ in the following for simplicity. We will also assume a local equilibrium of the soil water with the relative humidity $\mathrm{RH_s}$ of soil air, that is, $\mu_s = R_v \cdot T_s \cdot \ln(\mathrm{RH_s})$, where T_s is the soil temperature.

We obtain the total binding energy of water $\mathrm{BE_{water}}$ to the soil matrix due to adhesive and capillary forces by integrating the term $\mu \cdot \rho_{\mathrm{water}} \cdot \mathrm{d}\Theta$ in Eq. (7) from zero water content to the actual soil moisture content of the soil at depth z. Note that μ is highly dependent on Θ. The expression for $\mathrm{BE_{water}}(z)$ is then

$$\mathrm{BE_{water}}(z) = \int_0^{\Theta(z)} \mu_s(\Theta) \cdot \rho_{\mathrm{water}} \cdot \mathrm{d}\Theta \tag{9}$$

When we integrate Eq. (9) over the depth of the rooting zone D, we get an expression for the total binding energy associated with a given soil water distribution $\Theta(z)$ within the rooting zone

$$\mathrm{BE_{water}} = \int_0^{D} \mathrm{BE_{water}}(z)\mathrm{d}z \tag{10}$$

Note that $\mathrm{BE_{water}}$ is generally negative. This represents the fact that the bound water is in a lower state of energy and work needs to be done in order to remove it from the soil matrix.

The link to the total soil water content w_s (Eq. 6) is given by

$$w_s = \int_0^{D} \rho_{\mathrm{water}} \cdot \Theta(z)\mathrm{d}z \tag{11}$$

The potential energy $\mathrm{PE_{water}}$ relates to the vertical distribution of soil moisture within the soil column in relation to the soil surface. It is expressed as

$$\mathrm{PE_{water}}(z) = \rho_{\mathrm{water}} \cdot g \cdot \Theta(z)(z_{\mathrm{WT}} - z) \tag{12}$$

and

$$\mathrm{PE_{water}} = \int_0^{D} \mathrm{PE_{water}}(z)\mathrm{d}z \tag{13}$$

where z is defined as the depth from the surface and z_{WT} is the depth of the water table. The term z_{WT} is an arbitrary additive term (with no consequences on the following discussions) that defines the depth at which the potential energy is zero. We have chosen the depth of the water table z_{WT} as our reference depth simply because then the equilibrium distribution results in a value of the total energy TE_{water} of 0 (see below).

The total energy TE_{water} of the water stored in the soil column is given by

$$\mathrm{TE}_{water} = \mathrm{BE}_{water} + \mathrm{PE}_{water} \quad (14)$$

The three terms BE_{water}, PE_{water}, and TE_{water} have the units J m^{-2}. Note that one can get the expressions also by considering a modified chemical potential which includes the effect of an external (gravitational) field (Kondepudi and Prigogine 1998). Then, the modified chemical potential μ^* would be written as

$$\mu^*(z) = \mu_s(z) + g(z_{WT} - z) \quad (15)$$

When integrated and multiplied by the mass of water $\Theta(z)\rho_{water}$ at depth z, the two separate terms in Eq. (15) represent Eqs. (9) and (12), respectively.

The thermodynamic equilibrium state with respect to the soil water distribution is reached when TE_{water} is at a minimum for a given amount of soil water w_s (see also Soil Moisture Redistribution section below). This state is reached when no vertical (or horizontal) gradient in the modified chemical potential exists (i.e., local thermodynamic equilibrium at every depth z in the soil column), that is,

$$\mu_s(z) = \Psi_m(z) = -\rho_{water} \cdot g(z_{WT} - z) \quad (16)$$

Precipitation

Precipitation is clearly the most important driver of the surface water balance. As precipitation enters the surface system, liquid water enters in free form with a chemical potential near 0. However, falling raindrops are associated with a certain amount of kinetic energy KE_{rain}. This kinetic energy is dissipated into heat and possibly results in some deformation work at the surface at impact. The rate of dissipation can be calculated from the kinetic energy of the falling raindrops, which is assumed to be completely converted into dissipative heating at impact. The rate of kinetic energy dissipation D_{rain} is estimated from the rainfall amount P, the terminal velocity of the raindrops v_t (which depends on raindrop diameter), and the density of water by

$$D_{rain} = 1/2\rho_{water} \cdot P \cdot v_t^2 \quad (17)$$

Using a typical annual surface temperature of T_s = 288 K, the associated entropy production is estimated as

$$\sigma_{rain} = D_{rain}/T_s \quad (18)$$

Estimates of entropy production for three typical annual rainfall rates and for the extreme cases of drizzle and heavy rain are given in Table 2. While dissipative heating by falling raindrops can play an important role in the atmosphere (where the heating rates can be as much as 2–4 W m^{-2} (Pauluis et al. 2000, Pauluis 2005), the dissipative heating at the surface is less than 10^{-3} W m^{-2} for most regions of the world, and may exceed this value only in tropical regions of high, mostly convective rainfall.

Table 2 Estimation of dissipation rates (in W m^{-2}) (first number) and entropy production (in mW $m^{-2}K^{-1}$) (second number) due to the impact of falling raindrops at the surface for different rainfall intensities and annual rainfall amounts

Annual rainfall (mm)	Drizzle (droplet diameter ≈ 0.5 mm, terminal velocity ≈ 2 m s^{-1})	heavy rain (droplet diameter ≈ 5 mm, terminal velocity ≈ 9 m s^{-1})
100	6.4×10^{-6}	1.3×10^{-4}
	2.2×10^{-5}	4.5×10^{-4}
500	3.2×10^{-5}	6.5×10^{-4}
	1.1×10^{-4}	2.3×10^{-3}
1000	9.6×10^{-4}	2.0×10^{-3}
	3.3×10^{-4}	6.9×10^{-3}

Soil Moisture Change

Wetting of the soil matrix, that is, increasing the overall amount of soil water w_s by Δw_s, involves the irreversible binding of free water with a chemical potential $\mu_{water} = 0$ from the infiltrated water to a bound state with $\mu_s < 0$, thereby producing heat and exporting entropy. The released heat is called the heat of immersion (Hillel 1998). In other words, wetting decreases the total binding energy BE_{water} energy (Eq. 10). The released heat is then given by $\Delta Q = -(\mathrm{BE}_{water}(w_s + \Delta w_s) - \mathrm{BE}_{water}(w_s))$.

We can obtain a rough estimate of the magnitude of heating by assuming that the soil reaches the permanent wilting point ($\mu_s = \Psi_m = -1.5 \times 10^6$ J m^{-3}) before each rainfall event and that it is saturated after the rainfall. Then the difference in chemical potential is $\Delta\mu_s = 1.5 \times 10^6$ J m^{-3}. If we assume a plant available water (PAW) value of PAW = 100 mm, that is, the drawdown from saturation to the permanent wilting point is associated with the removal of 100 mm of water, the associated dissipative heating is approximately $P/\text{PAW} \cdot \Delta\mu_s/\rho_{\text{water}}$, where P is precipitation.

The estimated rates for three annual rainfall amounts are given in Table 3. Actual rates of entropy production would require a more detailed analysis that would need to take into account the interval between rainfall events, the levels of drying, and, naturally, soil characteristics that shape the relationship between the matric potential and soil wetness.

In case of soil drying, work needs to be performed against the surface tension for removing the water from the soil matrix that is of the same magnitude as the heat released by wetting.

Table 3 Estimation of heat of immersion (dissipative heating) and entropy production associated with soil wetting for three annual wetting rates, using a temperature of T_s = 288 K, plant available water of 100 mm, and a drawdown to the permanent wilting point after each rainfall event

Annual wetting (mm)	Heat of immersion (W m^{-2})	Entropy production (mWm$^{-2}K^{-1}$)
100	4.7×10^{-4}	0.002
500	2.4×10^{-3}	0.008
1000	4.7×10^{-3}	0.024

Evapotranspiration

Evapotranspiration represents a phase change of liquid water to the gaseous phase. In local thermodynamic equilibrium the two phases have equal chemical potentials, but heat is needed (i.e., the latent heat of vaporization L) to expand the molar volume from its liquid to gaseous value. This consists of an entropy flux into the system of $L \cdot E/T_s$ that results in an increase of entropy as water changes from the liquid to the gaseous phase (with E being the evapotranspiration rate and T_s the surface temperature of the liquid body from which the water is evaporated). This process alone does not result in entropy production since it is reversible (i.e., if the vapor would condense, it would release the same amount of heat that would be associated with an entropy export of $L \cdot E/T_s$).

Entropy is being produced once the saturated air at the surface is mixed with the unsaturated air of the atmospheric boundary layer (with a temperature T_a and a relative humidity RH_a). The entropy production associated with this mixing can be directly calculated from Eq. (1) by

$$\sigma_{\text{evap}} = -\rho_{\text{water}} \cdot R_v \cdot E \cdot (\ln(\text{RH}_a) - \ln(\text{RH}_s)) \quad (19)$$

Since RH_s is near saturation, $\ln(\text{RH}_s) \approx 0$ in Eq. (19). Equation (19) can be used to calculate the amount of entropy production for three annual evapotranspiration rates at different relative humidities (Table 4).

However, water is usually not in a free, unbound state, but needs to be extracted from the soil, and, in the case of transpiration, lifted up to the canopy. The work required to extract soil water is equivalent to the released heat of immersion when the soil is wetted. Hence, the rates are equivalent to the estimated heating rates shown in Table 3. Since evaporation of bound water requires work to be done against the surface tension, the latent heat of vaporization is somewhat higher compared to evaporation from a free surface (see also Tributsch et al. 2005).

For plant transpiration, we also need to account for the work done by vegetation in lifting the freed soil water from a depth z in the soil to a height h of the canopy. The resulting rate dW_{lift}/dt associated with a continuous transpiration rate T is

$$dW_{\text{lift}}/dt = \rho_{\text{water}} \cdot g(z + h)T \quad (20)$$

For instance, if a tropical rainforest extracts soil moisture from 5 m depth and transpires this at a canopy

Table 4 Estimation of entropy production (in mW m^{-2} K^{-1}) associated with evaporating free water into unsaturated air of the atmospheric boundary layer of different relative humidities RH for different annual evapotranspiration amounts

Annual evapotranspiration (mm)	RH = 60%	RH = 70%	RH = 80%	RH = 90%
100	0.7	0.5	0.3	0.2
500	3.7	2.6	1.6	0.8
1000	7.4	5.2	3.2	1.6

height of 60 m at a rate of $T = 1500\,\text{mm year}^{-1}$, the resulting rate at which work is performed is $dW_{lift}/dt = 3 \times 10^{-2}\,\text{W m}^{-2}$. The associated flow of water in the plant tissues would result in frictional dissipation, thereby producing entropy. This is neglected here.

Not considered here is the slight increase of RH as a result of the mixing of the evaporated water (saturated vapor) with the unsaturated air of the atmospheric boundary layer, which also results in entropy production. This effect would seem to be important to consider when global balances would be estimated since this, after all, is the process which pushes the hydrologic cycle back toward thermodynamic equilibrium.

A more detailed analysis of entropy production rates of land surfaces in different geographic settings would require information on the temporal evolution of relative humidity and evapotranspiration rates. This has been neglected here. Table 4 presents the entropy production rates for different, constant values of evapotranspiration and relative humidity. From Table 4 it is clear that both variables strongly affect the entropy produced by evapotranspiration.

Runoff

The flow of water associated with runoff results from the conversion of potential energy of soil moisture in the soil column into kinetic energy. We can estimate the amount of entropy produced by runoff by considering the energies at an initial stage and assume that the potential energy from the surface is converted into kinetic energy, which is subsequently dissipated into heat by friction. If we assume that the difference in potential energy between the top of the soil column and sea level is $\rho g \Delta z$, the amount of kinetic energy KE_{runoff} that can be generated (a generation rate of kinetic energy $G_{\text{KE,runoff}}$) by depleting this potential energy gradient for a given runoff rate R is

$$D_{\text{runoff}} = G_{\text{KE,runoff}} = R \cdot \Delta z \cdot \rho_{\text{water}} \cdot g \qquad (21)$$

with g being the gravitational acceleration ($g = 9.81\,\text{m s}^{-2}$), and the associated entropy production rate given by

$$\sigma_{\text{runoff}} = D_{\text{runoff}}/T_s \qquad (22)$$

Examples of entropy production rates for different runoff rates and elevation differences are given in Table 5. The rate of entropy production by runoff and subsequent river basin discharge is relatively small in comparison to the components of the global entropy budget (Table 1), except for the case of high rainfall rates and elevation gradients, for example, in mountainous areas with steep orography.

Table 5 Estimation of entropy production (in $\text{mW m}^{-2}\,\text{K}^{-1}$) associated with runoff generation and subsequent dissipation of kinetic energy for different values of annual runoff and elevation differences Δz

Annual runoff (mm)	$\Delta z = 100\,\text{m}$	$\Delta z = 1000\,\text{m}$
100	0.01	0.11
500	0.05	0.54
1000	0.11	1.08

Soil Moisture Redistribution

In addition to soil water addition by precipitation and removal by evapotranspiration and runoff generation, soil moisture is also redistributed within the soil. In other words, while the previous processes change the overall value of the soil water content w_s, pure redistribution alters the amount of the total energy associated with the soil moisture distribution, but leaves w_s unaffected.

The entropy produced by soil water redistribution can be estimated as follows. Let us consider an initial soil water distribution within a soil column. This distribution is characterized by a certain value of binding energy $BE_0 = BE(t = 0)$ and a potential energy $PE_0 = PE(t = 0)$. Here, only redistribution of moisture below the field capacity within the rooting zone is considered. Any movement of moisture outside the rooting zone is treated as runoff, which has been considered above in the section runoff. The flow of water is characterized by its associated kinetic energy KE_{redist}. This energy is generated by the reduction of the total energy TE, ultimately to its lowest possible value. After sufficiently long time, the motion will come to a rest due to frictional dissipation, that is, all kinetic energy is dissipated into heat, and BE + PE is at a minimum value. We hence have

$$\mathrm{BE}_0 + \mathrm{PE}_0 + \mathrm{KE}_{\mathrm{redist}}(0) = \mathrm{BE}(t) + \mathrm{PE}(t) + \mathrm{KE}_{\mathrm{redist}}(t) + Q_{\mathrm{redist}}(t) = \min(\mathrm{BE} + \mathrm{PE}) + \mathrm{Q}_{\mathrm{redist}} + \mathrm{KE}_{\mathrm{redist}}(\infty) \quad (23)$$

with Q_{redist} being the cumulative amount of heat produced by dissipation and immersion, and $\mathrm{KE}_{\mathrm{redist}}(0) = 0$ and $\mathrm{KE}_{\mathrm{redist}}(\infty) = 0$. If the state of thermodynamic equilibrium is approximately reached over a time scale τ_{redist}, the associated entropy production is

$$\sigma_{\mathrm{redist}} = Q_{\mathrm{redist}} / (\tau_{\mathrm{redist}} \cdot T_s) \quad (24)$$

The amount of entropy production associated with redistribution should be less than the more extreme case of soil moisture change considered above (Table 2). Hence, soil moisture redistribution for a given total w_s should result only in comparatively small rates of entropy production.

One could estimate entropy production associated with drying, wetting and redistribution of moisture in the soil column in more detail with a diffusion model for soil moisture, based, for example, on the Richard's equation (Hillel 1998). In such a setup one could also account for the frequency of wetting, which would likely prevent the soil moisture distribution to always reach its state of thermodynamic equilibrium.

Discussion

Limitations

This chapter provided a discussion of the irreversible nature of the processes involved in land surface hydrology and some rough, first-order estimates of the associated rates of entropy production. These very simple estimates certainly can be substantially improved, for instance by using simulation models of land surface hydrology. Such models include a fuller representation of the hydrological processes of the land surface and a coupling to realistic temporal forcing in different geographic settings. This would allow us to obtain estimates about the relevance of each component in different areas and seasons. This would seem to be relatively straightforward to do with present-day land surface schemes.

A shortcoming of our estimates is that other processes associated with water exchange that would produce entropy, were not considered here. Examples include (i) freeze-thaw dynamics in cold regions, where the phase transition from liquid to solid may be associated with irreversible phase changes, (ii) possibly associated work done by frost heaving, (iii) processes associated with snow and snowmelt as these involve phase changes as well, and (iv) possible interactions of water with other biogeochemical cycles.

Vegetation Effects

Despite the simplicity of the estimates performed in the previous section, evapotranspiration is clearly one of the dominant contributions to the entropy budget with respect to surface water exchange, especially the mixing of saturated surface air with the unsaturated air of the atmospheric boundary layer. Given the various ways by which vegetation affects water fluxes at the land surface, the entropy budget of the surface water balance is clearly affected. In particular, the following vegetation effects seem to have straightforward consequences on the thermodynamics of water exchange:

i. higher evapotranspiration rates in the presence of vegetation directly enhance entropy production rates;
ii. for a given evapotranspiration rate, the shift from evaporation to transpiration in the presence of vegetation would imply more work is done by lifting water to the canopy level. This would only marginally increase the overall entropy production, but would imply more work being done for a given rate of entropy production;
iii. higher evapotranspiration rates enhance soil drying, implying that more work is done in extracting soil moisture;
iv. for a given precipitation rate, the shift of the water budget to enhanced evapotranspiration results in reduced runoff, and consequently entropy production by this component would be reduced. However, since entropy production by runoff is small compared to evapotranspiration, it does not seem to reduce overall entropy production significantly;
v. hydraulic lift and redistribution of soil moisture through the root system of the vegetative cover (e.g., Caldwell et al. 1998) would enhance the relaxation to thermodynamic equilibrium of the soil

moisture distribution, that is, it acts to enhance the rate of entropy production associated with soil water redistribution.

Overall, these components would seem to raise the rate of entropy production associated with the surface water balance in the presence of terrestrial vegetation.

To provide a rough estimate of the magnitude by which vegetation affects the entropy budget of the surface water balance, we use the estimates from climate model simulations of Kleidon (2006). These simulations estimate that continental evapotranspiration is roughly doubled for the present day when compared to a "Desert World" void of terrestrial vegetation, mainly due to enhanced continental moisture recycling. This doubling in evapotranspiration rates would seem to roughly translate into a proportional increase in the associated rates in entropy production, although one would need to also include the effect that the relative humidity of the near-surface air would likely be higher, and therefore entropy production being reduced (see Table 4).

Implications

This thermodynamic perspective on land surface hydrology has potentially two important implications. First, gradients are a critical aspect in quantifying rates of entropy production. The necessary discrete representation of gradients and heterogeneity in land surface models necessarily involves the spatial and temporal aggregation, which is likely to "smear out" gradients and result in model biases. While these biases would seem difficult to quantify in terms of common variables (e.g., fluxes or concentrations), the effect on the entropy budget would seem straightforward in that rates of entropy production can only be underestimated. Entropy budget considerations would therefore seem to be a convenient and practical tool to quantify the appropriateness of model representations.

The calculation of entropy production rates by different processes could further be used in conjunction with the proposed MEP principle to make predictions about the dominance of different processes competing for water. An example would be the partitioning of precipitation into runoff and evapotranspiration, which might be optimal to maximize the catchment's entropy production.

On a more profound level, it would seem that entropy production is closely linked to hysteretic behavior and the memory of past hydrological conditions. Kleidon and Stieglitz (in preparation) introduce entropy production as a metric to measure the extent of hysteresis and memory with respect to the ground heat flux. The reasoning is as follows: in the absence of heat storage, and surface temperature responds and adjusts to equilibrium immediately to the forcing it is subjected to. In this case, the type of variability of the forcing is directly reflected in the variability of surface temperature. With an increased ability to store heat, past conditions play an increasing role in shaping the temperature response to the prevailing forcing, resulting in an increasingly frequency-dependent spectrum of surface temperature variations. At the same time, heat fluxes into and out of the soil heat reservoir take place at different temperatures, thereby producing entropy. Entropy production therefore reflects the sensitivity to past climatic conditions. An important implication of this is that the characteristic response of the heat storage to variability and change is linked to the rate of entropy production and its maximization.

The same line of reasoning applies to our case of soil water exchange (Fig. 2). For the above case of heat storage, surface temperature reflects the total amount of heat stored in the soil (at least to some extent). In our case of soil water exchange, the corresponding variable would be the total amount of the energy stored in the soil (TE). We would expect that increased rates of entropy production with water exchange would be associated with higher rates of evapotranspiration and higher rates of soil water variability, or more specifically, higher variability in TE. A higher variability in TE in turn would reflect more memory of past conditions. From this consideration we hypothesize that with increasing influence of vegetation on the exchange fluxes of water at the land surface we would expect: (i) higher rates of entropy production, possibly even maximizing it to a MEP state, mainly by enhancing evapotranspiration rates; (ii) higher variability in TE as soil is brought to drier conditions after rainfall events by enhanced evapotranspiration. This can also be illustrated by the extreme case of no evapotranspiration, for which the soil moisture content would vary between saturation and field capacity only, clearly resulting in a smaller

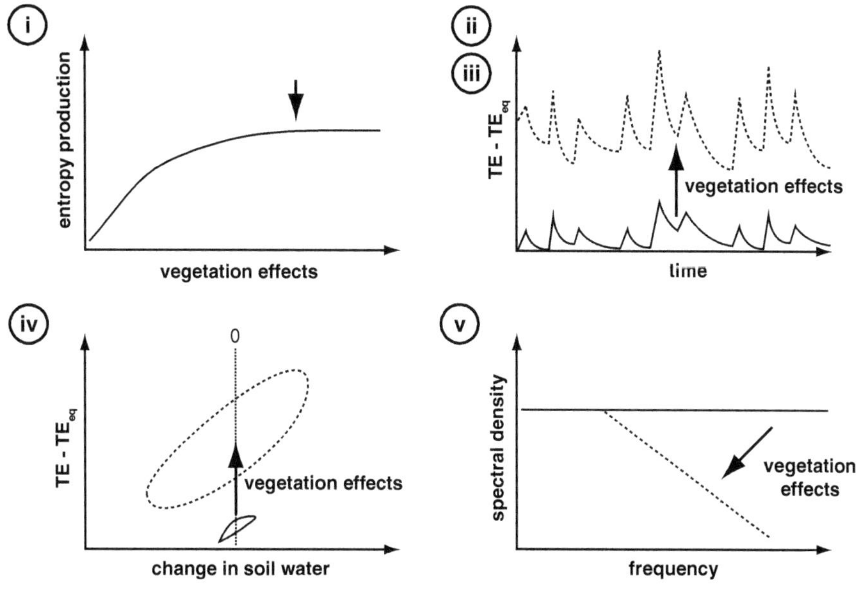

Fig. 2 Illustration of the five hypotheses described in the text of how entropy production can be used as a basis to characterize vegetation effects on the thermodynamics of surface water-exchange processes. Hypothesis (i) states that vegetation effects overall enhance the rate of entropy production associated with water-exchange fluxes at the land surface. Hypothesis (ii) is that because of enhanced evapotranspiration, the variability of total energy (TE) associated with soil moisture is enhanced, and the soil moisture distribution is further away from thermodynamic equilibrium, as measured by the distance of the actual value of TE to the total energy at thermodynamic equilibrium TE_{eq} (hypothesis (iii)). As the system is further from equilibrium, we would expect the effect of soil water change on total energy to depend more on previous conditions, therefore showing increased hysteretic behavior (hypothesis (iv)). Since increased dependence on previous conditions results in increased autocorrelation and memory, this would then lead to a more frequency-dependent spectrum of variability in total energy when subjected to noisy precipitation (hypothesis (v))

range of variability in TE; (iii) soil moisture that is on average further away from thermodynamic equilibrium; (iv) increased hysteretic behavior since the soil moisture distribution is further away from thermodynamic equilibrium (since processes are only reversible/do not reflect memory at thermodynamic equilibrium); and (v) a more frequency-dependent spectrum in TE variability.

These hypotheses could easily be tested in future work by using numerical simulation models of water movement in the soil.

Summary and Conclusions

We presented a first attempt to quantify the irreversible nature of the water-exchange fluxes associated with land surface hydrology. In this consideration, we included the irreversibility and entropy production rates associated with the most common components of the surface water balance: precipitation, soil moisture change, evapotranspiration, runoff, and soil moisture redistribution. Of these, evapotranspiration seems clearly to be contributing most significantly to the entropy production associated with the exchange of water at the surface, although runoff generated over large elevation differences could also add noticeable contributions in certain situations.

Based on these considerations, we formulated five hypotheses regarding the effect of vegetation on land surface hydrology, specifically to the state of soil moisture distribution in relation to its thermodynamic equilibrium distribution and how it relates to hysteretic behavior, hydrologic memory, and variability. If confirmed, these hypotheses would add significantly to our understanding of how the natural land surface responds to climate and climate variability and how this response is modified with land-cover change. It is therefore critical to further investigate and quantify the nonequilibrium thermodynamics of land surface hydrology. This chapter provides a first basis for doing so.

A logical next step would be to use the methodology established here and quantify the entropy budget in more detail, taking spatial and temporal changes into account. The considerations described here set the necessary foundations of such future work. These will help us in improving our understanding of the fundamental direction that drives land surface processes, what these mean for the response to climate variability and change, and how land-surface hydrology is affected by the terrestrial biosphere at a fundamental, but quantifiable level.

References

Caldwell MM, Dawson TE, Richards JH (1998) Hydraulic lift: consequences of water efflux from the roots of plants. Oecologia 113: 151–161

Campbell GS, Norman JM (1998) An introduction to environmental biophysics. Springer Publishers, New York, NY, 2nd edition

Hillel D (1998) Environmental soil physics. Academic Press, San Diego, 771pp

Kleidon A, Fraedrich K, Kunz T, Lunkeit F (2003) The atmospheric circulation and states of maximum entropy production. Geophys. Res. Lett. 30: 2223

Kleidon A (2004) Beyond Gaia: Thermodynamics of life and Earth system functioning. Clim. Ch. 66: 271–319

Kleidon A, Lorenz RD (2005) Non-equilibrium thermodynamics and the production of entropy: life, Earth, and beyond. Springer Publishers, Heidelberg

Kleidon A, Fraedrich K, Kirk E, Lunkeit F (2006) Maximum Entropy Production and the Strength of Boundary Layer Exchange in an Atmospheric General Circulation Model. Geophys. Res. Lett. 33: L06706

Kleidon A (2006) The climate sensitivity to human appropriation of vegetation productivity and its thermodynamic characterization. Glob. Planet. Ch. 54: 109–127

Kleidon A (2008) Energy balance. In: Jœrgensen SE, Fath BD (eds.) Global Ecology. Vol. 2 of Encyclopedia of Ecology, 5: 1276–1289, Elsevier, Oxford.

Kondepudi D, Prigogine I (1998) Modern thermodynamics, From heat engines to dissipative structures. Wiley, Chichester, 486pp

Lorenz EN (1955) Available potential energy and the maintenance of the general circulation. Tellus 7: 157–167

Lorenz RD, Lunine JI, Withers PG, McKay CP (2001) Titan, Mars and Earth: Entropy production by latitudinal heat transport. Geophys. Res. Lett. 28: 415–418

Martyushev LM, Seleznev VD (2006) Maximum entropy production principle in physics, chemistry and biology. Phys. Rep. 426: 1–45

Ozawa H, Ohmura A, Lorenz RD, Pujol T (2003) The second law of thermodynamics and the global climate system – A review of the Maximum Entropy Production principle. Rev. Geophys. 41: 1018

Paltridge GW (1975) Global dynamics and climate – a system of minimum entropy exchange. Q. J. Roy. Meteorol. Soc. 101: 475–484

Pauluis OM (2005) Water vapor and entropy production in the Earth's atmosphere. In: Kleidon A, Lorenz RD (eds) Non-equilibrium thermodynamics and the production of entropy: life, Earth, and beyond. Springer Verlag, Heidelberg, 107–120

Pauluis OM, Balaji V, Held IM (2000) Frictional dissipation in a precipitating atmosphere. J. Atmos. Sci. 57: 987–994

Pauluis OM, Held IM (2002) Entropy budget of an atmosphere in radiative-convective equilibrium. Part I: maximum work and frictional dissipation. J. Atmos. Sci. 59: 125–139.

Peixoto O (1992) Physics of climate. American Institute of Physics, New York

Roderick ML (2001) On the use of thermodynamic methods to describe water relations in plants and soil. Aust. J. Plant Physiol. 28: 729–742

Schneider ED, Kay JJ (1994) Life as a manifestation of the second law of thermodynamics. Math. Comput. Modeling 19: 25–48

Tesař M, Šír M, Lichner Ľ, Čermák J (2007) Plant transpiration and net entropy exchange on the Earth's surface. Biologia, Bratislava, 62/5:547–551

Tributsch H, Čermák J, Nadezhdina N (2005) Kinetic studies on tensile state of water in trees. J. Phys. Chem. B 109: 17693–1707

Ulanowicz RE, Hannon BM (1987) Life and the production of entropy. Proc. R. Soc. Lond. B 232: 181–192.

Winter Snow Supply in Small Mountain Watershed as a Potential Hazard of Spring Flood Formation

M. Hríbik, A. Majlingová, J. Škvarenina and D. Kyselová

Keywords Water equivalent of snow · Flood hazard · Water supply · Biosphere Reserve Polana · GIS

Introduction

The basic features, common for all Slovak rivers, are the preponderance of the flow rate in the spring season, low flow rate in summer and in the winter period with moderate increase in later fall months. Moreover the preponderance of the flow rate in the spring season is not a consequence of precipitation preponderance in this period. In the spring season the precipitation run is opposite and approaches to minimum values. Therefore, the regime of monthly precipitation and flow rate is expressed by long-term average.

Increase in spring outflows are caused by accumulated snow melting in the winter season. At the same time, there are still relatively low losses caused by evapotranspiration (Prazak et al. 1994, 1996; Strelcova and Mindas 2000; Tesar et al. 2006) and reduced to infiltration (Buchtele et al. 2006).

The floods present a dramatic demonstration of natural hazard. They have significant far-reaching consequences for humans and the environment. Simo (1972) introduces following causes of floods occurrence in Slovakia:

- Intensive storm rainfall (as well as long-time regional rain),
- Intensive snow melt (combined with rain mainly),
- Ice dams,
- Quick loosen of accumulated water in the reservoir, etc

The present chapter is oriented to the evaluation of flood risk potential as the consequence of an intensive snow melting, under conditions of a small mountain watershed. Generally, in the spring season, there is a gradual snow-cover melting here, with retention in higher sites. A dangerous situation usually occurs due to more than average warm advection from west or southwest trajectory, respectively. This is combined with warm square rain usually when the ground is not thawed out.

Pobedinskij and Krecmer (1984), who based an analysis of hydrological regimes in temperate Eurasian zone, note that there is usually no surface outflow on sites that are not covered by forest. Similar results for Slovak conditions are reported by Midriak (1992), who introduced the surface outflow lower than 1% from total annual precipitation for unbroken forest soil surface.

By spring snow melting and by downpours the outflow height as well as index on unforested areas is 2–4 times higher than that ones on forested sites (Pobedinskij and Krecmer 1984). In a given range, snow-melting retardation, surface outflow reduction and its transformation to underground outflow, and infiltration to substructural layers create a flood hazard and decrease a hazard of harmful erosion formation in forest areas. Forest soil, in difference with agriculturally cultivated soil, is not plain. Their surface is covered by cover humus and necrotic herbal organic material called litter. Other microclimatic insulation cover represents the forest stand itself, often refilled

M. Hríbik (✉)
Technical University in Zvolen, Faculty of Ecology and Environmental Sciences, T.G. Masaryka 24, 960 53 Zvolen, Slovakia
e-mail: vrchar@gmail.com

K. Strelcová et al. (eds.), *Bioclimatology and Natural Hazards*,
DOI 10.1007/978-1-4020-8876-6_10,

also with herbal and wood undergrowth in natural forests.

In relation to spring thaw flood formation, it is necessary to note that forest soils of mountainous sites are not very frozen throughout, if at all. This occurs in case of adequate high and continuously lasting snow-cover creation still before winter frosts coming (Petrik et al. 1986). This knowledge is mainly of hydrological significance, because by gradual melting, the snow-water supply can be absorbed by soil. The dangerous situation happens when black frosts occur in the later fall season. Then, coming snow-cover isolates frozen soil. In spring, such soils increase surface outflow from melted snow. As such, the risk of flood formation is increased.

If the climate changes lead to snow-cover changes it will also mean the changes in soils' freezing (Sály 1996). Soils of lower situated forests will continue to freeze deeper. By absence of an adequate snow supply, it can happen that mountainous soils will freeze as well. Moreover, from the hydrological point of view, the reduction of soils' releasing and absorption capabilities will occur.

However, forest cannot prevent flood waves absolutely, it can mitigate its run (Mindas et al. 2001).

The Characteristics of Natural Conditions in the Experimental Area

The experimental area is a small mountain watershed of Hucava stream in Polana Mountains. This situation is depicted in Fig. 1. Experimental area spreads from part of Polana Mountains ridge to limnigraph station near "Hrochotsky mlyn" (Hrochot millhouse). The description of the basic parameters and characteristics are presented in Table 1.

Fig. 1 Hucava mountain watershed localization within Biosphere Reserve Polana – hillshade of the digital relief model

The Characteristics of Geological and Soil Conditions

Polana is the highest volcanic massive in Slovakia with the maximum height of 1457.8 m above the sea level. As Ciesarik et al. (1977) and Kunca (2003) already noted, the mountain consists of two parts, each with different altitudes and morphology. One of them is an erosive caldera of Polana Mountain itself, together with the Lubietovsky Vepor promontory. The other is created by Detva Mountains. Based on the compactness, the caldera itself (result of explosion or raid of volcanic cone peak, respectively) represents a unique hydrological object, drained only by the Hucava stream (Fig. 1).

The whole massive is built with two products of the second andesites phase. Only in the middle part – in caldera, there are some granat ryolits. The scope of the massive and also the bottom of the volcanic complex is built with layers of tuffits of pyroxenic andesites that have a variable potency. In the upper strata this layer extends to tuffs. Tuffs are of different character from close-grained to coarse-grained, sometimes of till-boulder character. The ridge part of the Polana Mountains is composed nearly by pyroxenic andesites, only.

The mostly spread out soil types in Polana perimeter are saturated and unsaturated cambisols. Saturated soils occur in lower areas with the elevation up to 700–800 m above the sea level. There are also brown andosols here and typical andosols with an enclave representation.

The Characteristics of Climatic Conditions

The Hucava watershed is a region climatologically very differentiated with span of average annual temperatures 6.7–2.5°C. Average temperatures in July range from 16.5 to 11.5°C. The vegetation period (average daily temperature above 10°C) there

Table 1 Basic characteristics of Hucava watershed

Station	River	Area, $F(\text{km}^2)$	Forest cover (%)	Valley length, L(km)	H maximum (m)	H minimum (m)	Gradient (%)	Watershed shape (F/L^2)
Hrochot	Hucava	41.45	80	14.6	1458	522.5	6.41	0.19 transitive till feathery

in Hucava watershed lasts 65–155 days. Average annual precipitation range from 720 to 1200 mm and average precipitation in vegetation period are about 475–630 mm. The snow cover in studied area lasts 135–190 days on an average. Hucava watershed belongs, based on Koncek climate classification, into two climatic regions. First, moderately warm region (moderately warm, very humid, highlands district) in the Hrochot part of valley and in south sites of Bukovina part. Second, moderately cold region, moderately cold district in lower part of the caldera and on the north sites and also cold mountain district in the highest ridge parts of Polana Mountains. From the climatogeography point of view, this area is classified as cold subtype of mountain climate (north and the highest part of the massive) as well as moderate cold subtype of mountain climate in the south and lower parts of the caldera (Skvarenina and Mindas 2001).

The Characteristics of Hydrological Conditions

From the hydrological point of view, the studied area belongs to Hucava stream watershed. It is represented by right-side tributary of Slatina River and left-side part of Hron River watershed. Based on runoff classification by Dub, the implicit part of Polana caldera belongs to (B_2) middle-mountainous region (Simo 1972). Table 1 shows other hydrological characteristics.

The Characteristics of Forest Communities

The Polana massive crosses the southern spur of the prime Slovak climatic line, which is determined in forest communities spreading. On the lowest southern sites, the plan communities are sorted as the forest type, group of *Fageto–Quercetum*. That is followed by *Querceto–Fagetum* and in upper parts by *Fagetum pauper* and *Fagetum typicum*. On the main ridge, it crosses in form of slender strips of *Fagetum* and *Fageto–Abietum* to the southern-situated forest types, group of *Sorbeto–Piceetum*. On the north sides, there are extensive communities of *Abieto–Fagetum* and *Fageto–Abietum*. In south, expositions are dominated by beech stands and by mixed stands composed of beech. Fir and spruce dominate the north expositions. The original species composition is more preserved in higher parts and the composition of lower parts has been significantly changed. The south of the Polana Biosphere Reserve, in the composition of forest stands, is dominated by hornbeam against the oak. In the north, spruce dominates against beech and fir.

Methodology

Methodology of Terrain Monitoring

The measurements of snow-cover parameters were carried out in Polana Biosphere Reserve – in the Hucava watershed, covering 41.45 km^2, since winter season 2003/2004. The water equivalent of snow cover was obtained by mass method using the snow sampling tube (model VS-43). The snow-cover height was measured by communicable and stabile laths for snow measurements.

Basic physical parameters were monitored in 2-week intervals in 2004 and in 3-week intervals in 2005, 2006, and 2007. This was during the time of the highest snow supply occurrence in mountain locations. The monitoring was performed on Polana massive altitudinal transect. Testing areas were situated both, in open areas and in forest. Step by step, the measurements were carried out from Hucava stream mouth (525 m above sea level) up to Polana peak (1457.8 m above sea level) in each 100 altitudinal meters. In the forest, five measurements of snow-water equivalent and snow density were performed. There were three measurements performed in an open area. There were also 20 measurements of snow height undertaken, using snow-measurement laths. The measurements comply with the methodology standards used by Slovak Hydrometeorology Institute (Turcan 1973). The time constraints limited the number of measurements and also, more measurements were statistically of little significance. However, less measurements show negative results accuracy.

Methodology of Processing Terrain Monitoring Results in the GIS Environment

The analysis of spatial distribution of snow-cover water supply in area of Hucava small mountain watershed situated in the area of Polana Mountains, was performed using GIS tools in the IDRISI environment. Geostatistical and map algebra were used as analytical tools.

From the geostatistical analyses, linear regress dependencies (module REGRESS) among elevation as independent and snow-water equivalent as dependent variable, found out by terrain measurements in winter periods 2004–2007, were computed.

Using map algebra tools, based on computed regress dependencies (mainly regress and absolute coefficient), the maps of snow-cover water supply distribution in given area depending on elevation, were derived (Fig. 2).

Using this method, the obtained maps of water supply (m^3) distribution were then extracted (module EXTRACT): average, minimum, maximum, and total values of snow-cover water supplies in the area of the whole watershed as well as on its forested part and open areas (e.g., meadows).

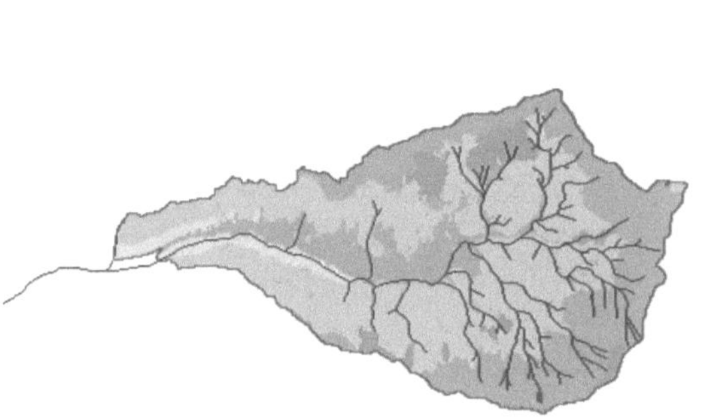

Fig. 2 The map of water supply distribution in snow cover – February 2004

Results

Snow-Water Equivalent

In accordance with winter aspect and elevation, winter precipitation fixed in snow cover in Hucava watershed represent from 15 to 40% of the total annual precipitation. The water equivalent of snow represents an important hydrophysical parameter defined as the height of water column in millimeters ($\mathrm{L\,m^{-2}}$) that comes from snow-cover melting at given place and depends on a snow height and its density. Figure 4 shows weighted means (weight is represented by the extent of area) of the snow-water equivalent for open areas, forest areas, as well as for the whole watershed. It points out the important fact that up to the top of the winter season (February, beginning of March, respectively), there are the highest water supplies in open areas. It is caused mainly by the absence of snow precipitation interception on the unforested area. Starting spring season, the situation of a snow-water supply is changing. The maximum value of snow-cover water equivalent is moving under crowns of forest stand. This is caused mainly by retarded snow melting in forest stand microclimate conditions. In situations when snow precipitation is wet and heavy, the snow is not caught by tree crowns, but cascades inside the forest where more compact snow mass with higher density is created. After its freezing it is more resistant to melting and lasts in the forest also after absolute loss of snow in opened areas. Figure 5 shows distribution of snow-cover water equivalent in millimeters ($\mathrm{L\,m^{-2}}$) in Hucava watershed during years 2004–2007. On the left, the situation presents the time of water supply culmination, while on the right, the situation in month of snow-cover loss in a given year is shown. This value graphically informs about the height of the water layer (in millimeters) that arises from melting snow in a given place. The figure also shows that snow supply approached its maximum in the winter of 2005/2006. During the winter of 2004/2005 the supply was lower, but more homogenously distributed along area of the watershed.

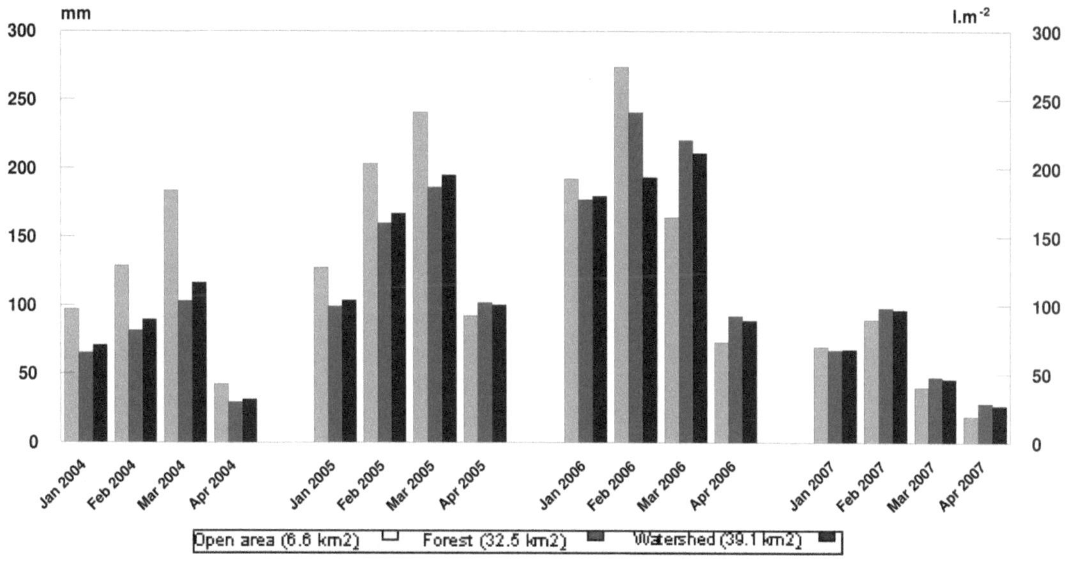

Fig. 3 Weighted means of snow-cover water equivalent (mm respective L m^{-2}) in Hucava watershed in period 2004–2007 concerning open areas, forests areas, and the total observed area

During the time of snow-cover loss, in years rich on snow (2004/2005 and 2005/2006), relatively high water values of snow right in forested ridge localities of Hucava watershed are found.

Snow-Cover Water Supply in Hucava Watershed

Based on processing data gained from snow-water equivalent measurements (4 years), the snow-cover water supply for watershed area was calculated using geostatistical methods. Results it can be concluded as follows (Fig. 3):

- Winters of 2004/2005 and 2005/2006 can be characterized as better than average, mainly in the winter season of 2005/2006 that marked out with total supply up to 9.65 millions m^3 of water fixed in snow.
- Owing to low water supply, the poorest was the winter of 2006/2007, when the deposit of water was represented only by 3.77 millions m^3 of water during the peak of the winter. This supply represented only one-third of the previous winter's water supply. Water supply in March was even lower, only one-fourth of the water supply was found in comparison with the same period of the previous year.

The Regime of Hucava Runoff in Relation to Snow Precipitation and Spring Flood Risk

The winter of 2005/2006 was marked with a high record-breaking snow-cover water supply and a long duration of snow cover. During this winter, the mutual progress of runoff (discharge), water equivalent, snow-cover height in dependency on meteorological situation (air temperature and precipitation total) was analyzed, (Fig. 6). The following coherences were shown:

- In the beginning of the winter season (since end of the November, when was registered accumulation beginning, up to January, 5th), the difference between snow cover in open area and in forest, mainly from the reason of high interception of snow in the tree crowns, was studied. The runoff height was influenced mainly by the total of fluid precipitation in lower sites of the watershed.
- Warmer and sunny periods with maximum temperatures over 0°C had an influence on significant decrease of snow cover total height mainly on meadows. The decrease was more gradual under the forest stand because of a specific microclimate of the forest stands. But process of snow accumulation continued up to end of March.

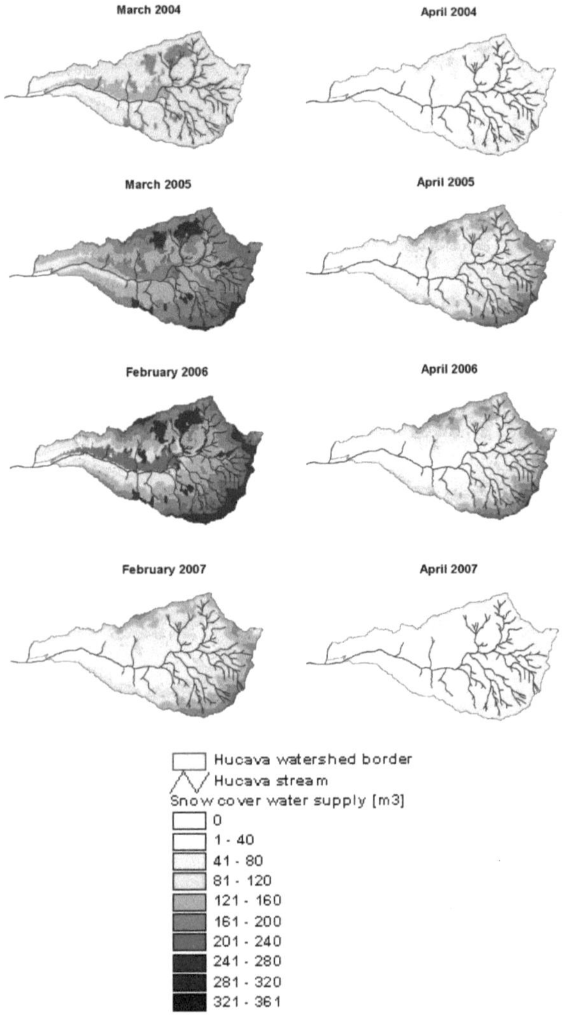

Fig. 4 Distribution of snow-cover water equivalent in millimeters ($L\,m^{-2}$) in Hucava watershed during the period of 2004–2007. On the left is depicted situation in time of a water supply culmination, on the right is depicted situation in month when snow cover ceased to exist in a relevant year

- The maximum supply of snow-cover water equivalent was recorded in open areas already at the beginning of March, meanwhile in the forest it was accumulated at the end of month. The run of water level was nearly steady up to the end of snow accumulation in the forest. This was impacted also by subnormal air temperatures during March.
- The melting process as a consequence of sublimation comes first in open areas and gradually continues since March, 5th. Owing to high forest cover percentage of Hucava watershed (80%) and low air temperatures, this fact did not show significant runoff height.
- The snow melting in forest started at the end of March after relatively high spring temperatures (daily maxima of air temperature above 10°C) together with abundant rain precipitation. The snow melting also occurred in forested parts of the watershed accompanied with significant increase in runoff (ca. 10 mm).
- The spring runoff was affected by a record water supply in snow cover, high air temperatures, and fluid precipitation. After such a significant increase (avalanche effect of snow melting), the gradual decrease began with typical spring shaken progress of water levels.

The temporal progress of snow cover and runoff in dependency on meteorological parameters was assessed by more authors, for example, Balazs et al. (1974), Ernstberger– and Sokollek (1984), and Miller (1977).

Ernstberger and Sokollek (1984) found, in middle-mountainous sites of Hessen conditions, that the forest cover has significant influence on snow-melting retardation (10 days) and runoff extremes decreasing in forested watersheds.

The influence of vegetation on snow-cover duration has been confirmed also in other Slovak experiments that are described in Hríbik (2007), (Hríbik 2007), Kantor et al. (2007), Kostka and Holko (2001), and Mindas (2003).

The Hydrological Regime of Hucava Stream

Figure 7 as well as the Table 2 reveal that Hucava watershed follows two different regimes of runoff. Here, the studied part of Hucava watershed "Hrochotsky mlyn," is characteristic of a middle-mountainous region stream, meanwhile for the remaining part, Hucava mouth is of a highland-to-lowland-stream character (draining mainly Detva foothills and part of Poniky highlands, representing mountains with lower altitude). The studied part has more attributes of middle-mountainous stream of snow – rain type. The maximum flow rate starts in April and is related to snow melting in the middle-mountainous parts of Polana caldera.

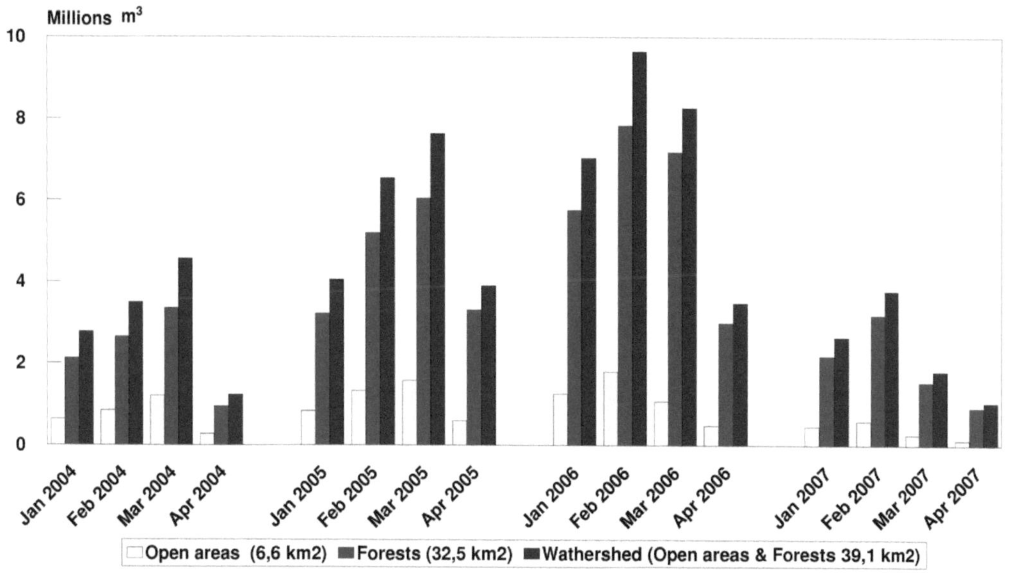

Fig. 5 Water supply in snow cover (million m^3) in watershed of Hucava during period 2004–2007

Hucava – mouth profile has runoff regime of highlands stream of the rain – snowy type. The portion of water supply accumulated in snow cover is significantly lower. The maximum flow rate starts a month before, it means in March (already in the end of February) after snow melting on highlands sides.

Middle-mountainous watershed of Hucava related with Polana caldera shows only small probability of spring flood formation. It is caused mainly by high forest cover percentage of the watershed that retards snow melting. Other reason confirming this fact is that significant quanta of winter supplies in ridge parts of Polana massive decreases only gradually from the end of April up to the start of May.

The segment of the watershed represented by Hucavamouth can lose its snow supply relatively quickly. This is caused by a quick accumulation of water level and flood formation in case of quick warming or by starting of abundant raining in spring season, respectively.

Conclusion

Snow cover represents significant ecological factor mainly in mountainous forest ecosystems. Processes of snow-cover water equivalent development and next snow melting has a significant influence on runoff and water balance in watershed in middle-mountainous conditions. The spatial and temporal distribution of snow in the watershed depends mainly on meteorological and topographic factors, and also on forest cover of the watershed (species composition, structure, and age of the forest stands). The interception of snow in tree crowns, as well as forest microclimate influence on snow cover under the tree crowns, retards snow melting and influences the progress of runoff.

This work brings results after 4 years of monitoring of snow-cover hydrophysical characteristics at elevation transect in Hucava mountain watershed in the Biosphere Reserve Polana. Experimental measurements were performed during culmination and snowmelting period at the mountain sites since the winter of 2003/2004 up to the winter of 2006/2007 and equally in the open and forest areas. The forest ecosystems in elevation interval 525–1457 m above sea level were observed.

Based on terrain measurements of snow-cover water equivalent, snow-supply spatial distribution analysis using GIS tools and mainly geostatistical (linear regression) and map algebra tools was performed.

Conclusions are concentrated in the following discussion:

From the water supply point of view, the winters of 2004/2005 and 2005/2006 can be characterized as

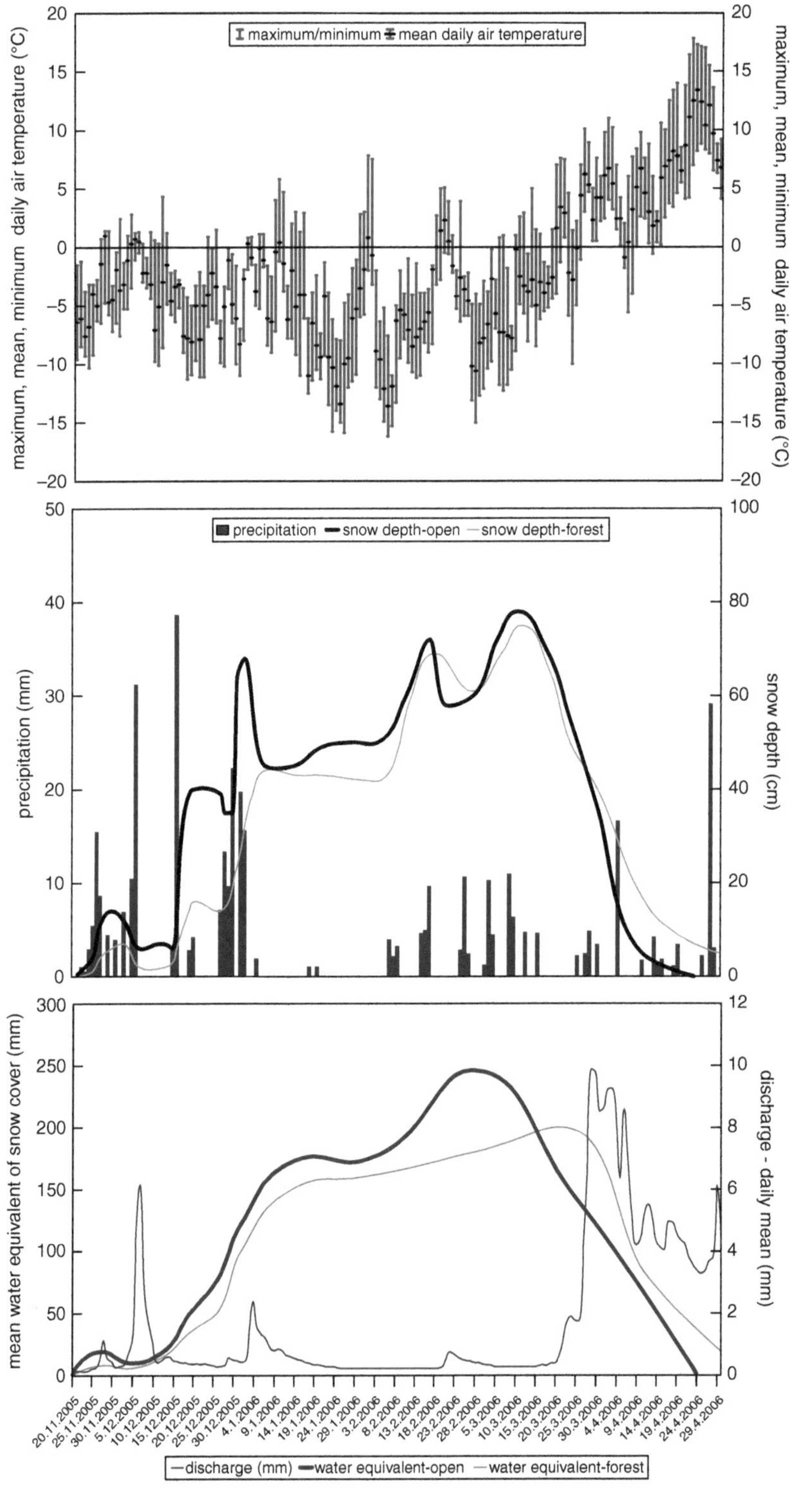

Fig. 6 Courses of basic hydrometeorological elements during the period of snow over accumulation and melting in winters of 2005–2006, at the mountain watershed of river Hucava Biosphere Reserve Polana Mountains

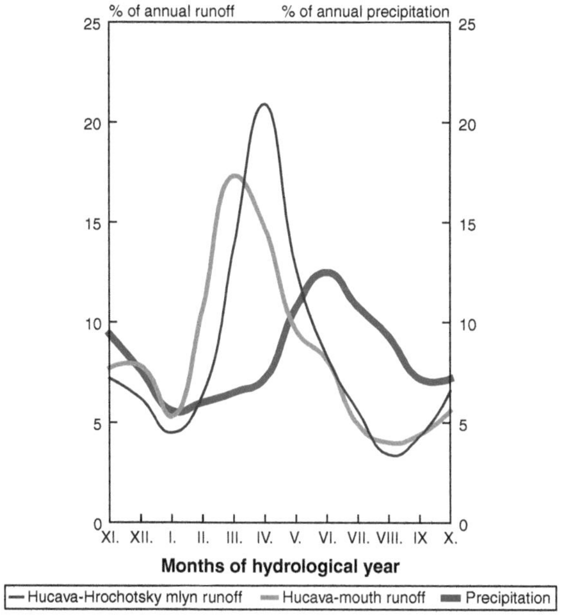

Fig. 7 The hydrograph of relative runoff and precipitation in the Biosphere Reserve Polana. The two types of runoff in the watershed are represented. One of them is Hucava – "Hrochotsky mlyn" runoff, measured in site where the stream has mountainous character (522 m above sea level). Another one is the Hucava-mouth runoff, measured in highlands – lowlands character site (300 m above sea level)

better than average. The highest water supply in snow cover was measured in turn of February and March 2006 representing 9.65 millions m^3 that represents up to 246 mm of water column on average. In the winter of 2004/2005 the supply was lower (7.65 millions m^3 with average water column of 195 mm), but more homogenously distributed along the area of watershed. In the same period of 2004, the total water supply in snow cover was about 3.5 millions m^3 (water column of 89 mm) and in 2007 the supply was represented by 3.78 millions m^3 (water column of 96 mm). This represents only one-third of volume in comparison with the year 2006. Even water supply volumes in March 2007 represented only one-fourth of those in 2006. The lowest values were found in April 2007, when the total water supply of the snow cover was only 1.06 millions m^3 (water column value of 27 mm). During the time of snow-cover loss, in years rich in snow (2004/2005 and 2005/2006) relatively high volumes of snow-water equivalent right in ridge localities of Hucava watershed can be found.

Based on these results, it can be stated that the influence of the forest on snow supply holding grows after culmination and after the start of snow-cover melting when the forest acts as a curtain and defends snow against melting. This fact confirmed the effect of forest which explains the more regular distribution of snow precipitation, and on the later retarded breakup of water on a land surface, and on underground waters. All these effects significantly decrease the potential risk of flood formation.

Acknowledgments Authors thank for grants to this project from VEGA MS SR No. 1/0515/08, 1/2357/05, 1/3283/06, 1/4393/07, 1/3528/06, from Slovak Research and Development Agency No. APVV-0022-07 and Europe Union from program Human Sources – project code 11230100453.

References

Balazs A, Liebscher HJ, Wagenhoff A (1974) Forstlich-hydrologische Untersuchungen in bewaldeten Versuchsgebieten im Oberharz – Ergebnisse aus den Abflußjahren 1949–1965 (In German). Reihe: Aus dem Walde, Heft 22. Mitteilungen aus der Niedersächsische Landesforstverwaltung. Hannover. M. & H.Schaper Vlg. Hannover, 265

Buchtele J, Buchtelova M, Tesar M (2006) Role of vegetation in the variability of water regimes in the Sumava Mts. Forest. Biologia, 19: 246–250

Ciesarik M, Mihalik A, Saly R, Tomlain A (1977) Excursion guidebook. Symposium of the International Pedological Society: "Soil as biotope factor of moderate and cold zone forests" (in Slovak). Zvolen, 5.09. 1977, ES VSLD, 34p.

Ernstberger H, Sokollek V (1984) Einfluß unterschiedlicher Vegetationsbestände auf Schneedeckenentwicklung und Schneeschmelzabfluß in unteren Mittelgebirgslagen. (In German). In: DVWK Mitteilungen 7, pp. 317–329

Hríbik M (2007) Beech and Spruce forest influence on snow and melting in period 2003–2007. (In Slovak). Acta Facultatis Forestalis. 49 (Suppl. 1): 98–107

Table 2 Average monthly and annual runoff of Hucava in period 1961–2000 (by SHMI Banska Bystrica)

Stream	Profile name	Watershed area (km^2)	Monthly and annual average runoff ($m^3.s^{-1}$)													
			XI	XII	I	II	III	IV	V	VI	VII	VIII	IX	X	IV–IX	$Q_{a1961-2000}$
Hucava	Hrochot	41.45	0.50	0.43	0.31	0.45	0.95	1.45	0.89	0.58	0.39	0.24	0.30	0.45	0.64	0.58
Hucava	mouth	72.22	0.65	0.66	0.45	0.91	1.48	1.26	0.82	0.69	0.42	0.34	0.37	0.48	0.65	0.71

Hríbik M, Škvarenina J (2007) Influence of spruce and beech forest stands (pole timber phases) on the formation of water storage in Biospherical reserve Polana (In Slovak). In: Roznovsky J, Litschmann T, Vyskot I (eds): "Forest Climate", Krtiny 11. – 12.4.2007. CD medium and Proceedings of Abstracts, p. 10

Kantor P, Karl Z, Šach F (2007) Storage and intensity of snow melting in young spruce stand in winter period 2005/2006 (In Czech) In: Roznovsky J, Litschmann T, Vyskot I, (eds): "Forest Climate", Krtiny 11.–12.4.2007, CD medium and Proceedings of Abstracts, p. 10

Kostka Z, Holko L (2001) Influence of vegetation cover change on hydrological regime of mountain watershed (in Slovak). National Climate Programme of the Slovak republic. 10: 82–92

Kunca V (2003) Critical loads of selected forest ecosystems in the Polana Biosphere Reserve. Scientific essay, 4/2003/A, Technical University in Zvolen, 72pp (in Slovak)

Midriak R (1992) Research of surface runoff and erosive soil losses in forest ecosystems. In: Ecological and ecophysiological research in forest ecosystems (in Slovak). Zvolen, Polana, pp. 32–36

Miller DH (1977) Water at the surface of the Earth. International Geophysics Series. Vol. 21. Academic Press, New York, San Francisco, London

Mindas J, Skvarenina J, Strelcová K (2001) Importance of forests in the landscape hydrological regime (In Slovak). Zivotné prostredie 35: 146–151

Mindas J (2003) Characteristic of snow conditions in forest stands of Polana middle mountainous region (In Slovak). Lesnicky casopis – Forestry Journal 49: 105–115

Petrik M, Havlicek V, Uhrecky I (1986) Forest bioclimatology (in Slovak). Priroda, Bratislava

Prazak J, Sir M, Tesar M (1994) Estimation of plant transpiration from meteorological data under conditions of sufficient soil moisture (in Czech). Journal of Hydrology 162: 409–427

Prazak J, Sir M, Tesar M (1996) Parameters determining plant transpiration under conditions of sufficient soil moisture (in Czech). Journal of Hydrology 183: 425–431

Pobedinskij AV, Krecmer V (1984) Function in forest in the field of soil and water protection (In Czech). State agricultural publisher. Praha. P. 256

Sály R (1996) The influence of global climate changes on the forests soils (in Slovak). In: Mindas J, Lapin M, Skvarenina J (eds.): Climate changes and forests of Slovakia In: National climate programme SR, Bratislava: MŽP SR. part 5. pp. 41–45

Simo E (1972): Ground water (In Slovak). In: Slovakia – Nature. Bratislava, Obzor. Pp. 283–341

Skvarenina J, Mindas J (2001) Climate. In: Bublinec E, Pichler V et al. Slovakian primeval forests – diversity and protection. IEF SAS Zvolen, p. 200

Strelcova K, Mindas J (2002): Beech transpiration in relation to changing environmental conditions (In Slovak). Technical University in Zvolen, Scientific studies 11/2000/A, p. 82

Tesar M, Sir M, Lichner L, Zelenkova E (2006) Influence of vegetation cover on thermal regime of mountainous catchments. Biologie (Suppl 61) 19: 246–250

Turcan J (1973) Introduction to determination snow supply in mountainous watershed (In Slovak). Vodohospodarsky casopis – Journal of Hydrology and Hydromechanics 21: 3–4

Mapping of Gumbel Extreme Value Distribution Parameters for Estimation of Design Precipitation Totals at Ungauged Sites

S. Kohnová, J. Parajka, J. Szolgay and K. Hlavčová

Keywords Extreme daily precipitation totals · Precipitation mapping · Maps of design precipitation totals

Introduction

N-year maximum precipitation totals are usually estimated for engineering hydrology in order to provide a hydrometeorological input for estimation of design flood. Recently, intensive efforts to develop complex statistical methods for estimating design rainfalls have been reported in the national meteorological offices around the world. Examples of such complex national studies on risk assessments of heavy precipitation include the German KOSTRA project (Barthels et al. 1997; Malitz 2005), the *Flood Estimation Handbook* (FEH 1999) in Great Britain, the Italian VAPI project (Ferrari 1994), the HIRDS system in New Zealand (Thompson 2002), and the Australian Guide to Rainfall and Runoff (Institutions of Engineers 1987).

Within the KOSTRA project (e.g., Barthels et al. 1997; Malitz 2005), the regionalization of N-year design rainfall values for N ranging from 0.5 to 100 years was achieved in two time frames using classical extreme value statistics and a complex climatological analysis, with a grid resolution of 8.5×8.5 km for the whole of Germany.

S. Kohnová (✉)
Department of Land and Water Resources Management, Faculty of Civil Engineering, Slovak University of Technology, Radlinského 11, 813 68 Bratislava, Slovakia
e-mail: silvia.kohnova@stuba.sk

The objective of the UK *Flood Estimation Handbook* (FEH 1999) was to develop a method for the estimation of rainfall depth–duration–frequency relationships for durations between 1 h and 8 days and for return periods of up to 1000 years for arbitrary locations in Great Britain. The Focused Rainfall Growth Extension method (FORGEX) was developed to estimate the growth curve (Reed et al. 1999; Faulkner and Prudhomme, 1998).

High Intensity Rainfall Design System (HIRDS) is a software package developed in New Zealand for estimating design rainfall values (Thompson 2002). The system involves mapping of the index rainfall (the median value of the annual maximum rainfall) and regional growth curves, respectively. Design rainfall values in HIRDS are computed for 10 standard durations ranging from 10 min to 72 h.

Regionalization of extreme precipitation totals has been of interest in Switzerland as well. Rainstorms were investigated by Geiger et al. (1986). This analysis resulted in the construction of several maps, which permitted the estimation of design rainfall values up to the duration of 5 days in any locality of the country.

Analysis of the annual maximum precipitation in the State of Washington (US) (Wallis et al. 2007) was carried out using index rainfall type methodology. The parameters of the Generalized Extreme Value (GEV) distribution function were estimated by L-moments, a modification of probability-weighted moments. In South Africa, the L-moment-based regionalization method (Hosking and Wallis 1997) was successfully applied to estimate design rainfalls.

For the estimation of design maximum daily precipitation totals at ungauged sites, interpolation methods are usually applied. The assessment of the mapping of climatological characteristics is described in

K. Strelcová et al. (eds.), *Bioclimatology and Natural Hazards*,
DOI 10.1007/978-1-4020-8876-6_11,

many studies, for example, Tabios and Salas (1985), Borga and Vizzacaro (1997), Dubois (1998), Goovaerts (2000), Weisse and Bois (2001), Loukas et al. (2001), and Wallis et al. (2007).

Rainfall frequency mapping for the whole territory of Greece was also described in Loukas et al. (2001). Rainfall depths for durations of 1–7 days were fit by the Gumbel (EV1) theoretical distribution, and the depth–duration–frequency relationships for each station were estimated and mapped for Greece using spline interpolation. (Weisse and Bois 2001, 2002) compared kriging and ordinary regression against the topography to model 10- and 100-year rainfall estimates for rainfall durations of 1–24 h in the French Alps. The tests included relationships between the rainfall and topography, which is important in a mountainous region. All these methods were compared by cross-validation. The comparison showed that topography is indeed an important parameter for short time steps. Watkins et al. (2005) conducted regional precipitation frequency analyses and spatially interpolated point intensity duration frequency (IDF) estimates at gauge sites for Michigan. Several interpolation and smoothing techniques were evaluated including a trend surface analysis, thin plate splines, inverse distance weighting (IDW), and several kriging algorithms. Ordinary block kriging was recommended as a practical and objective method for the index flood values and the developing of isopluvial maps.

In previous studies concerning the statistical analysis of extreme daily precipitation totals in Slovakia, the Gumbel and Pearson type III distributions were found to be the most appropriate in studies of Reinhartová (1967), Dzubák (1969), and Šamaj et al. (1982, 1985). The recent studies for example, Faško et al. (2000) and Gaál (2006), preferred the Pearson type III distributions or GEV for the estimation of daily precipitation totals in Slovakia.

Although the number of other suitable theoretical distribution functions for the analysis of extreme daily precipitation totals has increased during recent years, subsequent studies (Geiger et al. 1986; Loukas et al. 2001; Casas et a. 2007) have favored the use of the Gumbel distribution (EV1) which may be useful under particular circumstances, for example, when data series are short and the tail behavior is not very obvious. There are a large number of theoretical as well as "case study" evidences that the distributions of heavy precipitation amounts tend to be heavy-tailed (see e.g., Katz et al. 2002; Koutsoyiannis 2004a,b).

This chapter builds on the results derived in case studies of Parajka et al. (2002, 2004), in the upper Hron River basin, where we proposed a different approach for the mapping of design extreme daily precipitation totals. First, the daily precipitation measured at the climatological gauge stations was spatially interpolated using a user-defined interpolation method. As a result, grid maps of the mean daily precipitation totals were constructed for the selected study period. Next, a statistical estimation of the design values of the maximum daily precipitation totals was performed at each individual grid point. Here, based on the presented methodology, kriging and IDW interpolation methods were used for the interpolation of the daily precipitation totals in the pilot Hron River basin. According to the statistical tests and also due to the historical tradition in Slovakia, the Gumbel distribution was selected for the analysis. The performance of the estimated distribution function parameters using kriging and IDW interpolation methods was tested. Finally, derived maps of the 100-year precipitation totals were compared with the expert hand-drawn isohyets map of design 100-year maximum daily precipitation totals based on 50-year measurements at 557 stations in Slovakia and the Person III model, constructed by Faško and Lapin (Faško et al. 2000; Gaál et al. 2004).

Estimating N-Year Maximum Precipitation Totals

In order to estimate the N-year maximum daily precipitation totals the Swiss methodology according to Geiger et al. (1986) was applied. This methodology is based on the application of two extremal distribution functions; Gumbel EV1 and EV2.

The general formula for the probability density function F_x for EV1 is

$$F_x = \exp\left(-\exp\left(\frac{x-\mu}{c}\right)\right) \tag{1}$$

where x is the annual maximum daily precipitation total, and c and μ are the parameters of the distribution function EV1, calculated as follows:

$$\mu = \mu_x - c\mu_{yn} \tag{2}$$

$$c = \frac{\sigma_x}{\sigma_{yn}}\sqrt{\frac{n-1}{n}} \tag{3}$$

where

μ_x is the sample mean of x,
σ_x is the sample standard deviation of x,
n is the number of observations.

The values of μ_{yn} and σ_{yn} are available in a tabular form, depending on the number of observations (Geiger et al. 1986).

The probability density function equation for EV2 is

$$F_x = \exp\left(-\exp\left(\frac{z-\hat{z}}{c_z}\right)\right) \tag{4}$$

where

$$z = \ln x. \tag{5}$$

For the estimation of parameters c_z and $\hat{z}$ of the extremal EV2 distribution, the following equations are applied:

$$\hat{z} = \mu_z - c_z.\mu_{yn} \tag{6}$$

$$c_z = \frac{\sigma_z}{\sigma_{yn}}\sqrt{\frac{n-1}{n}} \tag{7}$$

where

μ_z is the sample mean value of z,
σ_z is the sample standard deviation of z,
n is the number of years of observation.

The Van Montfort test was used to select between the two extreme value distributions, the Gumbel EV1 and the EV2 (Geiger et al. 1986).

It is based on the estimation of the relationship between the $I_{(m)}$ values of the EV1 and the standardized values of $y_{(m+1/2)}$, where

$$I_{(m)} = \frac{\Delta x_{(m)}}{\Delta y_{(m)}} = \frac{x_{(m+1)} - x_{(m)}}{y_{(m+1)} - y_{(m)}} \tag{8}$$

$$y_{(m+1/2)} = -\ln\left(-\ln\frac{n-m+0,5}{n+1}\right) \tag{9}$$

where

$x_{(m)}$ is the value of the ordered extreme precipitation totals,
$y_{(m)}$ is the reduced variable,
m is the order of the x values,
n is the number of years of observation.

The testing variable is the correlation coefficient r, estimated as

$$r = \frac{S_{I,y}}{\sigma_I \sigma_y} \tag{10}$$

where

$S_{I,y}$ is the covariance of variables $I_{(m)}$ and $y_{(m+1/2)}$,
σ_I is the standard deviation of variable $I_{(m)}$, and
σ_y is the standard deviation of variable $y_{(m+1/2)}$.

The critical values of r_{krit} are estimated according to the length of observations and significance level α. In the study, we used r_{krit} on the significance level $\alpha = 10\%$.

The Mapping of Daily Precipitation Totals

A variety of methods, both deterministic and stochastic, can be applied to take point data (e.g., from gauges) and estimate the mapped variable over an area.

The general form of the equation for interpolation in the *z(x,y)* plane at point z_o (Meijerink et al. 1994) is

$$z_0 = \sum_{i=1}^{n} w_i z_i \tag{11}$$

where:

z_0 is the estimated value of the process at any point x_0 and y_0,
w_i is the weight of the sampling point i with coordinates x_i, y_i,
$z_i(x_i, y_i)$ is the observed value of the attribute at point x_i, y_i,
n is the number of sampling points considered.

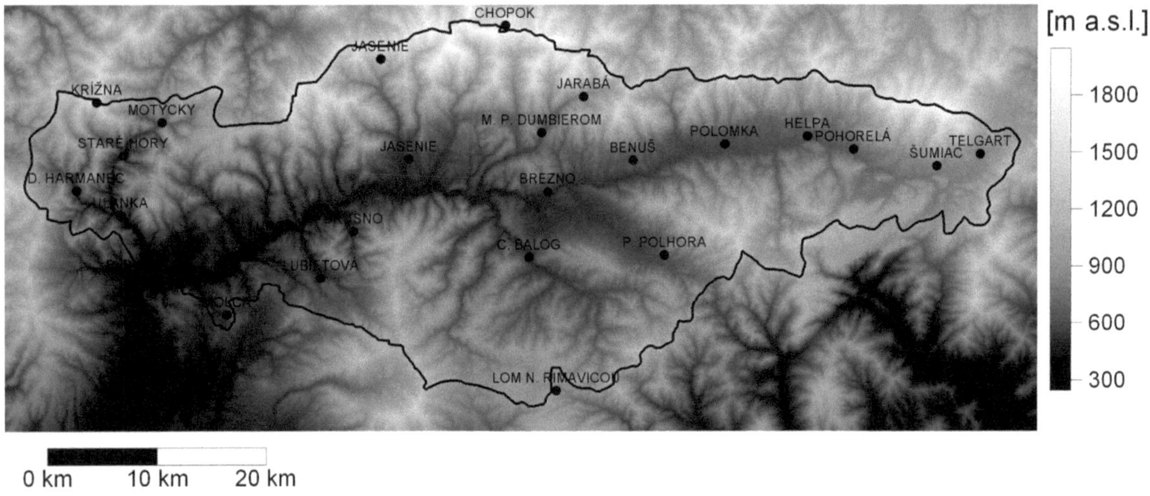

Fig. 1 The topography of the upper Hron basin and locations of precipitation stations

Here, in order to interpolate the mean daily precipitation, two interpolation algorithms were examined: the ordinary kriging and the IDW method. The IDW method is a deterministic approach that estimates the mapped variable at an ungauged site using a distance-weighted average of the data points. The weights w_i are determined as a decreasing function of the distance between the interpolated point x_0, y_0 and the points with measurements x_i, y_i (for details, see e.g., Tabios and Salas 1985); in this study the power parameter was set to be equal to 2. The kriging approach is a stochastic interpolation method that uses a statistical relationship (spatial correlation) between the values at sampled points to determine the weights w_i. In this approach, the weights w_i are assessed by minimizing the variance of the interpolation error (for details, see e.g., Isaaks and Srivastava 1989). The ability to assess an interpolation error is unique to stochastic interpolators and demonstrates the advantage of kriging over simpler deterministic interpolation methods.

Input Data

The upper Hron River basin to the Banská Bystrica profile was selected as the pilot basin. This region is located in central Slovakia. The area of the basin is 1763 km^2, and the elevations range from 340 m a.s.l. in the basin's outlet to more than 2000 m a.s.l. in the northern part of the catchment. The dataset includes daily precipitation totals measured at 25 precipitation stations and the period studied was from January 1961 to December 2000. The location of the stations with the topography and vertical zonality of the region is presented in Fig. 1.

Results of Mapping of the 100-Year Maximum Daily Precipitation Totals

For the interpretation of the spatial variability in the 100-year values of maximum daily precipitation totals, the following steps were taken: first, the mean daily precipitation totals were spatially interpolated. As a result, the grid maps (500 × 500 m) of the daily precipitation totals were constructed for the study period 1961–2000. Next, a statistical estimation of the design values of the maximum daily precipitation totals was performed at each individual grid point. The statistical estimation itself consists of the determination of the maximum daily precipitation totals for each year, the estimation of the parameters of the selected extreme value distribution and the calculation of the design values for the user-defined return periods. The Van Montfort test was then applied to select the appropriate EV1 or EV2 extremal distribution function. From the test results presented in Fig. 2 we can conclude that for approximately 88% of the upper Hron River basin, the EV1 (Gumbel distribution) was selected as appropriate and unified one and was used for the whole region.

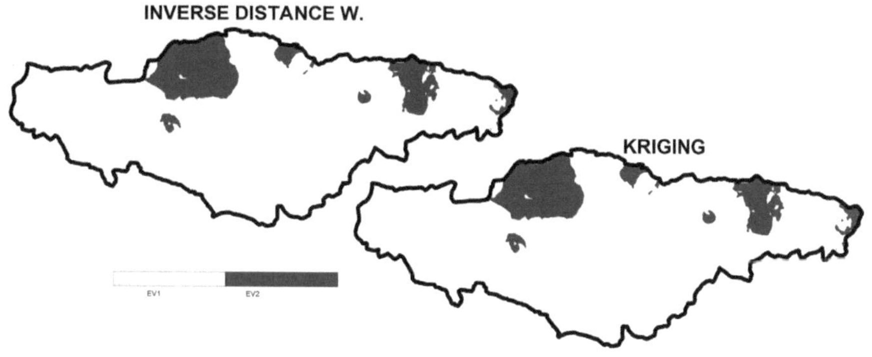

Fig. 2 Spatial interpretation of the EV1 and EV2 distributions according to the Van Montfort test

The EV2 distribution was the most suitable one for the region of Jaseniansky Creek and the Heľpa-Pohorelá region.

Using the Equations (2) and (3) the parameters of the EV1 distribution function were estimated. The comparison of the values of the μ and c parameters of the EV1 distribution function derived using the kriging and IDW interpolation method is presented in Fig. 3.

The cumulative distribution functions of the μ and c parameters over the upper Hron River basin area in Fig. 3 show that the values of parameter c are very similar using both kriging and the IDW method. The similar result was also observed by comparison of cumulative distribution functions of parameter c values over the upper Hron basin in different elevation zones (Fig. 4). For the parameter μ the IDW interpolation method gives slightly higher values in comparison to the kriging interpolation method in 70% of the basin area. Similar results can also be seen from Fig. 4, when comparing cumulative distribution functions of parameter μ values over the upper Hron basin in the elevation zones 500–1500 m. In the highest elevation zone, the IDW interpolation method underestimates the values of both EV1 parameters when compared to the kriging interpolation method.

In a subsequent step, the mapped μ and c parameters of the EV1 distribution function were used to construct a map of the 100-year maximum daily precipitation totals (Fig. 5).

Evaluation of Results and Conclusion

From the spatial distribution of the 100-year maximum daily precipitation totals in Fig. 5, it is evident that the highest values of around 100–120 mm are found in the Jaseniansky Creek basin. In the region of Starohorské Mountains and the spring region of the Hron River, the 100-year design maximum daily precipitation totals are around 90 mm. In the central part of the upper Hron River basin and in the Čierny Hron River basin the 100-year design precipitation totals are around 70 mm.

Subsequently, maps of the 100-year design precipitation totals derived in this study were compared with the expert hand-drawn isohyets map of the 100-year maximum daily precipitation totals

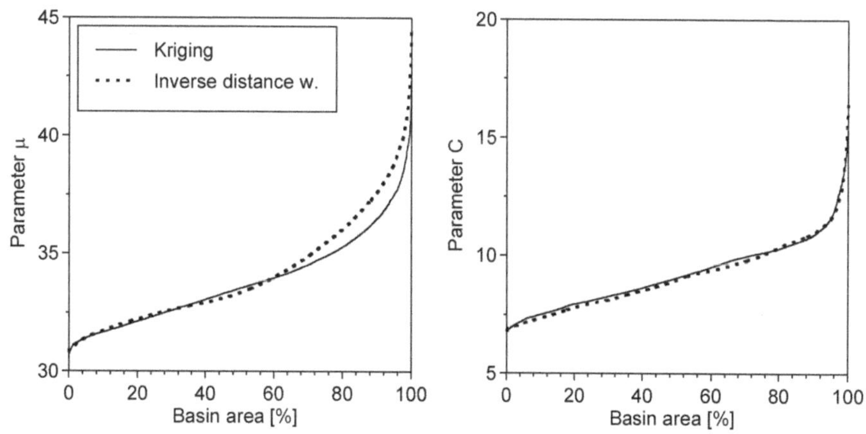

Fig. 3 Cumulative distribution functions of the μ and c parameters of the EV1 distribution function over the upper Hron basin

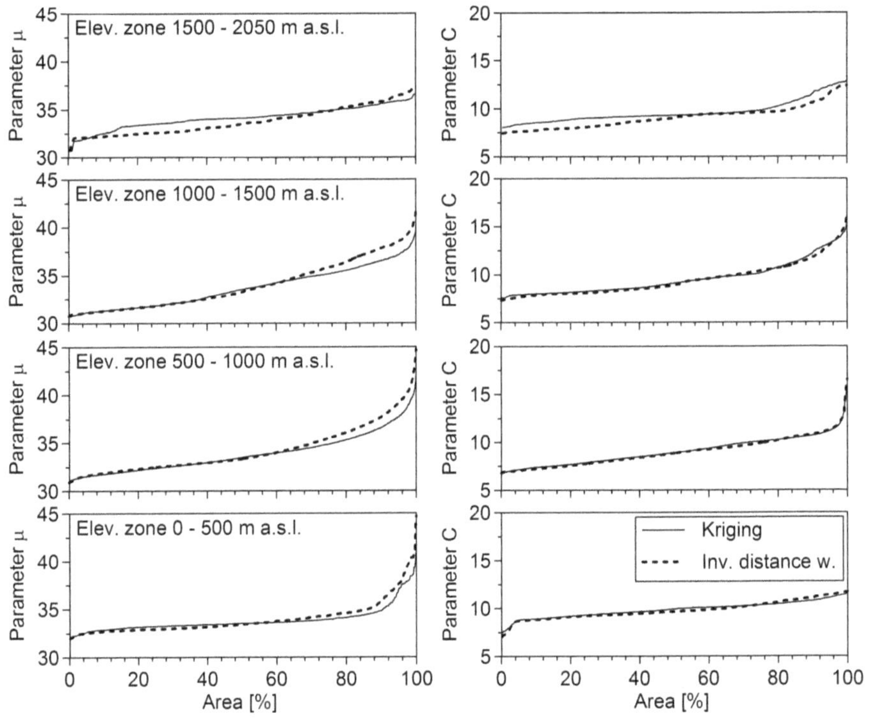

Fig. 4 Cumulative distribution functions of the μ and c parameters of the EV1 distribution function over the upper Hron basin in selected elevation zones

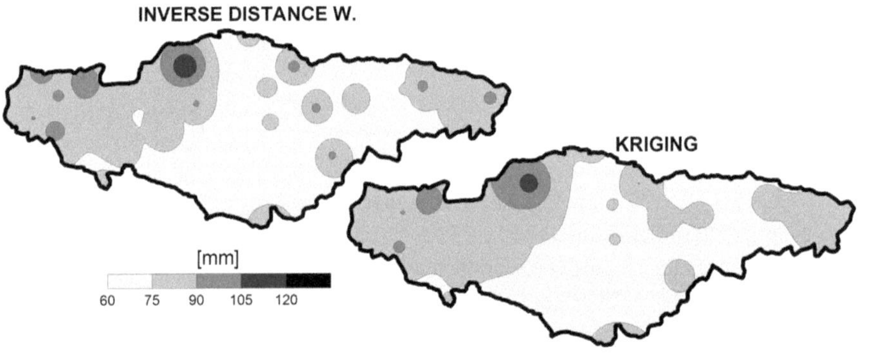

Fig. 5 Maps of 100-year maximum daily precipitation totals constructed using the kriging and IDW interpolation method

constructed by Faško and Lapin (Faško et al. 2000; Gaál et al. 2004), which can be considered to be more objective in a terrain with complex orography. A comparison of the relative catchment area belonging to a certain range of percentage differences between the maps of the 100-year maximum daily precipitation totals derived by these two approaches is presented in Table 1.

Differences from the expert-drawn map are less than ±30% in most of the basin's area. Differences in the category from 0 to ±10% amount to 29.7% of the basin's area for the kriging and 29.1% of the basin's area for IDW method. Differences in the range of (−10 to −20%) are on 47.2% of the basin's area for kriging, 43.2% of the basin's area for IDW. The differences

Table 1 Comparison of the catchment area (in %) belonging to a certain range of percentage differences between the maps of 100-year maximum daily precipitation totals derived using the kriging and IDW method and the expert-drawn isohyets map (Faško et al. 2000)

Category of relative differences (%)	Percentage of the basin's area (%)	
	Kriging interpolation method	IDWinterpolation method
Less (−30)	5.4	6.4
(−30)–(−20)	17.2	19.7
(−20)–(−10)	47.2	43.2
(−10)–0	24.5	21.6
0–10	5.2	7.5
10–20	0.6	1.4
20–30	0.0	0.1

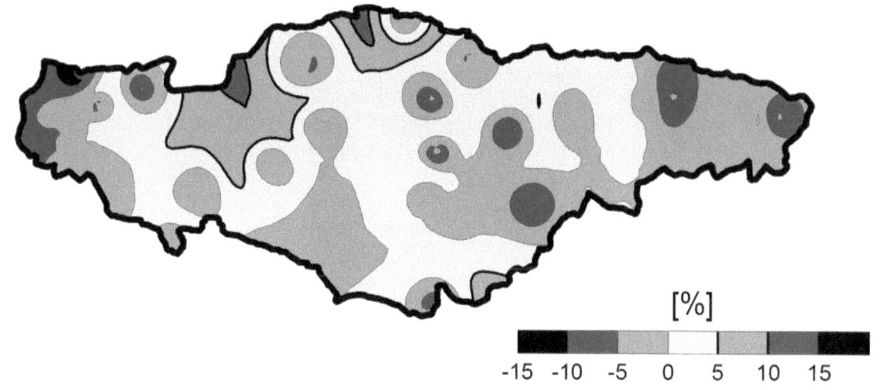

Fig. 6 Percentage differences between maps of 100-year maximum daily precipitation totals (derived in this chapter) interpolated by the kriging and IDW methods

higher than 20% take 19.6% of the basin's area for kriging and 26.1% of the basin's area for IDW. From the results we can conclude that the values of the 100-year maximum daily precipitation totals constructed by the expert hand-drawn isohyets map are higher than the values calculated using the approach presented in our study in approximately 70% of the catchment area. The differences compared to the expert-drawn map tend to be larger in mountainous areas with complex orography.

In Fig. 6 the percentage differences between the maps of the 100-year maximum daily precipitation totals (derived in this chapter) interpolated by the kriging and IDW methods are presented. The comparison showed very similar values of design precipitation in localities around the climate stations. Different results were achieved in locations without the direct measurement of precipitation; the highest differences were around 10–15%.

The benefits of the presented approach in comparison with classical approaches is the ability to map design maximum daily precipitation totals directly from an interpolated time series of annual maximum daily precipitation totals. Using this methodology it is possible to take into account the physical, regional, and seasonal characteristics of extreme precipitation totals in a region. On the other hand, this approach underestimates the values of design precipitation totals at localities with a lack of precipitation stations. This fact results especially from the underestimation of interpolated daily precipitation totals in the case of local extreme rainfall events.

Acknowledgments This work was supported by the European Commission's Sixth Framework Programme through a grant to the budget of the STREP Project HYDRATE, Contract GOCE 037024. The authors also gratefully acknowledge the Slovak VEGA Grant Agency for the support of grants 1/4024/07 and 1/4209/07 as well as the use of Slovak Hydrometeorological Institute data.

References

Barthels H, Malitz G, Asmus S, Albrecht FM, Dietzer B, Günther T, Ertel H, (1997) Starkniederschlagshöhen für Deutschland, KOSTRA. Deutscher Wetterdienst – Hydrometeorologie, Offenbach, Selbstverlag

Borga M, Vizzacaro A (1997) On interpolation of hydrologic variables: formal equivalence of multiquadratic surface fitting and kriging. Journal of Hydrology, 195 (1–4):160–171

Casas CM, Herrero M, Ninyerola M, Pons X, Rodriguez R, Riusb A, Redanoe A (2007) Analysis and objective mapping of extreme daily rainfall in Catalonia. International Journal of Climatology, 27:399–409

Dubois G (1998) Spatial interpolation comparison 97: foreword and introduction. Journal of Geographic Information and Decision Analysis, 2:1–10

Dzubák M (1969) To the estimation of distribution function of maximum daily precipitation totals. Journal of Hydrology and Hydroinformatics, 28(3):209–225 (in Slovak)

Faulkner DS, Prudhomme C (1998) Mapping an index of extreme rainfall across the UK. HESS, 2(2 -3):183–194

Faško P, Lapin M, Šťastný P, Vivoda P (2000) Maximum daily sums of precipitation in Slovakia in the second half of the 20th century. In: Images of Weather and Climate. Prace Geograficzne, fasc. 108. Cracow. 131–138

FEH – Flood estimation handbook (1999) Institute of Hydrology, Wallingford. vol. 1–5, 325pp

Ferrari E (1994) Regional rainfall and flood frequency analysis in Italy. In International Conference on "Developments in Hydrology of Mountainous Areas", Stará Lesná, Slovakia, 12–16 September 1994, 45–52

Gaál L (2006) Estimation methods of statistical properties of short-term to several-day design precipitation in Slovakia. PhD thesis, Faculty of Mathematics, Physics and Informatics, Comenius University, Bratislava, 220pp

Gaál L, Lapin M, Faško P (2004) Maximum 1- to 5-day precipitation totals in Slovakia. In.: Rožnovský, J.,

Litschmann, T. (ed): Proceedings of Seminary "Extreme Weather and Climate", Brno, 11. March 2004, ISBN 80-86690-12- 1, 15pp. on CD (in Slovak)
Geiger H, Stehli A, Castellazzi U (1986) Regionalisierung der Starkniederschläge und Ermittlung typischer Niederschlagsganglinien. Beitrage zur Geologie der Schweiz – Hydrologie, 33:141–193
Goovaerts P (2000) Geostatistical approaches for incorporating elevation into the spatial interpolation of rainfall. Journal of Hydrology, 228:113–129
Hosking JRM, Wallis JR (1997) Regional Frequency Analysis. Cambridge University Press, Cambridge, UK
Institutions of Engineers (1977) Australian rainfall and runoff. Flood analysis and design. Australia (IEA), Canberra, 149pp
Isaaks EH, Srivastava RM (1989) An introduction to applied geostatistics. Oxford University Press, New York, 561pp
Katz RW, Parlange MB, Naveau P (2002) Statistics of extremes in hydrology. Advances in Water Resources, 25:1287–1304
Koutsoyiannis D (2004a) Statistics of extremes and estimation of extreme rainfall: I. Theoretical investigation. Hydrology Science Journal, 49:575–590
Koutsoyiannis D (2004b) Statistics of extremes and estimation of extreme rainfall: II. Empirical investigation of long rainfall records. Hydrology Science Journal, 49:591–610
Loukas A, Vasiliades L, Dalezios NR, Domenikiotis C (2001) Rainfall-frequency mapping for Greece. Physics and Chemistry of the Earth (B), 26/9:669– 674
Malitz G (2005) Grundlagenbericht über Starkniederschlagshöhen in Deutschland, (Grundlagenbericht KOSTRA-DWD-2000). Deutscher Wetterdienst – Hydrometeorologie
Meijerink AMJ, De Brower HAM, Manaertz CM, Valenzuela CR (1994) Introduction to the use of geographic information systems for practical hydrology. ITC publication No. 23, The Netherlands
Parajka J, Kohnová S, Szolgay J (2002) Spatial interpolation of N-year maximum daily precipitation totals in the upper Hron River basin by stochastic interpolation methods. Acta Hydrologica Slovaca, 3(1):35–45 (in Slovak)
Parajka J, Kohnová S, Szolgay J (2004) Spatial interpolation of distribution function parameters of maximum daily precipitation totals in the upper Hron River basin. Acta Hydrologica Slovaca, 5(2): 258–265 (in Slovak)
Reed DW, Faulkner DS, Stewart EJ (1999) The FORGEX method of rainfall growth estimation II: Description. Hydrology and Earth System Sciences, 3:197–203
Reinhartová J (1967) Maximum daily pecipitation totals in Czech Republic. In: Meteorologické správy, 20(3–4):75–79. (in Czech)
Šamaj F, Valovič Š, Brázdil R, Gulčíková V (1982) Maximum daily pecipitation totals in Czechoslovakia. Meteorologické zprávy, 35, 3 (in Slovak)
Šamaj F, Valovič Š, Brázdil R (1985) Daily extreme pecipitation totals in Czechoslovakia during 1901–1980. Zborník prác SHMÚ, No 24. ALFA, Bratislava. (in Slovak)
Tabios GQ, Salas JD (1985) A comparative analysis of techniques for the spatial interpolation of precipitation. Water Resources Bulletin, 21
Thompson CS (2002) The high intensity rainfall design system: HIRDS (Abstract). In: International Conference on Flood Estimation, Berne, Switzerland
Wallis JR, Schaefer MG, Barker BL, Taylor GH (2007) Regional precipitation-frequency analysis and spatial mapping for 24-hour and 2-hour durations for Washington State. Hydrology and Earth System Sciences, 11(1): 415–442
Watkins Jr. DW, Link GA, Johnson D (2005) Mapping regional precipitation intensity duration frequency estimates. Journal of the American Water Resources Association, 41(1): 157–170
Weisse AK, Bois P (2001) Topographic effects on statistical characteristics of heavy rainfall and mapping in the French Alps. Journal of Applied Meteorology 40(4): 720–740
Weisse AK, Bois P (2002) A comparison of methods for mapping statistical characteristics of heavy rainfall in the French Alps: The use of daily information. Hydrological Sciences Journal, 47(5):739–752

Flood Prevention and Nature Conservation – Interdisciplinary Evaluation of Land Use Scenarios for an Agricultural Landscape

E. Richert, S. Bianchin, H. Heilmeier, M. Merta and Ch. Seidler

Keywords Landscape assessment · Landscape metrics · Hydrological modelling · Runoff generation · Mountainous catchments

Introduction

Flood prevention and nature conservation are often considered not to be compatible. The Weißeritz catchment (eastern Erzgebirge, Saxony, Germany) was heavily affected by floods in August 2002. In response to this event, the German Environmental Foundation (Deutsche Bundesstiftung Umwelt, DBU) funded the project 'Flood Prevention and Nature Conservation in the Weißeritz Area' (HochNatur) which aimed to design measurements integrating both flood prevention and nature conservation. Thus, land-use changes such as extensification of grasslands, transformation of arable fields into grasslands, partial afforestation and establishment of small landscape structures like hedgerows were in the focus of the project. In the following results on the evaluation of various land-use scenarios considering both flood prevention and nature conservation will be presented, using two sub-catchments with contrasting land use and biotope patterns as an example.

E. Richert (✉)
Technische Universität Bergakademie Freiberg,
Interdisciplinary Environmental Research Centre, Freiberg, Germany
e-mail: richert@ioez.tu-freiberg.de

Investigation Area

The Weißeritz catchment in the eastern Erzgebirge declines from about 800 m above sea level (a.s.l.) in the mountain ranges down to 200 m a.s.l. in the northern foreland. The two sub-catchments selected are the Weißbach (WB, 630–800 m a.s.l.) and Höckenbach (HB, 350–500 m a.s.l.) some 10 km apart and therefore there is no for the scenarios relevant interaction between the sub-catchments. The WB sub-catchment is characterized, apart from forests (24% of the total area), by grassland (42% of the area), which is mainly extensively used (33% of the area). Many endangered species and biotope types are present, especially along the rivulet corridor. In contrast, the HB sub-catchment is dominated by arable fields (69% of the area), whereas grasslands and forests occur on 6% and 13% only. Endangered species or biotope types are extremely rare.

Methods

A detailed survey of the present state of the two sub-catchment areas HB and WB with respect to landscape ecology and hydrology via systems analysis and modelling was performed using a method transferable to other mountainous regions. On this foundation different land-use scenarios were developed (Table 1) and evaluated both from the flood prevention and nature conservation perspective (Fig. 1). These scenarios considered land-use changes such as extensification of grasslands, transformation of arable fields into grasslands, partial afforestation (potential natural vegetation), conservation tillage and

K. Strelcová et al. (eds.), *Bioclimatology and Natural Hazards*,
DOI 10.1007/978-1-4020-8876-6_12,

Table 1 Compilation of the land-use scenarios analysed in the project 'HochNatur' (Flood Prevention and Nature Conservation in the Weißeritz Area), their abbreviations and the area percentages affected by land-use changes in comparison to the present state in the sub-catchments HB and WB

Scenario	Abbreviation	HB	WB
Present land use	Pres	–	–
Arable field into grassland	a-g	69.0	16.3
Conservation tillage	c_till	69.0	21.0
Partial afforestation (areas with quick runoff components)	p_aff	25.9	24.9
Extensification of grasslands	g_ext	n.a.	9.8
Nature conservation measures	Nat	83.7	53.8
Flood prevention measures	flood	82.2	51.8
Combination of nature conservation and flood prevention measures	comb	82.9	51.3

n. a. = scenario not analyzed

the establishment of small landscape structures like hedgerows (Richert et al. 2007).

The main focus of the ecological analysis was on high-resolution biotope mapping and the assessment of the present state and the developed scenarios (Bianchin et al. 2007; Richert et al. 2007). The assessment of the various biotope types was done with the help of three evaluation criteria, naturalness, substitutability, and rareness and endangerment. However, the assessment of biotope types only does not yield any information about their spatial distribution and structural composition of the landscape. Therefore an assessment for the whole landscape through landscape metrics was necessary to analyse the structural and biotope type diversity at the landscape level. For this analysis the Shannon–Weaver diversity index, the mean patch size index (Farina 2000) as well as the interdispersion–juxtaposition index were calculated (Bianchin et al. 2007). In order to compare the different land-use scenarios with each other and the current state a ranking system was used. Within a last step the results were weighted according to the percentage of the area with high conservation value.

For the analysis of the hydrological situation in the project area, three tightly coupled models were used. First, the expert system WBS FLAB – areas of equal runoff components – identified areas with fast runoff components (surface runoff, saturation overland flow, fast interflow) on the basis of landscape characteristics such as soil type, land use and slope angle (Merta et al. 2003, 2006, 2007b; Schulla and Jasper 2006; Zimmermann et al. 2001; Peschke et al. 1999). The results were used to parameterise the afterwards following runoff-precipitation models WaSiM-ETH (Schulla 1997) and SWMM (United States Environmental Protection Agency 2005), which were used to quantify the runoff of the respective sub-catchment.

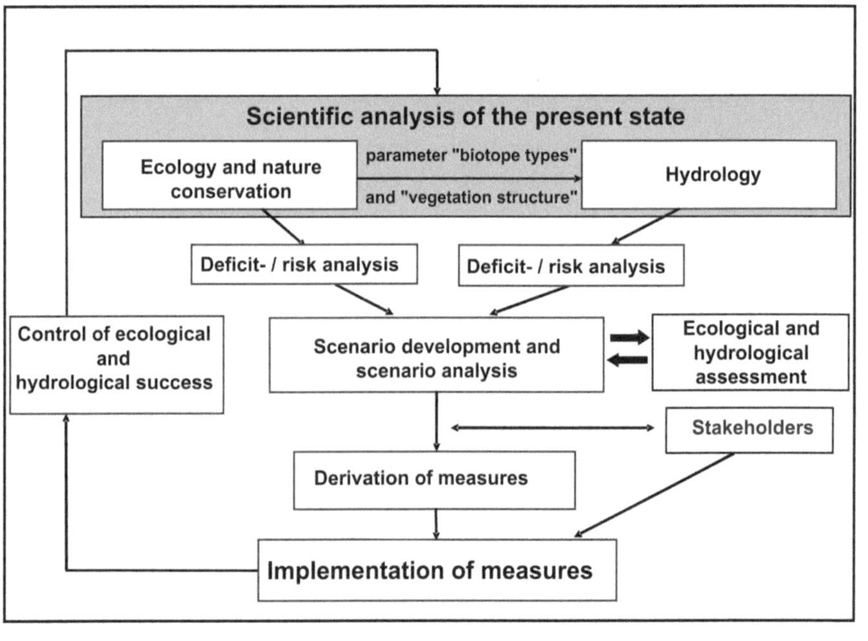

Fig. 1 Main steps in the project 'HochNatur'

Results and Discussion

Apart from the present state, 7 and 6 scenarios, respectively, were analysed for each sub-catchment (Table 1). The scenario 'extensification of grasslands' was not defined for the sub-catchment HB due to the small percentage of grasslands in this area. The scenarios 'nature conservation measurements' and 'flood prevention measurements' were elaborated nearly exclusively with respect to the relevant objective only.

The scenario 'combination of nature conservation and flood prevention measures' attempted to consider both aims as much as possible, on the basis of various guidelines (Richert et al. 2007). The assessment of the scenario 'arable fields with conservation tillage' according to landscape metrics was not meaningful since the borderlines and spatial distribution of the fields did not change with respect to the present state.

From the nature conservation point of view for both sub-catchments all scenarios resulted in higher scores compared to the present state (Figs. 2 and 4). The scenarios 'arable field into grassland' (a-g), 'combination of nature conservation and flood prevention measures' (comb) and 'nature conservation measures' (nat) all resulted in high scores (mostly > 50%). For these scenarios, a large amount of area was affected by land-use changes (Table 1). Due to the greater amount of area affected for the scenario 'arable field into grassland' in the HB (69%) than in the WB (16%) sub-catchment, scores are higher in the former. In comparison to the present state higher scores were also allocated to the scenarios 'flood prevention measures' (flood), but for both sub-catchments these scenario received the lowest scores of all. The extreme low scores for the WB sub-catchment are a result of the loss of biotope and landscape diversity and heterogeneity caused by, for example large-scale afforestation.

The reduction of area proportions with fast surface and sub-surface flow components was significantly correlated with the reduction of peak discharge in the river ($p < 0.05$). From the flood protection point of view especially the scenarios with large changes in land use (flood prevention measures, nature conservation measures, combination of nature conservation and flood prevention measures; 50–85%, Table 1) resulted in marked improvements in comparison to the present state, with the effects in the HB sub-catchment being more pronounced than in the WB sub-catchment (Figs. 3 and 4). The combination of flood prevention measures focused at the specific on-site problems will yield in high improvements (Merta 2007a). Moreover, these scenarios for flood prevention measures showed high scores from the nature conservation point of view in both sub-catchments also (Fig. 4). Similarly, the scenarios focused on nature conservation measures and on the combination of nature conservation and flood prevention measures markedly reduced the proportion of areas with fast surface and sub-surface flow components in both sub-catchments (Fig. 3). Conservation tillage, which does not change the proportion of arable fields in contrast to other scenarios, has a high relevance with respect to flood prevention and is as effective as conversion of arable fields into grassland. From the nature conservation point of view, the conservation tillage scenario slightly gains compared to the present state due to its soil protection effect.

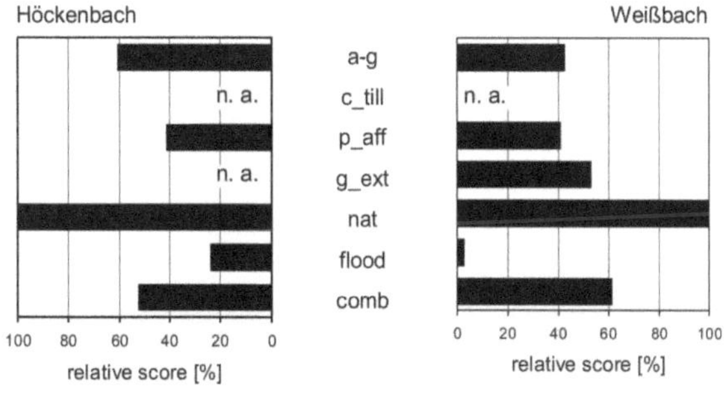

Fig. 2 Results from the evaluation of land-use change according to nature conservation criteria in relative scores (scenario 'present land use' = 0%, scenario 'nature conservation measures' = 100%) for the sub-catchments HB and WB (for abbreviations of scenarios refer to Table 1)

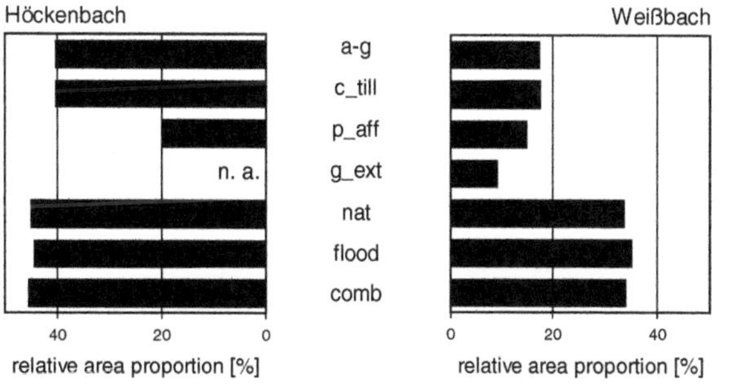

Fig. 3 Results from the hydrological assessment for the sub-catchments HB and WB. Numbers indicate the reduction of area percentages with fast surface and sub-surface flow components in comparison to the present state (for abbreviations of scenarios see Table 1)

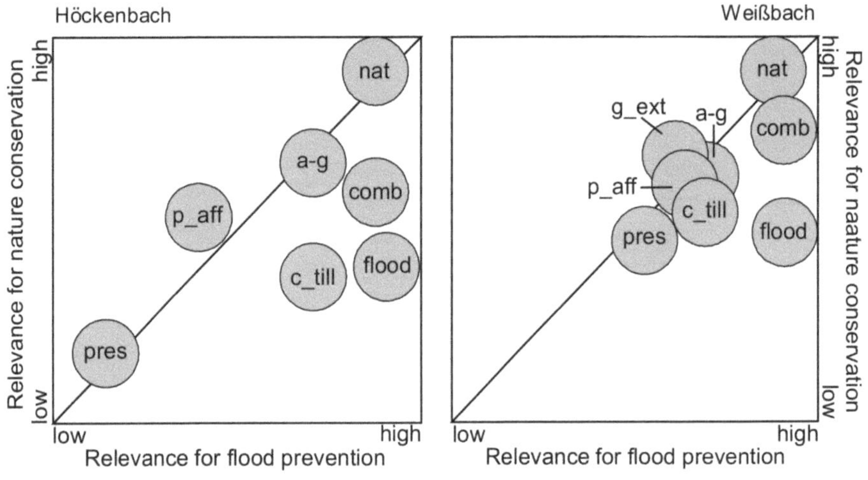

Fig. 4 Combined assessment of land-use scenarios with respect to flood prevention and nature conservation (for abbreviations of scenarios refer to Table 1)

In contrast to measures involving agricultural land use only, afforestation on sites with fast runoff components are less effective in reducing the proportion of area on which floods may be generated. Although the scenario for partial afforestation (p_aff) changed land use on 25% of the total sub-catchment area, especially in the WB sub-catchment there was a smaller reduction in the area proportion with fast surface and sub-surface flow components compared to measures on agricultural land (a-g, c_till) which affected 16% respectively 21% of the sub-catchment area only (Fig. 3). This is due to the very shallow soils on steep slopes in the upper ranges of the Erzgebirge with their low capacity for water storage. Therefore, in spite of the higher demand for agricultural land, the combined effect of partial afforestation on flood prevention and nature conservation is less than transforming arable fields into grasslands in the WB sub-catchment with its high abundance of endangered species and biotope types (Fig. 4).

Importantly, measures aimed at flood prevention synergistically interact with nature conservation (habitat and species diversity, connectivity), landscape conservation and aesthetics (tourism and recreation potential), and soil protection (erosion). Moreover, these measures contribute to a balanced regional hydrological budget, which can mitigate negative consequences of summers with low precipitation.

Conclusions

The integrative assessment of scenarios of land-use changes aimed at both flood prevention and nature conservation in a mountainous area has shown that through a wide spectrum of measures on varying proportions of the area of individual sub-catchments dramatic improvements for both objectives can be achieved. Even scenarios with measures directed exclusively to nature protection yielded in a reduction in the extent of areas with fast surface and sub-surface flow and reduced flood peaks in rivers (Merta et al. 2007b). Similarly, land-use management designed with respect to flood prevention had positive effects on nature protection. Highest effects were associated with land-use changes on large area proportions. However, also single measures like the establishment of hedgerows may be positive both from the nature conservation and flood prevention perspective (Richert et al. 2007). They affect especially the local habitat, for example by reducing soil erosion.

The analysed effects of land-use changes were the greatest in small-to-medium sized catchments and in the case of precipitation events with 5–50 years reoccurrence intervals (Merta et al. 2007b). In large catchments, the temporal and spatial multiplicity of processes in the different parts overlap and therefore it depends on these interactions whether the discharge of the whole catchment is influenced or not. Finally the flow processes in the river bed itself become more important.

The effect of land-use changes heavily depends on the specific conditions of the landscape such as the presence and spatial arrangement of habitat and landscape elements with high relevance for nature conservation, such as vegetation structure (density, height, root depth, etc.) with relevance for flood prevention (Merta et al. 2007b). Therefore results from the individual scenarios developed for the two sub-catchments cannot be transferred to other catchments. However, the methods developed for the assessment both from the nature conservation and flood prevention perspective can be transferred to other regions provided that necessary data such as a digital landscape model, type and distribution of land use and soil characteristics are available.

Acknowledgments This project was funded by the German Environmental Foundation (Deutsche Bundesstiftung Umwelt, DBU).

References

Bianchin S, Richert E, Heilmeier H, Merta M, Seidler Ch (2007) Landscape metrics as a tool for conservation – Assessment of scenarios for flood and nature protection. Landscape Online (submitted)

Farina A (2000) Landscape ecology in action. Kluwer, Dordrecht

Merta M, Hammer G, Seidler Ch, Richert E, Bianchin S (2007a) Analyse von Landnutzungsänderungen unter hochwasser- und naturschutzfachlichen Aspekten im Einzugsgebiet der Weißeritz (Osterzgebirge). Forum für Hydrologie und Wasserwirtschaft 20: 247–258

Merta M, Seidler Ch, Bianchin S, Heilmeier H, Richert E (2007b) Analysis of land use change in the Eastern Ore Mountains regarding both nature protection and flood prevention. Journal of Soil and Water Research 3(Special Issue 1):105–115

Merta M, Seidler Ch, Bianchin S (2006) A knowledge based system as decision support for landscape management in catchments. Biohydrology-Symposium: Impact of Biological Factors on Soil Hydrology, Prag

Merta M, Seidler Ch, Uhlenbrook S, Tilch N, Zillgens B, Kirnbauer R (2003) Das Wissensbasierte System FLAB als Instrument zur prozessbezogenen Raumgliederung von mesoskaligen Einzugsgebieten. In: Kleeberg H-B (ed.) Klima-Wasser-Flussgebietsmanagement – im Lichte der Flut. Beiträge zum Tag der Hydrologie am 20./21. März 2003 in Freiburg. Forum für Hydrologie und Wasserbewirtschaftung 1: 171–178

Peschke G, Etzenberg C, Töpfer J, Zimmermann S (1999) Das wissensbasierte System FLAB – ein Instrument zur rechnergestützten Bestimmung von Landschaftseinheiten mit gleicher Abflussbildung, IHI-Schriften 10

Richert E, Bianchin S, Hammer G, Heilmeier H, Matschullat J, Merta M, Seidler Ch (2007) Hochwasser- und Naturschutz im Weißeritzkreis (HochNatur). Final report to the Deutsche Bundesstiftung Umwelt (DBU), Osnabrück

Schulla J (1997) Wasserhaushalts-Simulations-Modell WaSiM-ETH – Anwenderhandbuch. ETH Zürich

Schulla J, Jasper K (2006) Model Description WaSiM-ETH. Internal report, IAC, ETH Zürich, 174pp. last update December 2006

United States Environmental Protection Agency (EPA) (2005) SWMM. Storm Water Management Model – User's Manual Version 5.0. EPA/600/R-05/040

Zimmermann S, Töpfer J, Peschke G (2001) A knowledge-based system to improve the processing of distributed precipitation-runoff models. In: "Runoff generation and implications for river basin management/modelling". Proceed. of the IAHS/ICT/ICSW Freiburg workshop October 9 – 12, Freiburger Schriften zur Hydrologie 13:175–182

Part III
FOREST BIOCLIMATOLOGY, NATURAL HAZARDS AND MODELLING

Risk Assessment of the Tatra Mountains Forest

P. Fleischer, M. Koreň, J. Škvarenina and V. Kunca

Keywords Tatra Mountains forest · Forest history · Insect outbreaks · Down-slop wind · Windfall · Ecological resistance

Introduction

Forest is the most outstretched natural element in the Tatra Mountains. Continuous forest belt covers an area of nearly 60,000 ha on both, Slovak and Polish sides. In addition to that, shrubby stands at timberline, formed by mountain pine (*Pinus mugo*), covers another 10,000 ha.

The Tatra Mountains is the most northern and also the highest part of the Carpathian mountain range. The mountain chain is oriented east–west and located almost in the centre of Europe. Steep peaks are surrounded by large basins. This 'Geographical island' constellation is one of the reasons for unique climate with frequent extreme meteorological changes; it has a notable impact on local vegetation, especially the forest. Altitudinal span (from 600 to 2600 m a.s.l.) provides space for presence of four altitudinal vegetation ranks. Together with wide spectrum of geoecological types, it forms conditions for unique biological and landscape diversity of this territory. On the other hand, tree-species composition is very simple due to glaciations of the Tatra Mountains during Pleistocene era, trees species spreading in Holocene and current climate conditions. Spruce-dominated stands occur mostly on the northern boreal forest instead of the central European mountain forest. This evaluation is confirmed by the following: natural occurrence of continental species like *Pinus cembra* and *Larix europea*, low percentage of *Abies alba* and broadleaf species, low occurrence of *Fagus sylvatica* and by absolute dominance of *Picea abies*.

Due to the 'island effect', strong winds occur frequently; the strongest winds reaching speed above 10 m/s and occurring more than 200 days a year. Prevailing wind direction is west–east.

Unique landscape of the high mountains and well-preserved countryside were the reasons for its declaration as a National Park in 1949. The forest has played an important role in protected area covering nearly 75% of the National Park. Relatively high degree of forest's naturalness provides habitats for existence of native and often rare and endangered species. Important is also hydrological function for water runoff regulation, especially after heavy summer storms and melting snow in spring. Climate conditions, favourable for establishing and developing local health centres and sanatoria, depend on forest's effect of reducing extreme weather conditions, as well as on concentration of fytoncides. The forest is not only an aesthetic background but it also offers space for recreation and sport. Multi-purpose function of the National Park relies on functional and stable forest ecosystems. Moreover, the Tatra forest also has an economical value. Nature-close forestry, which rehabilitates degraded status to more natural ones, has some commercial use. Sustainable use of the Tatra forest was approved in 1993 when bilateral biosphere reserve the Tatras was placed on the UNESCO Man and Biosphere list. Despite its legal protection, the status of the forest has remarkably deteriorated in last 25 years. Both aerial- and

P. Fleischer (✉)
Research Station of the Tatra National Park,
State Forest of TANAP, 059 60 Tatranská Lomnica, Slovakia
e-mail: fleischer@post.sk

K. Strelcová et al. (eds.), *Bioclimatology and Natural Hazards*,
DOI 10.1007/978-1-4020-8876-6_13,

ground-monitoring projects in the late 1980s confirmed increased defoliation, elevated concentration of heavy metals in plant tissues, increased precipitation acidity and changed soil properties. Forest degradation led to increased incidental felling caused by insect and wind. The situation of the forest was suddenly dramatized by a down-slope wind called Bora in November 2004. The area of 12,000 ha was completely destroyed with total volume of 2.3 million m^3. Windfall and consequent events are affecting current status and development of the Tatra forest.

Forest History

Forest vegetation in the Tatra Mountains territory is relatively young. Its formation has started after Pleistocene era, roughly 10,000 years ago. Based on samples of dozen pollen analyses in wider vicinity of the Tatra Mountains, one can state that formerly open tree vegetation is similar to taiga or even tundra type grown only in the foothills of the mountains. Stands were formed mostly by pines with mixture of cembra and larches. Good light conditions allowed spreading of junipers, willow and birch. At the end of younger Dryas (8800–8300 y BC), hydrological conditions favoured spruce and caused its strong expansion.

During Preboreal era (8300–6800 y BC), which was the beginning of continuous warming, an open pine-birch formation spread into the altitude 900–1000 m a.s.l. During second half of this era larch and cembra became more accommodated and pressed away spruce from lower altitude. This trend continued during warm and dry Boreal (6800–5500 y BC) when spruce stands were the prevailing forest type and spread up to 1700 m a.s.l. In neighbouring basins fir percentage increased, on the warmest sites also typical species for *Quercetum mixtum*, especially elm, lime and hazel occurred.

During warm and wet Atlantic era (5500–2500 y BC) alder in lower elevation spread into dominant spruce stands. Elm, lime, hazel and oak were only marginal. Dominant species in basins as well on the mountain slopes was spruce. Timberline at the 1800 m a.s.l. was formed by pine–spruce forest with an admixture of larch. Above timberline sparse Pinus mugo belt has developed in accordance with edaphic and relief conditions. Pine percentage slowly decreased and in some localities fir and beech occurred.

During the next, even warmer, era of Subboreal (2500–800 y BC) the significant decrease in percentage of broadleaves species on the behalf of fir occurred. Climate of Subatlanticum (800 y BC–till now) was stable and equal to the current conditions. During the whole era spruce and fir forest dominated, after systematic deforestation the percentage of pine and alder increased.

Current vast spruce stands are the consequence of postglacial vegetation development in the Tatra Mountains region. Character of the forest was also formed by climate conditions. These differ from comparable mountainous regions in Slovakia mostly by high solar radiation, higher summer temperature and lower summer precipitation, as expected in their altitude. Together with prevailing acid cambisoils and podzols they form a frame for very narrow spectrum for native forest's tree species inside taiga-like forest. Permanent presence of alder, birch or poplar is only at hydrophic sites in foothills of the Tatra Mountains. Notably shorter life cycle is the typical feature of birch, rowan, larch and pine which occur like pioneer species in early succession stage after large-scale wind disturbances. The special case of forest vegetation development is the occurrence of mixed larch and pine–spruce forests which are located on sites with frequent Bora effect.

The Tatra forest has been notably impaired by man since 15–16th century with an ambition to get open space for grazing and arable land. Logging was rather limited due to low population, undeveloped transport and market networks. Logging form was limited and even individually oriented; cutting in fact did not change the forest too much. Bigger demand for wood came during formation of the Tatra settlement in 18–19th century. Parallel reforestation of clear cut areas was applied. The largest areas were forested in early 1920s after windfalls in 1915 and 1919. Systematic reforestation continued after the formation of the National Park in 1949.

Forest Status at the End of the Last Century

Mortality is a natural part of life cycle of all living organisms and systems. Also in the boreal forest, large-scale dieback is assessed as a natural part of their living strategy, in fact, it is a prerequisite for natural regeneration. Until 1980s in the Tatra forest, large-scale windfalls and consequent insect calamities occurred regularly, together with fires, allowing natural regeneration.

Occurrence and size of such disturbances was causally conditioned. Since the second half of 1980s, forest injuries have showed new symptoms. Tree crowns at the most exposed localities turned to yellow and started to loose needles. Pollution was assessed as a reason for forest dieback and later, after some years of extreme weather, synergism of more direct and indirect factors was described as a driving factor.

In the early 1990s, based on ground and aerial monitoring, chemical composition of assimilating apparatus, the status of the forest in the Tatras was assessed as ordinary, but with the worst prognosis in Slovakia (Fleischer and Koreň 1995). Ten years later, based on 180 monitoring plots for defoliation assessment, we stated that average defoliation increased from 28% up to 33%. The most alarming was the fact that observed mortality was 20%. The most damaged trees were those with the best status in the previous census.

Status Analyses of Model Community

Current status and risk assessment we performed at the end of 1990s on forest community was ranked by previous preliminary analyses as the most vulnerable. Selected model community, larch–spruce forest, Lariceto-Piceetum, covers an area of 4500 ha and in the past there were established many research plots for its observation (Fleischer 1999).

As key factors determining forest status we assessed the following:

1. internal preconditions of the forest stand for further development with emphasis on the resistance to disturbing factors;
2. external factors causing changes in the forest's status (climate, pollution, forest management activities).

Internal Criteria

Internal criteria expressed the following:

Health Status

Health status we used in the assessment, comprised combination of defoliation and discolouration, pollen and seed germination, and heavy metals concentration in plant tissues. Mean defoliation in the upper story (biosociological rank 1 and 2) corresponded to average defoliation of the Tatra forest. Stronger defoliation we found was in the main end especially in the lower story of planted stands. Defoliation above 50% was not considered as a base for healthy development of the forest stands. We assessed only standing living trees. Standing dead trees were assessed separately; they are presented for three stand-layers in Table 1. Higher concentrations of heavy metals were found in vicinity of main roads as a consequence of heavy traffic.

Incidental Felling and Bark Beetle

Due to windy conditions on the south-oriented lee side of the Tatra Mountains, incidental felling caused by wind is recorded yearly, reaching the volume of several dozen thousands cubic metres of wood. An overview of incidental felling caused by both, wind and insect is presented in Fig. 1. The graph clearly confirms the dependence of $80,000\,m^3$ insect calamity from 2004 on $120,000\,m^3$ wind calamity from 2003. From 1989, the volume of annual incidental felling is equal to half of annual increment in the Tatra forest. Wind usually affects small areas, in 80% of the events less than 0.5 ha. In the past it was an appropriate size for natural regeneration or plantation, forming the future forest more vertically and more age-structured. At the present moment new gaps and edges are the main risk for insect outbreaks. Insect is swarming even in the altitude where it has not occurred before.

Naturalness of the Forest

We analysed naturalness of the forest stands according to presence of natural and anthropogenic signs, for ex-

Table 1 Average defoliation and mortality in Lariceto-Piceetum forest community according to biosociological rank (status in 1999)

Biosociological rank	Mean defoliation	Average mortality
1	31.5	2.7
2	32.4	31.7
3	42.3	51.0
4	48.4	Not recorded

Biosociological rank: 1 – plus trees, 2 – main story trees, 3 – trees from 1.3 m up to 1/2 of the main story tree height, 4 – trees up to 1.3 m.

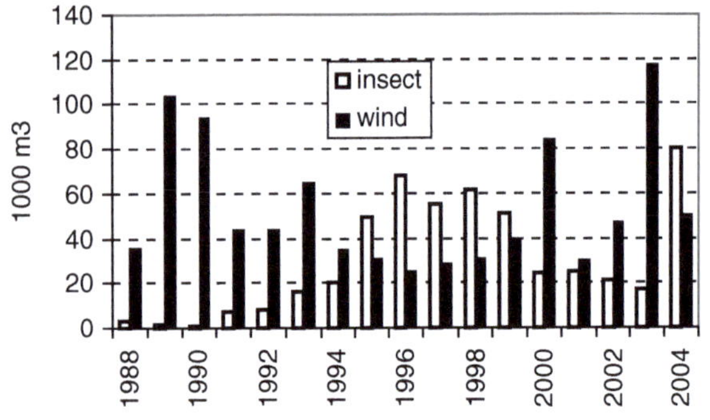

Fig. 1 Incidental felling volume (in m^3) according to a prevailing factor (wind, insect), LHC Tatry, 1988–2004 (before 19 November 2004)

ample tree and herbal species composition, dead wood, stamps and forest roads. In addition, historical drawings, photographs, maps and forestry archives were used.

According to occurrence of selected elements six categories of naturalness were chosen: (1) pristine forest – natural elements only, in past low-intensity selective cutting, (2) natural forest – native tree-species composition, partly changed spatial structure, (3) mostly natural forest – native species composition, moderately changed spatial structure, (4) slightly changed forest – both natural and anthropogenic signs are presented, natural are dominant, (5) moderately changed forest – only anthropogenic signs but nature-close outlook, (6) strongly changed forest – only anthropogenic signs and unnatural outlook, monoculture (Table 2; Fig. 2).

Table 2 Relative area of naturalness categories in Lariceto-Piceetum community

Degree of naturalness	Description	Area (%)
1	Pristine	3
2	Natural	13
3	Mostly natural	13
4	Slightly changed	23
5	Moderately changed	30
6	Strongly changed	18

Species, Age, Vertical and Spatial Structure of Forest

Species diversity as one of the traditional preconditions for higher ecological stability is loosing sense in naturally monospecies communities on the one hand. On the other hand, mixture of any, even single tree species – in this case larch, we need to accept this as existence of 'mixed' forest with expected higher resistance. In the model community more than 20% of the area covered pure spruce stands with zero percentage of larch. Nearly 40% were stands with 10% larch percentage. One-third of the area was covered by stands with 30% larch percentage. Only 5% of the area formed stands with more than 30% larch admixture. Higher heterogeneity, as a precondition for higher ecological stability, is guarantied by vertical and spatial differentiations. The most unsuitable one-layer storey covered more than 60% of the model community. Optimal, multi-storey structure covered 18%. Higher larch percentage and more diversified vertical structure played an important role for more suitable light and humus conditions, higher

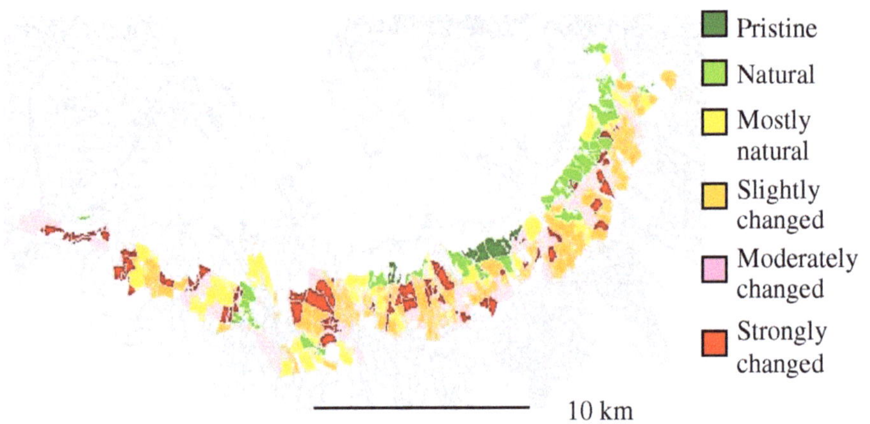

Fig. 2 Map of naturalness categories in Lariceto-Piceetum

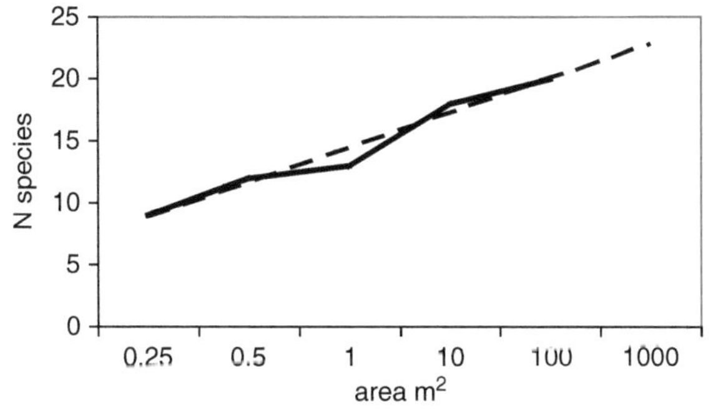

Fig. 3 Vegetation species number (per m^2) and linear trend, plot T20, 1350 m a.s.l.

ground-layer vegetation diversity and better natural regeneration. We found enormous seedlings mortality due to dense canopy closure.

High stand density was confirmed by tree numbers per hectare. We found on average 1082 individuals with DBH>8 cm. According to previous studies on forest stability (Korpel′ 1994) this number is 20% higher than accepted maximum. High numbers of trees meant also high wood volumes. According to Korepel' (1990) 250 m^3 is optimum and 500 m^3 is the critical value. We found 478 m^3 ha^{-1} as an average. In 62% of analysed cases we found more than 500 m^3 ha^{-1}. More than 85% of wood volume was concentrated in upper forest layers.

We confirmed close correlation between spatial distribution and naturalness. Using indexes by Pretzsch (1998) we confirmed long-term tendency of natural stands to form clusters, and man-made forest to keep regular distribution. Most of the analysed stands are age-classes stands, especially those which were affected by windfalls in 1920s. The areas of age-classes were nearly normally distributed along Gauss curve, confirming 100–150 year history of sylviculture in the Tatra forest.

Species homogeneity of stands affected the character of phytocenosis. In Fig. 3 we present an example of number of species per area. On average vegetation species number derived from 50 research plots was 10.3 and average vegetation cover was 67%. Mosses dominated in dense stands. Grass and herbal species dominated in more open, mixed and vertically heterogeneous stands.

Vegetation with high percentage of mesophytes and acidophilous mosses indicated moist, very acid soils with lack of nitrogen.

Geology, Soil and Humus Conditions

Model community was located on south-oriented slopes of the Tatra Mountains which are, from geological point of view, formed by moraines and polygenetic debris. Soils are oligotrophic, often podzolic cambisoils with different physical parameters depending on soil depth and particle size. Shallow, loam-sandy, often rocky soils are more suitable for larch. Deeper, more loamy and moistly soils offer opportunity for fir. We identify relatively thick humus layer (5–12 cm), moder, often mor with pH 3.3–3.7 on average. High humus percentage was also in mineral *A* horizon. Biological activity derived indirectly from cellulose test was very low, especially in mosses phytocenosis. Fulvoacids were more abundant than humin acids (Fleischer 1999).

External Factors

External factors were divided into the following categories:

Climate Conditions

Climate was analysed through mean and extreme diurnal, monthly and seasonal air temperature and precipitation. We stated important positive deviations from long-term temperature averages during growing season and often negative deviations from mean precipitation (Fig. 4a,b). Continuity of such conditions might

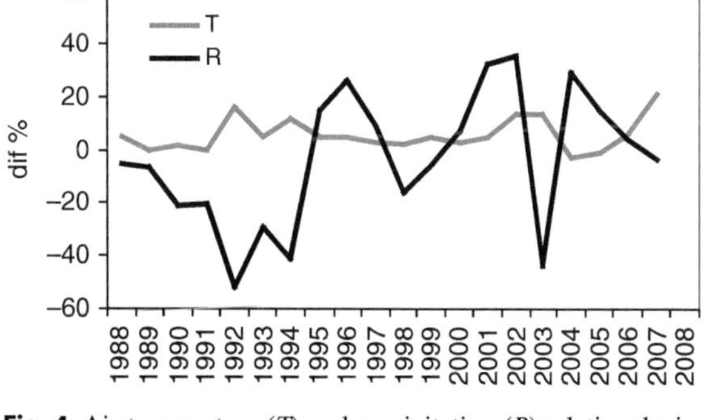

Fig. 4 Air temperature (T) and precipitation (R) relative deviation for years 1988–2007 from long-term values during growing season (V–VIII 1930–1960, T = 13.1°C, R = 413 mm)

lead, according to the CCCM scenario (Mind'áš and Škvarenina 2003), to the total collapse of analysed forest communities at height between 1000 and 1200 m a.s.l. during several decades mostly due to high interception (about 60% on average, occasionally 85%) and very sporadic natural regeneration. We found correlation between heath status and precipitation. During dry and warm periods, the health of the forest declined. Also mean defoliation was higher at rocky soils with weak retention capacity (Koreň et al. 1997).

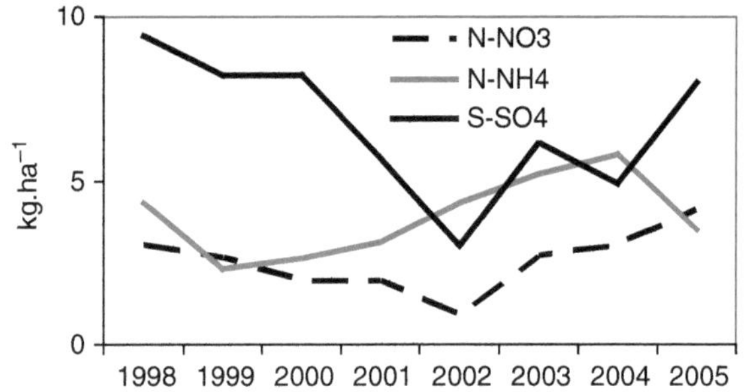

Fig. 6 Annual S–SO_4^{2-} and N–NO_3^-, N–NH_4^+ deposition in 1500 m a.s.l.

Pollution

Pollution impact was analysed by wet deposition with emphasis on pH, S and N content, nutrient leaching (Ca, Mg, K) from plant tissues and Al mobilisation in the soil. The average pH value of precipitation ranged between 3.1 and 4.5. Figure 5 shows annual pH average at two different heights. Annual H^+ input was 1–8 kg. Rain in higher altitudes is more acidic. Recent years confirmed decline of S input, but opposite is trend for N (both NO_3^-, NH_4^+) as shown in Fig. 6. We found higher acid lead in throughfall waters, coefficient of enrichment ranged between 2 and 7 (Fleischer 2005).

According to the critical load method (Kunca et al. 2003), current deposition exceeds soil buffer capacity and causes 10–30% mortality. Using the steady-state model Profile (Warfvinge and Sverdrup 1992) we assessed possible effects of atmospheric deposition on changes in soil chemistry which can influence the forest growth.

The research run at three forest sites in the Tatra Mountains with presence of Norway spruce, European larch, Arolla pine and Dwarf pine. This approach responded to the topical demand for describing the effects of acidification at the level of ecosystems. As the criteria different Bc/Al ratios were taken. In spite of evident reduction in sulphur emissions we found that the low-risk critical loads were exceeded for Norway spruce, European larch and Arolla pine at two sites. The critical deposition limits were low and calculated at interval from 1514 to 2511 eq ha^{-1} year^{-1} and the acid load values range from 1494 to 3424 eq ha^{-1} year^{-1}.

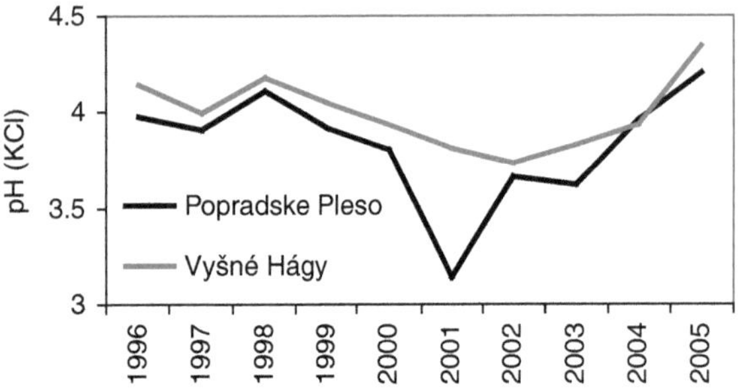

Fig. 5 Mean pH of precipitation on Popradske Pleso, 1500 m a.s.l. and Vysne Hagy, 1200 m a.s.l.

Air quality was analysed by O_3 concentration during growing season. The highest values occurred during warmer years. Recommended value 50 μg m^{-3} was exceeded continuously in the whole territory. We confirmed significant correlation between O_3 and altitude. Index AOT 40 was exceeded in 800 m a.s.l. twofold, in 1200 m a.s.l. threefold and at timber line (1500–1700 m a.s.l.) nearly fivefold. Necrotic ozone-like symptoms were recorded, for example on rowan and elder leaves (Fleischer et al. 2005).

Risk Assessment

A risk for further status and development of the forest community was understood as its resistance against known disturbing factors previously recorded. Based on the current knowledge we added some weight values to categorize the importance of each factor (Table 3).

Based on the final score we divided expected resistance into four categories. Spatial distribution of those categories fits well with the forest naturalness categories (Fig. 7). Model for the highest resistance was joined with the ecologically most stable virgin stands. We divided natural stands into more and less resistant according to the character of stand spatial structure. We

Table 3 Classification of factors on forest resistance

Factor	Weight	Note
Health status	?	– Fast dieback led to 'edge effect' + Slow defoliation allows humus decomposition and natural regeneration
Incidental felling	?	– Areas >0.5 ha suffer from extreme microclimate + Areas <0.5 ha supports natural regeneration
Insect	+	Spontaneous spreading in homogenous spruce stands
Naturalness	++	The highest possible resistance was assigned to natural stands
Stand structure	++	Age, species and vertical heterogeneity lead to higher stability
Stand texture	+++	Spatial distribution of natural and structured stands
Diversity	+	Presence of admixed species, for example larch, pine or fir increases physical stability of community
Humus	?	+ Buffering effect against acid deposition – Thick row humus layer limits natural regeneration
Soil – physical parameters	?	+ Rocky soils allows higher larch percentage, vertical structure – Loamy and moist conditions speed up biomass production, soft wood formation
Climate	+++	Extreme values increase vulnerability to insect attack Precipitation deficit in growing season Irregular large-scale windfalls
Pollution	+	Critical level for acid load exceeded Critical level for O_3 exceeded

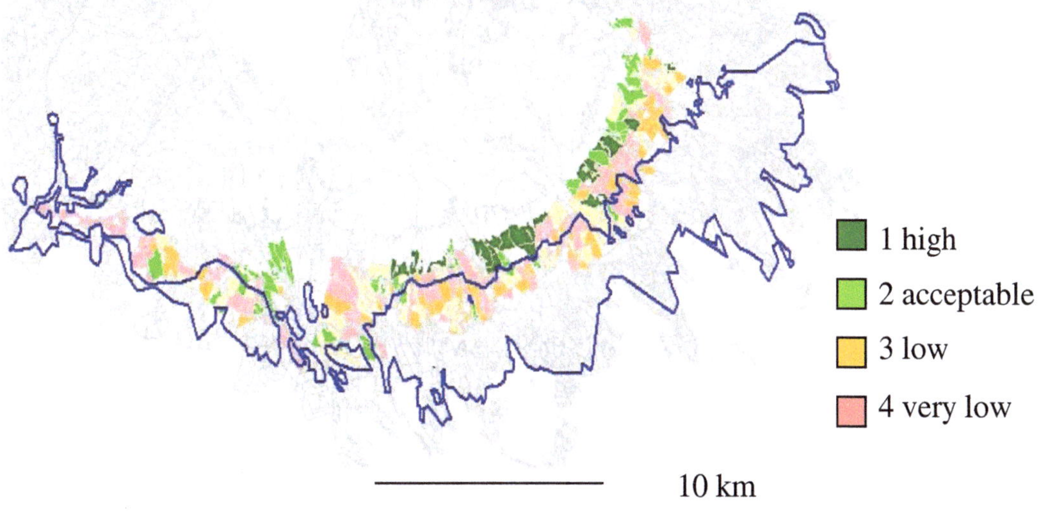

Fig. 7 Spatial distribution of expected resistance in Lariceto-Piceetum community and windfall 2004 border

categorised ecologically unstable spruce monocultures as the least resistant.

Expected resistance of the Lariceto-Piccetum community was as follows:

Forest categories	Resistance
Virgin forest	1
Natural forest, diversified	
Natural forest, undiversified	2
Semi natural forest, diversified	3
Semi natural forest, undiversified	4
Spruce monocultures	

Windstorm 2004

Verification of the expected resistance came unexpectedly soon. On 19 November 2004 northern wind with gusts over 200 km h^{-1} laid down forest stands on the area of 12,000 ha with wood volume 2.3 million m^3. The area affected by the windstorm was continuous. Cold, heavy air masses reached the earth surface at the altitude of 1200–1350 m a.s.l. and formed sharp contour-like line and completely devastated the forest. Only narrow belt at the altitude of 700–750 a.s.l. remained (Fig. 7). Most of the spruce communities with larch and pine were damaged. Since 1915, it was seventh large-scale windfall of this nature, affecting more or less the same territory. History of well-documented windfalls is shown in Fig. 8.

From this graph it is clear that the Bora 2004 caused more damage on wood than the sum of all previous events. It is correct to call the last event as the 'Large Bora' (Koreň 2005).

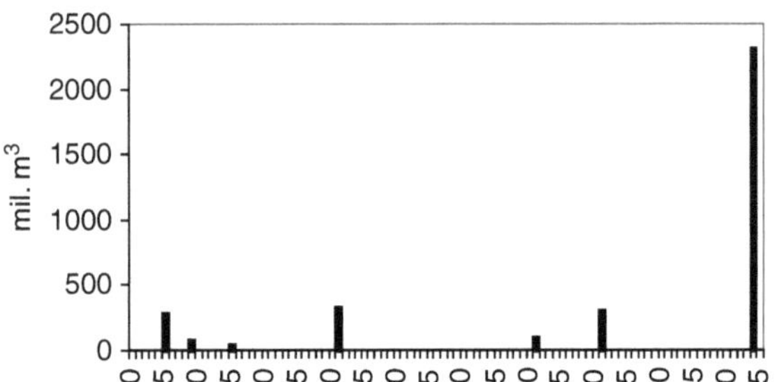

Fig. 8 History of the Bora wind in the Tatras and the volume of damaged forest

Current Status and Expected Forest Development

During the following years, the windfall status in the forest confirmed rather pessimistic prognosis derived from the mentioned criteria. Remnants of young stands, especially fir and spruce, started to die due to the extreme microclimate. Surrounding stand edges were attacked by bark beetle insect in unprecedented way. Besides common species like *Ips typographus* and*Pityogenes chalcographus*, formerly marginal species like *Ips amitinus* and *Ips cembrae* extended their populations to catastrophic scale. Increment index for *I. typographus* for 2006 versus 2007 was twofold, and for *P. chalcographus* even sixfold. Insect population profited not only from luxurious foot conditions but also extremely warm weather in 2006 and 2007 allowed more than usual massive swarming. Current insect attack on standing trees brings the most pessimistic prognosis. Large volume of flammable material after logging of coniferous trees and on-purpose left biomass, combined with high air temperatures, low air humidity lack of precipitation and increased risk of fires. In the summer 2005 after a series of extreme air temperatures exceeding 30°C, the largest fire in the history of the National Park broke out and burnt 250 ha of windfall and adjacent forest. There is evidence that further forest development will be driven by synergetic combination of negative factors and tendencies.

Objective for Further Forest Management

Forest management objectives of the Tatra National Park forest were defined in 1960s (Bezačinský and Greguš 1976) and updated in 1990s (Korpel' 1990; Koreň et al. 1999). According to these concepts some parts (virgin forest) should remain without human intervention. They should serve primarily as sites to study natural processes. Intervention in the forest, identical with current 'nature-close forestry' were planed and changed. Disturbed by both, human activity and natural disasters the forest lost the ability to provide expected landscape-ecological services. This attitude was approved in many papers (Korpel' 1990; Saniga 1996; Koreň 2006) and only slight modification for the forest development is needed.

Optimal general forest ecosystem rehabilitation should include

- conservation of fauna and flora gene pool by sensitive land use,
- production of high-quality drinking water,
- utilisation of countryside for education, health treatment, recreation and tourism. Especially important are strategies for forest management:
- biodiversity conservation, suitable stands with structures close to natural ones are ecologically applicable
- biodiversity rehabilitation, ecologically unstable stands with disturbed biodiversity are applicable, priority should be given to stands below 80 years with reduced risk for evoking calamities
- adaptation, which aims at minimising negative impact of the climate change on the living forest and thus less adaptable landscape systems. Wide span of impact is expected, from intact up to total devastation. Proposed measures are aimed at naturalisation of species composition and elevation of stand stability (Vladovič 2003; Voško et al. 1997). This means
  - forest stands are based on autochthonous species;
  - species and genetic diversity need to be as broad as possible to guarantee higher adaptation of phenotypes to changed ecological conditions;
  - due to low adaptation ability of spruce and absence of beech in the whole Tatra Mountains region and its foothills, it is important to increase percentage of broadleaf species, especially maple;
  - application of selective and shaded sylviculture forms in autochthonous stands, large utilisation of natural regeneration.

Conclusion

Based on the analysis of expected resistance and damaged forest in reality we can conclude that

1. There is no relation between expected and real resistance. Wind power was a driving factor in damaging the forest. Neither natural nor diversified strands could resist the wind speed over 230 km h^{-1}. Even tree species like birch, pine and alder did not survive. Wind broke many trees but majority of trees were uprooted despite the fact that that autumn 2004 was relatively dry. The least injured species was larch, due to short crown and small surface, and the needleless conditions in the late autumn.
2. Despite the fact that wind power exceeded all our expectations, we can state that the main reasons for such a large destruction were
   - forest density, and high volume per hectare,
   - weak spatial diversification of forest.
3. Only few islands of standing forest remained on the windfall. There were localities protected by geomorphologic structures (moraines, stream debris) but surprisingly there also were elevated sites oriented along windstorm direction (N–S). Special group represent former multi-layered stands, where young trees remained only partly damaged.

References

Bezačinský H, Greguš C (1976) Problems of forest management in the Tatra National Park. Studies on TANAP (in Slovak), No. 17, pp. 251–295

Fleischer P (1999) Current forest status in the TANAP as a bases for ecological stability assessment (in Slovak), Technical University Zvolen, Research Center of TANAP, T. Lomnica, 107p

Fleischer P, Koreň M (1995) Forest health conditions in the Tatra Biosphere Reserve. Ecology (Bratislava) 14, No.4, pp. 445–457

Fleischer P, Godzik B, Bicarova S, Bytnerowicz A (2005) Effects of pollution and climate change on forest of the Tatra Mts. In: Omasa, K. Nouchi, I. (eds.), Plant responses to air pollution and global change. Springer Tokyo 2005, pp. 111–121

Koreň M (2005) Windfall in the forest of the TANAP, reasons and consequences (in Slovak). Proceedings from seminar Actual problems in forest conservation 2005. Banská Štiavnica April, 28.–29. 2005, pp. 46–55

Koreň M (2006) Proposal for zoning, objectives and principles for forest management in Biela voda Nature Reserve (in Slovak). Studies on the TANAP, 8(41), pp. 441–456

Koreň M, Fleischer P, Turok J et al. (1997) Reasons for bark beetle outbreak in Javorina region, the Tatra National Park (in Slovak). Studies on TANAP, No.3 (36), pp. 113–187

Koreň M, Fleischer P, Ferenčík J, Slivinský J (1999) Degradation of mountain forest and its rehabilitation in the Tatra National Park (in Slovak). In: Proceedings from seminar Actual problems in forest conservation 1999. Banská Štiavnica April, 8.–9. 1999, pp. 37–46.

Korpeľ Š (1990) Research on forest development and structure of natural stands in the TANAP (in Slovak). In: Proceedings from conference 40 years of TANAP. Tatranska Lomnica, pp. 237–250.

Korpel′ Š (1994) Selective forestry models suitable for the High Tatra Mts conditions. Proceedings from the Conference Theory and practice in nature conservation in Tatra National Park (in Slovak). Tatranska Lomnica, pp. 125–140.

Kunca V, Škvarenina J, Fleischer P, Celer S, Viglaský J (2003) Concept of critical load applied in landscape ecology on an example of the Tatra Mts. Ekologia Bratislava, Vol. 22, Supplement 2/2003: 349–360

Mind′áš J, Škvarenina J (eds.) (2003) Forests in Slovakia and global climate change (in Slovak). EFRA Zvolen and FRI Zvolen, 128p

Pretzsch H (1998) Structural diversity as a result of sivlvicultural operations. Lesnictvi–Forestry, 44, 1998: 429–439

Saniga M (1996) Possibilities and reality for application of nature close forestry in the TANAP forest's (in Slovak). In: Koreň, M. (eds.): Fifty years of forest management in the TANAP. Proceedings from conference. Vysoke Tatry, June, 16–18, 1999, pp. 99–102

Vladovič J (2003) Regional basis and principles for ecological stability assessment of forest in Slovakia (in Slovak). Lesnícke štúdie 57. Bratislava. Príroda, 160p

Voško M et al. (1997) Prognosis and concepts of forestry development (in Slovak). Zvolen. FRI, 173p

Warfvinge P, Sverdrup HU (1992) Calculating critical loads of acid deposition with PROFILE – a steadystate soil chemistry model. Water Air Soil Pollution, 63: 119–143

Modeling Natural Disturbances in Tree Growth Model SIBYLA

M. Fabrika and T. Vaculčiak

Keywords Forest modeling · Injurious agent · Salvage cutting · Tree mortality · Factor analysis · Monte Carlo method

Introduction

Climatic changes belong to very important factors with significant influence on forest structure, stability, and production. During the last 10 years, frequency of natural disasters has been systematically increasing in forest ecosystems. However, not only their frequency has been increased, but also the amount of destroyed wood mass has risen. For instance, in November 2004, huge windstorm appeared in the High Tatras in Slovakia and completely destroyed spruce forests on an area of 2–5 km by 40–50 km. This windthrow caused a production loss in the magnitude of 90% of planned annual coniferous cutting in the Slovak Republic. Due to this increasing occurrence of disturbance models for different forest injurious agents have recently become popular in Slovakia. The development of such models is also important in connection with new trends in education system. E-learning methods successfully utilize strategic computer games and simulators. Our intention was to create a strategic game for the simulation of forest behavior, to place it on WEB (to make it available online), and to make it accessible to students and for e-learning process at Technical University of Zvolen. However, it is impossible to create "real simulations" of forest behavior without the model of natural disturbances. By now, only a few authors in Slovakia have been involved in the estimation and quantification of natural risks in forest ecosystems. First more complex papers have been concentrated with resistance potential of forest (Stolina 1980, 1992). Recent papers are mainly concentrated on insurance models or spatial distribution by GIS technology. From the newest models developed in Slovakia, we can mention the insurance model of Holecy (2000) and Mikusikova (2005), which has also been implemented in Germany (Holecy and Hanewinkel 2006) and spatial models by Hlásny et al. (2005), Turčáni and Hlásny (2007). However, the approach used in these models is not satisfactory for the purposes of forest simulators, where a more detailed model is needed. In this chapter, we present a new model of natural hazards that was developed for and implemented in the tree growth simulator SIBYLA (Fabrika 2005).

Material and Methods

We used data from forest inventory databases and management records covering the whole area of Slovakia (Fig. 1). Database consists of 89,707 forest stands. In total, 388,830 ha have been statistically investigated. The chosen sample represents approximately 19% of Slovak forest area. The stands were selected from all 47 Forest Eco-regions. Minimum and maximum number of forest stands selected from one Eco-region is 56 (295 ha), and 6,687 (21,188 ha), respectively. Forest inventory data included following information: Forest Eco-region, stand density, aspect, slope, site category, age, tree species (proportion in composition in percentage), mean diameter, mean height, and growing stock. From management records

M. Fabrika (✉)
Technical University in Zvolen, Faculty of Forestry,
T.G. Masaryka 24, 960 53 Zvolen, Slovakia
e-mail: fabrika@vsld.tuzvo.sk

K. Strelcová et al. (eds.), *Bioclimatology and Natural Hazards*,
DOI 10.1007/978-1-4020-8876-6_14, © Springer Science+Business Media B.V. 2009

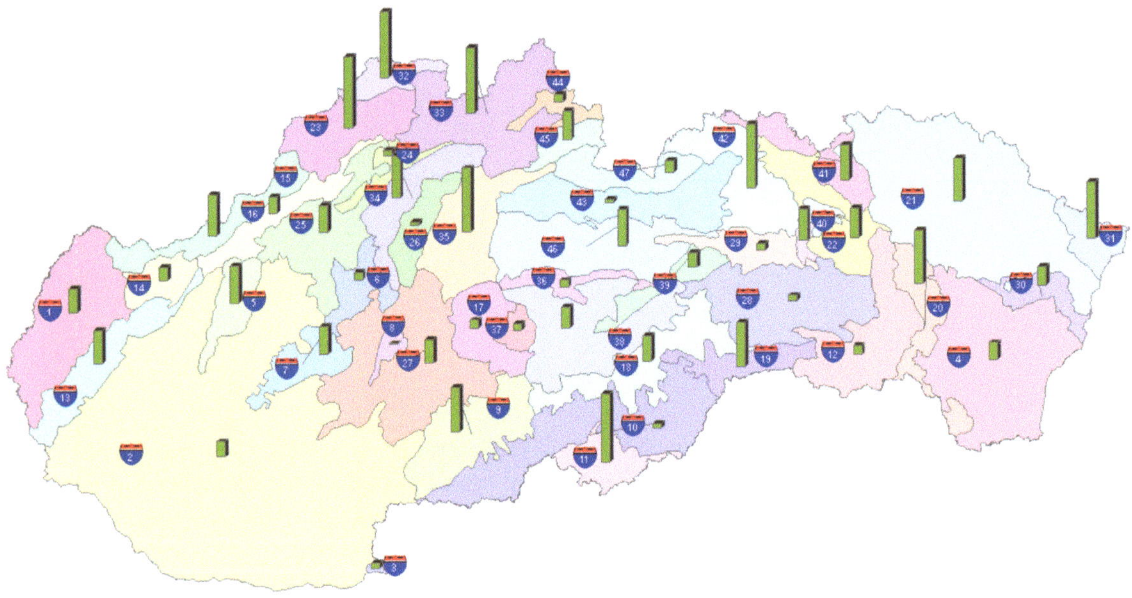

Fig. 1 Investigated area of all 47 Forest Eco-regions in Slovakia (vertical bar means sample size of Forest Eco-region)

the following data were used: tree species, year and amount of salvage cutting, and type of injurious agent. All information is related to a stand identifier. Four types of abiotic injurious agents were observed (wind, snow, icing, drought), three types of biotic agents (bark beetle and timber borer, defoliator and other insect, wood-destroying fungus), and three types of human-induced agents (air pollutants, fire, illegal cutting). Data were collected during 2000–2005, with different period of repeated collections from 1 to 6 years depending on individual stands. In total, we processed 26,246 stands damaged by wind, 993 by snow, 912 by icing, 3,048 by drought, 9,803 by stem insect, 504 by crown insect, 2,800 by fungi, 3,278 by air pollutants, 97 by fire, and 1,808 through illegal cutting. Considering tree species composition, data included 47,437 spruce stands, 5,719 fir stands, 8,034 pine stands, 20,048 beech stands, and 9,234 oak stands.

Risk for occurrence of salvage cuttings was calculated as

$$risk = \frac{\sum_{i=1}^{m} \sum_{k=1}^{l_i \in [1..6]} 1}{\sum_{j=1}^{n} \sum_{k=1}^{l_j \in [1..6]} 1} .100 \qquad (1)$$

where m is number of stands where salvage cutting occurred, n is the number of all stands, l_i is the number of years with salvage cutting in management records for stand i, l_j is the number of maximum years in management records for stand j. The risks were calculated for stands using the following cross-classification: Forest Eco-region (FER), tree species (SPECIES), site category (SITE), and type of injurious agent (AGENT).

In the next step, relative (5-years) amount of salvage cutting was calculated for stands

$$\%V_i = \min \left\{ \frac{\left(\sum_{k=1}^{l_i \in [1..6]} Vic_{ik} \right) \cdot \frac{5}{l_i}}{Vorig_i} ; 1 \right\} .100 \qquad (2)$$

where Vic_{ik} is the amount of salvage cutting of stand i in year k (in m^3) and $Vorig_i$ is the initial growing stock of stand i (in m^3) taken from forest inventory. Next, arithmetic means AVG(%V) and standard deviations SD(%V) were calculated for the same groups of stands defined by FER, SPECIES, SITE, and AGENT. The result of these calculations is the *classification table* (C-TAB), which includes Forest Eco-region, tree species, site category, type of injurious agent and for these groups risk of occurrence, mean relative amount of salvage cutting, and its standard deviation, together with mean stand variables: vegetation zone, aspect, slope, age, stand density, tree species percentage, site index, and relation between mean height and mean diameter (hd ratio). Together with C-TAB, *selection table* (S-TAB) was derived from the analyzed database. Selection table comprises information about occurrence of each type of injurious agent in groups of stands determined by Forest Eco-region and tree species.

Moreover, further modeling approaches were utilized as a modeling instrument:

- *factor analysis*, because salvage cuttings are activated by many factors (climate, geomorphology, stand stability, maturity, resistance, productivity, etc.),
- *regression analysis*, because salvage cuttings are dependent on many variables (altitude, aspect, slope, site, age, stand parameters, and others),
- *Monte Carlo method*, because salvage cuttings appear randomly as a result of unknown and complicated situation,
- *planar geometry*, because some injurious agents are propagated in typical spatial figures (ellipse, strip, circle, spot-propagation),
- *fuzzy-based rules*, because some injurious agents are propagated on the basis of uncertain rules.

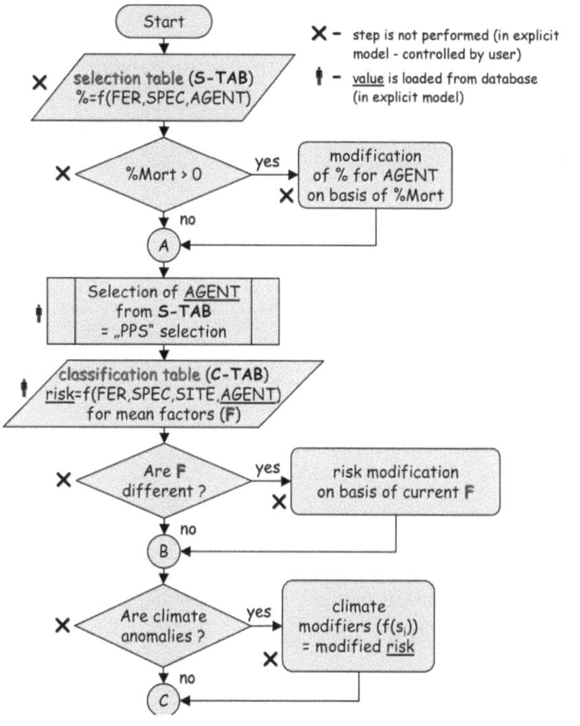

Fig. 2 Flowchart of the salvage cutting model (part 1)

Results

The model of natural disturbance is composed of two parts. The first part selects salvage cutting on a stand level; it specifies the type of injurious agent, its risk and total amount of cutting. This part is statistically oriented, since factor analysis, regression analysis, and Monte Carlo method are utilized. The second part determines salvage cutting on a tree level using fuzzy-based rules and planar geometry. The complete flowchart of the model is presented in Figs. 2, 3 and 4. The procedure is explained in following steps:

1. Frequencies of occurrence of all injurious agents are taken from S-TAB. Frequencies depend on Forest Eco-region and tree species. Table 1 presents an example for one Forest Eco-region and spruce. Next, frequencies are increased with multipliers if during the last 5-year period any dead trees occurred in the stand. This modification is applicable to bark beetles and timber borers, wood-destroying fungi and fire, because occurrence of these agents is higher if stand hygiene is worse. The multipliers are

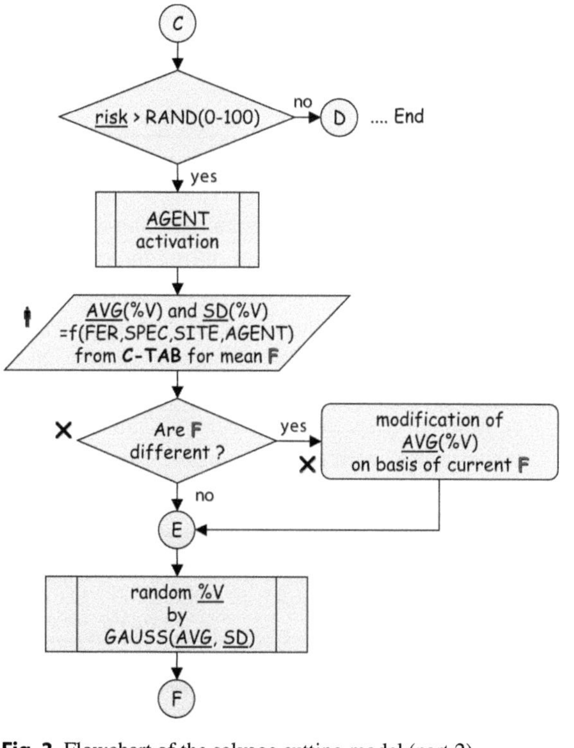

Fig. 3 Flowchart of the salvage cutting model (part 2)

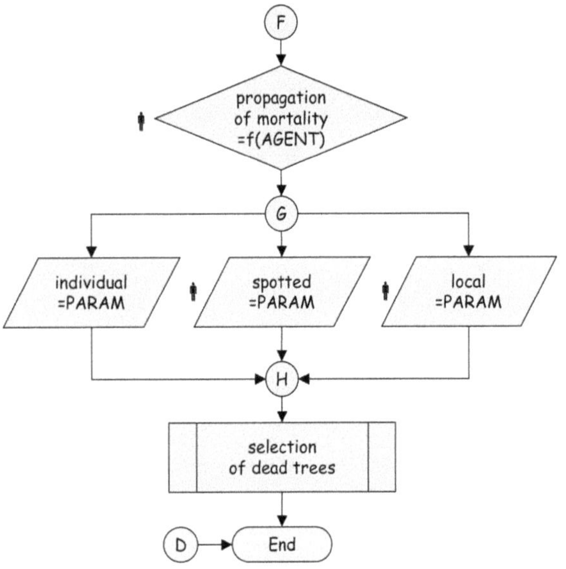

Fig. 4 Flowchart of salvage cutting model (part 3)

Table 1 Example from "Selection table" (S-TAB)

Forest Eco-region	Tree species	Injurious agent	Frequency
23 JAVORNIKY	Spruce	Windstorm	46.7
23 JAVORNIKY	Spruce	Bark beetles	31.3
23 JAVORNIKY	Spruce	Other	10.9
23 JAVORNIKY	Spruce	Snow damage	4.3
23 JAVORNIKY	Spruce	Illegal cutting	3.5
23 JAVORNIKY	Spruce	Drought	1.4
23 JAVORNIKY	Spruce	Defoliator	1.2
23 JAVORNIKY	Spruce	Fire	0.5
23 JAVORNIKY	Spruce	Air pollutants	0.3

derived from functions (Fig. 5) and they depend on the percentage of dead trees during the last growth period. In the next step, a particular injurious agent is selected by applying *probability proportional to size* (PPS): (1) cumulative frequencies are calculated, (2) random number between 0 and maximum cumulative frequency is generated, and (3) the injurious agent with equal or nearest cumulative frequency is selected.

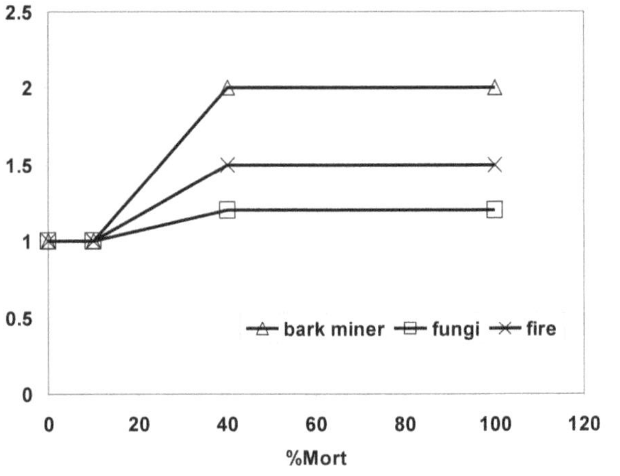

Fig. 5 Multipliers used to increase occurrence frequency of injurious agents

2. Mean risk (risk^{mean}) of injurious agent is retrieved from C-TAB. The risk depends on Forest Eco-region, tree species, site category, and injurious agent. Table 2 presents an example for one Forest Eco-region, spruce and windstorm. This risk is valid for mean conditions in cross-classified category that are specified by eight mean variables (x_j): vegetation zone (1), aspect (2), slope (3), age (4), stand density (5), percentage of tree species in species composition (6), site index (7), and hd ratio (8). For the analyzed stand, the actual risk is determined on the basis of the real stand conditions described by the same variables as above. Mean risk is modified using a model derived from *factor analysis*. The above-mentioned eight stand variables are combined into factors, while for each tree species and injurious agent different number of factors (k=2,. . .,4) are derived (Vaculčiak 2007). The factors F_i are linear combinations of the transformed stand variables into normal variables multiplied by their scores calculated in factor analysis:

$$F_i = \sum_{j=1}^{8} \left(score_{ij} . \frac{x_j - \bar{x}_j}{s_{x_j}} \right) \quad (3)$$

Table 2 Example from "Classification table" (C-TAB)

FER	SPECIES	SITE	AGENT	RISK	AVG (%V)	SD (%V)	Veg- zone	Aspect	Slope	Age	Stand density	Percentage	Site index	hd
23	Spruce	1	Windstorm	13.27	18.39	31.03	4.8	183	29	81	0.80	79	30	0.78
23	Spruce	2	Windstorm	28.04	15.76	19.64	5.0	164	39	89	0.78	90	31	0.76
23	Spruce	3	Windstorm	15.28	7.90	11.88	4.1	184	39	83	0.79	80	32	0.81
23	Spruce	4	Windstorm	18.52	14.28	19.20	4.0	212	44	85	0.78	74	31	0.83

Table 3 Example of factors table

SPEC	AGENT	FACTOR	Scores of linear combination							
			vegzone	aspect	slope	age	sdensity	percentage	siteindex	hd
spruce	windstorm	stability	−0.123243	0.057766	−0.224123	−0.738731	0.207547	0.174868	0.815942	0.793160
		diversity	−0.793107	0.023538	−0.173181	0.070574	0.162982	−0.799766	0.035486	−0.004895
		terrain	0.314044	0.006772	0.706223	0.081063	0.638186	−0.239916	−0.073781	0.088126
		aspect	0.080726	0.916622	0.236525	−0.051582	−0.329803	−0.076033	−0.092671	0.044231
fir	windstorm	stability	−0.072138	0.008514	−0.102457	−0.819394	0.025576	−0.062213	0.845124	0.724375
		diversity	0.140796	−0.003404	−0.284255	0.097013	0.710334	−0.804328	0.035255	0.181635
		terrain	0.838617	0.054434	0.743003	0.115097	−0.148218	−0.065398	−0.144667	0.042374
		aspect	0.062678	0.981793	−0.007793	−0.025703	0.134595	0.145238	0.027565	−0.041866
pine	windstorm	stability	0.002035	−0.077524	−0.014910	−0.799566	0.095944	0.042437	0.696902	0.848855
		terrain	−0.794081	−0.436846	−0.860241	0.077155	0.039856	0.258176	0.251750	0.021108
		density	−0.018982	0.359042	−0.050020	0.038754	0.922050	0.073460	0.145289	0.049189
		diversity	0.178664	0.118310	−0.004555	0.239430	−0.086423	−0.879039	0.465678	0.052959
beech	windstorm	stability	0.128001	0.033452	0.086098	0.803614	−0.029059	0.012653	−0.760635	−0.884057
		diversity	−0.529656	−0.087224	0.105628	0.186517	0.145024	0.907098	0.309903	0.020474
		density	0.197946	0.635618	−0.143354	0.151555	0.772599	0.078975	0.049473	0.079477
		terrain	0.606860	−0.010171	0.903248	0.106998	−0.041415	0.046732	−0.247284	0.067404
oak	windstorm	stability	0.070130	−0.001142	0.004954	0.814447	0.012914	0.100096	−0.644454	−0.884375
		diversity	0.886811	0.057240	0.342279	0.132466	0.107642	−0.810566	0.283735	0.032391
		density	−0.014329	0.046729	−0.591642	0.217322	0.798093	0.051661	0.395113	0.049584
		aspect	0.113164	0.988673	0.022911	−0.051688	0.073637	0.037037	−0.106577	0.000466

Table 3 shows an example of factors for windstorm and all tree species. Mean factors and real factors are calculated for mean and real conditions. Risk modification is based on multipliers, which are derived from occurrence of salvage cuttings depending on the factor value. For all factors, occurrences are derived from *regression analysis* using Weibull function. For mean factor (F^{mean}) and real factor (F^{real}) occurrences are calculated from the regression function separately (see example in Fig. 6a). Finally, real risk ($\mathrm{risk}^{\mathrm{real}}$) is calculated by

$$risk^{real} = risk^{mean}.\frac{\sum_{i=1}^{k} \frac{Weibull(F_i^{real})}{Weibull(F_i^{mean})}}{k} \quad (4)$$

3. Real risk for real condition is modified by climate anomalies

$$risk^{modif} = risk^{real}.b.\left(1 + \frac{\sum_{i=1}^{8} a_i.\left(\frac{s_i - s_i^{norm}}{0,5.(s_i^{\max} - s_i^{\min})}\right)}{\sum_{i=1}^{8} |a_i|}\right) \quad (5)$$

where a_i are dummy variables, s_i, the real values of site variables, and s_i^{norm}, the standard values of site variables obtained from regionalized climate (Fabrika et al. 2005). Site variables correspond to variables that are necessary for ecological site classification in model SIBYLA (Fabrika 2005). These are NO_x in air (1), CO_2 in air (2), nutrient supply in soil (3), number of vegetation days per year (4), annual temperature amplitude (5), mean temperature in vegetation season (6), soil moisture (7), and precipitation amount in vegetation season (8). Variables $s_i^{\min}$ and $s_i^{\max}$ are minimum and maximum values of site variables and represent ecological amplitude of a particular tree species (Kahn 1994). If dummy variable a_i is equal to 0, site difference between the reality and norm has no influence to risk disturbances; if a_i is equal to +1, site difference has a positive influence, and if a_i is equal to −1, site variable has a negative influence. Dummy variables are given in Table 4. Coefficient b is a multiplier sensitive to tree species and can obtain values between 0.5 and 1.5. Equation (5) is calculated only if the sum of absolute values of dummy variables in the denominator is more than 0, otherwise result of equation is 1.

4. Estimated risk is compared with random number from uniform distribution $U[0,100)$. If the risk is higher than the random number, the injurious agent is activated in the growth period.
5. Mean relative volume of salvage cutting AVG(%V) and standard deviation SD(%V) is retrieved from C-TAB (see example in Table 2). The volume is

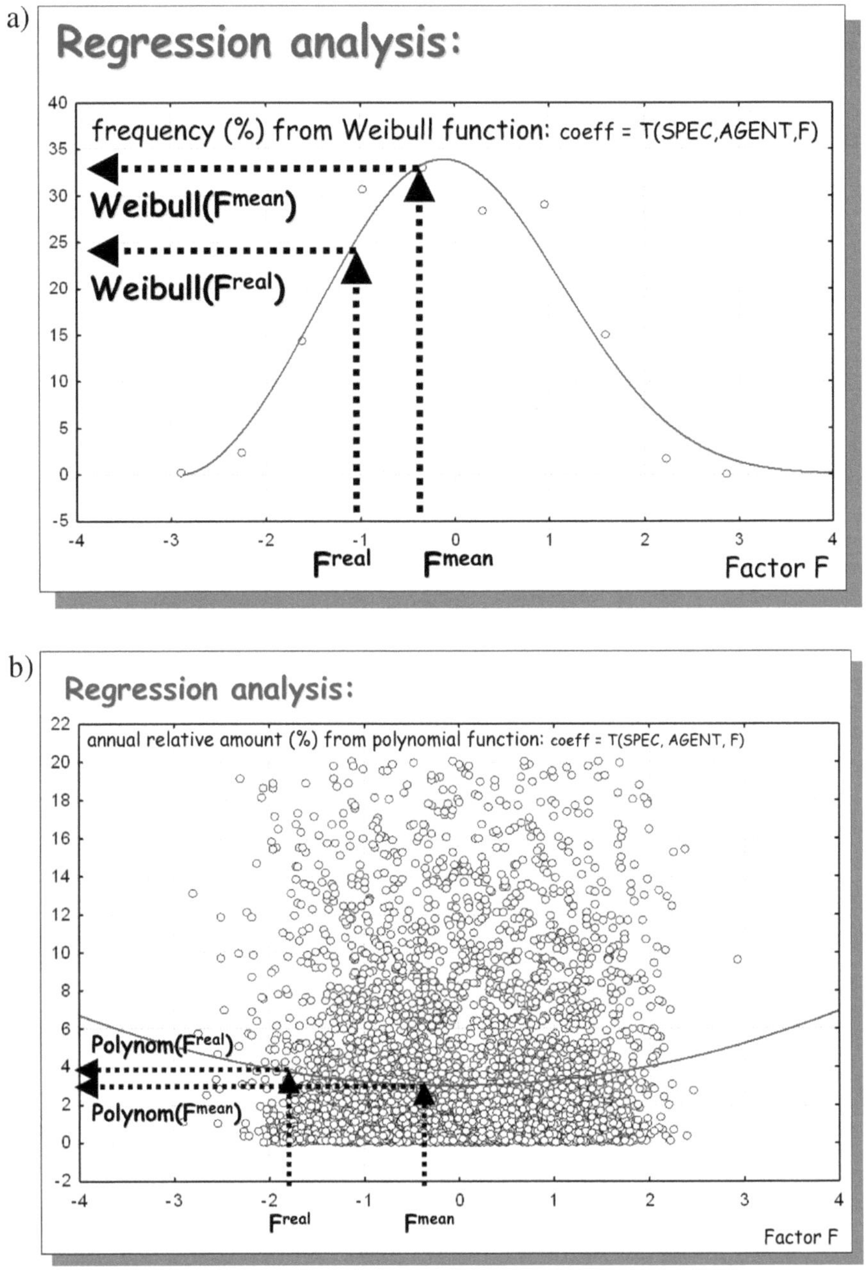

Fig. 6 Example of correction for: (**a**) risk of salvage cutting, (**b**) amount of salvage cutting

modified on the basis of mean factors and real factors as described in Step 2 using a polynomial function (example shown in Fig. 6b). The function was derived from *regression analysis*. The modification of volume is

$$AVG\,(\%V)^{real} = AVG\,(\%V)^{mean} \cdot \frac{\sum_{i=1}^{k} \frac{Polynom(F_i^{real})}{Polynom(F_i^{mean})}}{k} \tag{6}$$

Standard deviation is not modified because variance has been proven to be homoskedastic, without exchange depending to factor value.

6. In the final step of the statistical part of the model, random relative amount of salvage cutting %V in the stand is generated by applying the principle of Monte Carlo method. Random number is generated within Normal (Gauss) distribution defined by mean ($AVG(\%V)^{real}$) and standard

Table 4 Dummy variables for model of climate anomalies

Agent	a_1	a_2	a_3	a_4	a_5	a_6	a_7	a_8
Windstorm	0	0	0	0	0	0	0	0
Snow damage	0	0	0	−1	0	0	0	0
Icing damage	0	0	0	−1	0	0	0	0
Bark beetle	0	0	0	1	0	1	0	0
Defoliator	0	0	0	1	0	1	0	0
Wood-destroying fungi	0	0	0	1	0	1	0	1
Air pollutants	0	0	0	0	0	0	0	0
Fire	0	0	0	0	0	1	0	−1
Drought	0	0	0	0	0	1	0	−1
Illegal cutting	0	0	0	0	0	0	0	0
Other	0	0	0	0	0	0	0	0

deviation (SD(%V)). Generation is repeated in each growing period (1 period = 5 year). To obtain a plausible conclusion, we must repeat simulation many times. After a high number of simulations (more than 30) simulated mean converges to origin average with origin standard deviation. Therefore, single simulation simulates one possible amount of salvage cutting very well.

7. Individual trees are selected for salvage cutting. Selection depends on spatial propagation of mortality in the stand space. Propagation can be individual, spotted, or local. In the case of *individual* (*scattered*) propagation, trees are selected on the basis of their individual tree parameters (tree diameter, tree height, hd ratio, crown diameter, crown length, crown shape coefficient, tree vitality, bio-sociological tree status, tree competition pressure, tree quality, and score of existence). In this procedure, parameters are called selectors and are selected for each injurious agent separately. Afterwards, selectors are transformed by fuzzy functions and joined by OR/AND operators. Fuzzy functions are sensitive to individual parameters and hence, the parameters can be in positive or negative position to each other. Negative position is derived from the positive one by NOT operator. Trees with the highest final values of joined fuzzy functions are selected as dead trees for salvage cutting, while the number of trees is defined by %V. Individual propagation is automatically activated for snow damage, icing damage, drought, wood-destroying fungi, air pollutants, and illegal cutting. Default adjustment of the model is given in Table 5. In *spotted* propagation (*expanded from local point*), trees are selected on the basis of generated parameters of spotted elements. Random number of spots is generated from uniform distribution $U[1,5)$. Propagation is defined by the length of propagation period (usually 3 years). Size of spotted elements is defined by %V. Amount annual salvage cutting is expanded linearly. All trees inside the spotted elements are dead and prescribed for salvage cutting. Spotted propagation is automatically activated for bark beetles and timber borers. In the case of *local* propagation (*for specified sub-area*), trees are selected on the basis of generated parameters of a local area. The shape of the area is automatically selected from the pre-defined categories (circle, strip, ellipse) depending on injurious agent. The position of the area is randomly generated. The size of the local area depends on %V. All trees inside of the local area are dead and prescribed for salvage cutting. Local propagation is automatically activated for wind damages (in strips), defoliators (in ellipses), and fires (in ellipses).

Table 5 Default adjustment of the model for individual propagation of dead trees (+ stands for positive position, − stands for negative position)

		Snow	Icing	Fungus	Air pollutants	Drought	Illegal cutting
Selector	Tree diameter						
	Tree height						
		+					
	Crown diameter	+	+				
	Crown length		+				
	Crown shape coefficient						
	Tree vitality			−		−	
	Bio-sociological status				−		
	Competition pressure				+	+	
	Tree quality			−			
	Score of existence						
Link		AND	AND	OR	AND	OR	
Random		NO	NO	NO	NO	NO	YES

The model has been implemented as a software tool. The software is an integral component of SIBYLA growth simulator (Fabrika 2005). User can specify if (s)he wants to use implicit or explicit model for salvage cuttings. *Implicit model* is fully automatic. All necessary input data are retrieved from SIBYLA database and include information about the site, and about the stand. It is necessary to specify generalized data about site (Forest Eco-region, altitude, aspect, slope, and forest type). At the same time, it is important to select also detailed climate and soil data (i.e., concentration of NO_x and CO_2 in air, soil nutrient supply, number of days of vegetation season, annual temperature amplitude, mean temperature in vegetation season, soil moisture, and total amount of rainfall during vegetation season), if they are known. The software compares standard climate from generalized specification and real climate specified by user. Detected differences are used in modeling climate anomalies and their influence on the occurrence of salvage cuttings. If detailed climate and soil data are not specified, no climate anomalies will be accounted for in the model. Apart from site data, user should also specify stand data, namely information about age, stand density, tree species composition, site index, and hd ratio. Data are derived from forest structure, which is either directly specified by user or generated by structure generator. The number of dead trees, which occurred in the recent period is modeled by SIBYLA mortality model. Mortality is necessary for modification of occurrence frequencies of injurious agents. User can check the functionality of Monte Carlo generator for occurrence and amount of salvage cuttings in periods of simulation (see Fig. 7) before prognosis. *Explicit model* is fully controlled by user. User selects which injurious agent should be activated in the particular simulation period and inserts its risk and its amount of salvage cutting (specified by average and standard deviation). Furthermore, user needs to justify propagation of injurious agent on a tree level. Options are available from software dialog (see example in Fig. 8) together with fuzzy functions of individual tree selectors. All options are also saved in table DAMAGE in SIBYLA Microsoft Access Database. Explicit model is simplified and the model flowchart is reduced (see pictograms in Figs. 2, 3 and 4). The final model of disturbances has been integrated into PROPHESIER Unit and its subunit is called AGGRESSOR (Fig. 9). The final result of the model

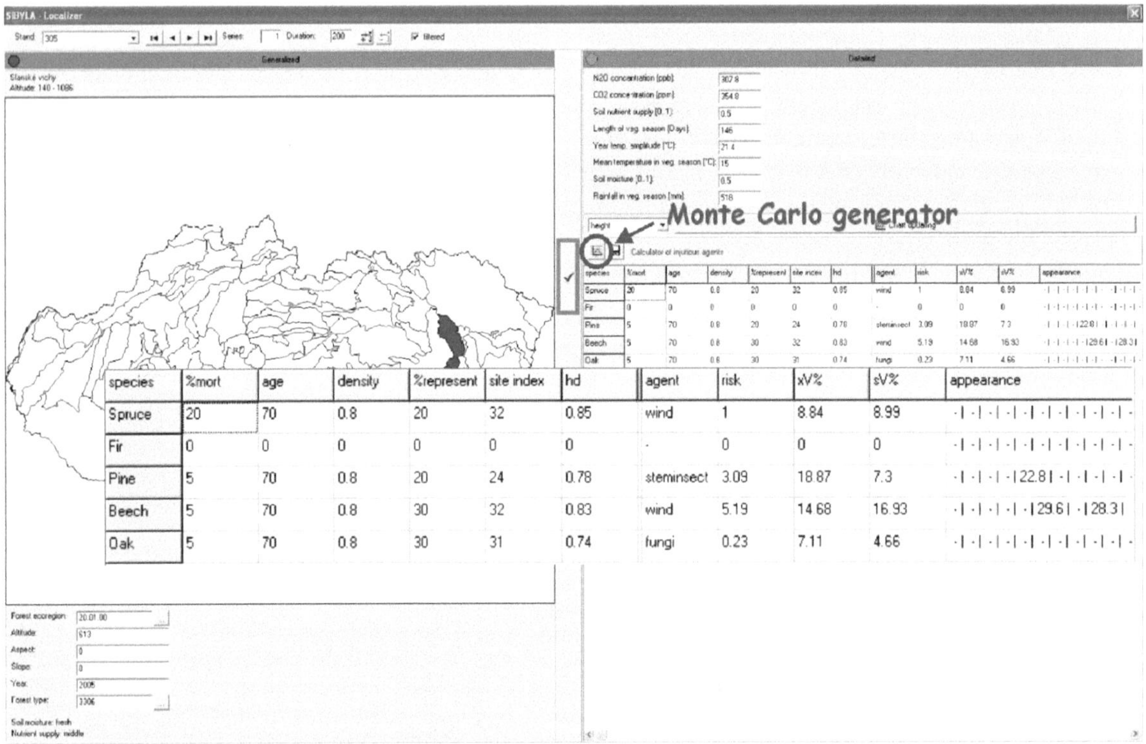

Fig. 7 Monte Carlo generator of salvage cuttings in Localizer unit

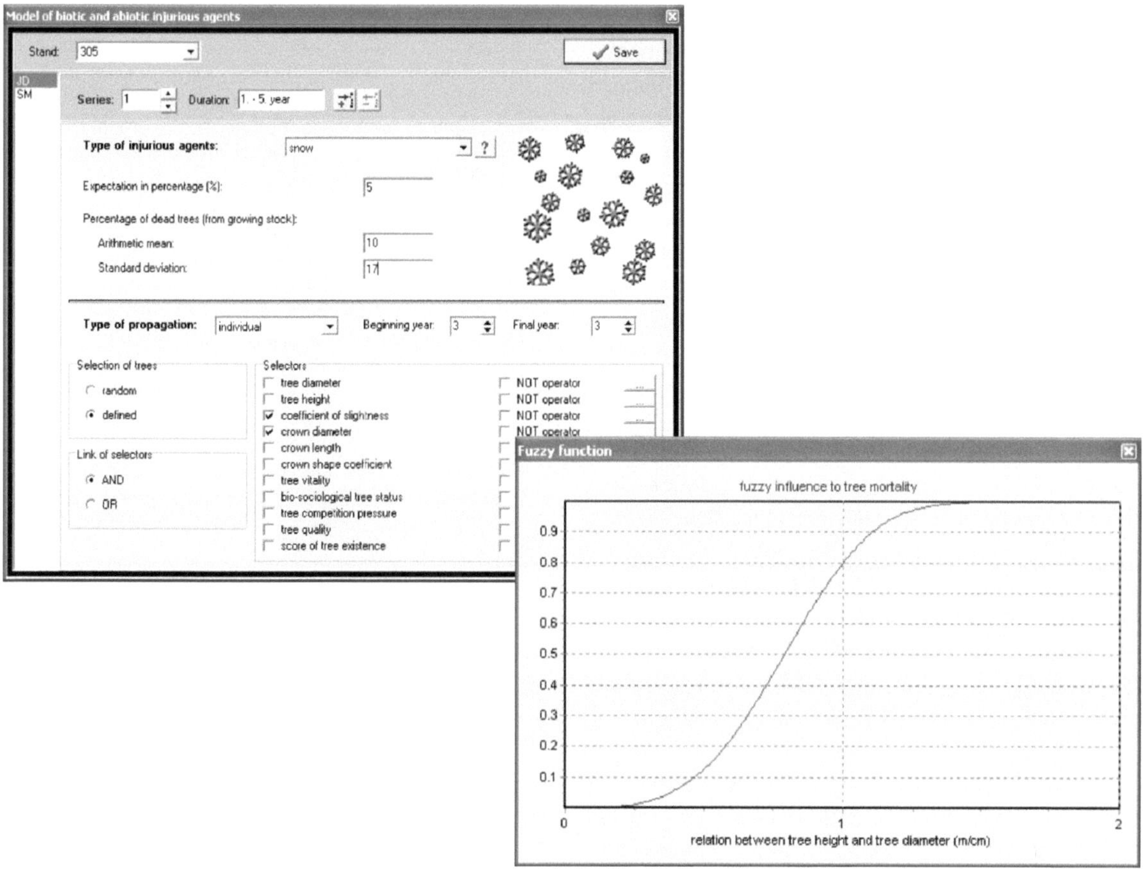

Fig. 8 Example of a dialog box for the explicit model fully controlled by user

Fig. 9 The position of Aggressor unit in SIBYLA software structure

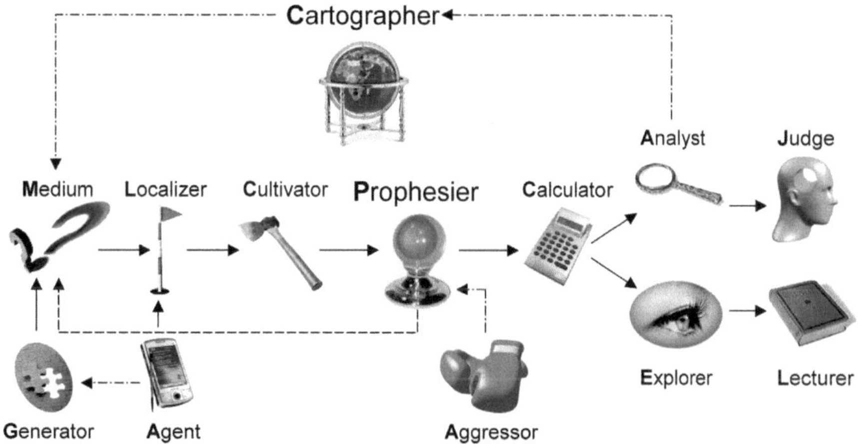

is the identification of dead trees with the transparent reason of mortality. It is possible to check reason of mortality in the SIBYLA database after the simulation or check it in the console (info panel) of virtual reality (if cursor goes over dead trees).

Conclusion

One cannot imagine forest development without influence of external factors either abiotic, biotic, or human-induced factors, of which many are often unpredictable (hardly predictable). This is the reason, why such factors usually do not belong to regular components of growth simulators. In this chapter, we presented the first version of the complex model simulating natural disturbances that was successfully implemented in tree growth simulator SIBYLA. This model responds to a wide range of injurious agents: wind, snow, icing, bark beetles, timber borers, defoliators, wood-destroying fungi, air pollutants, drought, fire, and illegal cutting. Injurious agents are activated by a great number of external variables: geomorphology parameters (altitude, aspect, slope), climate (temperature, precipitation, vegetation season), sanitation parameters (remaining dead trees), site parameters (Forest Eco-region, vegetation zone, site class), and stand parameters (age, stand density, tree species composition, site index, mean height and mean diameter). The model combines different modeling instruments: factor analysis, regression analysis, Monte Carlo method, planar geometry, and fuzzy-based rules. Salvage cuttings are simulated hierarchically from stand level (risk and total amount) to tree level (tree selection). The model cannot only help to find answers on a number of scientific questions, but is also applicable in e-learning process, and as a decision support in forest management. However, first the model needs to be evaluated on a scientific level, which is to be performed in the nearest future. We also have to solve some additional problems, which have not been implemented yet, for example, a sequential character and gradations of some injurious agents, or changes in tree increment and tree quality of living trees due to damages caused by injurious agents.

Acknowledgments The model of salvage cuttings was developed in the frame of research project VEGA No. 1/3531/06 – "Forest inventory and forecasting," KEGA No. 3/6080/08 – "Internet Forest Training Tool in Forestry E-learning" and 6FP EU Project CECILIA – "Central and Eastern Europe Climate Change Impact and Vulnerability Assessment."

References

Fabrika M (2005) Simulátor biodynamiky lesa SIBYLA, koncepcia, konštrukcia a programové riešenie [Simulator of forest biodynamics SIBYLA, conception, construction, and software solution], Habilitation work, Technical University of Zvolen, Zvolen, 238p

Fabrika M, Dursky J, Pretzsch H, Sloboda B (2005) Regionalization of climatic values for ecological site classification using growth simulator SIBYLA with GIS. In: Klein, Ch, Nieschultze, J, Sloboda, B (eds.), "Remote Sensing and Geographical Information Systems for Environmental Studies, Application in Forestry", Universität Göttingen, J.D. Sauerländer's Verlag Frankfurt am main: Vol. 138, 245–255

Hlásny T, Koreň M, Turčáni M, Jiřina M (2005) Neural network based system to analyse forest ecosystems health status at a regional scale, In: Priwitzer, T. (ed.), "Climate Change – Forest Ecosystems & Landscape", Proceedings from the international scientific conference and JRC workshop "Forest Monitoring from remote sensing at scales from global to local", Zvolen–Sielnica, 19–22 October 2005, 41–48

Holecy J (2000) Forest fire insurance problem in Slovakia and its solution by methods of elementary statistics. In: Development trends of processes management in wood processing industry and in forestry. Proceedings of the International Association for Technology Management–Wood/Technical University of Zvolen Publishers, Zvolen: 53–60

Holecy J, Hanewinkel M (2006) A forest management risk insurance model and its application to coniferous stands in southwest Germany, Forest Policy and Economics, 8:161–174

Kahn M (1994) Modellierung der Höhenentwicklung ausgewählter Baumarten in Abhängigkeit vom Standort, Forstliche Forschungsber, München, Vol. 141, 221p

Mikusikova B (2005) Modely poistenia lesného majetku proti riziku výskytu náhodných ťažieb [Forest property insurance models against risk of salvage cuttings occurrence], Dissertation work, Technical Univerzity of Zvolen, Zvolen, 80p

Stolina M (1980) Resistence Potential of Forest and Spruce Ecosystems. In: Stability of spruce Forest Ecosystems. International Symposium on MAB – UNESCO – IUFRO, Brno, 75–88

Stolina M (1992) Resistence Potential of Forest as Indicator of its Ecological Stability. In: Ecological Stability of Landscape. Federal Committee for Environment Prague – Kostelec nad Černými Lesy: 78–80

Turčáni M, Hlásny T (2007) Spatial distribution of four spruce bark beetles in north-western Slovakia. Journal of Forest Science, 53, 45–53

Vaculčiak T (2007) Modelovanie náhodných ťažieb v lesných porastoch poškodených vplyvom abiotických škodlivých činiteľov [Modeling salvage cuttings in forest stands damaged by abiotic injurious agents]. Acta Facultatis Forestalis, Suppl. 1, Zvolen – Slovakia, XLIX: 153–165

Insect Pests as Climate Change Driven Disturbances in Forest Ecosystems

T. Hlásny and M. Turčáni

Keywords *Lymantria dispar* (L.) · *Ips typographus* (L.) · Climate change · Spatial modelling · Outbreaks · Voltinism · Slovakia

Introduction

Climate change is generally agreed to have a profound impact on forest structure and its dynamics (Aber et al. 2001; Ayres and Lombardero 2000; Dale et al. 2000, 2001). As trees can live from decades to centuries, rapid changes of climate are also expressed through alterations of the disturbance regime (Franklin et al. 2002; He et al. 1999). All major disturbances are affected in a certain manner – fire, drought, introduced species, insect and pathogen outbreaks, hurricanes, windstorms, ice storms and landslides (Dale et al. 2001; Flanning et al. 2000; Aber et al. 2001; Hanson and Weltzin 2000). Although disturbances are an integral part of forests, when such disturbances exceed their natural range of variation the impact on forests may be extreme (Ayres and Lombardero 2000). Pest organisms have the ability to adapt much faster than their host trees, thereby increasing the likelihood of severe pest impacts (Docherty et al. 1997; Malcolm et al. 2001; Battles et al. 2006).

Pest organisms respond to environmental changes both directly and indirectly, through changes in forest structure and the decreased resistance of trees. A changing climate may also cause those pests that are currently of minor significance (i.e. not forming large-scale outbreaks) to become key species, thereby causing serious damage.

Insects are physiologically extremely sensitive to temperature, and even a small temperature increase may have a severe impact on forests (Lange et al. 2006). In addition to the pest's distributional ranges, climate also influences their voltinism (i.e. the annual number of generations) (Reynolds and Holsten 1994, Hansen and Bentz 2003; Lange et al. 2006). If climate warming extends the vegetation season, multi-voltinism is expected to shift to northern locations and higher altitudes; potentially causing further severe damage to forests.

The impact of climate variability on insect pests' abundance fluctuation is not at all clear (Kendall et al. 1998; Myers 1998; Liebhold and Kamata 2000). The most common hypotheses to explain population oscillations are based on density-dependent biotic interactions, that is predator–prey dynamics (Andersson and Erlinge 1977), maternal effects (Rossiter 1994), induced plant defences (Baltensweiler and Fischlin 1988), host–parasitoid interactions (Berryman 1996) and disease dynamics (Myers 1993). However, the concept of climatic release, as a cause of periodic population cycles, has been questioned due to statistical concerns and the lack of periodicity in the climatic deviations that are thought to trigger the outbreaks (Martinat 1987; Turchin and Berryman 2000).

An analysis of potential changes in the distribution of outbreak areas and fluctuation patterns of pests (both indigenous and introduced) is a key task in the assessment of the impact of global change on

T. Hlásny (✉)
Forest Research Institute, National Forest Centre,
T.G. Masaryka 22, 960 92 Zvolen, Slovakia; Czech University of Life Sciences, Faculty of Forestry and Wood Sciences, Department of Forest Protection and Game Management, Kamýcká 1176, Prague 6 – Suchdol 165 21, Czech Republic
e-mail: hlasny@nlcsk.org

K. Strelcová et al. (eds.), *Bioclimatology and Natural Hazards*,
DOI 10.1007/978-1-4020-8876-6_15,

forest ecosystems. The growing number of recent works on this topic is testimony to this issue's importance (Brasier 1996; Lonsdale and Gibbs 1996; Harrington et al. 2001; Gordon et al. 2001; Williams and Liebhold 2002; Logan et al. 2003; Woods et al. 2005; Esper et al. 2007). To contribute to this field, we focused on two key pests in spruce (*Picea abies* L.) and oak-beech ecosystems (*Quercus cerris* L., *Quercus petraea* Liebl. s.l., *Carpinus betulus*, *Quercus robur*, *Fagus sylvatica* L.) – *Ips typographus* (L. 1758) (hereinafter 'IT') and *Lymantria dispar* (L. 1758) (hereinafter 'LD'). These pest species act in diametrically opposed ways, responding to expected changes of climate differently (Logan et al. 2007; Netherer and Pennerstorfer 2001; Netherer et al. 2004). In particular, we focused on

- the impact of the expected temperature increase on changes in the number of IT generations per year (voltinism),
- the impact of the expected temperature increase on future spatial development of LD outbreaks, under the assumption of beech as the alternative host.

To meet these goals, we identify the locations where outbreaks will most probably occur, evaluate changes in voltinism in relation to climate conditions and assess their development under the climate change scenario.

Lifecycle of Investigated Pest Species

Eight-toothed bark beetle (IT) is the most important bark beetle pest in spruce stands. Its lifecycle and population dynamics have long been studied (Christiansen and Bakke 1988; Lieutier et al. 2004) even with various computer simulations (Byers 1993, 1996, 1999, 2000). The species is distributed across the whole of Eurasia, thus it has a very wide niche. It is one of the primary injurious agents in man-made spruce stands at medium latitudes in Europe (Turčáni and Novotný 1998), although large-scale outbreaks have also been reported in northern regions in nature-close forests (Christiansen and Bakke 1988). In normally functioning and balanced ecosystems, IT is not considered to be an aggressive or primary mortality agent of healthy trees. However, under favourable conditions it is able to attack healthy trees and may be the primary tree mortality factor (Hedgren and Schroeder 2004). Once IT populations reach a critical size, they are able to overwhelm almost every mature tree, healthy or otherwise. Outbreaks can last for many years and normally collapse only when every tree within reach has been killed, or when cold weather suppresses the populations (Raffa 1988).

The primary conditions for an outbreak are stand hazard, availability of windthrow and current beetle population size (Reynolds and Holsten 1994). Windstorms are outstandingly important precursors to outbreaks, because they quickly provide large quantities of breeding material in the form of broken or fallen branches, which eliminates intra- and inter-specific competition (Anderbrant 1990; Schopf and Köhler 1995). However, wood processing and control measures subsequent to such windstorms may reduce such impacts.

The emergence and migration of beetles of monovoltine generations were found to depend on geographical latitude. Northern populations emerged later and migrated less frequently before overwintering than those of southern origin (Forsse 1991). In a bivoltine situation, the overwintering generation was found to disperse more extensively (Furuta et al. 1996).

In warmer areas of Europe, IT is able to emerge from hibernation earlier; and two or – under favourable conditions – even three generations can develop in a single season. In northern Europe, lower temperatures normally constrain the beetles to a single generation (Annila 1969). Increasing temperatures are expected to shift the outbreaks to higher elevation (Baier et al. 2007) and to force more populations per year (Lange et al. 2006; Harding and Ravn 1985). Continued warming trends will increase the risk of spruce bark beetle outbreaks throughout the host's range (Logan et al. 2003).

Several studies have been performed to examine the influence of temperature on the development and reproductive cycles of IT (Vité et al. 1952; Annila 1969; Zumr 1982; Anderbrant 1986; Netherer 2003; Baier et al. 2007). IT limits proposed by Annila (1969) and Wermelinger and Seifert (1998), recently published by Lange et al. 2006), are generally accepted. This topic has also been discussed by Netherer (2003), who conducted extensive research in the High Tatra mountains. Table 1 states the stage-specific limits of IT development reported by these authors.

Baier et al. (2007) conducted extensive research on IT thermal limits to develop PHENIPS – a

Table 1 Specific developmental threshold temperatures T(°C) and heat sum requirements (D°) for IT. Variant A is by Annila (1969), and Wermelinger and Seifert (1998); variant B is by Netherer (2003)

Stadium	T(°C)		D°(dd)	
Variant	A	B	A	B
Flight of first generation	5	–	110.0	–
Egg	10.6	10.3	51.8	54.51
Larvae	8.2	4.07	204.4	246.64
Pupae	9.9	12.81	57.7	39.32
Immature adult	3.2	3.38	238.5	307.00

comprehensive phenology model of IT. The authors supposed that the onset of host tree infestation in spring is given by 16.5°C for flight activity coupled with a mean thermal sum of 140 degree days (dd) from the beginning of April 1st onwards. In contrast to other works, a nonlinear function was used to calculate the effective thermal sums, using threshold temperatures of 38.9 – 8.3°C. The discontinuation of reproductive activity appears at a day length of less than 14.5 h.

A thermal sum of 334.2 dd was found to be necessary to complete pre-imaginal development (egg to pupal stage), and 222.8 dd (i.e. two-thirds of the thermal sum for pre-imaginal development) is required for maturation feeding of the filial beetles (Wermelinger and Seifert 1998, Netherer 2003). Hence, a thermal sum of 557 dd is required for total development. For successful hibernation, the brood must complete pre-imaginal development (egg–pupae), that is it requires 60% of the thermal sum for total development before the onset of the cold period.

Gypsy moth (LD) is the most important defoliator of broad-leaved stands (mainly oak) in Eurasia (native) and North America (introduced). It is characterized by cyclical abundance fluctuations, causing defoliation in large areas across southern, eastern and central Europe, northern Africa, Asia Minor, Central Asia, the Middle East (Villemant and Fraval 1998) and the eastern part of North America (Doane and McManus 1981). Populations with flying females occur from Northern Asia to Japan (Liebhold et al. 2008). Contrary to the previous species, LD is strictly univoltine across all of its distributional territory, and it rarely acts as the primary tree mortality agent. The cyclicity of outbreaks seems to be regular across a wide span of natural conditions (Johnson et al. 2006), fluctuating from 3–4 years in southern Europe, 8–10 years in central Europe and 20–25 years in northern regions (Johnson et al. 2006). There are indications of altitudinal shifts of outbreaks in warm years (Csóka and Hirka 2006).

LD eggs are laid in August and the embryo immediately starts developing during the warm days of summer. In a month, the tiny larva is fully formed and ready to hatch. At this point, however, the larva goes into diapause, shutting down metabolic activities and becoming insensitive to the cold. LD can tolerate temperatures as low as −30°C provided such temperatures do not persist for several days (Doane and McManus 1981).

LD caterpillars are thermal conformers (Knapp and Casey 1986). Unlike IT, their growth is independent of temperatures between 25°C and 30°C.

Some newly hatched larvae may be wind-dispersed to new locations. This spring 'ballooning' is an important means of dispersal for both Asian and European races, and it is the primary natural means of dispersal for the European race. Although 'ballooning' has been recorded over distances of 50 km, it is usually effective from about 5–7 km from infestation sites (Doane and McManus 1981).

A specific point is that, under favourable conditions, the species feeds on beech (and occasionally other broadleaves) as alternative hosts. Formerly, this had been observed in isolated stands at lower elevations, mainly surrounded by infested oak stands. Recently, this has been observed on a larger scale, for example in Hungary and Slovakia (Hirka 2006; personal observation). Serious egg mass densities in beech stands were reported in 2004 in Hungary. The 2006 outbreak spread over thousands of hectares of beech stands of different ages at 500–700 m (Bakony Mountains, Bükk Mountains, Hungary). Observed defoliation was up to 50–70% over large areas. Therefore, the species must be considered an important climate change driven agent in both oak and beech stands.

Study Region

Both species have been investigated throughout Slovakia (49, 000 km^2). Forested land covers 41% of the country. Tree species' composition is made up predominantly of spruce (26.1%), fir (4%), pine (7.2%), beech (31.2%), oak (13.4%) and hornbeam (5.7%) (Moravčík et al. 2006). The altitudinal gradient varies between 94 and 2655 m a.s.l. The country is in the temperate and continental climatic zone. The mean annual temperature ranges between 5°C and 8.5°C, mean annual precipitation total is 740 mm. Diversified topography and land-use, ranging from agricultural lowlands in the south to high mountains in the north, are typical of the region. The geographic limits of key forest tree species and pests are within these limits. This facilitates the modelling of global warming-induced shifts of pest outbreaks and changes in population dynamics, as well as the generalisation of results across a broader region.

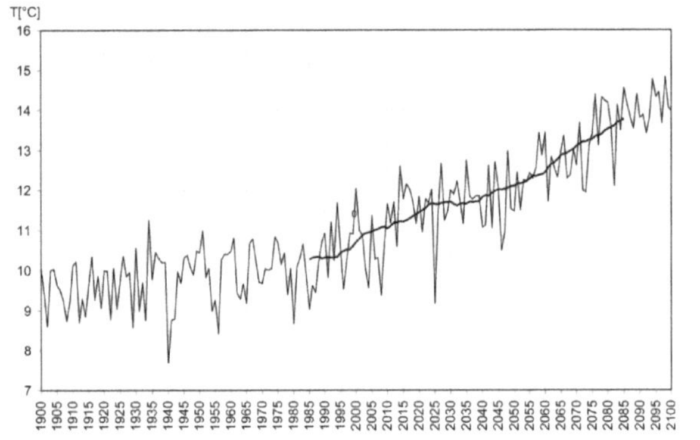

Fig. 2 Observed air temperature data (1900–2000) and modified outputs of CCCM 2000 climate model (2001–2100) (Lapin et al. 2001). The 30-year moving average is displayed for the 1985–2085 period

Data

We used the mean annual air temperature records for the 1951–1980 period as substantial climate data. This period is generally recognised as 'normal', without any trends or significant changes in the variability of main climate parameters: it is broadly used as the reference for climate change impact studies in the region. The available meteorological stations are irregularly distributed at 178 climatic stations across the country (Fig. 1).

Canadian Global Circulation Model (GCM) of the second generation (CCCM 2000) was used for regional downscaling (Lapin et al. 2001, 2006). The only climatic variable needed was mean annual air temperature and its future projection. Its course during the 1900–2100 period at Hurbanovo station (southern Slovakia) can be seen in Fig. 2. The references above discuss the downscaling procedure in more detail. The low spatial variability of air temperature data allows us to use this time series for the entire country (in the form of increments).

Utilized forestry data included species distribution and percentage in mixed stands. All data is spatially referenced through forest compartments. The distribution of investigated forest tree species is given in Fig. 3.

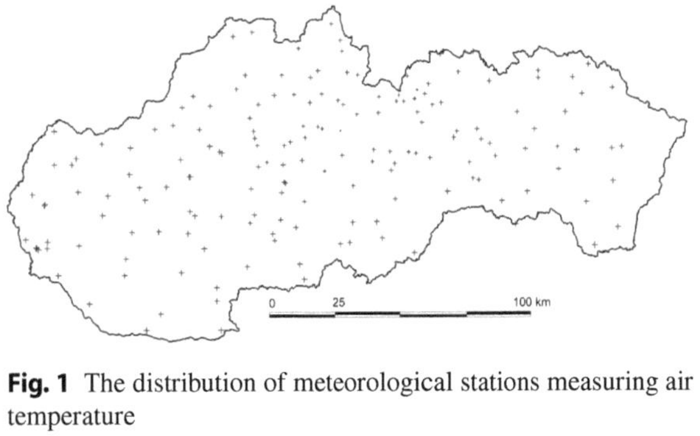

Fig. 1 The distribution of meteorological stations measuring air temperature

The data concerning LD was gathered from 1955 to 1982 (usually between May 10th and 15th), at 20 sites throughout Slovakia. The size of the forest stands in which the sites were located varied from several hectares to several hundred hectares. LD larvae were collected from each site by placing nets around 20 lower canopy branches (about 0.5 m long with about 100 leaves) of 20 *Quercus* spp. trees at the forest edge. Netted branches were cut and subsequently beaten over a sheet and fallen LD larvae counted (Patočka et al. 1962, 1999).

Methods

Changes in IT Voltinism

The sums of effective temperatures (SET) needed to meet the IT stage-specific developmental thresholds were calculated from the average air temperature data for the 1951–1980 period. As daily data was not available at all climatic stations, mean monthly values were centred to the middle of each month, and missing daily values were linearly interpolated. The sums of effective temperatures differed negligibly between A and B variants in Table 1, therefore we used variant A which also provided threshold for the flight of the first generation stage.

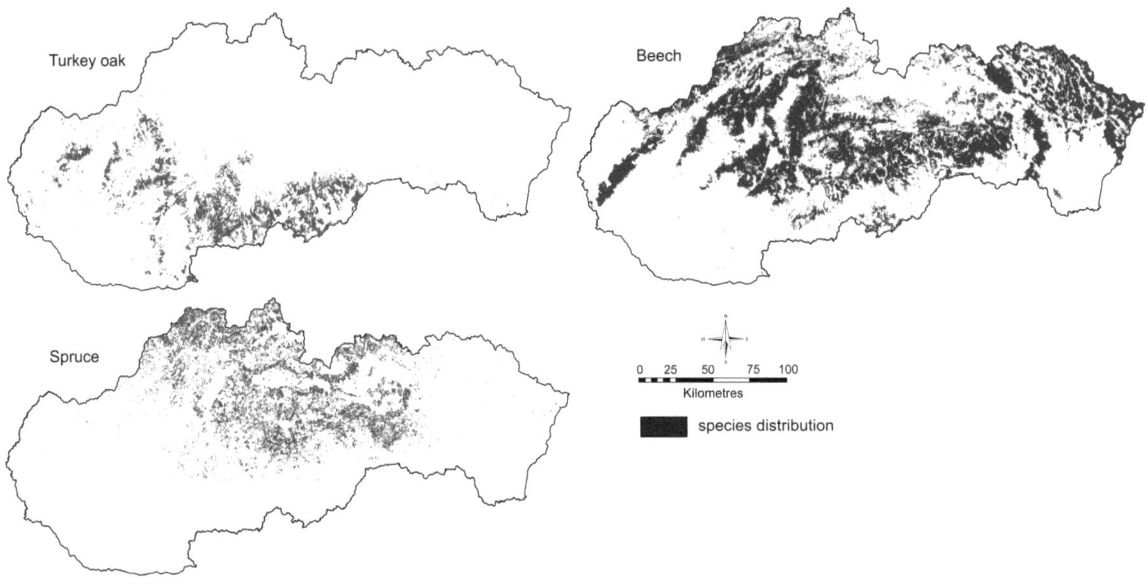

Fig. 3 The distribution of investigated forest tree species in Slovakia (stands with species occurence above 10%)

The strong linear correlation between SET and elevation ($r = 0.94$) allowed us, by using External Drift Kriging (e.g. Wackernagel 1998), to predict (interpolate) SET values at unrecorded locations over the whole study region. The technique is based on the combination of point distributed target variable (SET) with so-called auxiliary/predictor variable(s) (elevation), available at all grid-nodes in the considered spatial domain, assuming these are linearly correlated. To facilitate the time-demanding computations, we used a 180 m resolution grid for all the analyses.

Using this technique, we produced a series of surfaces bearing the sum values of effective temperatures accumulated at each grid cell, starting from the first day that met the criterion of minimal temperature for the flight of the first generation. Due to the 14.5 h day length limit (22nd August in the study region, Julian day 228), all calculations were restricted by this date. We only analysed the full generations (Immature Adult stage) which were expected to complete their development (SET = 557 dd). Other developmental stages were not considered. The methodology is summarised in the following steps:

1. Calculate the total number of dd for each meteorological station, beginning from the first day that meets the criterion of minimal temperature for the flight of the first generation, to August 22 (day length limit), provided that the temperature is not lower than required by the particular developmental stage.
2. Predict SET values at all unrecorded locations in the study region using External Drift Kriging.
3. Threshold the produced surfaces by multiples of 557 dd, in order to identify areas which allow for the full development of any particular IT generation.
4. Project the SET according to climate change scenario and evaluate changes in areas which allow for the full development of any particular IT generation.

Changes in Gypsy Moth Outbreak Ranges

We analysed expected changes in LD outbreak ranges in oak stands, as well as the future danger to beech stands, in the vicinity of identified oak stands outbreak spots. External Drift Kriging was used to produce the underlying temperature map and its future projections. Long-term data on LD abundance and environmental conditions were analysed using Canonical Correspondence Analysis (CCA) (McCune and Mefford 1999; ter Braak and Šmilauer 2002). CCA allows us to examine the patterns of the community structure of oak defoliator species (LD among them) in relation to a set of environmental variables, in order to design a regional model of species abundance–environment relationship. The particular steps of the proposed approach are

1. Investigate LD abundance – environment relationship by identifying the key explanatory variables and their respective importance.
2. Identify outbreak spots by means of a linear weighted combination of relevant environmental variables (as maps) standardized into the unit range.
3. Identify all beech stands in the vicinity of LD outbreak spots in oak stands, provided these are within the temperature limit allowing for LD occurrence.
4. Evaluate climate change impact on the size of LD outbreak areas in oak stands, as well as on the extent of potentially endangered beech stands.

Results

All the analyses have been carried out for four temporal scales, as presented in Table 2. These allow for the development of both medium-term and long-term strategies to prepare the forests for forthcoming changes.

Bark Beetle Voltinism

The sums of effective temperatures for particular developmental stages were calculated using the method described above. The surfaces of SET values were calculated for four temporal horizons (Table 3). All surface values were divided by 557 (SET required to complete the full development). This produced surface of real numbers ranging from 0 to approximately 5 (depending on the time scale) indicating the number of generations, which can potentially complete their development at a location. The maps illustrating the temperature increase induced shifts for the whole country are presented in Fig. 4. It is evident that the fifth IT generation was the highest identified. In fact, climatic conditions suitable for the development of the fourth and fifth generations appear far away from the current distributional ranges of spruce, even on the 2075 time scale.

The summary of the analysis is given in Table 3. We have calculated the proportion of the total country's area which provides climatic conditions that would allow for the full development of respective IT generations, as well as the percentage of area currently occupied by spruce.

Table 2 The 30-year moving average data of mean annual air temperature (T°C year) and increments to the 'normal' climate period (1951–1980)

Time scale	T° (year)	Increment to 1951–1980 (T°)]
1951–1980	9.94	0.00
2015 (2000–2030)	11.19	1.24
2045 (2030–2060)	11.98	2.03
2075 (2060–2100)	13.26	3.31

Gypsy Moth Outbreak Ranges

CCA was used to identify environmental variables controlling species abundance. The species–sites matrix consisted of average abundances across 27 years of sampling. The eight environmental variables used for the analysis were elevation, mean annual air temperature, soil moisture and the proportion of five oak species at a stand (*Q. robur*, *Q. petrea*, *Q. pubescens*, *Q. cerris*, *Q. rubra*). Statistically significant variables (p=0.01) were air temperature, soil moisture, *Q. petrea*, *Q. pubescens* and*Q. cerris*. An ordination plot suggested the pest's positive correlation with *Q. cerris* and air temperature. The temperature correlates with the species axis more significantly (r=0.811) than with *Q. cerris* (r=0.576). The weighted combination of these variables (as maps), rescaled to unit range according to Table 4, was used to identify those stands providing suitable conditions for LD outbreaks under both cur-

Table 3 The percentage of total country's area (1) and area currently occupied by spruce (2) that provides climatic conditions suitable for the full development of respective IT generations

Time scale/ Generation	First generation		Second generation		Third generation		Fourth generation		Fifth generation	
Variant	1	2	1	2	1	2	1	2	1	2
1951–1980 (+0°C)	98.72	97.59	51.54	1.01	90.07	52.92	0.00	0.00	0.00	0.00
2015 (+1.249°C)	100.00	98.57	68.78	7.43	95.74	78.42	20.88	0.00	0.00	0.00
2045 (+2.034°C)	100.00	100.00	78.04	20.48	97.29	88.01	35.61	0.12	0.00	0.00
2075 (+3.316°C)	100.00	100.00	89.76	51.95	98.64	96.77	54.68	1.44	0.77	0.00

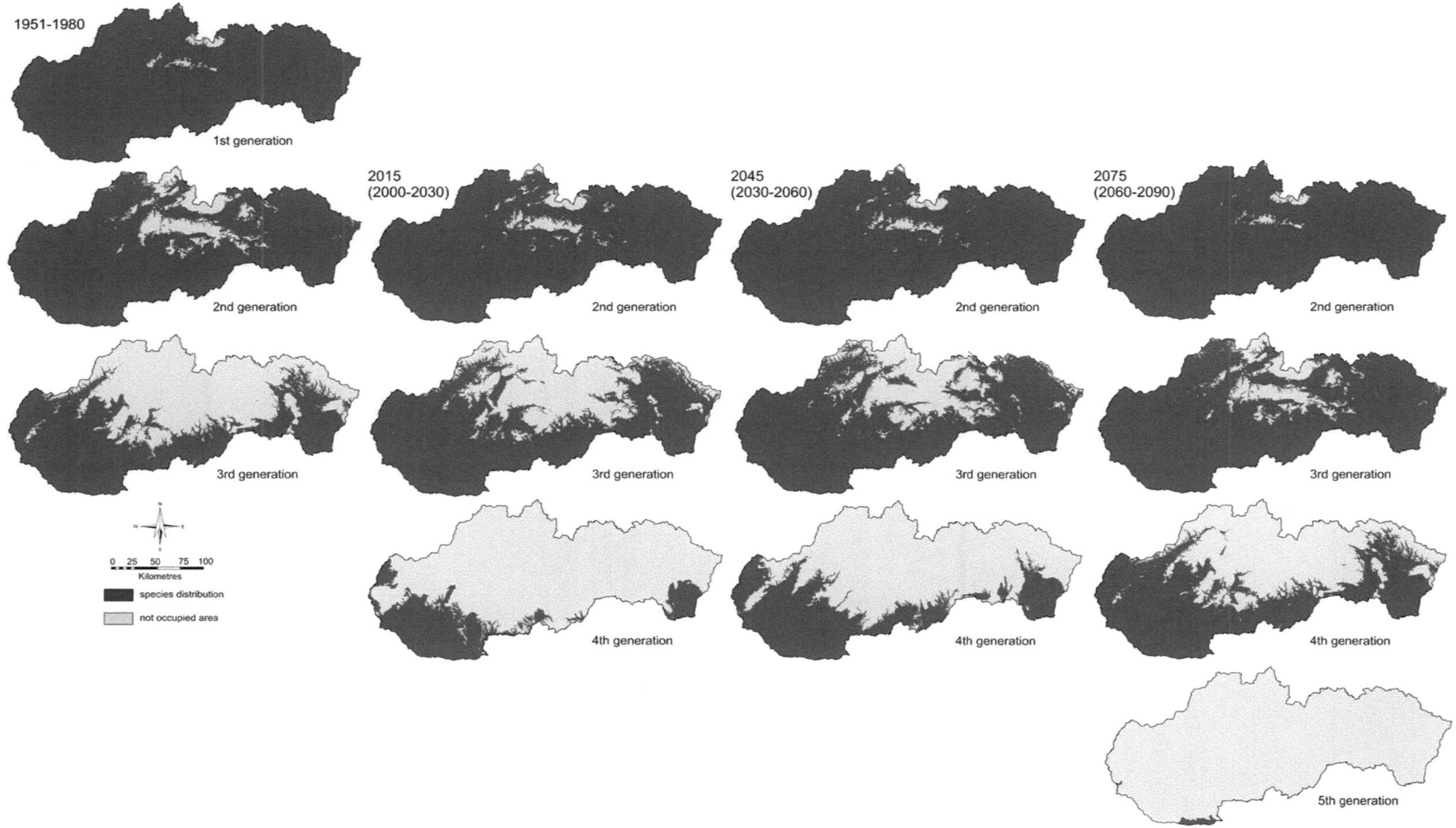

Fig. 4 The expected shifts of potential distributional ranges of particular bark beetle generation due to the increase of air temperature in Slovakia, irrespective of the distribution of the host Norway spruce

Table 4 Criteria used to predict potential LD outbreak areas in oak stands

	Q. cerris (%)	Temperature (°C)	Outbreak potential (suitability score)
Optimum	> 60	> 9.5	1
Suboptimum	30–60	8.75	0.5
Pessimum	< 30	< 8	0

rent and future climate. The respective weights were set to 0.4 (*Q. cerris*) and 0.6 (temperature).

In this way we obtained a surface indicating outbreak potential, taking on values ranging from 0 to 1. The arbitrary threshold of 0.7 was used to identify outbreak spots (according to previous knowledge about outbreak foci) (Fig. 5). This corresponded well with actual observations of defoliation. A quantitative analysis of their match has yet to be carried out.

Identified oak outbreak spots were considered as initial spots for pest spreading to beech stands. The parameters used in the prediction were distance from the oak outbreak spots and mean annual average air temperature (Table 5).

Table 5 Criteria used to assess the danger of potential LD outbreak spreading to beech stands

	Distance from oak outbreak spots (km)	Temperature (°C)	Outbreak potential (suitability score)
Optimum	< 2	> 9.5	1
Suboptimum	2–5	8.75	0.5
Pessimum	> 7	> 8	0

Finally, we calculated the extent of both potential oak outbreak areas and endangered beech stands for the four time scales mentioned above (Table 6). The map in Fig. 5 indicates the most critical areas and their growth under the respective climate change scenario. The results obtained are discussed below.

Table 6 The extent of predicted LD outbreak areas in oak stands; and area of beech forests within 7 km of oak outbreak areas in favourable temperature conditions

Time scale	Oak outbreak areas (ha)	Beech within 7 km from OOA (ha)
1951–1980 (+0°C)	17 149	83 925
2015 (+1.249°C)	44 640	193 288
2045 (+2.034°C)	48 768	212 378
2075 (+3.316°C)	49 273	216 623

Discussion and Conclusions

In this study we analysed how climate change can alter insect pest related forest disturbances in central Europe. The two species we focused on – *L. dispar* and *I. typographus* – are generally agreed to be sensitive to temperature, thus changes in their distributional ranges and population dynamics can be expected. Indications of this have already been observed in recent years worldwide. Many studies on climate change impacts suggest that the observed trends will accelerate. In this study we found that:

1. The area providing climatic conditions suitable for the full development of the second generation of IT within the current distribution range of spruce will almost double by 2075. Significant areas (20% of the current range of spruce distribution) with potential for the full development of the third generation are expected to appear around 2045. This will continue up to 50% in 2075. The fourth and fifth generations are not expected to occur in the current distributional range of spruce at all.
2. LD outbreak areas are expected to enlarge significantly in the near future. The outbreak areas will be more than double by 2015 compared to the 1951–1980 period. However, further growth will be limited by the distributional range of *Q. cerris*, therefore the extent of outbreaks will remain stable over future decades.
3. There are strong indications that pests will feed on beech as an alternative host, mainly in the vicinity of Turkey oak outbreak spots. Under the assumptions described above, the area of endangered beech stands will more than double by 2015, compared to the 1951–1980 period. Subsequently, the restricted growth of oak outbreak spots will keep this rate stable. Provided, the pest will spread regardless of the positions of Turkey oak outbreak spots, increasing temperature may enable it to grow unrestrictedly up to the upper distributional limit of other oak species (*Q. petraea* s. l.) and beech.

We focused on four temporal time scales: 1951–1980, 2015, 2045 and 2075. Thus, we can provide the fundamental data for the development of short-term, medium-term and long-term adaptation and mitigation strategies. Apart from increasing temperature, the main distinct feature of these time

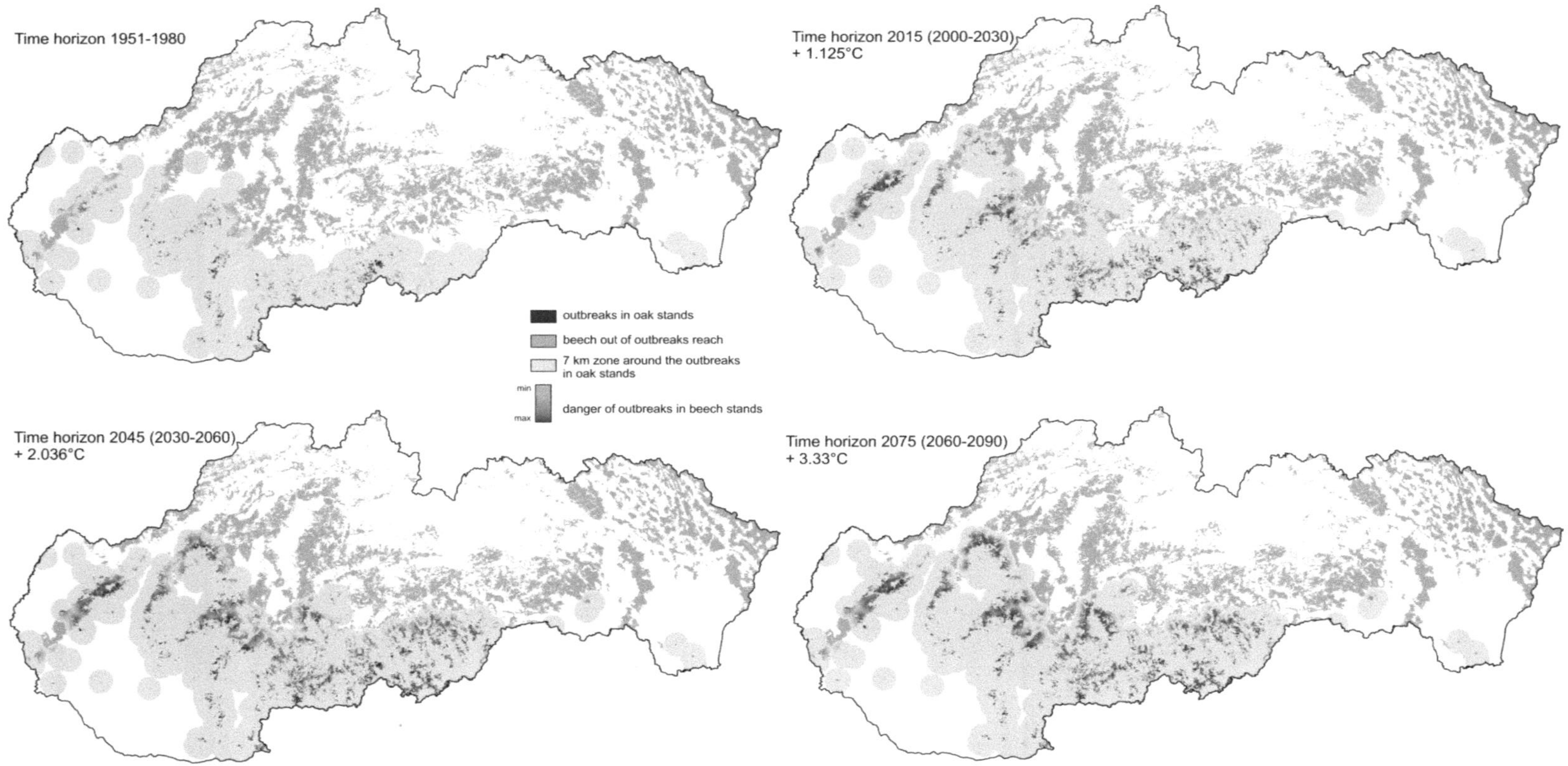

Fig. 5 Expected development of gypsy moth outbreaks in oak stands, and potential danger to beech stands in the vicinity of oak stands outbreak spots; in suitable temperature limit

scales is the potential distribution of target tree species. While in 2015 this remains similar to the status quo, 2045 and 2075 time scales move beyond this. In general, forest-management practices tend to adapt to forthcoming changes, therefore the impact may be expected to be less pronounced. However, it is difficult to currently assess the efficiency of applied measures.

In the case of bark beetle, the most endangered areas are those where the second and third generations appear, thereby dramatically altering the disturbance regime. Moreover, these are expected to occur in regions currently dominated by spruce. Trees in such regions are not adapted to more bark beetle generations per year, thus it may result in further extensive mortality of spruce stands.

Apart from temperature, moisture conditions play important role in IT population dynamics, as physiological condition of a potential host tree are crucial for the success of a bark beetle attack. Vital trees have higher resistance to prevent attacking bark beetles from successfully establishing broods. These mechanisms have been studied in detail under field and laboratory conditions (Baier 1996a,b; Rohde et al., 1996; Lieutier et al. 1997). Moisture probably influences the speed and success of attack and higher moisture may slightly delay effective dates of the beginning of IT generations.

IT shows variation in their reaction towards synthetic pheromone (Lieutier et al. 2004) which indicates their genetic variability. Probably, the gene pool in the current populations in central Europe still has a potential to adapt their voltinism to changing environment. These topics were not discussed in this chapter; however, they are expected to play an important rule in analysis of expected changes of IT population dynamics.

In the case of gypsy moth, outbreaks repeat every 10–11 years. Thus, a temperature increase between cycles may cause an unprecedented growth of outbreak areas and the intensity of defoliation. The last gradation occurred in 2004–2006 (Kunca et al. 2007). The defoliated area reached 48,000 ha. Compared to the previous outbreaks in 1992–1995, the defoliated area was 1.5 times larger (increasing from 32,000 to 45,000 ha). This means that the extent of the outbreak areas predicted for 2015–2045 were already observed in 2004–2006. In fact, there were evidently reported larger acreages than were really defoliated, due to the fact that some stands were defoliated repeatedly and thus counted more times. Nevertheless, the growing trend in the extent of outbreak areas is evident: for example, the total defoliated area during the 1983–1987 outbreak was only 8400 ha. The next outbreak is expected to occur in 2015, when the scenario indicates that at least 44,000 ha of oak stands will be defoliated.

Changes in gypsy moth outbreaks were evaluated only in relation to selected environmental parameters. Other factors, which are difficult to assess, are ecosystem relationships, such as host trees–gypsy moth, parasitoids–gypsy moth and predator–gypsy moth. Further research will also aim at shifts of host distributional ranges. Another aspect, which has yet to be analysed, are expected changes in the intensity of defoliation that is also forecasted to occur. In contrast, outbreak cycles should remain stable. Johnson et al. (2006) reported similar cycles as those observed in Slovakia also in southern countries with warmer climates – Hungary and northern Croatia. The south of Croatia is the first region where outbreaks are occurring in shorter cycles (the main period lasts 7–10 years and inter-outbreaks 2–3 years).

Acknowledgments This research was supported by the six FP project CECILIA (Central and Eastern Europe Climate Change Impacts and Vulnerability Assessment); and projects of the Ministry of Agriculture of the Slovak Republic 'Reconstruction of non-native stands endangered by changing conditions to more stable ecosystems' and 'Climate Change Impacts on the Forests of Slovakia' and by projects of the Ministry of Agriculture of the Czech Republic QH 81136 "Study and optimisation of real efficiency of control measures against Ips typographus in various fluctuation phases" and QH 71094 "The use of dendrochronology for reconstruction of fluctuation cycles of nun moth and gypsy moth in Central Europe".

References

Aber J, Neilson RP, McNulty S, Lenihan JM, Bachelet D, Drapek RJ (2001) Forest processes and global environmental change: Predicting the effects of individual and multiple stressors. BioScience, 51: 735–751

Anderbrant O (1986) A model for the temperature and density dependent reemergence of the bark beetle *Ips typographus*. Entomologia Experimentalis et Applicata, 40: 81–88

Anderbrant O (1990) Gallery construction and oviposition of the bark beetle *Ips typographus* (Coleoptera: Scolytidae) at different breeding densities. Ecological Entomology, 15: 1–8

Andersson M, Erlinge S (1977) Influence of predation on rodent populations. Oikos, 29: 591–597

Annila E (1969) Influence of temperature upon the development and voltinism of *Ips typographus* L. (Coleoptera, Scolitidae). Annales Zoologici Fennici, 6: 161–208

Ayres MP, Lombardero MJ (2000) Assessing the consequences of global change for forest disturbance from herbivores and pathogens. Science of the Total Environment 262: 263–286

Baier P (1996a) Auswirkungen von Vitalität und Brutbaum-Qualität der Europaischen Fichte, Picea abies, auf die Entwicklung der Borkenkäfer–Art *Ips typographus* (Coleoptera: Scolytidae). Entomology General, 21: 27–35

Baier P (1996b) Defence reactions of Norway spruce (Picea abies Karst) to controlled attacks of *Ips typographus* (L.) (Col Scolytidae) in relation to tree parameters. Journal of Applied Entomology, 120: 587–593

Baier P, Pennerstorfer J, Schopf A (2007) PHENIPS – A comprehensive phenology model of *Ips typographus* (L.) (Col. Scolytinae) as a tools for hazard rating of bark beetle infestation, Forest Ecology and Management, 249: 171–186

Baltensweiler W, Fischlin A (1988) The larch budmoth in the Alps. In: Berryman A (ed.) Dynamics of Forest Insect Populations: Patterns, Causes, Implications, Plenum, New York, pp. 331–351

Battles JJ, Robards T, Das A, Waring K, Gilless JK, Schurr F, LeBlanc J, Biging G, Simon C (2006) Climate change impact on forest resources, California Climate Change Center

Berryman AA (1996) What causes population cycles of forest Lepidoptera? Trends in Ecology and Evolution, 11: 28–32

Brasier CM (1996) *Phytophthora cinnamomi* and oak decline in southern Europe: Environmental constraints including climate change. Annales des Sciences Forestières, 53: 347–358

Byers JA (1993) Simulation and equation models of insect population control by pheromone-baited traps. Journal of Chemical Ecology, 19: 1939–1956

Byers JA (1996) An encounter rate model of bark beetle populations searching at random for susceptible host trees. Ecological Modelling 91: 57–66

Byers JA (1999) Effects of attraction radius and flight paths on catch of scolytid beetles dispersing outward through rings of pheromone traps. Journal of Chemical Ecology, 25: 985–1005

Byers JA (2000) Wind-aided dispersal of simulated bark beetles flying through forests. Ecological Modelling, 125: 231–243

Csóka G, Hirka A (2006) 2004- year of the gypsy moth in Hungary. In: Csóka G, Hirka A, Koltay A (eds.). Biotic damage in forests. Proceedings of the IUFRO WP. 7.03.10.) Symposium held in Mátrafüred, Hungary, 12–16 September 2004. pp. 271–275.

Christiansen E, Bakke A (1988) The spruce bark beetle of Eurasia. In: Berryman AA (ed.), Dynamics of Forest Insect Populations; Patterns, Causes, Implications, Plenum Press, New York, pp 479–503

Dale VH, Joyce LA, McNulty S, Neilson RP, Ayres MP, Flannigan MD, Hanson PJ, Irland LC, Lugo AE, Peterson CJ, Simberloff D, Swanson FJ, Stocks BJ, Wotton B (2001) Climate change and forest disturbance. BioScience 51(9): 723–734

Dale VH, Joyce LA, McNulty S, Neilson RP (2000) The interplay between climate change, forests, and disturbance. Science of the Total Environment 262(3): 201–204

Doane CC, McManus ME (eds) (1981) The gypsy moth: research toward integrated pest management. USDA Tech Bull 1584, Washington

Docherty M, Salt DT, Holopainen JK. (1997) The impacts of climate change and pollution on forest pests. In: Watt AD, Stork NE, Hunter MD (eds.), Forests and Insects, London: Chapman & Hall, pp. 229–247

Esper J, Buntgen U, Frank DC, Nievergelt D, Liebhold A (2007) 1200 years of regular outbreaks in alpine insects. Proceedings of the Royal Society, B 274: 671–679

Flannigan MD, Stocks BJ, Wotton BM (2000) Climate change and forest fires. Science of the Total Environment 262: 221–229

Franklin JF, Spies TA, Van Pelt R. et al. (2002) Disturbances and structural development of natural forest ecosystems with silvicultural implications, using Douglas-fir forests as an example. Forest Ecology and Management 155: 399–423

Forsse E (1991) Flight propensity and diapause incidence in five populations of the bark beetle *Ips typographus* in Scandinavia. Entomologia Experimentalis et Applicata, 61: 53–58

Furuta K, Iguchi K, Lawson S. (1996) Seasonal difference in the abundance of the spruce beetle (*Ips typographus japonicus* Niijima) (Col., Scolytidae) within and outside forest in a bivoltine area. Journal of Applied Entomology 120: 125–129

Gordon TR, Storer AJ, Wood DL (2001) The pitch canker epidemic in California. *Plant Dis.* 85(11): 1128–1139

Hansen EM. Bentz BJ, (2003) Comparison of reproductive capacity among univoltine, semivoltine, and re-emerged parent spruce beetles (Coeloptera: Scolytidae). The Canadian Entomologist, 135: 697–712

Hanson PJ,Weltzin JF. 2000. Drought disturbance from climate change response of United States forests. Science of the Total Environment 262: 205–220

Harding S, Ravn H, (1985) Seasonal activity of *Ips typographus* in Denmark. Z. Angew. Ent. 99: 123–131

Harrington R, Fleming RA, Woiwod IP (2001) Climate change impacts on insect management and conservation in temperate regions: can they be predicted? Agricultural and Forest Entomology, 3(4): 233–240

He HS, Mladenoff DJ, Crow TR (1999) Linking an ecosystem model and a landscape model to study forest species response to climate warming. Ecological Modelling, 114: 213–233

Hedgren PO, Schroeder LM (2004) Reproductive success of the spruce bark beetle *Ips typographus* (L.) and occurrence of associated species: a comparison between standing beetle-killed trees and cut trees. Forest Ecology and Management, 203(1–3): 241–250

Hirka A (2006) (ed.) A 2005. évi biotikus és abiotikus erdőgazdasági károk, valamint a 2006-ban várható károsítások [Biotic and abiotic forest damages in 2005 and forecasts for 2006.], Növényvédelem 42, 5 (in Hungarian)

Johnson DM, Liebhold AM, Bjørnstad ON (2006) Geographical variation in the periodicity of gypsy moth outbreaks. Ecography 148: 51–60.

Johnson DM, Liebhold AM, Bjornstad ON, McManus ML (2006) Circumpolar variation in periodicity and synchrony among gypsy moth populations. Journal of Animal Ecology, 74: 882–892

Kendall BE, Prendergast J, Bjørnstad ON (1998) The macroecology of population dynamics: taxonomic and biogeographic patterns in population cycles. Ecology Letters 1: 160–164

Knapp R, Casey MT (1986) Thermal ecology, behavior, and growth of Gypsy moth and eastern tent caterpillars. Ecology, 67(3): 598–608

Kunca A, Brutovský D, Findo S, Gubka A, Konôpka B, Konôpka J, Leontovyč, R, Longauerová V, Mindáš J, Novotný J, Pajtík J, Vakula J, Varínsky J, Zúbrik M, 2007. Occurrence of injurious factors in Slovakia in 2005 and their prediction for 2006. Forest Research Institute Zvolen. 89pp

Lange H, Økland B, Krokene P (2006) Thresholds in the life cycle of the spruce bark beetle under climate change. Interjournal for Complex Systems 1648

Lapin M, Damborská I, Melo M. (2001) Downscaling of GCM outputs for precipitation time series in Slovakia. Meteorologický časopis, 4(3): 29–40

Lapin M, Melo M, Damborská M, Vojtek M, Martini M (2006) Physically and statistically plausible downscaling of daily GCMs outputs and selected results. Acta Meteorologica Universitatis Comenianae, 34: 35–57

Liebhold A, Kamata N, (2000) Introduction: Are population cycles and spatial synchrony a universal characteristic of forest insect populations? Population Ecology, 42: 205–209

Liebhold AM, Turčáni M, Kamata N (2008) Inference of adult female dispersal from the distribution of Gypsy moth egg masses in a Japanese City. Agricultural and Forestry Entomology, Journal Summary 2007, 1–5

Lieutier F, Brignolas F, Sauvard D, Galet C, Yart A, Brunet M, Christiansen E, Solheim H, Berryman AA (1997) Phenolic compounds as predictors of Norway spruce resistance to bark beetles. USDA, Forest Service. General Technical Report NE 236: 215–216

Lieutier F, Day KR, Battisti A, Grégoire JC, Evans HF (eds.) (2004) Bark and Wood Boring Insects in Living Trees in Europe, A Synthesis 2004, XIV, Hardcover, Kluwer Academic Publishers Dordrecht/ Boston/ London.

Logan JA, Regniere J, Powell JA (2003) Assessing the impact of global warming on forest pest dynamics. Frontiers in Ecology, 1(3): 130–137

Logan JA, Regniere J, Gray DR, Munson AS (2007) Risk assessment in the face of a changing environment: Gypsy moth and climate change in Utah. Ecological Applications 17(1): 101–117

Lonsdale D, Gibbs JN (1996) Effects of climate change on fungal diseases. In: Frankland JC, Magan M, Gadd GM (eds.) Fungi and Environmental Change: Symposium of the British Mycological Society, Cranfield, England, pp. 1–19

Malcolm JR, Markham A, Neilson RP (2001) Can species keep up with climate change? Conservation Biology In Practice, 2(2): 24–25

Martinat PJ (1987) The role of climatic variation and weather in forest insect outbreaks. In: Barbosa P and Schultz J (eds.), Insect Outbreaks, Academic Press, New York, pp. 241–268

McCune B, Mefford MJ (1999) PC-ORD. Multivariate analysis of ecological data, version 4. MjM Software Design, Gleneden Beach, Oregon, USA.

Moravčík M et al. (2006) Report on Forestry in the Slovak Republic 2006 (Green Report). Bratislava, MP SR a NLC-LVÚ Zvolen

Myers JH (1993) Population outbreaks in forest Lepidoptera. American Scientist, 81: 240–251

Myers JH (1998) Synchrony in outbreaks of forest Lepidoptera: a possible example of the Moran effect. Ecology, 79: 1111–1117

Netherer S, Pennerstorfer J, Baier P, Schopf A, Führer E (2004) Modellierung der Entwicklung des Fichtenborkenkäfers, *Ips typographus* L., als Grundlage einer umfassenden Risikoanalyse. Mitt. Deut. Gesell. Allg. Ang. Ent. 14: 277–282

Netherer S, Pennerstorfer J (2001) Parameters relevant for Modelling the Potential Development of *Ips typographus* L. (Coleoptera, Scolitidae). Integrated Pests Management Reviews 6(3–4): 177–184

Netherer S (2003) Modelling of bark beetle development and ofsite and stand-related predispositions to*Ips typographus* (L.) (Coleoptera; Scolytidae). A contribution to risk assessment. Ph.D. thesis, Forstpathologie und Forstchutz der Universität fur Bodenkultur Wien

Patočka J, Čapek M, Charvát K (1962) The communities of Invertebrata on oaks in Slovakia. Biologické práce SAV, p. 98

Patočka J, Krištín A, Kulfan J, Zach P (eds.) (1999) Die Eichenschadling und ihre Feinde./ Institut fur Waldokologie der Slowakischen Akademie der Wissenschaften

Raffa KF (1988) The Mountain Pine Beetle in Western North America, In: Berryman AA (ed.) Dynamics of Forest Insect Populations, Plenum Press, New York

Reynolds KM, Holsten EH (1994) Relative importance of risk factors for spruce beetle outbreaks, Canadian Journal of Forest Research, 24: 2089–95

Rohde M, Waldmann R, Lunderstädt J (1996) Induced defence reaction in the phloem of spruce (Picea abies) and larch (Larix decidua) after attack by *Ips typographus* and *Ips cembrae*. Forest Ecology and Management, 86: 51–59

Rossiter MC (1994) Maternal effects hypothesis of herbivore outbreak. Bioscience, 44: 752–763

Schopf R, Köhler U (1995) Untersuchungen zur Populationsdynamik der Fichtenborkenkafer im Nationalpark Bayerischer Wald. Nationalpark Bayerischer Wald – 25 Jahre auf dem Weg zum Naturwald. Nationalparkverwaltung Bayerischer Wald, Neuschönau, 88–110

ter Braak CJF, Šmilauer P (2002) CANOCO Reference Manual and CanoDraw for Windows User Guide: Software for Canonical Community Ordination (version 4.5). Microcomputer Power (Ithaca NY, USA), 500pp

Turchin P, Berryman AA (2000) Detecting cycles and delayed density dependence: a comment on Hunter & Price (1998). Ecological Entomology, 25: 119–121

Turčáni M, Novotný J (1998) The importance of eight-toothed spruce bark beetle (*Ips typographus* L.) in Central Europe. In: McManus M. (ed.), Proceedings of U.S. Department of Agriculture Interagency Gypsy Moth Research Forum 1998. pp. 62–63

Vité JP (1952) Die holzzerstörenden Insekten Mitteleuropas. Göttingen: Musterschmidt, Wissenschafticher Verlag. pp. 68–84

Villemant C, Fraval A (1998) *Lymantria dispar* en Europe et en Afrique du Nord, INRA

Vité JP (1952) Die holzzerstörenden Insekten Mitteleuropas. Göttingen: Musterschmidt, Wissenschafticher Verlag, pp. 68–84

Wackernagel H (1998) Multivariate geostatistics: an introduction with applications, 2nd Edition, Springer Verlag, New York

Williams Dw, Liebhold Am (2002) Climate Change And The Outbreak ranges of two North American bark beetles. Agricultural and Forest Entomology, 4: 87–99

Wermelinger B, Seifert M (1998) Analysis of the temperature dependent development of spruce bark beetle *Ips typographus* L. (Coleoptera, Scolitidae). Journal of Applied Entomology, 122: 185–191

Woods A, Coates DK, Hamman A (2005) Is an unprecedented dothistroma needle blight epidemic related to climate change? Bioscience, 55(9): 761–769

Zumr V (1982) The data for the prognosis of spring swarming of main species of bark beetles (Coleoptera, Scolytidae) on the spruce (*Picea excelsa* L.). Z. Ang. Entomol. 93: 305–320

Genetic Background of Response of Trees to Aridification at the Xeric Forest Limit and Consequences for Bioclimatic Modelling

Cs. Mátyás, L. Nagy and É. Ujvári Jármay

Keywords Forest ecosystems · Changes of climatic environment · Aridity tolerance · Common-garden test results

Introduction: Xeric Limits and Genetics

Trees, as dominant components of forest ecosystems, are of high ecological importance in the temperate belt and receive much attention with regard to adaptation potential and future risks of diversity loss and extinction. Much of the climate change literature however is based on simulations and models, the genetic background of which is often deduced from results with annuals or other fast reproducing organisms. It should be remembered that lifespan plays a decisive role in the adaptation process and mechanisms functional in annuals or even shorter generation organisms might be irrelevant for trees when considering the timeframes of present climate change scenarios. Genetic analyses of forest trees demonstrate that their genetic system and diversity parameters are *diametrically different* from annual plants or animals (Hamrick et al. 1992).

The crucial problem of realistic interpretation of adaptation to climate change is however the missing of field observations, such as common-garden tests. In forestry, these tests have a very long tradition (provenance tests). Tests with trees are difficult to establish, laborious and time consuming to maintain and measure. Provenance testing of forest trees may be one of the most important contributions of forestry to biological sciences. They are unique because they have been established with natural-state populations and adapted to specific conditions. They are unique also because these tests have been established across continents, at many sites and maintained over decades. Important insights in evolutionary ecology, for example, on processes and patterns of adaptation, have been collected and utilised very early (see review of Langlet 1971). It is all the more surprising that much of the very extensive work of forest geneticists (e.g. Morgenstern 1996; Geburek and Turok 2005; Eriksson and Ekberg 2001; Mátyás 1997, 2000, Müller-Starck and Schubert 2001) failed to capture the attention of ecologists.

Another important field often missed when modelling and predicting responses to changes, is the production biology of forest trees (forest yield science). Large-scale assessments exist in forestry which analyse the response of forest stands to extant climate change effects and weather extremes (e.g. Briceno-Elizondo et al. 2006; Lapenis et al. 2005; Kramer and Mohren 2001). For instance, data show for large parts of Western Europe an unprecedented acceleration of forest growth in the recent warming decades, exceeding in some cases 50% (Spiecker et al. 1996). Interpreting these data might alleviate prediction difficulties of adaptive behaviour of tree populations. These shortcomings emphasise the importance of cross-disciplinary research (Mátyás 2006b).

Cs. Mátyás (✉)
Institute of Environmental Sciences, Faculty of Forestry, West Hungarian University, Ady Endre Str. 5, Sopron 9400, Hungary
e-mail: cm@emk.nyme.hu

K. Strelcová et al. (eds.), *Bioclimatology and Natural Hazards*,
DOI 10.1007/978-1-4020-8876-6_16,

Bioclimatic Modelling of Xeric Limits Needs Genetic Considerations

The determination of 'climatic envelopes' of forest tree species has been a long-time topic, in principle already since Alexander von Humboldt's observation of links between climates and vegetation physiognomy. Climatic demands of tree species and of forest ecosystems have however attained a sudden actuality in the context of adaptation to predicted climatic changes. Recent publications on bioclimatic modelling and predicted climate change–triggered vegetation shifts are abundant and have been considered also in the fourth report of IPCC (chapter on Europe, 2007).

Models and analyses deal however mostly with the shift of the thermal ('upper' or 'front') limits of distribution (e.g. for Europe: Savolainen et al. 2004; for North America: Wang et al. 2006; for North Asia: Lapenis et al. 2005; Rehfeldt et al. 2003). Migration at the front, that is the shift of vegetation is the most visible and illustrative response to climate change. 'Forward' colonisation is more sensitive to climatic changes than loss of vitality and retreat at the xeric limits, which is buffered by persistence and plasticity as explained later. No surprise that investigations at the xeric limits are seldom, mostly dealing with montane-Mediterranean conditions (e.g. Pigott and Pigott 1993; Peñuelas et al. 2001; Westphal and Millar 2004; Piovesan and DiFilippo 2005; Jump 2006; Fournier et al. 2006) and often lack a clear climatological foundation necessary for more general conclusions. The report of IPCC (2007) also misses to deal with the retreating rear limits according to their importance.

Bioclimatic modelling of distribution ranges is based on the concept that distributional patterns depend – among other factors – on the physiological tolerance limits to climatic effects. This generally recognised rule has to be extended by the statement that physiological tolerance is unquestionably determined by genetics. Tolerance can be defined as the ability of a genotype to maintain its fitness despite damage. It is also presumably genetically correlated with phenotypic plasticity, that is, with growth vigour across environments (Weis et al. 2000; Mátyás and Nagy 2005). Limits of tolerance are therefore genetically set and will determine the presence or absence of species (Fig. 1).

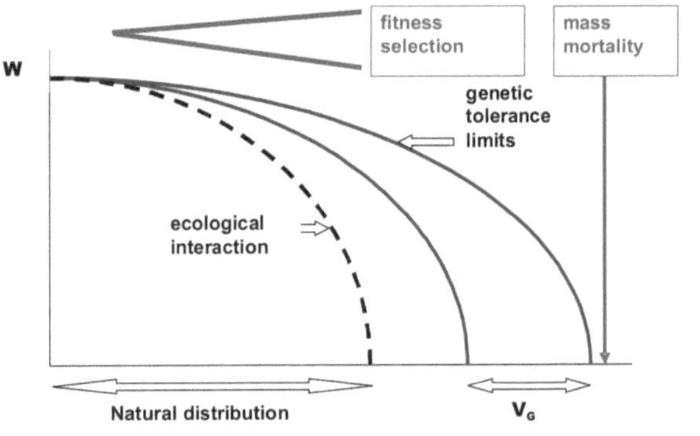

Fig. 1 Ecological-genetic model of fitness decline and mortality triggered by worsening of climatic (site) conditions. The genotypic variance of limits of tolerance (V_G) represents the basis of natural selection. Due to competitive or trophic interactions in the ecosystem, the natural distribution is usually stronger limited, than the genetically set critical tolerance, as marked by the dashed curve (Mátyás 2006a)

Thus, adaptive response to environmental stress is ultimately a genetic issue, and correct bioclimatic modelling is strongly dependent on genetically set tolerance limitations.

Quantitative Genetic Approach to Interpretation of Adaptive Behaviour of Forest Trees

The basic concept may be summed up as follows:

- effects of climate factors on quantitative traits may be effectively studied on zonal tree species, especially at the distributional limits;
- much of the intra-specific genetic differentiation among populations of dominant tree species with large distributional ranges, is linked to climatic adaptation and illustrates the functioning of climatic selection;
- given climatic conditions (at test sites) trigger differentiated responses, depending on the adaptedness of populations of different origin;
- limits of tolerance are genetically determined; the fitness of a population experiencing worsening environmental conditions declines gradually – depending on available genetic variability – to mass mortality when the genetic and ecological possibilities of adaptation are exhausted (Fig. 1);
- quantitative, adaptive responses (growth, phenology, health) measured in comparative tests may be

utilised to forecast the effects of climatic change, as the response of populations at the test site can be interpreted as a simulation of environmental changes (trading in space for time).

The Problem of the Xeric Forest and Tree Distribution Limits

Out of the complexity of biotic-ecological factors determining distribution limits, in this chapter the climatic and genetic ones are dealt with in detail, for the conditions of xeric (rear) limits on continental plains. The 'front' and 'rear' or upper and lower limits of distribution of widely distributed, zonal species differ by the key limiting climate factors, temperature and aridity, respectively. Thermal limits are determined by relatively accurately measurable temperature conditions, which makes modelling relatively reliable. Xeric (aridity) limits of distribution are determined by climatic aridity, modified by local soil water regime conditions. These limits are more difficult to trace than thermal limits. Beyond methodical difficulties, xeric forest tree limits have received much less attention in the past because of their peripheral situation, away from regions of mainstream European interest.

Xeric forest and tree limits at the planar border zone between closed forests and woodlands (forest steppe) are especially vulnerable and ecologically very important. Their importance is increasing with predicted climatic changes. The loss of closed forest cover implies the disruption of vital ecological services forests are providing. Beside certain Mediterranean and sub-Mediterranean regions (such as the Spanish Meseta or Aquitany in Southern France), this belt reaches from East-Central Europe (Moravian, Carpathian Basin) across the plains of southeast Europe (Romania, the Ukraine, South Russia) far into southern Siberia.

Xeric limits are fuzzy on flat terrain, where species and vegetation types occur mosaic-like and follow the pattern of small-scale topographic, water regime and soil variation. These distribution limits are strongly affected by biotic interactions, competition, diseases and pests, and various obstacles to regeneration.

It has to be emphasised that low-elevation distribution limits of tree species and forest types have been under continuous and strong human impact due to higher population density and easier accessibility as compared to upper limits. Therefore, present-day distribution patterns reflect long-lasting anthropogenic effects. Still, on a larger scale, present xeric limits at least indicate genetic-physiological limitations of climatic tolerance.

Ecological and Genetic Options to Adapt to Changes at the Xeric Limits

Early symptoms of climate change effects at the xeric limits, such as loss of vitality, sporadic mortality and forest health problems indicate the constraints of adaptability. There are both genetic and non-genetic mechanisms operating on the individual, population, species and ecosystem levels, balancing changes in environmental conditions. On species and ecosystem or landscape level, a non-genetic possibility of responding to large-scale changes in the environment is *migration through seed dispersal*, including species substitution (succession, immigration) provided there are suitable species available. Palaeoecological evidence is abundant on migration during the epochs of glacials and inter-glacials (Davis and Shaw 2001), and this is the underlying response mechanism modelled and described by practically all ecologically oriented future vegetation shift scenarios as well.

Extensive studies on *long-distance gene flow through pollen* have shed light also on this very effective mechanism of constant replenishment of genetic resources, which most probably contributes to the unexpectedly high diversity and adaptability of tree populations (Hamrick et al. 1992). It is self-evident that migration and gene flow are functional across the whole range of distribution. Both mechanisms have however limited importance at the xeric limits, because they *rather support the escape of species and genes instead of the persistence* in marginal situations.

Genetically set adjustment mechanisms sustain persistence both on population and individual level. On the level of populations, *natural selection* adjusts the average fitness of a population to changing conditions. The directed genetic change of the population's gene pool towards an optimum state is genetic adaptation in the strict sense. It is a well-accepted concept that the basic precondition for fast and effective genetic adaptation lies in sufficiently large variation, that is in sufficient genetic diversity (e.g. Booy et al. 2000; Beaulieu and Rainville 2004). Long-term genetic adaptability is therefore directly depending on the conservation or even reconstruction of broad adaptive genetic variance. The progress of selection will also depend on the intensity of selection pressure, as described by Fisher's theorem (Mátyás 2004). This progress may be counterbalanced by gene flow and migration.

Selection by climatic effects is certainly a key element among ecological factors. Although seemingly obvious, the role of selection in tracking changes is not uniformly seen. Some authors presume that climate selection is ineffective, plays a subordinate role or is at least very slow (Bradshaw 1991; Huntley 1991; Savolainen et al. 2004). The classic works of Turesson (1925) and Clausen et al. (1940) on annuals, and Langlet (1971) on trees support however the strong selective power of climate effects, backed by more recent papers such as Linhart and Grant (1996), Davis and Shaw (2001), Etterson and Shaw (2001), Jump and Peñuelas (2005).

On the individual (genotype) level, *phenotypic plasticity*[1] provides the ability to survive in a wide range of environments, without genetic change in the classic sense. Plasticity is the *environmentally sensitive production of alternative phenotypes by given genotypes* (DeWitt and Scheiner 2004). The term has been coined in zoology, where it has been applied in a relatively restricted manner for certain environmentally induced developmental or morphological phenomena. For plants, plasticity has to be interpreted relatively broadly (Bradshaw 1965). In our interpretation, derived from the practice of forest tree breeding, the ability of the genotype (clone), or of the population to maintain relative (usually superior) fitness across a wide range of environments is regarded as phenotypic plasticity or stability (Mátyás 2006a).

Plasticity implies that the phenotypic expression of genes is influenced by the environment, thus the organism may modify its responses within genetically set limits. It is especially effective in modular organisms such as trees, where the growth and developmental cycle may be strongly influenced by the environment. Phenotypic plasticity will set the limits of environmental heterogeneity within which a genotype or population can persist in its lifetime. In ecological literature, plasticity is often regarded as a non-genetic adaptation mechanism. It has to be emphasised that this trait is definitely heritable and also underlying climatic selection (Mátyás 2006a).

When analysing the chances to adapt to the rapidly changing climate, it is important to comprehend which of the described mechanisms will gain importance in the timeframe of expected scenarios, and which mechanisms might be even negligible, and to discern between adaptation options in the current and future generations.

Processes of Adjustment in the Extant (Currently Growing) Generation

Phenotypic plasticity provides the ability for individual, instant acclimation without any change in the inherited genetic resources of the population. Mainly for reasons of difficulty of experimental analysis, reaction norms and limits of adaptability set by phenotypic plasticity are rarely considered in connection with adaptation. It is an often underestimated issue both in forest genetics and ecology, in spite of the fact that *considering the speed and magnitude of predicted changes, phenotypic plasticity is the most important and practically only natural buffering mechanism* (Mátyás and Nagy 2005).

Natural selection is eliminating the genotypes of low fitness in the lifetime of a population. Although some studies on adaptation deal with selection processes and with changes in the genetic composition, few studies have been implemented in practice under conditions of severe change where populations are reaching their tolerance limits. In such extreme situations, the effectiveness of adjustment through selection ceases and mass mortality and even local extinction may follow (Lynch and Lande 1993; Mátyás and Yeatman 1992). Compared to plasticity, spontaneous natural selection plays therefore a smaller role in supporting persistence than generally assumed.

Processes of Adjustment in the Following Generations

The progeny generation of parents selected by ecological factors will be altered on ecosystem level by migration (including gene flow) and succession.

[1] Phenotypic plasticity is a trait which may be interpreted and measured also on population level as the average effect of individuals. Common garden experiments prove that significant differences exist between populations (see in following chapters).

Preconditions for effective adjustment in both cases are landscape connectivity, availability of propagules and suitable speed, tracking the pace of change.

Environmental signals might trigger also *genetic carryover* (imprinting, 'after-effects') effects. These are lasting changes of genetic regulation, which can be inherited (for a review, see Martienssen and Colot 2001).[2] Theoretically, spontaneous mutations might contribute to the replenishment of adaptive potential as well. These mechanisms are however too slow to become effective in a rapidly changing environment. There are also further genetic, migratory and evolutionary constraints to effective tracking (Loeschke 1987; Kremer et al. 1999; Mátyás 2006a).

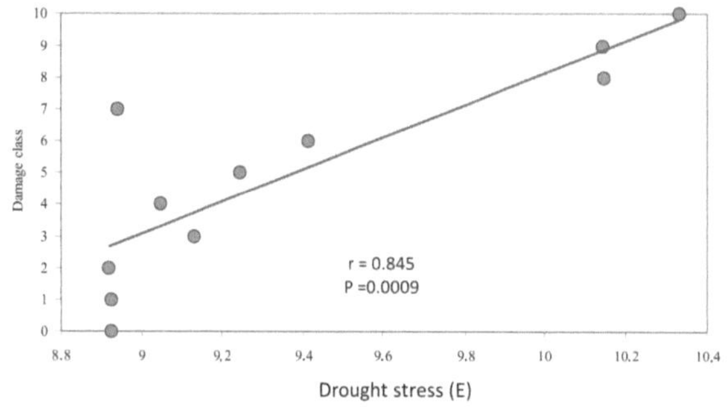

Fig. 2 Approaching the genetically set limits of tolerance: climatic summer drought stress (*E*) versus leaf damage classes of beech in permanent monitoring plots in Hungary. The figure shows the change in health status from healthy (class 0) to dead (class 10) along a climate gradient of increasing summer moisture stress (for explanation of '*E*' see text; analysis by G. Veperdi)

Selection at the Xeric Limit

Selective Pressure Is Determined by the Ecological-Climatic Position

The selective pressure increase when approaching the xeric limits has been seldom studied (see review of Kingsolver et al. 2001). Observations on tolerance are scarce even in forestry practice, as forest management strives to operate well above the tolerance limits to secure economic returns.

Having investigated the health condition of sampled trees in the permanent heath-monitoring network in Hungary, it turned out that the effect of selection pressure growing towards the xeric limits can be traced for climate indicator species such as beech or sessile oak. The position in relation to the distribution limit significantly determines vitality: the larger the distance from the limit, the better the health status.

Climatic distance from the xeric limit may be expressed for example by the average climatic moisture stress. A modified Ellenberg index was calculated as surrogate for climatic summer drought severity:

$$E = T_{\mathrm{Jul}} \times (P_{\mathrm{Jun}} + P_{\mathrm{Jul}} + P_{\mathrm{Aug}})^{-1} \times 100 \quad (1)$$

E index was calculated from the percentile ratio of mean July temperature (T_{Jul}) and sum of mean precipitation of summer months ($P_{\mathrm{Jun}} + P_{\mathrm{Jul}} + P_{\mathrm{Aug}}$). The index shows a very close correlation with leaf damage classes for both beech and sessile oak. Figure 2 shows the average damage classes for beech assessed in permanent monitoring plots in Hungary. It is visible that approaching the genetically set limits of tolerance, health status declines and reaches mortality at extreme locations.

Genetic Effects of Past Climatic Selection

The selective effect of local environment on the genetic pool of plant populations has been demonstrated for adaptive traits in a large number of studies (for review, see Linhart and Grant 1996). Genes coding fitness-related physiological processes such as certain metabolic enzyme loci should be under selection (Eanes 1999), leading to a 'selective sweep' if the effect is strong enough. However, few studies provide information on the concrete action of climate selection such as drought events on plants. Similarly, out of the very rich literature on intra-specific genetic variation of forest trees, papers dealing with adaptive response to climate/weather events are rather the exceptions.

To prove the hypothesis of increasing selection towards the tolerance limit (see Fig. 1), gene loci of adaptive/physiological relevance have to be studied. A systematic genetic inventory at gene loci coding key enzymes was carried out in sessile oak and

[2] Such effects were detected also in certain forest trees, for example in boreal populations of Norway spruce (*Picea abies*) (Skrøppa and Johnsen 2000). Similar effects are suspected in a progeny trial of the same species (Ujvári-Jármay and Ujvári 2006). The general significance of genetic carryover in adaptation of trees is so far unclear and should be treated with caution.

beech stands along climate severity gradients in Hungary (Borovics 2007). Results show a strikingly distinctive selective effect on allelic diversity. The effect depends on the adaptive role of the investigated gene or of the allele. In the majority of cases a frequency increase of certain, adaptively favourable alleles (gene variations) has been observed, which leads to a decrease of diversity and higher level of fixation towards the xeric limits, that is to a selective sweep. In a few cases, where heterozygotes seem to support tolerance, increasing heterozygosis was also observed. These results are all the more surprising as the sampled stands have been under prolonged human (silvicultural) influence.

As an example, the effective number of alleles (N_e) of the gene locus Skdh-A in sessile oak is presented in Fig. 3. The enzyme produced by this gene takes part in the synthesis of amino acids, of growth hormones and influences lignification. The cline in moisture conditions is represented by annual mean precipitation at the sampled locations. A strong positive correlation with rainfall ($r = 0.7528$, $p = 0.0012$) indicates unmistakeably the sweep of less-adaptive allelic variations towards the xeric limit. The reduction of allelic variation leads to the highly significant increase of the frequency of an obviously drought-tolerant homozygote genotype '33' at the Skdh-A gene locus at drought-stressed locations ($r = -0.797$, $p = 0.0004$, not shown). Climatic selection apparently selects for genotypes (individuals) carrying the adaptively most effective allele 3 in homozygote combination (Borovics 2007).

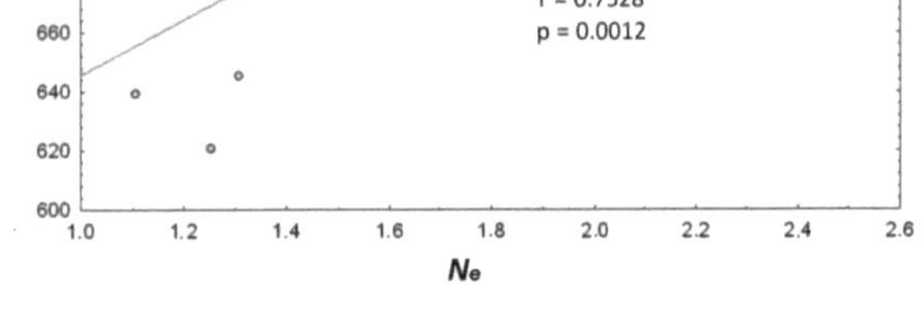

Fig. 3 Effective number (N_e) of alleles at the Skdh-A gene locus in sessile oak populations towards the xeric limit. Mean annual precipitation (Pyear) serves as surrogate for the ecological distance from the xeric limit (Borovics 2007)

These findings are in contradiction with certain current opinions on the genetic status of populations at the xeric limits. Some authors propose (e.g. Hampe and Petit 2005; Petit et al. 2002) that during Quaternary climate oscillations, many populations persisted in habitat islands (stable edges). These populations might be very old, and maintain a high level of between-population genetic diversity. Accordingly, marginal populations would generally harbour most of the species' genetic diversity.

This hypothesis – although permitting genetic drift (diversity loss) within individual populations – does not seem to hold for lowland xeric limit situations, where stable habitat refugia are seldom found. Persistence of trees in refugia means by definition that other populations not growing in exceptional ecological situations went extinct. Under planar conditions, refugia do not provide safe havens of diversity. These isolated populations show the effects of strong selection and diversity loss.

A further questionable opinion is the presumed decline of actual fitness as a result of diversity decline. The loss of adaptive diversity theoretically decreases the adaptive potential of the population. Its role in fitness loss is however somewhat overestimated. Narrowing genetic variation under selection pressure has by definition the function of increasing (instead of decreasing) fitness on population level. There is no 'optimum' diversity, and the effect of diversity loss on future competitiveness is therefore unpredictable (Whitlock in: Jump and Peñuelas 2005).

The observed tolerance loss at the xeric limits to pests and diseases is primarily not the consequence of diversity loss. The attack of insects and pathogens is usually connected either with the changed consumer life cycle (Woods et al. 2005) due to changed climate conditions, or the weakened physiological status of the trees (e.g. lower resin or sap pressure permitting the attack), or both.

Weather Extremes: The Real Drivers of Climatic Selection

Extremes and Mortality

It is self-evident that means of climatic parameters do not characterise correctly the severity of climatic selection at the xeric limits. Even under zonal conditions, the limit cannot be described properly in terms of average climate, as primary limiting conditions are the irregularly appearing extremes, often triggering damage by diseases and pests.

Spontaneous climatic selection is driven at the xeric limits by recurrent droughts. Therefore, if climatic means are used in analyses, they should be regarded rather as surrogates for extreme events.

For methodical reasons the analysis of causal relations between drought events and selection has to be confined to selected locations. In Fig. 4 severity and frequency of drought events are shown for a South Hungarian beech forest at the very limits of distribution. The drought index used calculates with the ratio of spring and summer precipitation and average temperature of the three summer months. Investigation of mortality frequency has shown that single drought events did not threaten the stability of populations. The recurrent drought period lasting for 5 years between 2000 and 2004 however resulted in very serious mortality in the investigated beech stands, in one case the population went extinct. It has been found that for beech, recurrent drought events of 4–5 years lead to irreversible mass mortality and local extinction (Berki et al. 2007). These cases illustrate in practice the process of retreating margins of distribution as proposed by the tolerance limit hypothesis (Fig. 1).

Expected Frequency Changes of Drought Events at the Xeric Limits

Drought events will happen in line with predicted climatic changes but their frequency and severity may change at a rate different from the average trends. Frequency change of drought events have been analysed for Hungary using the REMO climate model of the Max Planck Institute (MPI), Hamburg. The predicted frequency of drought years (precipitation decline exceeding 5% of the periodic mean) and of drought summers (precipitation decline exceeding 15% of the seasonal mean) are shown in Table 1. The anomalies are averages calculated for the territory of Hungary, related to the 1961–1990 period. The model indicates relatively modest increases of drought events for the scenario A2, at least for the first half of the twenty-first century. Still, the average precipitation decline will continuously progress, and the average precipitation loss of drought summers may reach 30% (related to the predicted averages). It has to be pointed out that the data refer to *average anomalies*, single drought events may be much more severe. It is highly remarkable in Table 1 that from 2050 onwards, the REMO model defines every second summer as drought event: 24 summers out of 50 years (Gálos et al. 2007)

The described shifts in drought frequency may cause catastrophic changes in lowland regions at the

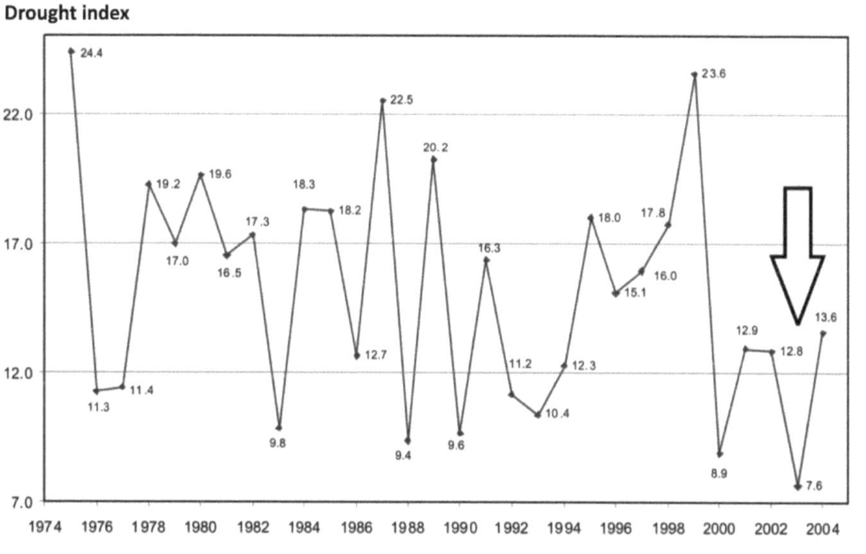

Fig. 4 Drought frequency and the initiation of mass mortality of beech at the location Fiad (south Transdanubia). Years with drought indices (*vertical axis*) below the 30-year average (<14.5) have been considered as drought events. Mass mortality started after the fourth year of consecutive drought (*arrow*) (after Berki and Rasztovics 2004)

Table 1 Frequency of recent and predicted drought events for Hungary, calculated with MPI's REMO climate model (from Gálos et al. 2007)

Period	Number of years	Mean of precipitation anomalies (%)	Mean of temperature anomalies (°C)
Drought years			
1951–2000	17	−12.42	+0.39
2001–2050	9	−16.52	+1.24
2051–2100	21	−19.07	+3.75
Drought summers			
1951–2000	15	−28.02	+0.95
2001–2050	9	−29.21	+2.00
2051–2100	24	−34.98	+2.86

Table 2 Present and predicted area of forest steppe climate in Hungary

	Present situation	Predicted for 2080
Area (million ha)	4.92	7.53
7.53 In percents of total territory	52.9	81.0

xeric limit. For the mildest climate change scenario, the following area change of non-forest, open woodland climate has been calculated for Hungary:

Table 2 indicates a predicted area increase for forest steppe climate by more than 50%, which implies that open woodlands could potentially replace a significant part of present-day closed forests. Part of the predicted change may be buffered by the plasticity and persistence of populations. Mass mortality is most probable at the rear edges, on sites with unfavourable water regime.

Estimation of Aridity Tolerance from Common Garden Test Results

There are two possibilities to elucidate the genetic background of tolerance:

- quantitative analysis of adaptive response in common gardens, or
- molecular genetic tracing of genetic regulation of adaptive traits.

Contemporary molecular genetics is investing great efforts to clarify the genetic mechanism regulating quantitative traits, however the breakthrough has not been reached yet. Compared to quantitative traits, variation at the molecular genetic (nearly exclusively neutral) loci is inconclusive (Savolainen 1994; Savolainen et al. 2004). This is true even for identified quantitative trait loci (QTLs). There might be hundreds of genes participating in the manifestation of quantitative traits. It seems that the identification of at least major gene groups responsible for these traits will be the task of coming decades (Neale and Wheeler 2004). Therefore the shortcut, simple method of direct analysis of quantitative responses in common gardens cannot be missed.

Transfer Analysis of Common Garden Data

The idea of transfer analysis, that is modelling of responses and forecasting responses to scenarios based on provenance test data, has been proposed originally by the author (Mátyás and Yeatman 1987; Mátyás 1994). The principle of this approach is the use of ecological variables to express the change of environment through transfer to the test site. Adaptive responses to changes can be interpreted, generalised and compared more easily if expressed as ecological distances. To observe tolerance and plasticity, populations (provenances) are assessed in different environmental conditions. Regression analysis can be applied to describe the change in fitness. The slope of the function represents the sensitivity to changes and possible limits of tolerance. Taking growth and health condition as proxy for fitness, the function is interpreted as the species' reaction norm of fitness to the variable investigated (precipitation, drought). Thus, growth and survival of populations adapted to a given site, transferred and tested in other environments as part of common garden tests, can be interpreted as a simulation of ambient changes at the original location. The transfer analysis validates the forecasting of adaptive response and of effects of environmental change (Mátyás and Nagy 2005; Rehfeldt et al. 2003). This approach has been applied by numerous researchers (e.g. Persson and Beuker 1996; Rehfeldt et al. 1999; Andalo et al. 2005; Wang et al. 2006).

In common-garden tests the response of populations is analysed under conditions often very different from their original climatic environment they are adapted to. The changes caused by the transfer to the test site are in extreme cases exceeding 5°C annual mean temperature, which is well beyond the predicted climate scenarios for medium latitudes (Fig. 9). Therefore field observations provide valuable concrete data on population responses, which are otherwise not obtainable from theoretical models. Traces of severe climate selection (mortality) are seldom observed in juvenile age in common gardens. One reason for this is the original concept of selecting sites for provenance tests: in hindsight it is a pity that extreme test locations provoking high mortality rates have been avoided for obvious reasons. The existing tests are therefore first of all informative in providing data on phenotypic plasticity.

Response to Changes of Climatic Environment

When testing a set of populations at a given site, a characteristic response pattern can be observed, where growth and vigour of populations originating from the vicinity of the test site tend to be the best and the performance of populations decreases with the ecological distance from the location of origin (Mátyás and Yeatman 1987, 1992). Equations describing these phenomena have been developed by a large number of authors, and response functions have been broadly utilised to define seed transfer rules and to delimit seed zones. Earlier models relied on describing genetic variation patterns on a geographical basis using latitude, longitude and elevation as independent variables to describe variation patterns for a given area (Shutyaev and Giertych 1997). The ecological relevance of these variables is ambiguous. With the growing availability of digital climatic data, multivariate analyses of provenance trials have provided convincing results on patterns of climatic selection and adaptation. Common garden tests of most tree species verify that populations originating from different climates show specific adaptation to local conditions and, accordingly, respond differently if grown under uniform conditions of a common garden. Height growth response shows significant correlations even with very general climatic parameters such as average annual mean temperature[3] at the location of origin (Mátyás and Yeatman 1987, 1992; Rehfeldt et al. 2003). Calculated regression functions show for most investigated species a very characteristic pattern of growth response, indicating an obviously clinal adaptation to the climate across the range.

Species with different life histories, such as Norway spruce and Scotch pine, show strikingly similar patterns. For example, the relation height at 15–16 years of age versus annual mean temperature at origin may be described by a typical response regression function, where in both cases a not very well expressed maximum appears around 6–7°C annual mean (Figs. 5 and 6). At the same time no clear trend is detectable in within-population variation, that is the magnitude of genetic variation in height between individuals remains roughly the same across the range. Expressed in standard deviation, a decreasing trend from the northern and central (optimum) populations towards the xeric limit (edge) is detectable only for Scotch pine.

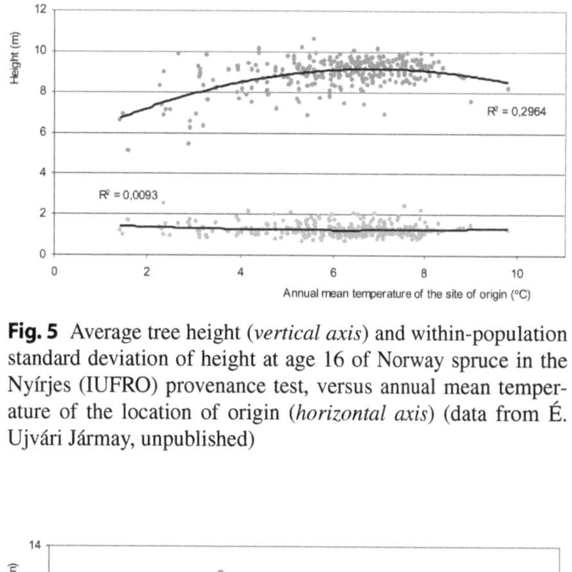

Fig. 5 Average tree height (*vertical axis*) and within-population standard deviation of height at age 16 of Norway spruce in the Nyírjes (IUFRO) provenance test, versus annual mean temperature of the location of origin (*horizontal axis*) (data from É. Ujvári Jármay, unpublished)

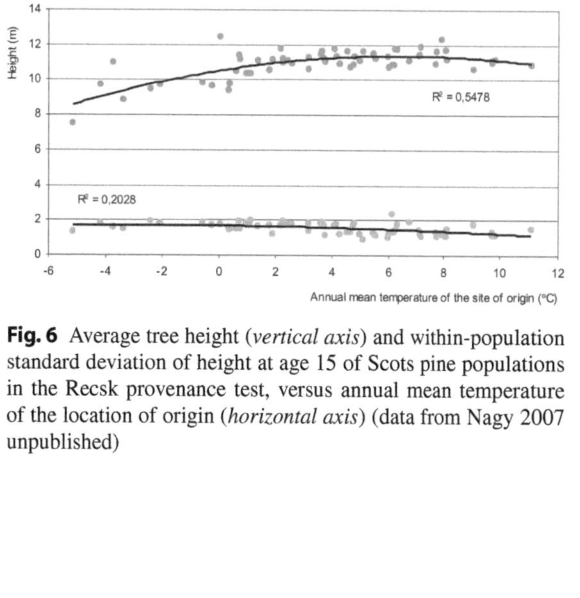

Fig. 6 Average tree height (*vertical axis*) and within-population standard deviation of height at age 15 of Scots pine populations in the Recsk provenance test, versus annual mean temperature of the location of origin (*horizontal axis*) (data from Nagy 2007 unpublished)

[3] Multivariate analyses using numerous climate parameters supply even better results. For the sake of simplicity, annual means are used in the following examples.

The effect of temperature conditions on height growth of populations has been studied in six Scots pine tests situated in the centre of European Russia (Mátyás and Nagy 2005). The climate there is continental, summers may show moisture deficit. In order to exclude the effect of site quality, data were standardized by expressing height at age 16 in percents of locally adapted populations (relative height, see Fig. 7). In the figure, transfer into cooler environments is shown by negative temperature sums. Tested populations were grouped according to their adaptedness into northern, central and southern groups. Regressions relative height versus temperature sum difference of the vegetation period at the test and original location were calculated from monthly averages. The comparison of the regressions show that the three groups behave very similarly and display a marked depression in height growth with increasing aridity (i.e. higher mean temperature) of test location. This means that if introduced to more arid conditions than they were adapted to, populations react with growth decline expressed in relative height. On the other hand, the transfer into cooler (=more humid) environments resulted in growth acceleration compared to the local, autochthonous populations (Mátyás and Nagy 2005).

Figure 7 illustrates that the simulation of climatic warming, that is the transfer into warmer environments, results in significant decline of productivity in the warmer part of the range, where moisture is in deficit in certain periods of the year.

Width of Adaptability: Phenotypic Plasticity

Analyses of field tests show remarkable width of adaptability and persistence (and, in consequence, an extended width of "local" adaptation) in the face of even drastic changes in thermal environment and, to a less extent, in moisture supply. This phenomenon indicates the substantial conservatism in the climatic adaptation of numerous tested tree species, which has an inherent genetic basis and may have been enhanced by evolution (Mátyás and Nagy 2005).

For illustration, data from three test sites of the international Norway spruce provenance experiment (Krutzsch 1974; Ujvári-Jármay and Ujvári 2006) are shown. Height at age 16 was measured for 300 identical populations of various origin at climatically widely differing test sites in Hungary (Nyírjes), south Sweden (Abild) and north Sweden (Lappkojberget, Persson and Persson 1992). Response regressions were calculated versus the mean annual temperature of the locations of origin of the populations for the three sites (Fig. 8). Comparing the three response functions, the first conclusion is that in spite of large environmental

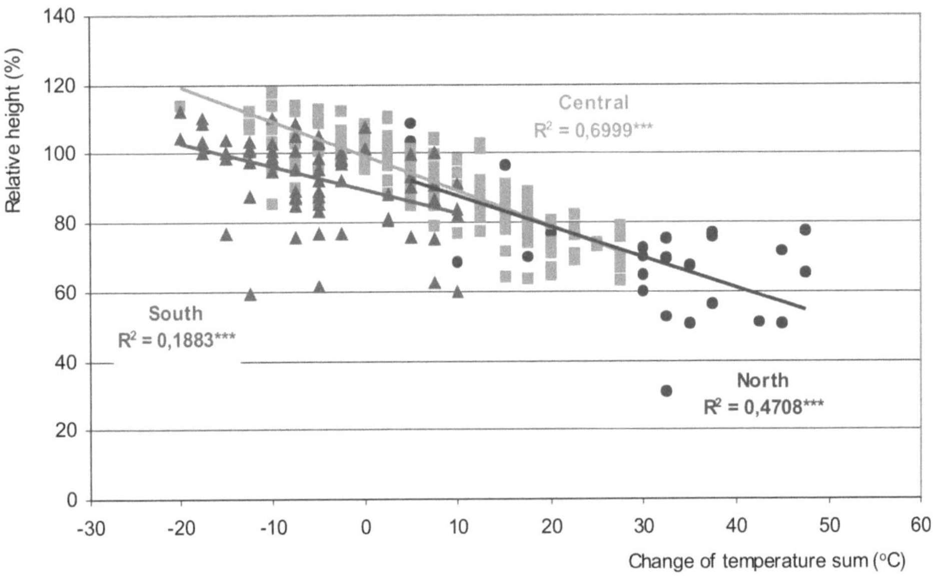

Fig. 7 Linear regressions of relative tree height versus change of temperature sum (degree days in °C) due to transfer by groups of provenances of Scots pine in six Russian tests (Mátyás and Nagy 2005)

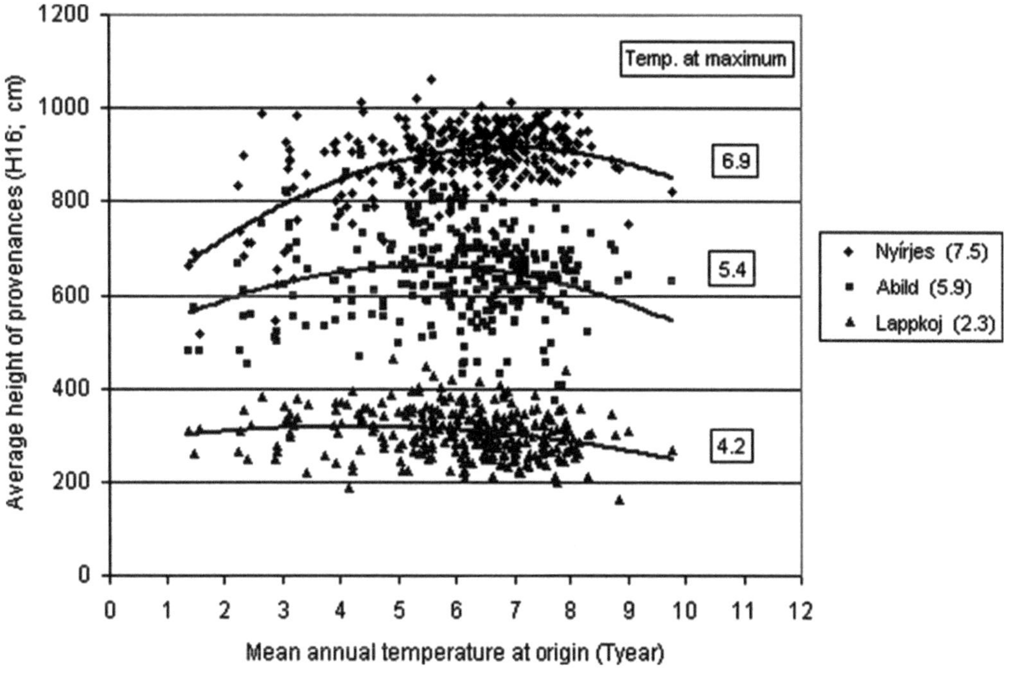

Fig. 8 Regression of 16-year average tree height (*vertical axis*) of identical Norway spruce provenances versus mean annual temperature (°C) at origin (*horizontal axis*) in three IUFRO provenance tests Nyirjes, Abild, Lappkojberget. Mean temperatures (°C) of the test sites are shown in the legend, maximum values of the curves in the graph (É. Ujvári Jármay, unpublished)

differences represented by the tested populations, the response does not indicate narrow (strictly local) adaptedness. It seems also that fitness differentiation increases towards more favourable environments, that is southward. This phenomenon is attributable to the fact that manifestation of genetic differences is favoured by better site conditions.

When comparing the temperature values at the maxima of the response functions (T_{max}) with the ones of the test sites (T_{loc}), it turns out that the 'fittest' populations at two milder locations originate from somewhat cooler environments, while at the harsh northern location, populations from milder environments perform somewhat better (Fig. 8, Table 3). According to Table 3, the difference between the annual mean temperature corresponding to the response regression maximum and the test site mean ($\Delta T = T_{max} - T_{loc}$) is negative at milder sites, indicating that populations from cooler climates perform better. The opposite result appears at the harsh northern site, where populations from milder environments outperform local ones. This surprising phenomenon was reported from some other boreal tests as well (e.g. Andalo et al. 2005).

The analysis has also shown that phenotypic plasticity of populations from certain geographic regions display characteristic differences, which may be determined by local microevolution. As an example, the height growth response of Norway spruce is shown for three selected regions of provenance: eastern Carpathians, the Beskids and Harz Mountains. Figure 9 shows

Table 3 Temperature and height data of Norway spruce experiments to Fig. 8

Experimental site data			Response function	
Test name	Mean annual temperature (T_{loc}, °C)	Mean height of test (cm)	Temperature at maximum (T_{max}, °C)	Mean height at T_{max} (cm)
Nyírjes (H)	7.5	890.3	6.9	916.1
Abild (S)	5.9	643.1	5.4	661.7
Lappkojberget (S)	2.3	305.4	4.2	319.1

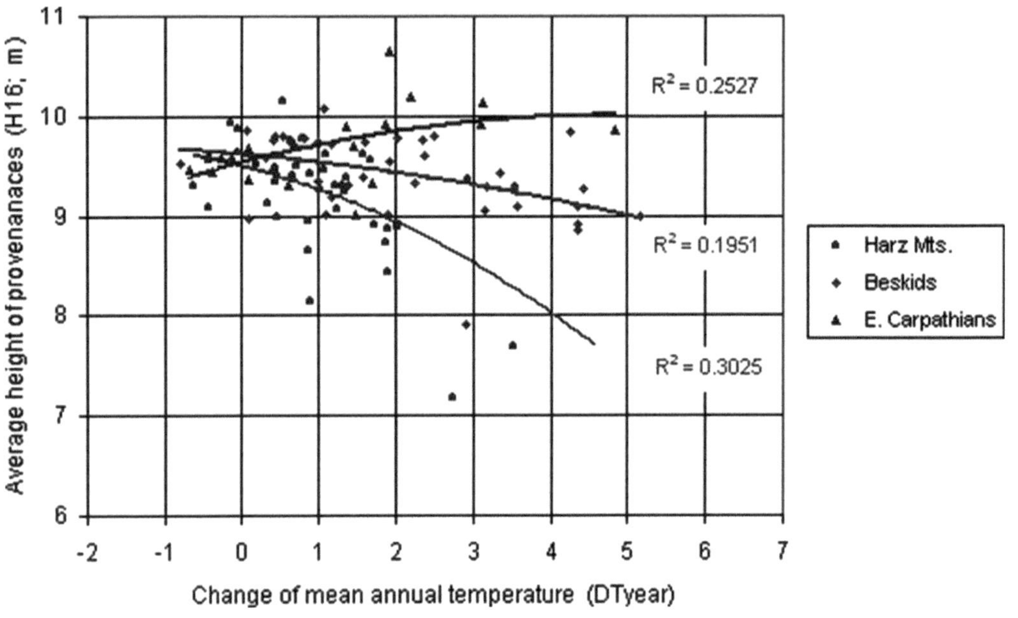

Fig. 9 Effect of mean annual temperature change (in °C) (*horizontal axis*, positive values stand for warming), on 16-year height of Norway spruce provenances. The three provenance groups Harz, Beskids, eastern Carpathians show different levels of plasticity (best: eastern Carpathians). All correlations are significant at the $p = 0.05$ level (É. Ujvári Jármay, unpublished)

that an increase of average annual mean temperature of up to 4°C has no significant effect on the 16-year height of east Carpathian provenances in the test. At the same time the populations from the Harz have reacted with a decline of roughly 15%. In cases, where the mean temperature of the test site Nyirjes was similar to the one at origin of the population ($\Delta T = 0$), the average height of provenances from all the three regions was similar ($H_{16} = 9.5$ m). The results support the higher plasticity of the eastern Carpathian populations. Similar experiences have been gained by Norway spruce breeders in northern Europe (Persson and Persson 1992).

Discussion

Adaptation Maintains Non-equilibrium State

Common sense would suggest that in any environment, the growth of the locally adapted population should represent the maximum, compared to populations adapted to other environments. This principle has been adopted instinctively by foresters already for centuries and represents the basic rule of reproductive material commerce and utilisation to the present day. As described before, the maximum of the response functions does not coincide with the climatic conditions of the test site, in other words, locally adapted populations do not perform best. This *adaptation lag* (Mátyás 1990) runs against the accepted principle of evolutionary optimisation and needs further examination.

The explanation that the lag may originate from the difference between the environments of natural regeneration under forest cover (the populations are assumed to be adapted to) and the grossly artificial conditions provided in a common-garden test (nursery-grown plants, site preparation, weed competition excluded, etc.) seems to be evident. Eriksson coined the term 'domestication fitness' (Eriksson and Ekberg 2001) to explain the difference to locally adapted fitness.[4] Apart from the fact that this effect arises rather from cultivation (cultivation fitness: Mátyás 2004) than from domestication, it does not explain fully the peculiarities of response shifts observed in various environments.

[4] Fitness is interpreted in this chapter as fitness to optimally utilise nutrient and energy resources of the local environment for growth, that is as vegetative fitness.

When looking for a parsimonious explanation, one should keep in mind that there is no reason to suppose the two adaptive mechanisms, (genetic) selection and phenotypic plasticity are acting alternatively or independently. On the contrary, they should be considered as jointly functioning forces, determining a certain position in the 'adaptive landscape'. Although lacking experimental proof, this concept surfaced already in literature (cited in Fournier et al. 2006): a question to be answered by common garden data. A logical explanation could be that *plasticity buffers the effect of natural selection*. It seems that the width of local adaptation is extended by phenotypic plasticity of genotypes towards less optimal environments. If environmental conditions improve (=transfer to milder sites), growth response will improve as well. This implies that genotypes growing under suboptimal conditions, 'camouflaged' by plasticity are genetically adapted to utilise more favourable site conditions than the actual ones. Consequently, reaction norms of such populations display shifts of the maximum performance towards better climates. One has to remember, however, that 'better' has a different meaning at opposite limits of distribution. Close to the xeric (aridity, lower) limits humidity is restraining, so lower temperatures (even at equal precipitation) mean more favourable conditions. At the thermal (northern, upper) limits, temperature sum is at minimum, and more southern, milder sites appear as more favourable.

Therefore, reaction norm of populations adapted to climates close to the xeric limit, exhibit an increased growth response northward of the original location. This effect is shown in Fig. 7. It indicates that *populations under climatic selection pressure adapt to local conditions simultaneously by genetic (natural) selection and by utilising phenotypic plasticity. With increasing distance from climatic conditions of the physiological/genetical optimum for a given species, populations toward the climatic limits of distribution display an increasing genetic adaptation lag, buffered by phenotypic plasticity.* Accordingly, autochthonous, local populations, considered as 'perfectly' adapted, are in reality under constant strain, and perform better in more favourable environments. As similar phenomena have been observed in numerous common garden experiments, on different species, cultivation or random effects such as gene flow are insufficient explanations. Taking growth performance as the measure of (vegetative) fitness, reproductive fitness and tolerance to diseases and pests are unconsidered and could be responsible for the described shift. Although plausible, this explanation is lacking experimental proof up to now. The migration hypothesis, that is that populations colonizing the site had no time yet to adapt locally, might hold at the thermal limits in the north, but not in the rest of the distribution area, and particularly not at the xeric limits. A parsimonious explanation is the assumption of *adaptive non-equilibrium.*

The proposed hypothesis of adaptive non-equilibrium means that within the distribution area of a (zonal) species, genetic adaptedness in the strict sense can be considered to be in an equilibrium state only in a narrow optimum zone. Approaching the thermal and xeric limits, the local populations get under increasing climatic stress due to the suboptimal functioning of genetic selection, which is buffered by phenotypic plasticity. Genetic diversity is then stabilised in a quasi-equilibrium state.

In contemporary ecology, the paradigm of non-equilibrium state of forest ecosystems is generally accepted. This refers to the regulation conditions at the level of species. It seems that *the paradigm of non-equilibrium state of natural systems may be valid also at the genetic level of adaptation to the (climatic) environment.*[5] Non-equilibrium state in genetic respect implies that in most cases, natural, adapted populations do not maintain a maximum vegetative fitness condition related to the local environment.

'Decoupling': Local Adaptive Optimum Disturbed by Changes?

The 'decoupling' of local populations from the climate they are adapted to is an accepted hypothesis applied for trees as well. The assumption is based on the equilibrium concept, that is that local populations are genetically well adapted to the local climate and any change will have detrimental effects. As a result, fitness loss will appear and extinction risk may increase across the whole range (see review of Jump and Peñuelas 2005).

[5] There are striking parallelities in the species-level and genetic-level regulation also in other respects, for example in the area size – diversity relation.

A corollary of the hypothesis is the immigration of pre-adapted individuals (Davis and Shaw 2001) or even of pre-adapted populations (Rehfeldt et al. 2003). Besides the fact that long-distance immigration of individuals or populations into existing ecosystems has a low probability, reference is made to the dominance of plasticity in adaptation as described before: tree populations experiencing climate warming will first utilise the available potential of phenotypic plasticity, which may lead even to growth acceleration, depending on the position within the range.

Consequences for Bioclimatic Modelling

The non-equilibrium concept, if valid, has important implications with regard to construction of climate envelopes and vegetation shift models. Climate envelope models are developed on the basis of the equilibrium hypothesis, assuming that local, autochthonous populations are optimally adapted to their environment. As a consequence, any change of the climatic conditions to the worse should trigger decline or even local extinction. The non-equilibrium concept proposes that these models predict responses too pessimistic: the genetic/physiological possibilities for persistence are not instantly exhausted under changing conditions, with the exception of populations at extreme, marginal sites. The non-equilibrium approach does not, however, consider the additional threat caused by unpredictable outbreaks of new diseases and pests (neither do the mentioned bioclimatic models), which is often observed if vitality and tolerance are constrained by climatic extremes.

Conclusions for Prediction and Bioclimatic Modelling of Adaptive Response

Asymmetry of Response

An important outcome of transfer analyses is the asymmetry of response. The effect of environmental change on populations in different parts of the distribution range is divergent as different climatic factors exert their selection pressure.

The reaction of indigenous tree populations to warming will differ according to climatic zones. In Europe, in the thermal-limited northern-boreal zone, the expected rise of temperature will lead to marked growth acceleration. At lower altitudes, in the temperate-maritime zone, growth will accelerate too, along with increasing or at least unchanged rainfall. In the sub-humid temperate-continental and sub-humid Mediterranean zones, however, even relatively minor temperature increases, coupled with growing drought stress, will trigger loss of compatibility, higher susceptibility to diseases and increased mortality. At the xeric limits warming leads to relatively fast growth and productivity loss, and selective mortality (Berki and Rasztovics 2004; Mátyás 2005). It should be noted that the described phenomena are generalisations. Substantial deviations may be caused by the genetic system of the species, the evolutionary-migratory past and regional or local climate effects. For example, there are indications that in certain regions of the boreal zone, where moisture stress is already present due to low precipitation, higher temperatures and increased drought stress may also lead to incremental decline (Lapenis et al. 2005).

Changes in Genetic Diversity Following Climatic Stress

Expectable genetic changes will be minor in the northern part of the distribution range despite the extreme speed of predicted (and already ongoing) changes. Improved growing conditions can be utilised through the plasticity potential of tree populations, without much migration or selection. As inherited plasticity will determine the response to changes, there is little room left for genetic adaptation. In temperate-Atlantic Europe, where moisture stress is predicted to stay low, populations will also be well buffered by their adaptability.

The situation is completely different along the xeric limit of main tree species, and at the limit of closed temperate forests. Here, natural selection becomes effective in the form of irregularly appearing health decline and mortality waves following weather extremes. The symptoms of pests (gradations) and diseases might be mistaken for primary causes (this was the case in many countries in recent decades). High selection rates will certainly exert a strong effect on the genetic resources of exposed populations, and if stress situations aggravate, it may lead to local population extinction, even for once well distributed, dominant species. This underlines the importance of management and conservation of forest genetic resources (Ledig and Kitzmiller 1992; Mátyás 2000).

Caveats for the Climatic Interpretation of Xeric Limits

To forecast climatic limits of genetically set tolerance has its constraints. The following have to be pointed out:

- It is a well-known ecological rule that actual distributions of species are regulated by complex, often hidden interactions which may modify tolerance limits. Genetically set (potential) limits may be *per definitionem* wider than realized actual ones.
- Due to the longevity and persistence of forest trees, the determination of fitness limits may be misleading. Short-term absence of seeding and reproduction may also mislead locally, as reproduction may happen anytime during the century-long lifetime of a mature tree, if suitable weather conditions favour it.
- The sequence of consecutive extreme weather events and linked biotic damages will concretely decide over survival or mortality at the fitness limits. Therefore *the use of climate (mean) data should be regarded only as surrogates for weather extremes*.
- The change of climatic environment affects also consuming and pathogenic organisms, the selection pressure by consumers may be rearranged. Forecasts in this respect are unreliable, especially because up to date negligible or unknown pests and diseases may appear. Environmental shifts may also lead to changing interactions between host and consumer.
- The limited precision of predicted precipitation changes in scenarios is of special significance in particular at the xeric limits which are extremely sensitive to relatively minor humidity variations.
- Significance of correlations is in itself no proof for a causal relationship, this has to be investigated and verified.

Conclusions for Mitigation and Management

No Mitigation Measures Needed?

Although genetic, migratory and evolutionary constraints are generally acknowledged (Namkoong 2001; Loeschke 1987; Davis and Shaw 2001; Mátyás 1990, 2006a), some scientists (e.g. Kelly et al. and others cited in Jump and Peñuelas 2005, also Hamrick 2004) claim that no measures will be needed to mitigate the effects of changing conditions because

- there is enough genetic variability in the populations, which might be further replenished by migration and gene flow;
- inter-annual fluctuation has the same magnitude as predicted changes of milder scenarios, so populations are prepared to adjust by pre-adapted individuals;
- regeneration is secured through the persistence, phenotypic plasticity and long life cycle of forest tree populations;
- palaeoecological data indicate that enough variation accumulates and is saved in refugia, and the selection pressure of recent climatic fluctuations was without effect;
- genetic adaptation may happen in relatively short periods, 2–3 generations.

Some of the opinions have been already answered in preceding chapters. The argument of limited consequences of vegetation and area shifts may be valid first of all in boreal regions with predominantly nature-close conditions, where human land use had no serious impact yet. In natural landscapes or national parks cyclical changes of vegetation do not pose serious threats as long as ecological space for retreat is available. In landscapes and regions transformed by humans there is however no room left for such fluctuations, especially not close to the lower distribution limits of a vegetation type or a species. At the xeric limits of distribution, migration or gene flow from better adapted populations is not happening.

Regarding inter-annual fluctuations, Tables 2 and 3 indicate that with increasing mean temperatures, severity of extremes will increase too: aridity stress will therefore increase, which will cause additional stress at the xeric limits.

Fast genetic adaptation is in contradiction with the accepted assumption of strong biological and ecological constraints. At the (zonal) xeric limits, an unlimited adaptation to declining environment is unthinkable, due to the evolutionary tradeoffs and constraints. This is proven by remarkable migrations and area shifts in the geological past.

Therefore the need of human intervention in mitigation has to be underlined (Hulme 2005; Mátyás 2006a). Due to ecological constraints to spontaneous adaptation, the policy of artificial translocation should be preferred instead of extensive enhancement of connectivity, at least with regard to tree species.

Consequences for Forest Management

The urgent necessity to put into practice the findings of quantitative genetics cannot be questioned. In addition, some aspects of forest management should not be overlooked when predicting responses and formulating mitigation strategies. Most of Europe's forests have been and still are under strong human influence, and are managed according to periodic management plans. Especially close to the xeric limit, the proportion of nature-close forests is low, regeneration is mostly artificial. For example, in Hungary, the rate of artificial regeneration is at present is over 70% on the Great Plain. The possibilities left for spontaneous processes, such as migration and succession are limited. Forest stand composition is primarily determined by forest policy and economic considerations. This means also that adjustments in species composition and in adaptive genetic potential may be achieved faster and more effectively compared to natural, spontaneous processes.

In drought stress climates, increment loss and higher incidence of diseases and pests will challenge the economics of forest operations, and will shift the emphasis towards the maintenance of ecological functions and conservation of stability and of genetic resources (Geburek and Turok 2005).

References

Andalo C, Beaulieu J, Bousquet J (2005) The impact of climate change on growth of local white spruce populations in Québec, Canada. For. Ecol. Manage., 205: 169–182

Beaulieu J, Rainville A (2004) Adaptation to climate change: genetic variation is both a short- and long term solution. The Forestry Chronicle, 81(5): 704–708

Berki I, Rasztovics E (2004) [Research in drought tolerance of zonal tree species, with special regard to sessile oak.] (in Hungarian with English summary) In: Mátyás Cs, Vig P (eds.), Erdő és klima – Forest and Climate IV. Sopron, Hungary, 209–220

Berki I, Móricz N, Rasztovics E, Vig P (2007) [Tolerance limits of beech.] (in Hungarian with English summary) In: Mátyás Cs, Vig P (eds.), Erdő és klima – Forest and Climate V. Sopron, Hungary, 213–218

Booy GR, Hendriks JJ, Smulders MJ et al. (2000) Genetic diversity and the survival of populations. Plant Biol., 2(4): 379–395

Borovics A (2007) Assessment of adaptive potential of beech and sessile oak by correlative analysis of allozymatic variation patterns and climate parameters. In: Mátyás Cs (Proj. leader), Climate uncertainty and threats to forest cover. Research report, in Hungarian, 89–98

Bradshaw AD (1965) Evolutionary significance of phenotypic plasticity in plants. Advances in Genet., 13: 115–155

Bradshaw AD (1991) Genostasis and the limits of evolution. Philos. Trans. Royal Soc., London, 333: 289–305

Briceno-Elizondo E, Garcia-Gonzalo G, Peltola H, Matala J, Kellomäki S (2006) Sensitivity of growth of Scots pine, Norway spruce and silver birch to climate change and forest management in boreal conditions. For. Ecol. Manage., 232: 152–167

Clausen J, Keck DD, Hiesey WW (1940) Experimental Studies on the Nature of Species. Vol I and II–IV (the additional volumes published in 1945, 1948, 1958) Carnegie Inst. Publ. Nr 520, Washington D.C.

Davis MB, Shaw RG (2001) Range shifts and adaptive responses to quaternary climate change. Science, 292: 673–679

DeWitt TJ, Scheiner SM (2004) Phenotypic variation from single genotypes. In: DeWitt, TJ, Scheiner SM (eds.), Phenotypic Plasticity; Functional and Conceptual Approaches. Oxford University Press, Oxford, 1–9

Eanes WF (1999) Analysis of selection on enzyme polymorphisms. Ann. Rev. Ecol. Syst. 30: 301–326

Eriksson G, Ekberg I (2001) Introduction to Forest Genetics. SLU Press, Uppsala, Sweden

Etterson JR, Shaw RG (2001) Constraint to adaptive evolution in response to global warming. Science, 294, 151–154

Fournier N, Rigling A, Dobbertin M, Gugerli F (2006) Faible differentiation génétique á partir d'amplification aléatoire d'RAPD, entre les types de pin silvestre d'altitude et de plaine dans les Alps á climat continental. Ann. Forest Sci., 63: 431–439

Gálos B, Lorenz Ph, Jacob D (2007) Will dry events occur more often in Hungary in the future? Env. Res. Letters 2, doi: 10.1088/1748-9326/2/3/034006

Geburek T, Turok J (eds.) (2005) Conservation and Management of Forest Genetic Resources in Europe, Arbora Publisher, Zvolen, Slovakia

Hampe A, Petit R (2005) Conserving biodiversity under climate change: the rear end matters. Ecol. Letters, 8: 461–467

Hamrick JL (2004) Response of forest trees to global environmental changes. For. Ecol. Manage., 197(1–3): 323–336

Hamrick JL, Godt JW, Sherman-Broyle SL (1992) Factors influencing levels of genetic diversity in woody plants. New Forests, 6: 95–124

Hulme PE (2005) Adapting to climate change: is there scope for ecological management in the face of a global threat? J. Appl. Ecol., 42: 784–794

Huntley B (1991) How plants respond to climate change – migration rates, individualism and the consequences for plant communities. Ann. Bot., London, 67: 15–22

IPCC WG II. (2007) Fourth assessment report for government and expert review. Alcamo J, Moreno JM, Nováki B (eds.) Chapter 12: Europe. Bruxelles, Belgium, 62p

Jump AS, Peñuelas J (2005) Running to stand still: adaptation and the response of plants to rapid climate change. Ecology Lett., 8: 1010–1020

Jump AS, Hunt JM, Peñuelas J (2006) Rapid climate change related growth decline at the southern edge of Fagus sylvatica. Global Change Biol., 12: 1–12

Kingsolver JG et al. (2001) The strength of phenotypic selection in natural populations. Am. Natur., 157(3): 245–261

Kramer K, Mohren G (2001) Long-term effects of climate change on carbon budgets of forests in Europe. Alterra Report, No. 194

Kremer A, Le Corre V, Mariette S (1999) Population differentiation for adaptive traits and their underlying loci in forest trees. In: Mátyás Cs (ed.), Forest Genetics and Sustainability. Kluwer, Dordrecht, 59–74

Krutzsch P (1974) The IUFRO 1964/8 provenance test with Norway spruce (Picea abies Karst.). Silvae Genet., 23: 58–62

Langlet O (1971) Two hundred years of genecology. Taxon, 20: 653–722

Lapenis A, Shvidenko A, Shepaschenko D, Nilsson S, Aiyyer A (2005) Acclimation of Russian forests to recent changes. Global Change Biol., 11: 2090–2102

Ledig FT, Kitzmiller JH (1992) Genetic strategies for reforestation in the face of global climate change. For. Ecol. Manage., 50: 153–169

Linhart YB, Grant MC (1996) Evolutionary significance of local genetic differentiation in plants. Ann. Rev. Ecol. Syst., 27: 237–277

Loeschke V (ed.) (1987) Genetic Constraints of Adaptive Evolution. Springer Verlag, Berlin

Lynch M, Lande R (1993) Evolution and extinction in response to global change. In: Kareiva PM, Kingsolver J (eds.), Biotic Interactions and Global Change. Sinauer Association, Sunderland, 234–250

Martienssen RA, Colot V (2001) DNA methylation and epigenetic inheritance in plants and filamentous fungi. Science, 293: 1070–1074

Mátyás Cs (1990) Adaptation lag: a general feature of natural populations. Invited lecture. Proc., WFGA-IUFRO Symp. Olympia, Wash. Paper no. 2.226, 10p

Mátyás Cs (1994) Modelling climate change effects with provenance test data. Tree Physiol., Victoria B.C. 14: 797–804

Mátyás Cs (ed.) (1997) Perspectives of Forest Genetics and Tree Breeding in a Changing World. IUFRO World Series Vol. 6. IUFRO, Vienna

Mátyás Cs (ed.) (2000) Forest Genetics and Sustainability. Kluwer, Dordrecht

Mátyás Cs (2004) Population, conservation and ecological genetics. In: Burley J, Evans J, Youngquist J (eds.), Encyclopedia of Forest Sciences. Elsevier Major Reference Works, Oxford, Vol 1, 188–197

Mátyás Cs (2005) Expected climate instability and its consequences for conservation of forest genetic resources. In: Geburek T and Turok J (eds.), Conservation and Management of Forest Genetic Resources in Europe. Arbora Publisher, Zvolen, Slovakia, 465–476

Mátyás Cs (2006a) Migratory, genetic and phenetic response potential of forest tree populations facing climate change. Acta Silvatica et Ligniaria Hung., 2: 33–46 (http://ASLH.NYME.hu)

Mátyás Cs (2006b) The missing link: synthesis of forest genetics and ecological research in view of challenges of environmental change. In: von Wühlisch G (ed.). Forest Genetics and its Contribution to Sustainability. Mitt. BFH, Nr 221, Kommissionsverlag, Hamburg, 1–14

Mátyás Cs, Nagy L (2005) Genetic potential of plastic response to climate change. In: Konnert M (ed.), Tagungsberichte, Forum Genetik und Wald 2004. Bavarian Centre f. For. Repr. Material, Teisendorf, 55–69

Mátyás Cs, Yeatman CW (1987) Adaptive variation of height growth of *Pinus banksiana* populations (in Hungarian with English summary). EFE Tud. Közl., (Scientific Proceeding of Sopron University, Hungary), 1–2: 191–197

Mátyás Cs, Yeatman CW (1992) Effect of geographical transfer on growth and survival of jack pine (*Pinus banksiana* Lamb.) populations. Silvae Genet., 43(6): 370–376.

Müller-Starck G, Schubert R (eds.) (2001) Genetic Response of Forest systems to Changing Environmental Conditions. Kluwer Academic Publishers, Dordrecht

Morgenstern, E.K. 1996. Geographic Variation in Forest Trees. UBC Press, Vancouver

Namkoong G (2001) Forest genetics – pattern and complexity. Can. J. For. Res., 31(4): 623–632

Neale DB, Wheeler NC (2004) Mapping of quantitative trait loci in loblolly pine and Douglas fir: a summary. Forest Genet. 11(3–4): 173–178

Peñuelas J, Lloret F, Montoya R (2001) Severe drought effects on Mediterranean woody flora in Spain. Forest Science, 47: 214–218

Persson B, Beuker E (1996) Distinguishing between effects of changes in temperature and light climate using provenance trials with *Pinus sylvestris* in Sweden. Can. J. For. Res., 26: 572–579

Persson A, Persson B (1992) Survival, growth and quality of Norway spruce (*Picea abies* (L.) Karst.) provenances at the three Swedish sites of the IUFRO 1964/68 provenance experiment. Swedish University of Agriculture Sciences, Department of Forest Yield Research. Report 29. 1–67

Petit R, Kremer A et al. (2002) Identification of refugia and postglacial colonisation routes of European white oaks based on chloroplast DNA and fossil pollen evidence. For. Ecol. Manage., 156: 27–40

Pigott CD, Pigott S (1993) Water as determinant of the distribution of trees at the boundary of the Mediterranean zone. J. Ecol., 81: 557–566

Piovesan G, DiFilippo AA (2005) Structure, dynamics and dendroecology of an old-growth *Fagus* forest in the Appenines. J. Veget. Sci., 16: 13–28

Rehfeldt GE, Tchebakova NM, Barnhardt LK (1999) Efficacy of climate transfer-functions – introduction of Eurasian populations of *Larix* into Alberta. Can. J. For. Res., 29: 1660–1668

Rehfeldt GE, Tchebakova NM, Milyutin LI, Parfenova EI, Wykoff WR, Kouzmina NA (2003) Assessing population responses to climate in *Pinus sylvestris* and *Larix* spp. of Eurasia with climate transfer models. Eurasian J. For. Res., 6(2): 83–98

Savolainen O, Bokma F, García-Gil R, Komulainen P, Repo T (2004) Genetic variation in cessation of growth and frost hardiness and consequences for adaptation of *Pinus sylvestris* to climatic changes. For. Ecol. Manage., 197: 79–89

Savolainen O (1994) Genetic variation and fitness: conservation lessons from pines. In: Loeschke V et al. (eds.), Conservation Genetics. Birkhaeuser Verlag, Basel, 27–36

Shutyaev AN, Giertych M (1997) Height growth variation in a comprehensive Eurasian provenance experiment of Pinus sylvestris L. Silvae Genet., 46: 332–349

Skrøppa T, Johnsen G (2000) Pattern of adaptive variation in forest tree species: the reproductive element as an evolutionary force in *Picea abies*. In: Mátyás Cs (ed.). Forest Genetics and Sustainability. Kluwer Academic, Dordecht, 49–58

Spiecker H, Mielikäinen K, Köhl M, Skovsgard JP (eds.) (1996) Growth trends in European forests. EFI Report 5, Springer Verlag, Berlin

Turesson G (1925) The plant species in relation to habitat and climate. Hereditas, 6: 147–236

Ujvári Jármay É, Ujvári F (2006) Adaptation of progenies of a Norway spruce provenance test to local environment. Acta Silvatica et Ligniaria Hung., 2: 47–56 (http://ASLH.NYME.hu)

Wang T, Hamann A, Yanchuk A, O'Neill GA, Aitken SN (2006) Use of response functions in selecting lodgepole pine populations for future climates. Global Change Biol., 12: 2414–2416

Weis AE, Simms EL, Hochberg ME (2000) Will plant vigor and tolerance be genetically correlated? Evol. Ecol., 14: 331–352

Woods A, Coates KD, Hamann A (2005) Is an unprecedented *Dothiostoma* needle blight epidemic related to climate change? BioScience, 55(9): 761–769.

Westphal RD, Millar CI (2004) Genetic consequences of forest population dynamics influenced by historic climate variability in the western USA. For. Ecol. Manage., 197(Special issue): 159–170

Seasonal Changes in Transpiration and Soil Water Content in a Spruce Primeval Forest During a Dry Period

F. Matejka, K. Střelcová, T. Hurtalová, E. Gömöryová and L'. Ditmarová

Keywords Transpiration · Soil water content · Spruce primeval forest · Mathematical modelling

Introduction

Transpiration covers approximately half of the annual precipitation total under humid temperate conditions in Europe (Denmead and Shaw 1962). The energetic equivalent of this amount of transpired water represents an important contribution to the energy balance of the Earth's surface. In response to water stress, plants regulate their transpiration by decreasing their stomatal conductance (Sperry 2000). This physiological control of transpiration plays an important role in processes of matter and energy exchange between vegetation and the atmosphere.

Many authors have analysed transpiration using the results of sap-flow measurements. Based on this methodical approach, the daily and seasonal variability of transpiration was analysed for various forest stands under non-limiting soil water conditions and under water stress (Cienciala et al. 1994, 1997; Čermák et al. 1976, 1982; Kučera et al. 1977, Morikawa et al. 1986; Granier and Loustau 1994). Al-Kaisi et al. (1989) described and quantitatively expressed the relationship between the leaf area index and transpiration. Jara et al. (1998) studied transpiration as a component of evapotranspiration.

F. Matejka (✉)
Geophysical Institute, Slovak Academy of Sciences, Dúbravská cesta 9, 845 28 Bratislava, Slovakia
e-mail: geofmate@savba.sk

The estimate of transpiration from results of the sap-flow measurements requires a scaling procedure that can be complicated sometimes. Further, the differences between sap-flow and transpiration can be neglected over periods longer than one day but they are important over shorter periods or if diurnal courses are studied (Cienciala 1992). Therefore, mathematical modelling of the water vapour transfer in the soil–vegetation–atmosphere system has become an alternative approach for the determination of transpiration rates. In recent years, considerable efforts have been made to improve the methods for modelling the transpiration. Recent mathematical models of the water exchange between vegetation and the atmosphere take into account the existence of two sources of water, which are the area of leaves and the soil surface (Choudhury and Monteith 1988; Iritz et al. 1999; Shuttleworth and Wallace 1985; Torula and Heikinheimo 1999; Wallace et al. 1990).

Soil drought may be a factor significantly affecting the transpiration rate and consequently the partitioning of energy in the energy budget of evaporating surfaces. Since this partitioning of energy determines the properties of the planetary boundary layer (Wilson and Baldocchi 2000), transpiration, reduced by water stress, may have a significant influence on the climate (Shukla and Mintz 1982). For these reasons, research on transpiration has become important, especially in the last decades when the frequency of extreme weather phenomena has risen (Karl et al. 1995).

Recently, several publications have appeared stating that dry air can be a similar stress factor as dry soil. Vapour pressure deficit is an important environmental factor which, together with soil moisture, affects the gas exchange between vegetation and the atmosphere (Calvet 2000; Gucci et al. 1996; Leonardi

K. Strelcová et al. (eds.), *Bioclimatology and Natural Hazards*,
DOI 10.1007/978-1-4020-8876-6_17,

et al. 2000; Habermann et al. 2003). A close statistical relationship exists between the vapour pressure deficit and the canopy resistance for the water vapour transfer (Granier et al. 2000; Xue et al. 2004). Consequently, it is easy to understand that evapotranspiration responds sensitively to changes in the vapour pressure deficit (Turner et al. 1984; Bunce 1996).

In spite of many results obtained on transpiration, the synergetic effect of the dry soil and dry air on the transpiration of forest stands have not been examined satisfactorily and further investigation into this topic is needed. The aim of this study is to quantify the response of daily totals of transpiration from a spruce primeval forest to changes in soil moisture under high evaporative demand of the atmosphere and to assess the impact of atmospheric drought on the water regime of the forest.

Material and Methods

The analysed data were obtained during the period May–September 2003 in a spruce primeval forest in the Biosphere Reserve Poľana (19° 28′, 48° 37′, 1347 m a.s.l.). This locality is situated in a humid, cool temperate region with abundant precipitation. The mean annual air temperature is about 4°C, ranging from occasional winter temperatures below −25°C to summer temperatures approaching 28°C. Mean daily temperatures for January and July are −5°C and 14°C. Mean monthly relative air humidity ranges between 72% in July and 83% in December. Precipitation (900–1100 mm per year) is unevenly distributed between seasons. The mean annual sum of the sunshine duration achieves 1700 h. There are 75 clear days and 130 cloudy days yearly in the area (Škvarenina et al. 2002).

There are andosols in the upper part of the mountains, with a high content of volcanic glass and thick humus horizon. Favourable physical and chemical features and high fertility of these soils permit the trees to achieve extraordinary dimensions and a high standing crop of the above-ground biomass.

The forest is 190 years old dominated by the Norway spruce (*Picea abies* [L.] Karst), beech (*Fagus sylvatica* L.) and rowan (*Sorbus aucuparia* L.) trees are sporadic. The forest type is *Sorbeto–Piceetum* and*Acereto–Piceetum.* During the analysed period, the experimental forest had the mean height of 25 m and the stand density was inhomogeneous due to the primeval forest structure and various age stages.

Table 1 Size of sample trees (*Picea abies* Karst.) used for sap-flow measurements

Sample tree	Stem perimeter (cm)	D.B.H. (cm)	Height of the tree (m)
1	190.5	60.8	24
2	193.5	61.8	25
3	202.5	64.6	21

Measurements of the global radiation, net radiation, air temperature and humidity were made above the spruce primeval forest. All sensors were sampled at intervals of 10 s, and 10 min averages were computed and stored on a data logger. The spatial and time variability of the soil water content was measured gravimetrically in three soil layers.

The sap flow of model spruce trees was estimated on three representative trees (Table 1) by direct, non-destructive and continuous measurements by the tree-trunk heat balance (THB) method with internal heating of xylem tissues and sensing of temperature (Čermák et al. 1976; Čermák and Kučera 1981).

The output from the thermocouples was registered by a data logger and the sap flow was obtained by a simple calculation based on the differential heat balance equation (Kučera et al. 1977). There were two measuring points at opposite sides of the trunk at 2 m height, to take account of possible variation of sap flow within the stem. The measuring points were insulated using 30 mm polyurethane foam covered by 0.5 mm aluminium shield.

The stand transpiration was calculated by scaling up the measured sap flow values to the whole stand.

Model Description

The model described here combines and extends the works of Bichele et al. (1980), Choudhury and Monteith (1988) and Wallace (1995). The soil–vegetation–atmosphere continuum is divided into three layers:

- a reference level in the atmosphere,
- the effective sink for momentum within the canopy,
- the soil surface.

The structure of such a layered soil–vegetation–atmosphere system is very similar to the approach used by Wang (2000) in the two-big-leaf model for calculating canopy photosynthesis.

The model is conceptually based on Darcy's law, the continuity equation and the idea that the rate of water uptake by roots from the soil is equal to the rate of water loss by transpiration, so the effect of plant water retention on transpiration is neglected. Obviously, such a presumption may be restrictive by modelling transpiration rates in a forest with a large water reserve within tree stems. However, errors caused as a result of this simplification tend to be compensated for over 24 h or longer time scales. Since daily totals of transpiration and their seasonal variability are required to be simulated in this study, neglecting water retention in trees can be accepted as a simplifying model assumption.

The soil block in the model is represented as a single material profile. The transport of water along the soil profile is not calculated but, due to evapotranspiration, an average soil water content w, soil water potential Ψ and soil hydraulic conductivity k are changing. Let us assume that the matrix potential is the main component of the soil water potential. According to the Darcy–Buckingham equation, the flux of water in soil q can be expressed as

$$q = -\frac{k}{\rho w\, g} grad\,\Psi \tag{1}$$

where ρ_w is the water density and g is the acceleration due to gravity. The values of the soil hydraulic conductivity k and the soil water potential Ψ are related according to the empirical equation (Wind 1972)

$$k = a(-\Psi)^{-b} \tag{2}$$

Let us suppose that the mean radius of roots is r and the mean distance between roots is $2d$. Let $\Psi(x)$ be the soil water potential at distance x from the root axis. Then, the rate of water uptake q' by a segment of the unit root length is

$$q' = -\frac{2\pi x a(-\Psi(x))^{-b}}{\rho w g}\frac{d(\Psi(x))}{dx} \tag{3}$$

Assuming that Ψ_R is the soil water potential at $x = r$ and Ψ_S the soil water potential at $x = d$ after integrating (3) and taking into account (2), we get

$$q'\int_r^d \frac{dx}{x} = -\frac{2\pi a}{\rho w g}\int_{\Psi_R}^{\Psi_S}(-\Psi)^{-b}d\Psi \tag{4}$$

The value of q' multiplied by the total root length of one plant l_r evidently gives the total amount of water taken up by one plant in the unit time interval. The root system, however, is often characterized by its total area S_R, rather than by its length. Using $S_R = 2\pi r l_r$, transpiration of N plants growing in 1 m^2, E_T can be expressed as

$$E_T = \frac{aNS_R}{\rho w\, g\, r\ln(d/r)}({\Psi_S}^b - {\Psi_R}^b) \tag{5}$$

The root potential Ψ_R cannot be measured directly; therefore, it is not a suitable model input. However, taking into account the neglected water holding capacity of trees, it can be eliminated using van Honert's relationship containing the plant resistance r_p and expressing transpiration rate as proportional to the difference between root and leaf water potentials Ψ_R–Ψ_L (Honert 1948):

$$E_T = \frac{1}{g}\frac{\Psi_R - \Psi_L}{r_p} \tag{6}$$

By expressing leaf water potential Ψ_R according to van Honert's relationship and substituting it in Eq. (5) we get

$$\frac{1}{-\Psi_S} = \frac{1}{-gr_pE_T - \Psi_L} + \frac{gr\ln(d/r)}{aNS_R}E_T \tag{7}$$

The parameter β expressed as

$$\beta = \frac{gr\ln(d/r)}{aNS_R} \tag{8}$$

integrates soil and vegetation characteristics and it quantifies the role of the root system in the process of the water transfer in the soil–vegetation–atmosphere system.

The transpiration rate E_T depends on the canopy resistance and on external factors according to the Penman–Monteith equation (Monteith 1965) expressed in terms used in Fig. 1 as

$$\mathrm{LE_T} = \frac{\Delta r_1 R_v + \rho cpD}{\Delta r_1 + \gamma(r_1 + r_c)} \tag{9}$$

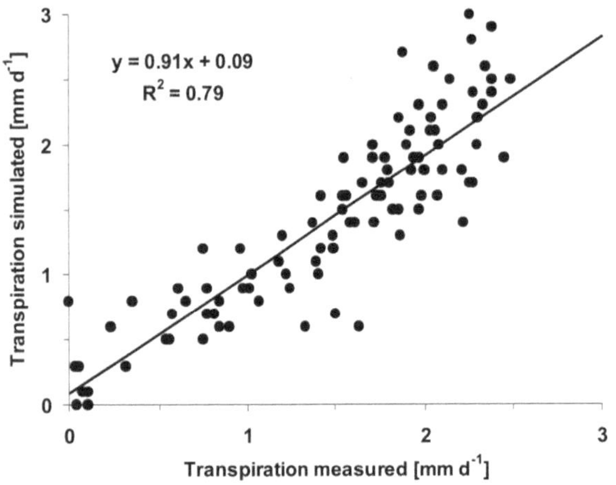

Fig. 1 Daily totals of the stand transpiration of the spruce primeval forest determined from the sap-flow measurements (*transpiration measured*) compared with the results of the model simulations (*transpiration simulated*)

where Δ is the temperature derivative of the saturated water vapour pressure, r_l is the boundary resistance, D is the vapour pressure deficit and r_c means the canopy resistance. The net radiation at the canopy R_v is determined as the difference of the net radiation above the canopy R_n and the net radiation at the soil surface R_s, provided that the net radiation at the soil surface is calculated according to Choudhury and Monteith (1988) by means of the relationship $R_s = R_n \exp(-\alpha''\mathrm{LAI})$. The value of $\alpha'' = 0.7$ is used in later calculations.

The leaf resistance r_{leaf} depends mainly on the radiation intercepted by leaves Q_L and on the leaf water potential Ψ_L (Choudhury and Idso 1985). This relationship was empirically determined using data obtained earlier in a spruce forest as follows:

$$r_{leaf} = r_0 \left(1 + \frac{n}{Q_L}\right) \exp(m\Psi_L) \tag{10}$$

where r_0 is the minimum leaf resistance and m and n are empirical constants. The total conductivity g_c of the canopy with the leaf area index L is related to the leaf conductivity g_L in individual canopy layers, according to the following equation (Choudhury and Monteith 1988):

$$g_c = \int_0^L gL\,dL \tag{11}$$

The global radiation $Q(L')$ in a homogeneous canopy below a leaf area index L' measured from the top of the canopy can be expressed using the extinction coefficient τ as follows:

$$Q(L') = Q\,\exp(-\tau L') \tag{12}$$

Then, the average amount of global radiation Q_L absorbed by the leaf unit area in a given layer is

$$Q_L = -\frac{\partial Q(L')}{\partial L'} = \tau\,Q\,\exp(-\tau\,L') \tag{13}$$

After the integration in (11) using the equations (10) and (13) and standing LAI (for the leaf area index) we finally obtain

$$r_c = r_0 \exp(-m\Psi_L)\left(LAI + \frac{1}{\tau}\ln\frac{1 + \frac{n}{Q\tau}}{1 + \frac{n}{Q\tau}\exp(-\tau\,LAI)}\right) \tag{14}$$

Equations (7), (8) and (9) constitute a system of three equations which can be solved for the three unknowns Ψ_L, r_c and E_T. In the next step of the model, the soil evaporation E_s is calculated by means of the Penman–Monteith equation using appropriate parameters of the soil surface as inputs to Eq. (9). These involve net radiation at the soil surface, the vapour pressure deficit and the aerodynamic resistance below the canopy. Values of the soil surface resistance r_s, which is now used instead of the canopy resistance in the denominator of Eq. (9), were calculated as a function of the soil moisture (Katerji and Perrier 1985). Daily and seasonal changes in the aerodynamic resistance were determined as a function of the roughness parameter, wind speed and thermal stratification. Finally, the evapotranspiration rate E is determined as the sum of the transpiration E_T and soil evaporation E_s rates.

Inputs to this model involve the hydrophysical parameters of the soil, including the soil moisture in the root zone, biometric characteristics of the stand and meteorological elements such as global radiation, net radiation, wind speed, air temperature and humidity. The outputs of the model provide the parameters of the water regime of the stand. By running the model with 1 h steps, the daily course of the stand transpiration was calculated. By summarizing the hourly sums of transpiration over the whole day, the daily totals of transpiration were determined. Finally, the changes in the soil water content in the root zone were estimated by the procedure which balanced precipitation, interception, water uptake by roots and soil evaporation from the soil layer of 0–60 cm depth.

Model Verification in Actual Weather Conditions

Using the mathematical model described above, actual transpiration, evaporation from the soil and finally the evapotranspiration were calculated with 1 h time steps over the period June–September 2003. Net radiation, wind speed, air temperature and air humidity measured above the forest served as the meteorological inputs to the model. The necessary soil and vegetation characteristics integrated in the parameter β from equation (8) were obtained during the process of model calibration that was based on an independent data set. To verify the model, the results of stand transpiration determined by scaling up the sap-flow measurements in the investigated spruce primeval forest during the period June–September 2003 were used.

Meteorological conditions during the period June–September 2003 were generally characterized by higher than normal values of global radiation and air temperature. The extremely high evaporative demands of the atmosphere triggered intensive evapotranspiration that resulted in a following rapid decrease of the soil water content. Consequently, periods of soil drought occurred during July, in the second half of August and in September. Thus, the model was verified under the conditions of intensive soil and atmospheric drought.

The daily totals of the stand transpiration simulated by the model were compared with corresponding measured data. The daily totals of the stand transpiration determined from the sap-flow measurements and the ones obtained from the model are comparable (Fig. 1), with a reasonably close correlation coefficient (r = 0.79), statistically significant at the level α = 0.05 and negligible systematic differences between the two data sets. The existing differences between the compared data are normally distributed with a standard deviation of 0.49 mm/day, corresponding to the probable error of the model equal to 0.32 mm/day. It is obvious that the model is able to simulate daily totals of the stand transpiration quite satisfactorily. Consequently, the seasonal courses of daily totals of the stand transpiration determined by the two methods are also quite similar (Fig. 2).

Seasonal changes in the soil water content in the root zone of the investigated spruce primeval forest were simulated during the vegetation period of 2003 by the additional soil block of the model which balanced precipitation, water uptake by roots and soil evaporation from the soil layer 60 cm deep. The results of the model simulations were compared with the direct measurements of soil water content. The compared data sets were closely related and the standard deviation of differences between measured and simulated values was comparable with the experimental data error.

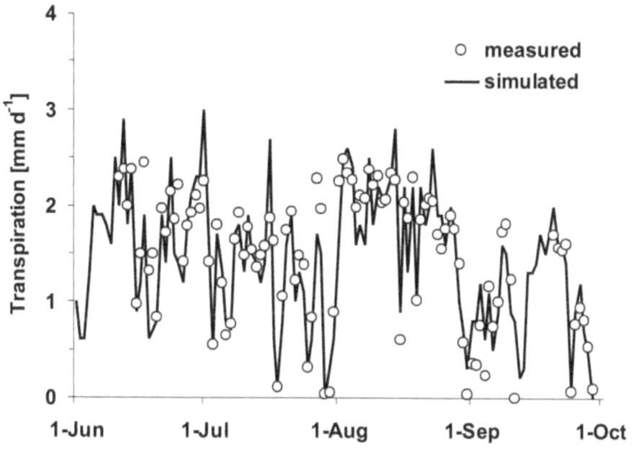

Fig. 2 Seasonal courses of the stand transpiration daily totals determined concurrently from the sap-flow measurements and simulated by the model

Errors in the model simulations are caused first of all by errors in the input data and by simplifying assumptions of the model. The model also does not account for internal plant processes that influence the behaviour of stomata and for the water storage in stems. Despite all this, the model can be applied to simulate daily sums of the stand transpiration soil moisture quite realistically and with acceptable accuracy. Similarly, the simulations of changes in the soil water content in the root zone can be performed by balancing precipitation, interception, water uptake by roots and soil evaporation from the soil layer 60 cm deep.

Sensitivity of the Model to Changes in Environmental Factors

The values of the stand transpiration calculated according to the model (Eqs. 7, 8, 9) depend on various soil, canopy and atmospheric characteristics. From theoretical and practical points of view, it is important to know how the transpiration rates respond to changes in the input data. A sensitivity analysis allows to quantify the partial effects of environment and to assess its relative importance.

Partial relationships of surface fluxes to different parameters of the soil–plant–atmosphere system were simulated by meteorological conditions corresponding to mid-day values on a bright summer day. LAI = 8 and 40% of the volume average soil water content was assumed, that is soil water was not limiting the evapotranspiration.

Regarding atmospheric factors, transpiration of the stand well supplied with soil water responds most sensitively to changes in global radiation, and less to vapour pressure deficit and air temperature. The detected dominant role of global radiation is related to the method of sensitivity testing.

In most tree species, and particularly in conifers, stomatal resistance of the leaf surface is strongly related to atmospheric water vapour deficit. This relationship was not explicitly considered by the design of the model that is used in this study. Nevertheless, the results of model simulations obviously indicate that the canopy conductance of the studied forest declines with increasing vapour pressure deficit approximately according to a hyperbolic function. There is an urgent need to validate the simulated relationship by experimental data and to clarify its theoretical background.

With the aim to evaluate the influence of changes in soil moisture on the stand transpiration of the spruce primeval forest, the relationship between transpiration and soil water content in the root zone was simulated for meteorological conditions typical for a bright summer day. To separate the influence of atmospheric factors on transpiration rates, the dependence of the ratio between actual and potential transpiration on the soil water content moisture was analysed. The effect of soil water on transpiration daily totals manifests itself if its value falls below 35% of volume.

In addition to atmospheric factors, the stand transpiration is influenced by plant characteristics, leaf area index, root system development and the response of stomata to changes in environmental factors. Model equations indicate that transpiration rates are strongly affected by the root–shoot ratio S_R/S_L. With the aim to quantify the influence of this parameter on transpiration, the model simulations of transpiration rates were performed with real data and then repeated with halved and doubled root–shoot ratio. The results of the simulations indicated that the transpiration rates in a hypothetical forest stand with a doubled root–shoot ratio was more intense and, consequently, drying of the soil was significantly faster in comparison with the real situation. On the other hand, the forest stand with a reduced root–shoot ratio saved the soil water.

Transpiration of the Primeval Forest Under High Evaporative Demands

The transpiration total during the period of June–September 2003 reached 158.9 mm. The daily mean of transpiration, averaged over the whole period, was 1.43 mm/day. The seasonal maximum of daily totals of transpiration of 3.05 mm/day fell on 1 July, when sufficient soil water was accompanied by high evaporative demands of the atmosphere. Transpiration represented the major part of evapotranspiration during the whole season.

Large monthly totals of global radiation, accompanied by high daily maximum temperatures, leading to a very large daily maximum atmospheric vapour pressure deficit, were typical for much of the analysed period. Daily maximum values of the vapour pressure deficit exceeding 20 hPa, extremely high values for the given locality, occurred over periods of several days. Conversely, a significant reduction in the precipitation was recorded in the first two months of the analysed period. During the period August–October, precipitations occurred more frequently, including irregular, intensive precipitation events. Consequently, the values of soil water content in the root zone varied significantly. Two periods occurred when soil water in the root zone approached the wilting point in the middle of July. Despite the significant reduction of soil water, the daily totals of transpiration remained relatively high. This was caused by the extremely high values of the vapour pressure deficit. The simultaneous effect of declining soil moisture and increasing vapour pressure deficit on transpiration has been supported also by results of the sensitivity tests of the mathematical model (Fig. 3).

To analyse the effect of high evaporative demands of the atmosphere on transpiration daily sums, the seasonal changes of transpiration were first simulated for real conditions. Then, the simulation was repeated using the data of the vapour pressure deficit reduced to one-half. All other model inputs remained unchanged. The comparison of the results of both simulations clearly manifested the effect of extremely high evaporative demands on daily totals of transpiration (Fig. 4). High vapour pressure deficit occurring in the

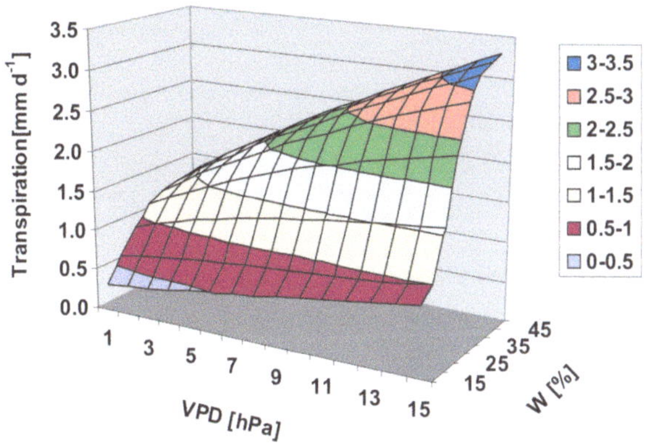

Fig. 3 The transpiration daily totals simulated for the primeval spruce forest related to changes in the vapour pressure deficit (VPD) and soil moisture in the root zone (*W*) expressed as a volumetric percentage

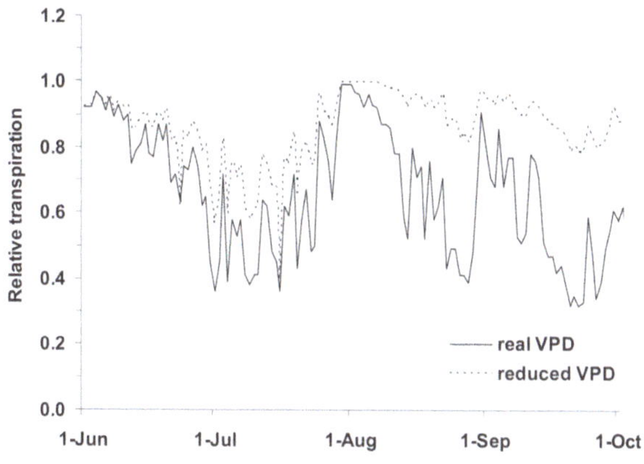

Fig. 5 The seasonal course of the relative transpiration simulated for the real situation (*real* VPD) compared with model simulation for the vapour pressure deficit reduced to one-half (*reduced* VPD)

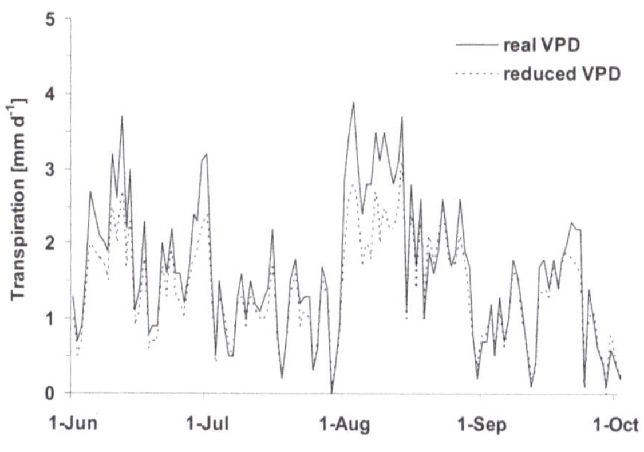

Fig. 4 The seasonal course of the transpiration daily totals simulated for the real situation (*real* VPD) compared with model simulation for the vapour pressure deficit reduced to one-half of the original values (*reduced* VPD)

analysed period accelerated the transpiration mainly in periods when the forest was sufficiently supplied with soil water. On the other hand, during the periods with dry soil, the differences between the transpiration affected by average vapour pressure deficit and under high evaporative demands of the atmosphere were negligible. Hence, the lack of soil water in the periods with extremely dry soil was the dominant factor resulting in the reduction of transpiration.

To quantify the water stress caused by a lack of soil water, relative transpiration as the ratio of actual to potential transpiration is commonly used. The potential transpiration was estimated in this case as drought-free transpiration (Cienciala 1992). For calculations of drought-free transpiration, the described mathematical model was used, with sufficiently high and constant values of the soil water in the simulation. All other inputs of the model remained unchanged. The water content of soil considered sufficiently wet was set to 40% of volume. The results of the model simulations of drought-free relative transpiration are graphically presented in Fig. 5, together with the seasonal course of relative transpiration simulated with reduced values of the vapour pressure deficit on the model input.

It is obvious that the actual transpiration of the investigated young spruce forest was significantly reduced by soil drought during the first two months of the analysed period. The mean ratio between the actual and drought-free transpiration calculated for the real conditions in the analysed period was 69.2%. When applying reduced vapour pressure deficit, this ratio exceeds 88%. It means that soil drought occurring simultaneously with high evaporative demand of the atmosphere reduced transpiration to one-half of the drought-free level within 1 month, despite the monthly precipitation total of 46.1 mm. Taking into account that the drought-free transpiration is very close to the potential transpiration, it can be concluded that the analysed primeval forest suffered from intense water stress in the prevailing part of the analysed period, especially in July and then in the second half of August and in September.

A lack of soil water accompanied by high evaporative demands of the atmosphere can affect not only the transpiration rates but also the dynamics of soil water in the root zone. To quantify the influence of dry air and dry soil on soil moisture, an extended form of the above-mentioned model was used. Seasonal changes in

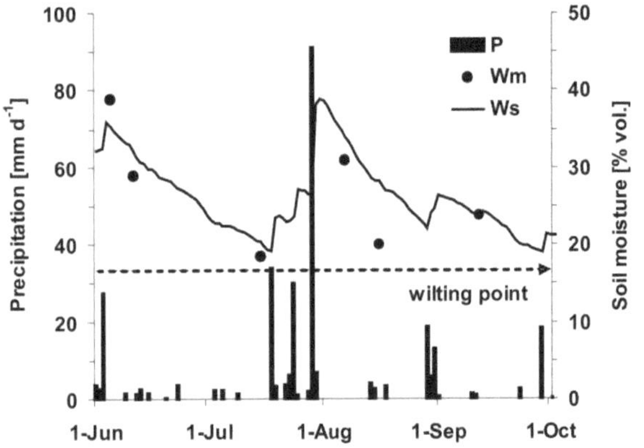

Fig. 6 Measured (W_m) and simulated (W_s) seasonal courses of soil water in the soil layer 0–60 cm with daily totals of the precipitation (*P*)

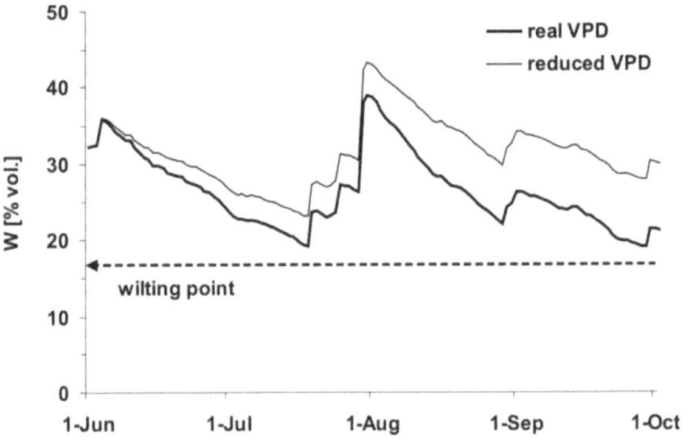

Fig. 7 The seasonal course of the soil water content (*W*) in the root zone simulated for the real situation (*real* VPD) and the vapour pressure deficit reduced to one-half (*reduced* VPD)

the soil water content in the root zone of the spruce primeval forest were simulated during the period of June–September 2003 by the extended model, which balanced precipitation, interception, water uptake by roots and soil evaporation from the soil layer 60 cm deep. The results of the model simulations were compared with the actual measurements of the soil water content (Fig. 6). The compared data are closely related and the standard deviation of differences between the measured and simulated values is comparable with the experimental data error. Then, the model was used as a tool for simulations of the soil water content in the root zone of the spruce primeval forest under high evaporative demands of the atmosphere. For this purpose, the seasonal changes in the soil water content in the root zone of the spruce primeval forest was simulated parallel for real environmental conditions and then using the data of the vapour pressure deficit reduced to one-half (Fig. 7). The results indicated that the water uptake by roots in the root zone of the forest stand under high evaporative demands of the atmosphere in the real situation was more intense and, consequently, drying of the soil was significantly faster in comparison with reduced vapour pressure deficit. However, the forest evaporating under lower evaporative demands significantly saved soil water.

It should be stressed that the halved values of the vapour pressure deficit used in the model simulations correspond well with the long-term averages of the vapour pressure deficit for the site of the analysed spruce primeval forest. Thus, the obtained results provided a possibility to interpret the simulated data alternatively as characterizing soil water regime and stand water balance, or as quantifying the impact of extremely high evaporative demands on the transpiration and soil water dynamics in the analysed forest. It can be concluded that the extremely high evaporative demands of the atmosphere affecting the transpiration of the analysed spruce primeval forest in the prevailing part of the analysed period were a significant risk factor for the stand water regime.

Conclusions

The designed and experimentally validated mathematical model can be considered as a suitable tool for the determination of transpiration daily sums above forest stands in a real situation and also for prediction of impacts of the changing environmental factors on transpiration of vegetation surfaces and on the seasonal variability of the soil water content in the root zone.

The results of the model simulations indicated that soil drought accompanied by dry air can affect substantially the transpiration and consequently the soil water dynamics in the root zone. It was shown that high evaporative demands of the atmosphere were able to compensate partially for the reduction in transpiration rates caused by reduced soil water in the root zone, but it was at the expense of the soil water content. On the other hand, the forest stand evaporating under lower evaporative demands significantly saved soil water.

From the aspect of natural hazards, the extremely high evaporative demands of the atmosphere affecting the forest transpiration during the period of a couple of days were a significant risk factor for the stand water regime. Under the investigated conditions of water stress, the forest stand was not damaged by the soil drought only because of sufficient amount of seasonal precipitation.

Acknowledgments This study was supported by Projects VEGA 2/5006/26, 1/3548/06 and by the Slovak Research and Development Agency under the contract No. APVV-51-030205, APVV-0468-06 and APVV-0022-07.

References

Al-Kaisi M, Brun LJ, Enz JW (1989) Transpiration and evaporation from maize as related to leaf area index. Agric. For. Meteorol., 48, 111–116

Bichele Z, Moldau H, Ross J (1980) Mathematical Modelling of Plant Transpiration and Photosynthesis under Soil Moisture Stress [in Russian]. Gidrometeoizdat, Leningrad, 222p

Bunce JA (1996) Does transpiration control stomatal responses to water vapour pressure deficit? Plant Cell Environ., 19, 131–135

Calvet JC (2000) Investigating soil and atmospheric plant water stress using physiological and micrometeorological data. Agric. For. Meteorol., 103, 229–247

Choudhury BJ, Idso SB (1985) Evaluating plant and canopy resistances of field grown wheat from concurrent diurnal observations of leaf water potential, stomatal resistance, canopy temperature and evapotranspiration flux. Agric. For. Meteorol., 34, 67–76

Choudhury BJ, Monteith JL (1988) A four-layer model for the heat budget of homogeneous land surfaces. Q. J. R. Meteorol. Soc., 114, 373–398

Cienciala E (1992) Assessment of transpiration estimates for Picea abies trees during a growing season. Trees, 6, 121–127

Cienciala E, Eckersten H, Lindroth A, Hällgren JE (1994) Simulated and measured water uptake by Picea abies under non limiting soil water conditions. Agric. For. Meterorol., 71, 147–164

Cienciala E, Kučera J, Lindroth A, Čermák J, Grelle A, Halldin S (1997) Canopy transpiration from a boreal forest in Sweden during a dry year. Agric. For. Meteorol., 86, 157–167

Čermák J, Kučera J (1981) The compensation of natural temperature gradient in the measuring point during the sap flow rate determination in trees. Biol. Plant., 23, 469–471

Čermák J, Palát M, Penka M (1976) Transpiration flow rate in a full grown tree of *Prunus avium* L. estimated by the method of heat balance in connection with some meteorological factors. Biol. Plant., 18, 111–118

Čermák J, Ulehla J, Kučera J, Penka M (1982) Sap flow rate and transpiration determination in full grown oak (Quercus robur L.) in floodplain forest exposed to seasonal floods, as related to potential evapotranspiration and tree dimensions. Biol. Plant., 24, 446–460

Denmead OT, Shaw RT (1962) Availability of soil water to plants as affected by soil moisture content and meteorological conditions. Agron. J., 54, 358–390

Granier A, Biron P, Lemoin D (2000) Water balance, transpiration and canopy conductance in two beech stands. Agric. For. Meteorol., 100, 291–308

Granier A, Loustau D (1994) Measuring and modelling the transpiration of a maritime pine canopy from sap-flow data. Agric. For. Meteorol., 71, 61–81

Gucci R, Massai R, Xiloyanis C, Flore JA (1996) The effect of drought and vapour pressure deficit on gas exchange of young kiwifruit (Actinidia deliciosa var. deliciosa) vines. Ann. Botany, 77, 605–613

Habermann G, Machado EC, Rodrigues JD, Medina CL (2003) Gas exchange rates at different vapor pressure deficits and water relations of 'Pera' sweet orange plants with citrus variegated chlorosis (CVC). Scientia Horticulturae, 98, 233–245

Honert TH (1948) Water transport in plants as a catenary process. Discuss Faraday Soc., 3, 146–153

Iritz Z, Lindroth A, Heikinheimo M, Grelle A, Kellner E (1999) Test of a modified Shuttlerworth – Wallace estimate of boreal forest evaporation. Agric. For. Meteorol., 98–99, 605–619

Jara J, Stockles JO, Kjelgaard JK (1998) Measurement of evapotranspiration and its components in a corn (*Zea Mays* L.) field. Agric. For. Meteorol., 92, 131–145

Karl TR, Knight RW, Plummer N (1995) Trends in high frequency climate variability in the twentieth century. Nature, 377, 211–227

Katerji N, Perrier A (1985) Determination of canopy resistance to water vapour and its various components: theoretical approaches and experimental verification.Agric. For. Meteorol., 34, 105–120

Kučera J, Čermák J, Penka M (1977) Improved thermal method of continual recording the transpiration flow rate dynamics. Biologia Plantarum 19, 413–420

Leonardi Ch, Guichard S, Bertin N (2000) High vapour pressure deficit influences growth, transpiration and quality of tomato fruits. Scientia Horticulturae 84, 285–296

Monteith JL (1965) Evaporation and environment. In: G.E. Fogg (ed.), The State and Movement of Water in Living Organisms. Academic Press, New York, 205–234

Morikawa Y, Hatori S, Kyiono Y (1986) Transpiration of a 31-year-old *Chamaecyparis odtusa* Endl. stand before and after thinning. Tree Physiol., 2, 105–114

Shukla J, Mintz Y (1982) The influence of land surface evapotranspiration on Earth's climate. Science, 215, 1498–1501

Shuttleworth WJ, Wallace JS (1985) Evaporation from sparse crops – an energy combination theory. Q. J. R. Meteorol. Soc., 111, 839–855

Škvarenina J, Střelcová K, Mind'áš J (2002) Bioclimatological and ecophysiological research in Biosphere Reserve Poľana. (In Slovak). In: Rožnovský, J., Litschmann, T. (ed.): XIV. Czech-Slovak Bioclimatological Conference, CD-ROM, ISBN 80-85813-99-8, 429–441

Sperry S (2000) Hydraulic constraints on plant gas exchange. Agric. For. Meteorol., 104, 13–23.

Torula T, Heikinheimo M (1999) Modelling evapotranspiration from a barley field in the growing season. Agric. For. Meteorol., 91, 237–250

Turner NC, Schulz ED, Gollan T (1984) The response of stomata and leaf gas exchange to vapour pressure deficits and soil water contents. Oecologia, 63, 338–342

Wallace JS (1995) Calculating evaporation: resistance to factors. Agric. For. Meteorol., 73, 353–366

Wallace JS, Roberts JM, Sivakumar SVK (1990) The estimation of transpiration from sparse dryland millet using conductance and vegetation area indices. Agric. For. Meteorol., 51, 35–49

Wang YP (2000) An improvement to the two-big-leaf model for calculating canopy photosynthesis. Agric. For. Meteorol., 10, 143–150

Wilson KB, Baldocchi DD (2000) Seasonal and interannual variability of energy fluxes over broadleaved temporate deciduous forest in North America. Agric. For. Meteorol., 100, 1–18

Wind PG (1972) A hydraulic model for simulation of non/hysteric vertical unsaturated flow of moisture in soils. J. Hydrol., 15, 227–246

Xue Q, Weiss A, Arkebauer TJ, Baenziger PS (2004) Influence of soil water status and atmospheric vapor pressure deficit on leaf gas exchange in field-grown winter wheat. Environ.Exp. Botany, 51, 167–179

Assessment of Water Deficiency in Forest Ecosystems: Can a Simple Model of Forest Water Balance Produce Reliable Results?

P. Baláž, K. Střelcová, M. Blaženec, R. Pokorný and Z. Klimánková

Keywords Drought · European beech · Norway spruce · Transpiration · Water balance · Water demands

Introduction

Increased frequency of drought events, as one of the main natural abiotic stress factors, has raised concerns in recent years about its potentially detrimental impact on forest ecosystems. Ongoing changes in global climate are associated with a predicted sharp increase in drought-induced damage to natural ecosystems (IPCC Third Assessment Report 2000). Increased frequency and intensity of dry events occurring in vegetation period is predicted also for the region of central Europe (Lapin et al. 2001). Detrimental impact of drought on tree species manifests itself through a range of symptoms including impaired growth (Hanson et al. 2001), defoliation and foliage yellowing (Solberg 2004; Zierl 2004), and increased sapling mortality (Helenius et al. 2002). It is also considered an important predisposition factor predictive of other harmful agents such as bark beetle (Schwenke 1996; Schopf and Köhler 1995; Zinecker 1957). Detrimental impact of drought is mostly associated with Norway spruce (*Picea abies* [L.] Karst.); however, other coniferous and broad-leaved tree species are prone to drought-induced changes as well (Hanson et al. 2001; Thomas et al. 2002; Kozlov and Niemelä 2003).

P. Baláž (✉)
Forest Research Institute, National Forest Centre,
T.G. Masaryka 22, 960 92 Zvolen, Slovakia
e-mail: balaz@nlcsk.org

In order to plan efficient measures precluding and restricting impact of drought, it is imperative first to understand how drought affects processes in forest ecosystems. To achieve this, we need an effective, universal, yet at the same time simple and multi-faceted tool for monitoring and quantification of intensity and frequency of drought-induced stress. The tool should take account of basic input and output components of forest ecosystem water balance; provide for species and developmental-stage-specific water demands; and last but not least be based on easily accessible input data interpolable for larger areas.

At present, there is a wide range of different drought-assessment methods and indices used; majority of them, however, are based on climatologic approach to water-deficiency assessment, often referring to a specific, long-term period or water demands of main agricultural crops. In addition, many of these methods make use of data averaged over longer time periods (weeks, months, etc.) and thus fail to reflect short-term variability of factors influencing drought frequency and intensity. More comprehensive review of the most frequently used drought indices is provided by Byun and Wilhite (1999), Heim (2002), and Narasimhan and Srinivasan (2005).

To date, large-scale assessment of frequency and intensity of drought in forest ecosystems is largely based on the utilisation and adjustment of various climatic and agricultural indices. Nonetheless, it is important to realise that drought is usually characterised by a non-dimensional variable (index), the value of which is difficult to interpret with regard to a particular tree species since the same value possesses different importance for different species depending on their natural water demands. Hence, for practical interpretation, models of forest ecosystem water

K. Strelcová et al. (eds.), *Bioclimatology and Natural Hazards*,
DOI 10.1007/978-1-4020-8876-6_18, © Springer Science+Business Media B.V. 2009

balance with a various degree of comprehensiveness are considered more appropriate. The aforementioned models have been tested, among others, by Bouten (1995), Jansson et al. (1999), Čermák and Prax (2001), Zierl (2001), Wellpott et al. (2005). The main disadvantage of this approach to forest stand water balance modelling is a comparatively high complexity of used models as they are very demanding for site conditions parametrization. Their application is therefore often limited to experimental plots, where a range of different climatic, physiologic and soil variables can be monitored. However, if we attempt to assess drought frequency and intensity over larger tracts of land, we need to ensure that all entering variables are interpolable for any assessed area and are based on easily accessible data. Such approach inevitably calls for a somewhat simplified 'version' of water-balance estimate, even at the expense of slightly compromised accuracy of final results.

The aim of the present study is to evaluate the applicability and effectiveness of assessment of frequency, intensity and duration of water deficiency based on a substantially simplified estimate of forest stand water balance taking due account of specific demands of particular tree species and their different developmental stages. This is done on the basis of two model stands and measurements of selected water balance components.

Material and Methods

Model Stands

Water balance estimate and assessment of water deficiency were carried out for two model forest stands: a spruce stand situated on the Bílý Kříž site and a mixed beech–dominant stand on the Hukavský Grúň site.

Bílý Kříž Spruce Stand

Bílý Kříž site is situated on the territory of the Czech Republic (in the immediate vicinity of the border with Slovakia), in the mountain range of Moravskoslezké Beskydy, at the altitude of 908 m (49°30′N, 18°32′E). It is overgrown by an even-aged Norway spruce (*Picea abies* [L.] Karst.) stand with an occasional admixture of silver fir (*Abies alba* L.) on the gently sloping south-east facing hillside with a natural gradient of 13.5°. Bedrock is formed by Mesozoic Godula sandstones atop. The soil type is classified as moderate Ferro-Humic Podzol with average depth of 60–80 cm and loamy/sand-loamy soil texture. The site lies in the region typical for moderately cold and wet climate. The stand constitutes an experimental research site of the Institute of Systems Biology and Ecology, Academy of Sciences of the Czech Republic.

Hukavský Grúň Mixed (Beech) Stand

Hukavský Grúň site carries a mixed mature forest stand with the predominance of European beech (*Fagus sylvatica* L.) and with the admixture of other common tree species such as Norway spruce (*Picea abies* [L.] Karst.), sycamore (*Acer pseudoplatanus* L.), common ash (*Fraxinus excelsior* L.) and aspen (*Populus tremula* L). For practical reasons, the stand is considered a pure beechwood for the purposes of this study. Accordingly, all measurements and calculations relate exclusively to beech only. The site is situated in the central part of the Slovak Republic, in the mountain range of Poľana, at the altitude of 860 m (48°37′N, 19°28′E). The stand itself lies on a moderately sloping north-east facing hillside with a natural gradient of 25°. Bedrock is formed by volcanic rocks most abundant of which are epiclastic volcanic sandstones. These rocks give origin to light soils with silty sand topsoil classified as Cambi-Eutric Andosols. The stand constitutes a permanent research plot for the National Forest Centre in Zvolen.

Selected characteristics of the aforementioned model stands are given in Table 1.

Table 1 Selected characteristics of the model stands

	Bílý Kříž	Hukavský Grúň
Stand age (years)	23	90
Species composition (%)	Spruce 99/fir 1	Beech 70/spruce 20/others 10
Mean tree height (m)	10.9	Beech 27.0/spruce 36.5
Mean breast-height diameter (cm)	13.0	Beech 24.0/spruce 44.7
Mean annual precipitation (mm)	1100	853
Mean annual temperature (°C)	4.9	5.8

Input Data

The study is based on the data sourced from a range of different research projects. Due to this fact, analyses for particular model stand were based on different time frame and modelling period. In case of Bílý Kříž stand, the data available covered the period from May through July 2004; in case of Hukavský Grúň, the data referred to May through September 1996. Still, actual modelling of forest water balance was conducted only for selected parts of the aforementioned periods because of discontinuous measurements of particular data caused by technical failure or measurement methods.

Meteorological Characteristics

In the study, the following meteorological characteristics were subject to investigation in model stands: above-canopy mean daily air temperature, total daily precipitation over an open area and canopy-reaching global daily radiation total.

In the Bílý Kříž stand, air temperature readings were taken every 30 min at 8 m height. The readings were further processed into daily averages. Intensity of global radiation (Kipp. Zonen Delf BV-CM5, NL) was measured 15 m above the canopy; daily totals were provided from 30-min averages. Daily precipitation totals were measured using an automated rain gauge (HOBO, AMET, CZ) installed on the adjacent open land.

In the Hukavský Grúň stand, air temperature readings were taken at 34 m stand height. Readings for both precipitation totals and canopy-reaching global radiation were recorded (equipment DELTA-T Devices Ltd., UK) at the height of 37 m in a 10 min interval. In addition, daily values (averages and totals) of particular meteorological characteristics were provided.

Other Variables

For the purpose of this study, a number of other variables were under investigation, namely average daily soil moisture (soil water content) in a particular soil profile; and daily stand transpiration total.

Soil moisture under the Bílý Kříž model stand was recorded in percent volume using Time Domain Reflectometry (TDR) technique (TRIME, IMKO, DE) at five points in the forest stand. The readings were taken on daily basis for three different soil layers: 1–17, 16–32 and 27–43 cm. Subsequent reading conversions were provided for average daily soil water content (mm) in the 0–43 cm depth.

Soil moisture under the Hukavský Grúň model stand was measured as soil water potential using tensiometres with 2.2 gauge ceramic cups (Soil moisture Equipment, U.S.A.) in weekly intervals. Tensiometer vacuum readings were taken using Marthaler manometer. Based on earlier results by Cassel and Klute (1986), we suggested the lowest measurable value of soil water potential at −850 hPa. Soil water potential was measured in weekly intervals in the 15, 30 and 50 cm depth profile at three points around each sample tree at the distance of 1–2 m from the tree.

The heat pulse velocity method (Sapflow Meter SF 300, Greenspan Technology, Australia) (e.g. Hatton et al. 1990) was used for the campaign measurements of sap flow rate on 10 sample trees at Bílý Kříž. The sap flux ($l\ h^{-1}$) obtained for the sample trees and values of respective sapwood areas were used for the calculation of mean specific sap flux for each stem diameter class ($l\ h^{-1}\ cm^{-2}$ of sapwood). These values together with summarised sapwood area of all trees in each stem diameter class (based on the site-specific equation, Pokorný 2000) were used for up-scaling to stand transpiration.

Transpiration of the Hukavský Grúň stand was estimated using up-scaling of sap flow readings taken at sample trees (Čermák and Kučera 1990; Čermák et al. 2004). Five sample trees were selected, and sap-flow rates for particular trees were estimated applying the tree-trunk heat balance method using direct, non-destructive and continuous measurements. Spatial arrangement of the measuring points conformed to Čermák et al. (1973, 1976 and 1982) and Kučera et al. 1977. The final mass flow was estimated using a simple calculation based on the differential heat balance equation (Kučera et al. (1977).

Sample trees at both sites were selected following the rules given by Swanson (1970),Čermák (1989), and Čermák and Kučera (1990). They recommended using a set of sample trees composed of at least three or more (6–15) trees.

Water-Deficiency Assessment

Drought Index

As mentioned earlier, many existing methods of drought assessment have to deal with the uncertainties of drought definition. From the hydrological viewpoint our approach is purposed to deal with water deficiency rather than with drought. On the contrary, the method chosen for this study assumes that particular tree species start to suffer from drought stress when unable to cover their water demands. Based on this fact we used the terms 'drought index' and 'drought intensity'. The method hence calculates with both the actual soil water availability and actual water demand of a particular forest stand. The frequency and immediate intensity of water deficiency are expressed by drought index (DI), which is given as a ratio between the actual (AT) and potential (PT) transpiration of a particular stand.

$$DI_n = 1 - \left(\frac{AT_n}{PT_n}\right) \quad (1)$$

where DI_n is n-day drought index, AT_n, the actual stand transpiration at the start of day n (mm day^{-1}) and PT_n, the potential stand transpiration at the start of day n (mm day^{-1}). Index values vary between 0 for ample soil water content (above the point of decreased availability) and 1 for zero availability (below the permanent wilting point).

Potential Transpiration

Potential stand transpiration represents transpiration unlimited by water deficiency. It means that PT value is dependent only on the values of selected meteorological characteristics and stand characteristics such as age, tree species composition and structure. In the study, PT value was estimated using simple and readily available input variables that are well interpolable over larger spatial units. Among the main meteorological characteristics influencing potential transpiration, just two characteristics – air temperature and global radiation – met this presumption. PT value was estimated as a function of reference (potential) evapotranspiration (PET) derived from the regression analysis of the relationship between potential transpiration and reference evapotranspiration:

$$PT = f(PET) \quad (2)$$

PET was estimated using Turc's equation (Turc 1961), which is

$$PET = 0.013\left(\frac{T_a}{15 + T_a}\right)(R_s + 50) \quad (3)$$

where PET is potential evapotranspiration (mm), T_a is air temperature (°C) and R_s is the total daily global radiation (cal cm^{-2} day^{-1}).

The reason for choosing the Turc's method of PET calculation is the availability of input variables. It is a method based exclusively on the influence of air temperature and radiation, which are easily available even on the regional scale. Potential stand transpiration was understood as total daily sap flow for days with soil moisture exceeding 25% (Bílý Kříž) or for days with soil water potential above −350 hPa (Hukavský Grúň). We assumed that differences in species-specific water demand would manifest themselves also through PT values obtained under comparable climatic conditions. On account of that, the correlation between PT and PET was estimated for both model stands (Bílý Kříž, Hukavský Grúň) separately.

Actual Transpiration

Actual n-day stand transpiration (AT_n) is the result of n-day potential stand transpiration (PT_n) and n-day soil water content (SWC_n). Because of the lack of relevant knowledge we based our AT estimate on the simple assumption that soil moisture above the point of decreased availability (PDA) allows tree species full satisfaction of their transpiration needs; in that instance, actual transpiration equals potential transpiration. If soil moisture drops below the PDA, the ability of trees to meet their transpiration needs linearly decreases to zero at permanent wilting point (PWP) (Fig. 1). Soil water limits for the Bílý Kříž stand were estimated using clay fraction percentage (Tlapák et al. 1992). Soil water limits for the Hukavský Grúň stand were supplied from Soroková (2001).

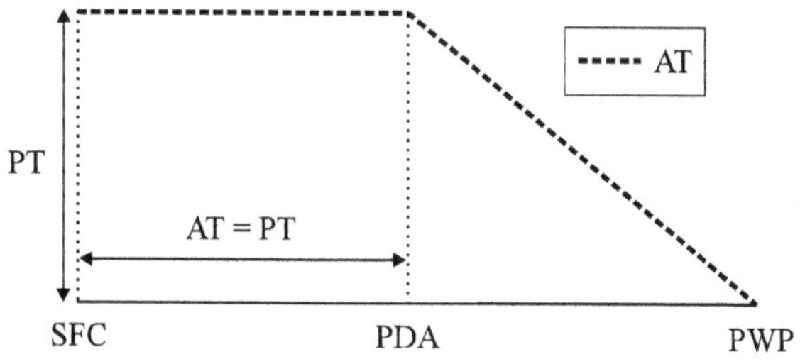

Fig. 1 Relationship between actual stand transpiration, its potential transpiration and soil moisture

Soil Water Content

Soil water content for a particular day (*n*-day) was estimated based on the following equation:

$$SWC_n = SWC_{n-1} + P_{n-1} - AT_{n-1} \qquad (4)$$

where SWC_{n-1} is $n-1$ day soil water content (mm), P_{n-1} is the $n-1$ day total precipitation after interception reduction (mm day^{-1}) and AT_{n-1} is the $n-1$ day actual stand transpiration (mm day^{-1}).

Soil water content is calculated only for the root development zone of the soil profile. For the estimate, this zone is considered homogenous with consistent physical properties. Rather than by zone properties, we were limited by the maximum depth of soil moisture measurements. The actual SWC fluctuates anywhere between PWP and soil field capacity (SFC). The amount of water exceeding SFC is classified as subsurface flow and soil depth penetration and as such was excluded from further calculations. Since SWC is based on the $n-1$ day values, for the first day of the assessment period we used known value of soil water content originating from soil moisture measurements. On the other hand, canopy storage capacity for both model stands was only estimated. Total stand interception was calculated from the canopy storage capacity and the amount of precipitation intercepted by the under story layer (herbaceous layer, litter, top humus layer). This approach was chosen because the model applied worked with the amount of precipitation influencing soil moisture in the root development zone rather than the throughfall amount. Earlier results by Intribus (1977) indicate that the amount of under-canopy interception can reach up to 27.7% of the total precipitation over an open area. Minďáš (1999) estimated the canopy storage capacity of beech at 1.6 mm and that of spruce at 2.3 mm. Since the model calculates with daily precipitation totals we assumed that during a precipitation day, canopy is on average able to dry out at least once; hence for average daily canopy interception we presumed double canopy storage capacity. Due to this fact, we estimated final average daily interception of 8 mm for the Bílý Kříž spruce stand and final average daily interception of 5 mm for the Hukavský Grúň beech stand.

Intensity of long-term drought stress was estimated using cumulative transpiration deficit (CTD); CTD value for the day '*n*' was given as:

$$CTD_n = PT_n - AT_n + CTD_{n-1} \qquad (5)$$

where CTD_n is cumulative transpiration deficit (mm), PT_n, *n*-day potential stand transpiration (mm day^{-1}), AT_n, the *n*-day actual stand transpiration (mm day^{-1}) and CTD_{n-1} is the $n-1$ day value of cumulative transpiration deficit (mm). Based on the assumption that following a period of ample transpiration water, an average tree is able to recover from drought stress, we introduced a 3-day sustained period of ample transpiration water as the limit annulling previous drought stress. The CTD value was upon three consecutive days with DI=0 habitually changed to zero.

Results

Potential Transpiration

For both model stands, only a limited number of days when sap flow, air temperature and radiation readings were taken and specific limit exceeding soil moisture recorded was available for the estimation of the correlation between the stand PT and reference PET (see the previous section 'Water-deficiency assessment'). In the case of Bílý Kříž spruce stand, 22 such days were available. Final linear function illustrating PET-based estimate of the particular model stand PT is given in Fig. 2.

The second stand at the Hukavský Grúň site provided better results as 117 days with comprehensive data were available; results are given in Fig. 2(b).

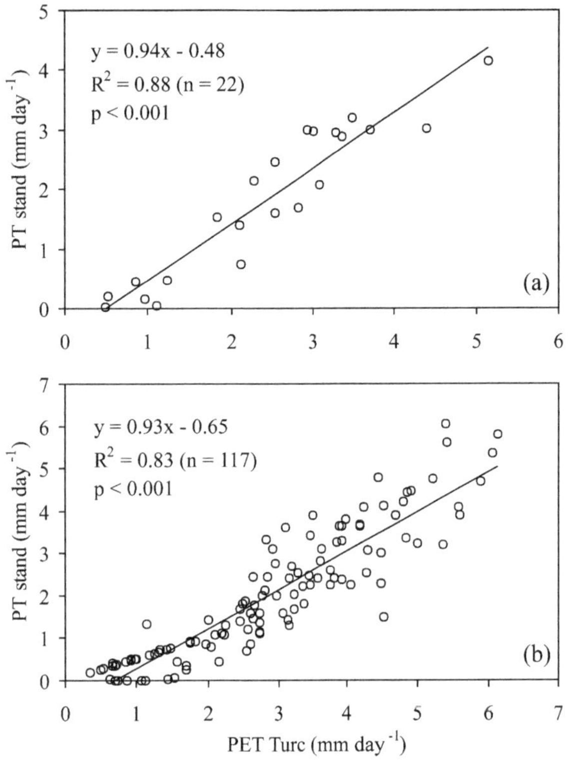

Fig. 2 Relationship between reference potential evapotranspiration by Turc (1961) and measured stand transpiration (PT) used for the estimate of potential transpiration of Bílý Kříž spruce stand (**a**) and Hukavský Grúň beech stand (**b**)

Drought Index

Values of drought index indicate differences in soil moisture regimes at chosen model stands referring to the assessment period. The Bílý Kříž spruce stand suffered from water deficiency for almost the entire length of the assessment period; DI value, however, was mostly relatively high (Fig. 3). It essentially means that the soil water content fluctuated between the point of decreased availability and wilting point for much of the assessment period.

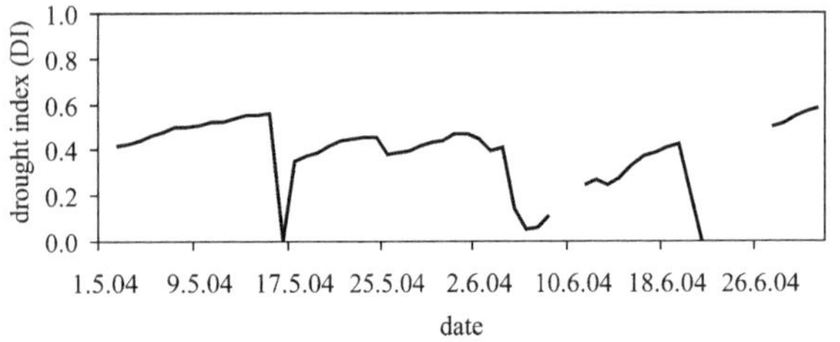

Fig. 3 Drought index (DI) for the Bílý Kříž spruce stand

Results from the Hukavský Grúň beech stand indicate ample soil moisture for the entire vegetation period of 1996. Here, drought index for much of the assessment period except for August was equal to zero. In other words, soil moisture for much of the assessment period except for August did not fall below the point of decreased availability. Even during August water deficiency, DI recorded was still below 0.3 (Fig. 4)

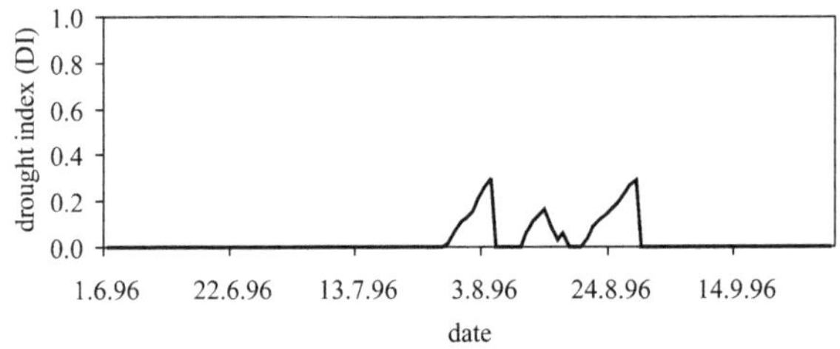

Fig. 4 Drought index (DI) for the Hukavský Grúň beech stand

Cumulative Transpiration Deficit

Values for cumulative transpiration deficit confirmed the results of formerly conducted assessment of in situ moisture regimes at both model stands. Results from the Bílý Kříž stand reflected the effect of permanent water deficiency (drought stress) occurring in the first half of the 2004 vegetation period (Fig. 5). CTD values showed a steady increase throughout the modelled period, reaching 33 mm at the end of the period. Detrimental effect of drought on physiological variables of this stand could thus be accumulated over the entire modelled period, because it was permanent. The Hukavský Grúň beech stand, on the contrary, benefited for much of the 1996 vegetation period (except

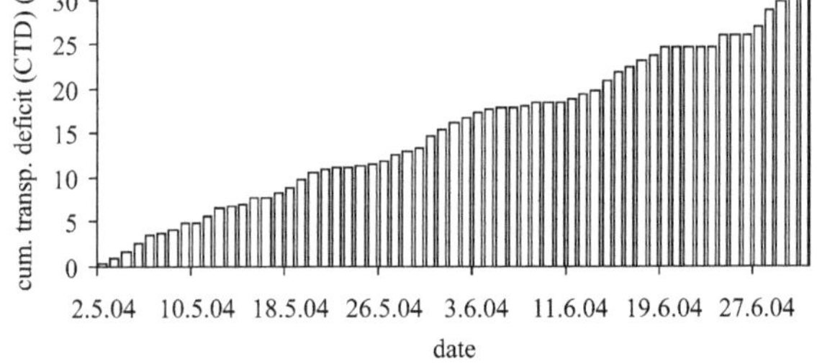

Fig. 5 Values of cumulative transpiration deficit for the Bílý Kříž spruce stand

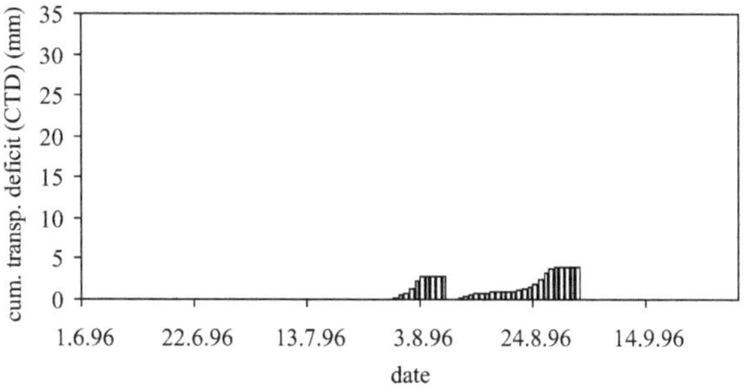

Fig. 6 Values of cumulative transpiration deficit for the Hukavský Grúň beech stand

for August) from a transpiration-satisfying level of soil moisture, as indicated by the CTD values (Fig. 6),. Transpiration deficit was temporary, and in August it was only limited; the CTD value, however, never exceeded 4 mm, largely thanks to the observed disruption in the water-deficiency period.

Model Validation

Functionality and correctness of the described forest stand water balance model was tested by the means of the comparison between the measured and computed parameters of water balance, namely actual stand transpiration and soil water content. Comparison of the measured and computed trends of soil water content is shown in Figs. 7 and 8. In the case of Hukavský Grúň beech stand, the comparison is merely informative, as soil moisture measurements were interpreted only in the form of soil water potential. The figure thus primarily serves to compare the conformity of the trends of measured and computed soil moisture values rather than to compare the conformity of its absolute values. Results of the regression analysis between measured and computed values of the selected water balance components are provided in Figs. 9 and 10. The total error of the model was evaluated by the root mean square error (RMSE). The values of RMSE are presented in the Table 2.

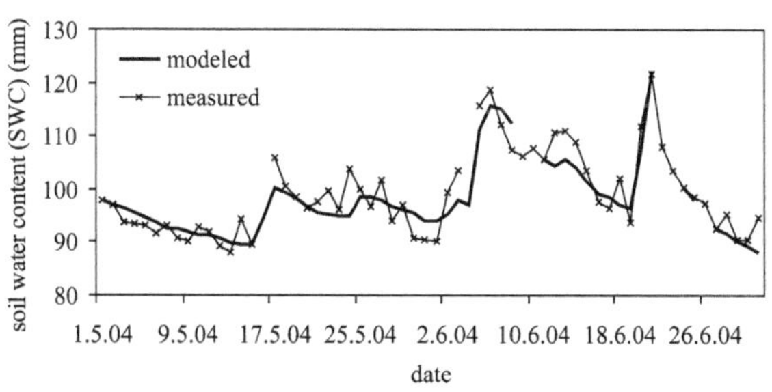

Fig. 7 Comparison of trends of measured and modelled values of soil water content in the Bílý Kříž spruce stand

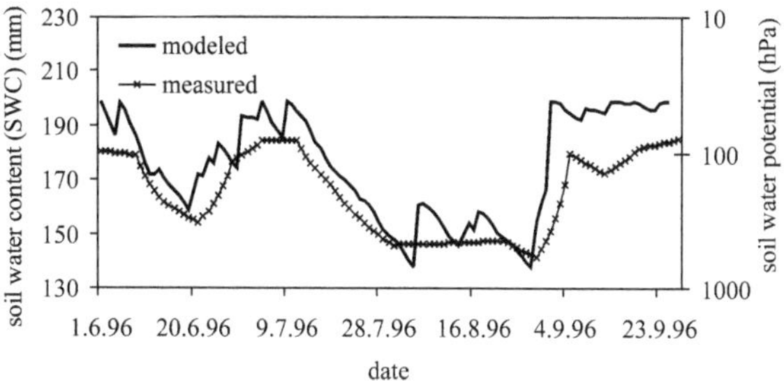

Fig. 8 Comparison of trends of modelled values of soil water content and measured values of soil water potential in the Hukavský Grúň beech stand. Values of soil water potential were interpolated from roughly weekly orchestrated measurements down to daily values

Results of the respective correlation analyses (Figs. 9 and 10) indicate a very tight correlation with a high level of statistical significance between the

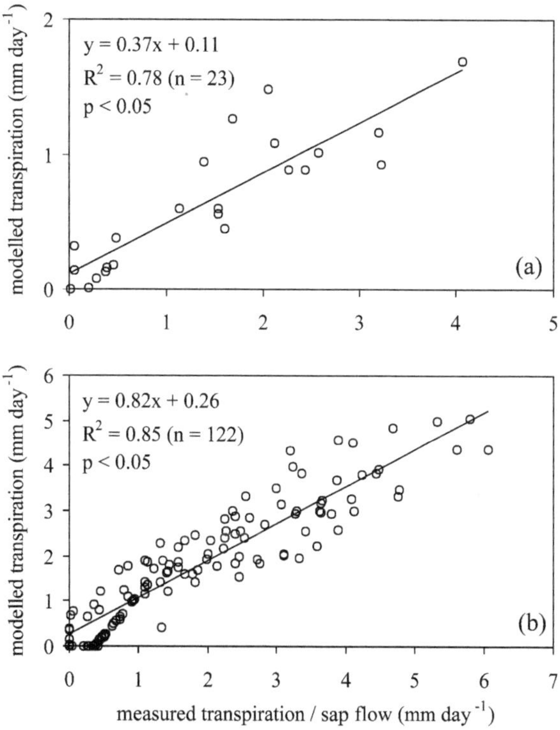

Fig. 9 Relationship between modelled transpiration (AT) and transpiration measured at Bílý Kříž spruce stand (**a**) the Hukavský Grúň beech stand (**b**)

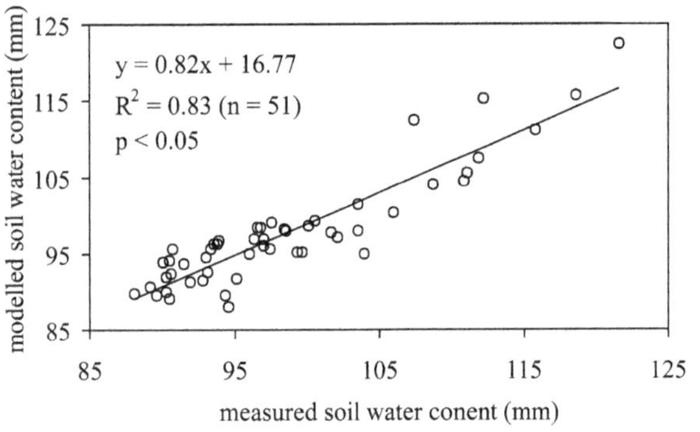

Fig. 10 Relationship between modelled and measured soil water content (SWC) at the Bílý Kříž spruce stand

measured and modelled values of actual transpiration for both model stands; in case of Bílý Kříž spruce stand, similar trend was observed between the measured and modelled values of soil water content. In spite of a tight correlation, the AT of Bílý Kříž spruce stand was modelled with a relatively high error – the RMSE is 1.09 mm (the measured values ranges from 0 to 4 mm). Similarly unfavourable result with high error shows also the RMSE of the model at SWC Bílý Kříž that is 23.7 mm (the measured values ranges from 88 to 121 mm). The RMSE of the modelled AT of Hukavský Grúň beech stand indicates much better prediction with an error of 0.59 mm (the measured values ranges from 0 to 6 mm).

Table 2 RMSE of modelled values in comparison to measured values of the actual transpiration and soil water content (mm)

Variable	Bílý Kříž	Hukavský Grúň
Actual transpiration	1.09	0.59
Soil water content	23.73	–

Discussion

Water-Deficiency Assessment

Assessment of water deficiency, climatic humidity, aridity and site moisture regime based on the relationship between potential and actual transpiration, and, if applicable, evapotranspiration, represents at present a very popular method of the estimation of these parameters (Lexer and Hönninger 1998; Schiller and Cohen 1998; Lai and Katul 2000; Zierl 2004; Wellpott et al. 2005 and others). Climatic water balance based on the difference between total precipitation and potential evapotranspiration as applied by Škvarenina et al. (2004) can also be classed within the mentioned group of methods. If there is no other limitation to actual tree species transpiration than insufficient soil water content during the vegetation period, drought index applied can be considered appropriate for the assessment of immediate water deficiency in a particular forest stand. As we are yet to learn which threshold DI values can trigger physiological responses in trees, for the meantime, we agreed DI > 0 to be drought stress generating values.

From the perspective of drought stress and its impact on species-specific physiological status, both the momentary intensity of soil water deficiency and its duration and intensity in the immediate past are of considerable significance. In our study, the duration of drought stress influence was described using cumulative transpiration deficit. Another similar cumulative drought characteristic referring to a specific time period is the transpiration deficit estimated as a difference between seasonal integrals of potential and actual transpiration (Ciencala et al. 1997). This approach, however, does not take into account possible recovery of drought-weakened trees resulting from replenishment of soil water above the easily available level. For this very reason, impact of drought over a specific time period might be overestimated if there were a few sufficiently wet episodes within this period. Practical application of the approach chosen in this study will however necessitate the identification of the minimum period with sufficient soil water content needed for trees to recover to the level preceding water deficiency. Different approach to drought stress assessment was applied by Zierl (2004). He constructed his approach around the presumption that forest stands have to a certain degree adapted to in situ soil moisture regimes; this view allowed him to characterise drought stress through the difference between the assessment period drought index and long-term average drought index. It is important to realise that such comparisons may also bear the risk of drought stress underrating, should previous years record several extremely dry periods with very low long-term averages. In our opinion, influence of natural adaptation mechanisms would be better included in the estimate of potential stand transpiration or the estimate of threshold limits of soil moisture availability to tree species (Fig. 1).

Potential Transpiration

Although Penman and Penman–Monteith equations and their modifications are presently the most preferred methods of potential stand transpiration computing (Ciencala et al. 1997; Schiller and Cohen 1998; Čermák and Prax 2001; Zierl 2001 and others), we based our estimate on Turc's equation. The reason was to minimise the number of climatic variables entering the model. Since the Turc's method does not utilise air moisture influenced by stand transpiration, it is particularly suitable for the modelling of site evaporative conditions (Ciencala et al. 1999). Differences in potential transpiration caused by species composition and age structure of a particular stand were projected into the function of stand PT and PET. We assumed it was possible to establish this function separately for each tree species and its different age classes. The approach is basically an alternative to coefficient using conversion of reference FAO Penman–Monteith PET to particular agricultural crop evapotranspiration (Allen et al. 1998). Since the functions we obtained in the study result from the analysis of data collected from a single stand for a single vegetation period, possibilities for their generalisation for pure spruce and beech stands of a particular age class are seriously limited. Possible complex species-specific PT-estimation research is supposed to lead to more realistic and representative non-linear functions establishment. Needless to say, availability of data for other variables creates basis for utilisation of more accurate standard methods of potential stand transpiration computing.

Water Balance Model

In an attempt to simplify the calculation of particular components of forest stand water balance, we decided to use, instead of more or less complex water balance models such as SWAT (Narasimhan and Srinivasan 2005), WAWAHAMO (Zierl 2001) and FORHYD (Bouten 1995), a simplified model consistent with the formula for soil water content calculation (Eq. 4).

We assumed interception-reduced precipitation to be the only source of water; accordingly, stand transpiration was the only recognised water output (except for water supply exceeding soil water capacity). Acceptance of the assessed soil profile for a homogenous layer allowed us to avoid a rather complicated estimation of sub-surface soil water movement. This move in return contributed to model simplification. Model adaptations did not compromise the calculation of selected water balance components; obtained results displayed a high level of correlation between the computed and measured values. Although the study showed a tight, highly significant correlation between measured and modelled values of actual transpiration and soil water content for the experimental stands (Figs. 9 and 10), the RMSE values indicate lower accuracy of AT and SWC prediction in the case of spruce stand Bílý Kříž. However, the obtained results suggest some potential of such a simplified approach in water-deficiency assessment and forest water balance modelling. Unfortunately, there was only a limited opportunity to test functionality of both the water balance model and water-deficiency assessment method in the process, mainly because of limitations imposed by data gathering method and data available for respective model stands. For the purpose of this study, data from a variety of different, mutually independent field investigations were used; however, for a more comprehensive analysis of model's reliability, we would need to work with data covering a wider range of stands and time periods.

Conclusions

In this chapter we attempted to outline possible application of a substantially simplified forest ecosystem water balance model for the assessment of drought stress frequency in forest stands. Simultaneously, we tested the appropriateness of this simplified model for obtaining an acceptable accuracy of results for selected water balance components. Due to missing physiological measurements we nevertheless failed to affirm the suitability of our approach for water-deficiency assessment in association with stress response in trees. On the other hand, the method is based on a widely used assessment of the rate between actual and potential transpiration and stand transpiration deficit. Results obtained in the course of the investigation provide for the following conclusions:

- Correlation analysis between the modelled and measured transpiration and soil water content of the model stands indicated the ability of the simplified model of forest stand water balance to deliver acceptable results. On the other hand, the total model accuracy analysis shows unfavourable values for the Bílý Kříž spruce stand, what can result also from the a low number of data for analysis, not only from the model simplicity. In spite of this, the presented results, especially from the Hukavský Grúň, show the potential of this method to be used as an effective tool for the assessment of frequency and intensity of water deficiency in cases when input data are insufficient for more complex and accurate methods of water balance assessment.
- The applicability of the proposed simplified approach in water balance modelling can be finally considered only after future analyses based on more comprehensive measurements of water balance components and selected variables indicative of drought stress in tree species.
- Drought index and cumulative transpiration deficit indicate significant level of water deficiency experienced by Bílý Kříž spruce stand in the 2004 vegetation period. On the contrary, Hukavský Grúň beechwood showed no signs of water stress throughout much of the 1996 vegetation period; partial signs of water deficiency were only temporary and restricted to August only.

Acknowledgments We acknowledge financial support from the grants No. 1/3524/06 of the Scientific Grant Agency of the Ministry of Education of Slovak Republic and the Slovak Academy of Sciences, from Slovak Research and Development Agency No. APVV-0468-06, APVV-0022-07, No. 526/03/H036 of the GA CR and No. 1P05OC027 of the Ministry of Education of CR, Youth and Sports of the Czech Republic. Research Intention of ISBE: AV0Z60870520. We also thank Zuzana Kmeťová for translation.

References

Allen RG, Pereira LS, Raes D, Smith M (1988) Crop evapotranspiration: guidelines for computing crop water requirements. FAO Irrigation and Drainage Paper NO. 56, Food and Agricultural Organization, Rome, Italy, 300

Bouten W (1995) Soil water dynamics of the Solling spruce stand, calculated with the FORHYD simulation package. Ecol. Model, 83, 67–75

Byun HR, Wilhite DA (1999) Objective quantification of drought severity and duration. J. Climate, 12 (9), 2747–2756

Cassel DK, Klute A (1986) Water Potential: Tensiometry. In: Methods of soil analysis, Part 1 – Physical and Mineralogical Methods, 2nd Edition, Klute, A. (ed.). ASA, Inc./SSSA Madison, 563–596

Čermák J, Deml M, Penka M (1973) A new method of sap flow rate determination in trees. Biol. Plantarum, 15, 71–78

Čermák J, Palát M, Penka M (1976) Transpiration flow rate in a full-grown tree of *Prunus avium* L. Estimated by the Method of Heat Balance in Connection with Some Meteorological Factors. Biol. Plantarum, 18, 111–118

Čermák J, Úlehla J, Kučera J, Penka M (1982) Sap flow rate and transpiration determination in full-grown oak (*Quercus robur* L.) in Floodplain Forest Exposed to Seasonal Floods, as Related to Potential Evapotranspiration and Tree Dimensions. Biol. Plantarum, 24, 446–460

Čermák J (1989) Solar equivalent leaf area: an efficient biometrical parameter of individual leaves, trees and stands. Tree Physiol., 5, 269–289

Čermák J, Kučera J (1990) Scaling up transpiration data between trees, stands and watersheds. Silva Carelica, 15, 101–120

Čermák J, Prax A (2001) Water balance of Southern Moravian floodplain forest under natural and modified soil water regimes and its ecological consequences. Ann. For. Sci., 58, 15–29

Čermák J, Kučera J, Nadezhdina N (2004) Sap flow measurements with some thermodynamic methods, flow iteration within trees and scaling up from sample trees to entire forest stands. Trees – Struct. Funct, 18, 529–546

Ciencala E, Kučera J, Lindroth A, Čermák J, Grelle A, Halldin S (1997) Canopy transpiration from a boreal forest in Sweden during a dry year. Agr. Forest Meteorol., 86, 157–167

Ciencala E, Kučera J, Lindroth A (1999) Long-term measurements of stand water uptake in Swedish boreal forest. Agr. Forest Meteorol., 98–99, 547–554

Hanson PJ, Todd DE Jr., Amthor JS (2001) A six-year study of sapling and large-tree growth and mortality responses to natural and induced variability in precipitation and throughfall. Tree Physiol., 21, 345–358

Hatton TJ, Catchpole EA, Vertessy RA (1990) Integration of sapflow velocity to estimate plant water use. Tree Physiol., 6, 201–209

Heim RR Jr. (2002) A review of twentieth-century drought indices used in the United States. B. Am. Meteorol. Soc., 83(8), 1149–1165

Helenius P, Luoranen J, Rikala R, Leinonen K (2002) Effect of drought on growth and mortality of actively growing Norway spruce container seedling planted in summer. Scand. J. For. Res., 17, 218–224

Intribus R (1977) Water balance components in lowland broadleaf forests. Closing report VI-2-1/20. VÚLH Zvolen

IPCC The Third Assessment Report, Summary for Policymakers, (2000) (www.ipcc.ch)

Jansson P-E, Cienciala E, Grelle A, Kellner E, Lindahl A, Lundblad M (1999) Simulated evapotranspiration from the Norunda forest stand during the growing season of a dry year. Agr. Forest Meteorol., 98–99, 621–628

Kozlov MV, Niemelä P (2003) Drought is more stressful for northern populations of Scots pine than low summer temperatures. Silva Fenn., 37(2), 175–180

Kučera J, Čermák J, Penka M (1977) Improved thermal method of continual recording the transpiration flow rate dynamics. Biol. Plantarum, 19, 413–420.

Lai Ch-T, Katul G (2000) The dynamic role of root-water uptake in coupling potential to actual transpiration. Adv. Water Resour., 23, 427–439

Lapin M, Dambborská I, Melo M (2001) Dowscaling of GCM outputs for precipitation time series in Slovakia. Meteorol. J. 4(3), 29–40

Lexer MJ, Hönninger K (1998) Defining the physiological amplitude of alpine tree species using the combined network of forest inventory, soil and meteorological data. Écologie, 29(1–2), 383–387

Minďáš J (1999) Quantitative and qualitative characteristics of beech-fir ecosystem rainfall pattern. Dissertation. Technical University Zvolen, p. 153

Narasimhan B, Srinivasan R (2005) Development and evaluation of Soil Moisture Deficit Index (SMDI) and Evapotranspiration Deficit Index (ETDI) for agricultural drought monitorin. Agr. Forest Meteorol., 133(1–4), 69–88

Pokorný R (2000) Sap flux simulation and tree transpiration depending on tree position within stand of different densities. Phyton., 40(4), 157–162

Schiller G, Cohen Y (1998) Water balance of *Pinus halepensis* (Mill.) afforestation in an arid region. Forest Ecol. Manag., 105, 121–128

Solberg S (2004) Summer drought: a driver for crown condition and mortality of Norway spruce in Norway. Forest Pathol., 34, 93–104

Schopf R, Köhler U (1995) Untersuchungen zur Populationsdynamik der Fichtenborkenkäfer im Nationalpark Bayerischer Wald. 25 Jahre auf dem Weg zum Naturwald. Berichte über die wissenschaftliche Beobachtung der Waldentwicklung. 88–111

Schwenke W (1996) Gründzüge des Massenwechsels und der Bekämpfung des Großen Fichtenborkenkäfers, *Ips typographus* (L.) (Col., Scolytidae). Anz. Schädlingskd. Pfl., 69, 11–15

Škvarenina J, Križová E, Tomlain J (2004) Impact of the climate change on the water balance of altitudial vegetation stages in Slovakia. Ekológia (Bratislava), 23(2) Supplement, 13–29

Soroková M (2001) Soil moisture regime in forest stands with different species composition. Dissertation. TU Zvolen, p. 143

Swanson RH (1970) Sampling for direct transpiration estimates. J. Hydrol., 9, 72–77

Wellpott A, Imbery F, Schindler D, Mayer H (2005) Simulation of drought for a Scots pine forest (*Pinus silvestris* L.) in the southern upper Rhine plain. Meteorol. Z., 14(2), 143–150

Thomas FM, Blank R, Hartmann G (2002) Abiotic and biotic factors and their interactions as causes of oak decline in Central Europe. Forest. Pathol., 32, 277–307

Tlapák V, Šálek J, Legát V (1992) Water in agricultural landscape. 1st Edition. (in Czech) Brázda Publishers in cooperation with the ME CR, Praha, p. 318

Turc L (1961) Estimation of irrigation water requirements, potential evapotranspiration: a simple climatic formula evolved up to date. Ann. Agron., 12, 13–49

Zierl B (2001) A water balance model to simulate drought in forest ecosystems and its application to the entire forested area in Switzerland. J. Hydrol., 242, 115–136

Zierl B (2004) A simulation study to analyze the relations between crown condition and drought in Switzerland. Forest Ecol. Manag., 188, 25–38

Zinecker E (1957) Der große Fichtenborkenkäfer (*Ips typographus* L.) in seiner Abhängigkeit vom Standort. Anzeiger f. Schädlingskunde, 30–31, 99–103

Forest Fire Vulnerability Analysis

J. Tuček and A. Majlingová

Keywords Vulnerability · Risk · Forest fire · GIS · SDSS – spatial decision support system

Introduction

After a strong wind disturbance in the High Tatras Mountains National Park in November 2004, the problem of forest fire prevention has attracted much attention from our society. Under our conditions, forest fire is an undesirable phenomenon, damaging not only the forest ecosystem but also, if the fire spreads behind the forest borders, endangering the property and lives of people.

The current problem of forest fire occurrence is evidently going to be an increasingly relevant phenomenon in the future. This arises from the latest reports of climatologists related with the progress of global climatic changes that can already be felt locally. According to these forecasts, we can expect more frequent and extensive fire occurrence under our conditions. The background of this fact is the progressive increase of the average annual air temperature that will lead to longer periods of drought. The drought periods are the most dangerous periods owing to the frequent occurrence of fire initiation and propagation. The most common reason for fire initiation is humans and their careless manipulation of fire, mainly in this period. Forest fires very often occur as a consequence of the fire's gradual transition from urban and agricultural sites to forest. The most common source of this transition is unmanaged grass burning, which is still a component of agricultural sites' management, regardless of all warnings.

Present forest fire research points out the fact of vulnerability of the area to fire initiation and next propagation. Vulnerability is characterized by hazards – factors that, in given conditions, are demonstrated to be dangerous, according to their final effect on the condition of a given area.

For the identification and then evaluation of individual hazards, the most effective tool for this purpose was chosen – a spatial decision support system. In general, the spatial decision support system includes the decision support system (DSS) itself and a geographical information system (GIS), mainly using its graphical and analytical functionality.

Problem

The Slovensky raj National Park is situated in the north-eastern part of the Slovenske Rudohorie Mountains near the Low Tatras Mountains. Figure 1 shows the localization of the experimental area (ESA) in the area of Slovakia. The climatology character shows that

Fig. 1 Localization of the experimental area

J. Tuček (✉)
Technical University in Zvolen, Faculty of Forestry,
T.G. Masaryka 24, 960 53 Zvolen, Slovakia
e-mail: tucek@vsld.tuzvo.sk

K. Strelcová et al. (eds.), *Bioclimatology and Natural Hazards*,
DOI 10.1007/978-1-4020-8876-6_19, © Springer Science+Business Media B.V. 2009

the preponderance of the area belongs to a moderately cold region with an average annual temperature of 5–6°C. The geological ground here consists of limestone and dolomite that allow the creation of karstic formations. The area of the Slovensky raj National Park is well known, mainly due to its numerous canyons and ravines. The predominant soil types here are rendzinas, pararendzinas and lithosols (80–90% of the area). Forest covers about 75% of the area. The most represented tree species (Leskovjanska 1999) is spruce (50%), followed by beech (30%).

The extent of forest fires in Slovakia is much smaller than the extent of fires in Mediterranean countries like Spain, Greece or Portugal but, owing to the growing intensity of their occurrence, there is a need to concern ourselves with them under our conditions as well. Prognoses of future climate changes refer to the fact that global climate change will also cause climate warming in our region.

Meteorological factors represent a significant factor for fire initiation. Due to meteorological situations, under our conditions, the highest risk periods are the spring season (March–May) and the summer season, with the months with the highest air temperature (July and August). From the point of view of the danger of meteorological factors, the parameter of most risk is relative air moisture, which has an influence on current vegetation moisture (trees, herbs, grass) as well as on the moisture of soil and its horizons. In particular, the upper soil layer is very risky, being formed from litter (leaves or needles). This layer, together with the grass and herbal component, becomes the prime source of fuel in the case of fire initiation. Therefore, grass burning is considered to be very dangerous in a period of drought. A fire is easy to start burning, but controlling and extinguishing it is very difficult, particularly in the case of such a fuel type. This is confirmed by the annually growing number of spring fires caused by grass burning. We should note that this type of fire causes not only material damages that can be calculated right on the site but also environmental damage. To this group belongs, for example the disturbance of important biotopes and the forest ecosystem, or contamination of the environment by chemical matters in the case of ecologic unfriendly extinguishing substance use. Sometimes, damage occurs with the highest price – injury or death.

The fire evidence, fire causes probes and the processing of documentation about fire, fire statistic surveys and fire analyses are regulated by the Instruction of the President of the Fire and Rescue Forces of the Slovak Republic No. 25/2005. The evidence includes such information as operational data about procedures related to fire fighting (date of fire initiation, time of announcement, time of alarm announcement, time of departure, fire site arrival, beginning of extinguishing activities, fighting and departure to the fire station (together with transportation distance)), descriptive data relating to the damaged area or object as well as data about the fire type, established reasons and calculation of the resulting damages. In the case of the occurrence forest fires, the age of the forest stand, its total area, the area that was destroyed and damaged by fire, species composition, representation and stem density are registered.

There is a problem with the fire evidence, because it is not unique. As to the experimental area, the fire evidence regulation has been changed during the period that was analysed, for the first time in 1998 and later in 2000 (the Instruction of the President of the Fire and Rescue Forces of the Slovak Republic No. 22/1998 and No. 12/2000). Besides, after a change of social background, the evidence, also including fires with a smaller extent, was described in more detail in the second part of the surveyed period. Conversely, the discipline of some subjects managing the forests has become significantly worse and a lot of data about disturbed forests were missing from the evidence.

The problem of forest fire vulnerability evaluation in the Slovensky raj National Park was solved in the frame of the WARM project. Particular results have already been published by Tucek et al. (2003), Tucek (2004, 2006), Skvarenina et al. (2003) and Holecy et al. (2003).

Methodology

The Statistical Analysis of Forest Fire Occurrence in the Experimental Area

Forest fire vulnerability can be described by means of probability $p(t)$ that reports the assumed disturbance of the forest (based on its species composition) in the age (t) during a common year. Probabilities for particular tree species were derived from empiric functions of frequency distribution, based on data about burned out forest areas, and divided into age classes, by processing records about fires in forest stands of the experimental area for a 25-year period (1976–2001).

Owing to the species composition of the forest stands in the experimental area, we had a sufficient set sample for coniferous trees, but there were only a few burned sites of broadleaves. As the degree of forest fire vulnerability in the area, the population proportion of the particular tree species' areas destroyed by fire in an average year was chosen.

$$p(t) = \frac{h(t)}{H(t)}, \tag{1}$$

where h is the sum of all areas of individual tree species disturbed by fire,

$$h = \sum h_i \tag{2}$$

and H is the sum of all areas of individual tree species in the experimental area,

$$H = \sum H \tag{3}$$

The variability of destroyed individual tree species' area with the extent of 1 ha, measured by this quantity, was shown to be extensively high. Some tree species, respectively tree species groups, are more susceptible to disturbance by fire than others. To simplify the problem, we tested the statistical significance of differences between obtained relative frequencies (probabilities) of 1 ha per 1 year's disturbance for all tested tree species by means of a null hypothesis.

$$\mathrm{H_0} : p_1 - p_2 = 0, \text{ it means } p_1 = p_2.$$

The testing characteristic (z) was calculated by the formula:

$$z = \frac{|p1 - p2|}{\sqrt{\frac{p_1(1-p_1)}{H1} + \frac{p_2(1-p_2)}{H2}}}, \tag{4}$$

where (p_1) and (p_2) are population proportions of disturbed hectares of compared tree species or tree species groups, and (H_1) and (H_2) are disturbed areas of both compared tree species or tree species groups.

We compared the testing characteristic (z) with the critical value of standard normal distribution (z_α) for a mutual test with a level of significance $\alpha = 0.05$. The null hypothesis ($\mathrm{H_0}$) was rejected at the significance level of α when ($z > z_\alpha$). Tree species for which it was not possible to reject the null hypothesis ($\mathrm{H_0}$) for a very significant correspondence of their vulnerability were merged into groups, in which vulnerability was compared with other tree species or tree species groups.

In the sampling area, we used the data of given tree species stands destroyed by fire, according to their age, for the calculation of empiric distribution function values $F_n(t)$ that report the forest stand disturbance probability by means of given age classes (t). To exclude an influence of the different number of hectares in individual age classes, $F_n(t)$ functions were composed of relative frequencies of expected areas of disturbance (f_i) in every age class (i). The areas of all age classes were homogenously distributed in the sampling area:

$$F_n(t) = \sum_{i \le t} \frac{^\wedge f_i}{^\wedge f}, \tag{5}$$

where

$$^\wedge f = \sum_{i=1}^{k} {^\wedge f_i} \tag{6}$$

The symbol n represents the number of expected areas of disturbance (ha) on homogenously distributed areas of individual age classes N (ha), and f represents the expected relative frequency of the destroyed areas in the selection ($f = n/N$). The symbol (k) represents the total number of age classes of the tree species or group of tree species, into which the selection (N) area is divided. In this case, $k = 15$.

For the description of the forest stand disturbance probability dependency in relation to its age (t), the Weibull probability distribution function was used:

$$F(t) = 1 - e^{-c.t^\gamma}. \tag{7}$$

The estimation of parameters (c) and (γ) of this distribution were performed by the method of quartiles, using the following formulae:

$$\gamma = \frac{1.572\,53}{\ln q_{75} - \ln q_{25}} \tag{8}$$

and

$$c = \frac{0.287\,67}{q_{25}^{\gamma}}. \qquad (9)$$

Symbols q_{25} and q_{75} represent the quartiles of the analysed empirical distribution functions $F_n(t)$, particularly describing the distribution of destroyed areas, for individual tree species groups with the same vulnerability. The analysis procedure was based on null hypothesis (H_0) testing of the correspondence of the functions $F_n(t)$ for particular tree species groups with the same vulnerability.

$$H_0: \quad F_n(t) - F(t) = 0.$$

Symbol $F(t)$ represents the distribution function of the equilibrated distribution values $W(c, \gamma)$.

For null hypothesis testing (H_0) of the correspondence of the functions, the Kolmogorov–Smirnov test was used for one selection, as it is recommended in Klein et al. (1997) and in Triola (1998). The procedure is based on a comparison of the absolute value maximum difference $F_n(t) - F(t)$ with the critical value $(n)_{a/2}$ for a mutual test of good correspondence. In all tests, a significance level of $\alpha = 0.05$ was used.

The probability of forest stand disturbance in individual age classes (t) in each tree species group equals the value $\Delta F(t)$:

$$\Delta F\,(t) = F(t) - F(t-1). \qquad (10)$$

Then, the values of expected disturbance probability values $p(t)$ – probabilities for tree species of pine, spruce, larch and broadleaves were estimated:

$$p\,(t) = u.\Delta F\,(t)\,.f. \qquad (11)$$

The tables of probabilities for individual tree species classified by age were later used in the Arc View GIS environment for the calculation of vulnerability of each forest stand in the experimental area, using scripts. Vulnerability, as a new attribute, was included in the database describing the forest stand. As a result, a map of the geographical distribution of fire vulnerability of forests was produced.

The Weibull distribution application for the purposes of description concerning the forest land management risk was also recommended by Kouba (2002) and Kouba and Kasparova (1989). The importance of description concerning the risk accompanying the forest management is also presented by Sisak and Pulkrab (2001).

In the second type of analysis, the influence of relevant geographic factors (elevation, slope, aspect, the nearest road distance, the nearest settlement and urbanized area distance) was tested on the fire occurrence. This was performed based on populated proportions' comparison of the analysed factor values on areas disturbed by fire and the whole experimental area. The differences in fire occurrence between factor value groups were next tested by means of a quotient test.

As a base, the mentioned probability $p(t)$ calculation has been used already, which informs us about the expected forest disturbance in relation to its age (t) in the whole experimental area. This could be consecutively iteratively revised according to considered geographical factors. The algorithm comes from the existence of the file of burning dependent probabilities $P(B \mid F_{x,y})_t$ of the forest stand (B) in the presence (x) of geographical factors $(F_{x,y})$ with possible (y) existing states. The first step of the algorithm is composed from the point assessments assignment $P(B \mid F_{1,y})_t$ of the first factor influence $(F_{1,y})$ in relation to all its conditions (y) in the experimental area. The second step is represented by the calculation of the a posteriori probabilities $P(F_{1,y} \mid B)$ that represent the fact that a burned area of 1 ha will belong to the area with the right characteristics (y) using the Bayes formula. The third step is the calculation of the dependent probabilities vector $p'(t)$ informing about the expected forest disturbance $p(t)$, but revised by factor influence $(F_{1,y})$ in the whole experimental area.

The algorithm continues with the second iteration repeating, here describing three steps using $p'(t)$ values instead of $p(t)$ as inputs and $P(B \mid F_{2,y})_t$ values instead of $P(B \mid F_{1,y})_t$ to obtain detailed geographical information about the fire occurrence relation and the second considered factor $(F_{2,y})$, expressed by dependent probability file by vector $p''(t)$. The number of iterations is not limited and the algorithm allows as many factors as are required to be taken into consideration. As it has already been said, the tests were performed to obtain information about the influence of elevation, slope, aspect and the nearest road and settlement distance. Then, the map of forest vulnerability in accordance with the main forestry factors (a priori probabilities) was corrected, using the results of the geographical factors' influence analysis. Every cell of the analysed raster was classified by reclassification process to the derived categories. Based on forestry factors, a priori probabilities were gradually revised (multiplied) by dependent probabilities, taking into consideration particular geographical factors for every cell.

Vulnerability Assessment in the EMDS Environment

In the analyses, the digital relief model of the study area with a 10 m cell resolution was used for the derivation of the slope and aspect surface. In addition, the digital vector layer of the forest stands' borders was used, related with the database containing detailed data about the forest and vector layer of the road network (including forest roads, tourist and cycle routes) and the vector layer of urban areas (settlements, dispersed houses).

As input data for the analysis process, the probabilities described in the above chapter were used and calculated for characteristics: elevation, slope, aspect, the nearest road and urban area distance.

The analysis process in EMDS is composed of three basic steps (Tucek, Majlingova 2007).

Dependency Network – Knowledge-Base Building

For the building of the dependency network – a formalized knowledge base, based on the decision rules – the NetWeaver environment, a subsystem of EMDS, was used.

The first step of dependency network building is the definition of the goals – groups of hazards that are input into the assessment. In this case, these goals were represented by hazards relating to the forest (species composition and age), geographical factors and the distance of the locality from the nearest road and settlement. To individual groups of factors, relevant calculated links representing viewed parameters (e.g. by terrain, elevation, slope, aspect) were added. For the faster visual control of the result, the transformation of origin probability values to fuzzy values (assigning values from unique interval [0; 1]) was performed in the NetWeaver environment. The values of probabilities' multiplication (vulnerability of the area) were then classified into five degrees of danger.

Assessment Process Performing

For the application of the assessment process, it is also necessary to create (except for the dependency network) input data describing the experimental area conditions (in the case of this analysis in raster representation), which is the applied knowledge included in the dependency network. Vector to raster conversion was performed in the basic environment of EMDS. Its interface is similar to the Arc View environment, because EMDS is an applicable extension for Arc View.

Result Representation and Output Production

A raster representation was used in the process of analysis. In the results, the image of each cell of the raster contains the fire danger (vulnerability) fuzzy value of the area. To simplify this raster, it was reclassified into five regular intervals – degrees of danger (fire vulnerability of the area) – based on the multiplication of hazard values of individual factors' groups.

Map composition and printing can be produced in the EMDS environment itself.

Results and Discussion

Results of Fire Occurrence Statistical Analysis

For statistical analysis purposes and in correspondence with the methodology, a database of the forest, agricultural and urban fires in the experimental area during the period 1976–2001 was created.

Various fire sources were registered, such as the manipulation of fire in a natural environment, children playing with fire, harvest waste burning, old grass and wood burning, home waste burning, smoking, electric wiring breakdown, thunderbolt, intentional setting of fire, self-ignition and railway traffic – trains.

In total, there were 328 fires recorded, consisting of 103 forest fires, 114 agricultural fires and 111 urban fires (built-up areas, respectively, buildings and objects) in the experimental area within the 25-year-long survey period.

As regards forest fires, the sample set of recorded forest fires contains two maximums of fire occurrence (Fig. 2). One of them is during spring (April) and the second during summer and autumn (August and October).

The relationships among geographical, forestry and meteorological factors and the probability of forest fire occurrence can serve as the base for the next rules or knowledge formulation about the problem. It is necessary to mention that not all the analyses that could be carried out were done based on the collected experimental material. In the chapter, the two main results of our investigation that are related to two groups of significance are presented. The first group includes essential forest mensuration field data (species composition and age). The second group includes geographical factors – site elevation, slope, aspect and the distance from the nearest road or settlement.

The first step was the calculation of the population proportion of burned out areas from particular age classes of a sample.

Having tested the differences between the obtained population proportions, five tree species were distinguished: respectively, their groups are pine, spruce, fir, larch and a group of broadleaved trees. Results obtained for tree species (groups), are shown in Table 1. The observed data pointed out obviously higher fire occurrence rates in younger forest stands than in older ones.

In the next step, values of empiric distributional functions of the probability distribution $F_n(t)$ were calculated. Then, the empiric functions correspondence test with theoretical probability distribution using Weibull distribution was performed. In all cases, the results for individual tree species (groups) showed very high correspondence (level $\alpha = 0.05$). The probability of forest stand disturbance in the age classes (t) for every tree species (group) then equalled $\Delta F(t)$. The probabilities of forest stand disturbance by fire are introduced in Table 2.

In the second and third group of conditions, the relations between fire occurrence and basic geographical and anthropogenic conditions were evaluated. In correspondence with the methodology, it was necessary to perform an analysis of relative frequencies of analysed factor occurrence in the areas disturbed by fire in comparison with the whole forested part of the experimental area. To realize the procedure, two groups of problems needed to be removed – the localization of

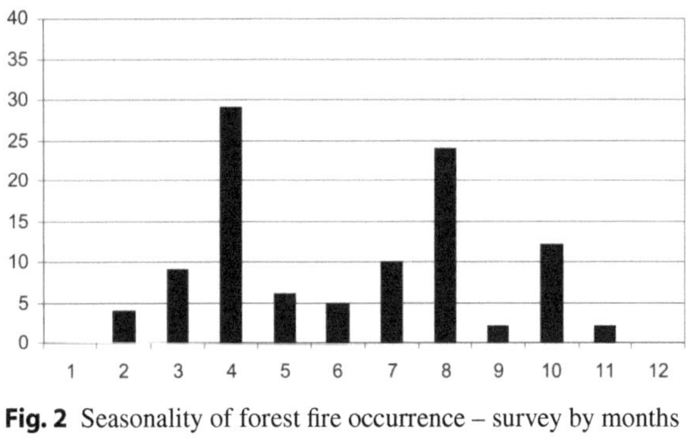

Fig. 2 Seasonality of forest fire occurrence – survey by months

Table 1 Probability of tree species' (groups') area disturbance by fire

Tree species (group)	Total area (ha)	Burned area (ha)	Fire rate occurrence probability
Pine	21,295.7	29.3	0.001377186
Spruce	106,028.3	83.5	0.000787769
Larch	12,572.0	7.9	0.000629969
Fir	16,204.8	5.7	0.000351664
Broadleaved	51,673.6	13.9	0.000268189
Total	207,774.4	140.3	0.000675400

Table 2 Probabilities calculated for particular tree species in relation to relevant age class

Age (t)	Broadleaves $p(t)$	Spruce $p(t)$	Fir $p(t)$	Larch $p(t)$	Pine $p(t)$
10	0.001068	0.003096	0.000639	0.000603	0.005032
20	0.000714	0.002002	0.000595	0.001804	0.003425
30	0.000549	0.001531	0.000540	0.002489	0.002672
40	0.000435	0.001217	0.000487	0.002445	0.002152
50	0.000350	0.000987	0.000438	0.001876	0.001763
60	0.000285	0.000811	0.000393	0.001163	0.001460
70	0.000234	0.000673	0.000352	0.000592	0.001218
80	0.000193	0.000562	0.000315	0.000249	0.001023
90	0.000160	0.000472	0.000281	0.000087	0.000862
100	0.000133	0.000399	0.000251	0.000025	0.000730
110	0.000111	0.000338	0.000224	0.000006	0.000620
120	0.000093	0.000287	0.000200	0.000001	0.000528
130	0.000078	0.000245	0.000178	0.000000	0.000451
140	0.000066	0.000210	0.000159	0.000000	0.000386
150	0.000055	0.000180	0.000142	0.000000	0.000331
160	0.000047	0.000154	0.000126	0.000000	0.000285
170	0.000039	0.000133	0.000112	0.000000	0.000245
180	0.000033	0.000115	0.000100	0.000000	0.000211

individual fires and the proposition of the procedure to obtain data about the occurrence of evaluated factors on the fire site.

For individual geographical factors, pixel sets with particular values for the area destroyed by fire (burned) and the whole experimental area were created. For the area destroyed by fire, there was a set with 7660 pixels; for the forested part of the experimental area, the set contained 317,568 pixels. That represents an area of 20,047 ha for the forested area and 478 ha for the fire-destroyed area, at a 25 m cell resolution. An example of elevation proportion distribution of occurrence values is introduced in Figs. 3a and 3b.

In the next analyses, the population proportion values of factors and not their absolute frequencies were required for each class. Therefore, the data about the analysed factor occurrence frequencies were exported to the Microsoft Excel environment, in which the required calculations were performed. The results of elevation, slope, aspect and distance from the nearest road and settlement (building) frequency calculation are introduced in Figs. 4, 5, 6, 7 and 8.

The categories of analysed factors' values, confirmed by statistical testing, are introduced in Table 3. The results show that, from the fire point of view, the vulnerability of the area is significantly higher on south-eastern, southern and south-western expositions (aspect 60°–160°), on sites with lower elevation (less than 775 m), on sites with lower terrain slope (less than 15°), on sites at a distance of less than 350 m from the nearest road and, finally, at a distance of less than 1000 m from the nearest settlement (building). The presumption that the fire risk increases with a higher terrain slope was not confirmed.

In correspondence with the procedure described in the methodology, the probability of forest fire occurrence can be considered as an a priori probability in view of the forest characteristics (age and species composition). This precision is possible using the influence of the relevant geographical factors based

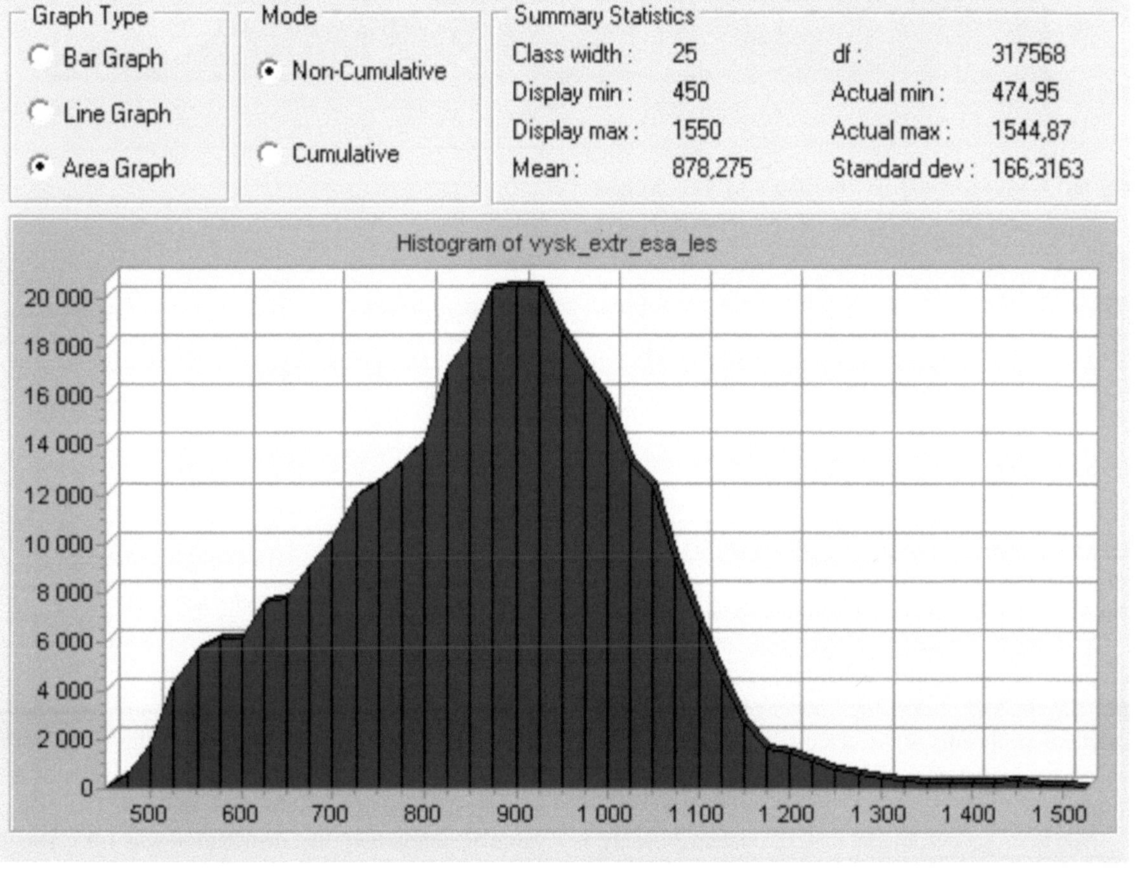

Fig. 3a Frequency histogram – forested area of the experimental area, elevation

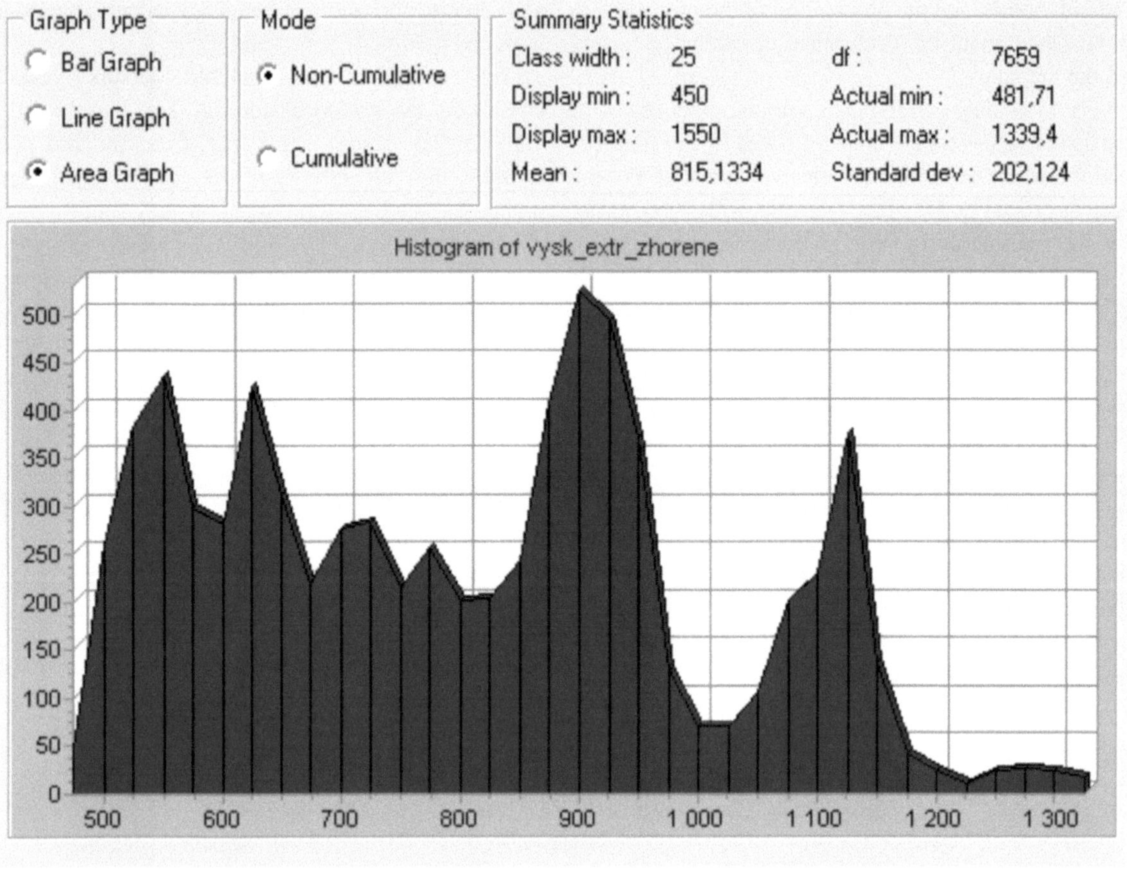

Fig. 3b Frequency histogram – fire-destroyed area, elevation

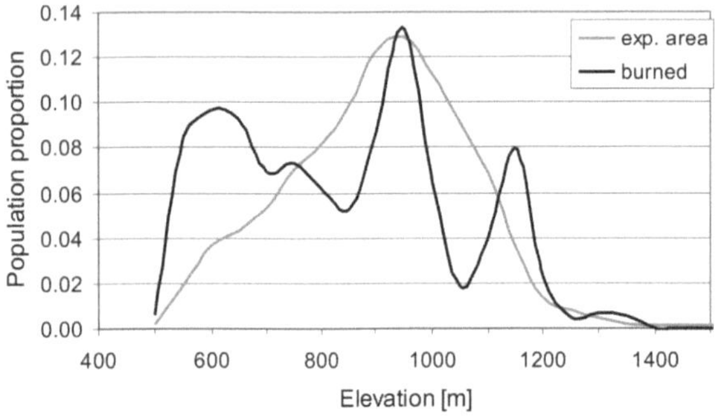

Fig. 4 Population proportion of individual factors – elevation

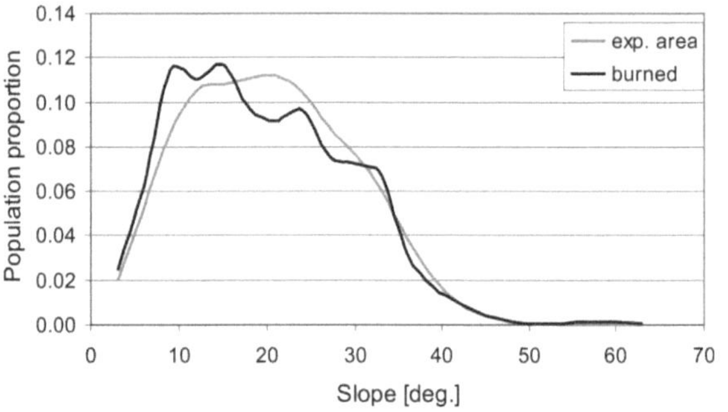

Fig. 5 Population proportion of individual factors – slope

on posteriori probability calculation using the formula of Bayes and next the dependency probability calculation. Results of the calculations for dependency categories, as was already mentioned, are introduced in Table 4.

Due to its geographic factors, the statistical analysis of the study area burnability, in this case, fire vulnerability, was performed based on fire occurrence rates by mutual comparison and testing whether their differences are significant or not, using a null hypothesis. The null hypothesis was rejected at the significance level of ($\alpha = 0.05$) when $z > z_\alpha$. The results of null hypothesis testing are shown in Table 5. The critical value of standard normal distribution was 1.96. The input data for null hypothesis testing are given in Tables 3 and 4.

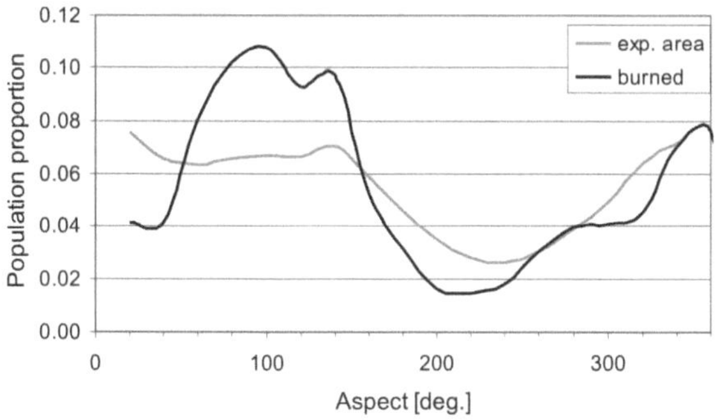

Fig. 6 Population proportion of particular factors – aspect

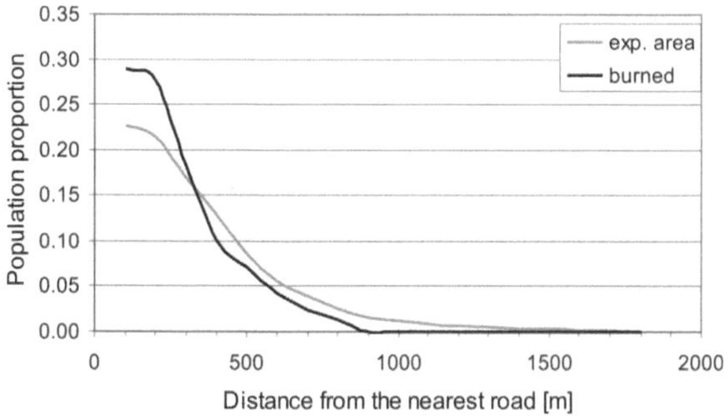

Fig. 7 Population proportion of particular factors – the distance from the nearest road

Based on the fact that the geographical distribution of all the assessed factors, such as forest characteristics, geographical characteristics of relief and anthropogenic factors' characteristics (distances computed from roads and urbanized areas), were available in digital form, there was a possibility to calculate the vulnerability value for each cell of the analysed area using derived relations and their parameters.

First, we used the programming tools of the tabular calculator Microsoft Excel to calculate the probability

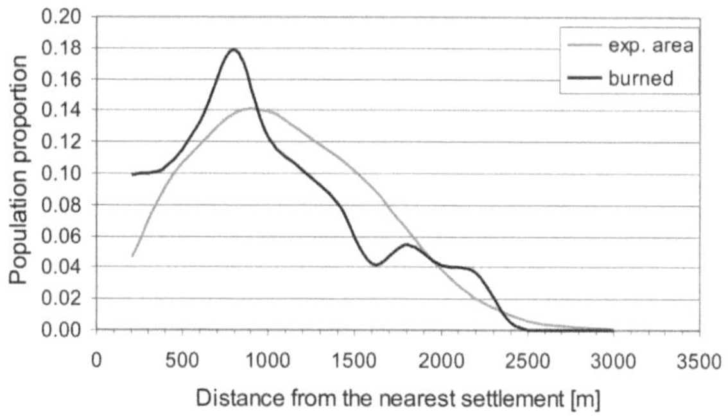

Fig. 8 Population proportion of particular factors – the distance from the nearest settlement

of each forest stand disturbance (vulnerability) based on its age and species composition. For that reason, the values of probability distribution were used, settled by Weibull distribution for each tree species and age class, and these were applied to forest stands based on their description. For mixed species composition in the forest stand, there were the probability values of the tree species of the composition weighted by their percentage rate in the forest stand. To the calculated value of the forest stand vulnerability, the new attribute (column) of the stand description table was added.

The probabilities of fire disturbance of the forest, calculated in this way, were used for the calculation of the fire vulnerability of the experimental area, using GIS and SDSS. These probabilities could also be used for the valuation of the forest stand for insurance purposes (Holecy 2004).

Vulnerability Assessment Results

The assessment process performance was carried out using the decision rules implemented into the

Table 3 Categories of the analysed factors' values determined based on the statistical test

Factor	Category of the factors' values	Area of the ESA	Burned area	Relative frequency
Aspect	60–160°	6 575.8	213.4	0.032439368
	160–60°	13 471.4	265.4	0.019700275
Elevation	450–775 m	5 297.7	215.8	0.040730437
	>775 m	14 749.5	262.9	0.017826415
Slope	0–15°	7 513.1	205.2	0.027321400
	> 15°	12 534.1	273.5	0.021821219
Distance from the nearest road	0–350 m	13 658.8	387.2	0.028344301
	>350 m	6 388.4	91.4	0.014305038
Distance from the nearest settlement	0–1 000 m	10 715.8	304.9	0.028456228
	>1 000 m	9 331.4	173.6	0.018603941

Table 4 Dependent proportions for analysed factors

Factor	Category of the factors' values	Relative frequency	Posterior frequency	Dependent frequency
Aspect	60–160°	0.032439368	0.445702470	1.358434837
	160–60°	0.019700275	0.554297529	0.824970276
Elevation	450–775 m	0.040730437	0.450799056	1.705633961
	>775 m	0.017826415	0.549200734	0.746502289
Slope	0–15°	0.027321400	0.430752383	1.149592698
	> 15°	0.021821219	0.569247616	0.910504824
Distance from the nearest road	0–350 m	0.028344301	0.809029176	1.187422736
	>350 m	0.014305038	0.190970823	0.599278425
Distance from the nearest settlement	0–1000 m	0.028456228	0.637222186	1.192120105
	>1 000 m	0.018603941	0.362777814	0.779377094

Table 5 Results of geographic factors' significance analysis using null hypothesis testing

Factor	z	z_α	H_0 result
Aspect	5.114214071	1.96	Rejected
Elevation	7.828403501	1.96	Rejected
Slope	2.411365928	1.96	Rejected
Distance from the nearest road	2.858177847	1.96	Rejected
Distance from the nearest settlement	4.62626007	1.96	Rejected

dependency network in the EMDS environment. The output of this analysis had the form of a raster (map).

As a base for the classification of fire hazards into vulnerability degrees, the results from the fire occurrence statistical analysis in the experimental area were used. The method of their application has already been described in the methodology. As one of the basics, the structure of the dependency network – the knowledge base built in the NetWeaver environment – is shown (Fig. 9).

The advantage of this approach to the vulnerability assessment, using the EMDS environment against classic processing and using the map algebra tools of the GIS environment is that SDSS allows more effective analysis processing, such as from the time and cost points of view. The possibility of the flexible deactivation or reactivation of individual factors (data links) or their groups (goals) entering the assessment process, the addition of a new factor group or a new link to an existing group, as well as the use of simply definable assessment based on fuzzy values (fuzzy logic principle) can be considered as very effective features.

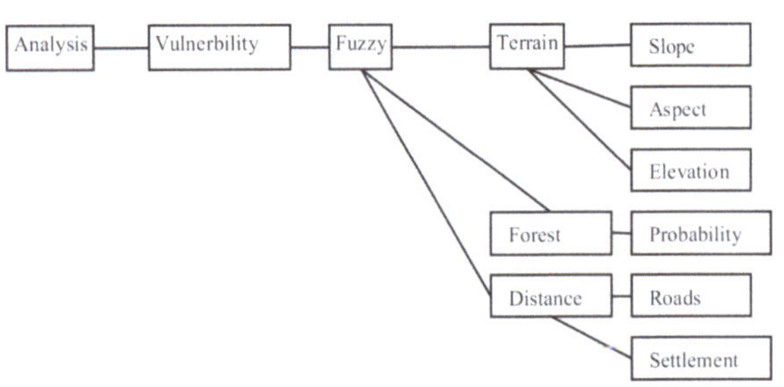

Fig. 9 Dependency network features and their relationships

Figure 10 represents the visualization of the vulnerability results of the experimental area to forest fire based on the evaluation of stand, geographical and distance factors.

Table 6 shows the results of the experimental area percentage rate assigned to the particular vulnerability degrees.

From the results presented in the table, from the total surface area of the experimental area (26,546.25 ha), the highest area proportion takes degrees 3, 2 and 1. A relatively big area also takes degree 4. There is also the highest degree – 5 – that takes 2% of the area. From the cell content of the result raster, it is possible to

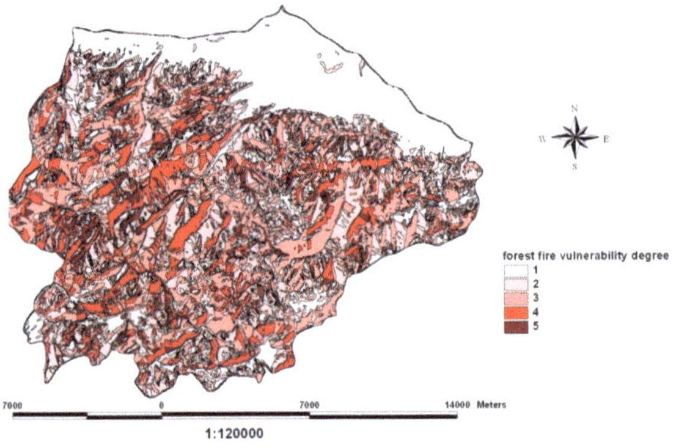

Fig. 10 Result of forest fire vulnerability analysis of the experimental area

Table 6 Percentage rate of vulnerability degrees for the experimental area

Vulnerability degree	Description	Percentage rate of the whole area
1	Very low danger	24
2	Low danger	27
3	Medium danger	31
4	High danger	16
5	Very high danger	2

extract vulnerability values for individual forest stands, too. This fact allows us to obtain a survey of individual forest stands' vulnerability and to calculate the average or another method of aggregated data.

In this analysis, the assessed factors (hazards) belong to the group of static factors, representing factors which do not change significantly in a short time. However, meteorological factors were not included in the analysis. It could be taken for granted that these dynamic factors could influence the fire risk degree significantly, mainly during the days with high air temperatures, during longer periods of drought, without precipitation, when the fire load of the forest and of the whole area grows enormously. This time, localities assessed as higher risk degree (3, 4) areas based on static factors' (vulnerability) assessment will be mainly endangered. In such a period, it is necessary to perform fire monitoring at regular time intervals, to tighten the control over the keeping of safety rules, to restrict works performed by forest workers and to regulate or forbid the entry of tourists to tourist routes in endangered localities.

Conclusion

In the chapter, the described approach to the analysis of the area vulnerability to forest fire for the area of the Slovensky raj National Park is only the first step in forest fire risk analysis. The risk analysis in this case is composed of two groups of factors. One of them is here described as the vulnerability of the forest stand to fire disturbance. The second one is represented by the meteorological factor that significantly influences the total risk value.

The advanced approach to the forest fire vulnerability evaluation, using a spatial decision support system based on decision rules and analytical and graphic display functionality, is more effective in terms of time and invested effort, as opposed to the analysis processing using only GIS tools (map algebra).

Sufficient data sources for the needs of modelling and new data derivations for the study area territory have been prepared. The analysis of forest and other fire data offers useful information for knowledge-base building.

In the future, we would like to extend the approach to forest fire risk analysis introduced here, using current data about weather conditions.

Acknowledgments This work was supported by the Science and Technology Assistance Agency under the contract no. APVT-51-037902.

References

Holecy J, Skvarenina J, Tucek J, Mindas J (2003) Fire risk insurance model for forest stands growing in the area of Slovak Paradise. In: Forest Fire in the Wildland-Urban Interface and Rural Areas in Europe: An Integral Planning and Management Challenge, Institute of Mediterranean Forest Ecosystems and Forest Products Technology, Athens, Greece, 15–16 May 2003, pp. 161–172

Holecy J (2004) Mathematical Model of Fire Insurance for Forests in Slovakia. Scientific Study, TU Zvolen, 65pp

Klein R, et al (1997) Adaptation to Climate Change: Options and Technologies, An Overview Paper. Technical Paper FCCC/TP/1997/3, United Nations Framework Convention on Climate Change Secretariat, Bonn, Germany, 33pp. http://www.unfccc.int/resource/docs/tp/tp3.pdf

Kouba J (2002) Das Leben des Waldes und seine Lebensunsicherheit [Forest life and its temporal uncertainty], German Journal of Forest Science, Vol 121: 211–228. [In German]

Kouba J, Kasparova I (1989) Steuerungs theorie des Ausgleichs prozesses zum Normal Wald auf Grund der Stochastischen Prozesse [Steering theory of compensatory processes to normal forest based on stochastic process], Res. Report, Czech Agricultural University, Prague, Czech Republic, 64pp. [In German]

Leskovjanska A (1999) Proceedings of papers from the 7th meeting of Slovak Bioclimatology Society by Slovak Academy of Sciences. Editorial: Management of the Slovensky raj National Park in Spisska Nova Ves and SBS, 1999. [In Slovak]

Sisak L, Pulkrab K (2001) Damaging factors and their influence on forestry in the Czech Republic. In: Proc. IUFRO Division 4 Conference "The Economics of Natural Hazard in Forestry", Solsana, Catalania, Spain, 7–10 June 2001, Padua University Press, Italy, pp. 125–132

Skvarenina J, Mindas J, Holecy J, Tucek J (2003) Analysis of the natural and meteorological conditions during two largest forest fire events in the Slovak Paradise National Park. In: Forest Fire in the Wildland-Urban Interface and Rural Areas in Europe: An Integral Planning and Management Challenge, Institute of Mediterranean Forest Ecosystems and Forest Products Technology, Athens, Greece, 15–16 May 2003, pp. 29–36

Triola MF (1998) Elementary Statistics, 4th edition, Benjamin Cummings Publishing Company, Redwood City, California, 1998, 284pp

Tucek J, Majlingova A (2007) Forest Fires in the Slovensky Raj National Park: Geoinformatics Applications. Scientific Study, TU Zvolen, Zvolen, ISBN 978-80-228-1802-5, 137pp. [In Slovak]

Tucek J (2004) Geoinformatics application in forest fire research – an example of the WARM project. In: 1st Goettingen GIS and RS Days, Proceedings, Part Environmental Studies, Georg August University Goettingen, 7–8 October 2004, pp. 368–380

Tucek J (2006) Influence of the geographical and forestry parameters of forest stand to the forest fire occurrence. In: Revitalization and Protection of Ecological Systems Disturbed by Natural Disasters, International Conference Zvolen, 3 March 2006, Technical University in Zvolen, Zvolen, ISBN 80-228-1573-X, p. 5

Tucek J, Skvarenina J, Mindas J, Holecy J (2003) Catalogue describing the vulnerability of landscape structures in the Slovak Paradise National Park. In: Proceedings of International Workshop on Forest Fire in the Wildland-Urban Interface and Rural Areas in Europe: An Integral Planning and Management Challenge, Institute of Mediterranean Forest Ecosystems and Forest Products Technology, Athens, Greece, 15–16 May 2003, pp. 73–84

The Paradigm of Risk and Measuring the Vulnerability of Forest by Natural Hazards

J. Holécy

Keywords Fire occurrence hazard · Forest management risk · Model of Forest vulnerability · Vulnerability assessment · Vulnerability ranking · Risk-free soil expectation value · Vulnerable value of forest

Introduction

Forest ecosystems are very sensitive to natural hazards of any kind. However, the concept of vulnerability concerning the forest growing including its value still has been and remains not very clearly defined and vague. Current approaches to measuring vulnerability often lack any systematic transparent and understandable development procedures (Birkmann 2006). Also the concept of risk and its relationship to vulnerability itself are often explained in different, sometimes even contradictory manners. The concept of forest management risk has been thoroughly discussed and explained by von Gadow (2000), Hanewinkel (2002) and also by Hanewinkel and Oesten (1998), but their explanations of risk and its definitions do not include the dimension of vulnerability, at all. Also Sisak and Pulkrab (2001) inform about the negative impacts of particular natural hazards occurrence on the sustainable management of the Czech forestry, but their evaluation of these effects does not take in account the vulnerability of forest as the decisive factor of their magnitude. The impact of fire occurrence risk on the capital value of forest soil and the rotation period investigated successfully Martell (1980) and Van Wagner (1978), but without any mention concerning both the vulnerability of forest and its estimation. Halada et al. (2006) and Skvarenina et al. (2004) also refer to the concept of vulnerability, but they do not provide any more detailed classification of this term.

The objective of this chapter is to propose the general mathematical model of vulnerability concerning the management of a forest ecosystem. This model is formulated and demonstrated at the extensive case study aimed to the estimation and investigation of forest fire occurrence in the area of the Slovak Paradise National Park. Except for experimentally measuring the vulnerability of forest by fire here, the risk of forest management due to the fire occurrence hazard is quantified and evaluated, as well.

The Paradigm of Risk and Its Shifting

Definitions of Basic Concepts

The undertaken investigation uses different rather similar terms that, however, have to be defined precisely in the following. A hazard, according to Thywissen (2006), is the potential natural or social danger to human life or property. In the present chapter, this potential is described by the probability of forest stands destruction for a certain period by the occurrence of fire. One of the key terms that is analysed in this case study is risk. The distinction between risk and uncertainty, where the probability of hazard that occurs is unknown, is here taken in account. Risk is the expected failure of a forest management project linked

J. Holécy (✉)
Technical University in Zvolen, Faculty of Forestry, T.G. Masaryka 24, 960 53 Zvolen, Slovakia
e-mail: holecy@vsld.tuzvo.sk

K. Strelcová et al. (eds.), *Bioclimatology and Natural Hazards*,
DOI 10.1007/978-1-4020-8876-6_20,

to a known probability. In this sense, risk is defined as hazard quantitatively measured in probability terms. Expected loss is defined as risk expressed in physical units and can be calculated as the size of a forest area in hectares multiplied by the probability of its destruction. An economic risk is defined as an expected loss expressed in monetary terms and obtained as the mentioned probability multiplied by the endangered value of property.

Vulnerability represents the central element of our approach to the formulation of risk and the assessment of its magnitude also from the economic point of view. The thorough explanation of this concept from both the engineering and ecological points of view has been given by Holling et al. (2002). In the broader discussion is this term applied by Caballero and Beltran (2003) at ranking the resistance concerning settlements exposed to forest fire under different conditions. Under the concept of vulnerability in the present chapter, we understand the damage potential of a forest ecosystem that can be expressed in both the physical units and monetary terms. In other words, it is the capacity of forest to be destroyed at the given frequency of fire occurrence or the likelihood that a certain part of forest will be destroyed. In this sense, the resistance of forest is understood as the ability of forest to survive or absorb the occurrence of fire and to maintain its own functioning further. The resilience then informs about the ability of forest to recover when its certain part has been destroyed and to change its tree-species composition and age structure in order to reduce its vulnerability, as much as possible. Here presented definitions are in accordance with opinion of Kasperson et al. (2001). The resistance and resilience are two components of the adapting capacity of forest. The vulnerability and resilience represent two sides of the dynamic forest evolution in the presence of fire occurrence hazard. Vulnerability informs about the possible disturbance of deterministic development of forest and resilience informs about the ability of the disturbance elimination and returning the forest ecosystem to the equilibrium state for further sustainable functioning, again.

The evolution of forest has been and remains very complex process in which many biological, physical, chemical and also social factors can play decisive roles. The deterministic analytical prediction of a forest development under these circumstances very seldom brings the reliable results. The mutual interaction of vulnerability and resilience is significantly stochastic and partly unpredictable in its very essence. However,

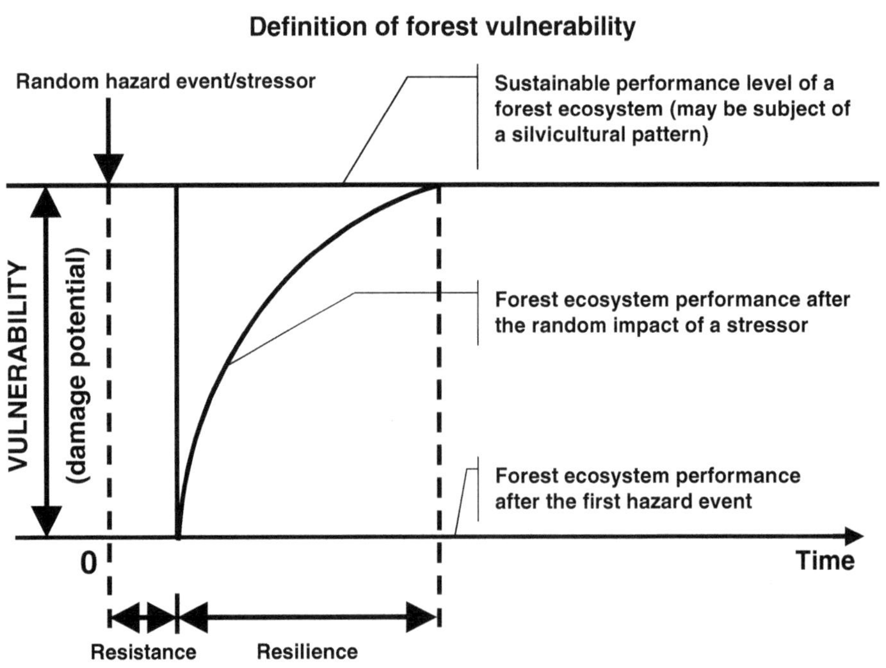

Fig. 1 The definition of forest ecosystem vulnerability in terms of its evolution under the pressure of the random occurrence of a stressor

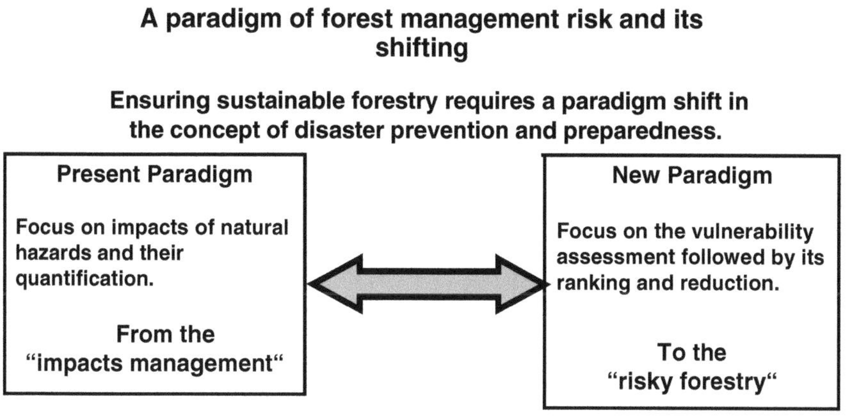

Fig. 2 The description of changing the paradigm of risk in forestry based on the knowledge improvement

we can observe size of destroyed area of forest, as well as to analyse frequency and intensity of occurrence concerning the particular kinds of hazard. Obtained probability distributions of a forest destruction under given conditions then inform about the inherent equilibrium among forces of destructive natural elements, the vulnerability of forest ecosystem and its corresponding resilience. The dynamic definition of vulnerability as the development of a forest ecosystem in the presence of a randomly occurring stressor is depicted in Fig. 1. The first occurrence of hazard event triggers the process of forest adaptation to the presence of a stressor and the forces of adapting capacity gradually change the structure of ecosystem in order to approach the level of its sustainable performance again. This definition follows the interpretation of vulnerability proposed by Bogardi and Birkmann (2006) and more precisely explained in the presentation of Bogardi (2006).

Based on this definition, the new more efficient approaches to the management of forestland could be developed and applied. Their essence is depicted in Fig. 2. The present commonly used paradigm of a forest management risk relies only on the quantitative assessment of hazard impacts, asserting the hazard prevention measures and seeking for the so-called security forestry. But, instead of lingering on the passive attitude to the occurrence of natural elements, the present paradigm of risk should be shifted to the new one based on the forest vulnerability assessment and ranking. The reason to change the present attitude of forest managers to risk is the obvious higher efficiency of the forest management resulting from the possibilities of a forest vulnerability reduction. However, the assertion of the new mentioned managerial attitude is much more demanding for a data collecting, including their processing and evaluation necessary for the improvement of our knowledge about risk.

Modelling and Ranking the Vulnerability of Forest

Data About Forest Fire Occurrence and Their Processing

To describe the random process of a forest growth under the pressure of fire occurrence hazard in a reliable way, there have been evaluated data about the forest fire occurrence in the whole territory of the Slovak Paradise National Park including its buffer zones for the period of years 1992–2001. These data were collected from the archives of four fire brigades and rescue corps offices situated in the districts of Spisska Nova Ves, Poprad, Roznava and Brezno who are responsible for the forest fire protection in the territory of a park in which the total of 2077.74 km^2 of forestland is situated.

For the purposes of evaluating the vulnerability of particular tree-species by fire, the population proportion of (n) burned out areas, ($\hat{p}$), from the total observed area of grown tree-species (N) during the period of 10 years in the sample was chosen:

$$\hat{p} = \frac{n}{N} \qquad (1)$$

Table 1 Particular groups of tree-species with the equal decennial fire occurrence rates (p̂) during the period of 1992–2001

Numbers of groups	Groups of particular tree-species	Total observed areas (ha)	Total destroyed areas (ha)	Real fire occurrence rates (p̂)
1.	Pine	2129.5701	29.3281	0.01377186
2.	Spruce	10602.8299	83.5258	0.00787769
3.	Larch	1257.2047	7.9200	0.00629969
4.	Fir	1620.4757	5.6986	0.00351664
5.	Broad-leaved	5167.3642	13.8583	0.00268189
Total		20777.4446	140.3309	0.06754001

These population proportions (p̂) were calculated for each growing tree-species and then mutually compared and statistically tested on the significance level of $(\alpha) = 0.05$, whether their differences could be regarded as so significant that they were not likely to occur by chance. The applied testing procedure was taken from Triola (1989). According to the results of these tests all tree-species were divided into groups with the same population proportions of destroyed hectares in the sample. The results of this division are presented in Table 1, where the mentioned estimated population proportions are referred to in terms of the decennial fire occurrence rates.

Modelling the Fire Occurrence Hazard

The essential problem of modelling the fire occurrence hazard concerning all growing tree-species in our research was the fact that the observed data had pointed out the obviously higher fire occurrence rates at the younger forest stands than at the older ones. Therefore, the description of forest vulnerability in the samples of growing tree-species related to their age is based on the Weibull probability distribution $W(c, \gamma)$ as proposed by von Gadow (2000), Kouba (2002) and Kouba and Kasparova (1989). This distribution is defined by its probability distribution function:

$$F(t) = 1 - e^{-c.t^{\gamma}} \tag{2}$$

where (t) denotes the assumed age of a forest stand.

The empirical distribution functions $F_n(t)$ created from the observed destroyed areas of particular tree-species in a sample arranged according to their age, were fitted by the corresponding assumed Weibull probability distribution functions $F(t)$. The parameters (c) and (γ) of the $W(c, \gamma)$ distributions were estimated by the method of quartiles described by Pacakova (2000) where

$$\gamma = \frac{1.57253}{\ln q_{75} - \ln q_{25}} \tag{3}$$

$$c = \frac{0.28767}{q_{25}^{\gamma}} \tag{4}$$

The goodness of fit between each particular $F_n(t)$ and $F(t)$ was tested as the null hypothesis (H_0):

$$H_0\colon F_n(t) - F(t) = 0 \tag{5}$$

The Kolmogorov–Smirnov critical values $D_{(n)\alpha}$ at this testing procedure taken from Klein et al. (1997) were used.

The hazard of fire at particular forest stands in relation to their age (t) was then estimated by probabilities $p(t)$ calculated by using the relation:

$$p(t) = k \cdot \Delta F(t) \cdot \hat{p} \tag{6}$$

Here k denotes the number of assumed age classes, $\Delta F(t)$ refers to the difference of two following values of the function $F(t)$. The symbol p̂ denotes the expected population proportion of areas destroyed by fire in the sample for the period of 10 years.

The results of the described statistical analysis and the calculation of values $p(t)$ including parameters c and γ of the $W(c, \gamma)$ distributions together with the corresponding population proportions p̂ for each analysed growing tree-species are presented in Table 2. The symbol h denotes the expected total of destroyed hectares when the areas of particular age classes are uniformly distributed.

Table 2 Probabilities $p(t)$ informing about the vulnerability of forest stands at growing particular tree-species in the Slovak Paradise

Tree-species	Spruce	Fir	Larch	Pine	Beech and Oak
Age	Probabilities of destruction				
(t)	$p(t)$	$p(t)$	$p(t)$	$p(t)$	$p(t)$
10	0.030959	0.006390	0.006025	0.050324	0.010576
20	0.020021	0.005951	0.018039	0.034253	0.007144
30	0.015310	0.005400	0.024888	0.026724	0.005489
40	0.012168	0.004870	0.024445	0.021525	0.004349
50	0.009870	0.004377	0.018760	0.017631	0.003503
60	0.008111	0.003927	0.011634	0.014598	0.002851
70	0.006729	0.003517	0.005919	0.012181	0.002338
80	0.005622	0.003146	0.002489	0.010225	0.001929
90	0.004724	0.002813	0.000868	0.008624	0.001598
100	0.003988	0.002512	0.000252	0.007301	0.001330
110	0.003380	0.002242	0.000061	0.006202	0.001110
120	0.002874	0.002000	0.000012	0.005284	0.000929
130	0.002452	0.001784	0.000002	0.004513	0.000780
140	0.002097	0.001590	0.000000	0.003862	0.000656
150	0.001797	0.001417	0.000000	0.003312	0.000553
160	0.001544	0.001262	0.000000	0.002846	0.000467
170	0.001329	0.001123	0.000000	0.002449	0.000395
180	0.001146	0.001000	0.000000	0.002110	0.000334
c	0.034396	0.010000	0.000407	0.029981	0.032062
γ	0.855000	1.027018	2.127214	0.879000	0.887222
h	83.525800	5.698600	7.920000	29.328100	13.858300
p	0.0078800	0.003520	0.006300	0.013770	0.002680

Measuring and Ranking the Vulnerability of Forest

Measuring the Resilience of Forest to the Occurrence of Fire

The shares of particular age classes that a forest ecosystem occupies in the absence of any kinds of hazards, are given by the vector (**b**) consisting the uniformly distributed elements having the same constant probability in any assumed future period. From the theoretical point of view, this age class distribution meets the requirements imposed on the deterministic development of a fully regulated forest.

Investigating the development of a forest and its resilience in probability terms means to admit that this random process evolves through time in a manner that is not completely predictable. It needs to realise, that the system to be modelled is described by a set of expected states. This set of states represents the totality of all possible conditions that the system could be observed in at any particular point in time. However, as the fully resilient can be regarded only one such a steady state that does not allow for any significant disturbance of the age class distribution a forest once has occupied. The theory of Markov processes provides several procedures how to obtain steady-state probabilities for such a randomly evolving systems. The mathematical tools for the estimation of these probabilities and their interpretation in case of both the ergodic and transient processes are discussed, for example by Hool (1981), and this approach enables to obtain information about the resilience of forest also in our case. For these purposes were used the transition probability matrices (**P**) describing the random development of growing particular tree-species. Transitions of forest stands from younger age classes to older ones at the all assumed tree-species were described by the fire occurrence probability matrix (**P**):

$$P = \begin{vmatrix} p_1 & 1-p_1 & 0 & \dots & 0 & \dots & 0 \\ p_2 & 0 & 1-p_2 & \dots & 0 & \dots & 0 \\ \cdot & \cdot & \cdot & & \cdot & & \cdot \\ \cdot & \cdot & \cdot & & \cdot & & \cdot \\ \cdot & \cdot & \cdot & & \cdot & & \cdot \\ p_k & 0 & 0 & & 0 & & 1-p_k \end{vmatrix} \tag{7}$$

The elements p_j are identical to the probabilities $p(t)$, except the value of p_k that equals 1. The steady-state probabilities as the elements of vectors (**a**) informing about the possible resilience of particular tree-species in the stress of forest fire occurrence were calculated by the formula derived by Kouba (1977):

$$a_i = \frac{\prod_{i=1}^{i-1} p_{i,i+1}}{\sum_{j=1}^{k} \prod_{i=1}^{j-1} p_{i,i+1}} \tag{8}$$

where $j = 1, 2, \ldots, k$ for all $i = 1, 2, \ldots, n$

$$\text{and } \prod_{i=1}^{0} p_{i,i+1} = 1.$$

The elements of vector (**a**) represent the point estimate of forest resilience or the revealed spent part of its adapting capacity under given conditions. The elements of vector (**b**) inform about the steady-state age class distribution which forest is allowed to occupy in the absence of any hazard. The elements of steady-state probability vectors (**a**) and (**b**) concerning growing analysed tree-species are depicted in Table 3.

Ranking the Vulnerability of Forest in Terms of its Resilience

The differences between the corresponding elements of vectors (**a**) and (**b**) point out the level of vulnerability by fire concerning growing particular tree-species. The level of vulnerability was evaluated directly by testing the null hypotheses about the statistically insignificant differences between vectors (**a**) and (**b**). The both vectors represent probability distributions of age classes of a forest under different growing conditions and so their elements must meet the following requirements:

$$\sum_{t=1}^{k} a(t) = 1 \text{ and } \sum_{t=1}^{k} b(t) = 1 \tag{9}$$

So, the null hypotheses (H_0) were formulated in terms of differences between the cumulative probability distribution functions $A_N(t)$ and $B(t)$ created from probabilities $a(t)$ and $b(t)$ at N grown hectares of a tree-species:

$$H_0 : A_N(t) - B(t) = 0 \tag{10}$$

As the age t of a forest stand is a continual variable, hypotheses (10) were tested by means of the

Table 3 The vectors of steady-state probabilities (**a**) informing about the age class structures of forest property at growing particular tree-species in the presence of fire occurrence risk. The elements of vector (**b**) describe the expected state in the absence of risk

Age classes of forest stand (decades)	Fully regulated forest $b(t)$	Forest in the presence of fire occurrence				
		Spruce $a(t)$	Fir $a(t)$	Larch $a(t)$	Pine $a(t)$	Beech $a(t)$
1	0.055556	0.061234	0.057510	0.060691	0.065702	0.057472
2	0.055556	0.059338	0.057143	0.060326	0.062395	0.056864
3	0.055556	0.058150	0.056803	0.059237	0.060258	0.056458
4	0.055556	0.057260	0.056496	0.057763	0.058648	0.056148
5	0.055556	0.056563	0.056221	0.056351	0.057385	0.055904
6	0.055556	0.056005	0.055975	0.055294	0.056374	0.055708
7	0.055556	0.055550	0.055755	0.054651	0.055551	0.055549
8	0.055556	0.055176	0.055559	0.054327	0.054874	0.055419
9	0.055556	0.054866	0.055384	0.054192	0.054313	0.055312
10	0.055556	0.054607	0.055228	0.054145	0.053845	0.055224
11	0.055556	0.054389	0.055090	0.054131	0.053451	0.055150
12	0.055556	0.054206	0.054966	0.054128	0.053120	0.055089
13	0.055556	0.054050	0.054856	0.054127	0.052839	0.055038
14	0.055556	0.053917	0.054758	0.054127	0.052601	0.054995
15	0.055556	0.053804	0.054671	0.054127	0.052398	0.054959
16	0.055556	0.053708	0.054594	0.054127	0.052224	0.054928
17	0.055556	0.053625	0.054525	0.054127	0.052075	0.054903
18	0.055556	0.053553	0.054464	0.054127	0.051948	0.054881
Total	1.000000	1.000000	1.000000	1.000000	1.000000	1.000000

Table 4 The values of cumulative probability distribution functions $A_N(t)$ and $B(t)$ set up from the elements of steady-state probability vectors (**a**) and (**b**)

Age classes of forest-stands (t)	Fully regulated forest $B(t)$	Forest in the presence of fire occurrence				
		Spruce $A_N(t)$	Fir $A_N(t)$	Larch $A_N(t)$	Pine $A_N(t)$	Beech $A_N(t)$
1	0.055556	0.061234	0.057510	0.060691	0.065702	0.057472
2	0.111111	0.120571	0.114653	0.121017	0.128097	0.114336
3	0.166667	0.178721	0.171456	0.180254	0.188355	0.170793
4	0.222222	0.235981	0.227952	0.238018	0.247003	0.226941
5	0.277778	0.292544	0.284174	0.294369	0.304388	0.282845
6	0.333333	0.348548	0.340148	0.349663	0.360762	0.338552
7	0.388889	0.404098	0.395904	0.404313	0.416312	0.394101
8	0.444444	0.459275	0.451463	0.458640	0.471186	0.449520
9	0.500000	0.514141	0.506847	0.512832	0.525499	0.504833
10	0.555556	0.568748	0.562075	0.566977	0.579344	0.560056
11	0.611111	0.623138	0.617165	0.621109	0.632795	0.615207
12	0.666667	0.677343	0.672131	0.675237	0.685915	0.670296
13	0.722222	0.731393	0.726988	0.729364	0.738754	0.725334
14	0.777778	0.785310	0.781746	0.783491	0.791355	0.780329
15	0.833333	0.839115	0.836417	0.837618	0.843753	0.835287
16	0.888889	0.892822	0.891011	0.891746	0.895977	0.890216
17	0.944444	0.946447	0.945536	0.945873	0.948052	0.945119
18	1.000000	1.000000	1.000000	1.000000	1.000000	1.000000

Kolmogorov–Smirnov test of goodness of fit. For ranking the observed vulnerability, the scale based on the P-value approach to testing these null hypotheses was proposed. The advantage of this approach is a possibility to evaluate vulnerability not only as the property of an evolving forest ecosystem but also as a quantity that is complementary to the adapting capacity of a forest. The obtained cumulative probability distribution functions $A_N(t)$ and $B(t)$ are compared in Table 4. The criteria of evaluation concerning vulnerability of forest also related to its corresponding resilience and resistance are arranged in Table 5.

Economic Analysis of Forest Land Management in the Presence of Fire Occurrence Risk

Economic Analysis of Forest Management Projects

For the purposes of evaluating change in capital value of forestland due to the fire occurrence hazard, the management projects of the following two groups of main commercial tree-species growing in this region have been investigated:

1. Spruce, fir, larch and pine forest stands representing the conifers growing.
2. Beech and oak forest stands, the characteristic for the broad-leaved growing.

Due to the presence of externalities, the rotation period (u) of all forests growing here, in the national park, approaches 180 years. This time is far beyond the economic optimum, but on the other hand, due to the production of valuable assortments in higher age classes, the forest management still has been remaining profitable.

The brief account of economics concerning the growing of particular tree-species from the economic point of view is presented in Tables 6, 7, 8, 9, 10, and 11. Except the values of expected stumpage, thinnings and costs values, these tables also inform about the levels of profitability concerning particular projects by using the net present value NPV(u) (Measure 1) and the risk-adjusted soil expectation value SEV(u) (Measure 2).

Determination of the Risk-Free Soil Expectation Value

Using the previous results, the risk-free soil expectation value $SEV_f(u)$ was determined as the NPV of a never-ending forest management project if fire occurs

Table 5 The criteria for the statistical testing vulnerability of particular tree-species by fire and its evaluation related to a revealed resilience

Test of vulnerability		Vulnerability		Resilience
P-value	Degree	Interpretation	Evaluation	Evaluation
$1 > P > 0.05$	0	Insignificant	Not vulnerable	Resistant
$0.05 > P > 0.01$	1	Statistically significant	Vulnerable	Very resilient
$0.01 > P > 0.001$	2	Highly statistically significant	Very vulnerable	Resilient
$0.001 > P > 0$	3	Extremely statistically significant	Endangered	Not resilient

Table 6 The NPV(u) and the SEV(u) of the spruce forest management project derived from the expected decennial cash flows ($r = 0.01$ p.a.)

Spruce age (t) (years)	Stumpage ST(t) (€.ha^{-1})	Thinnings TH(t) (€.ha^{-1})	Costs $C(t)$ (€.ha^{-1})	Measure 1 NPV(u) (€.ha^{-1})	Measure 2 SEV(u) (€.ha^{-1})
0	0	0	937	−937	
10	0	0	611	−1519	−31293
20	0	0	297	−1774	−12797
30	692	216	247	−1258	−5714
40	918	269	247	−1135	−3858
50	1404	343	247	−824	−2282
60	1726	356	247	−660	−1566
70	2030	352	247	−540	−1133
80	2985	446	247	−93	−177
90	3622	483	247	147	258
100	4442	513	247	422	690
110	4241	442	247	256	395
120	4836	447	247	368	540
130	4618	394	247	202	284
140	5074	389	247	232	314
150	6732	481	247	554	725
160	7228	467	247	556	707
170	8426	532	247	697	864
180	8125	0	247	446	541

Table 7 The NPV(u) and the SEV(u) of the fir forest management project derived from the expected decennial cash flows ($r = 0.01$ p.a.)

Fir Age (t) (years)	Stumpage ST(t) (€.ha^{-1})	Thinnings TH(t) (€.ha^{-1})	Costs $C(t)$ (€.ha^{-1})	Measure 1 NPV(u) (€.ha^{-1})	Measure 2 SEV(u) (€.ha^{-1})
0	0	0	1356	−1356	
10	0	0	505	−1837	−37840
20	0	0	247	−2049	−14780
30	1027	119	247	−1348	−6122
40	1364	227	247	−1200	−4082
50	1807	266	247	−996	−2760
60	2292	303	247	−793	−1881
70	2533	288	247	−770	−1617
80	3670	284	247	−339	−646
90	4672	316	247	−45	−78
100	5894	359	247	284	464
110	5746	418	247	75	116
120	6660	427	247	232	340
130	6428	387	247	4	6
140	7137	385	247	50	68
150	9536	491	247	498	652
160	10,242	477	247	485	617
170	11,832	397	247	614	762
180	11,669	0	247	326	395

Table 8 The NPV(u) and the SEV(u) of the larch forest management project derived from the expected decennial cash flows ($r = 0.01$ p.a.)

Larch Age (t) (years)	Stumpage ST(t) (€.ha^{-1})	Thinnings TH(t) (€.ha^{-1})	Costs $C(t)$ (€.ha^{-1})	Measure 1 NPV(u) (€.ha^{-1})	Measure 2 SEV(u) (€.ha^{-1})
0	0	0	1517	−1517	
10	0	0	392	−1891	−38,955
20	0	0	345	−2188	−15,782
30	1179	94	247	−1388	−6300
40	2390	380	247	−526	−1788
50	3384	479	247	98	271
60	3969	455	247	352	835
70	4144	401	247	307	644
80	4622	394	247	397	756
90	4681	339	247	255	447
100	5256	339	247	324	530
110	5226	305	247	141	217
120	6102	314	247	267	392
130	5946	280	247	47	67
140	6651	290	247	80	109
150	8948	345	247	482	631
160	9730	345	247	469	597
170	11,347	413	247	618	766
180	11,106	0	247	324	393

Table 9 The NPV(u) and the SEV(u) of the pine forest management project derived from the expected decennial cash flows ($r = 0.01$ p.a.)

Pine Age (t) (years)	Stumpage ST(t) (€.ha^{-1})	Thinnings TH(t) (€.ha^{-1})	Costs $C(t)$ (€.ha^{-1})	Measure 1 NPV(u) (€.ha^{-1})	Measure 2SEV(u) (€.ha^{-1})
0	0	0	1587	−1587	
10	0	856	337	−1094	−22,539
20	0	733	247	−675	−4867
30	1593	43	247	408	1852
40	1895	277	247	525	1786
50	2291	320	247	698	1935
60	2697	323	247	838	1989
70	3086	319	247	932	1957
80	3461	291	247	978	1860
90	3795	288	247	984	1723
100	4113	266	247	960	1571
110	4402	266	247	918	1415
120	4855	256	247	918	1347
130	5643	272	247	1006	1413
140	6391	282	247	1056	1429
150	7106	288	247	1077	1410
160	7797	290	247	1075	1367
170	7644	274	247	893	1107
180	7441	0	247	673	817

Table 10 The NPV(u) and the SEV(u) of the beech forest management project derived from the expected decennial cash flows ($r = 0.01$ p.a.)

Beech Age (t) (years)	Stumpage ST(t) (€.ha^{-1})	Thinnings TH(t) (€.ha^{-1})	Costs $C(t)$ (€.ha^{-1})	Measure 1 NPV(u) (€.ha^{-1})	Measure 2 SEV(u) (€.ha^{-1})
0	0	0	849	−849	
10	0	0	779	−1590	−32,764
20	0	0	247	−1803	−13,003
30	1321	141	247	−855	−3884
40	1947	190	247	−552	−1876
50	2371	277	247	−392	−1085
60	3498	325	247	163	386
70	3758	396	247	185	389
80	5564	510	247	979	1863
90	7191	590	247	1575	2759
100	8596	613	247	1971	3224
110	9707	640	247	2184	3369
120	10,587	621	247	2260	3316
130	11,144	600	247	2203	3095
140	11,513	557	247	2076	2809
150	11,592	522	247	1875	2455
160	11,400	464	247	1621	2061
170	10,875	399	247	1318	1634
180	10,361	0	247	985	1194

Table 11 The NPV(u) and the SEV(u) of the oak forest management project derived from the expected decennial cash flows ($r = 0.01$ p.a.)

Oak Age (t) (years)	Stumpage ST(t) (€.ha^{-1})	Thinnings TH(t) (€.ha^{-1})	Costs $C(t)$ (€.ha^{-1})	Measure 1 NPV(u) (€.ha^{-1})	Measure 2SEV(u) (€.ha^{-1})
0	0	0	1036	−1036	
10	0	0	685	−1687	−34, 765
20	501	113	247	−1371	−9888
30	602	151	247	−1408	−6394
40	2217	333	247	−252	−857
50	2662	364	247	−41	−114
60	3105	378	247	129	307
70	3521	407	247	261	548
80	3917	437	247	364	692
90	4292	450	247	437	765
100	4626	499	247	490	802
110	4929	479	247	508	783
120	5232	486	247	516	757
130	5529	323	247	466	655
140	6369	338	247	558	755
150	7152	346	247	609	797
160	7912	349	247	633	805
170	8671	328	247	636	788
180	9430	0	247	566	687

in particular age classes (j) with stationary probabilities p_j. According to the exact mathematical solution of the problem concerning the SEV$_f(u)$, calculation as provided by Reed (1984) who had used differential calculus, we have applied the numerical solution based on the theory of Markov chains as described by Kouba (1977), Martell (1980), Suzuki (1983) and Van Wagner (1978). For these purposes the above-mentioned transition probability matrix (**P**) was used again. The expected shares p_{ij} of 1 ha of forest originally planted and growing at the beginning of the next decade $(i+1)$ were given by the vector $\mathbf{p}^{(i)}$:

$$\mathbf{p}^{(i)} = \mathbf{p}^{(0)} \cdot \mathbf{P} \tag{11}$$

Vector $\mathbf{p}^{(0)} = [1, 0, 0, \ldots, 0]$ describes the initial state of the age class distribution when planting 1 ha of forest soil of a project in year (0). The expected shares of areas destroyed during the decade (i) in the age class (j), q_{ij} were derived as follows:

$$q_{ij} = p_{ij} \cdot g_j \tag{12}$$

The Gentan probabilities (g_j) as proposed by Suzuki (1983), for particular age classes (j), were calculated using the following equation:

$$g_{j+1} = \prod_{i=1}^{j} (1 - p_i) \cdot p_{j+1} \text{ for } g_1 = p_1 \tag{13}$$

The value g_j informs about the expected share of 1 ha of forest originally planted that is steadily destroyed during the decade (j) of its age within each assumed rotation period (u).

Similarly, as presented by Holecy and Hanewinkel (2006), the calculation of SEV$_f(u)$ was based on the expected age class distributions over time given by the areas of p_{ij} and q_{ij} evaluated with the costs and revenues in Tables 6, 7, 8, 9, 10 and 11. But, due to the more intensive site preparation, the costs of replanting after fire were the higher.

The algorithm of the SEV$_f(u)$ calculation ended, when the increment of the present value of the last assumed rotation had approached to less than 0.005 €. For all projects, the eight rotation periods were calculated to reach the desired level of precision.

The Costs of Forest Resilience

Under the stress of a fire occurrence, the adaptation process of a forest ecosystem evolves further approaching its steady-state equilibrium of age class distribution. The steady state decreases vulnerability of forest to the minimum level, ensures its next sustainable existence and resistance again, but both the necessary effort spent and the former higher revenue of forest management have already been lost forever. The cost of resilience CR(u) at the assumed rotation period of u years, in this sense, then can be determined as follows:

$$\mathrm{CR}(u) = \mathrm{RPSEV}(u) = \mathrm{SEV}(u) - \mathrm{SEV_f}(u) \tag{14}$$

The value of RPSEV(u) denotes the corresponding risk premium on the SEV(u) which is the second-possible interpretation of CR(u). The results informing about the values of SEV(u) and SEV$_f(u)$ including corresponding values CR(u) are arranged in Table 12

Table 12 The results of a forest soil valuation at the growing particular tree-species in both the absence SEV(u) and the presence SEV$_f(u)$ of a fire occurrence risk (u = 180 years, $r = 0.01$ p.a.)

Tree species	The measures of forest management projects			
	NPV(u) (€.ha^{-1})	SEV(u) (€.ha^{-1})	SEV$_f(u)$ (€.ha^{-1})	CR(u) (€.ha^{-1})
Spruce	446	541	390	151
Fir	326	395	27	368
Larch	324	393	136	257
Pine	673	817	345	471
Beech	985	1194	741	453
Oak	566	687	190	497

The Vulnerable Value of Forest Property

The valuation of forest vulnerability by fire belongs among the essential issues concerning the practical applications of a proposed procedure. In order to determine the vulnerable value of forest, the following three parameters of forest valuation had to be taken in account, calculated and combined:

1. The forest stand expectation value FEV(t) and the related SEV(u).
2. The salvage value SV(t) of a forest stand affected by fire.
3. The mentioned risk-free SEV$_f(u)$.

The loss a forest property suffers and thus the vulnerable forest value VFV(t) was then determined in the following way:

$$\mathrm{VFV}(t) = \mathrm{FEV}(t) - \mathrm{SV}(t) + \mathrm{SEV}(u) - \mathrm{SEV_f}(u) \quad (15)$$

The quantity t refers to the age of forest stand and the difference in value between the last two measures in the right-hand side of the equation refers to the RPSEV(u) informing about the corresponding risk premium on the SEV(u) when (u) denotes the length of assumed rotation period, at the forest management projects of growing particular tree-species.

The forest stand expectation value FEV(t) as expressed by using the Faustmann formula (1849) presented in terms of expected cash flows by Kilkki (1985) was calculated as follows:

$$\mathrm{FEV}(t) = \frac{\sum_{i=t}^{u} R_i (1+r)^{u-i} - \sum_{i=t}^{u} C_i (1+r)^{u-i} + \mathrm{SEV(u)}}{(1+r)^{u-t}} - \mathrm{SEV(u)} \quad (16)$$

Values R_i refer to the nominal values of the expected revenues during the ith period lasting 10 years that span the length of 1 assumed age class. Values C_i refer to the nominal values of the expected costs in the same period. The SEV(u) is the risk-adjusted soil expectation value calculated based on the NPV(u) of the underlying forest management project. The vulnerable forest value VFV(t) concerning all investigated tree-species obtained by the described procedure is depicted in Tables 13, 14 and 15.

The Definition of Risk in Terms of the Vulnerable Value of Forest

The exact definition of risk is necessary to evaluate economic effects caused by the changes of forest vulnerability that may occur as the results of new applied silvicultural patterns including more efficient fire protection measures. However, the vulnerability can be significantly increased due to the expected irreversibility or even the gradual propagation of a climate change. In this case, the change of probabilities $p(t)$ could become the most important factor of increasing the risk itself. The obvious functional relationship between the forest management risk $R(t)$ and the above-defined vulnerable forest value VFV(t) at known probability of a forest stand destruction can then be formulated as follows:

$$R(t) = p(t) \cdot \mathrm{VFV}(t) \quad (17)$$

Risk is hazard quantitatively expressed as the probability of forest fire occurrence multiplied by the vulnerability of forest expressed in monetary terms. The formulation (15) meets the definition of risk related to potential loss at a certain degree of vulnerability as it is explained by Rashed and Weeks (2003). In comparison with the simple product of hazard occurrence probability and the value of forest stand, this formulation enables to calculate more precisely the risk itself. The reason of a higher precision is the more detailed description and estimation of forest vulnerability, enlarged also by its economic dimension. The proposed procedure reduces the entropy of a forest land management as it is defined by Polster and Polsterova (2000) as well.

Results and Discussion

The evaluation of forest vulnerability by the proposed procedure has brought the results presented in Table 16. Pine and spruce proved to be the most vulnerable, even 'endangered' grown tree-species. The P-value test provides extremely strong evidences against the null hypothesis in the both cases and the expected steady-state age classes distributions at growing both tree-species inform about a threat to a sustainable forestry. They are also classified as 'not resilient', which informs about a necessity to support their fire protection to avoid their local extinction and a large-scale deforestation under given conditions.

Larch seems to be 'very vulnerable', but still has been 'resilient' to the occurrence of fire, as well. Although the P-value test for this tree-species gives a very strong evidence against the null hypothesis, the digression of its expected steady-state vector from that desired for the fully regulated forest does not threaten the sustainable development in its very essence. This fact informs about the ability of larch to cope the fire occurrence hazard and to approach the new fire resistant age class distribution, again. Even though it will take a longer time.

Table 13 The calculation of vulnerable forest value VFV(t) for spruce and fir forests ($u = 180$ years, $r = 0.01$ p.a.)

Tree-species	Spruce			Fir		
Age (t) (years)	FEV(t) (€.ha^{-1})	SV(t) (€.ha^{-1})	VFV(t) (€.ha^{-1})	FEV(t) (€.ha^{-1})	SV(t) (€.ha^{-1})	VFV(t) (€.ha^{-1})
0	0	0	151	0	0	368
10	1371	0	1522	1446	0	1814
20	2200	0	2351	2196	0	2564
30	2769	411	2509	2739	129	2979
40	3103	905	2349	3209	534	3043
50	3414	1614	1951	3608	1095	2881
60	3675	2121	1705	4005	1527	2846
70	3950	2362	1740	4404	1761	3011
80	4258	2741	1668	4861	2087	3141
90	4494	2842	1803	5370	2208	3530
100	4715	3125	1741	5897	2452	3813
110	4925	2740	2336	6432	2174	4625
120	5236	2926	2461	6957	2344	4980
130	5573	2641	3082	7527	2134	5761
140	6004	2773	3382	8201	2256	6314
150	6486	3536	3100	8948	2891	6425
160	6917	3673	3395	9656	3019	7004
170	7408	4487	3072	10453	3720	7101
180	7879	4409	3621	11422	4013	7777

Table 14 The calculation of vulnerable forest value VFV(t) for larch and pine forests ($u = 180$ years, $r = 0.01$ p.a.)

Tree-species	Larch			Pine		
Age (t) (years)	FEV(t) (€.ha^{-1})	SV(t) (€.ha^{-1})	VFV(t) (€.ha^{-1})	FEV(t) (€.ha^{-1})	SV(t) (€.ha^{-1})	VFV(t) (€.ha^{-1})
0	0	0	257	0	0	471
10	1615	0	1872	1710	0	2181
20	2258	0	2515	1401	0	1872
30	2917	371	2803	1096	267	1300
40	3432	747	2942	1522	608	1385
50	3685	1262	2680	1733	1072	1132
60	3855	1608	2504	1919	1395	995
70	4070	1745	2582	2121	1529	1063
80	4367	1977	2647	2348	1755	1064
90	4702	2018	2941	2630	1799	1303
100	5133	2189	3201	2946	1960	1457
110	5609	1894	3972	3318	1700	2089
120	6172	2001	4429	3729	1799	2402
130	6785	1792	5250	4195	1617	3049
140	7499	1863	5893	4692	1684	3479
150	8278	2362	6173	5229	2143	3558
160	9077	2453	6881	5816	2225	4062
170	9959	3100	7116	6463	2724	4209
180	10859	3296	7820	7194	2822	4843

Table 15 The calculation of vulnerable forest value VFV(t) for beech and oak forests (u = 180 years, r = 0.01 p.a.)

Tree-species	Beech			Oak		
Age (t) (years)	FEV(t) (€.ha^{-1})	SV(t) (€.ha^{-1})	VFV(t) (€.ha^{-1})	FEV(t) (€.ha^{-1})	SV(t) (€.ha^{-1})	VFV(t) (€.ha^{-1})
0	0	0	453	0	0	497
10	953	0	1406	1123	0	1620
20	2038	0	2491	2069	0	2566
30	2649	214	2889	2506	73	2929
40	3169	409	3213	2946	179	3264
50	3688	669	3472	3232	337	3391
60	4166	843	3776	3512	460	3549
70	4640	920	4173	3807	519	3785
80	5086	1057	4482	4100	604	3993
90	5452	1093	4812	4391	629	4259
100	5769	1199	5023	4698	695	4500
110	6093	1052	5494	4983	609	4870
120	6421	1127	5747	5320	652	5165
130	6804	1024	6234	5685	590	5591
140	7251	1080	6624	6267	622	6142
150	7792	1382	6863	6894	795	6596
160	8429	1444	7438	7578	828	7247
170	9196	1777	7872	8330	1018	7809
180	10,114	1845	8722	9184	1056	8624

Table 16 The mutual evaluating vulnerability and resilience of particular tree-species in a sample according to the results of testing the null hypothesis H_0: $D(t) = 0$

Age classes of forest (decades) (t)	Differences between distribution functions $A_N(t)$ and $B(t)$ $D(t) = A_N(t) - B(t)$				
	Spruce $D(t)$	Fir $D(t)$	Larch $D(t)$	Pine $D(t)$	Beech $D(t)$
1	0.005678	0.001955	0.005136	0.010146	0.001916
2	0.009460	0.003542	0.009906	0.016986	0.003225
3	0.012055	0.004790	0.013588	0.021688	0.004127
4	0.013759	0.005730	0.015795	0.024780	0.004719
5	0.014766	0.006396	0.016591	0.026610	0.005067
6	0.015215	0.006815	0.016329	0.027428	0.005219
7	0.015210	0.007015	0.015424	0.027423	0.005212
8	0.014830	0.007018	0.014196	0.026742	0.005076
9	0.014141	0.006847	0.012832	0.025499	0.004833
10	0.013193	0.006520	0.011422	0.023788	0.004501
11	0.012027	0.006054	0.009997	0.021684	0.004096
12	0.010677	0.005465	0.008570	0.019248	0.003629
13	0.009171	0.004765	0.007142	0.016532	0.003111
14	0.007533	0.003968	0.005713	0.013577	0.002551
15	0.005781	0.003084	0.004285	0.010419	0.001954
16	0.003933	0.002122	0.002857	0.007088	0.001327
17	0.002002	0.001092	0.001428	0.003608	0.000674
18	0.000000	0.000000	0.000000	0.000000	0.000000
Maximum $D(t)$	0.015215	0.007018	0.016591	0.027428	0.005219
Critical $D(n)_{0.05}$	0.004171	0.010669	0.012112	0.009306	0.005974
Critical $D(n)_{0.01}$	0.004999	0.012786	0.014516	0.011153	0.007160
Critical $D(n)_{0.001}$	0.005987	0.015314	0.017387	0.013359	0.008576
N	2129.5701	10602.8299	1257.2047	1620.4757	5167.3642
Degree of significance	3	0	2	3	0
Vulnerability ranking	Endangered	Not vulnerable	Very vulnerable	Endangered	Not vulnerable
Resilience ranking	Not resilient	Resistant	Resilient	Not resilient	Resistant

Fir together with beech and oak are classified as statistically 'not vulnerable' and 'resistant' to the fire occurrence under given circumstances, because *P*-value points out the insufficient evidence against the null hypothesis. The detected digressions of their steady-state age class distributions from the fully regulated one are obviously only due to chance fluctuations.

The annual risk premiums $R(t)$ expressing the risk of a forest owner at all investigated tree-species are shown in Table 17. They vary from €0.16 at 0 years old larch forest stand to €10.97 at 10 years old pine stand obviously according to the values of stumpage at particular age and detected vulnerability induced by a different levels of hazard. Striking is the fact, that a 'not vulnerable' fir is a little bit more risky growing than a 'very vulnerable' larch. These results point out the fact that the economic impact of vulnerability measured by risk premiums $R(t)$ can differ from the ecological vulnerability measured by digressions from a fully regulated forest management $D(t)$.

The obtained results can be concluded as follows:

1. The assessment of vulnerability carried out by the proposed procedure provides also with the complementary information concerning both the resistance and resilience of forest to a fire occurrence.
2. The vulnerability of the most endangered tree-species obviously can be substantially reduced by the conversion of pure spruce and pine forest stands to more resilient mixed forests by changing their unfavourable tree-species composition to the less vulnerable one by the higher admixture of resistant broad-leaved tree-species.
3. The calculated risk premiums $R(t)$ represent very valuable information especially for decision-making concerning capital investments of forest owners and landscape planners including protection and prevention measures of fire fighting and rescue corps. Risk premiums $R(t)$ can directly serve as the relevant information for the purposes of forest fire insurance within the whole area of the Slovak Paradise.

The proposed procedure of ranking the vulnerability of forest by a fire occurrence hazard based on its statistical evaluation can be regarded only as an attempt to clarify the relations between the random evolution of forest and its possible sustainable management when forest has to cope the presence of natural elements occurrence. The acquired knowledge about a forest management risk can increase the performance of a forest growth simulator proposed by Fabrika and Dursky (2005). Also, this procedure can be efficiently applied at ranking the vulnerability of forest by fire by using the tools of a map algebra as presented by Tucek and Majlingova (2007).

The applied vulnerability assessment and forecasting methods are based on the assumption that

Table 17 Annual risk premiums $R(t)$ for growing particular tree-species at the fire occurrence hazard in the Slovak Paradise

Age (t)	Spruce $R(t)$ (€.ha^{-1})	Fir $R(t)$ (€.ha^{-1})	Larch $R(t)$ (€.ha^{-1})	Pine $R(t)$ (€.ha^{-1})	Beech $R(t)$ (€.ha^{-1})	Oak $R(t)$ (€.ha^{-1})
0	0.47	0.23	0.16	2.37	0.48	0.53
10	4.71	1.16	1.20	10.97	1.49	1.71
20	4.71	1.53	1.50	6.41	1.78	1.83
30	3.84	1.61	1.51	3.48	1.59	1.61
40	2.86	1.48	1.43	2.98	1.40	1.42
50	1.93	1.26	1.17	2.00	1.22	1.19
60	1.38	1.12	0.98	1.45	1.08	1.01
70	1.17	1.06	0.91	1.30	0.98	0.88
80	0.94	0.99	0.83	1.09	0.86	0.77
90	0.85	0.99	0.83	1.12	0.77	0.68
100	0.69	0.96	0.80	1.06	0.67	0.60
110	0.79	1.04	0.89	1.30	0.61	0.54
120	0.71	1.00	0.89	1.27	0.53	0.48
130	0.76	1.03	0.94	1.38	0.49	0.44
140	0.71	1.00	0.94	1.34	0.43	0.40
150	0.56	0.91	0.87	1.18	0.38	0.36
160	0.52	0.88	0.87	1.16	0.35	0.34
170	0.41	0.80	0.80	1.03	0.31	0.31
180	0.41	0.78	0.78	1.02	0.29	0.29

estimated transition probabilities $p(t)$ are stationary (i.e. do not change in time). This allows for the reliable description of a forest ecosystem through time. In this sense, the estimated steady-state probabilities can be regarded as the limits of possible vulnerability under given conditions, unless the probabilities $p(t)$ change.

Acknowledgments The research works described in the present chapter were supported by using the funds of the granted research projects VEGA 1/3528/06 and 1/2382/05. The collection of data was funded by the APVT project 51-037902, as well. The mentioned analyses were carried out at the Department of Forest Economics and Administration belonging to the Technical University of Zvolen and at the Ecological and Forestry Research Agency (EFRA), Slovakia. Authors gratefully thank the all mentioned agencies for their significant support.

References

Birkmann J (2006) Measuring vulnerability to promote disaster-resilient societies: Conceptual frameworks and definitions. In: Measuring Vulnerability to Natural Hazards: Towards Disaster Resilient Societies. Birkmann, J. (Ed.). Tokyo, United Nations University Press. pp. 9–54

Bogardi JJ (2006) Human security and risk management within integrated river basin management. In: Towards Integrated River Basin Management. Casta – Papiernicka, The Slovak Water Research Institute. p. 243.

Bogardi JJ, Birkmann J (2006) Vulnerability assessment: The first step towards sustainable risk reduction. In: Disasters and Society – From Hazard Assessment to Risk Reduction. Malzahn, D., Plapp, T. (Eds.), Karlsruhe, Universität Karlsruhe. pp. 75–82.

Caballero D, Beltran I (2003) Concepts and ideas of assessing settlement fire vulnerability in the W-UI zone. In: Forest Fires in the Wildland-Urban Interface and Rural Areas in Europe. Xanthopoulos G. (Ed.). Athens, Mediterranean Agronomic Institute of Chania. pp. 47–54.

Fabrika M, Dursky J (2005) Tree growth simulators. EFRA (Ecological & Forestry Research Agency), Zvolen (in Slovak). 112pp.

Faustmann M (1849) Berechnung des Wertes welchen Waldboden sowie noch nicht haubare Holzbestände főr die Waldwitschaft besitzen. Allgemeine forst und Jagd-Zeitung, vol. 25, pp. 441–455.

Gadow K v. (2000) Evaluating risk in forest planning models. Silva Fennica, vol. 34(2): 181–191.

Halada L, Weisenpacher P, Glasa J (2006) Reconstruction of the forest fire propagation case when people were entraped by fire. Forest Ecology and Management, vol. 234S: p. 127.

Hanewinkel M (2002) Comparative economic investigations of even-aged and uneven aged silvicultural systems: a critical analysis of different methods. Forestry, vol. 75(4): 473–481.

Hanewinkel M, Oesten G (1998) Ökonomischer Modellvergleich risikobeinflusster Alterklassen und Plenterwald betriebsklassen. Allgemeine Forst Zeitschrift für Wald und Umweltvorsorge 53:427–430.

Holecy J, Hanewinkel M (2006) A forest management risk insurance model and its application to coniferous stands in southwest Germany. Forest Policy and Economics, vol. 8(2): 161–174.

Holling CS, Gunderson LH, Peterson G (2002) Sustainability and panarchies. In: Panarchy: Understanding Transformations in Human and Natural Systems. Gunderson, L. H.– Holling, C. S. (Eds.). Washington, D.C., Island Press. pp. 63–102.

Hool J (1981) Markov processes. In: Forest Operations Analysis Techniques in Planning and Control. Corcoran, T., Heij, W. (Eds.). Kyoto, Japan, 27pp.

Kasperson RE, Kasperson JX, Dow K (2001) Vulnerability, equity, and global environmental change. In: Global environmental risk. Kasperson, J. X., Kasperson, R. E. (Eds.). Tokyo, United Nations University Press and Earthscan Publications Ltd. 574pp.

Kilkki P (1985) Timber Management Planning. Joensuu, University of Joensuu. 159pp.

Klein T, Bahyl V, Vacek V (1997) Basics of Probability and Mathematical Statistics. Publishing House of the Technical University of Zvolen, Zvolen. 224p. (in Slovak).

Kouba J (1977) Markov chains and modeling the long-term development of the age structure and production of forests. Proposal of a New Theory of the Normal Forest. Scientia Agriculturae Bohemslovaca, vol. 26(3): 179–193.

Kouba J (2002) Das Leben des Waldes und seine Lebensunsicherheit [Forest Life and its Temporal Uncertainty] German Journal of Forest Science, vol. 121: 211–228.

Kouba J, Kasparova I (1989) Steuerungs theorie des Ausgleichs prozesses zum Normal Wald auf Grund der Stochastischen Processe. Res. Report, Czech. Agricultural University, Prague, Czech Republic. 64pp.

Martell DL (1980) The optimal rotation of a flammable forest stand. Canadian Journal of Forest Research, 10: 30–34.

Pacakova V (2000) Applied Insurance Statistics (Aplikovana poistna statistika). Bratislava, Elita Publishing Company. 248pp.

Polster P, Polsterova H (2000) Use of information entropy to define a mixed forest. Journal of Forest Science, vol. 46(6): 298–304.

Rashed T, Weeks J (2003) Assessing vulnerability to earthquake hazards through spatial multicriteria analysis of urban areas. International Journal of Geographical Information Science, vol. 17(6): 547–576.

Reed WJ (1984) The effects of the risk of fire on the optimal rotation of a forest. Journal of Environmental Economics and Management, vol. 11: 180–190.

Sisak L, Pulkrab K (2001) Damaging factors and their influence on forestry in the Czech Republic. pp. 125–132. In Proc. IUFRO Division 4 Conference "The Economics of Natural Hazards in Forestry", June 7–10, Solsona, Catalonia, Spain, Italy, Padua University Press. 164pp.

Skvarenina J, Krizova E, Tomlain J (2004) Impact of the climate change on the water balance of altitudinal vegetation stages in Slovakia. Ekológia (Bratislava), vol. 23, Supplement 2/2004: 13–19.

Suzuki T (1983) Gentan-Wahrscheinlichkeit, Vorhersage Modelle fur die Entwicklung des Normalwaldes und fur die Plannung des Holzaufkommens. In: Beitrage zur biometrischen Modellbildung in der Forstwirtschaft von T. Suzuki unter mitwirkung von B. Sloboda und J. Saborowski. Schriften aus der Forstlichen Fakultat der Universitat Goettingen, Band 76: 7–22.

Thywissen K (2006) Core terminology of disaster reduction: A comparative glossary. In: Measuring Vulnerability to Natural Hazards: Towards Disaster Resilient Societies. Birkmann, J. (Ed.). Tokyo, United Nations University Press. pp. 9–54.

Triola MF (1989) Elementary Statistics. 4th ed. Redwood City, California, Benjamin Cummings Publishing Company. 784pp.

Tucek J, Majlingova A (2007) Forest Fires in The Slovak Paradise. Zvolen, Technical University of Zvolen. 172pp.

Van Wagner CE (1978) Age class distribution and the forest fire cycle. Canadian Journal of Forest Research, vol. 8: 220–227.

Part IV

SOIL AND AGRICULTURE BIOCLIMATOLOGY, NATURAL HAZARDS AND RESPONSES

Responses of Soil Microbial Activity and Functional Diversity to Disturbance Events in the Tatra National Park (Slovakia)

E. Gömöryová, K. Střelcová, J. Škvarenina, J. Bebej and D. Gömöry

Keywords Windthrow · Wildfire · Spruce stands · Soil properties · Microbial activity

Introduction

Disturbance events may be human-induced or natural, both typically causing changes in ecosystem properties and functions (Wright and Coleman 2002). Many observations show an increasing frequency of risks related to natural hazards in most parts of Europe. Natural hazards (flood, drought, erosion, fire, storm damage, etc.), driven by weather extremes, are increasing as a consequence of climate change (Thürig et al. 2005, Anonymous 2007). Wind and fire can damage not only individual trees or group of trees but also forest stands at various spatial scales and at various level of their intensity. Empirical studies have documented changes of the ecosystem properties (nutrient cycling, hydrology, soil organisms, vegetation dynamics) especially at the large-scale disturbance-affected areas as a consequence of a complete destruction of tree canopy and soil perturbation. Windthrow areas are usually cleared. Consequently, microclimatic conditions become changed because of a greater input of precipitation, solar radiation and heat to the soil surface as well as a more intensive air circulation. Soil is affected as well: as the trees are removed, nutrient losses from the ecosystem are observed. Recently, many windthrow areas are left without clearing. However, there are only a few studies about the effects of these two management alternatives on soil properties (Thürig et al. 2005).

Forest fires (prescribed burning or wildfire) can affect many physical, chemical and biological properties of soil, the severity of these effects depending on the scale and intensity of the fire. Wildfires are generally considered to have negative effects on soil. They cause significant removal of organic matter, deterioration of structure and porosity, considerable loss of nutrients through volatilization, leaching and erosion, and marked alteration of both quantity and specific composition of microbial and soil-dwelling invertebrate communities (Certini 2005). These changes in soil depend on both the temperatures at different soil depths and the degree of heating, depending on the magnitude of energy transferred from the fire to soil and duration of this transfer, soil composition (including moisture), structure (porosity) and the fuel loads (Neary et al. 1999).

Soil microorganisms play the key role in various biogeochemical cycles and are responsible for the cycling of organic compounds. Because of their sensitivity, they are the early indicators of environmental changes. After disturbance events, changes in soil conditions are expected to affect the activity, structure and the function of microbial communities (Bååth et al. 1995, Papatheodorou et al. 2004). The effects of fire on soil microorganisms biomass and activity have been documented in many studies (Certini 2005, Fioretto et al. 2005, Guerrero et al. 2005, Hart et al. 2005, Mabuhay et al., 2006), however, the information about the changes in their diversity on pre- and post-fire plots is scarce. At the same time, there is a lack of knowledge about soil

E. Gömöryová (✉)
Technical University in Zvolen, Faculty of Forestry, T.G. Masaryka 24, 960 53 Zvolen, Slovakia
e-mail: egomory@vsld.tuzvo.sk

K. Strelcová et al. (eds.), *Bioclimatology and Natural Hazards*,
DOI 10.1007/978-1-4020-8876-6_21, © Springer Science+Business Media B.V. 2009

microorganisms on windthrow-affected plots with different management regimes.

In November 2004, spruce stands in the Tatra National Park (TANAP) of the area of 12,000 ha were affected by a large-scale windthrow and 1 year later (in July 2005) a wildfire broke out on a part (220 ha) of this area (Fleischer et al. 2007). On a part of the windthrow-affected area, all windthrow debris was left on site as it had fallen to study of the bio-, geo-chemical cycles and processes in such ecosystems.

The overall objective of this study was to examine the response of soil microbial biomass, activity and functional diversity to disturbance events.

Materials and Methods

Study Area

The experimental area is located in the TANAP in northern Slovakia. Four research plots, 100 ha each, were established by the Research Station of the TANAP (Fleischer et al. 2007) on areas treated in different ways: an area where timber was extracted (EXT), a non-extracted site (NEX), a burnt site (FIR) and in a reference intact forest (REF). On the extracted plot, the forest was destroyed by wind and the fallen trees were extracted. A large part of this plot has been overgrown by weeds, especially *Calamagrostis villosa*. At the non-extracted site, fallen trees have not been extracted and the plot was left to natural development. Ground-layer vegetation is slightly changed compared to closed forest, but light-demanding grasses and herbs do not predominate. On the burnt plot, fire completely destroyed the above-ground humus layer; however, mineral horizons were not immediately affected by fire. A part of this plot is covered by an ash layer, whereas on other parts, there is bare mineral soil. Ground-layer vegetation is scarce. The plots with a slope of 5–10% (only at REF 10–20%) are situated at an elevation of 1000–1260 m a.s.l. and oriented to south or southeast. The soil type is a Dystric Cambisol on moraine. More detailed information about soil profiles (thickness of the layers, texture) was published by Mičuda et al. (2005). Above-ground humus layer on the study plots (except burnt site, where it was burnt down) is up to 10 cm thick; mor is the dominant humus form. Before windstorm, the experimental area was covered by an old-growth forest of Norway spruce (*Picea abies* (L.) Karst) with an admixture of larch (*Larix decidua* Mill.).

Soil Sampling and Analyses

Soil samples (3–10) were taken from the mineral A-horizon (depth of 0–10 cm) on each study plot in July 2006, October 2006 and April 2007. Sampling was done in the central part of each plot along a line at approximately 10 m intervals from different places at each date.

After coarse material and plant roots were removed, the samples were stored at 4°C prior to analyses. Soil moisture was determined gravimetrically by oven-drying fresh soil at 105°C overnight. Soil pH was measured in water and 1 M KCl suspension (20 g soil plus 50 ml water and KCl, respectively). The content of soil organic carbon (SOC) was determined using Tyurin's method.

Basal soil respiration (BR) was measured by estimating the amount of CO_2 evolved during incubation of soil in a closed container for 24 h (Alef 1991). CO_2 was trapped in a solution of 0.05 M NaOH and titrated against 0.05 M HCl using phenolphthalein as indicator. For substrate-induced respiration (SIR), soil samples were amended by 1% (w/fresh weight) glucose and CO_2 evolved was determined as described above after 5 h. We decided for this amount of glucose on the basis of preliminary tests with different glucose levels, where 1% glucose induced most consistent respiration rates. Soil microbial biomass carbon (Cmic) was determined according to Islam and Weil (1998). Microwave energy was applied to moist soil to disrupt microbial cells. The flush of C released was then measured using Tyurin's method. N-mineralization (Nmin) was determined using the laboratory anaerobic incubation procedure described by Kandeler (1993). Soil samples (5 g) under water-logged conditions were incubated at 40°C for 7 days to prevent nitrification, and $NH_4 - N$ was measured by a colorimetric procedure. Catalase activity (Acat) was determined by measuring the volume of oxygen liberated after incubation of 10 g soil with 3% H_2O_2 for 10 min according to the method of Khazijev (1976).

Within each treatment, three soil samples were randomly selected at each date for the study of the functional diversity of soil microorganism employing BIOLOG EcoPlates (BIOLOG Unc., Hayward, CA, cf. Garland 1996). Microtitration plates with 31 different carbon sources were inoculated with 150 μl of soil extract (in 0.85% NaCl) at 27°C for 8 days. In addition to substrates, wells contain tetrazolium-violet dye to be reduced to purple formazan if microorganisms utilize the substrates. The presence of functional groups of microorganisms, which means the ability to digest a particular substrate, was indicated by the development of violet colour. The optical density of the colour in a particular well was taken as a measure of the abundance of the respective functional group. Optical density was measured as absorbance using a multi-optical reader (SUNRAY TECAN) at 590 nm each day of the microplate incubation. Absorbance limit of 0.2 was considered as the evidence for the presence of the functional group of microorganisms metabolizing a particular substrate in the soil. Richness in functional groups was calculated simply as the number of substrates exhibiting metabolic activity in a particular soil sample. Diversity of functional groups was assessed by Hill's diversity indices (Hill 1973):

$$N_2 = 1/\Sigma p_i^2 \qquad (1)$$

where p_i is a frequency of the ith functional group determined on the basis of the absorbance in a particular well.

Statistical Analyses

All the results are expressed on the oven-dry basis. Statistical analyses were done using the statistical package SAS/STAT® (SAS 1988). Two-way covariance analysis was performed on the data to detect the variability of soil microbial activity indicators among treatments and sampling dates (date and treatment were considered factors with fixed effects, soil moisture was used as a continuously distributed covariate; as organic C and pH were collinear with moisture, they were not used as predictors; we did not suppose temporal autocorrelation at the scale of 3–6 months and sampling was not repeated at the same points within a plot, so that date was considered an independent factor rather than using a repeated-measure model). Two-way analysis of variance was done for soil chemical properties (soil moisture, organic C, pH). For pair-wise comparisons of means Duncan's tests were used (procedure GLM).

To identify the patterns of microbial community composition, principal component analysis based on the abundances of functional groups (measured by optical density for individual BIOLOG EcoPlate substrates) was performed employing the procedure PRINCOMP.

Results

SOC, Soil Acidity and Soil Water Content

The SOC content exhibited a high variability within treatments (Fig. 1a), especially at the reference site. The average SOC was similar for all treatments except reference site where the highest average SOC was observed.

Soil acidity was also similar on disturbed plots with average values 4.15–4.23 for pH – H_2O and 3.20–3.48 for pH-KCl, respectively. The reference site exhibited

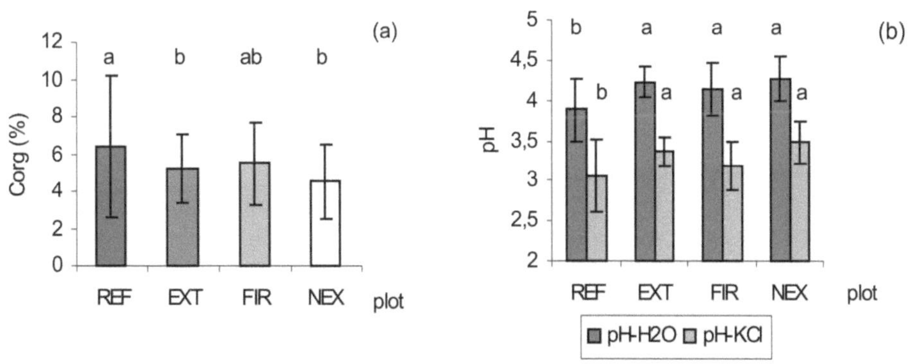

Fig. 1 (**a** and **b**) Mean values (±standard deviation) of the SOC and soil acidity at individual plots (Duncan's tests: the plots with the same letters do not differ significantly)

Table 1 Analysis of variance/covariance of soil variables (F-test)

Source	d.f.	F-value							
		Moisture	Respiration	SIR	Cmic	Nmin	Catalase	Richness	Diversity
Date	2	3.76*	7.80***	24.04***	1.49	9.50***	1.85	12.48***	8.94***
Treatment	3	1.50	12.25***	3.99*	4.82**	4.96**	7.33***	2.85	3.30*
Date × Treatment	6	7.16***	20.47***	5.18***	1.35	2.30*	11.06***	0.69	0.68
Moisture	1		30.79***	10.69**	3.18	18.46***	17.88***	0.26	0.72

*** $P > 0.999$, ** $P > 0.99$, * $P > 0.95$, d.f. – degrees of freedom

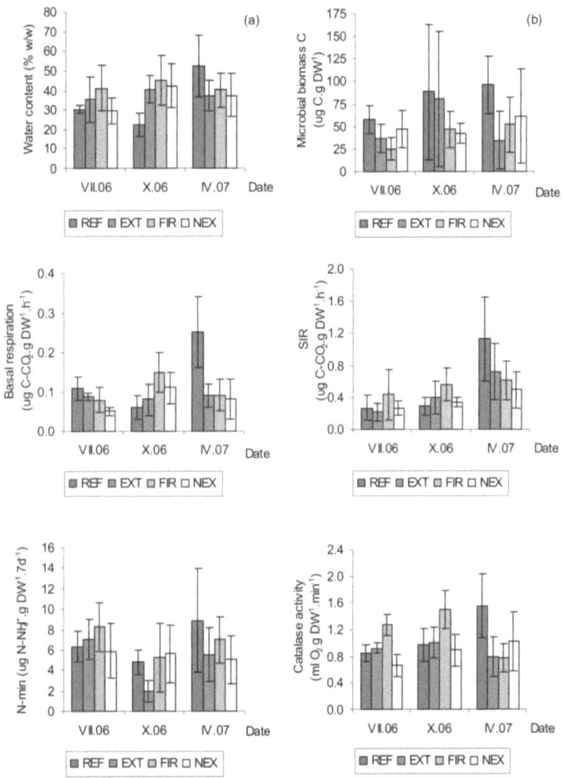

Fig. 2 (**a, b, c, d, e** and **f**) Mean values (±standard deviation) of the soil water content, soil microbial biomass carbon, basal respiration, SIR, N-mineralization and catalase activity at individual plots in different sampling dates

significantly higher acidity (pH – H_2O, 3.88; pH-KCl, 3.07) than the other treatments (Fig. 1b).

The gravimetric water content in the 0–10 cm of soil profile showed similar patterns in July and October 2006 (Fig. 2a). The lowest water content average was found at the reference site, the highest value at the burnt plot. In April 2007, however, the highest average water content was observed at the reference site. For the other treatments, the trend was very similar with highest value at the burnt site. At all plots, except reference site, average soil water content was higher in autumn than at other sampling dates as opposed to the reference site, where the lowest average water content was found in October 2006 and the highest value in the spring 2007. Generally, no significant differences were found in soil water content between plots (Tables 1 and 2).

Table 2 Duncan's pair-wise tests of the differences of means among treatments

Plot	Moisture	Respiration	SIR	Cmic	Nmin	Catalase
REF	35.99 a	0.15 a	0.65 a	87.80 a	6.75 a	1.21 a
EXT	38.42 a	0.09 b	0.51 ab	54.49 b	4.18 b	0.90 b
FIR	42.36 a	0.11 b	0.57 a	46.02 b	6.40 a	1.16 a
NEX	38.40 a	0.09 b	0.40 b	51.53 b	5.39 ab	0.92 b

Note: The treatments with the same letters do not differ significantly ($P > 0.95$)
Units used: moisture (% w/w), respiration (μg C – CO_2 gDW^{-1} h^{-1}), SIR (μg C – CO_2 gDW^{-1} h^{-1}), Cmic (μg C gDW^{-1}), Nmin (μg N – NH_4^+ gDW^{-1} $7day^{-1}$), catalase (ml O_2 gDW^{-1} min^{-1})

Microbial Biomass and Activity

The microbial biomass carbon exhibited a very high variability within treatments (Fig. 2b), especially in autumn 2006 and spring 2007. Cmic at the reference site (87.80 μg Cg^{-1}) was significantly higher (Table 2) compared to the other plots (46.02 – 54.49 μg C g^{-1}). In summer Cmic was mostly lower than in autumn and spring (except the extracted plot), however, the differences in averages between dates were not significant (Tables 1 and 3).

Basal respiration shows the same pattern in the summer 2006 and the spring 2007 with the significantly highest value at the reference site and the lowest one at the non-extracted site (Fig. 2c). In autumn, the situation was completely different, when the highest basal respiration was found at the FIR plot and the lowest at the REF site (Table 2). During our observations, basal

Table 3 Duncan's pair-wise tests of the differences of means among sampling dates

Date	Moisture	Respiration	SIR	Cmic	Nmin	Catalase
VII.06	33.27 b	0.08 b	0.29833 b	41.63 a	6.88 a	0.93 a
X.06	37.41 ab	0.10 b	0.39762 b	64.97 a	4.37 b	1.10 a
V.07	41.81 a	0.13 a	0.73903 a	60.78 a	6.61 a	1.04 a

Note: The dates with the same letters do not differ significantly ($P > 0.95$).
Units used: Moisture (% w/w), respiration ($\mu g\ C - CO_2\ gDW^{-1}.h^{-1}$), SIR ($\mu g\ C - CO_2\ gDW^{-1}\ h^{-1}$), Cmic ($\mu g\ C\ gDW^{-1}$), Nmin ($\mu g\ N - NH_4^+\ gDW^{-1}\ 7day^{-1}$), catalase (ml $O_2\ gDW^{-1}\ min^{-1}$).

respiration was significantly higher in the spring than in the summer and the autumn (Table 3).

SIR shows a similar trend in the autumn and the spring (Fig. 2d). Significant differences in SIR were found between treatments and also between sampling dates (Table 1). Significantly the lowest SIR was observed at the NEX site, the highest at the REF and FIR plots (Table 2).

N-mineralization belongs to parameters with considerable within-treatment variation (Fig. 2e). Significantly the lowest N-mineralization ($4.18\ \mu g\ N - NH_4^+\ gDW^{-1}\ 7day^{-1}$) was observed on the EXT and NEX plots (Table 2). In the summer 2006 and the spring 2007 the values were significantly higher compared to the autumn 2006 (Table 3).

As shown in Fig. 2f, similar trend was found for catalase activity with the highest value at the FIR plot and the lowest at the NEX plot in the summer and autumn 2006. However, generally the differences between dates were not significant (Table 1). At the REF and FIR plots, significantly the highest catalase activity was observed (1.21 and 1.16 ml O_2 $gDW^{-1}min^{-1}$, respectively) (Table 2).

Microbial Diversity

The analysis of variance of the richness and diversity of functional groups of microbes has shown that there are significant differences among dates (Table 1). Significantly the lowest richness was observed in the summer 2006, whereas more groups were active during the autumn 2006 and spring 2007. The diversity showed the same temporal pattern, the increase at later dates seems to be caused by increased richness rather than evenness. The plots where the trees were not extracted generally exhibited significantly higher diversity of functional groups than the others (Fig. 3a, b).

Principal component analysis of the composition of the microbial community has shown that there is no uniform response of functional group composition to the management regime (Fig. 4). On the other hand, there seems to be a temporal shift in the representation of functional groups, although the pattern is not completely consistent among plots. The composition of the microbial community was initially quite uniform at all plots. With time, it shifted in quite uniform direction and partially diverged. The clearest temporal shift was observed at the non-extracted plot. At the burnt and extracted plots, the unidirectional shift in community

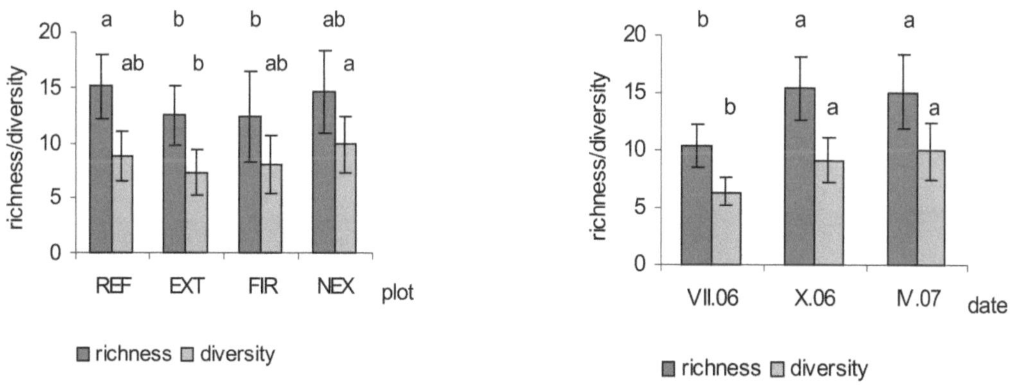

Fig. 3 (**a** and **b**) Mean values (±standard deviation) of the richness and diversity of functional groups at individual plots and at different sampling dates (Duncan's tests: the plots with the same letters do not differ significantly)

composition is still visible, although much less pronounced. On the other hand, the least changes seem to have occurred in the old-growth non-disturbed forest.

Discussion

Soil microorganisms compared to higher organisms have a high surface-to-volume ratio, which means a much more intense exchange of matter and energy with their environment. Therefore, soil microbiota tends to respond quickly to environmental stress. Changes in size, composition and activity of microbial communities can frequently be observed before detectable changes in soil physical and chemical properties occur (Nielsen and Winding 2002).

We suppose that after windthrow and fire in the TANAP environmental conditions at the affected area have changed in various ways due to the different amount and deployment of organic material on the ground. At the extracted and burnt plots fallen stems were logged; at the burnt plot also humus layer was completely destroyed. At the non-extracted plot fallen trees with all debris were left on the ground, so the most of soil surface is shaded.

Analysis of soil samples showed that disturbed plots did not differ in organic carbon content and soil acidity; however, the reference site compared to disturbed plots had higher amount of SOC and was more acid. Our measurements were performed quite shortly after disturbances and no older records are available for the investigated sites. The time elapsed is too short to expect significant changes in chemical soil properties between plots. Probably, differences between the old-growth forest and disturbed plots have already existed before the disturbance. However, it was not possible to find another non-affected stand with similar conditions near to disturbed plots.

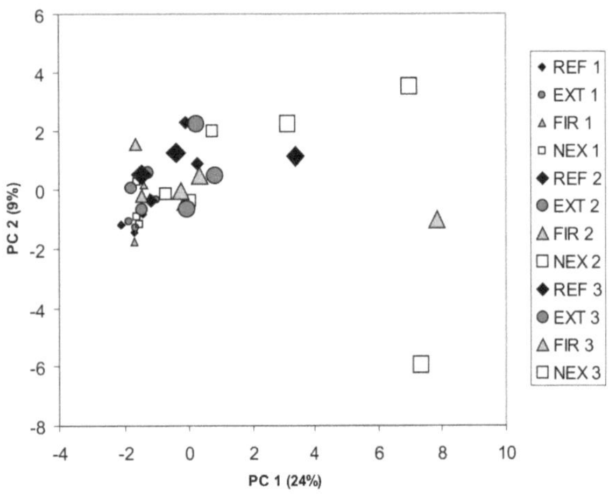

Fig. 4 Principal component analysis based on the abundances of functional groups of microorganisms at the investigated sites. The sizes of symbols corresponds to sampling dates

There were no significant differences in soil water content between treatments. The different pattern of soil moisture distribution between plots at various times of year can be explained by transpiration and interception of standing trees and different evaporation at the open area. Under forest canopy, higher accumulation of water after spring snowmelt in soil is usually observed as a consequence of a better infiltration. During the vegetation period the losses of water through transpiration of mature trees can be considerable and may represent more than 30% of total precipitation. Interception in spruce stands can account also up to 34–38% of the total precipitation. On the other hand, at the open area, the evaporation is by 50% higher than under the forest canopy. Soil moisture in the topsoil at the open area might reflect the changes due to the coming precipitation to soil surface earlier than under a forest stand (Petrík et al. 1986, Pichler 2006, Pichler et al. 2006). As a consequence, a different annual course of soil moisture is observed in a forest compared to an open area.

At our plots, all microbial characteristics exhibited high spatial variability, especially microbial biomass C- and N-mineralization. Hargreaves et al. (2003) found that the spatial variation in chemical determinants, for example pH and exchangeable cations, were much smaller than that of microbial biomass C. In our previous studies, microbial characteristics were also much variable in space than, for example soil water content, SOC or soil acidity (Gömöryová 2004, Gömöryová et al. 2006). This is not surprising because soil contains many microhabitats with suitable conditions for survival and reproduction of soil microbes. Bacteria and fungi may be spatially aggregated in soil at various scales, forming hotspots of microbial activity.

The present study has generally shown that soil microbial biomass and activity decreased at disturbed plots, except burnt site, where potential respiration, N-mineralization and catalase activity were significantly higher than at the other disturbed plots. No significant differences in soil microbial characteristics were observed between extracted and non-extracted site.

There are only few studies about the changes of soil microorganisms biomass or activity in windthrow area with different management (Wright and Coleman 2002). The conditions at the extracted plot can partially be compared to clear-cut plots, where all stems are logged as well. Much more observations are from the plots affected by wildfire or prescribed burning. Microbial biomass most often decreases as a result of these disturbances (especially fire) and this effect can last for many years. Mabuhay et al. (2006) found that 1 week after fire microbial biomass carbon represented only 3–5% of the average biomass carbon of the unburnt area. Even 13 months after the occurrence of fire, the biomass carbon was still around 36–50% of the biomass in the unburnt area. Smith et al. (2007) demonstrated that the heat of wildfire had a direct effect on soil organisms, killing a large part of microbes, but they cannot exclude other, indirect causes of the reduction of microbial biomass, such as changes in nutrient supply due to loss of plant cover. They also found that higher mortality of soil organisms occurred with greater soil moisture compared to dry conditions at the same temperature. On the other hand, Boyle et al. (2005) observed that tree removal alone had only a modest impact on soil microbial community 8 years after treatment. Wright and Coleman (2002) compared pre- and post-disturbance amounts of soil microbial biomass carbon and they did not find significant differences or patterns in response to these disturbance events. Litter decomposition and soil respiration decreased after disturbances, but microbial biomass was probably not affected. On the other hand, soil temperature, soil water content and soil structure are likely directly affected by these events. They found that the extracted-plot soil respiration was significantly higher than at the non-extracted plot; probably as a consequence of the increased solar radiation reaching soil surface, increased soil temperatures and more extreme soil wet/dry cycles. Pietikainen and Fritze (1995) assumed that the absence of litter fall should be compensated by the logging residues on the forest floor and the decomposition of dead roots in soil. Therefore the initial effect of clear-cutting can be beneficial to microorganisms. Heat during fire has a sterilizing effect on soil but later soil is gradually recolonized from propagules surviving in soil or introduced from neighbour areas.

At our plots, soil microoganisms responded on disturbance events by reducing their biomass and activity. Surprisingly, the fire plot exhibited relatively high microbial activity, while soil microbial biomass carbon was low. Andersson et al. (2004) stated that changes in microbial activity due to environmental changes can be independent of possible changes in microbial biomass. The fire caused an increase in microbial activity and soil respiration can be explained by higher pH, N availability, amount of readily decomposable organic material in soil and temperature, while the decrease of soil respiration was attributed to low soil moisture. In our case the absence of covering vegetation and especially the lack of a thick above-ground humus layer result likely in a broader variation in soil temperature and moisture, which may have either negative or also positive effects on microbial activity. Due to burning of humus layer, also release of mineral nutrients into the mineral horizon and its enrichment can be expected, resulting in a higher microbial activity.

We expected that soil microorganisms at the EXT and NEX plots will respond to different management in different ways. However, there were practically no differences in microbial biomass and activity between them, in concordance with what Wright and Coleman (2002) found. Attiwill and Adams (1993) also demonstrated that forest soils are resistant to major changes in patterns of nitrogen mineralization following disturbance by natural events such as windthrow and fire. We collected soil samples from the mineral horizon, because at the fire plot the humus layer was destroyed. It is possible, that the differences between the EXT and NEX plots exist mainly in the top litter sub-horizon, whereas for the mineral A-horizon, the time after disturbances has been too short to develop significant changes in microbial activity.

Microbial communities provide useful data for studying environmental events. Molecular methods, PLFA analysis, carbon source utilization methods and plate counts are used to study microbial diversity on plots affected by disturbances (Bååth et al. 1995, Staddon et al. 1998, Kang and Mills 2004, Mabuhay et al. 2006, Smith et al. 2007). Especially the study of functional diversity of microbial communities has been found to be very sensitive to environmental changes (Zak et al. 1994), therefore has been evaluated as early indicator of the impacts of management on soil biological properties (Bending et al. 2000).

We found that plots where fallen trees have been removed exhibited generally lower richness and diversity of functional groups than the others. At the extracted plot, one or two herb species predominated, and at the burnt plot, vegetation was also very poor, so a lower functional richness and diversity of soil microbes on these plots can be associated just with lower diversity of organic substrates available for decomposition, originating from root exudates, dead biomass, etc. However, there was no uniform temporal pattern of response of functional group composition to the management regime. It seems that at the reference plot microbial communities did not changed in such extent as at disturbed plots. There are many studies dealing with temporal variation in soil microbial properties, demonstrating seasonal cycles of changes in community structure and activity that are related to climate and vegetation factors. For example, Papatheodorou et al. (2004) found that bacterial diversity, richness and evenness as well as the mean oxidation of almost all BIOLOG substrate groups, were affected significantly by the seasonal fluctuations of soil temperature and moisture and that the diversity decreased from summer to winter on their study plots. Kang and Mills (2004) observed clear temporal trends in community structure throughout the 2-years study that coincided with the development of grasses in the field.

The plots affected by fire show usually lower microbial diversity than intact plots (Mabuhay et al. 2006, Smith et al. 2007) and these changes may last for 5 years after burning (Staddon et al. 1998). Fire alters the soil microbial community structure in the short-term through heat-induced microbial mortality. Over the long-term, fire may modify soil communities by altering plant community composition through plant-induced changes in the soil environment (Hart et al. 2005). Staddon et al. (1998) found that no differences between treatments (clear-cut and burnt plots) were detected when individual samples were analysed. However, when pooled samples from organic and mineral horizons were used, lower microbial diversity was observed at the burnt plot. Many studies showed that there is a large variability of the individual variables even at homogeneous plots as a consequence of heterogeneous environments at the microbial level (Staddon et al. 1998, Clegg et al. 2000, Smith et al. 2007). Smith et al. (2007) using molecular methods found that changes in bacterial community composition resulting from various treatments were evident; a greater impact of fire than of harvesting on soil microbial community was observed. Kang and Mills (2004) suppose that soil bacterial communities may more rapidly go through successional series than communities of other organisms, because they can modify the environment to more favourable conditions for other groups of microorganisms and for higher organisms. Development of a diverse plant community might be expected to provide the soil microbes with a broader selection of organic energy sources, resulting in a more diverse bacterial community.

Our study revealed a temporal shift in the representation of functional groups. A short observation period does not allow to distinguish, whether the development of microbial community composition is cyclic or unidirectional. Nevertheless, ecological conditions change in different ways and to different extent depending on the type of disturbance, provoking different trajectories of the secondary succession of plant communities. Future observations could reveal how the changes of ground vegetation will be reflected in the richness and diversity of functional groups of microorganisms.

Acknowledgments Technical assistance of Ž. Brnáková and V. Kriššák is heartily appreciated. This study was supported by the grant no. APVV-0468-06, APVV-0022-07, APVV-0456-07, VEGA 1/0703/08 and 1/0515/08.

References

Alef K (1991) Methodenhandbuch Bodenmikrobiologie. Aktivitäten, Biomasse, Differenzierung. Ecomed, Landesberg

Andersson M, Michelsen A, Jensen M, Kjøller A (2004) Tropical savannah woodland: effects of experimental fire on soil microorganisms and soil emissions of carbon dioxide. Soil Biology and Biochemistry, 36: 849–858

Anonymous (2007) Concluding message from the International Scientific Conference on Bioclimatology and Natural Hazards. BioClimatology and Natural Hazards Proceedings, Zvolen and Pol'ana, 17.–20. September 2007

Attiwill PM, Adams MA (1993) Nutrient cycling in forest. New Phytologist 124:561–582

Bååth E, Frostegård Å, Pennanen T, Fritze H (1995) Microbial community structure and pH responses in relation to soil organic matter quality in wood-ash fertilized, clear-cut or burnt coniferous forest soils. Soil Biology and Biochemistry 27:229–240

Bending GD, Putland C, Rayns F (2000) Changes in microbial community metabolism and labile organic matter fractions as early indicators of the impact of management on soil biological quality. Biology and Fertility of Soils 31:78–84

Boyle SI, Hart SC, Kaye JP, Waldrop MP (2005) Restoration and canopy type influence soil microflora in a ponderosa pine forest. Soil Science Society of America Journal, 69:1627–1638

Certini G (2005) Effects of fire on properties of forest soils: Oecologia, 143: 1–10

Clegg CD, Ritz K, Griffiths B S (2000) %G + C profiling and cross hybridisation of microbial DNA reveals great variation in below-ground community structure in UK upland grasslands. Applied Soil Ecology 14:125–134

Fioretto A, Papa S, Pellgrino A (2005) Effects of fire on soil respiration, ATP content and enzyme activities in Mediterranean maquis. Applied Vegetation Science, 8:13–20

Fleischer P, Giorgi S, Miglieta F, Schulze D, Valentini R (2007) Large-scale forest destruction by November 2004 windstorm in the Tatra Mts – reasons, consequences and ecological research. In: Střelcová K, Škvarenina J, Blaženec M (Eds).: BioClimatology and Natural Hazards Proceedings, Poľana: 56

Garland JL (1996) Analysis and interpretation of community-level physiological profiles in microbial ecology. FEMS Microbial Ecology, 24:289–300

Gömöryová E (2004) Small-scale variation of microbial activities in a forest soil under a beech (*Fagus sylvatica* L.) stand. Polish Journal of Ecology, 52:311–321

Gömöryová E, Gregor J, Pichler V, Gömöry D (2006) Spatial patterns of soil microbial characteristics and soil moisture in a natural beech forest. Biologia, Bratislava, 61/Suppl. 19:329–333

Guerrero C, Mataix-Solera J, Gomez I, Garcia-Orenes F, Jordan MM (2005) Microbial recolonization and chemical changes in soil heated at different temperatures. International Journal of wildland fire, 14:385–400

Hargreaves PR, Brookes PC, Ross GJS, Poultin PR (2003) Evaluating soil microbial biomass carbon as an indicator of long-term environmental change. Soil Biology and Biochemistry, 35:401–407

Hart SC, DeLuca TH, Newman GS, MacKenzie MD, Boyle SI (2005) Post-fire vegetative dynamics as drivers of microbial community structure and function in forest soils. Forest Ecology and Management, 220(1–3):166–184

Hill MO (1973) Diversity and evenness: Aunifying notation and its consequences, Ecology 54:427–432

Islam KR, Weil RR (1998) Microwave irradiation of soil for routine measurements of microbial biomass carbon. Biology and Fertility of Soils, 27:408–416

Kandeler E (1993) Bestimmung der N-Mineralisation im anaeroben Brutversuch. S. 160–161. In: Schinner F, Öhlinger R, Kandeler E, Margesin R (Eds).: Bodenbiologische Arbeitsmethoden. Springer Verlag, Berlin, Heidelberg, New York

Kang S, Mills AL (2004) Soil bacterial community structure changes following disturbance of the overlying plant community. Soil Science, 169:55–65

Khazijev FCh (1976) Fermentativnaja aktivnost' počv. Metodičeskoje Posobje, Moskva, (in Russian)

Mabuhay JA, Nakagoshi N, Isagi Y (2006) Soil microbial biomass, abundance and diversity in a Japanese red pine forest: first year after fire. Journal of Forest Research, 11(3): 165–173

Mičuda R, Šimonovičová A, Ďuriš M, Šimkovič L, Lancuch P, Hanajík P, Dlapa P (2005): Soil-ecological characteristics of localities and evaluation of some soil properties in the High Tatra Mountains after windthrow. Phytopedon 4:12–18

Neary DG, Klopate CC, DeBano LF, Ffolliot PF (1999) Fire effects on belowground sustainability. Forest Ecology and Management, 122:51–71

Nielsen MN, Winding A (2002) Microorganisms as Indicators of Soil Health. National Environmental Research Institute, Roskilde, Denmark, Technical Report No. 388

Papatheodorou EM, Argyropoulou MD, Stamou GP (2004) The effects of large- and small-scale differences in soil temperature and moisture on bacterial functional diversity and the community of bacterial nematodes. Applied Soil Ecology, 25:37–49

Petrík M, Havlíček V, Uhrecký I (1986) Forest bioclimatology. Príroda, Bratislava (in Slovak and Czech)

Pichler V (2006): Beech-stand density as a tool for the regulation of soils hydric and environmental functions. Technical University in Zvolen, 51pp (in Slovak)

Pichler V, Gregor J, Vál'ka J, Capuliak J, Homolák M (2006) Forest stand (*Fagus silvatica*, L.) density reduction as a GCC – adaptive forestry tool. Forestry Journal, 52:89–97

Pietikainen J, Fritze H (1995) Clear-cutting and prescribed burning in coniferous forest: comparison of effects on soil fungal and total microbial biomass, respiratory activity and nitrification. Soil Biology and Biochemistry, 27:101–109

SAS (1988): SAS/STAT® User's guide, Release 6.03 Edition. SAS Institute, Cary

Smith NR, Kishchuk BE, Mohn WW (2007) Effects of wildfire and harvest disturbances on forest soil bacterial community. Applied and Environmental Microbiology, 74:216–214

Staddon WJ, Duchesne LC, Trevores JT (1998) Impact of clear-cutting and prescribed burning on microbial diversity and community structure in a Jack pine (*Pinus banksiana* Lamb.) clear-cut using Biolog Gram-negative microplates. World Journal of Microbiology and Biotechnology, 14: 119–123

Thürig E, Palosuo T, Bucher J, Kaufmann E (2005) The impact of windthrow on carbon sequestration in Switzerland: a model-based assessment. Forest Ecology and Management, 210: 337–350

Wright CJ, Coleman DC (2002) Responses of soil microbial biomass, nematode trophic groups, N-mineralization and litter decomposition to disturbance events in the southern Appalachians. Soil Biology and Biochemistry, 34: 13–25

Zak JC, Willig MR, Moorhead DL, Wildman HG (1994) Functional diversity of microbial communities: a quantitative approach. Soil Biology and Biochemistry, 26: 1101–1108

Capacities of Modelling to Assess Buffer Strip Efficiency to Reduce Soil Loss During Heavy Rainfall Events

M. Kändler, I. Bärlund, M. Puustinen and C. Seidler

Keywords Erosion · Modelling · Buffer strips · ICECREAM · Erosion2D

Introduction

The loss of soil material and nutrients from agricultural land to waters pose on the one hand a problem to farmers in form of soil degradation and on the other hand to surface water quality by enhancing the eutrophication process. Efforts undertaken, both concerning conservation practices and policy measures, show for example for the U.S. (Uri 2001) that erosion was reduced by 32% from 1982 to 1992, but the costs for the remaining efforts are still estimated to be up to $37.6 billion annually. One of the measures to abate off-site damage caused by erosion is to introduce buffers strips or zones on the edge of fields to trap soil material detached by rain and transported by surface runoff. The effectiveness of these buffer strips to reduce soil loss from fields will be an important part of management practice evaluation required by the Water Framework Directive to reach good ecological conditions of surface water bodies. Many studies have dealt with construction and quality of buffer strips/zones (for summary see, e.g. Correll 2005) and their modelling (e.g. Muñoz-Carpena et al. 1999, Lowrance et al. 2000) but the dynamic relationship between the field output above the buffer strip and the buffer strip efficiency to trap suspended sediments remains a topic where little modelling effort has taken place. The effect of buffer strip length was tested for fields with uniform slope less than 3% by Rankinen et al. (2001) using the ICECREAM model. The reduction varied between 33% and 91% for winter wheat but the length of the strip did not play a major role on these low slopes. This interannual variability was mainly appointed to different hydrological conditions. In a recent study the erosion model Erosion3D was applied to a small catchment in Saxony and the efficiency of a grassed waterway was estimated to be 22.6% (Schob et al. 2006).

In this study the buffer strip was located at the lower edge of the field and is characterised by grass vegetation. The current buffer strip description of two field scale erosion models was tested against measured data from a Finnish research field at the river Aurajoki (southwest Finland, Fig. 1). The calibration strategy of the ICECREAM (Tattari et al. 2001) and Erosion2D (Schmidt et al. 1996) models to simulate buffer strip efficiency and to evaluate their potential with regard to water protection requirements is presented. In ICECREAM the buffer strip consists of perennial grass, in Erosion2D the grass cover of the buffer strip is characterized by its influence on surface runoff and erosion processes. Special emphasis was laid on the effect of field slope geometry and buffer strip length on buffer strip efficiency.

Material and Methods

Investigations on the influence of different management practices (conventional vs. conserving cultivation) on surface runoff, soil erosion and nutrient transport were carried out on an experimental field in southwest Finland at the River Aurajoki in the years

M. Kändler (✉)
International Graduate School Zittau, Markt 23,
D-02763 Zittau, Germany
e-mail: kaendler@ihi-zittau.de

K. Strelcová et al. (eds.), *Bioclimatology and Natural Hazards*,
DOI 10.1007/978-1-4020-8876-6_22,

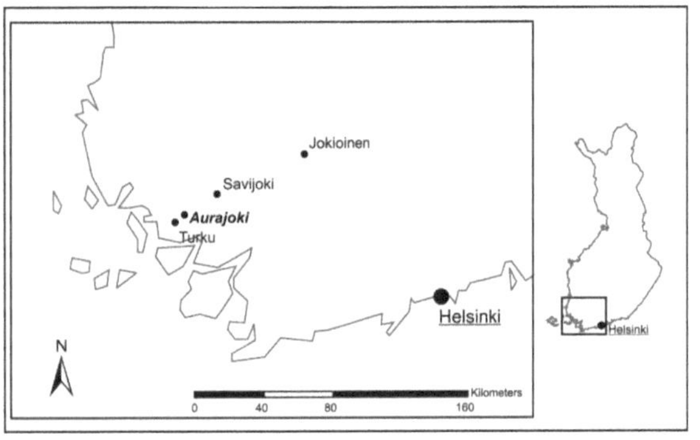

Fig. 1 Location of the experimental field (Aurajoki) and the climatic stations (Savijoki, Jokioinen, Turku)

1990–2002 (Puustinen et al. 2005). The field (Fig. 2) was divided into 12 plots, 9 with a length of 51 m and 3 with a length of 36 m. The width of all plots was 9 m, the average slope ca. 7% (Puustinen 1994). The soil type can be classified as heavy clay with over 60% clay and ca. 35% silt. Real precipitation events from Turku Airport were used for November 2000 (Fig. 3b). Due to the measurement method at the Aurajoki experimental field several runoff events were recorded as one. For modelling this meant constructing

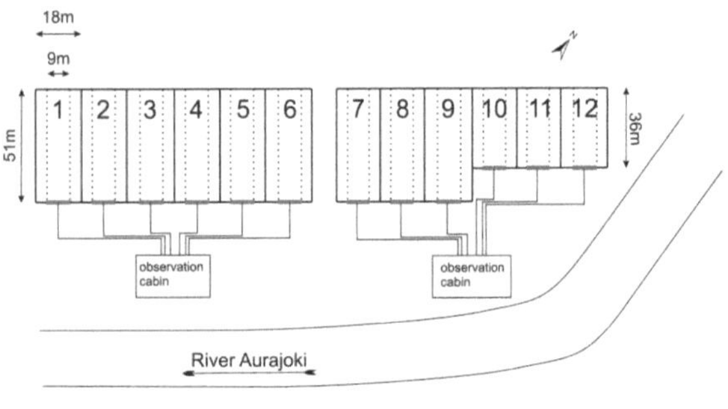

Fig. 2 The plan view of the Aurajoki experimental field

the sum of simulated daily values for these periods. In November 2000, these periods were 4–5.11 (sampled 6.11.), 6–12.11. (sampled 13.11.), and 18–19.11. (sampled 20.11.). The precipitation event in May 2000 (Fig. 3a) was constructed using a real 1-day event from May 2007 in Saxony, but it was added to a dry period in the Turku Airport climate data. This event was used for the hypothetical slope geometry study only. The different soil moisture conditions in May and November were considered in the models: in ICECREAM by the continuously simulated soil moisture content and in Erosion2D by adjusting the initial soil moisture content.

Erosion2D is an event-related model that needs precipitation values in a 10 min time-step. At the Aurajoki test field no precipitation was measured, that is why the precipitation intensity was calculated from daily values using the breakpoint method. Data from a pluviograph in the Savijoki catchment (Fig. 1) was digitalised by hand to get intensity values. These values were used to create the necessary precipitation intensity from daily values measured at Turku airport.

Two different models were used. The model Erosion2D is an event-related, physically based model for simulation of surface runoff and erosion on field slopes. A parameter catalogue (PK) has been developed (Michael et al. 1996) for practical model applications. The infiltration model has been developed by Schmidt (1993) and is based on precipitation data on a 10 min time-step. Particle detachment is described by an approach of physical forces that on the one hand hold the soil particles together (adhesion, coherence, friction, gravity) and on the other hand separate them (impact of rainfall and surface runoff) (Schmidt et al. 1996). The field scale nutrient transport model ICECREAM is likewise physically based, but it is

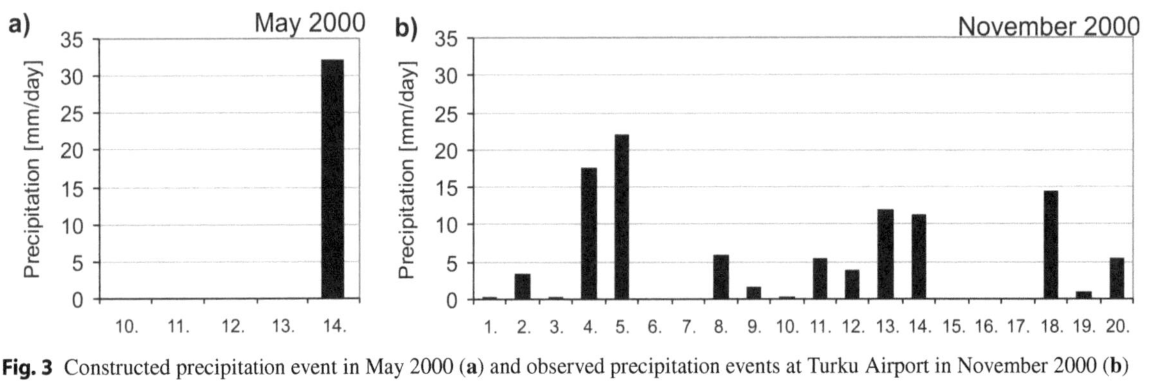

Fig. 3 Constructed precipitation event in May 2000 (**a**) and observed precipitation events at Turku Airport in November 2000 (**b**)

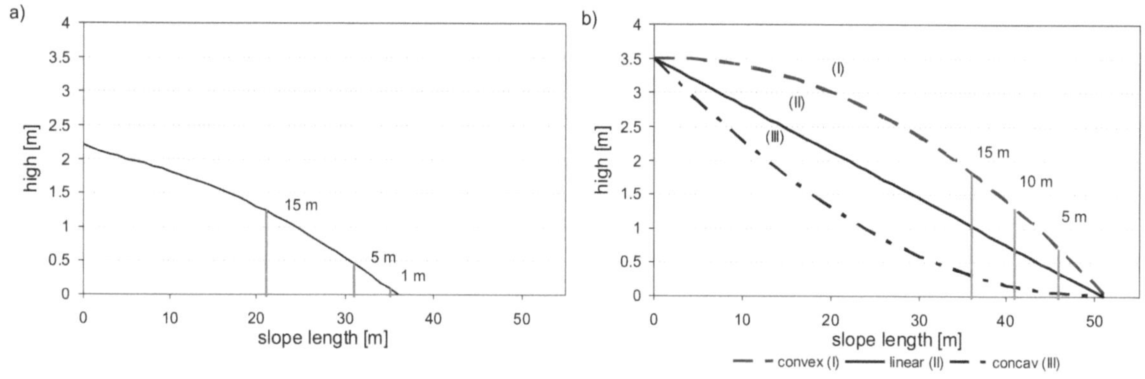

Fig. 4 The original slope geometry of the Aurajoki research field plots 10 and 12 and the buffer strip options (**a**) and the constructed slopes on plot 3 and the buffer strip options (**b**)

characterised by empirical formulation of several processes. It is based on the CREAMS and GLEAMS models developed in the U.S., but has been further developed to suit better the Finnish environmental conditions (Posch and Rekolainen 1993, Tattari et al. 2001). The ICECREAM model operates on daily, long-term basis and computes surface runoff using the SCS Curve Number method and a modified USLE for soil erosion.

For the parameterisation and calibration exercise we used data from the plots 12 (grass) and 10 (winter wheat) for November 2000. The first buffer strip exercise (buffer strip lengths 1, 5 and 15 m) concerning the testing of buffer strip length impact was thus conducted on the plot length 36 m (Fig. 4a). This was based on the calibration against measured data. The slope geometry exercise with a concave, convex and linear slope was constructed using the average slope of the Aurajoki experimental field plot 3 (plot length 51 m, Fig. 4b). The buffer strip length which was used here was 5, 10 and 15 m.

The Erosion2D model was calibrated for each of the four individual events in November 2000 (for winter wheat and grass separately) on the basis of the measured values for surface runoff and soil erosion. The calibrated variables were the *Skin factor* (surface runoff) and the *Erosion resistance* (soil erosion) with given *Surface roughness* (Tables 1 and 2). The ICECREAM model cannot be calibrated on event basis. The precipitation intensity parameter P1 in the erosion equation, the roughness coefficient (Manning's *n*, a parameter to which, e.g. sediment yield in the model REMM, Riparian Ecosystem Management Model (Graff et al. 2005) was found to be most sensitive) and the SCS Curve Number (CN2) were calibrated on annual basis.

Results

The result consists of three parts: first, a sensitivity analysis was performed for the initial soil moisture content in Erosion2D; second, the calibration result is presented as well as the simulated effect of a buffer strip on this real field slope and finally, two erosion events connected to heavy rainfall events and the corresponding effect of buffer strips on three constructed slopes of varying geometry (concave, linear and convex) are depicted.

Due to the fact that no soil moisture measurements were taken at the experimental site, this value could only be estimated. In order to explore the influence of

Table 1 Parameter values for the buffer strip (grass) in Erosion2D (derived from measured values); PK: derived from the parameter catalogue values of Erosion2D for buffer strips; *: estimated

	06.11.00	13.11.00	17.11.00	20.11.00	PK
Erosion resistance (kg m s^{-2})	0.01030	0.00143	0.00410	0.00170	1.00
Surface roughness (s $m^{-1/3}$)	0.032	0.032	0.032	0.032	0.300
Initial soil moisture* (Vol.–%)	44.5	44.5	45.0	45.5	45.5
Degree of coverage (%)	95	95	95	95	93
Skin factor (–)	1.1	1.0	1.4	3.2	10

Table 2 Parameter values for winter wheat in Erosion2D (derived from measured values); *: estimated, **: not utilised, for comparison only

	06.11.00	13.11.00	17.11.00	20.11.00	PK**
Erosion resistance (kg m s^{-2})	0.01000	0.00140	0.00370	0.00160	0.00417
Surface roughness (s m$^{-1/3}$)	0.025	0.031	0.028	0.030	0.015
Initial soil moisture* (Vol.–%)	44.5	44.5	45.0	45.5	
Degree of coverage (%)	5	5	5	5	4
Skin factor (–)	1.1	1.0	1.4	3.2	2.5

the initial soil moisture content on the simulation result a sensitivity analysis was performed (Kändler 2006) using the parameterisation (bulk density, particle size distribution) of plot 3 (Fig. 2). The initial soil moisture content has a huge influence on both surface runoff and soil loss (Fig. 5). Changing the value by 0.5% (38.5–39.0%) can change the result by a factor of 4 in one case but a change of 3% (41.0–44.0%) has nearly no influence at all. Due to this, we decided to use initial soil moisture values from the upper region for further investigations.

The calibration result shows that Erosion2D can be calibrated to reproduce the measured surface runoff events in November 2000 for both crop types (Fig. 6), whereas ICECREAM tends to underestimate some of the events for grass (Fig. 6a).

ICECREAM could not be calibrated to the measured soil loss values for grass (Fig. 7a), it underestimated the daily measured values heavily. In this model it is not possible to change the grass properties due to single events, the erosion resistance depends on crop growth status. In order to calibrate Erosion2D, however, the *Skin Factor* and *Erosion resistance* had to be substantially modified if compared to the parameter catalogue (Table 1). The *Skin factor* is used for describing macroporous flow in the model. With this factor it is possible to calibrate the surface runoff (higher values reduce surface runoff). The *Erosion resistance* specifies the resistance of the soil against particle displacement. For winter wheat the result is more variable (Fig. 7b), here both models have difficulties in depicting all four events.

The parameters gained from the calibration of the grass plot, served for the description of the buffer strip in the models. Their effect on surface runoff and soil loss was compared with a field growing winter wheat without a buffer strip (Fig. 8). For Erosion2D both calibrated values from the grass field and parameter catalogue values (PK) were utilised to characterise the buffer strip.

The buffer strip has a larger influence on soil loss than on surface runoff for both models. The parameters for grass that were derived from the measured values, do not have influence on the surface runoff in Erosion2D, since they have the same value for the correction factor and initial soil moisture content as for winter wheat. The values of the PK for Erosion2D have a large influence on the computed surface runoff. However, it was not possible to reproduce the measured data with these values. According to the ICECREAM model the

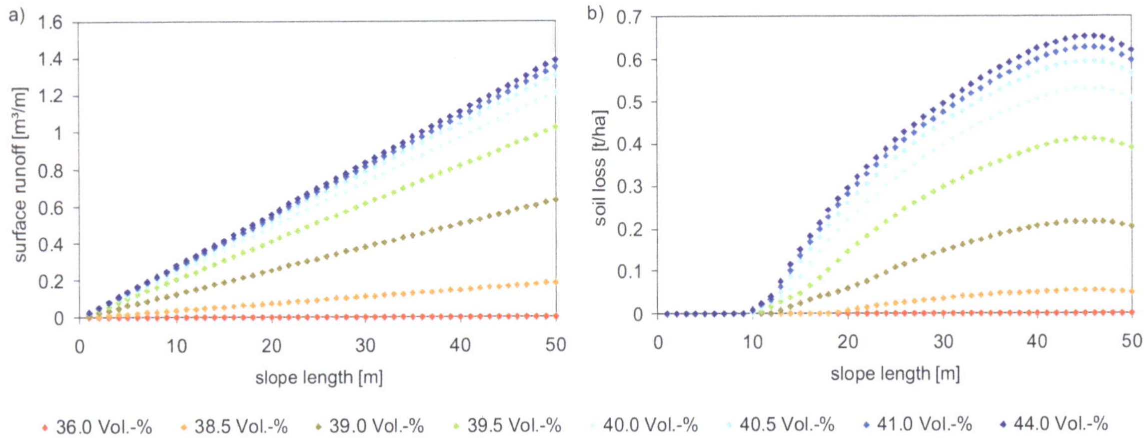

Fig. 5 The effect of initial soil moisture content on surface runoff (**a**) and soil loss (**b**) for plot 3 using Erosion2D

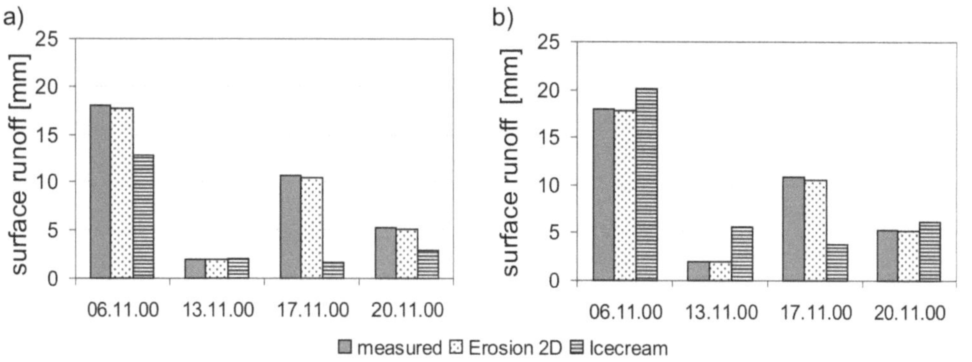

Fig. 6 Measured and calibrated surface runoff for the grass (**a**) and winter wheat plots (**b**) for both models in November 2000

surface runoff would reduce up to 33% using a buffer strip. Soil loss reduces due to a 1 m wide buffer strip in ICECREAM by less than 10%, for 5 m in both models by 20–40% and for 15 m between 30% and 70% (Erosion2D) and about 70% with the ICECREAM model. Using the PK value for Erosion2D leads to a 100% reduction on certain dates.

The effect of different slope geometry (convex, concave, linear) on simulated surface runoff and soil loss in the two models was tested using two heavy rainfall events (Fig. 9). Additionally, the capacity of buffer strips with different lengths (5, 10, 15 m) was investigated for these two events (Table 3).

It is to be recognised that, in both models, geometry does not have any or only negligible influence on surface runoff. In addition, surface runoff as simulated by ICECREAM is larger in the event in November than in May due to the higher initial soil moisture content. Surface runoff is in May in ICECREAM clearly smaller than in Erosion2D. Relative reduction of surface runoff due to the buffer strips is in Erosion2D higher than in ICECREAM (maximum 55% vs. maximum 21%, Fig. 10a). It is apparent that changing the infiltration capacity due to the grass in buffer strips is more efficient in the Erosion2D approach (higher *Skin factor*) than changing the SCS Curve Number CN2 as in ICECREAM.

Values of soil loss in the May event are higher in Erosion2D than in ICECREAM and, probably thus, the reduction of soil loss due to buffer strips is in ICECREAM more efficient than in Erosion2D (Table 4, Fig. 10b).

Compared to the linear slope, the convex slope increases soil loss and the concave slope reduces it. With respect to the concave slope both models react in a similar way. For the convex slope ICECREAM increases soil loss on a convex slope compared to a linear one much more than Erosion2D. It seems that ICECREAM reacts more sensitive to the high slopes of a convex hill slope towards the end of the field than Erosion2D.

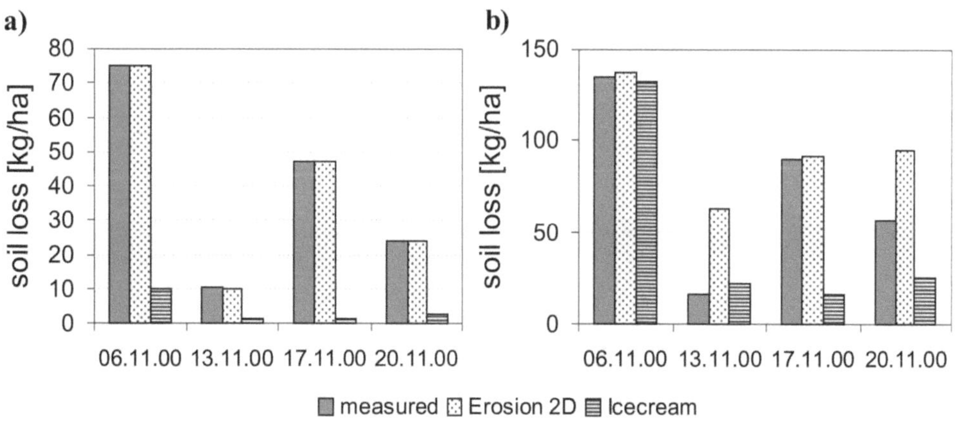

Fig. 7 Measured and calibrated soil loss for the grass (**a**) and winter wheat plots (**b**) for both models in November 2000

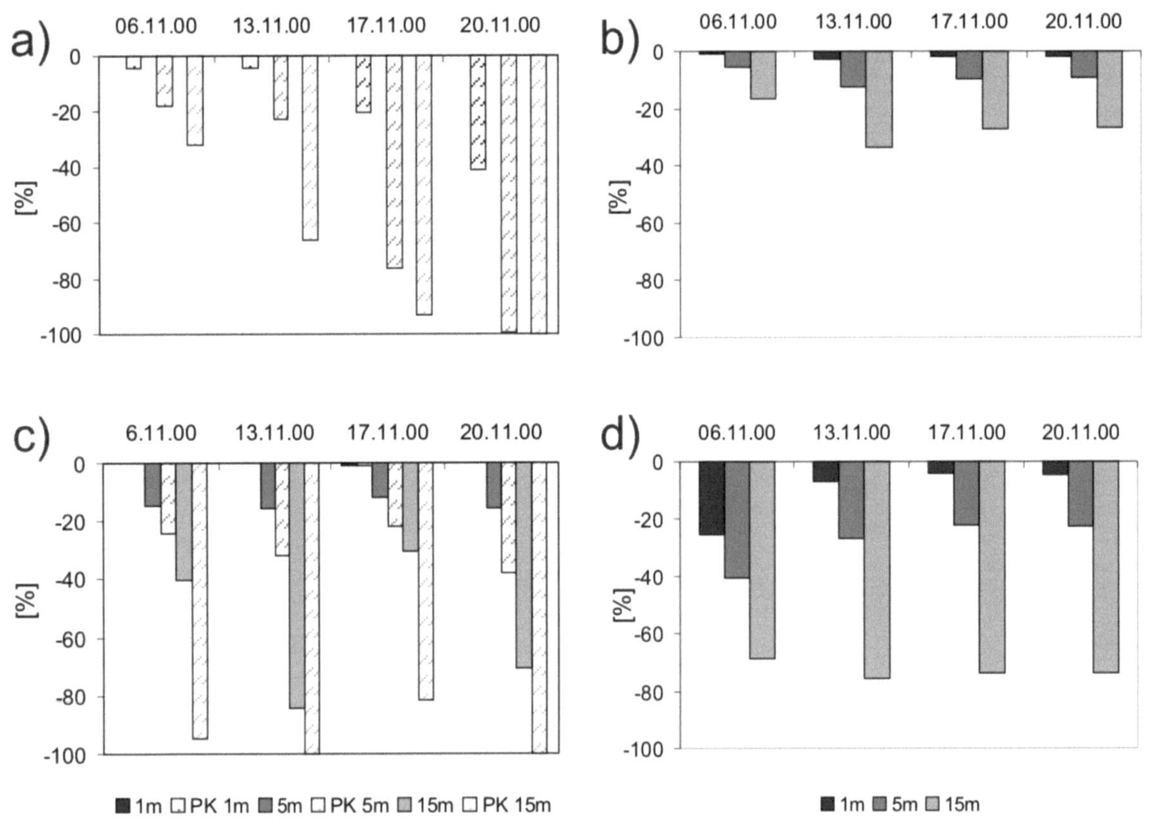

Fig. 8 Effect of buffer strip length on surface runoff in Erosion2D (**a**) and ICECREAM (**b**) and soil loss in Erosion2D (**c**) and ICECREAM (**d**) compared with winter wheat without buffer strip (PK: parameter catalogue of Erosion2D for buffer strips); Note: using calibrated values for Erosion2D leads to 0.0% reduction in surface runoff in (**a**)

It is obvious that buffer strip efficiency increases with buffer strip length for both models and all slope geometries (Fig. 10). The buffer strip on a convex slope is most efficient in both models, even though the reduction on the linear slope for ICECREAM in the May event is of same magnitude. For the concave slope long buffer strips are needed to reduce soil loss, but the absolute soil loss values without buffer strips are even smaller than values with the most efficient buffer strip on a linear slope. The buffer strip in Erosion2D is not as sensitive to the initial soil moisture content as ICECREAM, which has the same buffer strip efficiency for the linear and concave slopes under dry conditions (May). For wet conditions (November) efficiency is in the same order as for Erosion2D.

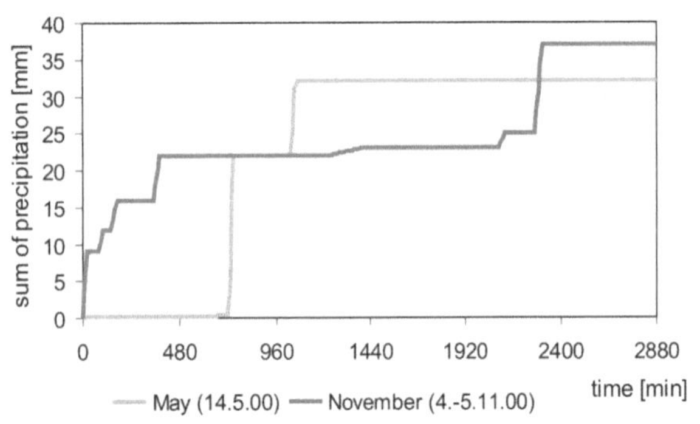

Fig. 9 Heavy rainfall events for the study on 14.5.2000 and 4–5.11.2000

Table 3 Simulated surface runoff (mm) for the two events on different slope geometries; relative deviation (%) of convex and concave to the result of the linear slope without buffer strip

	Strip	Convex		Concave		Linear	
14.5.00	(m)	(mm)	(%)	(mm)	(%)	(mm)	(%)
ICECREAM	0	8.1	−0.28	8.1	−0.07	8.1	–
Erosion2D	0	19.7	0	19.7	0	19.7	–
4–5.11.00							
ICECREAM	0	18.1	−0.07	18.1	−0.02	18.1	–
Erosion2D	0	17.2	0	17.2	0	17.2	–

Table 4 Simulated erosion (kg ha^{-1}) for the two events on different slope geometries; relative deviation (%) of convex and concave to the result of the linear slope without buffer strip

	Strip	Convex		Concave		Linear	
14.05.2000	(m)	(kg ha^{-1})	(%)	(kg ha^{-1})	(%)	(kg ha^{-1})	(%)
ICECREAM	0	227	+84	45.9	−63	123	–
Erosion2D	0	521	+31	199	−50	397	–
4–5.11.00							
ICECREAM	0	1051	+945	35.7	−64	101	–
Erosion2D	0	172	+55	25	−77	111	–

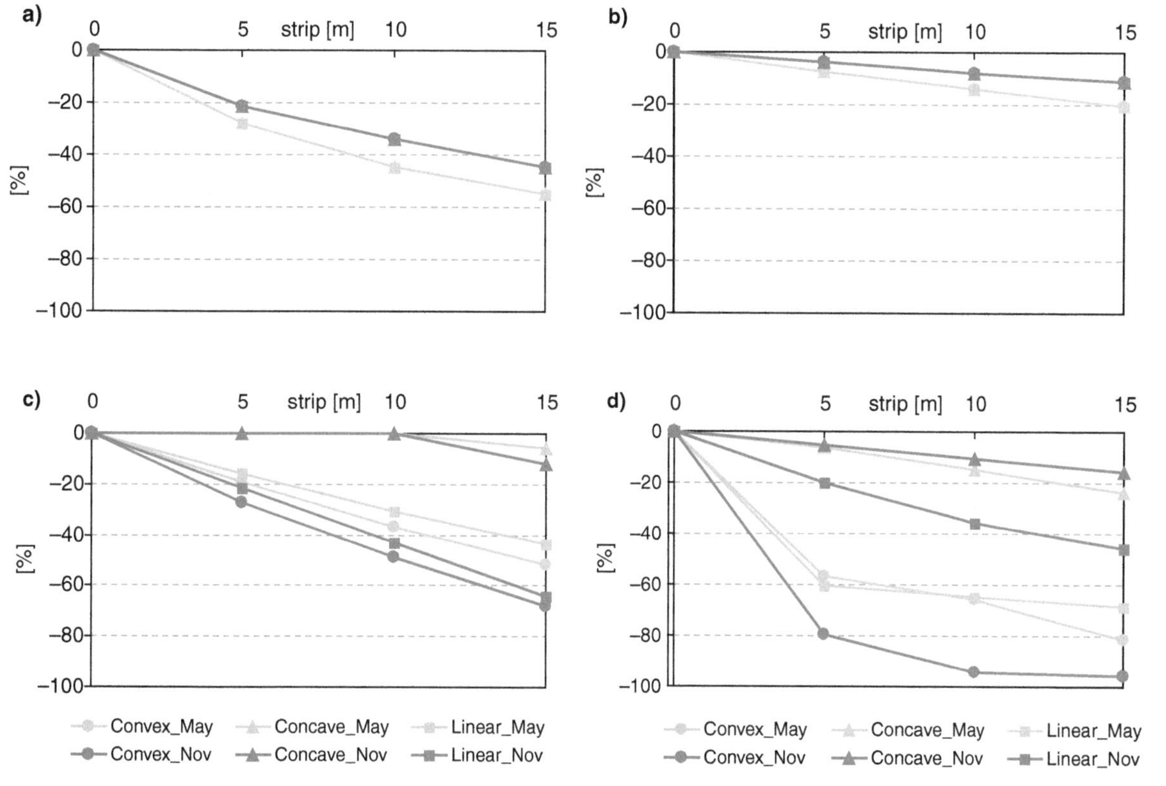

Fig. 10 Reduction of surface runoff in Erosion2D (**a**) and ICECREAM (**b**) and reduction of soil loss in Erosion2D (**c**) and ICECREAM (**d**) due to the usage of buffer strips with different length on different slopes for two rainfall events in May (14.5.2000) and November (4–5.11.2000), respectively

For the ICECREAM model buffer strip efficiency is higher for reducing soil loss than surface runoff. The Erosion2D model has the same rates for reducing surface runoff and soil loss (excluding the concave slope) under dry conditions (May), whereas soil loss is slightly higher (excluding concave slope) under wet conditions (November).

Discussion and Conclusions

The buffer strip has a larger influence on soil loss than on surface runoff in both models. Puustinen et al. (2005) indicate for grass after 10 years of measurement an average reduction for surface runoff of 10% and for soil erosion of 55% compared with winter wheat. The long-term average reduction of soil loss due to a 14 m wide buffer strip at Aurajoki experimental field was found to range between 58% and 67% (Puustinen 1999). In another study in south-western Finland (Uusi-Kämppä and Yläranta 1992) it was found that in a 1-year study a 10 m wide grass buffer strips reduced total solid load in runoff by 23.6%. This research field at Jokioinen (Fig. 1) consisted of a flat cropland area growing spring cereals and a steep buffer zone with an average slope of 16%. There grass buffer strips were effective in autumn but not in spring. Another study described that strips with a length of 10 m are able to reduce sediment discharge from 90% to 99% depending on the age of the grass (Van Dijk et al. 1996).

For the selected events only Erosion2D with, partially substantial, event-specific calibration could reproduce the measured values. Since, however, the model had to be re-calibrated for each event separately this procedure seems unsuitable for practical applications. On the other hand, the ICECREAM model delivers buffer strip scenarios without laborious calibration with similar values for surface runoff.

In addition, it has to be taken into account that the November 2000 event took place in late autumn where in Finland soils are saturated and the vegetation is not very dense anymore for permanent grass. Thus, the surface runoff can be of same magnitude or even higher from plots with permanent grass than from plots with winter wheat. The ICECREAM model could not be calibrated for sediment loss under grass for this event, since the dying of the grass cover in the model is not as strong as it should be. Actually both models assume that the efficiency of grass to retain water, but most of all sediment, is more efficient than what is the situation in certain periods in the northern conditions. Above all, the result highlights the discrepancy between event-based measurements and the desire to validate the models for credible simulation of long-term effects.

Abu-Zreig et al. (2004) found out for small slopes (2.3% and 5%) that efficiency of vegetated filter strips increased with length of filter strip. This could also be confirmed by this investigation. It was also stated by Abu-Zreig et al. (2004) that efficiency decreases for longer strips and that there will be no difference in efficiency for length from 10 to 15 m. This could not be confirmed by this model application. It was also found by Abu-Zreig et al. that efficiency related to surface runoff is smaller than with respect to erosion. This could also be confirmed reasonably.

As pointed out, for example by Rose et al. (2003) special attention should be paid on the deposition processes connected to the buffer strip. Is, for example in ICECREAM the sedimentation on low slopes (concave profile) dependent on the season, that is the stage of grass cover development of the buffer strip in the model.

All results depend strongly on specific conditions (age of grass, cover density, slope angle, geometry, ...) which have to be considered when modelling and planning buffer strips.

Acknowledgments This work was funded by the German Academic Exchange Service (D/05/26050) and the Academy of Finland project 'Process knowledge transferability and application in modelling' (reference number 206491).

References

Abu-Zreig M, Rudra RP, Lalonde MN, Whiteley HR, Kaushik NK (2004) Experimental investigation of runoff reduction and sediment removal by vegetated filter strips. Hydrological Processes 18 (11): 2029–2037

Correll DL (2005) Principles of planning and establishment of buffer zones. Ecological Engineering 24: 433–439

Graff CD, Sadeghi AM, Lowrance RR, Williams RG (2005) Quantifying the sensitivity of the Riparian Ecosystem Management Model (REMM) to changes in climate and buffer characteristics common to conservation practices. Transactions of the ASAE 48(4): 1377–1387

Kändler M (2006) Vergleichende Analyse der Modellansätze von ICECREAM und EROSION 2D bezüglich der hydrologischen Komponenten. [Comparative Analysis of modelling approaches in ICECREAM and EROSION 2D concerning hydrological components] diploma thesis. International Graduate School Zittau

Lowrance R, Altier LS, Williams RG, Inamdar SP, Sheridan JM, Bosch DD, Hubbard RK, Thomas DL (2000) REMM: The Riparian Ecosystem Management Model. Journal of Soil and Water Conservation 55(1): 27–34

Michael A, Schmidt J, Schmidt WA (1996) Parameter catalogue to EROSION2D. Sächsische Landesanstalt für Landwirtschaft, Sächsisches Landesamt für Umwelt und Geologie, Freiberg

Muñoz-Carpena R, Parsons JE, Gilliam JW (1999) Modeling hydrology and sediment transport in vegetative filter strips. Journal of Hydrology 214: 111–129

Posch M, Rekolainen S (1993) Erosivity factor in the Universal Soil Loss Equation estimated from Finnish rainfall data. Agricultural and Food Science in Finland 2: 271–278

Puustinen M (1994) Effect of soil tillage on erosion and nutrient transport in plough layer runoff. Publications on the Water Research Institute. National Board of Waters and the Environment, Finland. No. 17

Puustinen M (1999) Viljelysmenetelmien vaikutus pintaeroosioon ja ravinteiden huuhtoutumiseen. Suomen ympäristö 285. Helsinki, 116p. (in Finnish)

Puustinen M, Koskiaho J, Peltonen K (2005) Influence of cultivation methods on suspended solids and phosphorus concentrations in surface runoff on clayey sloped fields in boreal climate. Agriculture, Ecosystems and Environment 105: 565–579

Rankinen K, Tattari S, Rekolainen S (2001) Modelling of vegetative filter strips in catchment scale erosion control. Agricultural and Food Science in Finland 10: 99–112

Rose CW, Yu B, Hogarth WL, Okom AEA, Ghadiri H (2003) Sediment deposition from flow at low gradients into a buffer strip – a critical test of re-entrainment theory. Journal of Hydrology 280: 33–51

Schmidt J (1993) Teilmodell zur Simulation der Infiltration. Zwischenbericht zum BMFT-Vorhaben 03392233B

Schmidt J, von Werner M, Michael A (1996) EROSION2D – Ein Computermodell zur Simulation der Bodenerosion durch Wasser. Hrsg. Sächsische Landesanstalt für Landwirtschaft; Sächsisches Landesamt für Umwelt und Geologie Dresden-Pillnitz und Freiberg

Schob A, Schmidt J, Tenholtern R (2006) Derivation of site-related measures to minimise soil erosion on the watershed scale in the Saxonian loess belt using the model EROSION 3D. Catena 68: 153–160

Tattari S, Bärlund I, Rekolainen S, Posch M, Siimes K, Tuhkanen HR, Yli-Halla M (2001) Modeling sediment yield and phosphorus transport in Finnish clayey soils. Transactions of ASAE 44(2): 297–307

Uri ND (2001) A note on soil erosion and its environmental consequences in the United States. Water, Air and Soil Pollution 129: 181–197

Uusi-Kämppä J, Yläranta T (1992) Reduction of sediment, phosphorus and nitrogen transport on vegetated buffer strips. Agricultural Science in Finland 1(6): 569–575

Van Dijk PM, Kwaad FJPM, Klapwijk M (1996) Retention of water and sediment by grass strips. Hydrological Processes 10(8): 1069–1080

The Influence of Climate Change on Water Demands for Irrigation of Special Plants and Vegetables in Slovakia

V. Bárek, P. Halaj and D. Igaz

Keywords Climatic change · Irrigation · Vegetables · Special plants

Introduction

There exist proofs that the global warming, which is happening since the industrial revolution, is the consequence of the greenhouse emissions increase caused by the human activity. The development of computer models, besides the increasing evidences about the rising temperatures, and the more frequent weather variation are in accordance with the scientist's predictions about the climate change. The modelling tools also showed that in the twenty-first century temperatures can increase hereafter and influence the nature and the human being. The key event for scientists dealing with the climate was the establishment of the Intergovernmental Panel of Climate Change (IPCC) by the UNO in 1988. The IPCC has connected hundreds of scientists: those who have evaluated the studies and other relevant information with those who know more about climatic changing and that have handed in reports. The global climate change caused by the increasing of anthropogenic emission of the greenhouse gases is one of the most important environmental problems probably in the present human history. The world up to the year 2100, will suffer a dramatic rise of the greenhouse gases and the atmospheric CO_2 concentration in the atmosphere. For the most serious consequence of the development, besides the global warming from 2.0 to 2.5°C up to the year 2100, the IPCC is considering the general atmosphere circulation change with shift of the frontal zones and climate regions on the one hand and the rate of the climate change which exceeds all up to now climate changes on the other hand. An extreme scenario estimates warming up to 5.8°C in the opposite of the global average temperatures. It could be a reality after uncontrolled rise of the fossil fuels on the whole Earth, under the condition of the high economic progress only on the base of the energy from the fossil fuels and fast growth of the population up to 15.1 milliards by the year 2100 (Harrison et al. 1995).

For the area of agriculture it is necessary to highlight the specific bioclimatic uniqueness of the agricultural plants and animals in the Slovak Republic. Each ecosystem, each plant or animal exists in given abiotic conditions. Its genetic code contains information with the most successful alternative of existence for surviving in the given climate conditions. It is adapting not only for the mean climate parameters but also for their variability (characteristic variation from the absolute minimum to the absolute maximum and for combination of more probable climate conditions). If an ecosystem or an individual is removed from this system to other climatic conditions it is necessary to last certain period for adaptation and this period is considered as terminated only when it has a new equilibrium in the ecosystem. Some species are never adapted to the new environmental conditions and then, the previous equilibrium is never restored. Almost the same process will occur when the climate conditions have been changed in the given site (Špánik et al. 1996, Takáč and Zuzula 2000).

In the lowland Slovak areas, the new time horizon for the year 2075 expects that the daily average air

V. Bárek (✉)
Department of Landscape Engineering, Faculty of Horticulture and Landscape Engineering, Slovak University of Agriculture, Hospodárska 7, 949 76 Nitra, Slovakia
e-mail: viliam.barek@uniag.sk

K. Strelcová et al. (eds.), *Bioclimatology and Natural Hazards*,
DOI 10.1007/978-1-4020-8876-6_23,

temperature sums will increase about 32% for large vegetation period (LVP) and about 55% in the northern parts of Slovakia. In the southern part where the lowest locations in Slovakia are the photosynthetically active, radiation increase for the LVP is about 10% and in the highest agricultural exploited areas about 25%. In the southern lowest part of the Slovakia, there is a total increase of the photosynthetically active radiation QPAR to about 72 kWh m^{-2} during main vegetation period (MVP). That represents an increase of about 17%. In the highest locations it is 115 kWh m^{-2}, which is an increase of about 58%. The precipitation sums will probably increase about 27 mm during the MVP in the southern area of Slovakia and about 202 mm in the northern part of Slovakia. Evapotranspiration will be changed probably only slightly or nothing to the year of 2075. On the southern area of Slovakia the evapotranspiration will probably increase about 6% up to the year 2075 and on the north of Slovakia probably up to a 20% (Sobocká et al. 2005).

Changes of the temperature, rainfall totals, and also other factors of the environment often change the time course of the plants living displays, that is the phenophases beginning and by this also the phenophase intervals longitude and whole vegetation periods longitude of the individual plants. The presumption for this scenario is predicted by the basic indicators of the agro-climatic relationships of the LVP as its daily temperature sums increase (about 32–55%), photosynthetic active radiation increase about 10–25%, and vapour increase on the south of Slovakia up to 20% (Šiška et al. 2004). For vegetation periods, limited by the physiologic important temperatures to the time horizons of the year 2075, the earlier beginning and later finishing of the phenophases are valid. Up to the year 2075 it is assumed an increase of the production biomass potential about 10% in the south of Slovakia, about 25% in the north of Slovakia. It is assumed a shift of the fully rentable cornflake cultivar up to 500 m a. s. l. and the rentable cornflake cultivar up to 800 m a. s. l. (Šiška et al. 2004).

On base of the last decades, elaborated hydrological balances and hydrological balance prognosis based on climate change scenarios there is expected the decreasing of all three elements of water resources – surface water, groundwater and soil water. On the base of the total hydrologic situation evaluation for the last 5 years in Slovakia it is possible to state that the discharge extremeness has partly increased, while the average values assessed for the main Slovak river basins for the last years have not changed significantly from the long-time averages. The years 1996–2000 belonged to the period with largest floods and these were the floods in the river systems as well as the storms-floods flooding relatively small areas generally (Lapin et al. 2001, Lapin 2004). The long-term mean discharges in Slovak rivers have been decreasing, except the Danube after the year 1980. Concerning to the long-term monthly discharges, the notable decreasing had been observed in the central and eastern part of Slovakia, almost in the all months with the exception of May and June.

In the western part of Slovakia the summer and autumn months are observed decreasing of discharges in comparison with the past. The winter months have higher discharge in comparison with the long-term averages. The locality of Žitný ostrov belongs to one of the most important regions of Slovakia from the groundwater resources point of view.

Inasmuch as the groundwater regime of locality Žitný ostrov is in the tight relationship with the Danube's discharge regime and therefore groundwater of the territory is very sensitive to the potential consequences of expected climate change. In general it is possible the results obtained from remade hydrologic scenarios for Slovakia to summarize as it follows (Lapin et al. 2001, Lapin and Melo 2002): winter discharge increasing varied from 10% to 40% in the north and from 20% to 50% in the centre and from 30% to 80% in the south. The winter discharges can be even exceptionally higher. The winter discharge rises up to the farther time horizon of the year 2075.

The aim of the chapter is to calculate and analyse the calculation results of consumptive water usage V_c values, using the climate scenarios for Slovakia, for special plants and vegetables according to technical standards (STN no. 83 0635 – Irrigation water demand calculation) with modelled values of potential evapotranspiration ET_0 according to FAO methodology for horizons of the years 2010, 2030 and 2075 for scenarios $CCCM_{prep}$ and $GISS_{prep}$.

Materials and Methods

The irrigation amount rate represents the water amount that is necessary for the plant growth during vegetation period to refill the natural water content in soil and to recover all water losses in irrigation canals re-calculated in the unit area. The irrigation water amount quantity – M_Z – depends mainly on the consumptive water usage, the precipitation available during the vegetation period, soil water storage at the beginning of the vegetation period, the volume of capillary rise from groundwater table and on the efficiency of irrigation.

Total Consumptive Water Usage (V_c) Calculation According to Slovak Technical Standard STN No. 83 0635

The total water consumptive usage V_c is the amount of water that is necessary for evaporation, which assures the supposed development and the agricultural plant growth in given climate conditions while ensuring other growth factors during the whole vegetation period.

In irrigation practices in Slovakia, there is a consumptive water usage assessment methodology elaborated by Sláma (1971) that is based on dependence of the consumptive water usage on the vapour pressure deficit and biological consumptive water usage curve coefficient. Then consumptive water usage V_c (mm) is defined from the relation:

$$V_c = k_b.S_d \tag{1}$$

where k_b is biological consumptive water usage curve coefficient, and S_d is vapour pressure deficit of daily values sums in a considered period (mm).

The k_b values for common crops have been assessed by Sláma (1971) for time period responding to a sum of average day temperatures in the vegetation period's individual phases. Time intervals had been divided into temperature groups and those enable to define vegetation period phases of a given plant in which sum of average daily temperatures extended up to 100°C or 200°C. The beginning of the vegetation period in winter crop and more years fodder crops is assessed from the last day of 10-day period in which the sum of the average day air temperature have reached 50°C. For other plants it is given by a day of sowing or outplanting. Recommended values are presented in Tables 1 and 2. Crops without explicit recommended value are assessed according to sort or cultivate relative crop. Table 3 provides values of consumptive water usage V_c for the common crops of localities in southern Slovakia.

Table 1 Biological consumptive water usage curve coefficients

Crop	Sum of average daily temperatures (°C, temperature groups)												
	0–100	100–200	200–300	300–400	400–500	500–600	600–700	700–800	800–900	900–1000	1000–1100	1100–1200	1200–1300
Early Potato	0.37	0.36	0.38	0.44	0.57	0.60	0.67	0.80	0.80	0.75	0.50		
Early Kohlrabi	0.50	0.51	0.56	0.62	0.66	0.67	0.75	0.65					
Early Cauliflower	0.71	0.73	0.71	0.66	0.90	0.96	1.02	0.99	1.11	1.15	1.16	1.20	1.08
Early Cabbage	0.70	0.67	0.64	0.78	0.85	0.88	0.90	0.80	0.82				

Table 2 Biological consumptive water usage curve coefficients

Crop	Sum of average daily temperatures (°C; temperature groups)									
	0–200	200–400	400–600	600–800	800–1000	1000–1200	1200–1400	1400–1600	1600–1800	1800–2000
Potato	0.50	0.50	0.50	0.58	0.68	0.67	0.61	0.71	0.74	0.57
Cabbage	0.42	0.49	0.48	0.58	0.76	0.82	0.79	0.82	0.79	0.85
Celery	0.73	0.73	0.62	0.65	0.77	0.90	0.88	1.01	1.07	
Pulses	0.37	0.41	0.48	0.67	0.63	0.73	0.71	0.72	0.73	0.74
Cauliflower	0.71	0.73	0.71	0.66	0.90	0.96	1.02	0.99	1.11	1.15
Hops	0.82	0.84	0.67	0.65	0.67	0.67	0.80	0.80	0.77	0.75
Fruit	0.23	0.26	0.28	0.32	0.34	0.37	0.40	0.43	0.46	0.48
Grapes	0.18	0.22	0.24	0.27	0.30	0.32	0.34	0.38	0.42	0.43

Table 2 (continued)

Crop	Sum of average daily temperatures (°C; temperature groups)									
	2000–2200	2200–2400	2400–2600	2600–2800	2800–3000	3000–3200	3200–3400	3400–3600	3600–3800	3800–4000
Potato	0.44									
Pulses	0.63	0.57								
Cauliflower	1.16	1.20	1.08							
Hops	0.78	0.84	0.78	0.85	0.69					
Fruit	0.52	0.54	0.57	0.60	0.63	0.65	0.63	0.58	0.51	0.41
Grapes	0.45	0.48	0.50	0.46	0.36	0.32	0.25	0.18		

Table 3 Consumptive water usage V_c for the common crops for localities in southern Slovakia

Crop	Southern Slovakia		Crop	Southern Slovakia	
	Vegetative period	$V_c(m^3\ ha^{-1})$		Vegetative period	$V_c(m^3 ha^{-1})$
Apple	1.4.–30.9.	6500	Cauliflower	15.6.–30.9.	3500
Pear	1.4.–30.9.	6000	Potato	20.4.–20.9.	3000
Apricot	1.4.–31.8.	5500	Celery	15.5.–31.10.	4000
Cherry	1.4.–31.7.	4500	Pulses	1.5.–30.9.	2000
Black cherry	1.4.–31.7.	4000	Kohlrabi	1.4.–31.10.	3000
Plum	1.4.–30.9.	5500	Peach	1.4.–30.9.	6500
Redcurrant	1.4.–15.7.	5000	Hops	1.4.–20.8.	3600
Blackcurrant	1.4.–15.7.	4500	Table grapes	1.4.–31.8.	3600
Raspberry	1.4.–31.8.	4500	Wine grapes	1.4.–31.8.	3400
Strawberry	1.4.–30.7.	5000	Gooseberry	1.4.–31.7.	4500

Methods of Potential Evapotranspiration ET_c Calculation on Basis of Agrometeorological Data

Among the main factors that influence evaporation and transpiration may be assigned weather, crop characteristics, management form and environmental aspects. Radiation, air temperature, air humidity and wind speed belong among the main inputs of the weather which influence evapotranspiration. Actual evapotranspiration depends also on the kind of crop, its variety and phenological development. Reference evapotranspiration represents amount of water evaporated and transpired by well-maintained, well-watered grass canopy. Evapotranspiration differs from reference evapotranspiration if the soils cover, the crop canopy properties and the aerodynamic resistance of the crop are different from the grass.

Potential evapotranspiration ET_c, that is crop evapotranspiration under standard conditions (Allen et al. 1998, Hansen 1984) is evapotranspiration of a healthy, well-manured crop that is grown in a large area under optimal soil water content reaching full production in given climate condition. So the standard condition means sufficient soil water content and excellent agronomical conditions.

The water amount needed for evapotranspiration loss compensation is marked as a moisture requirement of the crop and basically equals to potential evapotranspiration. Irrigation need represents the difference between the moisture requirement of the crop and effective precipitation. Actual evapotranspiration ET_a, that is evapotranspiration under non-standard condition (Allen et al. 1998) is evapotranspiration from crop grown under farming management and natural conditions, which defers from standard conditions. During the second half of the twentieth century several methods of evapotranspiration assessment have been developed. However, many of them have only local validity.

Penman–Monteith Method of Calculation ET_0 According FAO

Reference evapotranspiration ET_0 (mm day^{-1}) was derived from original Penman–Monteith equation and from aerodynamic resistance equations and canopy resistance (Allen et al. 1998):

$$ET_0 = \frac{0,408\Delta(Rn - G) + \gamma \frac{900}{T+273} u2(es - ea)}{\Delta + \gamma(1 + 0,34u2)} \quad (2)$$

where R_n is net radiation on canopy surface ($MJ\,m^{-2}\,day^{-1}$), G the soil heat flux ($MJ\,m^{-2}\,day^{-1}$). T is average daily air temperature at the height of 2 m (°C), u_2 the wind speed at the height of 2 m ($m\,s^{-1}$), e_s the saturated water vapour pressure (kPa), e_a the water vapour pressure (kPa), $e_s - e_a$ the vapour pressure deficit (kPa), Δ the curve slope of dependency between saturated water vapour pressure and air temperature ($kPa\,°C^{-1}$) and γ the psychrometric constant ($kPa\,°C^{-1}$).

There are used standard climatic measurements of solar radiation, air temperature, air humidity and wind speed in the equation. Alternative computing techniques are developed for this method in case of data missing from some needed meteorological parameters. Estimation of reference evapotranspiration in FAO version requires following meteorological data:

- mean daily air temperature (T), maximum air temperature (T_{max}) and minimum (T_{min}) air temperature,
- water vapour pressure (e_a),
- net radiation (R_n) and
- wind speed at height of 2 m(u_2).

If the air humidity data are not available in sufficient quality we may use the fact that the dew-point temperature is the temperature to which the air needs to be cooled to make the air saturated. Then the actual vapour pressure (e_a) is the saturation vapour pressure at the dew-point temperature $e^o(T_d)$. Then we may use assumption that dew-point temperature T_d is close to minimum daily air temperature T_{min}, then

$$e_a = e^o(T_{min}) = 0.611 \exp\left[\frac{17.27T_{min}}{T_{min} + 237.3}\right] \quad (3)$$

Instruments for measuring radiation balance require professional treatment and therefore are not installed in agrometeorological stations. Radiation balance and long wave radiation are possible to calculate from global radiation or from duration of sunshine, air temperature and water vapour pressure. Short wave radiation R_s is possible to calculate according to Hargreaves equation (Hargreaves et al. 1985):

$$R_s = 0.16\sqrt{(T_{max} - T_{min})}R_a \quad (4)$$

where R_a is radiation on upper bound of atmosphere ($MJ\,m^{-2}\,day^{-1}$) and it is calculated according Eq. (5):

$$Ra = \frac{24(60)}{\pi} Gscdr\,[\omega s \sin(\varphi)\sin(\delta) + \cos(\varphi)\cos(\delta)\sin(\omega s)] \quad (5)$$

where G_{sc} is solar constant $= 0.0820\,MJ\,m^{-2}\,min^{-1}$, ω_s, the hour angle (rad), ϕ, the latitude (rad) and δ, the solar declination (rad).

Hargreaves equation is according to recommendation of FAO the alternative equation for case that some data of radiation, water vapour pressure and wind speed are missing (Allen et al. 1998):

$$ET_0 = 0.0023(T + 17.8)(T_{max} - T_{min})^{0.5} R_a \quad (6)$$

where ET_0 is reference evapotranspiration ($mm\,day^{-1}$).

Crop Evapotranspiration ET_c

If we use crop coefficient approach for crop evapotranspiration assessment where the ET_c is calculated by multiplying the reference crop evapotranspiration ET_o by a crop coefficient K_c:

$$ET_c = K_c \cdot ET_o \quad (7)$$

Differences in evaporation and transpiration between field crops and reference grass canopy may be expressed by simple crop coefficient K_c or we may divide it into two coefficients: basic crop coefficient K_{cb} and soil evaporation coefficient K_e, that is

$$K_c = K_{cb} + K_e \quad (8)$$

Simple crop coefficient is used in most of the applications. Dual coefficient is used for more exact estimation of evaporation, especially for research or water quality modelling, etc. Crop coefficient K_c integrates characteristics effect, which differs field crop from reference grass. Individual field crops have different crop coefficients K_c. Change of crop characteristics during vegetative period also influence value of K_c. As the evaporation is an integrated part of evapotranspiration, K_c also includes evaporation effect.

Evapotranspiration of the crops well supplied by water differs from reference evapotranspiraton ET_0 in consequence of differences in albedo, crop height, aerodynamic properties as well as leaf and stomata properties and is higher 15–20% by tall crops. As the crop grows its height and leaf area is changing as well as crop cover of soils surface. Thereby, also K_c is changing during vegetative period.

The crop vegetative period can be divided into four phenological stages (Fig. 1). The initial stage lasts from sowing to the period until the green vegetation does not cover approximately 10% of soil surface. The second phenological stage (stage of intensive crop growing) lasts from 10% covering until plants cover the whole soil surface. The middle phenological stage lasts from absolute covering to the beginning of maturity (leaf yellowing) and the last stage from the beginning of maturity to the harvest or leaf falling.

Table 4 presents values of simple crop coefficient K_c according to individual phenological stages of chosen crops for calculation according FAO methodology.

Crop evapotranspiration is calculated by multiplying reference evapotranspiration ET_0 with crop coefficient K_c expressing the difference in evapotranspiration between appropriate crop and reference of grass surface. The option is to use simple coefficient or combined dual coefficient expressing individual differences in evaporation and transpiration between both surfaces is influenced by accuracy requirements what meteorological data are available and what time step is going to be used for calculation.

In simple terms, crop coefficient K_c is combined crop transpiration effect and evaporation from the soil grouped into one coefficient integrating differences in crop transpiration and evaporation between appropriate crop and reference surface. As the evaporation from the soil can change daily due to precipitation and irrigation, the simple crop coefficient expresses only average multi-day crop evapotranspiration used for weekly and longer time periods, although the calculation is made on daily basis, for example by sprinkle irrigation.

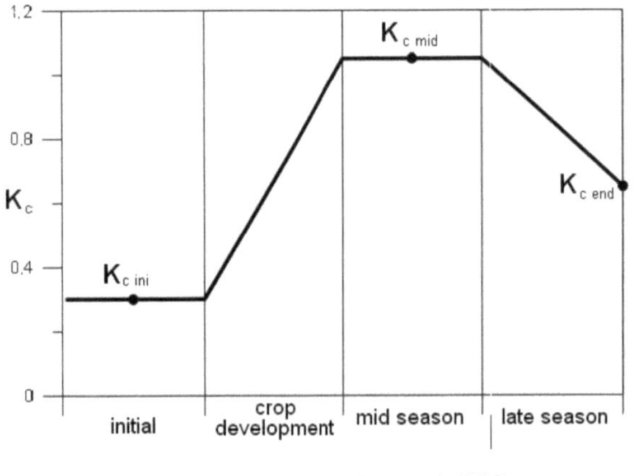

Fig. 1 Crop coefficient curve K_c (Allen et al. 1998)

Table 4 Values of simple crop coefficient curve K_c for individual phenological stages of selected crops of subhumid climate (RHmin ≈ 45%. $u_2 \approx 2\,\mathrm{m\,s^{-1}}$) (Allen et al. 1998)

Crop	K_{c-ini}	K_{c-mid}	K_{c-end}
Cauliflower	0.3	0.9	0.3
Crucifer	0.3	0.9	0.4
Faba bean dry	0.4	0.9	0.6
Celery	0.25	0.95	0.2
Potato	0.3	0.5	0.3
Deciduous orchard	0.2	1.2	0.6
Grapes	0.2	1.2	0.6
Hops	0.25	0.8	0.1

The assessment of irrigation water demand for the special crops and vegetables, its quantitative and time expression for planning, design and operation purposes are very difficult problems. But there is a necessity for negotiation at the base of the most suitable way to make possible the construction and operation of irrigation objects that use the state-of-art knowledge system. Seriousness for resolving an irrigation problem is escalating mainly by the irrigation operation. Very often, it is necessary to negotiate not only with different climate conditions, which keep changing with time as a consequence of climate change, but also with the specific biological demands of the irrigated crops, due to demands of the maximal exploitation and of possibilities of the use of an intensive agro-technology with the soil relationship with respect to technical factors and so on; the narrowest relationship had been assured at the maximal possible measure between the biological and economic irrigation relation, which determines the effect of almost whole irrigation application.

For basic crop evapotranspiration assessment, there exists several methods differing by the included meteorological elements and with respect to crops by development stage and crop status characteristics. In Slovakian conditions, crop evapotranspiration assessment is done according to consumptive water usage. V_c represents water amount supplied to a crop during a vegetation period demanded for evapotranspiration coverage for an optimal agriculture crop development and growth assurance in given climate conditions while assuring other growth factors during the whole vegetation period.

With climate scenarios results in Slovakia, the chapter is focused on assessment of consumptive water usage values using the older methodology based on the biological consumptive water usage curve coefficient for horizon of the year 2075 for special crops and vegetables cultivated in Slovakia as a base for suggestion and irrigation regulation and also for the water resources capacity design for changed conditions. The further part of the chapter will be oriented on the ET_c consumptive water usage assessment according to FAO methodology using the modified scenarios ($GISS_{mod}$ and $CCCM_{mod}$) for time horizons of the years 2010, 2030 and 2075. The General Circulation Models GISS (Goddard Institute for Space Studies) and CCCM (Canadian Climate Center Model) provide the average monthly air temperature and precipitation due to current atmospheric concentration of carbon dioxide and its doubling, which is predicted for 2075. The simulations had been performed for Hurbanovo locality, which has the climate station that has the longest observing series in Slovakia (from the year 1871). This city lies in the southern Slovakia and its climate relationships are characterized for irrigation areas of southern Slovakia (latitude: 47°52′N, longitude: 18°12′E and altitude 124 m a. s. l.). The similar calculations of the consumptive water usage for the special crops – fruit trees, hops, wine grapes and some vegetable types in the time horizons of the year 2075 have not yet been performed in Slovak conditions.

Consumptive Water Usage V_c in Changed Climate Conditions

The key assumption for the consumptive water usage (V_c) assessment according to methodology of Sláma (1971) is based on consumptive water usage dependence on the saturation deficit. Input data – daily mean air temperature characterizing climate conditions for $1xCO_2$ scenario were used for Hurbanovo climate station (time series 1951–1980). From these obtained values according to physical correlations has been calculated the water vapour pressure (e) expressing the instant content of the water vapour into air (hPa). It has been kept count of the specific air humidity (q) as a ratio of water vapour density to total humid air density. The specific air humidity was re-calculated on basis of coefficients provided by Lapin et al. (2001) for horizon of the year 2075. These data were origin of vapour pressure deficit calculations for consumptive water usage assessment of horizon of the year 2075. WATGEN as a sub-model of software DSSAT (version 3.1.) was used for the generation of daily climate datasets for climate change conditions of $2xCO_2$.

The sum of vapour pressure deficit was calculated for temperature groups equal to 200°C (respectively 100°C) for climate change conditions of $2xCO_2$ The beginning of temperature groups depends on the start date of the vegetative period. The relevant biological consumptive water usage curve coefficient value was assigned for each temperature group from Table 1. The total value of consumptive water usage V_c is calculated as a products sum of particular biological consumptive water usage curve coefficient and relevant vapour pressure deficit sum (Eq. (1)) for given temperature groups with sum of temperatures 200°C (respectively 100°C).

Evapotranspiration ET_c in Changed Climate Conditions

Daily climate data series (global radiation, air temperature, rainfall) for climate station Hurbanovo have been processed into input datasets for numerical simulation with model DAISY. Software DAISY (Hansen et al. 1990) represents one-dimensional numerical modelling tool for crop production assessment, soil water dynamics and nitrogen dynamics for different strategies of crop production management. Particular phenomena included in model are described by transport and transformation processes of water, heat, carbon and nitrogen.

Snow accumulation and melting, interception, evaporation of soil and vegetative cover surfaces, plant roots water uptake, transpiration and vertical soil water flow represent the hydrological processes included in the model. Heat processes include heat transmission and heat-induced volume changes by processes of freezing and melting. The model requires the following input datasets: climate data, soil and crop data as well as the type crop production management. The meteorological data required for transformation and transport of mass and energy are represented by global radiation, air temperature and rainfall.

Soil parameters required by model are soil moisture retention curve, hydraulic conductivity and specific water capacity. Richard's equation is used for soil water flow modelling in unsaturated zone. Upper boundary conditions create either soil water flow in upper soil layer (infiltration rate exceeds snowmelt, rainfall or irrigation rate intensity or water is evaporated from soil surface) or soil water tension of soil surface (water is accumulated on soil surface) or soil water flow in specific depth (Richard's equation is not solved for whole profile depth because of freezing or melting). The lower boundary condition may be assumed either soil water tension in given depth (groundwater-level depth is known) or groundwater flow in given depth. Richard's equation is numerically solved by method of finite differences. Soil water uptake by roots of plants is described by steady radial flow that depends on root's diameter and density of root system, soil moisture near the plant roots and soil water tension.

We realized series of simulation by DAISY model for representative period 1966–1985 (without assumption of climate change influence) and for climate change scenarios $CCCM_{prep}$, $GISS_{prep}$ for horizons of the years 2010, 2030 and 2075 for climate station Hurbanovo. As meteorological input data (global radiation, air temperature and rainfall) we used 20 years time series with daily time step. The long-term tendencies of basic meteorological elements development (mean daily air temperature and annual rainfall totals) are shown in Figs. 2 and 3.

Simulation results have been evaluated for representative soil profile with respect to dominant soil class. The soil profile properties were described by the following parameters: soil water retention curve. hydraulic conductivity, clay particles content (according to international classification system), quartz and other minerals content, litter content, initial

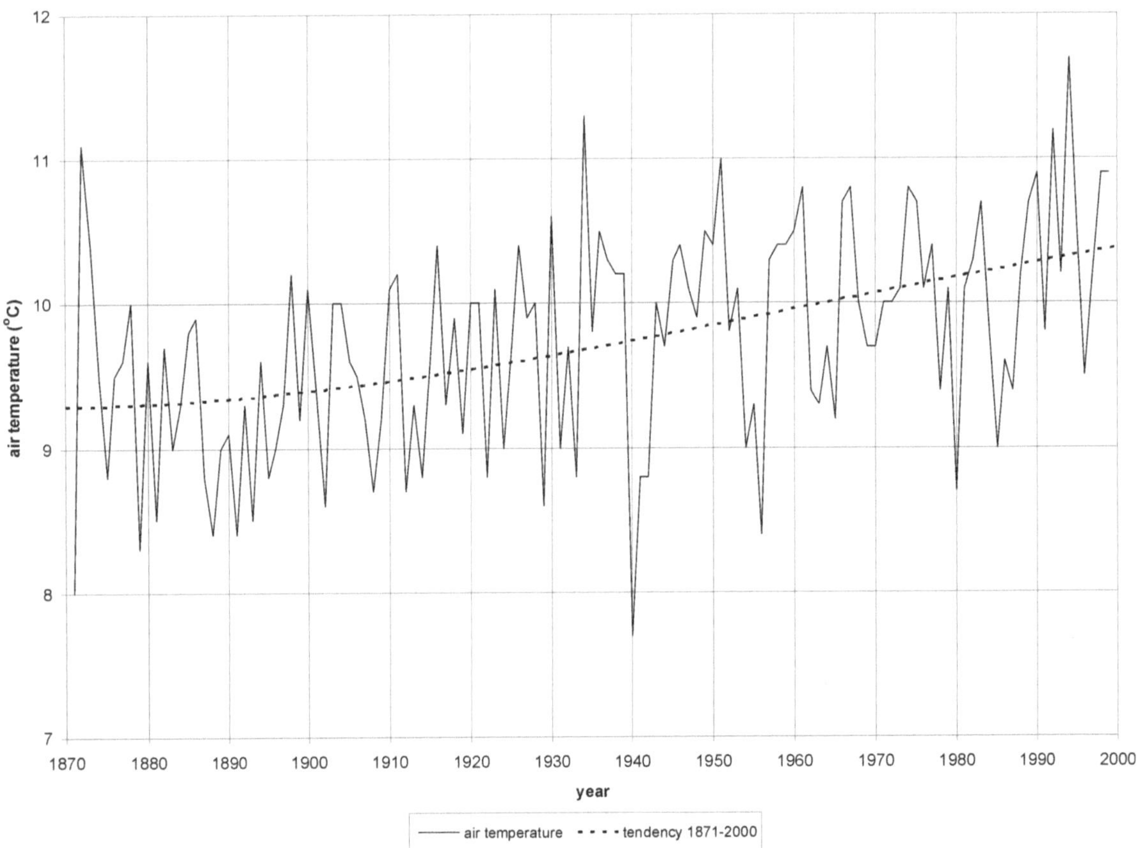

Fig. 2 Mean annual air temperature (°C), climate station Hurbanovo, period 1871–1999

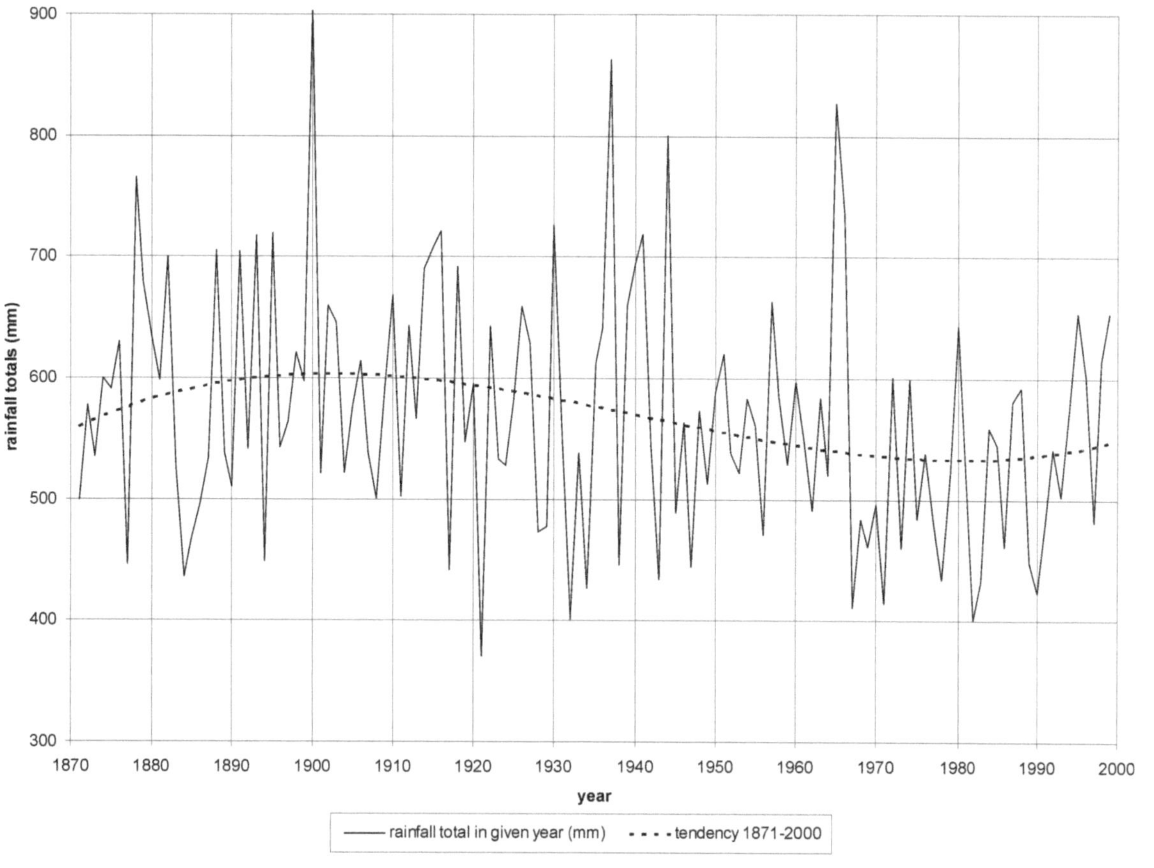

Fig. 3 Annual rainfall totals (mm), climate station Hurbanovo, period 1871–1999

value of soil water potential, temperature and moisture of soil. The groundwater level occurrence in root zone has not been assumed.

The results of simulations with daily time step and for given layers of soil profile and their comparison with representative period are described by elements of soil water balance, namely evapotranspiration, cumulative infiltration and soil water storage in particular layers.

The $CCCM_{prep}$ and $GISS_{prep}$ daily scenarios in the form of time series of monthly 10 years averaged data characteristics were prepared as T and R coefficient deviations for horizons of the years 2010, 2030 and 2075. The representative period from 1 January 1966 to 31 December 1985 has been selected because of the availability of high-quality data from climate station Hurbanovo. We especially used daily data of the representative period for scenarios of T, G and R (daily maximum, minimum and mean air temperature, daily sum of global radiation and daily rainfall totals) obtained for horizons of the years 2010, 2030 and 2075. Then modelled scenarios are valid for periods 2002–2021, 2022–2041 and 2066–2085.

Because the DAISY model has been calibrated only for main crops in Slovak conditions and calibration is long time-consuming process, potential evapotranspiration (ET_0) for special plants and vegetables have been calculated according to FAO methodology. We used simple crop coefficients for calculation (Table 3). The selected procedure provides the results with adequate accuracy advisable for optimization of irrigation operation purposes.

Results and Discussion

The first part of the chapter is focused on consumptive water usage V_c assessment for horizon of the year 2075 by method of consumptive water usage with biological consumptive water usage curve coefficient, that is recent valid Slovak methodology with input data

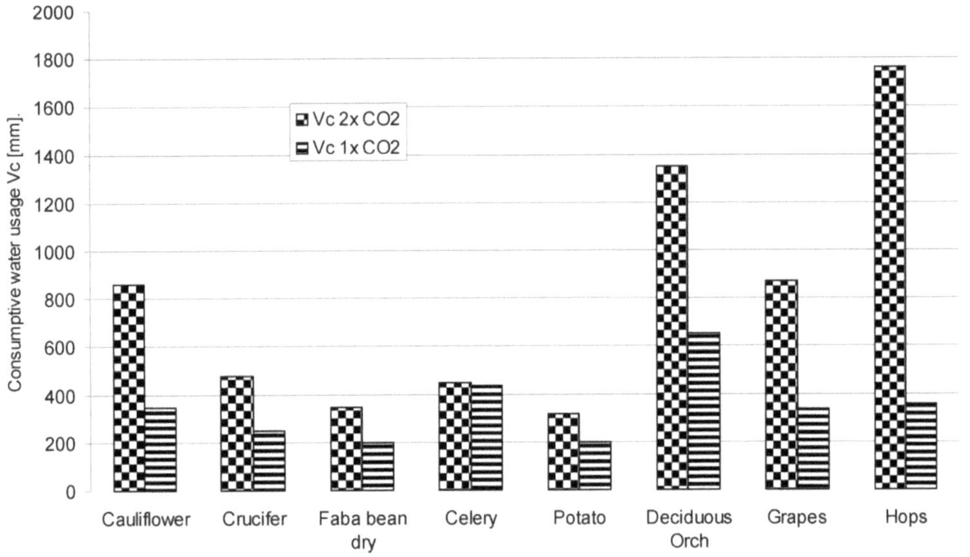

Fig. 4 Results of consumptive water usage V_c calculation for horizon of the year 2075 according to valid Slovak technical standard STN no. 83 0635

represented by vapour pressure deficit for horizon of the year 2075. The results are documented in Fig. 4.

Analysis of Fig. 4 (concentration $2xCO_2$) shows indisputable limitation of recently used method in irrigation practice for climate change scenarios in horizon of the year 2075 in Slovakia because the biological consumptive water usage curve coefficients distort the results of V_c calculations. This fact is markedly evident for crops with long vegetative period. The reason in fact is that the consumptive water usage assessment comes out from only meteorological characteristic – vapour pressure deficit. Because of practical inapplicability of this way of obtaining results we did not continue using the method.

The other part of chapter introduces the results of potential evapotranspiration modelling (ET_0) by DAISY model for scenarios $CCCM_{prep}$ and $GISS_{prep}$ for horizons of the years 2010, 2030 and 2075. At the next step we calculated crop potential evapotranspiration according to FAO methodology. The results were compared with consumptive water usage V_c that is recently used in Slovakia. The results are shown in Fig. 5.

On basis of modelled results for climate change scenarios we may state increasing of potential evapotranspiration of crops for both climate change scenarios. The scenario $CCCM_{prep}$ shows markedly higher values of ET_0. There is an increase in ET_0 for particular crops ranging from 9% to 57% in comparison with consumptive water usage calculated according to valid Slovak Technical Standard STN no. 83 0635. The increase in ET_0 for $GISS_{prep}$ scenario is within the interval 3–51%. Mean potential evapotranspiration value of vegetables will increase about 23% and for special plants about 34%. The highest increment value of ET_0 according to both scenarios is for grapes −51% and the lowest for cauliflower −4%.

Summary and Conclusions

The methods for the determination of the surface water resources quantity come out from the assumption of hydrologic processes stationarity. For the last 20 years water resources reduction has been happening and according to climate change scenarios this reduction will continue. It will lead to increase in number of water-deficit basins. The rising demand for water will require more integrated approach to water resources management. Under some conditions farmer's activities can be endangered without purposive and the right-time-operated irrigation devices. Irrigation can be understood as a controlling and stabilizing factor of the agricultural system, decreasing the influence of the casual elements. Estimate of irrigation water need for agricultural plants, its quantitative and time reference for planning, design and operation purposes are very

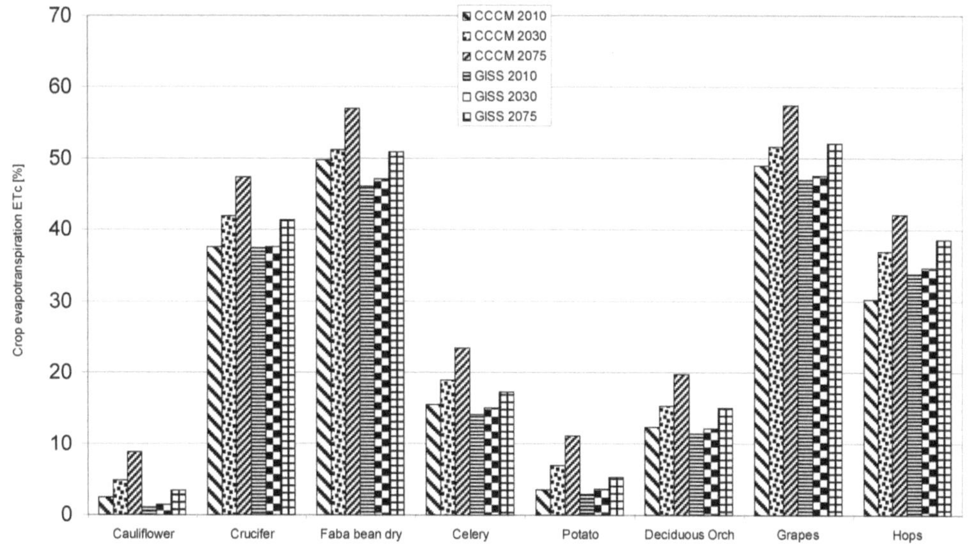

Fig. 5 Comparison of consumptive water usage V_c for selected crops according to STN no. 83 0635 with modelled values of potential evapotranspiration ET_0 according to FAO methodology for horizons of the years 2010, 2030 and 2075 for scenarios $CCCM_{prep}$ and $GISS_{prep}$

complicated and complex problems. However, it is necessary to understand this theme in the full scope to allow construction and operation of top-level irrigation objects on the basis of the recent information.

In Slovakian conditions, the classical approach is based on total consumptive water usage assessment (V_c) determination that represents water quantity needed for transpiration and evaporation. Its value covers agricultural plant water demand for given climate conditions during the whole growing period when all other factors are in optimal state.

The analyses of calculations results of consumptive water usage V_c for selected crops using the climate scenarios for Slovakia according to technical standard (STN 0000 no. 83 0635 – Irrigation water demand calculation) and modelled values of potential evapotranspiration ET_0 according to FAO methodology for horizons of the years 2010, 2030 and 2075 have shown increasing of potential evapotranspiration of special crops for both climate change scenarios ($CCCM_{prep}$ and $GISS_{prep}$). The increasing of ET_0 for $GISS_{prep}$ scenario is within the interval 3–51%. Mean potential evapotranspiration value of vegetables will increase about 23% and for special plants about 34%.

Results of analyses also have shown indisputable limitation of recent irrigation practice used for the calculation of water demands for special plants and vegetables for climate change scenarios in horizon of the year 2075 in Slovakia. The increase in ET_0 for particular crops ranges from 9% to 57% in comparison with consumptive water usage calculated according to valid Slovak Technical Standard (STN no. 83 0635). The author's results have shown that biological consumptive water usage curve coefficients provide inaccurate results of V_c calculations for irrigation water demands of special plants and vegetables. The obtained results emphasize need of revision of technical standard required for precise planning, design and controlling irrigations and also for water resources management services in changed climate conditions.

Acknowledgments The chapter was supported by Slovak Grant Agency Project (VEGA 1/3470/06) and by the Science and Technology Assistance Agency under the contract of APVT 51-019804.

References

Allen R et al. (1998) Crop evapotranspiration – Guidelines for computing crop water requirements. FAO Irrigation and drainage paper 56, Rome. 1998. ISBN 92-5-104219-5

Hansen S (1984) Estimation of potential and actual evapotranspiration. Nordic Hydrol 15: 205–212

Hansen S, Jensen HE, Nielsen NE, Svendsen H (1990) DAISY: Soil Plant Atmosphere System Model. NPO Report No. A 10, The National Agency for Environmental Protection, Copenhagen: 272

Hargreaves GL, Hargreaves GH, Riley JP (1985) Agricultural benefits for Senegal. River Basin. Journal of Irrigation and Drainage Engineering 111: 111–124

Harrison PA, Butterfield RE, Downing TE (1995) Climate Change and Agriculture in Europe: Assessment of Impacts and Adaptation. Research Report No. 9. Environmental Change Unit. University of Oxford: 41

Lapin M, Melo M (2002) Time series scenarios of ten climate characteristics for horizon 2001–2090 according to models CCCM2000 a GISS98. In: Proceedings of Abstracts of XIV. Czechoslovak Bioclimatological Conference: Bioclimate–Environment–Economy. Lednice: 4 (In Slovak)

Lapin M, Melo M, Damborská I (2001) Scenarios of Set of Several Physically Consistent Climate Elements. National Climate Program of the SR, VI. 2001, vol. 11. (In Slovak)

Lapin M (2004) Climate Change and Their Possible Impact on Hydrological Cycle – Scenarios of Changes Till the Year 2100 www.dmc.fmph.uniba.sk (In Slovak)

Šiška B, Takáč J, Igaz D (2004) May we expect the yield distribution changes as an influence a climate change in Danubian Lowlands? In: Conference proceedings: Climate Change – Weather Extremes. Organisms and Ecosystems. International Bioclimatological Workshop. Viničky 2. – 26. 8. 2004. CD ROM. (In Slovak)

Sláma V (1971) Biological Consumptive Water Usage Curve. Research Final Report R VI- 4/1-04. Praha. (In Czech)

Sobocká J, Šurina B, Torma S, Dodok R (2005) Climate Change Impact on Soil Properties and Processes in Slovakia. Research Institute of Soil Science and Soil Conservation, Bratislava (In Slovak)

Špánik F, Šiška B, Repa Š (1996) Impact Assessment of Climate Change on Agriculture and Appropriate Adaptive Measures: National Climate Program of SR. III. 1996. vol. 4. (In Slovak)

Takáč J, Zuzula I (2000) Climate Change Adaptation Measures for Agriculture in the Slovak Republic. National Climate Program of SR. 2000. vol. 9. (In Slovak)

STN no. 83 0635 Irrigation water demand calculation

Climate Change Impact on Spring Barley and Winter Wheat Yields on Danubian Lowland

J. Takáč and B. Šiška

Keywords Climate change · Spring barley · Winter wheat · Yield · Modelling

Introduction

Global warming is probably one of the most important environmental problems of mankind in its history. Disproportion between stability of the Earth climate system as a natural source on the one hand and energetic and water needs of mankind on this source on the other hand contributes to the increase of potential risk in the agricultural sector too. Possible climate change impact could have crucial influence on agricultural production and consequent socio-economic impacts.

Several climate change impact studies were elaborated for agricultural sector in Danubian lowland region during last decade of years. Except for works focused on drought effects of climate change (Škvarenina et al. 2004, Bárek 2006) most of studies were focused on crop yield modelling and testing of adaptive measures to reduce negative climate change impact. Results are strongly affected by level of growth simulation models as well as general circulation models outputs (Eitzinger et al. 2004) and therefore still new results and relations are found.

In the Slovak Republic, climate change impacts on agriculture were evaluated in the frame of several programmes of U.S. Country Study Program (Špánik et al. 1997, Takáč and Heldi 1996) and National Climate Program of the Slovak Republic (Špánik et al. 1996, Šiška and Mališ 1997, Šiška and Špánik 1999, Takáč 2001).

Today the simulation models are practically the only complex tool to estimate crop response on the climate change without carrying out expensive experiments. Winter wheat as a strategic crop was the most simulated crop in the Slovakia (Takáč and Heldi 1996, Šiška and Mališ 1997). Spring barley was not evaluated so frequently as compared with winter wheat. Analyses were focused also on evaluation of the possible acclimation effects (Šiška 1997). All yields were simulated as nutrient non-stressed yields.

New generation of climate change scenarios is available for agroclimatic modelling in conditions of Slovak republic since 2006 (Lapin et al. 2006). This generation of scenarios was already used for simulation of maize yields in climate change condition (Samuhel and Šiška 2007, Šiška and Samuhel 2007). The aim of this chapter is to evaluate possible climate change impact on spring barley and winter wheat yields according to new generation of GCM in two variants of emissions scenarios – SRES A2 and SRES B2. Nutrient and irrigation level were tested as possible adaptive measures to reduce negative impacts of climate change in condition of Danubian lowland – the most productive agricultural area of Slovakia.

Materials and Methods

Evaluation of the climate change impacts on spring barley and winter wheat yields was based on simulations by agroecological model DAISY. DAISY is a one-dimensional model simulating water, energy, nitrogen and soil organic matter content

J. Takáč (✉)
Soil Science and Conservation Research Institute, Gagarinova 10, 827 13 Bratislava, Slovakia
e-mail: takac@vupu.sk

K. Strelcová et al. (eds.), *Bioclimatology and Natural Hazards*,
DOI 10.1007/978-1-4020-8876-6_24,

balance. Crop development and yield is possible to simulate in dependence on crop rotation and various management strategy. DAISY simulates plant growth and development, including the accumulation of dry matter and nitrogen content in different plant parts. The main plant-growth processes considered in DAISY are photosynthesis, respiration, partitioning of assimilates, stress factors and leaf and root development. DAISY allows for building complex management scenarios (Hansen et al. 1990, Hansen 2000, Abrahamsen and Hansen 2000).

Global radiation, air mean temperature and precipitation for $2xCO_2$ climate were generated by general circulation model CCCM (Lapin et al. 2001). Meteorological data for reference period of years representing Danubian Lowland came from climatic station Hurbanovo. Crop yields were simulated for the reference period of years 1966–1985 in variants for emission scenarios SRES A2 and B2.

Crop parameters were set up according to the experimental data from field trials of Research Institute of Irrigation located in Most near Bratislava during years 1981–1987. Confirmation of the model was based on the results from experimental plots in Lehnice farm as well as on the data from Stationary Experiment of Research Institute of Irrigation from the year 1990 to 1994 (Takáč and Košč, 1995). Rainfed and irrigated variants were evaluated.

As a mineral fertilizer, $60\,kg\,N\,ha^{-1}$ was applied into spring barley crop. Mineral nitrogen was applied in two terms: $40\,kg\,N\,ha^{-1}$ before sowing in March and $20\,kg\,N\,ha^{-1}$ after sowing in May.

In winter wheat crop $150\,kg\,N\,ha^{-1}$ as a mineral fertilizer was applied. Mineral nitrogen was applied in three terms: $30\,kg\,N\,ha^{-1}$ before sowing in October, $50\,kg\,N\,ha^{-1}$ in March and $70\,kg\,N\,ha^{-1}$ in May.

Irrigation was applied automatically after lowering of soil water content below 50% of available water capacity for both evaluated cereals. Thirty millimetres of irrigation water was simulated. Interval between irrigations was set up to 10 days. This restriction assumption was chosen because limited irrigation water supply is expected. Nitrogen non-stressed full-irrigated variant was not simulated as this one was evaluated in the past (Šiška et al. 2004).

DAISY model calculates photosynthesis rate using a light saturation response curve. The effect of CO_2 concentration was included to the DAISY parameterization according to light-saturated photosynthesis rate

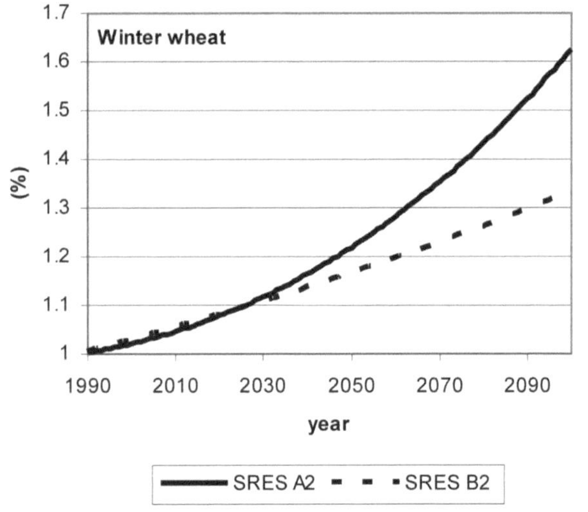

Fig. 1 Efficiency of photosynthetically active radiation as influenced by CO_2 concentration in winter wheat crop up to year 2100 – emission scenarios SRES A2 and SRES B2

Fm ($g\,CO_2\,m^{-2}\,h^{-1}$) and initial light use efficiency ϵ [$(g\,CO_2\,m^{-2}\,h^{-1})/(W\,m^{-2})$].

Efficiency of photo synthetically active radiation for winter wheat (Fig. 1) and spring barley (Fig. 2) crops was recalculated in dependence upon CO_2concentration in the atmosphere (Cure and Ackock 1986) for emission scenarios SRES A2 and SRES B2 (IPCC 2001).

Representative horizons of soil profiles of Danubian lowland were defined according to texture, parameters

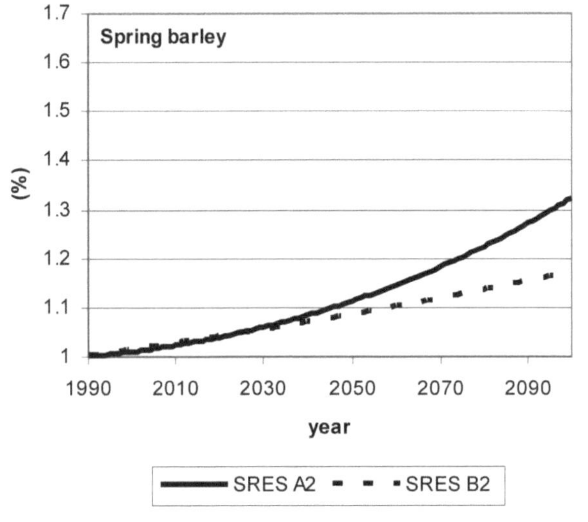

Fig. 2 Efficiency of photosynthetically active radiation as influenced by CO_2 concentration in spring barley crop up to year 2100 – emission scenarios SRES A2 and SRES B2

of retention curves, hydraulic conductivity, humus content and C/N ratio. Danubian Lowland was characterized with medium soil profile.

Results

Spring barley as well as winter wheat grain yields were favourably affected by soil properties and climatic conditions in the reference period of 1966–1985. They were limited first of all by available soil water (Table 1).

Due to increasing air temperature the flowering and harvest of spring barley and winter wheat should be accelerated by about 15 days in the decade 2061–2070 as compared to the reference period of 1966–1985 according to the scenario SRES B2 and by 12 days according to the scenario SRES B2.

Increase of CO_2 concentration and consequent increase of photosynthesis rate will positively affect the yields of spring barley, especially towards more distanced time horizons. On the other hand, course of meteorological elements generated by general circulation models cause some negative effects which frequently led to the fall of simulated yields. Generally, the yield increase in conditions of climate change will not correspond to the theoretical level of efficiency of photosynthetically active radiation for both spring barley and winter wheat crops that was calculated in dependence upon CO_2 concentration in the atmosphere for emission scenarios SRES A2 and SRES B2.

Results of crop yield simulations are influenced by interactive effect of factors taking into account. Lack of water was expressed in yield decline. Spring barley as well as winter wheat yields are also affected by duration of growing season and possible absorption of photosynthetic active radiation. Due to global warming the growing season will move towards months at the beginning of year that are characterized by lower radiation inputs.

Based on the simulation results, fertilization effect of CO_2 on spring barley and winter wheat top dry matter yield in rainfed conditions is evident according to both emission scenarios. Because the role of CO_2 concentration on photosynthesis rate in DAISY model is dominant as compared with other factors influencing formation of yield, most significant increase of biomass yield was found in simulations according to SRES A2. According to the scenario SRES B2 the effect of CO_2 on top dry matter yields in Danubian Lowland is insufficient to compensate the negative effect of the other environmental factors (Figs. 3, 4, 5 and 6).

Top dry biomass yield will be formed mainly by straw yield in future climate. Simulated harvest index dropped in range from 2 to 12 per cent on average as compared with reference period of years (Figs. 7 and 8). Grain yields would be probably reduced by high temperatures during ripening. Transport of assimilates from other parts of the plant into grains is not so effective because of accelerating effect of high temperatures on ripening of cereals (Šiška 1997). If we compare rainfed and irrigated variants, variability of yields is significantly smaller in irrigated ones (Figs. 3, 4, 5 and 6). Irrigation would be an effective measure to reach stabile yield of cereals in future climate.

Soil water content seems to be the main limiting factor of spring barley yields. Spring barley top dry matter yield increased by 27 per cent and grain yields

Table 1 Average simulated grain yields of spring barley and winter wheat (tons ha^{-1}) on Danubian Lowland in the period 1966–1985

Crop	Rainfed	Irrigated
Spring barley	4.22	5.79
Winter wheat	5.17	7.29

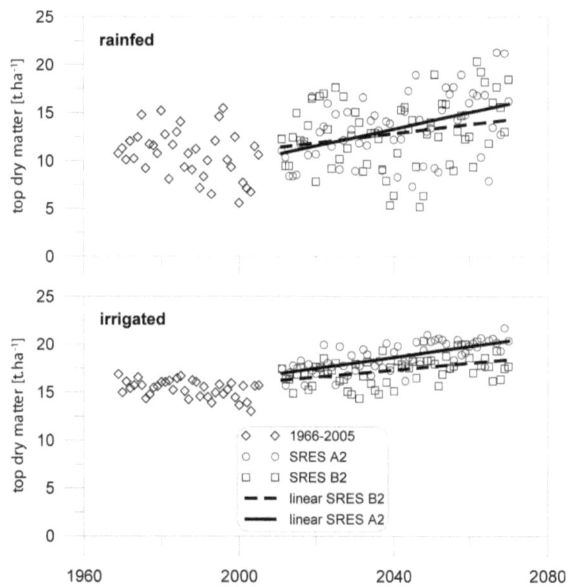

Fig. 3 Simulated rainfed and irrigated top dry matter yields (grain and straw) of winter wheat (tons ha^{-1}) in Danubian Lowland in the period 1966–2005 and according to the scenarios SRES A2 and SRES B2

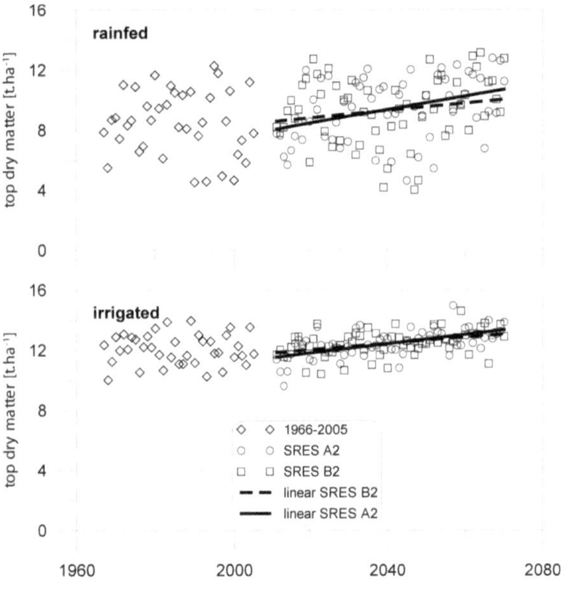

Fig. 4 Simulated rainfed and irrigated top dry matter yields (grain and straw) of spring barley (tons ha^{-1}) in Danubian Lowland in the period 1966–2005 and according to the scenarios SRES A2 and SRES B2

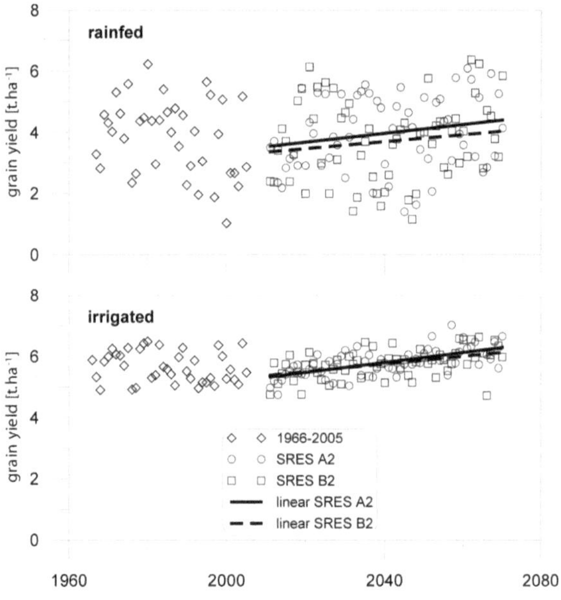

Fig. 6 Simulated rainfed and irrigated grain yields of spring barley (tons ha^{-1}) in Danubian Lowland in the period 1966–2005 and according to the scenarios SRES A2 and SRES B2

by 24 per cent in water and nutrient non-stressed conditions (Šiška et al. 2004). Yields are influenced by CO_2 concentration first of all in the $2xCO_2$ climate. Effect of CO_2 on spring barley yields was suppressed in rainfed variants with limited irrigation and variants with nitrogen fertilization, mainly according to the scenario SRES B2, which assumes less raise of CO_2 concentration.

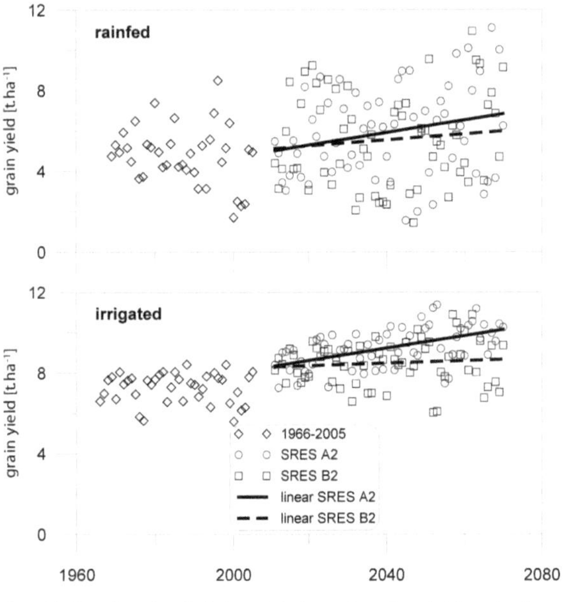

Fig. 5 Simulated rainfed and irrigated grain yields of winter wheat (tons ha^{-1}) in Danubian Lowland in the period 1966–2005 and according to the scenarios SRES A2 and SRES B2

Irrigation is the unavoidable condition of the water regime optimization and agricultural production stabilization.

Increase of irrigation demands of spring barley by about 3–20 per cent in dependence on the soil properties, region, scenarios and time horizon was found. According to the simulations the irrigation season of cereals will start by about 14 days earlier in time horizon 2070 (Takáč 2001). On the other hand, irrigation efficiency of spring barley and winter wheat decreases according to the scenarios on an average. Nevertheless the maximum values of irrigation efficiency of spring barley and winter wheat in dry years increased to $4.1-4.3$ kg m^{-3} and $4.3-4.6$ kg m^{-3}, respectively.

According to the statistical analyses of simulated yields there was found significant interactive effect of irrigation and fertilization on spring barley and winter wheat grain yields. Yields of spring barley increased by 53 per cent according to SRES A2 and 45 per cent according to SRES B2 on average as compared with rainfed variants. Yields of winter wheat increased by 88 per cent according to SRES A2 and 35 per cent according to SRES B2 on average as compared with rainfed variants.

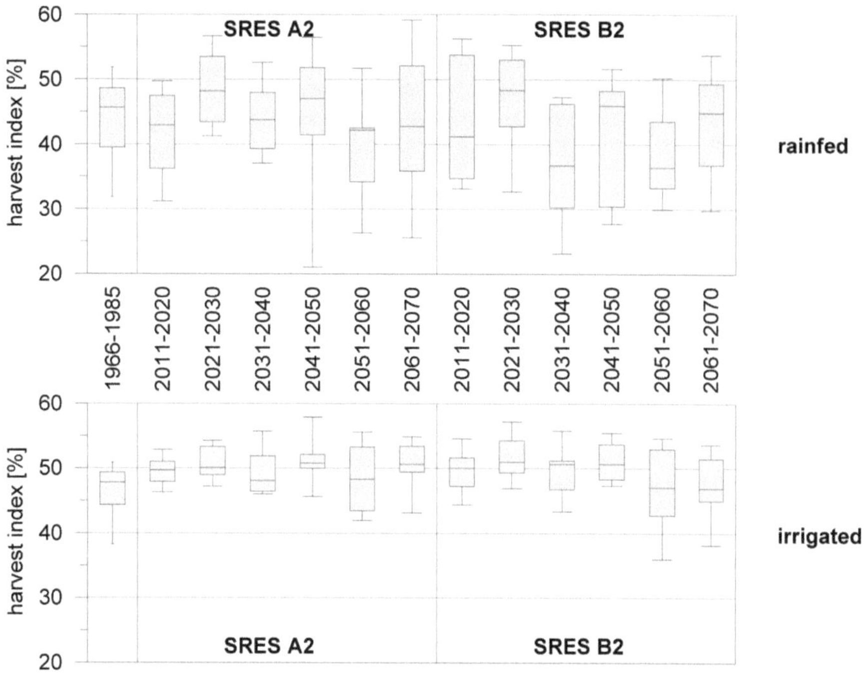

Fig. 7 Statistical characteristics of winter wheat harvest index (per cent) in Danubian Lowland in the period 1966–1985 and according to the scenarios SRES A2 and SRES B2

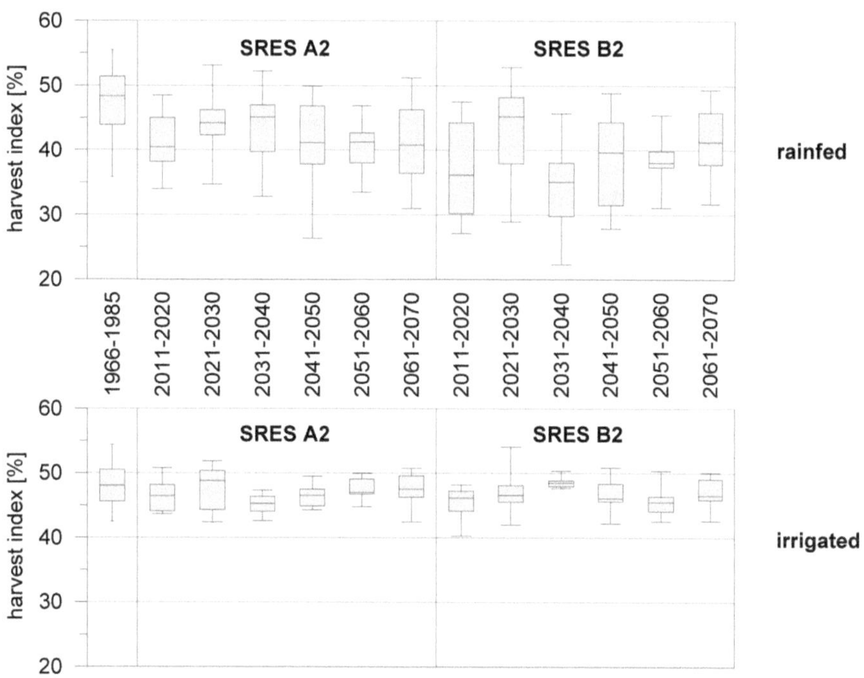

Fig. 8 Statistical characteristics of spring barley harvest index (per cent) in Danubian Lowland in the period 1966–1985 and according to the scenarios SRES A2 and SRES B2

Conclusion

Climate change impacts on yield of spring barley and winter wheat were evaluated according to agroecological model DAISY. Simulations confirmed the acceleration of spring barley and winter wheat development due to temperature increase. Simulated yields were influenced first of all by CO_2 concentration defined by emissions and climate change scenarios.

Course of meteorological elements generated by general circulation models cause some negative effects which frequently led to the fall of simulated yields.

Yield increase in conditions of climate change will not correspond with the theoretical level of efficiency of photosynthetically active radiation for spring barley crop that was calculated in dependence upon CO_2 concentration in the atmosphere for emission scenarios SRES A2 and SRES B2.

Soil water content was the main limiting factor of spring barley and winter wheat yields. Irrigation is an important factor of spring barley and winter wheat yields stabilization in conditions of the climate change. Despite the fact that shortage of water does not allow fully to utilize positive effect of CO_2 concentration on yield formation of spring barley and winter wheat the 10-day irrigation interval was found as the sufficient interval for yield stabilization.

Acknowledgments This study was made with a help of grant project VEGA 1/4427/07: Design of new agroclimatic regionalization of plant production in condition of changing climate in Slovakia and aAV/1109/2004: Climate change and drought in SR and from the study of project 2004 SP 20/06K 0A 03/000 00 10: Actual climate change and its impact to human society and Project No. 037005 Cecilia.

References

Abrahamsen P, Hansen S (2000) DAISY: An Open Soil – Plant – Atmosphere System Model. Environmental Modelling & Software 15: 313–330

Bárek V (2006) Climate change and irrigation. In: Study XXIII, Slovak biocimatogical society, Zvolen (in Slovak)

Cure JD, Ackock B (1986) Crop responses to carbon dioxide doubling. A literature survey. Agriculture and Forestry Meteorology, 38: 127–145

Eitzinger J, Trnka M, Hösch J, Žalud Z, Dubrovský M (2004) Comparison of CERES, WOFOST and SWAP models in simulating soil water content during growing season under different soil conditions. Ecological Modeling vol 171, Issue 3, 15: 223–246

Hansen S (2000) DAISY a flexible soil – plant – atmosphere system model. Equation section 1. The Royal Veterinary and Agricultural University, Copenhagen, pp 1–47

Hansen S, Jensen HE, Nielsen NE, Svendsen H (1990) DAISY – A Soil Plant System Model. Danish simulation model for transformation and transport of energy and matter in the soil-plant-atmosphere system. Copenhagen, The National Agency for Environmental Protection, pp 1–272

IPCC (2001) Climate Change 2001: The Scientific Basis. Contribution of Working Group in to the Third Assessment Report of the Intergovernmental Panel on Climate Change (IPCC). Houghton JT, Ding Y, Griggs DJ, Noguer M, van der Linden PJ, Xiaosu D (eds) Cambridge Univ. Press, UK, 944pp

Lapin M, Damborská I, Drinka R, Gera M, Melo M (2006) Scenarios of climatic daily values for Slovakia until 2100. Meteorological Journal 9(3–4): 149–156

Lapin M, Melo M, Damborska I (2001) Scenarios of several physically plausible climatic elements. MoE SR Bratislava, NCP SR 11: 5–30 (in Slovak)

Samuhel P, Šiška B (2007) Parametrization of crop simulation model "CERES-MAIZE" in Nitra – Dolná Malanta. Journal of Environmental Engineering and Landscape Management: Research Journal of Vilnius Gediminas Technical University 15(1): 1–5

Šiška B (1997) Supposed impact of CO_2 concentration on spring barley yields on Danubian lowland region. Acta horticulturae et regio tecturae 2: 107–120 (in Slovak)

Šiška B, Mališ J (1997) Supposed changes in production of winter wheat in consequence of climate change in Danubian lowland up to year 2075. MoE SR Bratislava, NCP SR 7: 84–92 (in Slovak)

Šiška B, Samuhel P (2007) Modelling climate change impact on maize (*Zea mays* L.) yields in conditions of Danubian lowland. Meteorological Journal 10(2): 81–84

Šiška B, Špánik F (1999) Supposed changes in phenological relation of winter wheat and spring barley as a consequence of climate change impacts in Danubian lowland up to year 2075. Meteorological Journal 2(3): 35–40 (in Slovak)

Šiška B, Takáč J, Igaz D (2004) Can we expect change of yield variability of cereals on Dunabian lowland as a climate change impact? In: Climate change – weather extremes – organisms and ecosystems [CD-ROM] Slovak Bioclimatological Society 16pp (in Slovak)

Škvarenina J, Krížová E, Tomlain J (2004) Impact of the climate change on the water balance of altitudinal vegetation stages in Slovakia. Ecology 23, Supplement 2: 13–29

Špánik F, Šiška B, Kostrej A, Liška E, Takáč J (1997) Climate Change Impacts on Plant Production and Adaptive Measures. Final Report. Slovak Republic's Country Study, AgTU (VŠP) Nitra, 1–41pp

Špánik F, Šiška B, Repa Š (1996) Climate change impacts on agriculture and adaptive measures. MoE SR Bratislava, NCP SR 4: 93–109 (in Slovak)

Takáč J (2001) Climate change impacts on water balance in agricultural landscape. MoE SR Bratislava, NCP SR 10: 16–26 (in Slovak)

Takáč J, Heldi A (1996) Possible Impacts of Climate Change On Irrigation Management. Final Report. SR's Country Study. MoE SR, SHMÚ, VÚZH Bratislava, 1–25pp. (in Slovak)

Takáč J, Koščo V (1995) Unsaturated Zone and Agriculture. Phare Project No. PHARE/EC/WAT/1 Danubian Lowland Ground Water Model. Final Report, MoE SR, VÚZH, Bratislava 64pp. (in Slovak)

Emissions from Agricultural Soils as Influenced by Change of Environmental Factors

J. Horák and B. Šiška

Keywords N_2O emissions · DNDC model · IPCC methodology · Sensitivity analysis

Introduction

Global climate change is probably one of the most important environmental problems of mankind during its history. Soils play an important role as a source of N_2O emissions in this process. Especially from ecological aspect this is taking a big account on individual evaluation of N_2O losses from the soils to the atmosphere (Bielek 1998).

Biogeochemical dynamics of carbon (C) and nitrogen (N) in ecosystem can be significantly affected by increasing N-fertilizers consumption which is the most significant source of N_2O emissions from agricultural soils in Slovakia. Different ways of farming management (tillage, depth of fertilizers application, crop residuals incorporated into the soil, multicrop systems) as well as environmental factors (air temperature, precipitation, soil pH) influence nitrogen balance and consequently N_2O emissions via microbial processes of nitrification and denitrification. Both farming management and environmental factors are interacted with climate-change influences because every change in farming management, climate, or soil condition can change biochemical and geochemical processes and N_2O emissions too (Li et al. 2004).

Total N_2O emissions are divided into direct and indirect emissions. Direct N_2O emissions from crop with fertilizers are of natural origin as a consequence of microbial processes nitrification and denitrification. They depend on N inputs from synthetic fertilizers, animal wastes from livestock production, plant residuals, and symbiotic fixation of leguminous. Indirect N_2O emissions are the result of processes of atmospheric ammonia and NO_x deposition and transformation of N from leaching and run off (Bouwman 1990, cited in IPCC 1996).

Nitrification is the aerobic microbial oxidation of ammonium ions to nitrite via NH_2OH, and then to nitrate

$$NH_4^+ \rightarrow NH_2OH \rightarrow NO_2^- \rightarrow NO_3^- \qquad (1)$$

N_2O is also emitted in the course of denitrification, the anaerobic microbial (mainly bacterial) reduction of nitrate successively to nitrite and then to the gases NO, N_2O, and N_2:

$$NO_3^- \rightarrow NO_2^- \rightarrow NO \rightarrow N_2O \rightarrow N_2 \qquad (2)$$

Both processes can simultaneously occur in soils, although the rates of the two processes depend on soil aeration and microsite availability (Van Cleemput and Baert 2002).

N_2O emissions in agriculture sector represent approximately 75% of all N_2O emissions in Slovakia. Despite the fact that N_2O emissions from agricultural sources decreased from 17,000 tons in year 1990 to 10,000 tons in year 1995 in the first half of 1990s (Šiška and Igaz 2005), it is important to develop strategies, which effectively decrease N_2O emissions.

N_2O emissions from agricultural soils evidently correlate with consumption trend of N-fertilizers in Slovakia. Quick decrease of N-fertilizers consumption in Slovakia after 1990 positively influenced N_2O

J. Horák (✉)
Department of Biometeorology and Hydrology,
Slovak Agricultural University, Hospodárska 7, 949 01 Nitra, Slovakia
e-mail: jan.horak@uniag.sk

K. Strelcová et al. (eds.), *Bioclimatology and Natural Hazards*,
DOI 10.1007/978-1-4020-8876-6_25, © Springer Science+Business Media B.V. 2009

emissions from agricultural soils. Economic growth will very probably bring consistent increase of N-fertilizers consumption in the following years which will affect increase of N_2O emissions from agricultural soils to the atmosphere. From the point of view of mitigation strategies there are new precise evaluation methods needed to reduce N_2O soil emission potential in context of environmental effects (Bielek 1998).

There are several ways for evaluation of N_2O emissions. IPCC methodology (1996) is used as the most common. Because the IPCC methodology (1996) does not take into account regional differences N_2O emissions are of high rate of uncertainties (20–200%). These differences can play an important role in such heterogeneous country as Slovakia is.

Modeling is an effective way to reduce uncertainties in estimation of N_2O emissions and is also very effective from the point of view of adaptive strategies proposal. Several models, like denitrification-decomposition (DNDC), CASA, CENTURY, EXPER-N, exist that can simulate greenhouse gas emissions. DNDC model was used for modeling of N_2O emissions from agricultural soils in this study.

Model can predict nitrogen (N) and carbon (C) cycle in these soils. DNDC and IPCC estimates were compared. Sensitivity analysis of DNDC model was made from the point of view of influences of farming management and environmental factors.

Materials and Methods

Study area is situated north-east from Nitra, which is the part of Danubian lowland, 160–180 m above the see level. Average year precipitation is 538 mm and average air temperature is 9.8°C (Climatic normal, 1961–1990).

There were two methods for estimation of N_2O emissions from agricultural ecosystem compared in this study:

1. IPCC methodoloy
2. simulation of N_2O emissions by DNDC model

Both methods were also subject of intercorrelation by Pearson correlation coefficient. DNDC model was consequently subject of sensitivity analysis considering changing farming management and changing environmental factors.

IPCC Methodology

N_2O emissions according to IPCC methodology depends upon the amount of nitrogen from fertilizers, plant residuals, and symbiotic fixation applied into the soil. The emission factor for inorganic N-fertilizer applied to the soils is 0.0125, that is, one assumes that $1.25\% \pm 1\%$ of total N applied to a field as mineral N-fertilizer is lost in the form of N_2O to the atmosphere (IPCC and Greenhouse Gas Inventory Workbook 1997).

Equation for calculating of N_2O emissions from the soil

$$E(\mathrm{kg\,N_2O-N\,ha^{-1}\,yr^{-1}}) = 1 + 0.0125\mathrm{N_f} \quad (3)$$

where

E = annual N_2O emissions from the soil,
N_f = amount of N in fertilizers ($\mathrm{kg\,N_2O - N\,ha^{-1}\,yr^{-1}}$)

Presented equation is based on direct proportion of acceleration of N_2O emissions from the soil depending on N-fertilizers amount. This equation depicts that there is an average loss of 1 kg N–N_2O in a year from 1 ha of unmanured agricultural soils (Bouwman, cit in Bielek 1998).

IPCC methodology requires national statistics of fertilizers consumption, livestock population, and plant residue management. It does not require data like area of agricultural used land, soils, meteorological data, fertilizers type or other details of agricultural management (Li et al. 2001).

DNDC Modeling

DNDC model was created by Institute for the Study of Earth, Oceans, and Space on New Hampshire University. DNDC model for estimating N_2O emissions is process oriented on computer simulation of soil carbon and nitrogen. The model consists of two components. The first component, consisting of the soil climate, crop growth, and decomposition submodels, predicts soil temperature, moisture, pH, redox potential (Eh), and substrate concentration profiles driven by ecological driver (e.g., climate, soil, vegetation, and anthropogenic activity). The soil climate submodel calculates vertical profiles of soil temperature, moisture, and soil redox potential driven by meteorological data and soil properties. The crop growth submodel calculates crop growth and its influence on soil environmental factors such as soil moisture, dissolved organic carbon (DOC), and available nitrogen concentrations. The decomposition submodel then generates vertical concentration profiles of substrates (e.g., DOC, NH_4^+, NO_3^-). The second component, consisting of the nitrification, denitrification, and fermentation submodels, predicts NO, N_2O, N_2, CH_4, and NH_3 fluxes based on the modeled soil environmental factors. Description of DNDC model is available on http://www.dndc.sr.unh.edu/.

DNDC Model Inputs

DNDC model requires inputs like meteorological data, soil properties (e.g., texture, pH, bulk density), vegetation (e.g., crop type), and management (e.g., tillage, fertilization, manure amendment, planting, harvest, etc.).

Soil

Soil type genetic properties including horizons and basic description of morphogenetic characters were found from soil profile examination into the depth of 1.75 m. Mechanical, chemical, and physical properties were established for each diagnostic horizon. Sandy loam is a dominant soil type with humus content (2.149% Hm). Soil organic carbon range from 0.96% to 1.31% and soil pH range from 4.59 to 5.39 (Chlpík and Pospíšil 2004).

Database includes detailed information on soil type, texture, soil pH, SOC, bulk density, etc. Observed area is an upland crop field with sandy loam soils with parameters that are given below:

– Bulk density ($g\,cm^3$)	1.4000
– Soil pH	4.9900
– Initial organic C content at surface soil (kg C/kg)	0.0135
– Clay fraction	0.0900
– Initial NO_3^- concentration at soil surface (mg N/kg)	4.0500
– Initial NH_4^+ concentration at soil surface (mg N/kg)	0.4050

Farming Management

Field trial during years 2000–2004 consists of crops in rotation as follows:

- 2000 – sugar beet (*beta vulgaris*)
- 2001 – barley (*Hordeum vulgare*)
- 2002 – maize (*Zea Mays*)
- 2003 – winter wheat (*Triticum aestivum*)
- 2004 – sunflower (*Helianthus annuus*)

Data on plowing, planting, harvest, and application of N-fertilizers were set up in relation to individual needs of different crops and actual weather conditions in years 2000–2004 (Fig. 1).

Meteorological Data

Daily maximum and minimum air temperatures in °C as well as daily precipitation in millimeters during

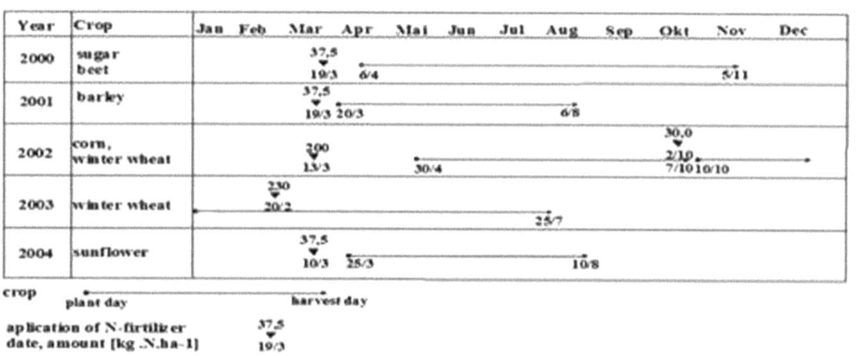

Fig. 1 Scheme of farming management

years 2000–2004 were measured on agrometeorological station of Department of Biometeorology and Hydrology, Slovak Agricultural University in Nitra located near the research area.

Sensitivity Analysis of DNDC Model

Sensitivity analysis of DNDC model was made for sunflower crop (*H. annuus* L.) in 2004 according scenarios as listed in Table 1.

Tested farming management:

- application of different types of mineral and organic fertilizers,
- depth of fertilizers incorporating,
- percentage of crop residue incorporated to the soil.

Tested environmental factors:

- changing soil types,
- soil organic carbon (C),
- soil pH,
- air temperature,
- atmospheric precipitation.

Table 1 Actual farming management, environmental factors, and alternative scenarios used for sensitivity analysis of DNDC model in sunflower vesture (*H. annuss* L.)

Actual farming management (2004)	
Crop residue	Alternative scenarios
10%	0%, 50%, 90%
Fertilizers	Alternative scenarios
37.5 kg/ha	0 kg/ha, 37.5, urea
Depth of fertilizers incorporating	Alternative scenarios
5 cm	15 cm, 25 cm
Actual environmental factors (2004)	
Air temperature (year average)	Alternative scenarios
10.1°C	−2°C, −4°C, +2°C, +4°C
Atmospheric precipitation (year average)	Alternative scenarios
561 mm	−20%, −10%, +10%, +20%
Soil types	Alternative scenarios
Sandy loam	Loamy sand (LS), loam (L), silty loam (SL), clay loam (CL), clay (C)
Soil organic carbon (C) content	Alternative scenarios
1.35%	0.5%, 1%, 2%, 3%
Soil pH	Alternative scenarios
4.99	5.5, 6.5, 7.5

Results

Estimation of N_2O Emissions According to IPCC Methodology

N_2O emissions from agricultural soils of the research area estimated according to IPCC methodology were in range 1.47–3.88 kg N_2O ha^{-1} yr^{-1} with average 2.43 kg N_2O ha^{-1} yr^{-1} in dependence upon amount of applied N-fertilizers during years 2000–2004 (Fig. 2). Among evaluated crops (sugar beet, barley, maize, winter wheat, sunflower) the highest N_2O emissions – 3.88 kg N_2O ha^{-1} yr^{-1} were found in maize and winter wheat crops in years 2002 and 2003.

Estimation of N_2O Emissions by DNDC Modeling

N_2O emissions estimated by DNDC model were in range 0.9–2.58 kg N_2O ha^{-1} yr^{-1} with average 1.7 kg N_2O ha^{-1} yr^{-1} during years 2000–2004 (Fig. 2). The highest N_2O emissions were found to be 2.58 kg N_2O ha^{-1} yr^{-1} from maize in 2002 and 2.31 kg N_2O ha^{-1} yr^{-1} from winter wheat in 2003.

Correlation and Comparison of Both Methods

N_2O emissions according to IPCC methodology and DNDC modeling were correlated by Pearson correlation coefficient (Rimarčík 2007).

A very close correlation was found between N_2O emissions estimated according IPCC methodology and that by DNDC (R = 0.95). However, the close correlation is also influenced by small number of analyzed years.

N_2O emissions according to IPCC methodology as compare with DNDC modeling emissions are higher by 25% in average. N_2O emissions are closely related to the amount of fertilizers applied. N-fertilizers of 230 kg was applied in years 2002 and 2003 while in years 2000, 2001 and 2004 only 37.5 kg was applied.

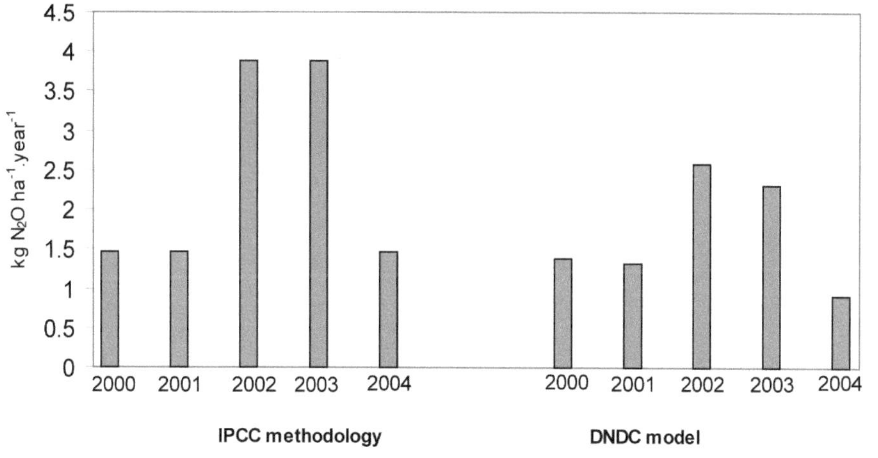

Fig. 2 N_2O emissions estimated according IPCC methodology and using DNDC model

Sensitivity Analyses Results of DNDC Model

Alternative Scenarios of Farming Management

Highest N_2O emissions were found for urea fertilizer (Fig. 3). Urea as a liquid fertilizer has a high emission potential because of evaporating to the atmosphere. Percentage of crop residuals incorporated to the soil after harvest also influenced N_2O emissions, but not so significantly as the type of applied fertilizer (Fig. 4). Depth of fertilizer incorporation into the soil did not influenced N_2O emissions significantly (Fig. 5).

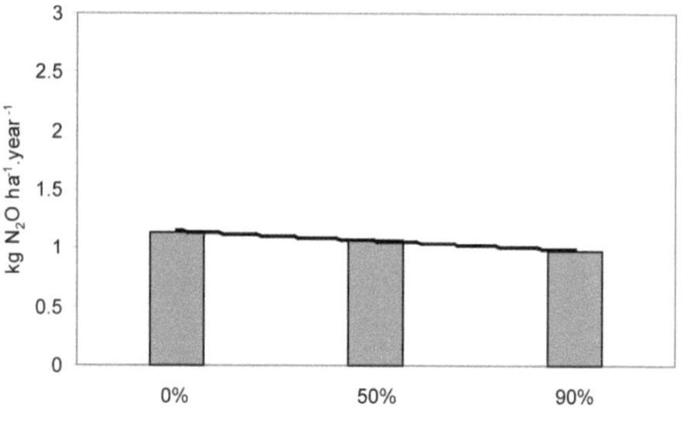

Fig. 4 Alternative scenarios of crop residue

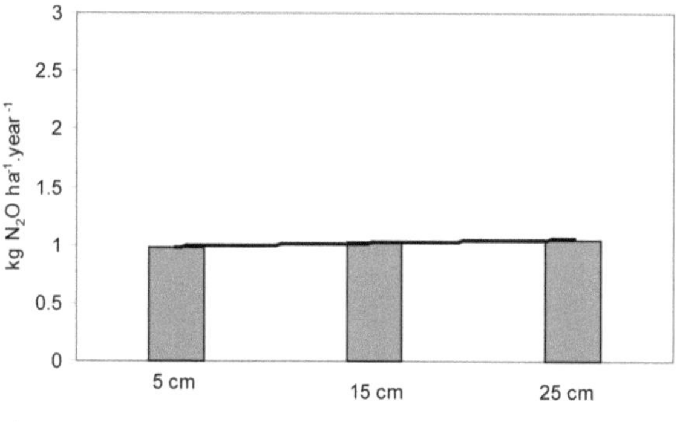

Fig. 5 Alternative scenarios of depth of fertilizers incorporation

Alternative Scenarios of Environmental Factors

According to computer simulations of different alternative scenarios of environmental factors the N_2O emissions were most sensitive to different soil type, air temperature, and soil organic carbon content (C). N_2O emissions exponentially increased from heavy soils to the light soils (Fig. 6). Light soils allow better move of water in the soil profile. Due to quick drying of these soils the nitrogen as an easy water-soluble compound can be emitted in form of N_2O to the atmosphere. Exponential dependence of N_2O emissions on linear increasing of air temperature was found. Both evaporation and temperature reflect radiation balance

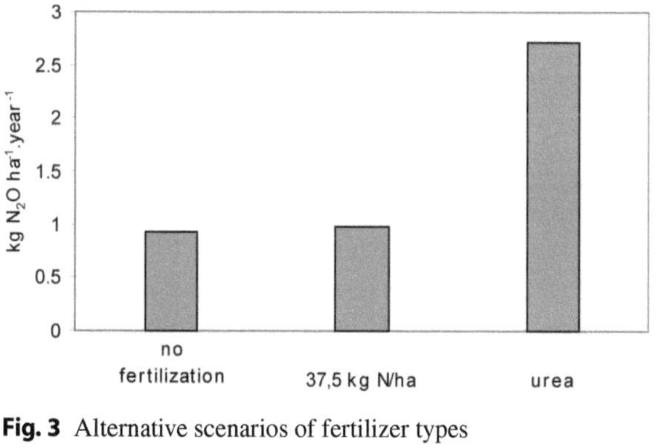

Fig. 3 Alternative scenarios of fertilizer types

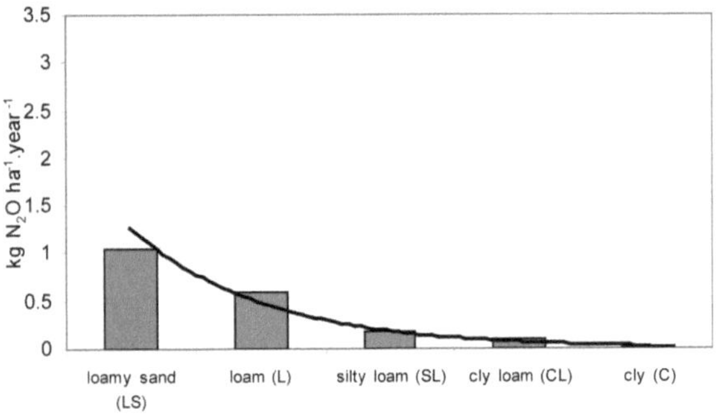

Fig. 6 Alternative scenarios of soil types

conditions and so evaporation rate naturally correlated with air temperature (Fig. 7). Increase of soil organic carbon (C) also caused exponential increase of N_2O emissions (Fig. 8). Atmospheric precipitation and soil pH did not significantly influenced flows of N_2O emissions (Figs. 9 and 10).

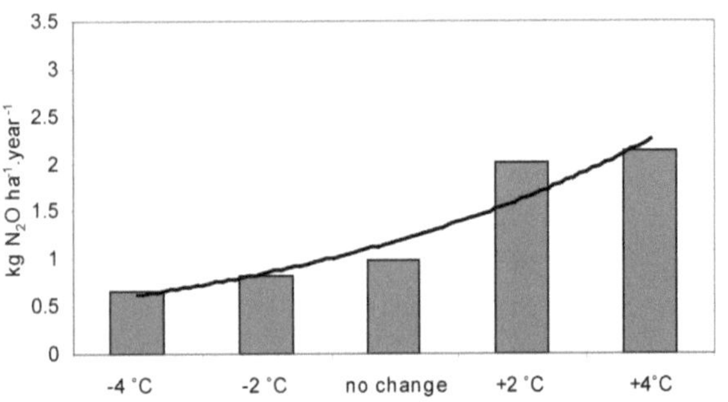

Fig. 7 Alternative scenarios of air temperature

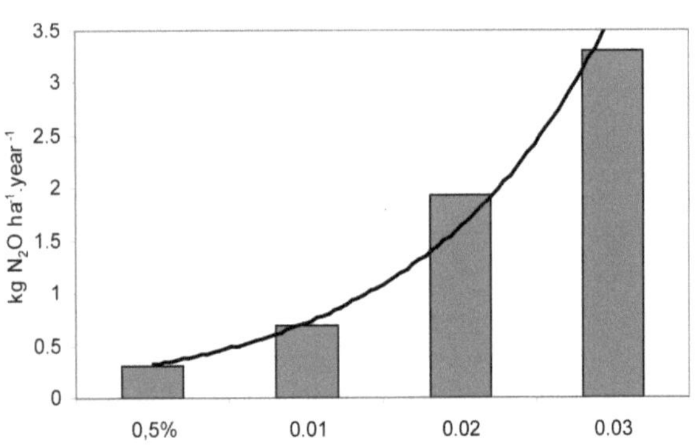

Fig. 8 Alternative scenarios of soil organic carbon (C)

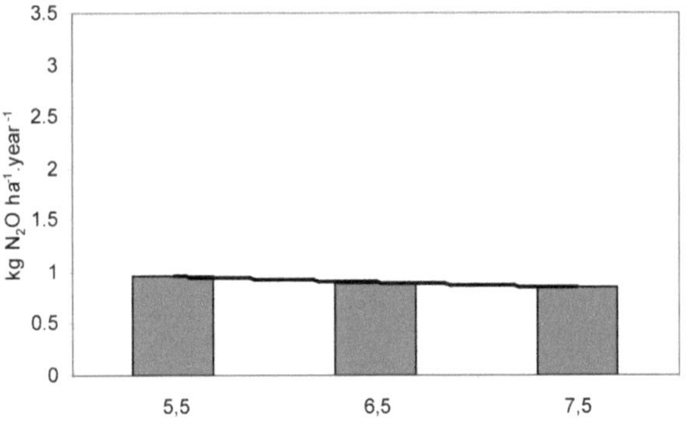

Fig. 9 Alternative scenarios of soil pH

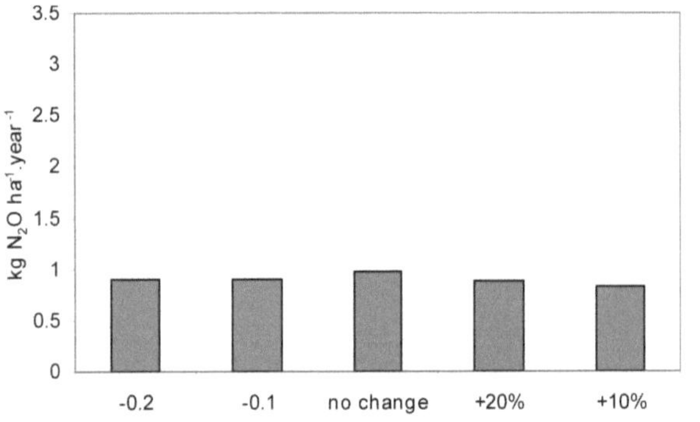

Fig. 10 Alternative scenarios of atmospheric precipitation

Conclusion

There was found very close correlation between N_2O emissions estimated according IPCC methodology and by using DNDC ($R = 0.95$) and therefore can be considered as relevant.

N_2O emissions are closely related to the amount of applied fertilizers.

N_2O emissions estimated according IPCC methodology are higher by 25% on average as compared with DNDC simulations.

Results of sensitivity analysis show that different farming management practices as well as environmental factors can significantly influence N_2O emissions from agricultural soils. Local soil properties (soil type, soil organic carbon C, soil pH), or meteorological conditions (air temperature, atmospheric precipitation, etc.) play important role and so modeling of N_2O emissions is necessary.

On regional scale the modeling gives a possibility to use geographic information system (GIS) to support more precise estimation of N_2O emissions through the DNDC model.

Acknowledgments This project was supported by grant agency of Slovak republic – VEGA 1/4427/07: New Agroclimatic Regionalization in condition of changing climate.

References

Bielek P (1998) Nitrogen in agricultural soils of Slovakia. Bratislava

Chlpík J, Pospíšil R (2004) Spatial characteristic of mechanical and chemical soil properties on experimental base of Slovak agricultural University in Nitra – Kolinany. Acta fytotechnica at zootechnica 1(7): 28–35 (in Slovak)

IPCC, Greenhouse Gas Inventory Workbook (1997) (Revised 1996 Guidelines for National Gas Inventories), vol 2, Bracnell, England

Li C, et al. (2001) Comparing a process-based agro-ecosystem model to the IPCC methodology for developing a national inventory of N_2O emissions from arable lands in China. Nutrient Cycling in Agroecosystem 60: 159–175

Li C, et al. (2004) Modeling greenhouse gas emissions from rice-based production systems: Sensitivitiy and upscaling. Global Biogeochemical cycles vol 18, GB1043, doi:10.1029/2003GB002045

Rimarčík M (2007) Statistics for practice: pp. 170

Šiška B, Igaz D (2005) Emissions of N_2O from agricultural sector during years 1990–2003 in Slovakia. In: Jubilee Scientific Conference – State-of-the-Art and Problems of Agricultural Science and Education. Agricultural University Plovdiv, Scientific Works, vol L, book 6 pp. 359–364

Van Cleemput O, Baert LA (2002) Key compound in N loss processes under acid conditions. Plant Soil 76: 233–241

Index

MIX
Papier aus verantwortungsvollen Quellen
Paper from responsible sources
FSC® C105338

If you have any concerns about our products,
you can contact us on
ProductSafety@springernature.com

In case Publisher is established outside the EU,
the EU authorized representative is:
Springer Nature Customer Service Center GmbH
Europaplatz 3, 69115 Heidelberg, Germany

Printed by Libri Plureos GmbH
in Hamburg, Germany

For EU product safety concerns:
IN TECH d.o.o., Prolaz Marije Krucifikse Kozulić 3, 51000 Rijeka, Croatia,
info@intechopen.com or visit our website at intechopen.com.

Meet the editor

Associate professor Dr. Doblin Sandai is a molecular medical mycologist. He graduated from the Institute of Medical Sciences, University of Aberdeen, Scotland, UK. Currently, he is working as a university lecturer at the Infectomics Cluster, Advanced Medical and Dental Institute, Universiti Sains Malaysia Bertam, Penang, Malaysia. His research is centered on virulence factors, proteins, and gene regulation, as well as developing alternative inhibitors from natural products to combat infections and resistance of the pathogenic yeast *Candida albicans*. Associate professor Sandai is involved in teaching and supervising postgraduate students. He has published over 40 articles, including books and peer-reviewed journals related to carbon source attribution to the pathogenicity of *C. albicans*. He is an international member of the American Society of Microbiology and the International Society of Human and Animal Mycology.

Contents

Preface

Candida albicans and other medically relevant *Candida* species are mainly commensal yeasts commonly found on mucosal surfaces and in the gastrointestinal and genitourinary tracts of humans. These yeasts are opportunistic pathogens that are mostly benign, but can be infectious when the host immune system becomes impaired or an environmental niche becomes available. The infections caused by *Candida* species are generally known as candidiasis, which can be further classified into superficial infections such as oral thrush, vaginal candidiasis, and invasive candidiasis. Invasive candidiasis is one of the most common nosocomial invasive infections with a high mortality rate, even with first-line antifungal treatments. This book will mainly focus on *C. albicans* and is divided into four sections, namely emerging fungal diseases, adaptation and resistance, interaction of cell hosts and other microbes, and new therapeutics and management.

Section 1 starts with an overview of the emerging and emerged pathogenic *Candida* species. These chapters will offer a historic perspective of *C. albicans* and other non-albicans (NAC) that cause serious disease and will review the current knowledge of emerging NAC pathogens with useful graphics and current references. Also, the articles consulted great variations of *Candida* species were tested and in MIC concentration from the development of new natural products, displayed the best performing plants that are effective in control of oral and systemic candidiasis, cheap and accessible for the population.

Section 2 highlights the unique metabolic adaptation of *C. albicans* that relates to the successful assimilation of available carbon sources during infections and resistance. Also, it has been proven that host carbon sources, blood serum, high temperature, acidic environments, and morphogenesis have a direct impact on mannoprotein content. Therefore, these changes might be involved in the adherence, drug resistance, and virulence of *C. albicans*. The proteomics data obtained were performed via comparative analysis of the proteins found in cell walls, whose primary location and function have been reported in different cellular compartments.

Section 3 covers important issues regarding the common role of phagocytes in antifungal immunity, defects in other immune cells, and immunologic pathways that give rise to susceptibility to infection with *Candida* and other fungi. Cell-mediated immunity is essential for antifungal immunity. This is evidenced by the predisposition to severe fungal infection in infants with severe combined immunodeficiency, a life-threatening condition manifested by low, absent, or severely dysfunctional T cells. Also, the mechanism of interaction between *C. albicans* and diverse microorganisms, including bacteria and yeast-like fungi from the genus *Candida*, is discussed in these chapters.

Section 4 is a special section on the therapeutics and management of fungal infections and new treatment strategies that involve the use of nanotechnology to overcome challenges associated with antifungal therapy. This review focuses on various means that have been employed to effectively counter problems supplementary to *C. albicans* by using nanotechnology.

This book is an updated overview of collected data on *C. albicans* research and review information. Clinicians and researchers should promptly manage pre- and post-candidiasis before proceeding to more sophisticated intensive investigation. Lastly, proper infection control principles should be kept in mind at all times to avoid invasion by this opportunistic pathogenic yeast.

Doblin Sandai
Infectomics Cluster
Advanced Medical and Dental Institute
SAINS@BERTAM
Universiti Sains Malaysia
Bertam, Penang, Malaysia

Section 1

Emerging Fungal Diseases

Chapter 1

Emerging Pathogens of the Candida Species

Bo Yang and Reeta Rao

Additional information is available at the end of the chapter

http://dx.doi.org/10.5772/intechopen.80378

Abstract

In recent years, opportunistic and nosocomial fungal pathogens have been dominated by yeasts of the genus *Candida*. Most of the research has focused on *Candida albicans* since it is the most prominent etiological agent. There are numerous publications that describe the biology, virulence factors, morphology, immunity, genomics, diseases, and laboratory aspects of *Candida albicans*. In this chapter we offer a historic perspective of *C. albicans* and focus on other non-albicans candida (NAC) that cause serious disease. We review the current knowledge of emerging NAC pathogens with useful graphics and current references. This chapter is laid out as an overview and is geared for students seeking basic information and may be superficial for an infectious diseases clinician.

Keywords: *C. albicans*, *C. tropicalis*, *C. glabrata*, *C. parapsilosis*, *C. guilliermondii*, *C. auris* and *C. krusei*, non-albicans candida (NAC), fungal pathogen

1. Introduction

By the end of the twentieth century, hospital-acquired fungal infections were on the rise. Infections by yeasts of the genus Candida have become one of the most common causes of bloodstream infections [1]. The increase in fungal infections is typically attributed to the longer survival of immuno-compromised individuals as well as the increase in the number of people in long term health care facilities undergoing, immunosuppressive therapy, long-term catheterization, broad-spectrum antibiotic use among others. This alarming increase in nosocomial fungal infections has alerted clinicians and scientist that yeasts, previously thought innocuous and relegated to plant pathology or industrial use were capable of causing serious illness.

While infections caused by Candida species are typically superficial and restricted to the urogenital or mucosal oral cavities, they are also capable of entering the bloodstream leading to deep-tissue infections [2].

The predominant yeasts in bloodstream infection remain restricted to the genus Candida [3] most of which, belong to the CTG clade, where the CTG codon is translated as Serine rather than Leucine [4]. Although the recent rise in the number of these infections [5] is mainly associated to *C. albicans*, non-albicans candida (or NAC) related diseases are also increasingly reported in different parts of the world [6]. The relative rates of infection among all Candida infections are shown in **Figure 1**. There are at least a dozen *Candida* species that can be pathogenic for humans, but more than 90% of reported invasive infections are associated with *C. albicans*, *C. glabrata*, *C. parapsilosis*, *C. krusei*, and *C. tropicalis* [7] .

The definition of a new or "emerging" pathogen is subjective at best. For example, how many independent isolations are required before an emerging pathogen is established as an infectious agent? Indications of emerging infections typically consist of case reports. The incidence of yeast infections is likely under-reported because it is dependent on clinical diagnosis and their desire to investigate, then confirm the novelty of the etiological agent, write a compelling report, and to withstand the critique and rigor of publishing the report. A close partnership between a research scientist and a clinical physician is critical for the most rigorous reports.

Furthermore, a single report describing several cases of infection by a novel microbe does not necessarily indicate that a new infection has emerged. Since these temporally separated clinical samples must be confirmed as being independent and not affected by laboratory practices, personnel and/or environmental factors [8]. For example, recent retrospective studies of blood stream infection caused by the genus *Candida*, reported *C. albicans*, *C. tropicalis*, *C. glabrata*, *C. parapsilosis*, *C. guilliermondii*, and *C. krusei*. This list closely resembles the list of known pathogenic Candida species in 1963 [3] suggesting that the emergence of other yeasts

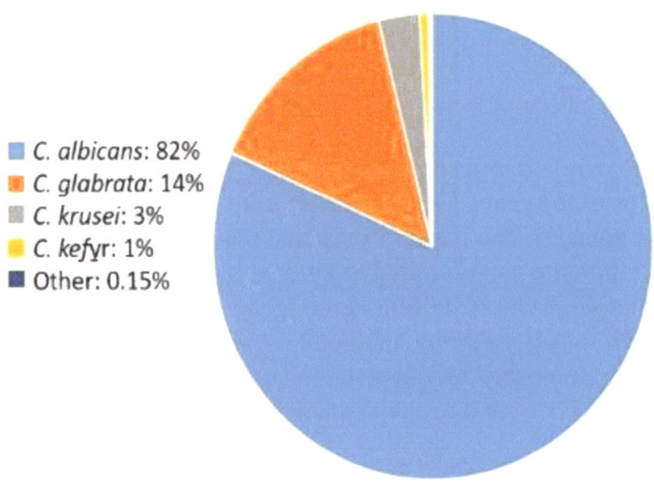

Figure 1. Relative rates of infection with Candida species.

as potential bloodstream isolates is a reflection of the changes in medical and laboratory practices since the early 1960s. Another important variable is that the advances in genomic tools coupled with the ability of contemporary blood culture systems and procedures to support the growth of the unusual yeast isolates have greatly accelerated the frequency of isolation and identification of novel (non-*C. albicans*) isolates as important pathogen.

For the purposes of this chapter we limit the definition of emerging pathogens to those that have recently appeared within a population or those whose incidence or geographic range is rapidly increasing or threatens to increase in the near future or those caused by previously undetected or unknown infectious agents.

2. Established pathogens of the Candida species

2.1. *Candida albicans*

The first description of a yeast infection was of thrush, by Hippocrates in the fifth century B.C. [9]. Since its first microscopic detection in thrush swabs by Langenbeck, subsequently Berg and Gruby [9] in 1839, *Candida albicans* has been confirmed to be the primary etiologic agent of thrush and cause of many other forms of mucosal disease. In fact, *C. albicans* is most frequently isolated species from human yeast infections [3]. Here we touch upon some of key features that make *C. albicans* a successful pathogen. The various morphological forms of *C. albicans* (**Figure 2**) have been associated with the shift from commensal or pathogenic states. The change from yeast to hyphae is thought to help cell adhesion and facilitate tissue infection, macrophage evasion and biofilm development [10].

C. albicans is a diploid polymorphic yeast (**Figure 2**) of human mucosal surfaces, a fungus that commonly found in the gastrointestinal (GI), respiratory, and urogenital tracts. It is generally commensal but able to turn into an opportunistic pathogen in immunocompromised or immuno-deficient individuals. It is the major species causing invasive candidiasis (46.3%), followed by *C. glabrata* (24.4%) and *C. parapsilosis* (8.1%) [11]. Systemic infections of *C. albicans* often cause death with a mortality rate of ~40% [12]. One reason leading to over growth of *C. albicans* is its ability to respond to a myriad of environmental imbalances such as changes in nutrition, temperature and pH [13]. Other important factors that increase the risk of *C. albicans* infection are prolonged treatment with broad spectrum antibiotics, surgical procedures, various diseases such as diabetes, trauma, and other genetic disease or congenital malformation. [2].

Candida albicans belongs to the CTG clade of fungal species and translates canonical leucine codons, CTG, to serine [14]. It prefers to use glucose as the carbon and has multiple approaches to transport and metabolize glucose. Nutritional changes alter virulence of *C. albicans* both during systemic infection as well as mucosal vaginitis, suggesting that alternative carbon sources within host environments are important during *C. albicans* infections [15].

Post-transcriptional mechanisms underlying this transition include mRNA stability, alternative transcript localization, and translation and influence *C. albicans* virulence processes. Below we highlight some key pathways but for details refer to book Candida and Candidiasis [16].

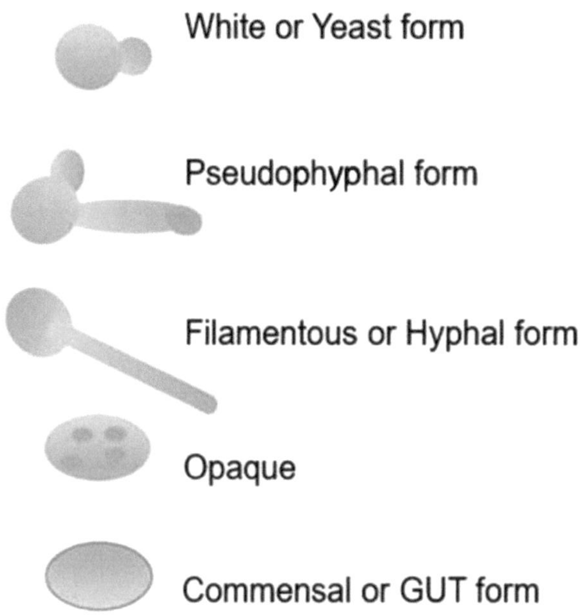

Figure 2. *Candida albicans* polymorphology.

At the transcriptional level, *C. albicans* fibrin Sac6 modulates morphogenesis and oxidative stress responses [17]. As mentioned above, metabolic adaptation a key virulence determinant is involved in the susceptibility of *C. albicans* to antifungal drugs as well as stress resistance and innate immune responses [18]. *C. albicans* typically infects epithelial cells through two specific mechanisms, active penetration and endocytosis. Before the infection, *C. albicans* transfer between commensal and invasive states through distinct genetic pathways to regulate the expression of hypha-specific and/or phase-specific genes. And these genes express proteins to regulate directly or indirectly to the pathogenesis and virulence of *C. albicans* [19]. The gastro-intestinally induced transition (GUT) highlights how these pathways are used [20]. Superficial candida infections require the interaction between fungal cell surface proteins or pathogen associated molecular patters (or PAMPs) with host innate immune cells system or pathogen recognition receptors (or PRRs) (**Figure 3**). For example, cell-wall β-glucans can stimulate monocyte reprogramming as one of the main immunological responses in hosts [2].

When interacting with macrophages, the SPS system is stimulated in the nutrient poor host environment and is critical for resistance of *C. albicans* to macrophages. It consists of three components; an amino acid transporter, SSY1, a membrane associated protein PTR3 and a chymotrypsin-like serine endoprotease, SSY5. Under conditions of carbon deprivation, signaling through the Stp2 transcription factor triggers the use of amino acids as carbon source which helps neutralize the acidic environment of the phagosomes [21]. *C. albicans* employs several mechanisms to evade immune detection (**Figure 3**, [22, 23]).

C. albicans has significant phenotypic and genetic diversity. It contains a diploid genome of 14.4 megabases arranged within eight chromosomes [24]. The heterozygosity and heterozygous

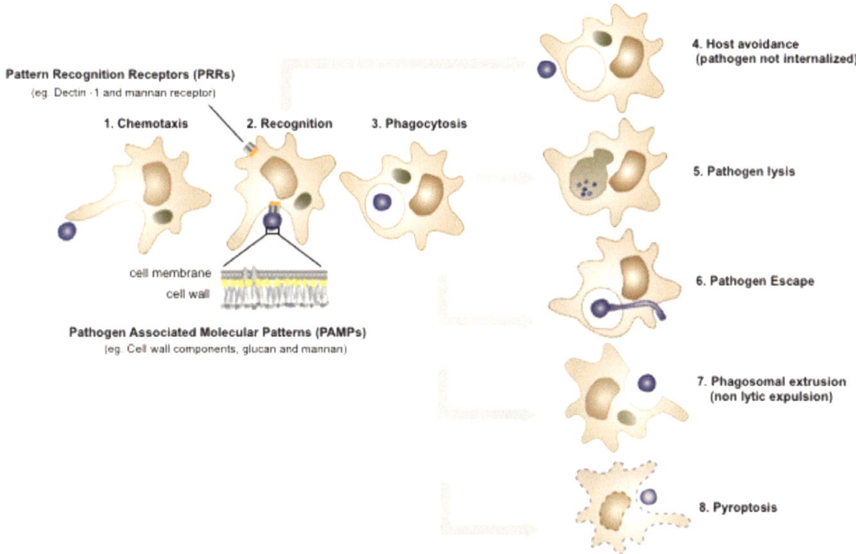

Figure 3. Host-Pathogen interactions. 1. Immune cells chemotax towards the pathogen, 2. fungal cell wall components pathogen associated molecular patterns (PAMPs) interact with macrophage via Pathogen Recognition Receptors(PRRs) such as Dectin-1 and Mannose Receptor that recognize β-(1,3)-glucan and mannan respectively, 3. Phagocytosis, 4. Host avoidance, 5. Pathogen lysis. 6. Pathogen escape, 7. Phagosomal extrusion where pathogen is expelled without lysis and 8. Pyroptosis.

of genome is thought to be related with *C. albicans* virulence and the genomic instability is crucial in its pathogenesis [2]. Differences specific to strains may contribute on the interaction of *C. albicans* and host [24].

Detection and identification of various yeasts has been a challenge. These yeasts can be distinguished morphologically on CHROMagar or spider media (**Figure 4**). However, genome sequencing is the most reliable method for species identification. In addition, detection of microsatellites also represents a reliable method for molecular typing and genetic analysis of *Candida*. A recent *Candida* distribution study conducted in a hospital, reported a clonal population including 62 identified genotypes among the tested isolates [25]. Beyond that, multilocus sequence typing (MLST) is another valuable method to understand the epidemiology of systemic *Candida* infections [2]. Here the DNA of seven housekeeping genes is sequenced. The results showed that MLST of *C. albicans* isolates are highly reproducible and sensitive. Comparative studies using MLST database of *C. albicans* are available online (http://calbicans.mlst.net/). These studies allowed further stratify the geographic isolation of *C. albicans*. For example, the most common MLST cluster within the *C. albicans* species is defined as clade 1. While, clade 2 is mainly located in the United Kingdom, clade 4 includes isolates from the Middle East and Africa, clade 11 includes isolates from continental Europe, and clades 14 and 17, where various gene clusters are regrouped include isolates from the Pacific region [26].

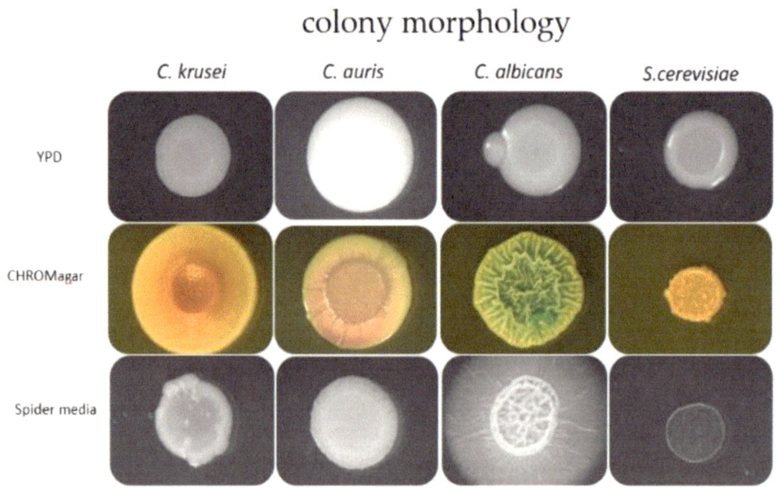

Figure 4. Colony morphology of *Candida krusei, Candida auris, Candida albicans* vs. *Saccharomyces cerevisiae.*

Pathogen profile: Diploid, belongs to the CTG clade, genome sequence available, antifungal resistance is moderate, molecular laboratory tools available.

2.2. *Candida glabrata*

Candida glabrata is often the second most common cause of candida infections after *C. albicans*. Historically, it has been considered nonpathogenic within the normal flora of healthy individuals without causing serious infection in humans. During the late 1990s, *C. glabrata* genetics was the most advanced among the non-albicans Candida (NAC) species due to its haploid status, its classical codon usage which allows direct usage of *S. cerevisiae* tools), and its high frequency of isolation in hospitals [27]. *C. glabrata* infections can be mucosal or systemic and often occur in immunocompromised persons or people with diabetes [28].

In contrast to most *Candida* species, *C. glabrata* is not dimorphic and exists as small blastoconidia under all environmental conditions both as commensal and pathogenic. In animal models *C. glabrata* is relatively nonpathogenic suggesting that it has few virulence attributes [28]. However, within the host environment, *C. glabrata* spreads rapidly, and is difficult to treat because it is resistant to many azole antifungal agents. Therefore, *C. glabrata* infections have a high mortality rate in compromised hospitalized patients [28].

C. glabrata has haploid genome, in contrast to the diploid genome of *C. albicans* and some NAC species [29]. It is distinguishable from *C. albicans* by its small-subunit rRNA [28]. *C. glabrata* only use glucose and trehalose. Such unique sugar utilization among *Candida* species can be applied to identify yeast to the level of genus and species. Now commercial kits (API 20C, Uni-Yeast-Tek, and YeastIdent) are available to identify *C. glabrata* in mixed cultures [28].

Both *C. glabrata* and *C. albicans* are commensal suggesting that similar host mechanisms such as suppressing expression of pathogenic determinants may be at play to effectively prevent infection of both microorganisms. However, the relatively low virulence of *C. glabrata* in animal models compared to *C. albicans* suggests genes controlling the virulence of *C. glabrata* may be different from those in *C. albicans* [28].

C. glabrata isolates are often associated with high resistance to the azole class of antifungal agents and less susceptibility to most other antifungal agents including amphotericin B [30]. Several mechanisms of azole resistance of *C. glabrata* have been identified, such as increased P-450-dependent ergosterol synthesis, an energy-dependent efflux pump of fluconazole and possibly via a multidrug resistance-type transporter [28].

Pathogen profile: Haploid, does not belong to the CTG clade, genome sequence available, antifungal resistance is high, several molecular laboratory tools available.

3. Emerging pathogens of the Candida species

3.1. *Candida auris*

Fluconazole-resistant *Candida* has been identified as a serious public health threat (www.cdc.gov). Among these, *Candida auris* has simultaneously and independently emerged on three continents in several countries as a multi drug resistant fungal pathogen with high mortality [31]. Phylogenetic analysis and polymorphism typing indicate that *C. auris* strains are clonal suggesting likely transmission from an environmental source.

Candida auris, as the name suggests, was first isolated from the drainage of the external ear a Japanese patient [31] and 15 Korean patients in 2009 [32]. Identification of it is challenging with standard microbiologic techniques. It frequently exhibits multidrug-resistance. Ever since these initial cases, *C. auris* has become an emerging global health threat causing massive invasive infections and outbreaks in healthcare facilities. Cases have already been identified in India, South Africa, Kuwait, the United Kingdom, Venezuela, Brazil, the United States, Colombia, Pakistan, Spain, Germany, Israel, Norway, and Oman [33].

Phylogenetically, *C. auris* is closer related to *Candida haemulonii* and *Candida ruelliae* [31] while distantly related to other more common pathogens *C. albicans* and *C. glabrata* [34]. Four distinct clades have been identified from geographic separate origins, suggesting almost simultaneous emergence of populations [35]. Susceptibility to antifungal reagents and survival from phagocytosis is largely different among four clades (CDC report).

Risk factors for *C. auris* infection appear to be similar to infections from Candida in general. *C. auris* is reported to grow at temperatures ranging from 35 to 42°C. It forms pink colonies on chromogenic media. *C. auris* does not form pseudohyphae (**Figure 4**) but capable of forming biofilms [36] and adhering to catheter material, although to a lesser degree as compared to *Candida albicans* [37]. Some *C. auris* strains produce phospholipase and proteinase, which may account for the variability in pathogenicity in a murine model [37].

The genomes of several isolates have been sequenced and they appear to parse into four distinct clades by geographic region [35]. Clades were separated by thousands of single-nucleotide polymorphisms, but within each clade isolates were typically clonal. Various mutations in *ERG11*, the gene encoding for lanosterol 14-alpha-demethylase and induced upon prolonged growth with antifungal drugs were shown to be associated with azole resistance in each geographic clade.

Again, while whole genome sequencing is the most reliable method for species identification, PCR and real-time PCR assays have shown excellent accuracy and have been effective for diagnosis, to rapidly identify *C. auris* [38]. The development of new antifungal medications with activity against *C. auris* will be vital to controlling *C. auris* as therapeutic options are already limited. Also aggressive infection control measures are critical to reducing the spread of *C. auris* [33].

Pathogen profile: Haploid, belongs to the CTG clade, Genome sequence available, antifungal resistance is high, several minimal molecular laboratory tools available.

3.2. *Candida krusei*

Candida krusei was first discovered in 1839 by Langenbeck from a patient with typhus, 75 years later Castellani proposed the suggestion that *C. krusei* may cause disease in humans [39]. Since then, it has been generally considered as a commensal in warm-blooded animals with low pathogenicity and virulence. In humans, *C. krusei* is generally considered to be a transient commensal and is infrequently isolated from mucosal surfaces. However, since 1960s there has been an increase in the number of reports of *C. krusei* as a human pathogen.

In contrast to most other ovoid shaped Candida spp., *C. krusei* cells are generally elongated in a feature similar to *C. kefyr* (formerly known as *C. pseudotropicalis*) among clinically important Candida spp. *C. krusei* has various colony morphologies. It has a multilayered cell wall consisting of an outer irregular coat of flocculent material, an electron-dense zone, a granular layer, a less granular layer, a thin layer of dense granules and another sparsely granular layer outside the cell membrane. The mannan component of the *C. krusei* cell wall has been shown to be different from other Candida spp. in containing (1–2) and (1–6) side chains in the ratio of 3:1 as being lightly branched [40]. Such differences may account for the variable behavior of *C. krusei* in biological fluids such as saliva and bronchial lavage fluid comparing with other Candida spp.

C. krusei has two basic morphological forms, yeast and pseudohyphae and both are often present simultaneously in growing cultures and not easily separated. *C. krusei* grows at a 37°C but can withstand temperature up to 45°C. *C. krusei* can grow in vitamin-free media even though most common Candida spp. require biotin or additional vitamin for growth. *C. krusei* ferments and assimilates glucose only as carbohydrate [39].

Like *C. auris*, *C. krusei* can adhere to abiotic surfaces but not to the same extent as *C. albicans*. Although adhesion to host surfaces is essential for colonization and invasion, *C. krusei* is able to colonize readily to inert surfaces such as implants and catheters by virtue of its cell surface hydrophobicity. Less pathogenic species—*C. parapsilosis*, *C. pseudotropicalis* (now *C. kefyr*) and *C. glabrata*—usually produce significantly less biofilm than the more pathogenic *C. albicans*, but *C. krusei* produced the most extensive biofilm on the surfaces of polyvinyl chloride catheter disks regardless of the growth medium. This could demonstrate the very high cell surface hydrophobicity, and adherence of *C. krusei* to inert plastic surfaces, which may then have other species, facilitated biofilm development [39]. *C. krusei* does not adhere to buccal epithelial cells whereas *C. tropicalis, C. parapsilosis, C. guilliermondii,* and *C. kefyr* do.

The susceptibility to lysozyme, an antimicrobial enzyme produced in phagosomes has been used as a method to assess the microbial virulence. Such tests indicate that the susceptibility to lysozyme of *C. krusei* > *C. parapsilosis* > *C. tropicalis* > *C. guilliermondii* > *C. albicans* > *C. glabrata,* the latter being the most resistant to lysozyme [41]. Interesting when pre-incubated in sucrose-supplemented media, *C. krusei* becomes highly sensitive to lysozyme as compared to *C. albicans*.

Pathogen profile: Diploid, does not belong to CTG clade, genome sequence available, antifungal resistance is high, minimal molecular laboratory tools available.

3.3. *Candida kefyr*

C. kefyr was first found in dairy products such as fermented milk, cheese, and yoghurt [42]. It was first isolated from a *C. kefyr* sample in 1909, named *Saccharomyces fragilis* at first, then *C. pseudotropicalis* and was reclassified as *Kluyveromyces marxianus* [43].

C. kefyr is rarely associated with disease [44] representing ~1% isolates of *Candida* spp. from clinical specimens [7, 45] (**Figure 1**). The first case of invasive *C. kefyr* (*C. pseudotropicalis*) was identified in a 58-year-old female with metastatic adenocarcinoma of the breast [46]. It has been reported to colonize oral cavities, gastrointestinal tract, and urinary tract. All infected patients were immunocompromised and had several potential risk factors [43] . There are only a few published cases of invasive *C. kefyr* infections.

These emerging pathogens of the Candida species themselves are typically not more virulent than *C. albicans*. It is generally thought that their conversion from commensalism to parasitism is largely determined by the host immune status [39]. In some cases these pathogens are resistant to multiple antifungal agents. For example, >90% of the recently emergent *C. auris* isolates are resistant to fluconazole, >30% are resistant to amphotericin B, and >5% to echinocandins, and >40% are resistant to classes of antifungal agents, while 4% are resistant to all three classes of antifungals available [35].

Pathogen profile: Ploidy is not determined, does not belong to CTG clade, Genome sequence available, antifungal resistance is moderate, few molecular laboratory tools available.

4. Other Candida species

Candida lusitaniae was firstly isolated from warm-blooded animals and was shown to cause opportunistic infections in humans in 1979 [47]. It is distinguished from other Candida species by its resistance to Amphotericin B however resistance profile *in vivo* is similar to other Candida species. Like most pathogens of the Candida species, *C. lusitaniae* has similar ability to colonize individuals but can opportunistically infected in immune-compromised patients [48].

Pathogen profile: Haloid, belongs to CTG clade, genome sequence available, antifungal resistance is moderate—high, few molecular laboratory tools available.

Candida tropicalis is another prevalent NAC pathogen in Candida species. In immunocompromised mice and human patients, *C. tropicalis* isolates appeared to have increased virulence. Secreted aspartyl proteinase 5 and 9 (Sap5 and Sap9) antigens are expressed by *C. tropicalis*. Invasive *C. tropicalis* infections were found more frequently in acute leukemia or bone marrow transplants patients may indicate that polymorphonuclear leukocytes are the first defense line against of *C. tropicalis* [49]. Overexpression of *ERG11* gene mutations in *C. tropicalis* likely causes resistance to azoles.

Pathogen profile: Diploid, belongs to CTG clade, genome sequence available, antifungal resistance is moderate, several molecular laboratory tools available.

Candida dubliniensis is a species of chlamydospore- and germ tube-positive yeast, primarily recovered from HIV-infected individuals and AIDS patients. It has been shown to grow well at temperatures ranging between 30 and 37°C but not 42°C. *C. dubliniensis* is unable to express beta-glucosidase activity [50].

Pathogen profile: Diploid, belongs to CTG clade, genome sequence available, antifungal resistance is moderate, several molecular laboratory tools available.

Candida parapsilosis has increased in significance and prevalence over the past 2 decades. The infections are mainly associated with prosthetic devices and catheters, especially in the nosocomial spread. Risk factors of *C. parapsilosis* infections include the hydrolytic enzymes secretion, prosthetics adhesion, and biofilm formation [51].

Pathogen profile: Diploid, belongs to CTG clade, genome sequence available, antifungal resistance is moderate-high, several molecular laboratory tools available.

Candida guilliermondii is the sixth frequently isolated Candida species, an emerging pathogen in Latin America that rarely causes invasive candida infections. However, it has been reported to exhibits reduced susceptibility to fluconazole [52] thus further study of the anti-drug mechanism is required.

Pathogen profile: Haploid, belongs to CTG clade, genome sequence available, antifungal resistance is high, several molecular laboratory tools available.

Candida lypolytica (also known as *Yarrowia Lipolytica*) isolates formed narrow, multi-branched, true hyphae on cornmeal-Tween 80 agar [53]. *C. lipolytica* is a weakly virulent pathogen that is most clearly vascular catheter-related. It is sensitive to Amphotericin B and Ketoconazole *in vitro*.

Pathogen profile: Haploid, does not belongs to CTG clade [54], genome sequence partially available, antifungal resistance is moderate [55], few molecular laboratory tools available.

Candida rugosa rarely causes invasive infections; however, recently, isolates have been shown to be an increasing cause of fungal infections especially in Latin America. Besides, *C. rugosa* appears decreased susceptibility to fluconazole with various patterns following geographic regions [56].

Pathogen profile: Haploid, belongs to CTG clade, genome sequence is not available, antifungal resistance is high, few molecular laboratory tools available.

Acknowledgements

This work was supported by Worcester Polytechnic Institute.

Conflict of interest

There is no conflict of interest concerning this chapter.

Author details

Bo Yang and Reeta Rao*

*Address all correspondence to: rpr@wpi.edu

Worcester Polytechnic Institute, Worcester, MA, USA

References

[1] Banerjee SN, Emori TG, Culver DH, Gaynes RP, Jarvis WR, Horan T, Edwards JR, Tolson J, Henderson T, Martone WJ. Secular trends in nosocomial primary bloodstream infections in the United States, 1980-1989. National Nosocomial Infections Surveillance System. The American Journal of Medicine. 1991;**91**(3B):86S-89S

[2] Dadar M, Tiwari R, Karthik K, Chakraborty S, Shahali Y, Dhama K. *Candida albicans*—Biology, molecular characterization, pathogenicity, and advances in diagnosis and control—An update. Microbial Pathogenesis. 2018;**117**:128-138. DOI: 10.1016/j.micpath.2018.02.028

[3] Hazen KC. New and emerging yeast pathogens. Clinical Microbiology Reviews. 1995;**8**(4):462-478

[4] Mühlhausen S, Kollmar M. Predicting the fungal CUG codon translation with Bagheera. BMC Genomics. 2014;**15**(1):411. DOI: 10.1186/1471-2164-15-411

[5] Sobel JD, Vazquez J, Lynch M, Meriwether C, Zervos MJ. Vaginitis due to *Saccharomyces cerevisiae*—Epidemiology, clinical aspects, and therapy. Clinical Infectious Diseases. 1993; **16**(1):93-99. DOI: 10.1093/clinids/16.1.93

[6] Oriol A, Ribera JM, Arnal J, Milla F, Batlle M, Feliu E. *Saccharomyces cerevisiae* septicemia in a patient with myelodysplastic syndrome. American Journal of Hematology. 1993;**43**(4):325-326

[7] Pfaller MA, Diekema DJ, Gibbs DL, Newell VA, Ellis D, Tullio V, Rodloff A, Fu W, Ling TA. Group global antifungal surveillance, results from the ARTEMIS DISK global antifungal surveillance study, 1997 to 2007: A 10.5-year analysis of susceptibilities of Candida species to fluconazole and voriconazole as determined by CLSI standardized disk diffusion. Journal of Clinical Microbiology. 2010;**48**(4):1366-1377. DOI: 10.1128/JCM.02117-09

[8] Wickes BL, Hicks JB, Merz WG, Kwonchung KJ. The molecular analysis of synonymy among medically important yeasts within the genus Candida. Journal of General Microbiology. 1992;**138**:901-907. DOI: 10.1099/00221287-138-5-901

[9] Ainsworth GC. Introduction to the History of Medical and Veterinary Mycology. Cambridge, New York: Cambridge University Press; 1986

[10] Jiyoti V-G, Ana T. Post-transcriptional gene regulation in the biology and virulence of *Candida albicans*. Cellular Microbiology. 2016;**18**(6):800-806. DOI: 10.1111/cmi.12593

[11] Andes David R, Nasia S, Baddley John W, Barbara A, Lisa B, Allison F, Susan H, Loreen H, Carol K, Marshall LG, Vicki M, Thomas P, Trish P, Randall W, Tim H, Tom C, Pappas Peter G. The epidemiology and outcomes of invasive Candida infections among organ transplant recipients in the United States: Results of the Transplant-Associated Infection Surveillance Network (TRANSNET). Transplant Infectious Disease. 2016;**18**(6):921-931. DOI: 10.1111/tid.12613

[12] Kearney T, Gunsalus W, Tornberg-Belanger SN, Matthan NR, Lichtenstein AH, Kumamoto CA. Manipulation of host diet to reduce gastrointestinal colonization by the opportunistic pathogen *Candida albicans*. mSphere. 2016;**1**(1). DOI: 10.1128/mSphere.00020-15

[13] Pfaller MA, Diekema DJ. Epidemiology of invasive candidiasis: A persistent public health problem. Virulence. 2007;**2**:119-128

[14] Santos MA, Tuite MF. The CUG codon is decoded in vivo as serine and not leucine in *Candida albicans*. Nucleic Acids Research. 1995;**23**(9):1481-1486

[15] Ene Iuliana V, Adya Ashok K, Silvia W, Brand Alexandra C, MacCallum Donna M, Gow Neil AR, Brown Alistair JP. Host carbon sources modulate cell wall architecture, drug resistance and virulence in a fungal pathogen. Cellular Microbiology. 2012;**14**(9):1319-1335. DOI: 10.1111/j.1462-5822.2012.01813.x

[16] Calderone RA, Clancy CJ. Candida and Candidiasis. Washington, DC: ASM Press; 2012

[17] Zhang B, Yu Q, Wang Y, Xiao C, Li J, Huo D, Zhang D, Jia C, Li M. The Candida albicans fimbrin Sac6 regulates oxidative stress response (OSR) and morphogenesis at the transcriptional level. Biochimica et Biophysica Acta (BBA)—Molecular Cell Research. 2016;**1863**(9):2255-2266. DOI: 10.1016/j.bbamcr.2016.06.002

[18] Alistair J, Brown P, Brown GD, Netea MG, Gow NAR. Metabolism impacts upon Candida immunogenicity and pathogenicity at multiple levels. Trends in Microbiology. 2014;**22**(11):614-622. DOI: 10.1016/j.tim.2014.07.001

[19] Liu H. Co-regulation of pathogenesis with dimorphism and phenotypic switching in *Candida albicans*, a commensal and a pathogen. International Journal of Medical Microbiology. 2002;**292**(5):299-311. DOI: 10.1078/1438-4221-00215

[20] Anne Neville B, d'Enfert C, Bougnoux M-E. *Candida albicans* commensalism in the gastrointestinal tract. FEMS Yeast Research. 2015;**15**(7):6-9. DOI: 10.1093/femsyr/fov081

[21] Pedro M, Lorenz Michael C. The SPS amino acid sensor mediates nutrient acquisition and immune evasion in *Candida albicans*. Cellular Microbiology. 2016;**18**(11):1611-1624. DOI: 10.1111/cmi.12600

[22] Erwig LP, Gow NAR. Interactions of fungal pathogens with phagocytes. Nature Reviews Microbiology. 2016;**14**:163. DOI: 10.1038/nrmicro.2015.21

[23] da Silva Dantas A, Lee KK, Raziunaite I, Schaefer K, Wagener J, Yadav B, Gow NA. Cell biology of *Candida albicans*-host interactions. Current Opinion in Microbiology. 2016; **34**:111-118. DOI: 10.1016/j.mib.2016.08.006

[24] Braunsdorf C, LeibundGut-Landmann S. Modulation of the fungal-host interaction by the intra-species diversity of *C. albicans*. Pathogens. 2018;**7**(1):11

[25] Guerrero HT, Espinosa IM, Ibarra MG, García MA. Distribution of Candida species and molecular typing of *C. albicans* isolates in a Mexico City tertiary care hospital from 2011 to 2013. Open Journal of Medical Microbiology. 2016;**06**(02):14. DOI: 10.4236/ojmm.2016.62010

[26] Odds FC, Bougnoux M-E, Shaw DJ, Bain JM, Davidson AD, Diogo D, Jacobsen MD, Lecomte M, Li S-Y, Tavanti A, Maiden MCJ, Gow NAR, d'Enfert C. Molecular phylogenetics of *Candida albicans*. Eukaryotic Cell. 2007;**6**(6):1041-1052. DOI: 10.1128/ec.00041-07

[27] Bendel CM, Hostetter MK. Correlation of adhesion and pathogenic potential in yeast. Pediatric Research. 1991;**29**(4):A167-A167

[28] Fidel, Jr PL, Vazquez JA, Sobel JD. Candida glabrata: Review of epidemiology, pathogenesis, and clinical disease with comparison to *C. albicans*. Clinical Microbiology Reviews. 1999;**12**(1):80-96

[29] Whelan WL, Simon S, Beneke ES, Rogers AL. Auxotrophic variants of Torulopsis-Glabrata. FEMS Microbiology Letters. 1984;**24**(1):1-4

[30] Vazquez JA, Dembry LM, Sanchez V, Vazquez MA, Sobel JD, Dmuchowski C, Zervos MJ. Nosocomial *Candida glabrata* colonization: An epidemiologic study. Journal of Clinical Microbiology. 1998;**36**(2):421-426

[31] Kazuo S, Koichi M, Yayoi H, Yayoi N, Katsuhisa U, Hideyo Y. *Candida auris* sp. nov., a novel ascomycetous yeast isolated from the external ear canal of an inpatient in a Japanese hospital. Microbiology and Immunology. 2009;**53**(1):41-44. DOI: 10.1111/j. 1348-0421.2008.00083.x

[32] Kim M-N, Shin JH, Sung H, Lee K, Kim E-C, Ryoo N, Lee J-S, Jung S-I, Park KH, Kee SJ, Kim SH, Shin MG, Suh SP, Ryang DW. *Candida haemulonii* and closely related species at 5 University Hospitals in Korea. Identification, Antifungal Susceptibility, and Clinical Features, Clinical Infectious Diseases. 2009;**48**(6):e57-e61. DOI: 10.1086/597108

[33] Sears D, Schwartz BS. *Candida auris*: An emerging multidrug-resistant pathogen. International Journal of Infectious Diseases. 2017;**63**:95-98. DOI: 10.1016/j.ijid.2017.08.017

[34] Munoz JF, Gade L, Chow NA, Loparev VN, Juieng P, Farrer RA, Litvintseva AP, Cuomo CA. Genomic basis of multidrug-resistance, mating, and virulence in *Candida auris* and related emerging species. bioRxiv. 2018. DOI: 10.1101/299917

[35] Lockhart SR, Etienne KA, Vallabhaneni S, Farooqi J, Chowdhary A, Govender NP, Colombo AL, Calvo B, Cuomo CA, Desjardins CA, Berkow EL, Castanheira M, Magobo RE, Jabeen K, Asghar RJ, Meis JF, Jackson B, Chiller T, Litvintseva AP. Simultaneous emergence of multidrug-resistant *Candida auris* on 3 continents confirmed by whole-genome sequencing and epidemiological analyses. Clinical Infectious Diseases. 2017;**64**(2):134-140. DOI: 10.1093/cid/ciw691

[36] Oh BJ, Shin JH, Kim M-N, Sung H, Lee K, Joo MY, Shin MG, Suh SP, Ryang DW. Biofilm formation and genotyping of *Candida haemulonii*, *Candida pseudohaemulonii*, and a proposed new species (*Candida auris*) isolates from Korea. Medical Mycology. 2011;**49**(1):98-102. DOI: 10.3109/13693786.2010.493563

[37] Larkin E, Hager C, Chandra J, Mukherjee PK, Retuerto M, Salem I, Long L, Isham N, Kovanda L, Borroto-Esoda K, Wring S, Angulo D, Ghannoum M. The emerging pathogen *Candida auris*: Growth phenotype, virulence factors, activity of antifungals, and effect of SCY-078, a novel Glucan synthesis inhibitor, on growth morphology and biofilm formation. Antimicrobial Agents and Chemotherapy. 2017;**61**(5):4-8. DOI: 10.1128/ AAC.02396-16

[38] Kordalewska M, Zhao Y, Lockhart SR, Chowdhary A, Berrio I, Perlin DS. Rapid and accurate molecular identification of the emerging multidrug-resistant pathogen *Candida auris*. Journal of Clinical Microbiology. 2017;**55**(8):2445-2452. DOI: 10.1128/JCM.00630-17

[39] Samaranayake YH, Samaranayake LP. *Candida krusei*: Biology, epidemiology, pathogenicity and clinical manifestations of an emerging pathogen. Journal of Medical Microbiology. 1994;**41**(5):295-310. DOI: 10.1099/00222615-41-5-295

[40] Kogan G, Pavliak V, Sandula J, Masler L. Novel structure of the cellular mannan of the pathogenic yeast *Candida krusei*. Carbohydrate Research. 1988;**184**:171-182

[41] Tobgi RS, Samaranayake LP, MacFarlane TW. In vitro susceptibility of Candida species to lysozyme. Oral Microbiology and Immunology. 1988;**3**(1):35-39

[42] Andrighetto C, Psomas E, Tzanetakis N, Suzzi G, Lombardi A. Randomly amplified polymorphic DNA (RAPD) PCR for the identification of yeasts isolated from dairy products. Letters in Applied Microbiology. 2000;**30**(1):5-9. DOI: 10.1046/j.1472-765x.2000.00589.x

[43] Khan Z, Ahmad S, Al-Obaid K, Joseph L, Chandy R. *Candida kefyr* as a cause of bloodstream infection and adjunctive role of biomarkers in its diagnosis. Journal de Mycologie Médicale. 2015;**25**(1):71-75. DOI: 10.1016/j.mycmed.2014.04.002

[44] Yavuz S, Kocabay S, Cetinkaya S, Akkaya B, Akkaya R, Yenidunya AF, Bakici MZ. Production, purification, and characterization of metalloprotease from *Candida kefyr* 41 PSB. International Journal of Biological Macromolecules. 2017;**94**(Pt A):106-113. DOI: 10.1016/j.ijbiomac.2016.10.006

[45] Pfaller MA, Diekema DJ. Epidemiology of invasive candidiasis: A persistent public health problem. Clinical Microbiology Reviews. 2007;**20**(1):133-163. DOI: 10.1128/CMR.00029-06

[46] Morgan MA, Wilkowske CJ, Roberts GD. Candida pseudotropicalis fungemia and invasive disease in an immunocompromised patient. Journal of Clinical Microbiology. 1984;**20**(5):1006-1007

[47] Pappagianis D, Collins MS, Hector R, Remington J. Development of resistance to amphotericin B in *Candida lusitaniae* infecting a human. Antimicrobial Agents and Chemotherapy. 1979;**16**(2):123-126

[48] Merz WG. *Candida lusitaniae*: Frequency of recovery, colonization, infection, and amphotericin B resistance. Journal of Clinical Microbiology. 1984;**20**(6):1194-1195

[49] Kothavade RJ, Kura MM, Valand AG, Panthaki MH. *Candida tropicalis*: Its prevalence, pathogenicity and increasing resistance to fluconazole. Journal of Medical Microbiology. 2010;**59**(Pt 8):873-880. DOI: 10.1099/jmm.0.013227-0

[50] Sullivan D, Coleman D. *Candida dubliniensis*: Characteristics and identification. Journal of Clinical Microbiology. 1998;**36**(2):329-334

[51] Trofa D, Gacser A, Nosanchuk JD. *Candida parapsilosis*, an emerging fungal pathogen. Clinical Microbiology Reviews. 2008;**21**(4):606-625. DOI: 10.1128/CMR.00013-08

[52] Pfaller MA, Diekema DJ, Mendez M, Kibbler C, Erzsebet P, Chang SC, Gibbs DL, Newell VA. *Candida guilliermondii*, an opportunistic fungal pathogen with decreased susceptibility to fluconazole: Geographic and temporal trends from the ARTEMIS DISK antifungal surveillance program. Journal of Clinical Microbiology. 2006;**44**(10):3551-3556. DOI: 10.1128/JCM.00865-06

[53] Walsh TJ, Salkin IF, Dixon DM, Hurd NJ. Clinical, microbiological, and experimental animal studies of *Candida lipolytica*. Journal of Clinical Microbiology. 1989;**27**(5):927-931

[54] Muhlhausen S, Kollmar M. Molecular phylogeny of sequenced saccharomycetes reveals polyphyly of the alternative yeast codon usage. Genome Biology and Evolution. 2014;**6**(5):3226. DOI: 10.1093/gbe/evu093

[55] Zhao Y, Chan JF-W, Tsang C-C, Wang H, Guo D, Pan Y, Xiao Y, Yue N, Chen JH-K, Lau SK-P, Xu Y, Woo PC-Y. Clinical characteristics, laboratory identification, and in vitro antifungal susceptibility of Yarrowia (Candida) lipolytica isolates causing Fungemia: A multicenter, prospective surveillance study. Journal of Clinical Microbiology. 2015; **53**(11):3639-3645. DOI: 10.1128/JCM.01985-15

[56] Pfaller MA, Diekema DJ, Colombo AL, Kibbler C, Ng KP, Gibbs DL, Newell VA. Candida rugosa, an emerging fungal pathogen with resistance to azoles: Geographic and temporal trends from the ARTEMIS DISK antifungal surveillance program. Journal of Clinical Microbiology. 2006;**44**(10):3578-3582. DOI: 10.1128/JCM.00863-06

Chapter 2

Antifungal Activity of Brazilian Medicinal Plants against *Candida* Species

Vagner Rodrigues Santos and
Elizete Maria Rita Pereira

Additional information is available at the end of the chapter

http://dx.doi.org/10.5772/intechopen.80076

Abstract

Due to the resistance of *Candida* sp. to the usual antifungal, the demand for active principles found in the plants has been the target of diverse studies around the world. There are few *in vivo* and human studies on the antifungal activity of medicinal plants in the mouth. Native and imported medicinal plants, used by the Brazilian population for traditional medicine use, are the subject of study in this chapter. Thirty-eight Brazilian plants were related to information on species, family, name, used part, and medical indication of popular use. All the species mentioned had their extracts tested *in vitro* against *C. albicans*, *C. tropicalis*, *C. krusei*, *C. parapsilosis*, among other species that occur more frequently in the mouth. In the articles consulted, there is a great variation in *Candida* species tested and in minimum inhibitory concentration. The *in vitro* studies serve as information for the continuity of studies on the best performing plants, validate the popular belief about the use, and provide subsidies for the development of new products that are effective in the control of oral and systemic candidiasis and that are cheap and accessible for the population.

Keywords: Brazilian medicinal plants, *Candida* sp., oral candidiasis, minimum inhibitory concentration

1. Oral candidiasis—by Elizete Maria Rita Pereira/Vagner Rodrigues Santos

The most common fungal pathogens detected from the oral cavity are *Candida* sp. and their transition of harmless commensals to pathogenic microorganisms is often related to decrease immunity. Often, candidiasis occurs in a localized and superficial manner in the oral cavity, but it can be systemic or invasive and even lead to death in immunocompromised individuals [1, 2]. Like other pathogenic fungi, *Candida* spp. present dimorphism in the yeast and pseudohyphatic forms. The hyphal form is associated with epithelial cell invasion and thus causing tissue damage [3]. *Candida* spp. has other virulence factors such as adherence, production of tissue-damaging hydrolytic enzymes such as proteases, and the biofilm formation in host tissue and in medical devices [4]. Several predisposing factors, local and systemic, may result in the transition from yeast Candida to the hyphal form (pathogenic). Local factors include the use of prostheses, corticosteroid inhalers, and xerostomia, while systemic factors include immunosuppressed states as human immunodeficiency virus—HIV for example. Psoriasis, recently, was described as a predisposing factor for oral candidiasis [5].

The type of *Candida* spp. most commonly isolated from the oral cavity is *C. albicans*, which occurs in both healthy and diseased individuals [1, 6, 7]. Nonalbicans species were also isolated from the oral cavity of immunocompromised patients, such as *C. glabrata, C. krusei, C. parapsilosis, C. tropicalis, C. guilliermondii*, and *C. dubliniensis* [8]. Protective immune response can be induced by Candida in the host that allows its own survival. Immunocompetent adults, usually, present acquired immunity underlying the fungus that prevents the progression of oral colonization to symptomatic infection. Integrity of the mucosa is important for oral health because normally it prevents the invasion of microorganisms, as well as macromolecules, which may be antigenic [9].

1.1. Main types, diagnosis, and treatment of oral candidiasis

Candidiasis is an acute or chronic infection produced by *Candida* sp., often, limited to the skin and mucous membranes, but can produce a severe systemic disease in immunocompromised patients [3]. It may exhibit various clinical patterns such as the four primary oral forms: pseudomembranous candidiasis (thrush), acute erythematous oral candidiasis, chronic erythematous oral candidiasis, and chronic hyperplastic candidiasis (**Table 1**). These primary forms of candidiasis are also associated with lesions called secondary forms of candidiasis, such as angular cheilitis, median rhomboid glossitis, and *Candida* sp.-associated prosthesis stomatitis [10].

Diagnosis of oral candidiasis is established by identification of clinical signs and symptoms in conjunction with the presence of Candida organisms in the examination of an injury smear, biopsy examination containing hyphal form in the epithelium (Schiff's Periodic Acid), or positive culture and serological tests [11–13].

Generally, the drugs of choice for localized, uncomplicated candidiasis in patients with normal immune function are topical antifungal agents (**Figure 1**). These agents can achieve elevated levels of concentration in the oral epithelium [14]. Azoles act inhibiting the lanosterol 14-α-demethylase (enzyme involved in ergosterol biosynthesis) activity 14-α-demethylase (enzyme involved in ergosterol biosynthesis) and disrupt the cell membrane. The resistance to azole, generally, can be observed in HIV/AIDS patients receiving treatment for pre-HAART oral or esophageal candidiasis, for example. The resistance mechanisms of *C. albicans* include mutations that result in increased expression of efflux pumps (CDR1P, CDR2P, and MDR1P) and mutations in the target ERG11 drug [15].

Polyenes act by direct binding to ergosterol within the membranes of fungal cells, therefore, inducing the leakage of cytoplasmic content and leading to the death of microorganism.

Primary forms of oral candidiasis	Characteristics
Pseudomembranous candidiasis (oral thrush)	Confluent white plaques with epithelial desquamation and accumulation of keratin, fibrin, necrotic tissue, and fungal hyphae [5, 11] • Ability to easily remove white plaques with gauze [5] • Most commonly found in immunosuppressed individuals [5]
Acute erythematous oral candidiasis	• Known as "sore mouth antibiotic," due to its association with prolonged use of broad-spectrum antibiotics [11] • It presents as erythematous patches, commonly on the palate, but may also present on the buccal mucosa or on the back of the tongue [5]
Chronic erythematous oral candidiasis	• These are atopic lesions associated with angular cheilitis and prosthetic stomatitis [5] • Prevalence in HIV-positive and AIDS patients [11]
Chronic hyperplastic candidiasis (candida leukoplakia)	• Homogeneous type of adherent white or erythematous nodular/speckled type that cannot be easily erased [5, 11] • It is present bilaterally in the commissure regions of the buccal mucosa [11]

Table 1. Primary forms of oral candidiasis [10].

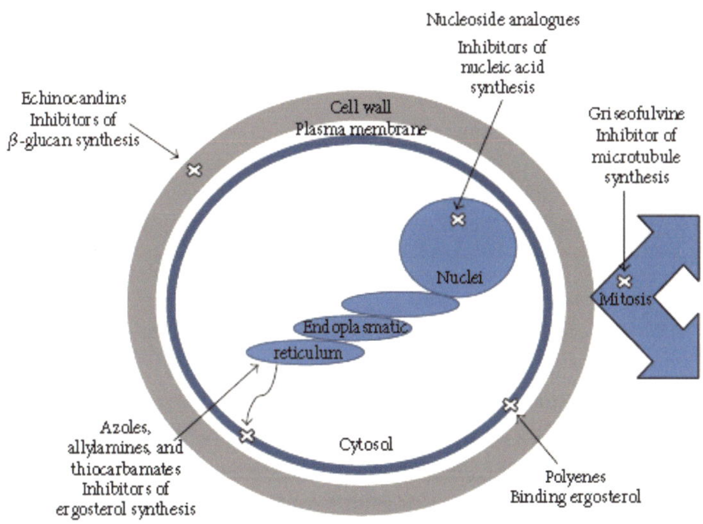

Figure 1. Primary targets of several antifungal agents by Garcia-Cuesta et al. [14].

Formulations of nystatin or amphotericin B are used for 4 weeks [11, 14]. Resistance was observed in case reports of cancer patients on chemotherapy and those who have received long-term prophylactic therapy. The mechanism of Candida resistence to the polyenes are not yet known, but seem to involve changes in the cell membrane composition [15].

Echinocandins are noncompetitive inhibitors of the β-1,3 glucan synthase encoded by the FKS1 gene of *C. albicans*, which leads to the formation of fungal cell walls with impaired structural integrity, leading to osmotic lysis. Resistance is associated with acquired or intrinsic

FKS1 point mutations [15]. Flucidosycin is a pyrimidine analogue, which is transported to fungal cells by cytosine permeases. After this, 5-fluorouracil and phosphorylation of 5-fluoro-deoxyuridine monophosphate are deaminated. This nucleotide acts by inhibiting thymidylate synthase and interferes with DNA synthesis [14]. Mutations in the cytosine-permease FCY2 gene or the cytosine-deaminase gene FCY1 are the most common causes of drug resistance. To avoid this increased resistance, flucytosine is almost always administered to patients in conjunction with amphotericin B [15]. Griseofulvin, first isolated from griseofulvin, first isolated from *Penicillium griseofulvum*, inhibits the fungal mitosis to act by disrupting the production of microtubules in the spindle and cytoplasm [14].

Antifungal systemic agents are indicated in fungemia when found in low immunity or immunodeficiency, high agranulocytosis, cancer patients or patients with intravenous catheters [11]. Worldwide, an increase in the number of antifungal resistant yeasts is recognized. An important factor in contribution to human candidiasis is the ability of *Candida* species to form drug-resistant biofilms [4]. The antifungal resistance can occur by different mechanisms, such as the reduction of the intracellular accumulation of the drug, decreased affinity of the target by the drug, and neutralizing the effect of the drug. Depending on the mode of action of the antifungal compounds, the mechanism of resistance will be different [14].

The search for new antifungal agents and the characterization of new targets that are more appropriate and efficient have been proposed [4]. Potential alternative therapies include the use of new active principles obtained from different general sources, such as natural products, in particular, the plants that contain several components that are important sources of biologically active molecules [14].

2. Brazilian medicinal plants tested against *Candida* spp.—by Vagner Rodrigues Santos

2.1. Introduction

The search for therapeutic applications of medicinal plants and their derivatives has grown in the past years throughout the world. Several studies have been carried out in order to evaluate new biological properties from the biodiversity. The discovery of new antimicrobial components is of great relevance, particularly for dentistry, since bacterial and fungal infections of the oral cavity are a relatively common problem: *Candida albicans* is an opportunistic yeast commonly identified in denture stomatitis and other oral candidosis clinical forms [12].

These are examples of infectious conditions of the mouth, and the resistance to antimicrobials in clinical cases has stimulated the search for natural agents as alternative treatments for the mouth infectious conditions. In Brazil, local communities use plants and their extracts for different medicinal purposes and take advantage of the availability of these plants and the low cost for product preparation. Plants have been used as antimicrobial, anti-inflammatory, wounds scarring, and antihemorrhagic agents, just to mention a few [16].

Medicinal plants continue to be widely used in rural and urban areas of Brazil. However, the intense miscegenation of crops over the last few centuries has more popularized the use of exotic native Brazilian plants and plants imported from other countries in popular medicine, especially in the southcentral part of the country. Most of these species were introduced by the Europeans and Africans, and are usually used according to the traditions of their places of origin [15, 16]. The growth of the pharmaceutical industry during the second half of the last century also distanced the Brazilian population from traditional medicine based on native plants. In the mid-1970s, for example, commercial pharmacies had lost their importance as the pharmaceutical industry completely dominated the drug market. This period was also marked by intense repression of mysticism, including the traditional use of medicinal plants. These facts are aggravated also by the continuous destruction of the rich Brazilian ecosystems, a process initiated with the exploration of Brazilwood by the Portuguese. As a consequence, remedies prepared with native plants, especially those of Amerindian origin, are now little known or used [16].

The Brazilian territory has about 20% of the world's biodiversity, including plants, which serve as raw materials for the production of herbal medicines and other products. The great cultural and ethnic diversity of Brazil is responsible for the knowledge transmitted over generations on the management and use of medicinal plants [17]. The high frequency of infections by the *Candida* species, as well as the occurrence of resistance to the usual antifungal, either in the hospital environment or in domestic use, as well as the increasing number of immunocompromised patients puts us in check and leads us to search for new active principles originated from medicinal plants that are effective in the control of microorganisms [18].

Several plants have been studied in Brazil based on popular use, mainly by rural communities [16].

The antifungal activity observed in some plants may be related to the presence of flavonoid glycosides and tannins, components that have antimicrobial and anti-inflammatory properties [19]. There is a growing interest in the use of tannins as antimicrobial agents. The activity of tannins against bacteria and yeasts can be measured by their action on the membranes, since they can cross the cell wall, composed of polysaccharides and proteins, and bind to its surface [20].

Studies with natural products generate difficulties regarding the comparison of results. This situation is due to the different presentations of the products used as tincture, ethanolic extract, aqueous extract, essential oil, among others, as well as the various methodological criteria employed [21] and also the different forms of phytotherapic presentation, among them, oral solutions, gel, and tea by decoction. The greater or lesser biological activity of the essential oils has been shown to be dependent on the composition of their chemical constituents, such as citral, pinene, cineole, caryophyllene, elemeno, furanodiene, limonene, eugenol, eucalyptol, and carvacrol. These constituents are responsible for the antiseptic, antibacterial, antifungal, and antiparasitic properties [22].

The mode of extraction of the active principles can influence significantly the antimicrobial activity. Biosynthesis of the constituents of a plant is strongly affected by the environment, harvest and postharvest, rainfall, temperature, luminosity, and humidity [23].

The mechanisms of action of medical plant extents on *Candida* spp. are still poorly studied. Several mechanisms of action have been proposed from the rupture of the cellular membranes, which several mechanisms of action have been proposed from the rupture of the cellular membranes, which seems to interrupt the cell cycle through the synthesis of proteins and alteration of the yeast DNA [24].

The most common microbiological methods for testing plant-derived products such as extracts, resins, and essential oils are agar diffusion tests and liquid-liquid tests such as macrodilution and microdilution [25]. The techniques of application of the plant antimicrobial substance in the diffusion method are by means of disc, stainless steel, or glass cylinders and agar perforation. The agar diffusion test, also called plaque diffusion, is a physical method in which a microorganism is challenged against a biologically active substance in solid culture medium and relates the size of the growth inhibition zone of the challenged microorganism [25, 26]. The application of the diffusion method is limited to fast-growing microorganisms, which are aerobic or anaerobic. The evaluation is comparative against a reference biological standard (positive control), and the zone or halo of inhibition of growth is measured starting from the circumference of the disc or well, to the margin where there is growth of microorganisms [27]. According to the size of the halo, the microorganisms can be classified as: sensitive, when the diameter of the zone of inhibition is greater or no more than 3 mm less than the positive control; moderately sensitive, halo greater than 2 mm, but smaller than the positive control of more than 3 mm; and resistant, diameter equal to or less than 2 mm. As a positive control, a standard antimicrobial is used, and as a negative control, the solvent is used for the dissolution of the extracts [28–32]. The recommended incubation conditions are 35–37°C for bacteria for 24 to 48 hours and for fungi from 25 to 27°C for 48 to 72 hours [33–39]. These tests serve to define the minimum inhibitory concentration that quantifies the lowest concentration of the product capable of inhibiting the growth of microorganisms [40–46] (**Table 3**).

Table 2 shows the species, families, popular names, and used parts of plants for the various applications in traditional medicine.

Herbs	Family	Local popular name	Used source	Medical use	Ref.
Allium sativum L.	Liliaceae	Garlic	Bulb	Antimicrobial, healing, antioxidant, antitumor	[47, 48]
Anacardium humile L.	Anacardiaceae	Cajuzinho-do-cerrado, little cuckoo	Shells, sheets, pulp	Antifungal, anti-inflammatory, hypoglycemic antioxidant, antimicrobial antiparasitic	[49, 50]
Anadenanthera colubrina (Vell) Brenan	Fabaceae	Angico branco, white Angico	Shells, resin	Healing, anti-inflammatory, antimicrobial	[51, 52]
Annona crassiflora Mart.	Annonaceae	Araticum	Shells, sheets, fruits	Antimicrobial cytotoxicity	[53, 54]

Herbs	Family	Local popular name	Used source	Medical use	Ref.
Arrabidaea chica (Hum. & Bonpl.) B. Verlot	Bignoniaceae	Crajiru	Sheets, shells	Anti-inflammatory, Antimicrobial Antihypertensive, antitumoral	[55, 56]
Azadirachta indica A. Juss	Meliaceae	Neem, nim	Oil, flowers, leaves, seeds, bark	Antimicrobial, insecticide, antimalarial	[57, 59]
Baccharis dracunculifolia DC	Asteraceae	Rosemary, broom	Sheets, flowers, stalk	Antimicrobial antioxidant, antitumoral, healing	[38, 60, 61]
Baccharis trimera (Less.) DC	Asteraceae	Carqueja	Flowers sheets oil	Antioxidant, antihepatotoxic, anti-inflammatory	[62, 63]
Calendula officinalis L.	Asteraceae	Calendula	Flowers	Anti-inflammatory, healing, antimicrobial	[64, 65]
Ceiba speciosa (A.St-Hil) Ravena	Malvaceae	Paineira	Shells, sheets	Antiematism, antihypertensive, antimicrobial	[66, 67]
Centaurium erythraea Rafn	Gentianaceae	Centaurea	Shells, sheets	Digestive, emetic, febrifuge, hepatic, antioxidant, anti-inflammatory	[49, 68]
Chrysobalanus icaco L.	Chrysobalanus	Ajiru	Sheets	Antimicrobial, anti-inflammatory, antitumoral	[71]
Coriandrum sativum L.	Apiaceae	Coriander, coentro	Sheets, seeds	Antibacterial, antioxidant, hepatoprotective, anticonvulsivant	[69, 70]
Croton campestris (A. St-Hill.)	Euphorbiaceae	Canopy, velame	Oil, barks, root	Anti-inflammatory, antimicrobial, antioxidant	[49, 72]
Curatella americana L.	Dilleniaceae	Sambaiba	Sheets	Antimicrobial, anti-inflammatory, antiulcerogenic, antihypertensive	[73, 74]

Herbs	Family	Local popular name	Used source	Medical use	Ref.
Dalbergia ecastophyllum (Linn.) Taub.	Leguminosae	Rabo-de-bugio	Resin, sheets	Antitumoral, antimicrobial, antioxidant, anti-inflammatory	[75, 76]
Drimys winteri (J.R.Forst & G. Forst)	Winteraceae	Casca d'anta	Bark	Antifungal, antibacterial, antioxidant	[77, 78]
Eugenia dysenterica ex DC Mart.	Myrtaceae	Cagaita	Leaves, barks	Antidiarrhoeic, antileukemic	[79, 80]
Eugenia uniflora L.	Myrtaceae	Pitanga	Leaves	Diarrhea, fever, diabetes, inflammation, headache	[81, 82]
Equisetum arvense L.	Equisetaceae	horsetail	Sheets, bark	Antioxidant, anti-inflammatory, antimicrobial, antitumoral	[83, 84]
Glycyrrhiza glabra L.	Fabaceae	Licorice Alcacuz	Root, rhizome	Antioxidant, anti-inflammatory, antiosteoporotic	[85, 86]
Hymenaea courbaril L.	Leguminosae	Jatobá	Sap, peel	Antimicrobial, anti-inflammatory, bronchitis, antidiarrheal	[87, 88]
Jacaranda cuspidifolia Mart.	Bignoniaceae	Rosewood Jacarandá	Barks, leaves, resin	Antimicrobial anti-inflammatory, antitumor	[89, 90]
Lafoensia pacari (A.St-Hill).	Lythraceae	Mangava brava	Stem bark	Anti-inflammatory analgesic	[91, 49]
Lippia sidoides Cham.	Verbenaceae	Rosemary-pepper, Alecrim-pimenta	Leaves, barks	Antinociceptive, anti-inflammatory antimicrobial	[60, 92]
Malva sylvestris L.	Malvaceae	Mauve, malva	Sheets, flowers	Cough, anti-inflammatory healing	[93, 94]
Maytenus salicifolia Mart Ex Reissek	Celastraceae	Holy Thorn, Espinheira santa	Sheets	Antiseptic, dyspepsia, antiulcer	[95, 96]
Melaleuca alternifolia Cheel	Myrtaceae	Melaleuca	Essential oil	Antiseptic anti-inflammatory, antifungal	[87, 97, 98]

Herbs	Family	Local popular name	Used source	Medical use	Ref.
Mentha piperita L.	Lamiaceae	Peppermint, H ortelã--pimenta	Sheets	Expectorant, carminative, anti-inflammatory antimicrobial	[55, 92]
Myroxylon peruiferum L.f.	Fabaceae	Cabreúva	Bark, fruits	Anti-inflammatory, anti-headache, antifungal	[87, 99]
Psidium guajava L.	Myrtaceae	Guava tree goiabeira	Leaves, fruits	Antioxidant, antimicrobial, anti-inflammatory	[81, 100]
Punica granatum L.	Punicaceae	Pomegranate Romã	Bark, peel, pericarp, leaves, juice	Antioxidant, anti-inflammatory, antimicrobial, anticarcinogenic	[20, 101, 102]
Ricinus communis L.	Euphorbiaceae	Castor mamona	Aerial parts	Antidiabetic, antifertility, anti-inflammatory, antimicrobial, antioxidant	[103, 104]
Sapindus saponaria var. drummondii (Hook. & Arn.) L. Benson	Sapindaceae	Soapberry	Leaves, fruits, barks	Diuretics, expectorants antifungal, antioxidant	[105, 106]
Schinus terebinthifolius Raddi	Anacardiaceae	Aroeira	Fruits, leaves, stem bark, essential oil	Antioxidant, anti-inflammation antimicrobial, antifungal, antiulcer	[69, 107]
Stryphnodendron adstringens (Mart) Coville, 1910	Fabaceae	Barbatimão	Bark leaves	Anti-inflammatory, cicatrizant, antimicrobial	[108–110]
Vismia guianensis (Aubl.) Pers.	Clusiaceae	Sealing wax Pau-de-lacre,	Resin, sheets, stalk	Anti-inflammatory, antifungal	[87, 111, 112]
Ziziphus joazeiro Mart.	Rhamnaceae	Juazeiro	Leaves, fruits, bark, root.	Anti-inflammation, antimicrobial, healing	[87, 113]

The table lists the native species and those imported or brought to Brazil.
*Ref.—References.

Table 2. Relation of Brazilian medicinal plants tested *in vitro* against *Candida* species.

Herbs	Active compounds	Microorganism	MIC: µg//mL	Ref.
Allium sativum L.	Quercetin, cyanidin, allistatin, allicin, ajoene	*C. albicans*	0.125	[47, 48]
		C. glabrata	0.312	
		C. tropicalis	1.56	
		C. parapsilosis	12.5	
Anacardium humile	Tannins, saponins, flavonoids amentoflavone	*C. albicans*	1.50	[114, 49, 50]
Anadenanthera colubrina (Vell)	Tannins, flavonoids	*C. albicans*	0.031	[115, 51, 52]
Annona crassiflora(Mart.)	Antioxidant, tannins	*C. albicans*	2.0	[54, 116]
		C. tropicalis	0.25	
		C. krusei	0.5	
Arrabidaea chica	Isoscutellarein, 6-hydroxyluteolin, hispidulin, scutellarein, luteolin, apigenin; anthocyanins, coumarins, flavonoids, saponins, tannins, triterpenes	*C. albicans,*	0.007/0.015	[55, 56, 102]
		C. dubliniensis		
		C. parapsilosis		
		C. tropicalis		
		C. krusei		
		C. guilliermondii,		
		C. utilis		
		C. lusitaniae		
		C. glabrata		
		C. rugosa		
Azadirachta indica	Nimonol, mahmoodin, naheedin	*C. albicans*	1000/500	[58, 117, 118]
Baccharis dracunculifolia	Artepillin C, baccharin, kaempferide, drupanin, p-coumaric acid, culifolin, caffeic acid phenethyl ester, chlorogenic acid, kaempferol, pinocembrin, naringenin, chrysin	*C. albicans,*	0.350	[24, 38, 92, 119]
		C. glabrata.	0.43	
		C. albicans,	20–320	
		C. tropicalis,		
		C. stellatoidea,		
		C. krusei		
Baccharis trimera	Flavonoids, phenolic acids, quercetin, luteolin, nepetin or eupafolin, apigenin, hispidulin, phytoalexin	*C. albicans*	2.0	[120, 121]
Calendula officinalis L.	Quercetin, hyperosides, α-cadinol, gamma-cadinene, 1,2,3-cadinatriene, α-muurolol	*C. albicans*	11.0 a 30	[64, 65]
		C. parapsilosis		
		C. dubliniensis		
		C. glabrata		
Ceiba speciosa (A.St-Hil) Ravena	Quercetin, ruthin, kaferol, gallic acid, chlorogenic acid, elagenic acid, caffeic acid	*C. albicans*	0.2	[66, 67]

Herbs	Active compounds	Microorganism	MIC: μg//mL	Ref.
Centaurium erythraea Rafn	Erytaurin, gentiopicrin, erytro-centaurin	*C. albicans,*	10.5	[49, 68]
Coriandrum sativum L.	Decanal, trans-2-decenal, 2-decen-1-ol, cyclodecane, mono- and sesquiterpene hydrocarbons	*C. albicans*	0.007	[60, 92, 122]
		C. tropicalis	20	
		C. stellatoidea	32	
		C. krusei	20	
Chrysobalanus icaco	Pomolic acid	*C. albicans,*	1.56	[71]
		C. tropicalis	6.25	
Croton campestris L.	Spathulenol, borneol,	*C. albicans*	5.25	[49, 123, 124]
	B-caryophyllene,			
	1,8-cineole			
Curatella americana L.	Tannins, 4-O-methyl—catechin; epicatechin-3-O-gallate; 4-O-methyl-catechin-3-O-gallate	*C. albicans*	15.6	[73]
		C. tropicalis	31.3	
		C. parapsilosis	31.3	
Dalbergia ecastaphyllum	Luteolin, quercetin,	*C. albicans,*	64	[125, 126]
	biochanin A,	*C. glabrata,*		
	Dalbergin, liquiritigenin,	*C. tropicalis*		
	rutin			
Drimys winteri	Polygodial,	*C. albicans*	0.015	[92, 127]
	caffeic acid,			
	3-caffeoylquinic acid			
Eugenia dysenterica ex DC Mart.	Caryophyllene,	*C. albicans,*	20	[69, 92]
	bicyclogermacrene,	*C. tropicalis,*	32	
	spathulenol,	*C. stellatoidea,*	32	
	Caryophyllene oxide	*C. krusei.*	20	
		C. albicans	0.250	
Eugenia uniflora Linn	Castor oil,	*C. krusei*	250	[79, 80, 128]
	isoquercetin,	*C. famata,*	125	
	quercetin	*C. guilliermondii*	500	
		C. tropicalis	125	
Equisetum arvense L.	Camptothin A,	*C. albicans*	0.250	[129, 81, 82, 130]
	Eugiflorins D1 and D2,	*C. albicans*	1000	
	afzelin, myricitrin, quercetin, myricetin, betulinic acid, centelloside C	*C. tropicalis*	31.2	
		C. krusei	31.2	
		C. parapsilosis	125	
Glycyrrhiza glabra L.	Flavonoids, phenolic acids, alkaloids, phytosterols, tannins, and triterpenoids	*C. albicans*	0.78–3.12	[131, 132]

Herbs	Active compounds	Microorganism	MIC: µg//mL	Ref.
Hymenaea courbaril L.	Triterpenic saponins, glycyrrhizin, glabridin	*C. glabrata*	8	[133]
Jacaranda cuspidifolia Mart.	Terpene,	*C. albicans*	1.25	[87]
	phenolic,	*C. glabrata*	0.625	
	salicylic acid	*C. krusei*	1.25	
		C. parapsilosis	1.25	
		C. tropicalis	0.625	
Lafoensia pacari St. Hil.	Saponins, coumarins, quinones, flavonoids, tannins, triterpenes, steroids, alkaloids	*C. albicans*	16	[90]
Lippia sidoides Cham.	Ellagic acid	*C. albicans,*	4.40	[60, 92]
Malva sylvestris L.	Isoborneol, bornyl acetate, α-humulene, α-fenchene	*C. albicans*	0.250	[92, 134]
Maytenus salicifolia Mart Ex Reiss	Mucopolysaccharides,	*C. albicans*	20	[69, 93, 94]
	mucilages,	*C. tropicalis*	32	
	flavonoids	*C. stellatoidea*	32	
		C. krusei	20	
Melaleuca alternifolia Cheel	Tannins, nepeticin, rigidenol, gliquidone, 11-Î ± –hydroxygliquidone, 16-b-hydroxypristymerin	*C. albicans*	50	[95, 96]
Mentha piperita L.	8-cineole, neomenthol, menthol, carvone, acetato de metila, trans-cariofileno e viridiflorol menthol, menthone	*C. albicans*	0.73	[87, 97, 98]
Myroxylon peruiferum	α-Copaene, safrole, δ-cadinene, cumarin, cabreuvina	*C. albicans*	0.500	[55, 92]
Psidium guajava	Phenolic, flavonoid,	*C. albicans*	1.25	[87, 99, 135]
	carotenoid, terpenoid	*C. glabrata*	1.25 µg	
	triterpene	*C. krusei*	0.625	
		C. parapsilosis	0.625	
		C. tropicalis	1.25	
Punica granatum	Tannins, piperidine alkaloids, polyphenols, oxalic acid, malic acid, ascorbic acid, estrone punicic acid, punicalagin	*C. albicans*	20	[136, 69, 81]
		C. tropicalis	32	
		C. stellatoidea	20	
		C. krusei	32	
		C. albicans	125	
		C. krusei	15,6	

Herbs	Active compounds	Microorganism	MIC: μg//mL	Ref.
Ricinus communis	Ricinoleic acid,	*C. albicans,*	125	[137–139]
	ricin, ricinin	*C. dubliniensis,*	15.6	
		C. parapsilosis	125	
		C. tropicalis	12.5	
		C. krusei	125	
		C. guilliermondii	100	
		C. utilis,	100	
		C. lusitaniae,	30	
		C. glabrata,	100	
		C. rugosa	30	
Sapindus saponaria var. Drummondii (Hook. & Arn.) L. Benson	Terpene-acetylated saponins	*C. albicans*	200/400	[140, 141]
	hederagenin			
Schinus terebinthifolius Raddi	α-Pinene, sabinene,	*C. albicans*	300/600	[105, 106, 142]
	z-salven, β- pinene,	*C. parapsilosis*	600	
	α-funebrene,	*C. glabrata*	300	
	limonene, myrcene,	*C. tropicalis*	300	
	alphaphellandrene			
Stryphnodendron adstringens (Mart) Coville, 1910	Gallic acid,	*C. albicans,*	4.25; 20	[49, 143, 144]
	procyanidin	*C. tropicalis,*	32	
	tannins,	*C. stellatoidea,*	32	
	delphinidin	*C. krusei*	20	
		C. albicans	7.8	
Vismia guianensis (Aubl.) Pers.	Vismiofuranoxantona, isojacareubina	*C. albicans*	4.25	[108, 110, 145]
	flavan-3-ol: epicatequina			
Zizyphus joazeiro Mart.	Betulinic acid, oleanolic acid, caffeine, amphibine D, jujubogenin	*C. albicans*	0.625	[87, 92, 112, 146]
		C. glabrata	1.25	
		C. krusei	0.625	
		C. parapsilosis	1.25	
		C. tropicalis	1.25	
		C. albicans	1.00	

* Ref—References.

Table 3. Minimum inhibitory concentration(MIC)—*in vitro* antimicrobial test of Brazilian medicinal plants against *Candida* species.

3. Oral candidiasis clinical trial studies

There are few clinical studies in humans on the efficacy of extracts from Brazilian plants in the treatment of oral candidiasis. More recently, human effectiveness of Brazilian green propolis derived from *B. dracunculifolia* on plaque control and gingivitis [49] has been shown for the prevention and control of oral mucositis and candidiasis [147, 148] and compared green propolis gel with benzydamine hydrochloride in cancer patients and irradiated in the head and neck regions [149]. Also, the Brazilian red propolis, derived from *D. ecastaphyllum*, both extract and gel, inhibited *C. albicans* in vitro, periodontopathogenic bacteria in vitro and in vivo, besides controlling periodontitis in rats [150]. The antifungal activity of Brazilian green propolis, originated by *Baccharis dracunculifolia*, was proven when tested against *Candida albicans* collected from HIV-patients mouth. The authors also demonstrated the more effectivity of propolis compared with usual antifungal tested [29].

4. Conclusions

In this chapter, there is information about the most used Brazilian species of plants against *Candida* species. The highest antimicrobial activities were obtained with diverse plant extracts. Some tests were done with the wild-type microorganisms collected from patients' mouths; however, most tests were done using standardized American Type Culture Collection (ATCC) samples. There is a certain difficulty in doing clinical tests on humans, especially when it comes to natural products. On the other hand, tested are antibacterial and anticandidal agents and could be used in the treatment of various oral diseases caused by multiresistant microbial agents. It is also clear from this study that the antifungal activity of these 38 medicinal plants was found with ethanolic, methanol, n-butanol, and chloroformic fractions. Studies have also revealed that the plants tested are not toxic at therapeutic doses with good antimicrobial properties. However, this study is an important step toward clinical evaluation in order to produce improved phytomedicine in the treatment of oral candidiasis for multiresistant *Candida albicans*.

Acknowledgements

The authors thank Ms. Lorena de Melo Santos for the English revision.

A. Plant image glossary index

Photo Credits (alphabetical order of Latin names): 1 Allium sativum, alho; 2 *Anacardium humile*, cajuzinho do campo. https://www.dicaparasaude.com/beneficios-do-cajui/; 3 *Anadenanthera colubrina* (Vell.) Brenan; Angico branco. https://sites.unicentro.br/wp/manejoflorestal/8598-2/; 4 *Annona sylvatica* (Araticum) http://independente.com.br/os-araticuns/; 5 *Arrabidea chica* (crajirú), https://manausalerta.com.br/pesquisa-analisa-acao-anti-inflamatoria-do-crajiru/, 6 *Azadirachta indica* (nem) -https://br.pinterest.com/pin/520095456944307200/, 7 *Baccharis dracunculifolia*, alecrim. http://www.ufrgs.br/fitoecologia/florars/open_sp.php?img=14749, 8 *Baccharis trimera* (carqueja) - http://plantaslujan-a.blogspot.com/2015/01/baccharis-trimera.html, 9 *Calendula officinalis*, calêndula. https://plantsam.com/calendula-officinalis/10 *Ceiba*

speciosa(https://www.ebay.co.uk/itm/50-seeds-of-Ceiba-speciosa-Chorisia-speciosa-bottle-of-fake-kapok-tree-R-/151774995083, 11 *Centaurium erythraea*, centaurea. http://naturalhealingroom.com/Centaury-Herb-Centaurium-Erythraea_p_1337.html, 12 https://en.wikipedia.org/wiki/Chrysobalanus_icaco; 13 *Coriandrum sativum*, coentro. https://commons.wikimedia.org/wiki/File:Starr_070906-8875_Coriandrum_sativum.jpg; 14 *Croton campestris*, https://www.flickr.com/photos/142712970@N03/38936756240; 15 *Curatella americana*, cajueiro-bravo-do-campo. https://pt.wikipedia.org/wiki/Cajueiro-bravo-do-campo; 16 Dalbergia ecastophyllum http://www.ufal.edu.br/unidadeacademica/ceca/pt-br/pos-graduacao/zootecnia/dissertacoes/talita-almeida-de-paula; 17 *Drimys winteri*, https://futureforests.ie/products/drimys-winteri; 18 *Equisetum arvense*, https://www.ebay.com/itm/Equisetum-Arvense-15-Fresh-Seeds-Herb-Medicinal-Plant-Field-Common-Horsetail-/132339802749; 19 *Eugenia dysenterica*, cagaita. http://www.viveiroipe.com.br/?mudas=cagaita; 20 *Eugenia uniflora*, pitangueira. https://www.fazfacil.com.br/jardim/pitangueira/; 21 Glycyrrhiza glabra https://www.indiamart.com/proddetail/glycyrrhiza-glabra-mulethi-extract-12502491912.html; 22 Hymenaea courbaril http://fredsonpaivareporter.blogspot.com/2017/01/hymenaea-courbaril-viveiro-da-patrulha.html; 23 Jacaranda cuspidifolia -https://br.pinterest.com/pin/247698666729355594/?lp=true; 26 *Malva sylvestris*, malva. http://johnstarnesurbanfarm.blogspot.com/2011/11/malva-sylvestris.

html; 27 *Maytenus salicifolia*, espinheira santa. http://www.ervanariamarcosguiao.com/product-page/espinheira-santa; 28 Melaleuca alternifolia; 29 Mentha piperita, 30 Myroxylon peruiferum, 31 Psidium guajava, 32 Punica granatum, http://www.medicinanatural.com.br/; 34 Sapindus saponaria, https://www.flickr.com/photos/mercadanteweb/10829969066; 35 Schinus therebintifolia, https://en.wikipedia.org/wiki/Schinus_terebinthifolia; 36 *S. adstringens* https://www.tudosobreplantas.com.br;37Vismiaguianensis, lacre. http://tropical.theferns.info; 38 *Z. joazeiro* http://www.naturezabela.com.br/2011/04/juazeiro-ziziphus-joazeiro.html; 23 Photos of the author Vagner Rodrigues Santos: (1) *A. sativum*, (13) *Coriandrum sativum*, (24) *Lafoensia pacari*, (25) *Lippia sidoides*, (33) *Ricinus communis*.

Author details

Vagner Rodrigues Santos[1,2,3,4]* and Elizete Maria Rita Pereira[4,5,6]

*Address all correspondence to: vegneer2003@yahoo.com.br

1 Clinical Pharmacology, UCA, Berkeley, USA

2 Oral Pathology, UFMG, Brazil

3 INRA, Jouy-em-Josas, France

4 Department of Clinic, Pathology and Surgery, Laboratory of Microbiology and Biomaterials/School of Dentistry, Universidade Federal de Minas Gerais/Belo Horizonte, Brazil

5 Medicine, Biomedicine, IEP-SCBH, Brazil

6 Biochemical and Molecular Pharmacology, Faculty of Medicine, UFMG, Brazil

References

[1] Manfredi M, Polonelli L, Aguirre-Urizar JM, Carrozzo M, McCullough MJ. Urban legends series: Oral candidosis. Oral Diseases. 2012;**19**:245-261

[2] Pietrzak A, Grywalska E, Socha M, Roliński J, Franciszkiewicz-Pietrzak K, Rudnicka L, et al. Prevalence and possible role of Candida species in patients with psoriasis: A systematic review and meta-analysis. Mediators of Inflammation. 2018 May 6;**2018**:9602362. DOI: 10.1155/2018/9602362. eCollection 2018

[3] Dabas PS. An approach to etiology, diagnosis and management of different types of candidiasis. Journal of Yeast and Fungal Research. 2013;**4**:63-74

[4] Sardi JCO, Scorzoni L, Bernardi T, Fusco-Almeida AM, Mendes-Giannini MJS. Candida species: Current epidemiology, pathogenicity, biofilm formation, natural antifungal products and new therapeutic options. Journal of Medical Microbiology. 2013;**62**:10-24

[5] Millsop JW, Fazel N. Oral candidiosis. Clinics in Dermatology. 2016;**34**:487-494

[6] Cannon RD, Chaffin WL. Colonization is a crucial factor in oral candidiasis. Journal of Dental Education. 2001;**65**:785-787

[7] Farah CS, Lynch N, McCullough MJ. Oral fungal infections: An update for the general practitioner. Australian Dent J. 2010;**55**(1 Suppl):48-54

[8] Anibal PC, Peixoto IT, Foglio MA, Höfling JF. Antifungal activity of the ethanolic extracts of Punica granatum L. and evaluation of the morphological and structural modifications of its compounds upon the cells of Candida spp. Brazilian Journal of Microbiology. 2013;**44**(3):839-848. DOI: 10.1590/S1517-83822013005000060. eCollection 2013

[9] Dineshshankar J, Sivakumar M, Karthikeyan M, Udayakumar P, Shanmugam KT, Kesavan G. Immunology of oral candidiasis. Journal of Pharmacy & Bioallied Sciences. 2004;**6**(Suppl 1):S9-S12

[10] Sawant B, Khanb T. Recent advances in delivery of antifungal agents for therapeutic management of candidiasis. Biomedicine & Pharmacotherapy. 2017;**96**:1478-1490

[11] Patil S, Rao RS, Majumdar B, Anil S. Clinical appearance of oral candida infection and therapeutic strategies. Frontiers in Microbiology. 2015;**6**:1391

[12] Ellepola AN, Morrison CJ. Laboratory diagnosis of invasive candidiasis. The Journal of Microbiology. 2005;**43**:65-84

[13] Rossie K, Guggenheimer J. Oral candidiasis: Clinical manifestations, diagnosis, and treatment. Practical Periodontics and Aesthetic Dentistry. 1997;**9**:635-641

[14] Garcia-Cuesta C, Sarrion-Pérez MG, Bagan JV. Current treatment of oral candidiasis: A literature review. Journal of Clinical and Experimental Dentistry. 2014;**6**:5:e576-e582

[15] Taff HT, Mitchell KF, Edward JA, Andes DR. Mechanisms of Candida biofilm drug resistance. Future Microbiology. 2013;**8**:1-19

[16] Brandao MGL. Plantas Uteis de Minas Gerais na obra dos Naturalistas. Belo Horizonte: Codigo Editora; 2010. 120pp

[17] Lopes RH, Macorini LF, Antunes KÁ, Espindola PP, Alfredo TM, da Rocha Pdos S, Pereira ZV, Dos Santos EL, de Picoli Souza K, Lorenzi H, Matos FJA. Plantas Medicinais do Brasil—Nativas e exóticas. São Paulo: Instituto Plantarum; 2002

[18] Freire JCP, Nóbrega MTC, Oliveira Júnior JK, Freire SCP, Dias-Ribeiro E. Herbal of antifungal activity on oral candidose: A literature review. SALUSVITA. 2016;**35**(4):537-546

[19] Pereira JV, Pereira MSV, Sampaio FC, Sampaio MCC, Alves PM, Araújo CRF, Higino JS. In vitro antibacterial and antiadherence effect of *Punica granatum* Linn extracts upon dental biofilm microorganisms. Brazilian Journal of Pharmacognsosy. 2006;**16**:88-93

[20] Vasconcelos LC, Sampaio MC, Sampaio FC, Higino JS. Use of Punica granatum Linn as an antifungal agent against antifungal activity of Higino JS 847 candidosis associated with denture stomatitis. Mycoses. 2003;**46**:192-196

[21] Freires Ide A, Murata RM, Furletti VF, Sartoratto A, Alencar SM, Figueira GM, de Oliveira Rodrigues JA, Duarte MC, Rosalen PL. *Coriandrum sativum* L. (Coriander) essential oil: Antifungal activity and mode of action on Candida spp., and molecular targets affected in human whole-genome expression. PLoS One. 2017;**9**(6):e99086. DOI: 10.1371/journal.pone.0099086. eCollection 2014

[22] Souza EL et al. Inhibitory action of some essential oils and phytochemicals on the growth of moulds isolated from foods. Brazilian Archives of Biology and Technology. 2005;**48**:245-250

[23] Tiwari R, Latheef SK, Ahmed I, Iqbal HMN, Bule MH, Dhama K, et al. Herbal immunomodulators: A remedial panacea for designing and developing effective drugs and medicines: Current scenario and future prospects. Current Drug Metabolism. 2018; **19**(3):264-301. DOI: 10.2174/1389200219666180129125436

[24] Mello AM, Gomes RT, Lara SR, Silva LG, Alves JB, Cortés ME, et al. The effect of Brazilian propolis on the germ tube formation and cell wall of *Candida albicans*. Pharmacology. 2006;**3**:352-358

[25] Soliman S, Alnajdy D, El-Keblawy AA, Mosa KA, Khoder G, Noreddin AM. Plants' natural products as alternative promising anti-Candida drugs. Pharmacognosy Reviews. 2017;**11**(22):104-122. DOI: 10.4103/phrev.phrev_8_17

[26] Besra M, Kumar V. In vitro investigation of antimicrobial activities of ethnomedicinal plants against dental caries pathogens. 3 Biotech. 2018 May;**8**(5):257. DOI: 10.1007/s13205-018-1283-2

[27] Hindler JA, Jorgensen JH. Procedures in antimicrobial susceptibility testingin. In: Mahon C, Manusels Jr. G. Text Book of Diagnostic Microbiology. Chapter 3. Philadelphia, United States: W.B. SaundersComp; 1995. 58-96pp

[28] de Paula AMB, Gomes RT, SW Kwasnicka, Dias RS, Cortés ME, Santos V R. Susceptibility of oral pathogenic bacteria and fungi tobrazilian green propolis extract. Pharmacologyonline. 2006;**3**:467-473

[29] Martins RS, Péreira ESJ, Lima SM, Senna MI, Mesquita RA, Santos VR. Effect of commercial ethanol propolis extract on the in vitro growth of *Candida albicans* collected from HIV-seropositive and HIV-seronegative Brazilian patients with oral candidiasis. Journal of Oral Science. 2002;**44**(1):41-48

[30] HMoreno PR, da Costa-Issa FI, Rajca-Ferreira AK, Pereira MA, Kaneko TM. Native Brazilian plants against nosocomial infections: A critical review on their potential and the antimicrobial methodology. Current Topics in Medicinal Chemistry. 2013;**13**(24): 3040-3078

[31] Giriraju A, Yunus GY. Assessment of antimicrobial potential of 10% ginger extract against *Streptococcus mutans*, *Candida albicans*, and *Enterococcus faecalis*: An in vitro study. Indian Journal of Dental Research. 2013;**24**(4):397-400. DOI: 10.4103/0970-9290.118356

[32] Clinical and Laboratory and Standards Institute (CLSI). Reference Method for Broth Dilution Antifungal Susceptibility Testing of Yeasts, Approved Standard. 2nd ed. CLSI document M27-S4. Wayne, PA, USA: CLSI; 2012

[33] Silva EJ, Herrera DR, Rosa TP, Duque TM, Jacinto RC, Gomes BP, et al. Evaluation of cytotoxicity, antimicrobial activity and physicochemical properties of a calcium aluminate-based endodontic material. Journal of Applied Oral Science. 2014;**22**(1):61-67. DOI: 10.1590/1678-775720130031.q

[34] Kirbag S, Erecevit P, Zengin F, Guvenc AN. Antimicrobial activities of some Euphorbia species. African Journal of Traditional, Complementary, and Alternative Medicines. 2013;**10**(5):305-309

[35] Elansary HO, Szopa A, Kubica P, Ekiert H, Ali HM, Elshikh MS, Abdel-Salam EM, El-Esawi M, El-Ansary DO. Bioactivities of traditional medicinal plants in Alexandria. Evidence-Based Complementary and Alternative Medicine. 2018, Jan 31;**2018**:1463579. DOI: 10.1155/2018/1463579. eCollection 2018

[36] Franca JR, De Luca MP, Ribeiro TG, Castilho RO, Moreira AN, Santos VR, et al. Propolis--based chitosan varnish: Drug delivery, controlled release and antimicrobial activity against oral pathogen bacteria. BMC Complementary and Alternative Medicine. 2014; **14**:478. DOI: 10.1186/1472-6882-14-478

[37] Gomes BP, Ferraz CC, Garrido FD, Rosalen PL, Zaia AA, Teixeira FB, et al. Microbial susceptibility to calcium hydroxide pastes and their vehicles. Journal of Endodontia. 2002;**28**(11):758-761

[38] Pereira CA, da Costa AC, Machado AK, Beltrame Júnior M, Zöllner MS, Junqueira JC, Jorge AO. Enzymatic activity, sensitivity to antifungal drugs and *Baccharis dracunculifolia* essential oil by Candida strains isolated from the oral cavities of breastfeeding infants and in their mothers' mouths and nipples. Mycopathologia. 2011;**171**(2):103-139. DOI: 10.1007/s11046-010-9353-y

[39] De Luca MP, Franca JR, Macedo FAFF, Grenho L, Cortes ME, Faraco AAG, Moreira AN, Santos VR. Propolis varnish: Antimicrobial properties against cariogenic bacteria, cytotoxicity, and sustained-release profile. BioMed Research International. 2014;**2014**:6. Article ID: 348647. DOI: 10.1155/2014/348647

[40] Ostrosky EA, Mizumoto MK, Lima MEL, Kaneko TM, Nishikawa SO, Freitas BR. Methods for evaluation of the antimicrobial activity and determination of minimum inhibitory concentration (MIC) of plant extracts. Revista Brasileira de Farmacognosia. Apr/June 2008;**18**(2):João Pessoa. DOI: 10.1590/S0102-695X2008000200026

[41] Zgoda JR, Freyer AJ, Killmer LB, Porter JR. Polyacetylene carboxylic acids from Mitrephora celebica. Journal of Natural Products. 2001;**64**(10):1348-1349

[42] Kunicka-Styczyńska A, Śmigielski K, Prusinowska R, Rajkowska K, Kuśmider B, Sikora M. Preservative activity of lavender hydrosols in moisturizing body gels. Letters in Applied Microbiology. 2015;**60**(1):27-32. DOI: 10.1111/lam.12346

[43] Åhman J, Matuschek E, Kahlmeter G. The quality of antimicrobial disks from nine manufacturers—EUCAST evaluations in 2014 and 2017. Clinical Microbiology and Infection. 2018 Jun 7. pii:S1198-743X(18)30456-7. DOI: 10.1016/j.cmi.2018.05.021. [in press]

[44] Dalyan CB, Topaç T, Ağca H, Sağlam S, Efe K, Ener B. Comparison of clinical laboratory standards institute (CLSI) and European committee on antimicrobial susceptibility testing (EUCAST) broth microdilution methods for determining the susceptibilities of Candida isolates. Mikrobiyoloji Bülteni. 2018;**52**(1):35-48. DOI: 10.5578/mb.63991

[45] de-Souza-Silva CM, Guilhelmelli F, Zamith-Miranda D, de Oliveira MA, Nosanchuk JD, Silva-Pereira I, Albuquerque P. Broth microdilution in vitro screening: An easy and fast method to detect new antifungal compounds. Journal of Visualized Experiments. Feb 14 2018;**132**:4. DOI: 10.3791/57127. Abstract

[46] Caamal-Herrera IO, Carrillo-Cocom LM, Escalante-Réndiz DY, Aráiz-Hernández D, Azamar-Barrios JA. Antimicrobial and antiproliferative activity of essential oil, aqueous and ethanolic extracts of *Ocimum micranthum* Willd leaves". BMC Complementary and Alternative Medicine. 2018 Feb 8;**18**(1):55. DOI: 10.1186/s12906-018-2122-z

[47] Li WR, Shi QS, Dai HQ, Liang Q, Xie XB, Huang XM, et al. Antifungal activity, kinetics and molecular mechanism of action of garlic oil against Candida albicans. Scientific Reports. 2016;**6**:22805. DOI: 10.1038/srep22805

[48] Khan S, Imran M, Imran M, Pindari N. Antimicrobial activity of various ethanolic plant extracts against pathogenic multidrug resistant Candida spp. Bioinformation. 2017 Mar 31;**13**(3):67-72. DOI: 10.6026/97320630013067. eCollection 2017

[49] Pereira EM, Gomes RT, Freire NR, Aguiar EG, Brandão Md, Santos VR. In vitro antimicrobial activity of Brazilian medicinal plant extracts against pathogenic microorganisms of interest to dentistry. Planta Medica. 2011 Mar;**77**(4):401-404. DOI: 10.1055/s-0030-1250354. Epub 2010 Sep 22

[50] Urzêda MA, Marcussi S, Silva Pereira LL, França SC, Pereira AM, Pereira PS, et al. Evaluation of the ypoglycemic properties of Anacardium humile aqueous extract. Evidence-Based Complementary and Alternative Medicine. 2013;**2013**:191080. DOI: 10.1155/2013/191080

[51] Lima RF, Alves ÉP, Rosalen PL, Ruiz ALTG, Duarte MCT, Góes VFFi, Medeiros ACD, Pereira JV, Godoy GP, Costa EMMB. Antimicrobial and antiproliferative potential of *Anadenanthera colubrina* (Vell.) Brenan. Evidence-Based Complementary and Alternative Medicine. 2014;**2014**:802696. DOI: 10.1155/2014/802696

[52] Barreto HM, Coelho KM, Ferreira JH, Dos Santos BH, de Abreu AP, Coutinho HD, et al. Enhancement of the antibiotic activity of aminoglycosides by extracts from *Anadenanthera colubrine* (Vell.) Brenan var. cebil against multi-drug resistant bacteria. Natural Product Research. 2016 Jun;**30**(11):1289-1292. DOI: 10.1080/14786419.2015.1049177

[53] Lage GA, Medeiros Fda S, Furtado Wde L, Takahashi JA, de Souza Filho JD, Pimenta LP. The first report on flavonoid isolation from *Annona crassiflora* Mart. Natural Product Research. 2014;**28**(11):808-811. DOI: 10.1080/14786419.2014.885518

[54] Formagio AS, Vieira MC, Volobuff CR, Silva MS, Matos AI, Cardoso CA, et al. Carvalho JE In vitro biological screening of the anticholinesterase and antiproliferative activities of medicinal plants belonging to Annonaceae. Brazilian Journal of Medical and Biological Research. 2015 Apr;**48**(4):308-315. DOI: 10.1590/1414-431X20144127

[55] Patel K, Patel DK. Medicinal importance, pharmacological activities, and analytical aspects of hispidulin: A concise report. Journal of Traditional and Complementary Medicine. 2016 Dec 10;**7**(3):360-366. DOI: 10.1016/j.jtcme.2016.11.003. eCollection 2017 Jul

[56] Siraichi JT1, Felipe DF, Brambilla LZ, Gatto MJ, Terra VA, Cecchini AL, Cortez LE, Rodrigues-Filho E, Cortez DA. Antioxidant capacity of the leaf extract obtained from *Arrabidaea chica* cultivated in Southern Brazil. PLoS One. 2013 Aug 29;**8**(8):e72733. DOI: 10.1371/journal.pone.0072733. eCollection 2013

[57] Sujarwo W, Keim AP, Caneva G, Toniolo C, Nicoletti M. Ethnobotanical uses of neem (*Azadirachta indica* a.Juss.; Meliaceae) leaves in Bali (Indonesia) and the Indian subcontinent in relation with historical background and phytochemical properties. Journal of Ethnopharmacology. 2016 Aug 2;**189**:186-193. DOI: 10.1016/j.jep.2016.05.014

[58] Gupta SC, Prasad S, Tyagi AK, Kunnumakkara AB, Aggarwal BB. Neem (*Azadirachta indica*): An Indian traditional panacea with modern molecular basis. Phytomedicine. 2017 Oct 15;**34**:14-20. DOI: 10.1016/j.phymed.2017.07.001

[59] Santos WL, Freire MGM, Bogorni PC, Vendramim JD, Macedo ML R. Effect of the aqueous extracts of the seeds of *Talisia sculenta* and *Sapindus saponaria* on fall armyworm. Brazilian Archives of Biology and Technology. 2008;**51**(2) Curitiba Mar/Apr. DOI: 10.1590/S1516-89132008000200018

[60] Freires IA, Bueno-Silva B, Galvão LC, Duarte MC, Sartoratto A, Figueira GM, et al. Rosalen PL the effect of essential oils and bioactive fractions on *Streptococcus mutans* and *Candida albicans* biofilms: A confocal analysis. Evidence-based Complementary and Alternative Medicine. 2015;**2015**:871316. DOI: 10.1155/2015/871316

[61] Veiga RS, De Mendonça S, Mendes PB, Paulino N, Mimica MJ, Lagareiro Netto AA, et al. Negrão V4, Marcucci MC. Artepillin C and phenolic compounds responsible for antimicrobial and antioxidant activity of green propolis and *Baccharis dracunculifolia* DC. Journal of Applied Microbiology. 2017 Apr;**122**(4):911-920. DOI: 10.1111/jam.13400

[62] Soicke H, Leng-Peschlow E. Characterisation of flavonoids from *Baccharis trimera* and their antihepatotoxic properties. Planta Medica. 1987;**53**(1):37-39

[63] Menezes AP, da Silva J, Fisher C, da Silva FR, Reyes JM, Picada JN, et al. Chemical and toxicological effects of medicinal *Baccharis trimera* extract from coal burning area. Chemosphere. 2016 Mar;**146**:396-404. DOI: 10.1016/j.chemosphere.2015.12.028

[64] Gazim ZC, Rezende CM, Fraga SR, Svidzinski TI, Cortez DA. Antifungal activity of the essential oil from *Calendula officinalis* L. (asteraceae) growing in Brazil. Brazilian Journal of Microbiology. 2008 Jan;**39**(1):61-63. DOI: 10.1590/S1517-838220080001000015

[65] Lovecka P, Lipov J, Thumova K. Macurkova a characterization of biologically active substances from *Calendula officinalis*. Current Pharmaceutical Biotechnology. 2017; **18**(14):1167-1174. DOI: 10.2174/1389201019666180226151910

[66] Dörr JA, Bitencourt S, Bortoluzzi L, Alves C, Silva J, Stoll S, et al. In vitro activities of *Ceiba speciosa* (A.St.-Hil) Ravenna aqueous stem bark extract. Natural Product Research. 2018 May;**24**:1-4. DOI: 10.1080/14786419.2018.1478823

[67] Vargas MS, Moreira FF, Cardoso MCK, Faoro D, Silva JSB, Machado MM. Análise do extrato aquoso das cascas de *Ceiba speciosa* (paineira) por cromatografia liquida de alta eficiência. anais do salao internacional de ensino, pesquisa e extensao. 2014;**6**(2):3-4

[68] Berkan T, Ustünes L, Lermioglu F, Özer A. Antiinflammatory, analgesic and antipyretic effects of an aqueous extract of *Erythraea centaurium*. Planta Medica. 1991;**57**:34-37

[69] Alves PM, Queiroz LM, Pereira JV, Pereira Mdo S. In vitro antimicrobial, antiadherent and antifungal activity of Brazilian medicinal plants on oral biofilm microorganisms and strains of the genus Candida. Revista da Sociedade Brasileira de Medicina Tropical. 2009 Mar-Apr;**42**(2):222-224

[70] Galvao LCC, Furletti VF, Bersan SMF, Cunha MG, Ruiz ALTG, et al. Antimicrobial activity of essential oils against *Streptococcus mutans* and their antiproliferative effects. Evidence-Based Complementary and Alternative Medicine. 2012. DOI: 10.1155/2012/751435

[71] Silva JP, Peres AR, Paixão TP, Silva AS, Baetas AC, Barbosa WL, Monteiro MC, Andrade MA. Antifungal activity of hydroalcoholic extract of *Chrysobalanus icaco* against oral clinical isolates of Candida species. Pharmacognosy Research. 2017 Jan-Mar;**9**(1):96-100. DOI: 10.4103/0974-8490.199772

[72] Araújo Monteiro P, Maccari Zelioli ÍA, de Oliveira Sousa IM, Ruiz ALTG, Vendramini-Costa DB, Foglio MA, et al. Chemical composition and antiproliferative activity of *Croton campestris* A.St.-Hil. Essential oil. Natural Product Research. 2017 Nov;**9**:1-4. DOI: 10.1080/14786419.2017.1399377

[73] Mendes de Toledo CE, Santos PR, Palazzo de Mello JC, Dias Filho BP, Nakamura CV, Ueda-Nakamura T. Antifungal Properties of crude extracts, fractions, and purified compounds from bark of *Curatellaamericana L.* (Dilleniaceae) against Candida species. Evidence-Based Complementary and Alternative Medicine. 2015;**2015**:673962. DOI: 10.1155/2015/673962

[74] Lopes RH, Macorini LF, Antunes KÁ, Espindola PP, Alfredo TM, da Rocha PS, et al. Antioxidant and hypolipidemic activity of the hydroethanolic extract of *Curatella Americana* L leaves. Oxidative Medicine and Cellular Longevity. 2016;**2016**:9681425. DOI: 10.1155/2016/9681425

[75] Daugsch A, Moraes CS, Fort P, Park YK. Brazilian red propolis--chemical composition and botanical origin. Evidence-Based Complementary and Alternative Medicine. 2008 Dec;**5**(4):435-441. DOI: 10.1093/ecam/nem057

[76] Santos, VR. Alternative medicine for the treatment of oral microbial diseasis. In: Sakagami H. Alternative Medicine, Chapter 07. Croatia: IntechOpen; 2012

[77] Doust AN, Drinnan AN. Floral development and molecular phylogeny support the generic status of *Tasmannia* (Winteraceae). American Journal of Botany. 2004 Mar;**91**(3): 321-331. DOI: 10.3732/ajb.91.3.321

[78] Meinhart AD, Damin FM, Caldeirão L, da Silveira TFF, Filho JT, Godoy HT. Chlorogenic acid isomer contents in 100 plants commercialized in Brazil. Food Research International. 2017 Sep;**99**(Pt 1):522-530. DOI: 10.1016/j.foodres.2017.06.017

[79] Galheigo MR, Prado LC, Mundin AM, Gomes DO, Chang R, Lima AM, et al. Antidiarrhoeic effect of *Eugenia dysenterica* DC (Myrtaceae) leaf essential oil. Natural Product Research. 2016;**30**(10):1182-1185. DOI: 10.1080/14786419.2015.1043633

[80] Vitek R, de Novais LMR, Torquato HFV, Paredes-Gamero EJ, de Carvalho MG, de Sousa PT Jr, et al. Chemical constituents and antileukemic activity of *Eugenia dysenterica*. Natural Product Research. 2017 Aug;**31**(16):1930-1934. DOI: 10.1080/14786419.2016.1261343

[81] Holetz FB, Pessini GL, Sanches NR, Cortez DA, Nakamura CV, Filho BP. Screening of some plants used in the Brazilian folk medicine for the treatment of infectious diseases. Mem Inst Oswaldo Cruz. 2002 Oct;**97**(7):1027-31

[82] Falcão TR, de Araújo AA, Soares LAL, de Moraes Ramos RT, Bezerra ICF, Ferreira MRA, de Souza Neto MA, Melo MCN, de Araújo RF Jr, de Aguiar Guerra ACV, de Medeiros JS, Guerra GCB. Crude extract and fractions from *Eugenia uniflora* Linn leaves showed anti-inflammatory, antioxidant, and antibacterial activities. BMC Complementary and Alternative Medicine. 2018 Mar 9;**18**(1):84. DOI: 10.1186/s12906-018-2144-6

[83] Gallo FR, Multari G, Federici E, Palazzino G, Giambenedetti M, Petitto V, et al. Chemical fingerprinting of *Equisetum arvense* L. using HPTLC densitometry and HPLC. Natural Product Research. 2011 Aug;**25**(13):1261-1270. DOI: 10.1080/14786419.2011.558015

[84] Gründemann C, Lengen K, Sauer B, Garcia-Käufer M, Zehl M, Huber R. *Equisetum arvense* (common horsetail) modulates the function of inflammatory immunocompetent cells. BMC Complementary and Alternative Medicine. 2014 Aug 4;**14**:283. DOI: 10.1186/1472-6882-14-283

[85] Simmler C, Guido FP, Chen SN. Phytochemistry and biological properties of Glabridin. Fitoterapia. Author manuscript; available in PMC 2014 Oct 1. Fitoterapia. 2013 Oct;**0**:160-184

[86] Najafi S, Koujan SE, Manifar S, Kharazifard MJ, Kidi S, Hajheidary S. Preventive effect of *Glycyrrhiza glabra* extract on oral mucositis in patients under head and neck radiotherapy: A Randomized Clinical Trial. Journal of dentistry (Tehran) 2017 Sep;**14**(5):267-274

[87] Costa MCMF, Silva AG, Silva APS, Lima VLM, Bezerra-Silva PC, Rocha SKL, Navarro DMAF, Correia MTS, Napoleão TH, Silva MV, Paiva PMG. Essential oils from leaves of medicinal plants of Brazilian flora: Chemical composition and activity against Candida species. Medicines (Basel). Jun 2017;**4**(2)27:8. DOI: 10.3390/medicines4020027

[88] Boniface PK, Baptista Ferreira S, Roland KC. Current state of knowledge on the traditional uses, phytochemistry, and pharmacology of the genus Hymenaea. Journal of Ethnopharmacology. 2017 Jul 12;**206**:193-223. DOI: 10.1016/j.jep.2017.05.024

[89] Ferreres F, Grosso C, Gil-Izquierdo A, Valentão P, Andrade PB. Phenolic compounds from Jacaranda caroba (Vell.) A. DC.: Approaches to neurodegenerative disorders. Food and Chemical Toxicology. 2013 Jul;**57**:91-98. DOI: 10.1016/j.fct.2013.03.012

[90] Yuan J, Gan T, Liu Y, Gao H, Xu W, Zhang T, et al. Composition and antimicrobial activity of the essential oil from the branches of *Jacarandacuspidifolia Mart*. Growing in Sichuan, China. Natural Product Research. 2018 Jun;**32**(12):1451-1454. DOI: 10.1080/14786419.2017.1346644

[91] Queiroz e Silva SM, Pinheiro SM, Queiroz MV, Pranchevicius MC, Castro JG, Perim MC, Carreiro SC. In vitro activity of crude extracts of two plant species in the Cerrado on yeast of the Candida SPP variety. Cien Saude Colet. 2012 Jun;**17**(6):1649-1656

[92] Bersan SM, Galvão LC, Goes VF, Sartoratto A, Figueira GM, Rehder VL, Alencar SM, Duarte RM, Rosalen PL, Duarte MC1. Action of essential oils from Brazilian native and exotic medicinal species on oral biofilms. BMC Complementary and Alternative Medicine. 2014 Nov 18;**14**:451. DOI: 10.1186/1472-6882-14-451

[93] Gasparetto JC, Martins CA, Hayashi SS, Otuky MF, Pontarolo R. Ethnobotanical and scientific aspects of *Malva sylvestris* L.: A millennial herbal medicine. The Journal of Pharmacy and Pharmacology. 2012 Feb;**64**(2):172-189. DOI: 10.1111/j.2042-7158.2011.01383.x

[94] de Souza GC, Haas AP, von Poser GL, Schapoval EE, Elisabetsky E. Ethnopharmacological studies of antimicrobial remedies in the south of Brazil. Journal of Ethnopharmacology. 2004 Jan;**90**(1):135-143

[95] Miranda RR, Silva GD, Duarte LP, Fortes IC, Filho SA. Structural determination of 3beta-stearyloxy-urs-12-ene from Maytenus salicifolia by 1D and 2D NMR and quantitative 13C NMR spectroscopy. Magnetic Resonance in Chemistry. 2006 Feb;**44**(2):127-131

[96] Magalhães CG, Silva GDF, Duarte LP, Takahashi JÁ, SantosVR, Figueiredo RC, Vieira Filho SA. *Maytenus salicifolia* Reissek (Celastraceae): Evaluation of the activity of extracts and constituents against *Helicobacter pylori* and oral pathogenic microorganisms. Revista Virtual de Química. 2016;**8**(5):1524-1536

[97] Agarwal V, Lal P, Pruthi V. Effect of plant oils on *Candida albicans*. Journal of Microbiology, Immunology and Infection. 2010;**43**(5):447-451

[98] Oro D, Heissler A, Rossi EM, Scapin D, Boff E. Antifungal activity of natural compounds gainst Candida species isolated from HIV-positive patients. Asian Pacific Journal of Tropical Biomedicine. 2015;**5**(9):781-784

[99] Akisue G. Secretions of *Myroxylon peruiferum* L.f. II. physical and chemical characterization of the balsam and qualitative analysis of various components. Revista de Farmácia e Bioquímica da Universidade de São Paulo. 1972;**10**(1):73-96

[100] Gutiérrez RM, Mitchell S, Solis RV. *Psidium guajava*: A review of its traditional uses, phytochemistry and pharmacology. Journal of Ethnopharmacology. 2008;**117**(1):1-27. DOI: 10.1016/j.jep.2008.01.025

[101] Paul S, Kalyani M, Kannan I. Antifungal activity, gas chromatographic-mass spectrometric analysis and in silico study of *Punica Granatum* peel extracts against fluconazole resistant strains of *Candida* species. Current Pharmaceutical Biotechnology. 2018;**19**(3):250-257. DOI: 10.2174/1389201019666180515104800

[102] Holfing JF, Anibal PC, Obando-Pereda GA, Peixoto IAT, Furletti VF, Foglio MA. Antimicrobial potential of some plant extract against Candida species. Brazilian Journal of Biology. 2010;**70**(4):1065-1068

[103] Leite VM, Pinheiro JB, Pisani MX, Watanabe E, de Souza RF, Paranhos Hde F, et al. In vitro antimicrobial activity of an experimental dentifrice based on *Ricinus communis*. Brazilian Dental Journal. 2014;**25**(3):191-196

[104] Pinelli LA, Montandon AA, Corbi SC, Moraes TA, Fais LM. *Ricinus communis* treatment of denture stomatitis in institutionalised elderly. Journal of Oral Rehabilitation. 2013;**40**(5):375-380. DOI: 10.1111/joor.12039

[105] Shinobu-Mesquita CS, Bonfim-Mendonça PS, Moreira AL, Ferreira IC, Donatti L, Fiorini A, et al. Cellular structural changes in *Candida albicans* caused by the hydroalcoholic extract from *Sapindus saponaria* L. Molecules. 2015;**20**(5):9405-9418. DOI: 10.3390/molecules20059405

[106] Fiorini A, Rosado FR, Bettega EM, Melo KC, Kukolj C, Bonfim-Mendonça PS, et al. *Candida albicans* protein profile changes in response to the butanolic extract of *Sapindus saponaria* L. Revista do Instituto de Medicina Tropical de São Paulo. 2016;**58**:25. DOI: 10.1590/S1678-9946201658025

[107] Santos MRG, Silva JHS, Caxito MLC. Brief review on the medicinal uses and antimicrobial activity of different parts of *Schinus terebinthifolius raddi*. International Journal of Pharmacy and Pharmaceutical Sciences. 2015;**7**(12):1-7

[108] Brandão MGL, Cosenza GP, Moreira RA, Monte-mor RLM. Medicinal plants and other botanical products from the Brazilian official pharmacopoeia. Brazilian Journal of Pharmacognsosy. 2006;**16**(3):408-420

[109] Santos VR, Gomes RT, Oliveira RR, Cortés ME, Brandão MGL. Susceptibility of oral pathogenic microorganisms to aqueous and ethanolic extracts of *Stryphnodendron adstringens* (barbatimão). International Journal of Dentistry. 2009;**8**(1):1-5

[110] Morey AT, de Souza FC, Santos JP, Pereira CA, Cardoso JD, de Almeida RS, et al. Antifungal activity of condensed tannins from Stryphnodendron adstringens: Effect on *Candida tropicalis* growth and adhesion properties. Current Pharmaceutical Biotechnology. 2016;**17**(4):365-375

[111] Hussein AA, Bozzi B, Correa M, Capson TL, Kursar TA, Coley PD, et al. Bioactive constituents from three Vismia species. Journal of Natural Products. 2003;**66**:858-860

[112] Chagas RCR. Estudo fitoquimico de Vismia brasiliensis (Chisy) (Clusianaceae). Tese: UFLA; 2009. 253 p

[113] Dantas FCP, Tavares MLR, Targino MS, da Costa AP, Dantas FO. *Ziziphus joazeiro* Mart.—Rhamnaceae: Características biogeoquímicas e importância no bioma Caatinga. Divulg Científica Tecnol IFPB. 2014;**25**(12):51-57

[114] Bailão EFLC, Devilla IA, Conceição E C, Borges LL. Bioactive compounds found in Brazilian cerrado fruits. International Journal for Molecular Science. 2015;**16**(10):23760-23783. DOI: 10.3390/ijms161023760

[115] Rocha RS, Kassuya CA, Formagio AS, Mauro Mde O, Andrade-Silva M, Monreal AC, et al. Analysis of the anti-inflammatory and chemopreventive potential and description of the antimutagenic mode of action of the *Annona crassiflora* methanolic extract. Pharmaceutical Biology. 2016;**54**(1):35-47. DOI: 10.3109/13880209.2015.1014567

[116] Mota GS, Sartori CJ, Miranda I, Quilhó T, Mori FA, Pereira H. Bark anatomy, chemical composition and ethanol-water extract composition of *Anadenanthera peregrina* and *Anadenanthera colubrina*. PLoS One. 2017 Dec 27;**12**(12):e0189263. DOI: 10.1371/journal.pone.0189263 eCollection 2017

[117] Santos VR, Chaim FA. Antimicrobial efficacy of Brazilian Azadirachta indica (neem) extract and mouthwash against oral pathogens microorganisms. Anais 29° Congresso Brasileiro de. Microbiología. 2017;**1**(1):122-122

[118] Raja Ratna Reddy Y, Krishna Kumari C, Lokanatha O, Mamatha S, Damodar Reddy C. Antimicrobial activity of *Azadirachta indica*(neem) leaf, bark and seed extracts. Internatinal Research Journal of Phytochemistry. 2013;**3**(1):1-4

[119] Kitamura H, Saito N, Fujimoto J, Nakashima KI, Fujikura D. Brazilian propolis ethanol extract and its component kaempferol induce myeloid-derived suppressor cells from macrophages of mice in vivo and in vitro. BMC Complementary and Alternative Medicine. 2018;**18**(1):138. DOI: 10.1186/s12906-018-2198-5

[120] Vieira ML, Johann S, Hughes FM, Rosa CA, Rosa LH. The diversity and antimicrobial activity of endophytic fungi associated with medicinal plant Baccharis trimera (Asteraceae) from the Brazilian savannah. Canadian Journal of Microbiology. 2014; **60**(12):847-856. DOI: 10.1139/cjm-2014-0449

[121] Duarte MC, Figueira GM, Sartoratto A, Rehder VL, Delarmelina C. Anti-Candida activity of Brazilian medicinal plants. Journal of Ethnopharmacology. 2005;**97**(2):305-311

[122] Laribi B, Kouki K, M'Hamdi M, Bettaieb T. Coriander (*Coriandrum sativum* L.) and its bioactive constituents. Fitoterapia. 2015 Jun;**103**:9-26. DOI: 10.1016/j.fitote.2015.03.012

[123] Fontenelle ROS, Morais SM, Brito EHS, Brilhante RSN, Cordeiro RA, Nascimento NRF, Kerntopf MR, Sidrim JJC, Rocha MFG. Antifungal activity of essential oils of Croton species from the *Brazilian Caatinga* biome (Article). Journal of Applied Microbiology. May 2008;**104**(5):1383-1390

[124] Oliveira-Tintino CDM, Pessoa RT, Fernandes MNM, Alcântara IS, da Silva BAF, de Oliveira MRC, et al. Anti-inflammatory and anti-edematogenic action of the *Croton campestris* A. St.-Hil (Euphorbiaceae) essential oil and the compound β-caryophyllene in in vivo models. Phytomedicine. 2018 Mar 1;**41**:82-95. DOI: 10.1016/j.phymed.2018.02.004

[125] Siqueira AB, Rodriguez LR, Santos RK, Marinho RR, Abreu S, Peixoto RF, Gurgel BC. Antifungal activity of propolis against Candida species isolated from cases of chronic periodontitis (Article). Brazilian Oral Research. 2015;**29**(1):1-6

[126] Santos Junior EM, Santos VR. Antimicrobial activity of mouthwash- brazilian red propolis content against oral microorganisms. Asian Academic Research Journal of Multidisciplinary. 2017;**5**(4):83-93

[127] Santos VR, Pinto LSFS., Araujo e Ribeiro BCB, Kubo I. Drug enhancement, synergism and antifungal activity of miconazole associated polygodial against *Candida albicans*. International Journal of Research—GRANTHAALAYAH. 2017;**5**(11):95-101

[128] Correia AF, Silveira D, Fonseca-Bazzo YM, Magalhães PO, Fagg CW, da Silva EC, et al. Activity of crude extracts from Brazilian cerrado plants against clinically relevant Candida species. BMC Complementary and Alternative Medicine. July 2016;**16**(1):11, Article number 203. DOI: 10.1186/s12906-016-1164-3

[129] Lazarini JG, Sardi JCO, Franchin M, Nani BD, Freires IA, Infante J, et al. Bioprospection of *Eugenia brasiliensis*, a Brazilian native fruit, as a source of anti-inflammatory and antibiofilm compounds. Biomedicine & Pharmacotherapy. 2018;**102**:132-139. DOI: 10.1016/j.biopha.2018.03.034

[130] Santos JFS, Rocha JE, Bezerra CF. Do Nascimento Silva MK, de Matos YMLS, de Freitas TS, dos Santos ATL, da Cruz RP, Machado AJT, Rodrigues THS, de Brito ES, Sales DL, de Oliveira Almeida W, da Costa JGM, Coutinho HDM, Morais-Braga MFB. Chemical composition, antifungal activity and potential anti-virulence evaluation of the *Eugenia uniflora* essential oil against Candida spp. Food Chemistry. 2018 Sep 30;**261**:233-239. DOI: 10.1016/j.foodchem.2018.04.015

[131] Pellati D, Fiore C, Armanini D, Rassu M, Bertoloni G. In vitro effects of glycyrrhetinic acid on the growth of clinical isolates of *Candida albicans*. Phytotherapy Research. 2009 Apr;**23**(4):572-574. DOI: 10.1002/ptr.2693

[132] PallagA FGA, OlteanuD CS, Baldea JT, MicleO VL, MarianE SO, Cenariu M, et al. *Equisetum arvense* L. extract induces antibacterial activity and modulates oxidative stress, inflammation, and apoptosis in endothelial vascular cells exposed to hyperosmotic stress. Oxidative Medicine and Cellular Longevity. 2018;**2018**:3060525. DOI: 10.1155/2018/3060525

[133] Najafi S, Koujan SE, Manifar S, Kharazifard MJ, Kidi S, Hajheidary S. Preventive effect of *Glycyrrhiza Glabra* extract on Oral mucositis in patients under head and neck radiotherapy: A randomized clinical trial. Journal of Dentistry (Tehran). 2017 Sep;**14**(5):267-274

[134] de Morais SR, Oliveira TL, de Oliveira LP, Tresvenzol LM, da Conceição EC, Rezende MH, et al. Essential oil composition, antimicrobial and pharmacological activities of Lippia siDOIdes Cham. (Verbenaceae) from São Gonçalo do Abaeté, Minas Gerais, Brazil. Pharmacognosy Magazine. 2016 Oct-Dec;**12**(48):262-270

[135] Schwarcz KD, Bajay MM, Macrini CM, Salazar VL, Souza AP, Pinheiro JB, et al. Microsatellite markers for the Cabreúva tree, *Myroxylon peruiferum* (Fabaceae), an endangered medicinal species from the Brazilian Atlantic Forest. Genetics and Molecular Research. 2014 Mar 26;**13**(3):6920-6925. DOI: 10.4238/2014.March.26.1

[136] Morais-Braga MFB, Sales DL, Carneiro JNP, Machado AJT, Dos Santos ATL, de Freitas MA, et al. *Psidium guajava* L. and *Psidium brownianum* Mart ex DC.: Chemical composition

and anti-Candida effect in association with fluconazole. Microbial Pathogenesis. 2016 Jun;**95**:200-207. DOI: 10.1016/j.micpath.2016.04.013

[137] Oliveira JR, de Castro VC, das Graças Figueiredo Vilela P, Camargo SE, Carvalho CA, Jorge AO, de Oliveira LD. Cytotoxicity of Brazilian plant extracts against oral microorganisms of interest to dentistry. BMC Complementary and Alternative Medicine. 2013 Aug 15;**13**:208. DOI: 10.1186/1472-6882-13-208

[138] Salles MM, Badaró MM, Arruda CN, Leite VM, Silva CH, Watanabe E, et al. Antimicrobial activity of complete denture cleanser solutions based on sodium hypochlorite and *Ricinus communis*: A randomized clinical study. Journal of Applied Oral Science. 2015 Nov-Dec;**23**(6):637-642. DOI: 10.1590/1678-775720150204

[139] Foss SR, Nakamura CV, Ueda-Nakamura T, Cortez DA, Endo EH, Dias Filho BP. Antifungal activity of pomegranate peel extract and isolated compound punicalagin against dermatophytes. Annals of Clinical Microbiology and Antimicrobials. 2014 Sep 5;**13**:32. DOI: 10.1186/s12941-014-0032-6

[140] Naz R, Bano A. Antimicrobial potential of *Ricinus communis* leaf extracts in different solvents against pathogenic bacterial and fungal strains. Asian Pacific Journal of Tropical Biomedicine. 2012;**2**(12):944-947

[141] Suurbaar J, Mosobil R, Donkor AM. Antibacterial and antifungal activities and phytochemical profile of leaf extract from different extractants of *Ricinus communis* against selected pathogens. BMC Research Notes. 2017 Dec 1;**10**(1):660. DOI: 10.1186/s13104-017-3001-2

[142] Tsuzuki JK, Svidzinski TI, Shinobu CS, Silva LF, Rodrigues-Filho E, Cortez DA, et al. Antifungal activity of the extracts and saponins from *Sapindus saponaria* L. Anais da Academia Brasileira de Ciências. 2007 Dec;**79**(4):577-583

[143] Johann S, Silva DL, Martins CVB, Zani CL, Pizzolatti MG, Resende MA. Inhibitory effect of extracts from Brazilian medicinal plants on the adhesion of *Candida albicans* to buccal epithelial cells(article). World Journal of Microbiology and Biotechnology. November 2008;**24**(11):2459-2464

[144] Bernardes NR, Araújo MH, Borges IFJC, Almeida FM, Amaral EP, Lasunskaia EB, et al. Nitric oxide production, inhibitory, antioxidant and antimycobacterial activities of the fruits extract and flavonoid content of *Schinus terebinthifolius*. Revista Brasileira de Farmacognosia. 2014;**24**:644-650

[145] de Freitas ALD, Kaplum V, Rossi DCP, da Silva LBR, Melhem MSC, Taborda CP, et al. Proanthocyanidin polymeric tannins from *Stryphnodendron adstringens* are effective against *Candida* spp. isolates and for vaginal candidiasis treatment. Journal of Ethnopharmacology. 2018 Apr 24;**216**:184-190. DOI: 10.1016/j.jep.2018.01.008

[146] Brito SM, Coutinho HD, Talvani A, Coronel C, Barbosa AG, Vega C, et al. Analysis of bioactivities and chemical composition of *Ziziphus joazeiro* Mart. using HPLC-DAD. Food Chemistry. 2015 Nov 1;**186**:185-191. DOI: 10.1016/j.foodchem.2014.10.031

[147] Noronha VR, Araujo GS, Gomes RT, Iwanaga SH, Barbosa MC, Abdo EN. Ferreira e Ferreira E, Viana Campos AC, Souza AA, Abreu SR, Santos VR. Mucoadhesive propolis gel for prevention of radiation-induced oral mucositis. Current Clinical Pharmacology. 2014;**9**(4):359-364

[148] Pina GM, Lia EN, Berretta AA, Nascimento AP, Torres EC, Buszinski AF, et al. Efficacy of propolis on the denture stomatitis treatment in older adults: A multicentric randomized trial. Evidence-based Complementary and Alternative Medicine. 2017;**2017**:8971746. DOI: 10.1155/2017/8971746

[149] Noronha VRAS, Abdo EN, Persio FPCL, Santos VR. Propolis gel versus benzydamine in preventing oral mucositis for patients irradiated in head and neck: A preliminary study. Oncology Journal | Cancer Reports and Reviews. 2017;**1**(2):24

[150] Santos VR, Oliveira CF, Abreu Rosa FH, Araujo ACS, Ribeiro TG, Faraco AAG, Lopes MTP, Castilho RO. A new red propolis mucoadhesive gel: Antimicrobial activity against some oral pathogen. Alternative and Integrative Medicine. 2017;**4**(3):129. DOI: 10.4172/2327-5162.S1.011. Abstract

Section 2

Adaptation and Resistance

Chapter 3

Carbon Sources Attribute to Pathogenicity in *Candida albicans*

Doblin Sandai, Yasser Tabana and Rosline Sandai

Additional information is available at the end of the chapter

http://dx.doi.org/10.5772/intechopen.80211

Abstract

This topic was to examine the impact of galactose or fructose upon the assimilation of secondary carbon sources by *Candida albicans*. *C. albicans ICL1* gene is repressed upon addition of 2% galactose or fructose to lactate- and oleic acid-grown cells. Further studies on Ca*FOX2*, Ca*FBP1* and Ca*MLS1* transcripts in response to galactose or fructose on assimilation of lactate and oleic acid resulted in repression of these genes. The Ca*ICL1* gene, which encode the glyoxylate cycles enzymes isocitrate lyase are required for growth on non-fermentable carbon sources. However, the enzyme CaIcl1 was not destabilized by galactose, but was degraded in response to fructose. In contrast, *S. cerevisiae* Icl1 has retained the molecular apparatus of protein degradation in response to either galactose or fructose. Screening of ubiquitination site by http://www.ubpred.org/ showed that *C. albicans* lacks ubiquitination site in gluconeogenic and glyoxylate cycles enzymes as compare to *S. cerevisiae*. Addition of a putative *S. cerevisiae* ubiquitination site carboxy terminus of CaIcl1 led to galactose- accelerated degradation of this protein in *C. albicans* cell via a ubiquitin-dependent process. In the other hand, CaIcl prior to addition of ubiquitination site was degraded upon exposure to fructose; addition of *S. cerevisiae* ubiquitination site to CaIcl1 further increased the speed of protein degradation.

Keywords: *Candida albicans*, galactose, fructose, glyoxylate cycles, isocitrate lyase, gluconeogenic, metabolic adaptation, protein degradation, ubiquitination

1. Introduction

Candida albicans is eukaryotic diploid (2N) sexual yeasts of the kingdom fungi [1]. *C. albicans* can divide asexually or can undergo parasexual reproduction, heterothallic or homothallic mating [2]. *C. albicans* genome sequencing project revealed the presence of sequences homologous to the *Saccharomyces cerevisiae MAT* (mating type) loci, *C. albicans MTL* (mating-type-like) loci [3]. Members of the genus *Candida* are very heterogeneous, can grow in at least three different morphologies; yeast, pseudo-hyphae and hyphae, such as *C. albicans* and *C. dubliniensis* [4, 5]. Further morphological forms exist during colony switching, such as white-opaque switching system involves switch between white domed colonies and opaque flat colonies [6], mating type-like locus a (*MTLa*) and *MTLα* cells must switch from white to opaque to become mating-competent [7]. Opaque cells secreted pheromones resulting in the formation of conjugation tubes, and subsequently, cell and nuclear fusion occur to form tetraploid (4N) cells. Mating products can be induced to undergo concerted chromosome loss to return to the diploid state [8, 9].

In order to proliferate in a wide range of environmental niches, pathogens must not only depend on certain virulence factors, it is also important to have a flexible metabolism; therefore, they can assimilate a variety of carbon sources that are scarce or the only available carbon source at a specific environmental niche. The primary and preferred sources of metabolic carbon for most organisms are carbohydrates; it is used for generating energy and producing biomolecules. Before entering the glycolytic pathway, most sugars are converted to fructose-6-phosphate or glucose-6-phosphate. ATP and NADH are produced from the conversion of hexose phosphates into the key metabolite pyruvate in glycolysis pathway. From there, two major strategies of energy production (fermentation and respiration) are carried out by cells. Although NAD+ is regenerated by both processes, respiration is more efficient than fermentation as it produces additional ATP through the oxidative phosphorylation and tricarboxylic acid (TCA) cycle. Glycolysis is the central, common pathway for both processes and it is critical for carbon assimilation; the pathway has been shown to be up-regulated during infections and important to the virulence in pathogenic bacteria, parasites, and fungi [10–12]. Glycolysis, gluconeogenesis, and the glyoxylate cycle are part of *C. albicans* central metabolism.

Central metabolic pathway such as glycolysis is strictly regulated, transcript of glycolytic enzymes is regulated in response to environmental conditions such as carbon source and availability, oxygen levels, and to cellular demands such as energy needs and metabolite concentrations. However, the regulators of glycolytic gene expression in most species have not been identified; for eukaryotes understanding of transcriptional control of glycolysis is mainly based on the non-pathogenic yeast *Saccharomyces cerevisiae* [13]. The transcription regulators (Gcr1 and Gcr2) are responsible for inducing the expression of the glycolytic genes in *S. cerevisiae* [14, 15], Gcr1p binds to CT boxes (*5'-CTTCC-3'*) upstream of the glycolytic genes and Gcr2p acts as a co-activator by forming a complex with Gcr1p [16]. Inactivation of either gene (*GCR1* and *GCR2*) result in growth defects during culture on glucose, due to the decreasing expression levels of the glycolytic genes [15]. However, on non-fermentable carbon sources, the mutant strains display wild type growth rates [17]. Another glycolysis-specific regulator in *S. cerevisiae* is Tye7 (also known as Sgc1); it is involved in the activation of several glycolytic genes, although not to the extent of Gcr1 and Gcr2, this activation is independent of *GCR1.* No growth defects under any carbon source regime are detected in *tye7* mutant strain [18].

C. albicans is an opportunistic fungus that is capable of metabolizing carbon sources through respiration and fermentation pathway similar to that of a typical eukaryotic cell. The culprit for more than half of all *Candida* infections is *C. albicans* [19]; this shows the importance of studying the metabolism of this pathogen for the development of effective antifungal treatments. *C. albicans* is a Crabtree-negative organism that lacks *GCR1/2* homologs and must control transcription of glycolytic genes differently as compared to *S. cerevisiae*. Tye7 and Gal4 are identified as two fungal-specific activators of the glycolytic pathway in *C. albicans*, severe growth defects were observed in the mutant strains (*tye7* and *gal4*) cultured on fermentable carbon sources when respiration was inhibited or oxygen was limited. Furthermore, chromatin immunoprecipitation coupled with microarray analysis (ChIP-CHIP) and transcription profiling showed these factors bind to and regulate expression of the glycolytic pathway genes. *C. albicans* mutant strains (*tye7* and *gal4*) showed attenuated virulence in *Galleria mellonella* infection model, therefore *TYE7* and *GAL4* genes are required for pathogenicity and virulence of *C. albicans* [20].

Gluconeogenesis is required for yeast cells to generate sugar phosphates for the synthesis of essential cellular components, during the growth on non-fermentable carbon sources. Under physiological conditions with two exceptions, most of the glycolytic reactions are reversible. In many yeasts including *S. cerevisiae* and *C. albicans*, due to unfavorable thermodynamic balances, pyruvate kinase and phosphofructokinase have to be bypassed, this is achieved by phosphoenol-pyruvate carboxykinase (Pck1, building phosphoenolpyruvate), and fructose-1,6-bisphosphatase (Fbp1, building fructose-6-phosphate). The enzymes of the glyoxylate cycle are necessary for gluconeogenesis, such as malate synthase (*MLS*) and isocitrate lyase (*ICL*). Moreover, in *S. cerevisiae,* in other to replenish TCA cycle intermediate (anaplerotic reactions), the succinate/fumarate transporter, Acr1p, has been shown to be a link between the glyoxylate cycle and the TCA cycle [21]. Not much is known about gluconeogenesis in *C. albicans*, whereas carbon sources/sugar phosphate may become available during lymphatic and blood infection, on skin after incorporation in macrophage, gluconeogenesis and the glyoxylate cycle may be depended for providing carbohydrates for cell wall biosynthesis [22].

The glyoxylate cycle is a "modified tricarboxylic acid (TCA) cycle," instead of the two decarboxylation steps of the TCA cycle the key enzymes of the glyoxylate cycle (isocitrate lyase, *ICL1* and malate synthase, *MLS1*) convert isocitrate and Acetyl-CoA into succinate and malate. *ICL1* splits the isocitrate, C_6-unit into succinate and glyoxylate, which in turn is condensed by malate synthase, *MLS1* with Acetyl-CoA generating free CoA-SH and malate. Malate is then processed malate dehydrogenase to continue the cycle and succinate is released as net product, which can be used to replenish the TCA cycle or to function as precursors for carbohydrate biosynthesis or amino acid biosynthesis. C_4-units (succinate) are generated by converting two Acetyl-CoA units generated by various catabolic processes. Therefore, the glyoxylate cycle enable cells to utilize fatty acids or C_2-units such as ethanol or acetate as sole carbon source [23].

Fatty acids are broken down in the mitochondria by catabolic process (beta oxidation) to generate two-carbon units (Acetyl-CoA), which can be oxidized to CO_2 and H_2O via the TCA cycle or used to generate hexose via the glyoxylate and gluconeogenesis, and NADH and $FADH_2$, which are co-enzymes used in electron transport chain. In yeast *S. cerevisiae,* β-oxidation and peroxisome biogenesis are regulated by transcription factors, *OAF1* and *PIP2*. However, such homolog is not present in *C. albicans.* Instead, *C. albicans* share a single homolog of transcription factors FarA/FarB with *A. nidulans*, *CTF1* which regulate genes for β-oxidation enzymes, glyoxylate cycle, and the gluconeogenesis in the presence of fatty acids. *CTF1* is important for growth on fatty acids and regulation of several genes encoding enzymes associated with β-oxidation, including *FOX2*. Deletion of *FOX2* (*fox2*Δ/*fox2*Δ (*fox2*Δ/Δ) strains) confers a mild attenuation of virulence [24, 25]. Therefore, *C. albicans* appears to share more similarity to that of filamentous fungi than to that of budding yeast at phenotypic and genotypic level in the regulation of alternative carbon assimilation pathways, but there are clearly *Candida*-specific adaptations in these regulatory networks [26].

Glyoxylate cycle is present in fungi but not in mammals, and β-oxidation of fatty acids develop in different directions between mammals and fungi. Therefore, in order to have a better understand on the importance of these pathways, physiological and virulence of mutant strains lacking of these genes encoding key enzymes the β-oxidation multifunctional protein (*FOX2*), glyoxylate cycle (isocitrate lyase, *ICL1*), and gluconeogenesis (fructose-1,6-bisphosphatase, *FBP1*) was studied. Deletion of *ICL1* or *FOX2* or *FBP1* confers attenuation to some degree in a mouse model of disseminated candidiasis. Therefore, confirming that alternative carbon sources are relevant nutrients in vivo, and *C. albicans* probably consume multiple carbon sources during infection. Finally, in vitro genomic and phenotypes analysis indicate that the regulatory networks that control alternative carbon metabolism in *S. cerevisiae* differ significantly from the paradigms developed in *C. albicans* [25, 27].

In the present of glucose, *C. albicans* undergoes yeast-to-hyphal transition which is crucial for virulence and invasion of host cells [28]. The hyphal form can diffuse through the tissues or form mycelial biofilms, and the yeast form is thought to disseminate easily via body fluids [29]. Glucose is an important carbon sources in central metabolism pathway, thus glucose sensing and response is closely regulated and highly evolved in most organisms. For example, *C. albicans* appears to have over 20 different hexose transporters that are expressed under different conditions and *S. cerevisiae* has at least 17 [30], it is presumed that this large number of hexose transporters is required because *S. cerevisiae* prefers to ferment glucose. *S. cerevisiae* demands a high influx of glucose because for each molecule of glucose metabolized, it yields only two molecules of ATP. In contrast, *C. albicans* also has many hexose transporters although it is presumed to prefer respiration and thus can yield up to 38 ATPs per glucose molecule metabolized [31]. This might be the reason why *C. albicans* can thrives in various niches, which likely release many different sugars as carbon sources.

Not much is known about how *C. albicans* senses and responds to sugars, therefore *S.* cerevisiae a paradigm for non-conventional yeasts is used as a model to elucidate sugar-sensing and signaling pathway that primarily regulates glucose transport. In yeast *S. cerevisiae*, glucose is detected by two glucose sensors in the cell membrane, which are Snf3 and Rgt2. The glucose sensors are orthologs of transporters with long cytoplasmic tails for intracellular signaling, which are missing from glucose transporters. The sensors cannot transport sugar but they control sugar acquisition by regulating the expression of genes encoding hexose transporters (*HXTs*) [32]. A single point mutation in the *S. cerevisiae* glucose sensors Snf3 (R229K) or Rgt2 (R231K) resulted in constitutive expression of genes encoding hexose transporters, because these mutations may convert the sensors into their glucose-bound (signaling) conformations [33]. *C. albicans* possesses glucose sensors, Hgt4 (orf19.5962) which is an ortholog of the Snf3 and Rgt2, Hgt4 generates an intracellular signal to induce expression of certain *HGT* genes encoding hexose transporters which is required for growth on glucose, for optimal virulence, and for filamentation [34].

Similar observation is observed between *S. cerevisiae* and *C. albicans*, *ICL1* and *PCK1*, whereby upon exposure to 2% glucose the transcription from DNA to mRNA are repressed. However, in contrast to *S. cerevisiae*, Icl1 and Pck1, which are degraded rapidly upon exposure to glucose. *C. albicans*, Icl1 and Pck1 remain stable [35]. This suggests that *C. albicans* undergoes catabolite inactivation on transcriptional level but not on the protein degradation level, which further implicate that the proteins (Icl1 and Pck1) are subjected to post-translational modifications. Post-translational modifications of proteins, such as acetylation, phosphorylation and ubiquitination, play important roles in generating and relaying signal in almost every cellular pathway as well as cellular adaptation of all organisms, which includes their growth, differentiation, division and development. These protein modifications are important for the regulation of the pre-existing proteins' functions, activities and stabilities, thereby controlling dynamic cellular processes [36].

Similar to that in *S. cerevisiae*, *C. albicans* polyubiquitin is encoded by the *UBI4* gene, which has three rather than five tandem repeats in a consecutive head-to-tail arrangement that is found in *S. cerevisiae* [37]. Decrease in the polyubiquitin encoded by the *UBI4* gene in *C. albicans* induces the growth of hyphae and pseudohyphae [38]. This was compatible with previous studies indicating that the ubiquitination via the E2 enzyme, Rad6 inhibits hyphal development in *C. albicans* [39]. Furthermore, polyubiquitin is presumed to contribute to stress responses in *C. albicans*, because downregulation of *UBI4* resulted in mild temperature sensitivity in stationary cells grown in glycerol, but not during growth on glucose [38]. Therefore, it has been implied that the *UBI4* gene is necessary for viability in *C. albicans* [38]. For example, recent studies show that ubiquitination sites are present in glyoxylate cycle and gluconeogenesis enzymes from *S. cerevisiae* but absent from their *C. albicans* homologs [35]. This indicates that evolutionary rewiring of ubiquitination targets upon glucose exposure; this implies *C. albicans* proteins remain functional, allowing it to continue metabolize alternative carbon sources [35]. This metabolic flexibility is presumed to be important during infection and colonization of dynamic host niches with variable carbon sources.

Lorenz and Fink [27] have proved the importance of key enzyme, *ICL1* of glyoxylate cycle to the virulence and pathogenicity of *C. albicans*. Deletion of the gene cause attenuated virulence in mice model. This highlights the importance of *ICL1* as a factor that contributes to the virulency and viability of *C. albicans* and provides a new target site for antifungal drugs test. For example, caffeic acid, rosmarinic acid, and apigenin were found to have antifungal activity against *C. albicans ICL1* when tested under glucose-depleted conditions [40].

Thus, it is important to have a better understanding and studying the mechanism involved, and the fitness attribute of the key enzymes in central metabolism of *C. albicans*. Recent studies by Sandai et al. [35] indicate that glucose trigger degradation of transcript in *C. albicans*, *ICL1* and *PCK1* but not the protein. This suggests that *C. albicans* have undergone evolutionary rewiring and lacks ubiquitination site as compared to *S. cerevisiae*. As such, this study was mainly focused on the effects of galactose and fructose on the central metabolism of *C. albicans*. Galactose is metabolized to the more metabolically useful glucose-6-phosphate by the enzymes of the Leloir pathway. This pathway is necessary as the initial enzymes of glycolysis are unable to recognize galactose. Of all the sugars found in nature, only D-fructose feeds directly into glycolysis, the central pathway of carbohydrate metabolism.

2. Effect of galactose or fructose on *C. albicans FOX2, FBP1, MLS1* and *ICL1* mRNAs

There are reports that state repression on transcriptome by glucose in *S. cerevisiae* [41]. For example, study by Yin et al. [42] using northern blotting and transcriptomic analyses showed that transcripts encoding the gluconeogenic enzymes (FBP1 and PCK1) are repressed by glucose in *S. cerevisiae*. Furthermore, previous work by Sandai et al. [35] reported that *ICL1* and *PCK1* are repressed by glucose in *C. albicans* and *S. cerevisiae*. To reconfirm this report and to compare it with the galactose and fructose responses of *C. albicans* more directly in this study, we first examined the responses of *S. cerevisiae* glyoxylate cycle (Sc*ICL* and Sc*MLS1*), β-oxidation (Sc*FOX2*) and gluconeogenesis (Sc*FBP1*) mRNAs using the following experimental approach.

S. cerevisiae ICL1 mRNA levels showed a dramatic decrease within 1 hour of galactose or fructose addition to the cells growing on lactate or oleic acid. Similarly, Sc*FBP1*, Sc*FOX2* and Sc*MLS1* mRNAs levels declined after galactose or fructose addition to the cells growing on lactate or oleic acid media. This strong repression occurred 1 hour after galactose or fructose addition. These confirmed that in *S. cerevisiae*, the *ICL1*, *FBP1*, *FOX2* and *MLS1* transcripts are strongly repressed by galactose and fructose [43].

Global transcriptional responses of *C. albicans* to low (0.01%), medium (0.1%) and high (1%) glucose concentrations by microarray analysis indicated that a total of 347 *C. albicans* gene were up-regulated and 344 genes were down-regulated in responses to at least one of the glucose concentration examined. There are 170 of these genes were up-regulated and 180 genes were down-regulated by 0.01% glucose, indicating about half of glucose-regulated genes are responsive to low glucose levels [42]. Therefore, it is concluded that *C. albicans* and *S. cerevisiae* is acquisitively sensitive to glucose. Hence, at the start of this study, an aim was to confirm the impact of galactose and fructose upon specific mRNAs that encode enzymes required for the assimilation of alternative carbon sources. The transcripts encoding the glyoxylate cycle enzyme isocitrate lyase (Ca*ICL1*), fructose-1,6-biphosphatase (Ca*FBP1*), malate synthase (Ca*MLS1*) and multifunctional enzyme of the peroxisomal fatty acid beta-oxidation pathway (Ca*FOX2*) were the main focus here.

C. albicans cells were grown to mid-exponential phase in media containing lactate or oleic acid as the sole carbon source using the same procedures [35]. Galactose or fructose was then added to a final concentration of 2%, samples taken for RNA analysis at various times thereafter, and the levels of the Ca*ICL1*, Ca*FOX2*, Ca*MLS1*, Ca*FBP1* and Ca*ACT1* mRNAs were measured. The relative expression of Ca*ICL1*, Ca*FBP1*, Ca*FOX2* and Ca*MLS1* (compared to the internal Ca*ACT1* control) was high in lactate- and oleic acid-grown cells compared to cells that were exposed to either galactose or fructose.

3. Role of ubiquitination in sugar phosphate-accelerated protein degradation in *C. albicans*

Research done by Sandai et al. [35] suggested that *S. cerevisiae* has retained the ability to degrade target proteins in response to glucose, but that *C. albicans* isocitrate lyase (*ICL1*) has

(A) C. albicans and S. cerevisae Fbp1
CaFbp1 MSGPVNSVSK--QMNVDTDIITLTRFILQEQQTVAPTATGELSLLLNALQFAFKFIAHNI 58
ScFbp1 MPTLVNGPRRDSTEGFDTDIITLPRFIIEHQKQFK-NATGDFTLVLNALQFAFKFVSHTI 59
CaFbp1 RRAELVNLIGVSGSANSTGDVQKKLDVIGDEIFINAMRSSNNVKVLVSEEQEDLIVFPG- 117
ScFbp1 RRAELVNLVGLAGASNFTGDQQKKLDVLGDEIFINAMRASGIIKVLVSEEQEDLIVFPTN 119
CaFbp1 GGTYAVCTDPIDGSSNIDAGVSVGTIFGVYKLQEGSTGGISDVLRPGKEMVAAGYTMYGA 177
ScFbp1 TGSYAVCCDPIDGSSNLDAGVSVGTIASIFRLLPDSSGTINDVLRCGKEMVAACYAMYGS 179
CaFbp1 SAHLALTTGHGVNLFTLDTQLGEFILTHPNLKLPDTKNIYSLNEGYSNKFPEYVQDYLKD 237
ScFbp1 STHLVLTLGDGVDGFTLDTNLGEFILTHPNLRIPPQKAIYSINEGNTLYWNETIRTFIEK 239
CaFbp1 IKK-------EGYSLRYIGSMVADVHRTLLYGGIFAY------PTLKLRVLYECFPMALL 284
ScFbp1 VKQPQADNNNKPFSARYVGSMVADVHRTFLYGGLFAYPCDKKSPNGKLRLLYEAFPMAFL 299
CaFbp1 MEQAGGSAVTIKGERILDILPKGIHDKSSIVLGSKGEVEKYLKHVPK-- 331
ScFbp1 MEQAGGKAVNDRGERILDLVPSHIHDKSSIWLGSSGEIDKFLDHIGKSQ 348
(B) C. albicans and S. cerevisae Fox2
CaFox2 MS-PIDFKDKVVIITGAGGGLGKYYSLEFAKLGAKVVVNDLGGALDGQGGNSKAADIVVD 59
ScFox2 MPGNLSFKDRVVVITGAGGGLGKVYALAYASRGAKVVVNDLGGTLGGSGHNSKAADLVVD 60
CaFox2 EITKNGGVAVADYNNVLD-GAKIVETAVKSFGTVHIIINNAGILRDSSIKKMTEKDFKLV 118
ScFox2 EIKKAGGIAVANYDSVNENGEKIIETAIKEFGRVDVLINNAGILRDVSFAKMTEREFASV 120
CaFox2 IDVHLNGAYAVTKAAWPYFQKQKFGRVVNTSSPAGLYGNFGQTNYSAAKSALLGFAETLA 178
ScFox2 VDVHLTGGYKLSRAAWPYMRSQKFGRIINTASPAGLFGNFGQANYSAAKMGLVGLAETLA 180
CaFox2 KEGDRYNIKANAIAPLARSRMTESILPPPILEKLGPEKVAPLVLYLSSAENEVTGQFFEV 238
ScFox2 KEGAKYNINVNSIAPLARSRMTENVLPPHILKQLGPEKIVPLVLYLTHESTKVSNSIFEL 240
CaFox2 AAGFYAQIRWERSGGVLFKPD-QSFTAEVVAKRFSEVLNFDDSGKPEYLKNQHPFMLNDY 297
ScFox2 AAGFFGQLRWERSSGQIFNPDPKTYTPEAILNKWKEITDYRD--KP-FNKTQHPYQLSDY 297
CaFox2 TTLTTEARKLPSNDASGAPKVTLKDKVVLITGAGAGLGKEYAKWFARYGAKVVVNDFKDA 357
ScFox2 NDLITKAKKLPPNEQGSVKIKSLCNKVVVVTGAGGGLGKSHAIWFARYGAKVVVNDIKDP 357
CaFox2 TKTVEEIKAAG--GEAWADQHDVASQAEEIIKNVIDKYGTIDVLVNNAGILRDKSFAKMS 415
ScFox2 FSVVEEINKLYGEGTAIPDSHDVVTEAPLIIQTAISKFQRVDILVNNAGILRDKSFLKMK 417
CaFox2 DQEWDQVQKVHLLGTFNLSRLAWPYFAEKKYGRIVNISSTSGIYGNFGQANYASAKAGIL 475
ScFox2 DEEWFAVLKVHLFSTFSLSKAVWPIFTKQKSGFIINTTSTSGIYGNFGQANYAAAKAAIL 477

```
CaFox2    GLSKTLAVEGARNNIKVNVVAPHAETAMTLTIFREQDK-NLYHADQVAPLLVYLGSEEVE 534
ScFox2    GFSKTIALEGAKRGIIVNVIAPHAETAMTKTIFSEKELSNHFDASQVSPLVVLLASEELQ 537
          *:***:*:***:.. * ***:********* *** *:.  * . * **:**:* * ***..

CaFox2    ------VTGETFEAGGGWIGNTRWQRAKGAVSHDEHTTVEFIRDNLKDITNFDSDTENPK 588
ScFox2    KYSGRRVIGQLFEVGGGWCGQTRWQRSSGYVSIKETIEPEEIKENWNHITDFSRNTINPS 597
                * *. ** **** *:*****. * ** *    * *..* . **.* .* **.

CaFox2    STTESSMAILSAVGGDDDDDDDDEEEEDEGDEEEEEEDEEEDDPVWRFNDRDVILYNIALG 648
ScFox2    STEESSMATLQAV-----------------QKAHSSKELDDGLFKYTTKDCILYNLGLG 639
          ** ***** * **                  :: ...* ** ..... .* ****. **

CaFox2    ATTKQLHYVYENDSDFQVIPTFGHLITFNSGKSQNSFAKLLRNFNPMLLLHGEHYLKVHK 708
ScFox2    CTSKELKYTYENDPDFQVLPTF-AVIPFMQATATLAMDNLVDNFNYAMLLHGEQYFKLCT 698
          .*:*.*.* ****.****:***  .*.*  ....  .. .*. ***  .*****.*.*. .

CaFox2    WPPPTEGAIKTTFEPISTTPK-GSNVVIVHGSKSVDNDSGEVIYSNEATYFIR--NCQAD 765
ScFox2    PTMPSNGTLKTLAKPLQVLDKNGKAALVVGGFETYDIKTKKLIAYNEGSFFIRGAHVPPE 758
          . *..*..** .*...   * * . ..:* * .. * . ..:* ** ..:***  :  .:

CaFox2    NKVYAERRS-FATNPFPAPK-RAPDYQVDVPISEDLAALYRLTGDRNPLHIDPNFAKGAK 823
ScFox2    KEVRDGKRAKFAVQNFEVPHGKVPDFEAEISTNKDQAALYRLSGDFNPLHIDPTLAKAVK 818
          ..*   .*: ** . * .*. : **:......  .* ******.** *******  .** *

CaFox2    FPKPILHGMCTYGLSAKVLIDKFGMFDEIKARFTGIVFPGETLRVLAWKESDDTVVFQTH 883
ScFox2    FPTPILHGLCTLGISAKALFEHYGPYEELKVRFTNVVFPGDTLKVKAWKQGS-VVVFQTI 877
          **.*****.** *.*** *....*  .:.*.*** .:***.**.* ***... .*****

CaFox2    VVDRGTIAINNAAIKLVGDKAKI 906
ScFox2    DTTRNVIVLDNAAVKLSQAKSKL 900
            .*..*..:***.**    *.*:
```

(C) *C. albicans* and *S. cerevisae* Icl1

```
CaIcl1    MPY----TPIDIQKEEADFQKEVAEIKKWWSEPRWRKTKRIYSAEDIAKKRGTLK-INHP 55
ScIcl1    MPIPVGNTKNDFAALQAKLDADAAEIEKWWSDSRWSKTKRNYSARDIAVRRGTFPPIEYP 60
          **     * *:  .*.:. ..***:****. ** **** *** ***  .***:  *..*

CaIcl1    SSQQADKLFKLLEKHDADKTVSFTFGALDPIHVAQMAKYLDSIYVSGWQCSSTASTSNEP 115
ScIcl1    SSVMARKLFKVLEKHHNEGTVSKTFGALDPVQISQMAKYLDTIYISGWQCSSTASTSNEP 120
          **  * ****:**** . : *** *******....*******.**:****************

CaIcl1    SPDLADYPMDTVPNKVEHLWFAQLFHDRKQREER-LTLSKEERAKT-PYIDFLRPIIADA 173
ScIcl1    GPDLADYPMDTVPNKVEHLFKAQLFHDRKQLEARSKAKSQEELDEMGAPIDYLTPIVADA 180
          .******************. ********* * . :.**  .  . **.* **.***

CaIcl1    DTGHGGITAIIKLTKMFIERGAAGIHIEDQAPGTKKCGHMAGKVLVPVQEHINRLVAIRA 233
ScIcl1    DAGHGGLTAVFKLTKMFIERGAAGIHMEDQTSTNKKCGHMAGRCVIPVQEHVNRLVTIRM 240
          *:****:**..****************:***.. .*********. ..*****:****:**

CaIcl1    SADIFGSNLLAVARTDSEAATLITSTIDHRDHYFIIGATNPEAGDLAALMAEAESKGIYG 293
ScIcl1    CADIMHSDLIVVARTDSEAATLISSTIDTRDHYFIVGATNPNIEPFAEVLNDAIMSGASG 300
          .***: *:*:.************:**** ******:*****: :: .:  .* .* * *

CaIcl1    NELAAIESEWTKKAGLKLFHEAVIDEIKNGNYSNKDALIKKFTDKVNPLSHTSHKEAKKL 353
ScIcl1    QELADIEQKWCRDAGLKLFHEAVIDEIERSALSNKQELIKKFTSKVGPLTETSHREAKKL 360
          :*** **.:* :.**************...  ***: *****.** **..***:*****

CaIcl1    AKELTGKDIYFNWDVARAREGYYRYQGGTQCAVMRGRAFAPYADLIWMESALPDYAQAKE 413
ScIcl1    AKEILGHEIFFDWELPRVREGLYRYRGGTQCSIMRARAFAPYADLVWMESNYPDFQQAKE 420
          ***: *::*:*:*:: *.*** ***:*****::**.*********:****  **: ****
```

```
CaIcl1    FADGVKAAVPDQWLAYNLSPSFNWNKAMPADEQETYIKRLGKLGYVWQFITLAGLHTTAL 473
ScIcl1    FAEGVKEKFPDQWLAYNLSPSFNWPKAMSVDEQHTFIQRLGDLGYIWQFITLAGLHTNAL 480
          **:*** .*************** ***..*** *:*:*** ***:*********** .**

CaIcl1    AVDDFSNQYSQIGMKAYGQTVQQPEIEKGVEVVKHQKWSGATYIDGLLKMVSGGVTSTAA 533
ScIcl1    AVHNFSRDFAKDGMKAYAQNVQQREMDDGVDVLKHQKWSGAEYIDGLLKLAQGGVSATAA 540
          **.:**.:::: *****.*.***.*:..**:*:******** *******:..***..***

CaIcl1    MGQQGVTEDQFKESKAKA 550
ScIcl1    MGTGVTEDQFKENGVKK 557
          ** *********..*
```

Legend:

Label	Score range	Sensitivity	Specificity
Low confidence	$0.62 \le s \le 0.69$	0.464	0.903
Medium confidence	$0.69 \le s \le 0.84$	0.346	0.950
High confidence	$0.84 \le s \le 1.00$	0.197	0.989

(D)

Strain	Number of high confident ubiquitination site			
	Icl1	Mls1	Fox2	Fbp1
C. albicans	0	0	1	0
S. cerevisae	2	0	0	0

Figure 1. Comparison of an amino acid pile-up of (A) CaMls1 and ScMls1. (B) CaFbp1 and ScFbp1. (C) CaFox2 and ScFox2. (D) CaIcl1 and ScIcl1. In which the high confidence ubiquitination sites are highlighted. (E) Compilation of high ubiquitination site in *C. albicans* and *S. cerevisiae* Icl1, Mls1, Fox2 and Fbp1.

lost the specific signal(s) that trigger this glucose-accelerated protein degradation. However, studies have not been done on the effect of other sugar phosphate such as galactose or fructose, both are important carbon sources that are absorbed directly into the blood from intestine. This studies focus on whether CaIcl1 protein has lost the specific signal that trigger protein degradation with the formation of ubiquitin-proteasome complex.

Ubiquitination play an important role in the glucose-accelerated degradation of gluconeogenic enzymes in *S. cerevisiae* [20] by acting as a degradation signal in *S. cerevisiae* [44]. This fact is supported by Eschrich et al. [22], in which ubc8 functions in the catabolite degradation of fructose-1,6-biphosphatase in *S. cerevisiae*. Therefore, consensus ubiquitination target sites were examined in CaIcl1 and ScIcl1, CaMls1 and ScMls1, CaFox2 and ScFox2, and CaFbp1 and ScFbp1 using ubpred (predictor of protein ubiquitination site, from http://www.ubpred.org/index.html) [45, 46] (**Figure 1**).

Based on this bioinformatics comparison, the ScIcl sequence containing strong consensus ubiquitination sites at amino acid 158 and 551, but there is a lack of high confidence ubiquitin target in CaIcl1, CaMls1, ScMls1, CaFbp1 and ScFbp1 (**Figure 1**). Interestingly, CaFox2 contain one strong consensus ubiquitination site at amino acid 588, but there is none high confidence ubiquitination target in ScFox2. This prediction was based on high level of confidence which is described in UbPred system containing score range $0.84 \le s \le 1.00$, 0.197 for sensitivity and 0.989 for specificity. However, focus of this study is on Icl1 because it is more important for *C. albicans* virulency. Deletion of *ICL1* gene in *C. albicans* attenuated virulence [27] while deletion of FOX2 confers mild attenuation of virulence of *C. albicans* [26].

The bioinformatics screening of ubiquitination site includes the hydrophobic nature of the ubiquitination target site for high confidence prediction (TEDQFKENGVKK), which is contrast to the low and medium confidence sites which contain acidic and basic residues in the putative ubiquitination site (NGVKK; FNWPKAMSVD) [45, 46]. Therefore, the presence of consensus ubiquitination sites in these proteins correlated with glucose-accelerated degradation [35].

4. Overview of carbon sources attribute to the pathogenicity to *C. albicans*

The effect of galactose or fructose on the expression levels of the CaIcl protein in *C. albicans* and to compare this response with the corresponding situation in *S. cerevisiae* were tested. The effects of glucose on *S. cerevisiae* fructose-1,6-bisphosphatase (FBPase) have been intensively studied and it was reported that the FBPase protein (ScFbp1) is rapidly degraded upon addition of glucose [47]. It is also reported by Hammerle et al. [48] that the levels of cytosolic malate dehydrogenase, fructose-1,6-biphosphatase, isocitrate lyase and phosphoenolpyruvate carboxykinase are all low in *S. cerevisiae* after glucose addition. However, the degradation of FBPase is not correlated with the prediction of ubiquitination target site presence in the protein as ScFbp1 has only one low confidence ubiquitination target site at amino acid 250 (**Figure 1B**). The reason is not known whether or not FBPase had undergone ubiquitin-proteasome machinery although Regelmann et al. [49] have found that there are two different degradation pathways: cytosolic ubiquitin-proteasome machinery, and the other dependent on vacuolar proteolysis. Genes that are essential for vacuolar degradation are unnecessary for proteasome-dependent degradation [49].

Therefore, as a starting point whether Icl1 decline in *S. cerevisiae* upon galactose or fructose addition. To achieve this, the *S. cerevisiae* which is tagged at its 3′-end with Myc_9 is obtained from Sandai et al. [35]. Cells were grown on lactate, and then subjected to 2% of galactose or fructose. Samples were harvested at various times point and the Icl1 proteins expression was determined by western blotting and normalized by beta Actin. Clearly galactose or fructose addition led to the degradation of ScIcl1. These results confirmed that in *S. cerevisiae,* Icl1 are degraded in response to galactose or fructose. Similar results are observed in study done by Sandai et al. [35], whereby ScIcl1 and ScPck1 are degraded in response to glucose.

Next was to determine the effect of galactose or fructose addition to *C. albicans* cell growing on lactate. *C. albicans* strains expressing Myc_3-tagged Icl1 was confirmed by western blotting. *C. albicans* cells were grown in media containing the non-fermentable carbon sources lactate as sole carbon sources and the 2% galactose or fructose was added while cells were in exponential phase. Cells were then harvested at various times, and their proteins extracted and were loaded equally onto the SDS-PAGE gels, and expression of the Myc_3-tagged CaIcl1 protein is detected with anti-Myc antibodies and normalized by beta-actin. The *C. albicans* Icl1 protein was expressed during growth on lactate and was not destabilized by the addition of 2% galactose. Indeed CaIcl1 proteins levels were not significantly different from the control (lactate) even after 4 hours. Likewise, similar result was obtained by Sandai et al. [35] in which

CaIcl1 proteins remained stable after addition of 2% glucose to lactate grown cells. However, interestingly CaIcl1 protein is destabilized after addition of 2% fructose to lactate grown cells. This result contradicts the finding that CaIcl1 has lost the specific signal(s) that trigger protein degradation. This might prove the points by Regelmann et al. [49] that except from ubiquitin proteasome pathways the protein can undergo degradation through vacuolar proteolysis. However, the exact mechanism of why fructose triggers such pathway is not known.

Testing was carried out to see whether *C. albicans* is able to degrade proteins in response to galactose or fructose, the *S. cerevisiae ICL1* gene is expressed in *C. albicans. C. albicans ICL1* allele was replaced with a tagged *S. cerevisiae ICL1*. Ca(ScIcl1-Myc$_3$) was then grown in lactate before subjected to 2% galactose or fructose. Western blot was performed to confirm the expression of the Myc$_3$-tagged ScIcl1 which will be normalized to housekeeping protein beta-actin. From the results, ScIcl1 protein expression remained stable in *C. albicans* cells grown on lactate. However, following addition of 2% galactose or fructose to *C. albicans* cells, ScIcl1 was degraded. This indicates that *C. albicans* has retained capacity to destabilize target proteins in response to galactose or fructose. In other words, ScIcl1 retained its molecular apparatus for protein degradation even if it is expressed in *C. albicans* cells.

Testing was also carried out to determine whether *C. albicans* Icl1 expression remains stable in response to galactose or fructose, the *C. albicans ICL1* gene was expressed in *S. cerevisiae*. *S. cerevisiae ICL1* allele was replaced with a tagged *C. albicans ICL1*. Sc(CaIcl1-Myc$_3$) was then grown in lactate before subjected to 2% galactose or fructose. Cells were harvested at different time points thereafter and western blot was performed to confirm the expression of the Myc$_3$-tagged CaIcl1 which is normalized to beta-actin. From the results, CaIcl1 protein expression remained stable in *S. cerevisiae* cells grown on lactate. However, following addition of 2% fructose to *S. cerevisiae* cells, CaIcl1 was degraded. This supports the previous result that *C. albicans* Icl1 although lacks ubiquitination site but is sensitive to fructose and might undergo a different proteolysis. *C. albicans* has lost capacity to destabilize target proteins in response to 2% galactose. In other words, CaIcl1 was not degraded and remained stable in *S. cerevisiae*.

S. cerevisiae Icl1 has two high confidence putative ubiquitination sites located at residues 551 and 158, whereas CaIcl1 has no such sites (**Figure 1**). Therefore, ubiquitination plays a role in glucose-accelerated protein decay in *C. albicans*; the addition of a ubiquitination site to CaIcl1 would confer glucose-accelerated degradation upon this protein Sandai et al. [35]. Therefore carboxy-terminal ubiquitin site from ScIcl (TEDQFKENGVKK) was added into CaIcl1, together with the Myc$_3$ tag into the wild type polyubiquitin containing *C. albicans* cells. The function of ubiquitination in regulation rewiring of *C. albicans* Icl1 in response to galactose or fructose was tested. Cells were grown in lactate and galactose or fructose was added to final concentration of 2%. Cells were harvested at different time points thereafter, protein extracted and quantified, and these subjected to western blotting. The CaIcl-Ubi-Myc$_3$ proteins are normalized to housekeeping protein beta-actin. The result obtained shows that CaIcl-Ubi-Myc$_3$ is degraded upon addition of galactose or fructose to non-fermentable carbon sources, lactate. The results confirm a finding in a research done by Sandai et al. [35] whereby additions of glucose to lactate grown DSCO4 cells degrade the protein. This has also proven that ubiquitination plays a significant role in regulation of central metabolism enzymes in both *C. albicans* and *S. cerevisiae*.

5. Conclusion and future perspective

The assimilation of carbon sources is fundamentally important for the growth of *C. albicans* and for the establishment of infections in the human host [10]. For most yeast, such as *S. cerevisiae*, glucose is generally a more favorable carbon sources and the chosen mode of metabolism is often fermentative via glycolytic pathway-Embden-Meyerhof fermentation. In the presence of excess glucose, *S. cerevisiae* utilize glycolytic pathways and produce ethanol. *S. cerevisiae* is Crab-negative yeast that has the ability to produce ethanol even in the absence of oxygen [50]. In contrast, *C. albicans* is Crab-positive yeast because it retains respiratory capacity in the presence of excess glucose [51].

C. albicans is glucose Crabtree positive yeast which retains respiratory activity even following exposure to glucose [51]. During growth on glucose, *ADH1* mRNA levels rise to maximum levels during late exponential growth phase and the decline to low levels in stationary phase [52]. The *ADH1* mRNA is relatively abundant during growth on galactose, glycerol, pyruvate, lactate or succinate, and less abundant during growth on glucose or ethanol. However, alcohol dehydrogenase levels do not correlate closely with *ADH1* mRNA levels. This locus may be controlled in both transcriptional and post-transcriptional levels, or other differentially regulated *ADH* loci may exist in *C. albicans* [51]. Interestingly, a significantly smaller proportion of glucose is fermented to ethanol by *C. albicans* than by *S. cerevisiae* [51].

S. cerevisiae is not able to assimilate both non-fermentable carbon sources and galactose/fructose at the same time because of repression. Hence, these yeasts have evolved different responses to galactose/fructose. Therefore in this analysis was done on regulation of carbon assimilation in *C. albicans* focusing on genes/enzymes involved in gluconeogenesis, glyoxylate cycle and β-oxidation. *ICL1, FOX2, MLS1,* and *FBP1* gene expression and the impact of galactose or fructose on the assimilation of non-fermentable carbon sources were analyzed. *C. albicans* responses were then compared to those of *S. cerevisiae* under equivalent conditions. The following conclusions can be drawn from these findings;

First, gluconeogenic, glyoxylate cycle and β-oxidation mRNAs are sensitive to galactose or fructose in both *C. albicans* and *S. cerevisiae*. This reconfirmed previous finding, whereby the transcriptome, such as *PCK1* and *FBP1* in *S. cerevisiae* [42], *ICL1* and *PCK1* in both *S. cerevisiae* and *C. albicans* [35], was degraded upon exposure to glucose. The dramatic decreases in *ICL, MLS1, FOX2* and *FBP1* mRNA levels in *C. albicans* after exposure to 2% galactose or fructose to lactate- and oleic acid-grown cells was observed. It is already known that *C. albicans* sensitive to glucose concentrations even as low as 0.01%, this value is within the physiological range of blood glucose (about 0.1%) [10, 12]. Therefore, *C. albicans* is able to respond to blood glucose levels during disseminated hematological infections. Interestingly, diabetic patients who often have elevated blood glucose levels have a higher risk of systemic *Candida* infections [53] and dietary glucose enhances *C. albicans* colonization and invasion [54]. Galactose and fructose are both important sugar phosphate to human and yeast. Galactose can be converted to glucose through Leloir pathway which utilized three enzymes in human liver [55] and five enzymes in yeast [56]. In mammary gland galactose covalently bound to glucose to formed disaccharide lactose [57]. Moreover individual with galactosemia, a rare inherited metabolic disorder that affects the body ability to metabolize galactose properly and have high content of galactose in both urine and blood. This increases the risk of developing *E. coli* sepsis and sometimes or rarely fungal sepsis, such as *Candida* species [58].

Second main observation was that the Icl1 proteins are stable in *C. albicans* following galactose exposure but not fructose. The addition of 2% galactose to *C. albicans* cells growing on lactate did not trigger the degradation of the Icl1 proteins, but the protein was degraded upon addition of 2% fructose, at least 4 hours examined. This is in contrast to the situation in *S. cerevisiae*, where the addition of 2% galactose or fructose triggered the rapid degradation of the Icl1 proteins. The estimated half-lives for these proteins in *S. cerevisiae* are more than 20 hours [46] indicating that these proteins are very stable. This probably represents a significant difference in the physiological responses of these pathogenic and benign yeasts to glucose. *C. albicans* is able to establish infections in complex niches, many which contain a rich mixture of alternative carbon sources [10]. The stability of the Icl1 protein in *C. albicans*, even in the presence of galactose and glucose [35] might suggest that this pathogen is capable to assimilate alternative carbon sources at the same time as galactose or glucose in these carbon-rich niches. The observation from addition of galactose and glucose is similar in *C. albicans*, this may be because galactose and glucose is reversibly convertible in Leloir pathway, and galactose is first converted to glucose to enter the glycolysis pathway.

Interestingly there is a finding that fructose increases phosphorylation/activation of hypothalamic AMP kinase causing phosphorylation/inactivation of Acetyl-CoA carboxylase, whereas glucose has the inverse effects [59]. This finding is interesting because acetyl-coenzyme A (Acetyl-CoA) is an essential cofactor in central metabolism, this molecule is the entry point to the tricarboxylic acid (TCA) cycle that generates energy, biomass, and intermediates for macromolecules [59]. This means that addition of fructose might repressed Acetyl-CoA in *C. albicans* which will in turn affects the TCA and glyoxylate cycle. However, this might not be the only reason for Icl1 protein decay in *C. albicans* in response to fructose. Glucose enters the glycolytic pathway via glucokinase-catalyzed phosphorylation and its further metabolism is subjected to control in phosphofructokinase (PFK) in the glycolytic pathway, which is considered to be rate-limiting step. In contrast, fructose is metabolized faster than glucose because it bypasses this step entering the glycolytic pathway at the level of the triose phosphates. This showed that fructose is a much efficient carbon sources compared to glucose or alternative non fermentable carbon sources such as lactate.

The third observation whether *C. albicans* has lost signal that trigger destabilization in response to galactose and retained molecular capability of destabilizing target proteins in response to fructose. *C. albicans ICL1* ORF was expressed in *S. cerevisiae* and vice versa *S. cerevisiae ICL1* ORF was expressed in *C. albicans*. The CaIcl1 in *S. cerevisiae* remained stable even after addition of 2% galactose but was degraded after addition of 2% fructose was tested. The finding supported the previous report. This is interesting because it proved that *C. albicans* Icl1 is sensitive to fructose but not to galactose. In contrast, ScIcl1 in *C. albicans* was degraded in response of 2% galactose or fructose.

The analysis of ubiquitination in *C. albicans* was done by exposing CaIcl1-Ubi-Myc_3 to 2% galactose or fructose. *C. albicans* Icl1 is destabilized by galactose because of the addition of carboxyl-terminal ubiquitination site. In previous finding the fructose destabilize CaIcl1 and addition of ubiquitin site accelerated the destabilization of the protein. Cells can no longer able to metabolize via the glyoxylate cycle following galactose and fructose addition due to the degradation of Icl1 protein. As a result, those cells would presumably be less able to cause infections and would be less able to compete for available nutrients against other microorganisms such as endogenous bacteria in the gastrointestinal (GI) tract compared to their wild type. It is likely to show less successful colonization, virulence and fitness due to this defect in its ability to assimilate both glucose and alternative carbon sources at the same time. Similarly, mutant *C. albicans* cell with deletion of *ICL1* attenuated virulence [27].

It is interesting to further study the mechanism involved in protein degradation in response to fructose. Why is only fructose and not glucose or galactose that triggers protein degradation of CaIcl1? And which pathways could CaIcl1 take for proteolysis? Is it vacuolar proteolysis or ubiquitin mediated proteolysis, such as ubiquitin conjugating enzyme Ubc8p [43] and E3 ubiquitin ligase, Gid complex [49, 60]? It is also interesting to explore other central metabolism enzymes that carry the same characteristic of CaIcl1 and also to determine the effect of exposure to other carbon sources such as malate and succinate.

Acknowledgements

The first author would like to sincerely thank Universiti Sains Malaysia for providing financial support for this project from the Research University Individual (RUI) grant no: 1001/CIPPT/812196.

Author details

Doblin Sandai[1]*, Yasser Tabana[2] and Rosline Sandai[3]

*Address all correspondence to: doblin@usm.my

1 Infectomics Cluster, Advanced Medical and Dental Institute, Universiti Sains Malaysia, Bertam, Penang, Malaysia

2 Faculty of Pharmacy and Pharmaceutical Sciences, University of Alberta, Edmonton, Alberta, Canada

3 Faculty of Languages and Communication, Sultan Idris University of Education, Perak, Malaysia

References

[1] Kurtzman C, Fell JW. The Yeasts—A Taxonomic Study. Elsevier Science; 1998:1076. ISBN 9780080542690

[2] Ene IV, Bennett RJ. The cryptic sexual strategies of human fungal pathogens. Nature Reviews. Microbiology. 2014;**12**(4):239-251. DOI: 10.1038/nrmicro3236

[3] Hull CM, Johnson AD. Identification of a mating type-like locus in the asexual pathogenic yeast *Candida albicans*. Science. 1999;**285**(5431):1271-1275

[4] Calderone RA. Candida and Candidiasis. Washington DC: ASM Press; 2002

[5] Sudbery P, Gow N, Berman J. The distinct morphogenic states of *Candida albicans*. Trends in Microbiology. 2004;**12**(7):317-324. DOI: 10.1016/j.tim.2004.05.008

[6] Slutsky B, Staebell M, Anderson J, Risen L, Pfaller M, Soll DR. "White-opaque transition": A second high-frequency switching system in *Candida albicans*. Journal of Bacteriology. 1987;**169**(1):189-197

[7] Butler G, Rasmussen MD, Lin MF, Santos MA, Sakthikumar S, Munro CA, et al. Evolution of pathogenicity and sexual reproduction in eight Candida genomes. Nature. 2009; **459**(7247):657-662. DOI: 10.1038/nature08064

[8] Bennett RJ, Johnson AD. Completion of a parasexual cycle in *Candida albicans* by induced chromosome loss in tetraploid strains. The EMBO Journal. 2003;**22**(10):2505-2515. DOI: 10.1093/emboj/cdg235

[9] Forche A, Alby K, Schaefer D, Johnson AD, Berman J, Bennett RJ. The parasexual cycle in *Candida albicans* provides an alternative pathway to meiosis for the formation of recombinant strains. PLoS Biology. 2008;**6**(5):e110. DOI: 10.1371/journal.pbio.0060110

[10] Barelle CJ, Priest CL, Maccallum DM, Gow NA, Odds FC, Brown AJ. Niche-specific regulation of central metabolic pathways in a fungal pathogen. Cellular Microbiology. 2006; **8**(6):961-971. DOI: 10.1111/j.1462-5822.2005.00676.x

[11] Costa M, Borges CL, Bailao AM, Meirelles GV, Mendonca YA, Dantas SF, et al. Transcriptome profiling of *Paracoccidioides brasiliensis* yeast-phase cells recovered from infected mice brings new insights into fungal response upon host interaction. Microbiology. 2007;**153**(Pt 12):4194-4207. DOI: 10.1099/mic.0.2007/009332-0

[12] Rodaki A, Young T, Brown AJ. Effects of depleting the essential central metabolic enzyme fructose-1,6-bisphosphate aldolase on the growth and viability of *Candida albicans*: Implications for antifungal drug target discovery. Eukaryotic Cell. 2006;**5**(8):1371-1377. DOI: 10.1128/EC.00115-06

[13] Chambers A, Packham EA, Graham IR. Control of glycolytic gene expression in the budding yeast (*Saccharomyces cerevisiae*). Current Genetics. 1995;**29**(1):1-9

[14] Clifton D, Weinstock SB, Fraenkel DG. Glycolysis mutants in *Saccharomyces cerevisiae*. Genetics. 1978;**88**(1):1-11

[15] Uemura H, Fraenkel DG. gcr2, a new mutation affecting glycolytic gene expression in *Saccharomyces cerevisiae*. Molecular and Cellular Biology. 1990;**10**(12):6389-6396

[16] Baker HV. GCR1 of *Saccharomyces cerevisiae* encodes a DNA binding protein whose binding is abolished by mutations in the CTTCC sequence motif. Proceedings of the National Academy of Sciences of the United States of America. 1991;**88**(21):9443-9447

[17] Clifton D, Fraenkel DG. The gcr (glycolysis regulation) mutation of *Saccharomyces cerevisiae*. The Journal of Biological Chemistry. 1981;**256**(24):13074-13078

[18] Nishi K, Park CS, Pepper AE, Eichinger G, Innis MA, Holland MJ. The GCR1 requirement for yeast glycolytic gene expression is suppressed by dominant mutations in the SGC1 gene, which encodes a novel basic-helix-loop-helix protein. Molecular and Cellular Biology. 1995;**15**(5):2646-2653

[19] Wisplinghoff H, Ebbers J, Geurtz L, Stefanik D, Major Y, Edmond MB, et al. Nosocomial bloodstream infections due to *Candida* spp. in the USA: Species distribution, clinical features and antifungal susceptibilities. International Journal of Antimicrobial Agents. 2014;**43**(1):78-81. DOI: 10.1016/j.ijantimicag.2013.09.005

[20] Askew C, Sellam A, Epp E, Hogues H, Mullick A, Nantel A, et al. Transcriptional regulation of carbohydrate metabolism in the human pathogen *Candida albicans*. PLoS Pathogens. 2009;**5**(10):e1000612. DOI: 10.1371/journal.ppat.1000612

[21] Bojunga N, Kotter P, Entian KD. The succinate/fumarate transporter Acr1p of *Saccharomyces cerevisiae* is part of the gluconeogenic pathway and its expression is regulated by Cat8p. Molecular & General Genetics. 1998;**260**(5):453-461

[22] Eschrich D, Kotter P, Entian KD. Gluconeogenesis in *Candida albicans*. FEMS Yeast Research. 2002;**2**(3):315-325

[23] Kornberg HL, Madsen NB. The metabolism of C2 compounds in microorganisms. 3. Synthesis of malate from acetate via the glyoxylate cycle. The Biochemical Journal. 1958; **68**(3):549-557

[24] Piekarska K, Mol E, van den Berg M, Hardy G, van den Burg J, van Roermund C, et al. Peroxisomal fatty acid beta-oxidation is not essential for virulence of *Candida albicans*. Eukaryotic Cell. 2006;**5**(11):1847-1856. DOI: 10.1128/EC.00093-06

[25] Ramirez MA, Lorenz MC. Mutations in alternative carbon utilization pathways in *Candida albicans* attenuate virulence and confer pleiotropic phenotypes. Eukaryotic Cell. 2007;**6**(2):280-290. DOI: 10.1128/EC.00372-06

[26] Ramirez MA, Lorenz MC. The transcription factor homolog CTF1 regulates {beta}-oxidation in *Candida albicans*. Eukaryotic Cell. 2009;**8**(10):1604-1614. DOI: 10.1128/EC.00206-09

[27] Lorenz MC, Fink GR. The glyoxylate cycle is required for fungal virulence. Nature. 2001;**412**(6842):83-86. DOI: 10.1038/35083594

[28] Hudson DA, Sciascia QL, Sanders RJ, Norris GE, Edwards PJ, Sullivan PA, et al. Identification of the dialysable serum inducer of germ-tube formation in *Candida albicans*. Microbiology. 2004;**150**(Pt 9):3041-3049. DOI: 10.1099/mic.0.27121-0

[29] Bendel CM, Hess DJ, Garni RM, Henry-Stanley M, Wells CL. Comparative virulence of *Candida albicans* yeast and filamentous forms in orally and intravenously inoculated mice. Critical Care Medicine. 2003;**31**(2):501-507. DOI: 10.1097/01.CCM.0000049954.48239.A1

[30] Fan J, Chaturvedi V, Shen SH. Identification and phylogenetic analysis of a glucose transporter gene family from the human pathogenic yeast *Candida albicans*. Journal of Molecular Evolution. 2002;**55**(3):336-346. DOI: 10.1007/s00239-002-2330-4

[31] Dumitru R, Hornby JM, Nickerson KW. Defined anaerobic growth medium for studying *Candida albicans* basic biology and resistance to eight antifungal drugs. Antimicrobial Agents and Chemotherapy. 2004;**48**(7):2350-2354. DOI: 10.1128/AAC.48.7.2350-2354.2004

[32] Johnston M, Kim JH. Glucose as a hormone: Receptor-mediated glucose sensing in the yeast *Saccharomyces cerevisiae*. Biochemical Society Transactions. 2005;**33**(Pt 1):247-252. DOI: 10.1042/BST0330247

[33] Ozcan S, Johnston M. Two different repressors collaborate to restrict expression of the yeast glucose transporter genes HXT2 and HXT4 to low levels of glucose. Molecular and Cellular Biology. 1996;**16**(10):5536-5545

[34] Brown V, Sexton JA, Johnston M. A glucose sensor in *Candida albicans*. Eukaryotic Cell. 2006;**5**(10):1726-1737. DOI: 10.1128/EC.00186-06

[35] Sandai D, Yin Z, Selway L, Stead D, Walker J, Leach MD, et al. The evolutionary rewiring of ubiquitination targets has reprogrammed the regulation of carbon assimilation in the pathogenic yeast *Candida albicans*. MBio Journal. 2012;**3**(6). DOI: 10.1128/mBio.00495-12

[36] Leach MD, Stead DA, Argo E, MacCallum DM, Brown AJ. Molecular and proteomic analyses highlight the importance of ubiquitination for the stress resistance, metabolic adaptation, morphogenetic regulation and virulence of *Candida albicans*. Molecular Microbiology. 2011;**79**(6):1574-1593. DOI: 10.1111/j.1365-2958.2011.07542.x

[37] Sepulveda P, Lopez-Ribot JL, Gozalbo D, Cervera A, Martinez JP, Chaffin WL. Ubiquitin-like epitopes associated with *Candida albicans* cell surface receptors. Infection and Immunity. 1996;**64**(10):4406-4408

[38] Roig P, Gozalbo D. Depletion of polyubiquitin encoded by the UBI4 gene confers pleiotropic phenotype to *Candida albicans* cells. Fungal Genetics and Biology. 2003;**39**(1):70-81

[39] Leng P, Sudbery PE, Brown AJ. Rad6p represses yeast-hypha morphogenesis in the human fungal pathogen *Candida albicans*. Molecular Microbiology. 2000;**35**(5):1264-1275

[40] Cheah HL, Lim V, Sandai D. Inhibitors of the glyoxylate cycle enzyme ICL1 in *Candida albicans* for potential use as antifungal agents. PLoS One. 2014;**9**(4):e95951. DOI: 10.1371/journal.pone.0095951

[41] Gancedo JM. Yeast carbon catabolite repression. Microbiology and Molecular Biology Reviews. 1998;**62**(2):334-361

[42] Yin Z, Wilson S, Hauser NC, Tournu H, Hoheisel JD, Brown AJ. Glucose triggers different global responses in yeast, depending on the strength of the signal, and transiently stabilizes ribosomal protein mRNAs. Molecular Microbiology. 2003;**48**(3):713-724

[43] Ting SY, Ishola OA, Ahmed MA, Tabana YM, Dahham MT, Agha MT, et al. Metabolic adaptation via regulation enzyme degradation in the pathogenic yeast *Candida albicans*. Journal de Mycologie Medicale. 2017;**27**:98-108. DOI: 10.1061/j.mycmed.2016.12.002

[44] Schule T, Rose M, Entian KD, Thumm M, Wolf DH. Ubc8p functions in catabolite degradation of fructose-1, 6-bisphosphatase in yeast. The EMBO Journal. 2000;**19**(10):2161-2167. Doi: 10.1093/emboj/19.10.2161

[45] Johnson ES, Bartel B, Seufert W, Varshavsky A. Ubiquitin as a degradation signal. The EMBO Journal. 1992;**11**(2):497-505

[46] Hitchcock AL, Auld K, Gygi SP, Silver PA. A subset of membrane-associated proteins is ubiquitinated in response to mutations in the endoplasmic reticulum degradation machinery. Proceedings of the National Academy of Sciences of the United States of America. 2003;**100**(22):12735-12740. DOI: 10.1073/pnas.2135500100

[47] Rost B. Review: Protein secondary structure prediction continues to rise. Journal of Structural Biology. 2001;**134**(2-3):204-218. DOI: 10.1006/jsbi.2001.4336

[48] Hammerle M, Bauer J, Rose M, Szallies A, Thumm M, Dusterhus S, et al. Proteins of newly isolated mutants and the amino-terminal proline are essential for ubiquitin-proteasome-catalyzed catabolite degradation of fructose-1,6-bisphosphatase of *Saccharomyces cerevisiae*. The Journal of Biological Chemistry. 1998;**273**(39):25000-25005

[49] Regelmann J, Schule T, Josupeit FS, Horak J, Rose M, Entian KD, et al. Catabolite degradation of fructose-1,6-bisphosphatase in the yeast *Saccharomyces cerevisiae*: A genome-wide screen identifies eight novel GID genes and indicates the existence of two degradation pathways. Molecular Biology of the Cell. 2003;**14**(4):1652-1663. DOI: 10.1091/mbc.E02-08-0456

[50] Piskur J, Rozpedowska E, Polakova S, Merico A, Compagno C. How did *Saccharomyces* evolve to become a good brewer? Trends in Genetics. 2006;**22**(4):183-186. DOI: 10.1016/j.tig.2006.02.002

[51] Niimi M, Kamiyama A, Tokunaga M. Respiration of medically important *Candida* species and *Saccharomyces cerevisiae* in relation to glucose effect. Journal of Medical and Veterinary Mycology. 1988;**26**(3):195-198

[52] Bertram G, Swoboda RK, Gooday GW, Gow NA, Brown AJ. Structure and regulation of the *Candida albicans* ADH1 gene encoding an immunogenic alcohol dehydrogenase. Yeast. 1996;**12**(2):115-127. DOI: 10.1002/(SICI)1097-0061(199602)12:2<115::AID-YEA889>3.0.CO;2-E

[53] Odds FC. Candida and Candidosis. Bailliere T, 2nd Edition. London. 1988

[54] Vargas SL, Patrick CC, Ayers GD, Hughes WT. Modulating effect of dietary carbohydrate supplementation on *Candida albicans* colonization and invasion in a neutropenic mouse model. Infection and Immunity. 1993;**61**(2):619-626

[55] Holden HM, Rayment I, Thoden JB. Structure and function of enzymes of the leloir pathway for galactose metabolism. The Journal of Biological Chemistry. 2003;**278**(45):43885-43888. DOI: 10.1074/jbc.R300025200

[56] Sellick CA, Campbell RN, Reece RJ. Galactose metabolism in yeast-structure and regulation of the leloir pathway enzymes and the genes encoding them. International Review of Cell and Molecular Biology. 2008;**269**:111-150. DOI: 10.1016/S1937-6448(08)01003-4

[57] Shatton JB, Gruenstein M, Shay H, Weinhouse S. Enzymes of galactose synthesis in mammary gland and mammary tumors of the rat. The Journal of Biological Chemistry. 1965;**240**:22-28

[58] Verma S, Bharti B, Inusha P. Association of fungal sepsis and galactosemia. Indian Journal of Pediatrics. 2010;**77**(6):695-696. DOI: 10.1007/s12098-010-0082-5

[59] Cha SH, Wolfgang M, Tokutake Y, Chohnan S, Lane MD. Differential effects of central fructose and glucose on hypothalamic malonyl-CoA and food intake. Proceedings of the National Academy of Sciences of the United States of America. 2008;**105**(44): 16871-16875. DOI: 10.1073/pnas.0809255105

[60] Menssen R, Schweiggert J, Schreiner J, Kusevic D, Reuther J, Braun B, et al. Exploring the topology of the Gid complex, the E3 ubiquitin ligase involved in catabolite-induced degradation of gluconeogenic enzymes. The Journal of Biological Chemistry. 2012;**287**(30):25602-25614. DOI: 10.1074/jbc.M112.363762

Chapter 4

The Cell Wall of *Candida albicans*: A Proteomics View

Elizabeth Reyna-Beltrán, César Isaac Bazán Méndez,
María Iranzo, Salvador Mormeneo and
Juan Pedro Luna-Arias

Additional information is available at the end of the chapter

http://dx.doi.org/10.5772/intechopen.82348

Abstract

Candida species are a natural commensal of humans and when the immune system is compromised it can cause candidiasis. *C. albicans* is the main etiological agent of candidiasis, representing nearly 60% of the total cases worldwide. The cell wall provides protection against several physical and chemical aggressions and is responsible for the different shapes displayed by *C. albicans*. The cell wall is not a static structure, but a dynamic one, having great plasticity to allow different cell morphologies, molecular remodeling and changes in the cell wall composition as a result from adaptation to the surrounding environment. It is mainly composed of chitin, β-glucan and mannoproteins. Therefore, the cell wall components are putative targets for the discovery and development of new drugs. The cell wall reprogramming in response to several conditions, including a host carbon source, blood, serum, high temperature, acidic environment, and morphogenesis, have a direct impact on the mannoprotein content and might be involved in adherence, drug resistance and virulence of *C. albicans*. In this chapter, we performed an analysis of the proteins that have been identified in the *C. albicans* cell wall by our group and others, which allowed the identification of proteins from different intracellular compartments.

Keywords: *Candida albicans*, cell wall, proteomics, review

1. Introduction

Fungal infections are a serious health concern worldwide. There has been estimated that more than 13% of the total human population has a fungal infection on skin, 300 million people are suffering a severe fungal infection, and 25 million are in great danger of dying or blindness as a result of an infection by fungi [1]. Infections caused by fungi provoke over 1.5 million deceases worldwide, mainly in people carrying HIV infections or suffering two or more pathological conditions [2]. Even though there are more than 150 species of *Candida*, only nearly 20 species are known to cause human infections, being *C. albicans* the most frequent causative agent of candidiasis and the leading fungal infection [3]. *C. albicans* is part of the normal microflora in humans and can be found in normal mucosae or cutaneous microflora of healthy individuals [2]. Under certain conditions, as when patients are being treated with antibiotics, immunosuppressive drugs, or chemotherapeutic agents, or are in surgical intensive care units for prolonged time, the commensal *C. albicans* yeast cells turn into pathogenic cells implicated in life-threatening invasive candidiasis [3]. Moreover, there have been increasing reports of *C. albicans* strains exhibiting a multidrug resistance phenotype to antifungal drugs. Therefore, *Candida* infection is a clinical problem worldwide due to the difficulty of treating systemic candidiasis.

2. The cell wall of *Candida albicans*

C. albicans has developed several molecular tools for sensing and evading the host immune system, which include fungal virulence factors involved in adherence, cell damage, and invasion [4]. Many of these factors are localized in the fungal cell wall, which constitute the main structure to be in contact with host cells and is essential for the fungal cell integrity. The cell wall is a remarkable structure because of its key role in protection against environmental stress conditions, including osmotic changes, dehydration, heat, cold, immune system response, or attack by other microorganisms [5–7], and it is also responsible for the cell shape and adhesion to host cells through adhesins.

The cell wall is mainly composed of proteins, glycans, and chitin [5–7]. Cell wall proteins are heavily mannosylated through *O*- and *N*-chemical bonds and function as structural elements, adhesins, cross-linking enzymes, molecular sensors of environmental changes, as well as in protection from environmental changes [5–8]. There are two classes of covalently bound fungal cell wall proteins (CWPs), the glycosylphosphatidylinositol (GPI)-dependent CWPs and Pir-CWPs [5–7], which can be isolated from cell walls by treatment with mild alkali (alkali-sensitive linkage, ASL). In the case of *Saccharomyces cerevisiae*, Cwp2 is a GPI cell wall protein harboring a Pir repeat involved in linking Cwp2 to β-1,3-glycan for increasing cell wall integrity [6, 8]. Proteins lacking homology to Pir proteins have been designated alkali-sensitive linkage cell wall proteins (ASL-CWPs), and are covalently linked to the cell walls of *C. albicans* and *S. cerevisiae* by mild-alkali sensitive chemical bonds [5, 6, 8]. Other proteins are linked to CWPs through disulfide bonds [9].

Most of the cell wall components have covalent linkages established to enhance the cell wall stability and, consequently, protection against environmental changes. One type of proteins described as main cross linkers of cell wall components are the members of the gas family that only contains members of the fungal kingdom. *C. albicans* possesses five genes encoding members of this family, the *PHR1*, *PHR2*, *PHR3*, *PGA4*, and *PGA5* genes [10]. These proteins are β-1,3-glucanases that also exhibit transglucosidase activity and are needed for cross linking β-1,3-glucan and β-1,6-glucan. *PHR1* and *PHR2* genes are the only members of the family that are responsive to changes in pH [10]. *PHR1* gene is able to complement the *S. cerevisiae GAS1/GGP1* null mutant, which has as main phenotypic characteristics, spherical shape, and affectation of normal budding pattern in the stationary phase of growth curve, indicating that the *PHR1* gene is the ortholog of *GAS1/GGP1* in *C. albicans* [11]. *PHR2* gene encodes a protein that shows 55.15 and 34.48% identity to Phr1 and Phr3, respectively, while Phr1 has 32.03% identity to Phr3, and interestingly, Phr3 does not have a GPI-anchoring site. *PGA4* and *PGA5* genes, the other members of the Gas family, encode proteins that share 25–35% identity to Phr1 and Phr2, and are not pH responsive.

Other proteins involved in the formation of covalent linkages are transglutaminases (TGases), a group of enzymes that are widely distributed in microorganisms, plants, and animals. They have been involved in a number of biological processes, such as differentiation, growth regulation, cellular adhesion, and tissue integrity. These proteins are enzymes with several functions, including posttranslational modifications, protein cross linking, incorporation of amine groups, and deamination. The best studied TGase function is cross linking, which involves a transamination reaction between the side chains of glutamine and lysine residues in target proteins to form N-ε-(γ-glutamyl) lysine amide bonds, rendering an increase in mechanical strength and resistance to proteolytic degradation [12]. Most of the transglutaminases are fully dependent of Ca^{2+} ions for their activity. However, there are transglutaminases from microorganisms that do not require Ca^{2+} for their function. Similarly, TGases from rodent intestinal mucosa are also independent of Ca^{2+} for their enzymatic activity [12, 13]. One of the best studied transglutaminases is the human TGase 2 that has a key role in the stabilization of the extracellular matrix and modulation of the interaction between fibronectin and integrin [14]. In addition, it has been involved in several human diseases of great impact in health worldwide, such as Alzheimer's, Huntington, and Crohn's diseases, as well as in fibrosis, cancer and other diseases [15]. In the case of *Chlamydomonas reinhardtii*, TGase has a key role in the assembly of cell wall through the cross linking of the cell wall components [16]. TGase activity has also been reported in *C. albicans* and *S. cerevisiae*, where it was found localized in the cell walls fraction of yeast cells and showed the capacity to establish covalent cross links between proteins. TGase activity from these microorganisms was inhibited by the specific inhibitor cystamine, which diminished the incorporation of several polypeptides into the cell wall, affected the regeneration of protoplasts, and inhibited the yeast-to-mycelium transition, revealing a role in the establishment of covalent cross links between cell wall proteins and the structural polymers chitin and/or glucan [17, 18]. In a recent work, the inhibition of TGase activity with cystamine inhibited the growth of yeast cells and induced autophagy. It also affected the yeast division pattern and the cell wall structure. Interestingly, the protein responsible for TGase activity in cell wall was identified as enolase by mass spectrometry [19].

3. Glucan

Glucan is the main component of the yeast cell wall, and along with chitin, these are assembled into a three-dimensional and dynamic network responsible for cell shape that renders protection against environmental insults. In the dimorphic yeast *C. albicans*, chitin abundance is approximately 1–2% of dry weight of the cell wall and is located adjacent to the cell membrane. β-1,3 and β-1,6-glucan represent 40% and 20% of dry weight of the cell wall, respectively, and are situated immediately after chitin. The outer layer of the cell wall is mainly composed by mannoproteins that define the cell surface properties [5, 20].

C. albicans glucan is a highly branched polymer. It is composed of linear D-glucose molecules bound through β-1,3- and β-1,6-glycosidic linkages. β-1,3-glucan forms a helical backbone as a single residue or a chain made of three strands attached by hydrogen bonds. β-1,6-glucan is responsible for the ramification due to cross linking with chitin and mannoproteins through a GPI-anchor bonded to the nonreducing end of β-1,3-glucan. Likewise, chitin and Pir proteins are attached to the nonreducing end of the β-1,3-glucan molecule [21]. It has been shown that during biofilm formation, the *C. albicans* glucan increases the resistance to several antifungal drugs, such as amphotericin B, fluconazol [22], and anidulafungin (an echinocandin) because of a mechanism of drug sequestration [23]. Biofilm lifestyle on medical devices is a high challenge on medicine because of the difficulty to treat patients with contaminated prosthetic materials.

The synthesis of β-1,3-glucan involves several steps and enzymes, and the major and best characterized is a plasma membrane complex named glucan synthase that uses UDP-glucose as universal donor and is stimulated by GTP; due to a transglycosylation reaction, the nascent linear chain is extruded into the cell wall where it can be further processed by several enzymes. The complex contains the catalytic subunit Fks/Gsc and the regulatory subunit Rho1 [24]. There has been identified three genes encoding the glucan synthase, *FKS/GSC, GSC1, GSL1* and *GSL2*. The main activity comes from *GSC1* gene and the incapability to obtain the null mutant highlight the importance of this gene [25]. Inhibition of glucan synthesis by blocking Fks1/Gsc1 activity to damage the cell wall is the main target of antifungal drugs. Echinocandins are the most important of these type of drugs. Most of the *C. albicans* clinical isolates are susceptible to these drugs; however, mutations in *FKS* genes result in echinocandin resistance [26], augmented chitin amount [27], and reduced expression of *FKS2* and *FKS3* genes [28]. *FKS2* and *FKS3* null mutants contain an increased amount of wall glucan and are more tolerant to echinocandin drugs indicating that both genes negatively regulate *FKS1* [29].

β-1,3-glucan undergoes some modifications after its synthesis, such as hydrolysis at the nonreducing end or within the glucan molecule, transference and bound to another chain or cross linking to β-1,6-glucan. Xog1/Exg1 is the major β-exoglucanase activity that is secreted to the cell wall and then to the milieu, and has a dual activity as it can also act as a glycosyltransferase [30, 31]. Xog1/Exg1 null mutant has no significant changes in viability, pathogenesis, or morphogenesis compared with the parental strain [32]. Bgl2 (GH17) is an endo-β-glucanase and 1,3-β-glucanosyltransferase in *C. albicans*. The null mutant of this enzyme has more sensitivity to chitin inhibitors and reduced virulence, but glucanosyltransferase activity remains in these null mutants suggesting the presence of other transferases [33]. *BGL2* is up-regulated during cell wall regeneration and when cells are challenged with fluconazole [34]. Interestingly, the absence of exoglucanases and glycosyltransferases augment the sensitivity to antifungal drugs during biofilm formation [22]. Another well-characterized glucanase is Eng1, an endoglucanase member of the family GH81; null mutants form cell chains because mother and daughter cells do not accomplish cytokinesis, indicating Eng1 has a key role in cell division [34].

The synthesis of β-1,6-glucan remains poorly understood. It requires Kre9, an *O*-glycosylated cell surface protein that belongs to the GH16 family. In the *KRE9* null mutant, the amount of β-1,6-glucan is undetectable and GPI-wall proteins are released to the extracellular medium and its virulence is attenuated [35]. Nevertheless, there have identified more genes involved in the synthesis of β-1,6-glucan [5].

4. Chitin

Chitin is a linear polymer of N-acetylglucosamine that corresponds to 1–2% of the dry weight of the yeast cell wall, and reaches up to 10% in hypha [36]. Chitin can also be deacetylated to produce chitosan by the enzymatic action of chitin deacetylases Cda1 and Cda2 [37], which has been reported in the formation of spores in *S. cerevisiae* under conditions of nitrogen starvation and absence of fermentable carbon sources [38]. The synthesis of chitin occurs as a transglycosylation reaction, where N-acetylglucosaminyl residues are transferred from UDP-N-acetylglucosamine (UDP-GlcNAc) to growing polysaccharide chains through β-(1–4) chemical bonds [39]. The *C. albicans* chitin is synthesized by chitin synthases organized into a family of four isoenzymes that are classified into three classes of chitin synthases, class I (Chs2 and Chs8), class II (Chs1), and class IV (Chs3) [39]. These enzymes synthesize chitin, which deposits at the tips during polarized growth of buds and hypha, and in septum. Chs1 protein is required for viability, cell shape, and septum formation. Under conditions of repression, the conditional mutants *Δchs1* can grow, but daughter buds are not separated from mother cells, forming yeast chains [40]. Class I chitin synthases participate in cell integrity during polarized growth in yeasts and hyphae, and contribute to the protection of nascent cell wall during polarized growth, as well as to the integrity of cells under stress affecting the cell wall [41]. Mutant *chs8* yeast cells showed the lack of long chitin microfibrils in septa of yeast and hyphae, while *chs3* mutants showed absence of short microfibrils [42].

The treatment of *C. albicans* with sub-minimum inhibitory concentration (MIC) levels of caspofungins has been reported to cause a compensatory increase in chitin content [43]. Walker et al. [44] showed an increase of chitin content in response to the use of echinocandins in several *Candida* species. The protein kinase C (PKC) and calcineurin signaling pathways were shown to be activated in isolates of *C. albicans, C. krusei, C. parapsilosis,* and *C. guilliermondii,* generating an increase in chitin and a reduced susceptibility to caspofungin. Thus, this is a mechanism of tolerance to caspofungin in *Candida* species [44].

5. Mannoproteins

The *C. albicans* cell wall mannoproteins correspond to approximately 40% of the cell wall content and are localized in the outermost layer of the cell wall, and consequently, they interact with host proteins and are the first line of defense against host response [45]. One of the main sugar residues in mannoproteins is mannose, which is linked to proteins during the *N*-, *O*-, and *C*-glycosylation processes in *C. albicans*. The mechanisms leading to these modifications of proteins in *C. albicans* are quite similar to those present in *S. cerevisiae*. There are excellent reviews on this issue in both microorganisms. In this section, we will briefly describe the two main glycosylation processes.

Most of the cell surface glycoproteins are extensively glycosylated and harbor both *N*-linked and *O*-linked oligosaccharides. The *N*-glycosylation of proteins begins with the attachment of the Glc_3Man_9-$GlcNAc_2$ core oligosaccharide by the dolichyl-diphosphooligosaccharide-protein glycosyltransferase complex (OST), where Ost1 is essential for the catalytic activity [46]. This core is attached to the asparagine residue in Asn-X-Ser/Thr sequons of proteins in the lumen of the endoplasmic reticulum, where X is any amino acid but proline. The nature of the X amino acid has a key role in the extent of mannosylation of the outer chain. Hyperglycosylation is inhibited by negatively charged amino acids, while it is enhanced by positively charged residues [47].

The synthesis of the Glc_3Man_9-$GlcNAc_2$ core begins with the synthesis of dolichol-phosphate-mannose (Dol-P-Man) by the dolichyl-phosphate-β-mannosyltransferase or dolichyl-phosphate-mannose synthase (DPMS) using the isoprenoid lipid dolichol-phosphate (Dol-P) and GDP-mannose as substrates, while the dolichol-phosphate-glucose (Dol-P-Glc) is synthesized by the dolichyl-phosphate-β-glucosyltransferase or dolichyl-phosphate-glucose synthase (DPGS or Alg5p) using Dol-P and UDP-glucose as substrates [46]. Both Dol-P-Man and Dol-P-Glc are translocated from the cytosolic face to the lumen of the endoplasmic reticulum by not yet described flippases, where they will be used during the synthesis of the Glc_3Man_9-$GlcNAc_2$ core oligosaccharide. The synthesis of the lipid-linked oligosaccharide Glc_3Man_9-$GlcNAc_2$-P-P-dolichol begins with the transfer of N-acetylglucosamine-phosphate (GlcNAc-P) group to dolichol phosphate (Dol-P) from UDP-GlcNAc by the UDP-N-acetylglucosamine dolichyl-phosphate N-acetylglucosamine phosphotransferase (Alg7) to produce the dolichyl-pyrophosphate-GlcNAc (Dol-PP-GlcNAc), a chemical reaction inhibited by tunicamycin. The second N-acetylglucosamine residue is added by the Alg7, Alg13, and Alg14 protein complex from UDP-GlcNAc to generate Dol-PP-$GlcNAc_2$. The first mannose residue is then added to Dol-PP-$GlcNAc_2$ by the chitobiosyldiphosphodolichol β-(1,4)-mannosyltransferase (Alg1) using GDP-Man, followed by the sequential addition of two mannose residues by the α-1,3/1,6-mannosyltransferase (Alg2p) to generate Dol-PP-$GlcNAc_2Man_3$, being the first mannose linked through an α-1,3 chemical bond and the second through an α-1,6 linkage. These two enzymes are assembled into a protein complex as well. Two mannose residues are then further linked to the α-1,3-mannose of Dol-PP-$GlcNAc_2Man_3$ by the GDP-Man:$Man_3GlcNAc_2$-PP-Dol α-1,2-mannosyltransferase (Alg11) to assemble into Dol-PP-$GlcNAc_2Man_5$. Subsequent reactions occur in the lumen of the endoplasmic reticulum after the Dol-PP-$GlcNAc_2Man_5$ translocation by the oligosaccharide translocation protein (Rft1). The first mannose residue added in the lumen is linked through an α-1,3-chemical bond to the α-1,6-mannose of Dol-PP-$GlcNAc_2Man_5$ by the Dol-P-Man:$Man_5GlcNAc_2$-PP-Dol α-1,3-mannosyltransferase (Alg3). Then, a mannose residue is α-1,2 linked to the α-1,3-mannose of Dol-PP-$GlcNAc_2Man_6$ to generate Dol-PP-$GlcNAc_2Man_7$ by the Dol-P-Man:$Man_5GlcNAc_2$-PP-Dol α-1,3-mannosyltransferase (Alg9). The next mannose residue is linked to the α-1,6-mannose of Dol-PP-$GlcNAc_2Man_7$ by a Dol-P-Man:$Man_7GlcNAc_2$-PP-Dol α-1,6-mannosyltransferase (Alg12), which shows 37% identity and 56% similarity to *S. cerevisiae* Alg12. The last mannose is linked to the α-1,6-mannose of Dol-PP-$GlcNAc_2Man_8$ by the Dol-P-Man:$Man_8GlcNAc_2$-PP-Dol α-1,2-mannosyltransferase (Alg9p) to get Dol-PP-GlcNAc$_2$Man$_9$. The final steps in the synthesis of the core oligosaccharide are the sequential additions of three glucose residues by the Dol-P-Glc:Man$_9$GlcNAc$_2$-PP-Dol α-1,3-glucosyltransferase (Alg6), Dol-P-Glc:Glc$_1$Man$_9$GlcNAc$_2$-PP-Dol α-1,3-glucosyltransferase (Alg8p) and Dol-P-Glc:Glc$_2$Man$_9$GlcNAc$_2$-PP-Dol α-1,2-glucosyltransferase (Alg10) [46].

The core oligosaccharide attached to proteins is further modified by the three α-glycosidases Cwh41, Rot2, and Mns1 in the endoplasmic reticulum before proteins are transported to Golgi apparatus. The mannosyl-oligosaccharide glucosidase Cwh41 cleaves the outer glucose residue bound through an α-1,2 chemical bond. Rot2 (glucan α-1,3-glucosidase subunit 2) removes the two glucose residues bound through α-1,3 linkages. Finally, Mns1 or mannosyl-oligosaccharide α-1,2-mannosidase removes the mannose residue bound through an α-1,2 linkage rendering the final mature Dol-PP-GlcNAc$_2$Man$_8$ oligosaccharide attached to proteins ready for further modifications in the Golgi apparatus [46].

The synthesis of the mannan outer chain in the Golgi apparatus is initiated by the initiation-specific α-1,6-mannosyltransferase or Och1, which adds a mannose residue through an α-1,6 chemical bond to the inner α-1,3-mannose of the GlcNAc$_2$Man$_8$ core bound to the asparagine residue in proteins. The elongation of the outer chain is first carried out by the mannan polymerase I complex (M Pol I), which adds up to 10 mannose residues trough α-1,6 linkages. M Pol I has several subunits including the mannan polymerase I complex Van1 subunit and the mannan polymerase complex subunit Mnn9. The addition of further mannose residues (up to 50 in yeast) is performed by the α-1,6-mannosyltransferase activity of the mannan polymerase II complex (M Pol II), which contains Mnn9, Anp1, Mnn10, Mnn11, and Hoc1 [48, 49].

After the action of these enzymes, the α-1,6 mannose polymer that constitutes the backbone of the outer mannan chain, is ready for the formation of branches, mainly made of mannose residues linked through α-1,2 chemical bonds, both α-1,2/α-1,3 mannose units, or α-1,2/β-1,2 mannose residues. The addition of the first α-1,2-mannose is performed by the α-1,2-mannosyltransferase Mnn2, and the subsequent α-1,2-mannose residues is carried out by Mnn5 in yeast. *C. albicans* has an ortholog of Mnn2 protein, which shows α-1,2-mannosyltransferase activity and adds the fourth and fifth mannose residues during *O*-glycosylation [50]; however, the mannosylation role during *N*-glycosylation has been assigned to Mnt4 and Mnt5/Ktr2 [50, 51]. The addition of the α-1,3 mannose units is performed by Mnn1 in yeast, but the corresponding enzymatic activity has not yet been identified in *C. albicans*. The addition of β-1,2 mannose residues is performed by Bmt1 and Bmt3 in *C. albicans*. Another modification in oligosaccharide branches is the phosphomannosylation, which occurs in the α-1,2-mannose and is catalyzed by Mnn6 in yeast, while in *C. albicans* is done by Mnt3 and Mnt5/Ktr2. This phosphomannan can also be further modified by the addition of β-1,2 mannose units by the catalytic activity of Bmt2, Bmt3, and Btm4 [50, 51].

The initial step in the *O*-glycosylation process, also known as *O*-mannosylation, is the attachment of an α-1,2 mannose unit to serine or threonine residues on proteins in the endoplasmic reticulum by Pmt1, Pmt2, Pmt4, Pmt5, and Pmt6 enzymes using as substrate Dol-P-Man. The elongation of the *O*-oligosaccharide (up to six mannose units in length) is carried out in the Golgi apparatus. The second and third α-1,2 mannose units are added by Mnt1/Kre2 and Mnt2 enzymes. This *O*-Man$_3$ oligosaccharide can also be modified by the addition of β-1,2 mannose residues by Bmt1 and Bmt3 proteins. The addition of more α-1,2 mannose units is performed by Mnt1 [50, 52].

6. Proteomics of the *Candida albicans* cell wall

C. albicans sequencing genome project has revealed that it possess 6218 open reading frames (ORFs). However, only 1686 ORFs (~27%) have been characterized so far, and 4380 putative OFR's are of unknown function [53]. In order to determine which genes are expressed to proteins, it is necessary to determine the proteome through mass spectrometry, which has become the workhorse in mining proteomes. The cell wall is the first structure to make contact with host, and consequently, several studies have been performed to identify the proteins contained within the cell wall and those involved in the interaction with host human cells. One of the first works to elucidate the high complexity of the *C. albicans* cell wall was carried out by Pitarch et al. [54], who analyzed a series of fractionations from isolated cell walls of either blastospore or hyphae by two-dimensional polyacrylamide gel electrophoresis (2D-PAGE) and matrix assisted laser desorption/ionization time-of-flight mass spectrometry (MALDI-TOF-MS). Unusual and conventional proteins were identified and grouped on classical CWPs, heat shock and chaperone, folding, elongation factors, glycosylation, fermentation enzymes, and miscellaneous with unknown functions. Proteins highly glycosylated were not successfully identified; nevertheless, identification was possible when PNGase F was used. Likewise, to minimize possible contamination with proteins from different cell compartments, biotinylation has been used to purify CWPs from yeasts and hyphae, prior to cell fractionation [55]. Proteins were purified by affinity chromatography, precipitated, separated by SDS-PAGE, and further identified by (MALDI-TOF-MS). Interestingly, some intracellular proteins were localized in the cell wall during blastospore formation. As an example, the thiol-specific antioxidant-like protein 1 (Tsa1) was found in the nucleus in blastospores, whereas during the hyphae formation, it was localized on the cell surface.

De Groot et al. [56] carried out a series of extractions using isolated cell walls, and the proteins in each fraction were submitted to anion-exchange chromatography. Proteins in fractions were then separated through SDS-PAGE and peptides, after proteinase digestion, were identified with a Quadrupole-TOF (Q-TOF) MS system. Covalently bound CWPs were identified and classified in several protein classes, including carbohydrate-active enzymes, adhesion, host defense, and proteins of unknown function. A total of 12 proteins were identified as GPI-linked proteins and two were classified as sensitive to mild-alkali extraction. In a similar way, peptides from proteins obtained through several chemical extractions of cell walls from blastospores and mycelium, separated through a C18 reverse phase column and identified with an ESI-quadrupole MS system allowed an increased number of identified proteins, both covalently and noncovalently bounded to the cell wall. In the former, most of the proteins were identified as GPI-anchored proteins. In the latter, they were classified in two groups, one named *bona fide* surface proteins, such as agglutinin-like sequence (Als) proteins, glucanases, chitinases, among others, and a group termed atypical cell wall proteins involved in cell cycle, metabolism and energy, protein synthesis, stress response, translation elongation, transport, and others, as well as those with unknown functions that had never been reported as part of the cell wall. In addition, MS analysis of SDS extracts obtained from cell walls allowed the identification of most of these atypical cell wall proteins and some others, suggesting that it is possible to identify proteins directly form cell wall extracts [57]. Cell surface proteins that were nonbound to the glucan network were identified from *C. albicans* biofilms, blastospores, and hyphae using 2D-fluorescence difference in gel electrophoresis (2D-DIGE) and MALDI-TOF/TOF-MS, finding 108 differentially expressed proteins. Totally, 25 proteins were exclusively found in the biofilm and were classified in different classes, including metabolic process, protein fate and synthesis, cell surface rescue defense and virulence, biogenesis of cellular components, protein metabolism, and uncharacterized [58].

Reference cultures that were metabolically labeled with ^{15}N for relative quantification using an ESI-Fourier transform (FT) ion cyclotron resonance mass spectrometer (ESI FT-ICR MS) have been used to determine the difference among the yeast and hyphal cell wall proteomes, and cultures under different stimulus, including fetal calf serum, N-acetylglucosamine (GlcNAc), and Iscove's modified Dulbecco's medium (IMDM). The latter stimulus showed the strongest hyphal induction [59]. Cell wall fractions were obtained as described in [56] and quantified 21 proteins. Proteins involved in cell wall remodeling and maintenance showed no significant variations in all conditions tested. Several proteins that are specifically linked to hyphal growth were also identified, including Hyr1, Sod5, Hwp2, Plb5 and Als3, as well as to yeasts only, such as Rhd3, Ywp1, and Sod4. Similarly, the effects of fluconazole on the secretome and cell wall proteome and integrity were studied [60], revealing that fluconazole not only affects the cellular membranes, but the cell wall proteome as well, as was shown in both planktonic and sessile cultures, and confirming that fluconazole inhibits the morphogenesis to hyphae, lowering the amount of cell wall proteins distinctive from hyphae, and increasing the number of cell wall repair associated proteins.

Sosinska et al. [61] demonstrated that pH induces modifications of the *C. albicans* cell wall proteome using a system that resembles the mucosal surface. When cells were cultured at pH 7.0, they were mainly characterized by hyphal growth, while at pH 4.0, only yeasts and pseudohyphae were obtained. Relative quantitation of cell wall proteins labeled with ^{15}N using ESI-FT-MS allowed the identification of 21 proteins covalently bounded to the cell wall, being most of them GPI-linked proteins with the exception of Mp65, Pir1, and Tos1. Nine of these proteins were overexpressed at pH 7.0, while 12 were overexpressed at pH 4.0. Additionally, the transglucosidase Phr2 was detected at pH 4.0, while Als1, Als3, Hyr1, Phr1, Rbt1, Sod5, and Tos1 were found at pH 7.0. Moreover, it has been observed that the carbon source provokes changes in the cell wall proteome and secretome [62]. Glucose, lactate, and glucose plus lactate were utilized to induce changes in the cell wall architecture. There was found an upregulation of proteins implicated in cell wall remodeling when cells were grown in lactate in comparison with cells cultured in glucose, modifying the elasticity of the cell wall. Thus, the cell wall structure is modified in response to stress, antifungals, and osmotic changes. In the same way, the cell wall from cells grown as biofilms and lactate, incorporated proteins involved in adhesion, consequently, improving biofilm formation, and adhesion on plastic surfaces.

There has also been shown that incubation of *C. albicans* at 42°C triggers the expression of stress-related proteins [63] as previously described [60], increases both the phosphorylation the Mkc1 MAP kinase (involved in the cell wall integrity) and the chitin amount [64]; on the contrary, the amount of Cht3 protein is diminished, impeding cell separation after cytokinesis and showing more sensitivity to Calcofluor white, Congo red, and SDS, suggesting that high temperature induces cell wall stress. Mannoproteins covalently linked to the cell wall either by disulfide or alkali-labile chemical bonds were determined after SDS-treatment, deglycosylation, SDS-PAGE, and reverse phase chromatography coupled to an ESI-LTQ-Orbitrap XL MS system [65]. A total of 10 proteins lacking a GPI-anchor were identified; nevertheless, these proteins possess a signal peptide and were classified as cell wall proteins by *in silico* tools. On the other hand, 16 proteins with a GPI-anchor were identified following extraction with β-mercaptoethanol, which were previously identified after HF-pyridine treatment [56]. Moreover, a group of proteins lacking a signal peptide and characteristic of other cell wall compartments, but not from the cell wall, were identified.

Hernández et al. [66] developed an in vitro model to study the yeast-to-hypha transition of *C. albicans* that allowed the determination of the cytoplasmic proteome of hypha that can be used as a reference. This study described the isolation of the cytoplasmic fraction from yeasts and hyphae for 2D-PAGE analysis and spot identification by MALDI-TOF/TOF. There were identified 66 proteins, but only 43 were found involved in diverse metabolic routes; 20 proteins belong to metabolic pathways of amino acids, carbohydrates, nucleotides, lipids and fatty acids, 23 corresponded to enzymes of glycolysis, gluconeogenesis, pentose and tricarboxylic acid pathways, 13 were involved in transcription and protein synthesis, 8 proteins have functions related to cell rescue, virulence and defense, and only 2 with unknown functions.

The cell wall proteome of *C. albicans* ATCC10231 cells cultured with 10% fetal bovine serum (FBS) has been recently published [67]. The morphogenesis induced with FBS at 30 and 37°C revealed that 285 proteins were differentially expressed from a total of 1177. At 30°C, 152 proteins were up-regulated and clustered in several processes, including signal transduction, cell wall biosynthesis, metabolism, stress response, DNA replication, transcription, RNA processing, among others, while 62 were down-regulated and grouped in cell wall proteins, metabolism, stress response, RNA processing, translation and posttranslational modification and transport. At 37°C, 18 proteins were up-regulated and found associated to cell wall biosynthesis, metabolism, translation and posttranslational modifications, and transport, whereas 53 were down-regulated and found with a role in signal transduction, cell wall, metabolism, stress response, transcription, RNA processing, and others. All these data indicate that morphogenesis induced with FBS at 30°C provokes more changes than those observed at 37°C. Thus, FBS induces morphogenesis through several signaling transduction pathways, the inhibition of glycolysis and enhancement of oxidative phosphorylation.

A gel-free method to identify cell surface proteins from alive *C. albicans* cells incubated for 5 min with trypsin and DTT, a technique named cell shaving or surfomics, has been reported [68]. Peptides were separated by RP-HPLC and analyzed in a MALDI-TOF/TOF system. Identified proteins are implicated in cell wall organization and biogenesis, cell rescue, defense and virulence, cytoskeleton organization, transport, metabolism, protein fate and those with unknown function. Using this approach, surfomics differences among yeasts, hyphae and biofilms were determined [69]. A total of 131 proteins covalently and noncovalently bound to the cell wall were identified and 35 of them had never been reported; 27 of these proteins were present in all tested conditions; however, 22 are of unknown functions. Three were found in the biofilm only, 26 on hyphae and 38 on yeast. Proteins were classified in GPI-anchored, cell wall biogenesis and maintenance, ligand binding, unknown function, and noncell wall function or description. Under this similar method, proteins from planktonic and hyphae cells using an RP-ESI-LTQ-Orbitrap were determined [70]. A total of 943 proteins were found; 423 were shared between yeast and hyphae, 15 were solely detected in yeast and 505 in hyphae. Proteins were classified in cell wall organization and biogenesis, GPI-anchored, cell surface, pathogenesis, stress, plasma membrane, others, and unknown function, being the latter the largest group. About 14 out of 20 of the most abundant proteins were present in both growth styles, 6 were cell surface proteins, 3 involved in stress response, one from pathogenesis, and 4 from other classes. The normalized spectral abundance factor (NSAF) was utilized to calculate the relative protein abundance. Marín et al. [71] challenged *C. albicans* cells with normal human serum (NS) and heat inactivated serum (HIS); they were able to detect human serum proteins covering the cell surface and fungal cell wall proteins. A total of 214 human and 372 *C. albicans* unique proteins were identified, of which 371 proteins were found in NS, 134 in HIS, and 133 proteins were found in both conditions. Hyphae induction with human serum resulted in the identification of 12 proteins that were not found in any other morphogenesis induction. Moreover, 147 out of 372 proteins were classified as cell wall surface proteins, 60 possessed signal peptide, and 23 were GPI-anchored proteins, being glycerol 3-phosphate-dehydrogenase 2 (Gpd2) and pH-regulated antigen 1 (Pra1) the most relevant, both participating in complement evasion, and seven proteins that bind to plasminogen, resulting in immune protection and evasion.

Our group recently performed the ^{14}C-lysine labeling of proteins of the cell wall using them as endogenous substrates by the TGase present in this cellular fraction. Labeled proteins were then sequentially extracted with 2% SDS, and chitinase, separated by 10% SDS-PAGE and the proteins in radioactive areas were identified by tandem mass spectrometry (MS/MS) [19]. Most of the radioactive labeled proteins were identified in the SDS-solubilized fraction (1048 proteins), 37 proteins in the fraction solubilized with zymolyase, and 41 proteins were released with chitinase. Only 24 proteins were extracted in common with the three treatments used. We found the S7A, S13, and S16A 40S ribosomal proteins; 60S ribosomal protein L30; the glycolytic enzymes enolase 1, glyceraldehyde-3-phosphate dehydrogenases 1 and 2, pyruvate kinase, fructose-bisphosphate aldolase, phosphoglycerate kinase 1, phosphoglycerate mutase, and pyruvate decarboxylase; the heat shock proteins of 70 kDa SSA1 and SSB1; the mitochondrial outer membrane protein porin 1 and ADP/ATP carrier protein; the plasma membrane ATPase PMA1, ADP/ATP carrier protein, and galactose transporter-related Hgt7; the elongation factor 1-α 1; the cell wall agglutinin-like protein 1 (Als1) and 2 (Als2), 1,3-β-glucanosyltransferase PGA4, and chitinase 2. About 16 of these proteins have evidence in the cell wall, being Eno1, Pgk1 and Als1 reported as immunodominant proteins [8]. Phosphoglycerate kinase 1 (Pgk1) was shown to confer poor immune protection against *C. albicans* in an infected mouse model [72]. Hsp70 protein is involved in protein folding and translocation, has antigenic properties, and could play a role in the biosynthesis, secretion, and assembly of other components of the cell wall [73]. The full list of cell wall proteins identified from the sequential extractions with 2% SDS, zymolyase, and chitinase has been reported [19].

Asc1, a protein component of the 40S ribosomal subunit and a signal transducer, was found in fractions extracted with both SDS and zymolyase, and has been described as a protein that has a key role in the process of cell-cell adhesion and cell adhesion to substrate. The deletion of *Asc1* showed that this gene could be related to the signaling pathway of the cAMP-dependent protein kinase Tpk, which has a role in the processes of morphogenesis under differential environmental conditions [61]. Other identified proteins in the SDS cell wall extract were SOD1, SOD2, while SOD5 was found in the extract obtained with zymolyase. These proteins belongs to a family of at least six members in *C. albicans*, whose main function is the transformation of superoxide radicals into hydrogen peroxide that can be subsequently converted into water by catalase enzyme. SOD1 is cytosolic, SOD2 is localized in mitochondria, and SOD5 is a GPI-anchored protein localized in the cell wall [74]. SOD5 has been described as essential for the infection of *C. albicans* in a mouse model, although the deletion of *SOD5* gene showed no effect in *C. albicans* killing by macrophages. It has also been implicated in the degradation of reactive oxygen species generated by the host to evade surveillance of the innate immune system, resistance to miconazole through biofilm formation, and defense against host insults. Other protein that was also extracted with SDS and zymolyase was the extracellular glycosidase CRH11, a glycosylphosphatidylinositol-anchored protein involved in cell wall organization [75]. Tsa1-B was exclusively found in the fraction of proteins extracted with chitinase treatment. Tsa1-B is a peroxiredoxin, a thiol-specific peroxidase that plays a protective role against reactive oxygen species, organization of the cell wall [76], and is essential for the yeast-to-mycelium transition [77].

The analysis of all SDS-extracted proteins [19] in the PANTHER classification system according the cellular component revealed that 518 proteins (43.2%) correspond to cell part, two proteins (0.2%) belong to the extracellular region, 254 proteins (21.2%) are part of macromolecular complexes, 119 (9.9%) are membrane proteins, one (0.1%) belongs to the nucleoid, and

Nuclear proteins	U6 snRNA-associated SM-like protein LSM1 and LSM4; RuvB-like helicases 1 (RVB1) and 2 (RVB2); replication factor C subunit 3 RFC3; nuclear pore complex protein NUP50; histones H2AZ, H2A.1, H2A.2, H3.3, and H4; histone deacetylase complex subunit SAP30L; high mobility group protein DSP1; transcription elongation factors SPT5 and SPT6; mediator of RNA polymerase II transcription subunits 10 (NUT2) and 4 (MED4); DNA-directed RNA polymerase II subunits RPB2, RPB7 and RPB8; single-stranded telomeric DNA-binding/mRNA-binding protein GBP2; DNA-directed RNA polymerase I subunit RPA49; negative cofactor 2 transcription regulator complex subunit HFL2; DNA-directed RNA polymerase III core subunit RPC19 and RPC40; RNA polymerase-associated protein LEO1 homolog; transcription factor IWS1; snoRNP complex protein SIK1; nucleolar protein 58 NOP58; karyopherin β (importin subunit β) and karyopherin β-3 (importin subunit β); importin subunit α; restriction of telomere capping protein 3; U3 small nucleolar RNA-associated protein 22 UTP22; decapping protein 1, isoform a	U6 snRNA-associated SM-like protein LSM1 and LSM4; RuvB-like helicases 1 (RVB1) and 2 (RVB2); replication factor C subunit 3 RFC3; nuclear pore complex protein NUP50; histones H2AZ, H2A.1, H2A.2, H3.3 and H4; histone deacetylase complex subunit SAP30L; high mobility group protein DSP1; transcription elongation factors SPT5 and SPT6; mediator of RNA polymerase II transcription subunits 10 (NUT2) and 4 (MED4); DNA-directed RNA polymerase II subunits RPB2, RPB7 and RPB8; single-stranded telomeric DNA-binding/mRNA-binding protein GBP2; DNA-directed RNA polymerase I subunit RPA49; negative cofactor 2 transcription regulator complex subunit HFL2; DNA-directed RNA polymerase III core subunit RPC19 and RPC40; RNA polymerase-associated protein LEO1 homolog; transcription factor IWS1; snoRNP complex protein SIK1; nucleolar protein 58 NOP58; Karyopherin β (Importin subunit β) and Karyopherin β-3 (Importin subunit β); Importin subunit α; restriction of telomere capping protein 3; U3 small nucleolar RNA-associated protein 22 UTP22; decapping protein 1, isoform a

Proteasome	Proteasome endopeptidase complex subunits PRE8 (proteasome subunit α type 2) and PRE9 (proteasome subunit α type 4); proteasome subunit β, proteasome regulatory particle base subunit RPN2 (26S proteasome non-ATPase regulatory subunit 1); proteasome regulatory particle lid subunit (26S proteasome non-ATPase regulatory subunit 8); proteasome regulatory particle base subunit RPN10 (26S proteasome non-ATPase regulatory subunit 4); proteasome regulatory particle lid subunit RPN5 (26S proteasome non-ATPase regulatory subunit 12); proteasome core particle subunit beta 1 PRE3; proteasome endopeptidase complex PRE10 (proteasome subunit α type 3); proteasome endopeptidase complex PRE5 (proteasome subunit α type-related); proteasome assembly chaperone 2; proteasome core particle subunit β4; proteasome core particle subunit β6; proteasome regulatory particle base subunit RPN1; proteasome endopeptidase complex PRE6; proteasome endopeptidase complex PUP2; 26S proteasome non-ATPase regulatory subunit 13 (proteasome regulatory particle lid subunit); proteasome endopeptidase complex SCL1 (proteasome subunit α type-6).	Proteasome endopeptidase complex subunits PRE8 (proteasome subunit α type 2) and PRE9 (proteasome subunit α type 4); proteasome subunit β, proteasome regulatory particle base subunit RPN2 (26S proteasome non-ATPase regulatory subunit 1); proteasome regulatory particle lid subunit (26S proteasome non-ATPase regulatory subunit 8); proteasome regulatory particle base subunit RPN10 (26S proteasome non-ATPase regulatory subunit 4); proteasome regulatory particle lid subunit RPN5 (26S proteasome non-ATPase regulatory subunit 12); proteasome core particle subunit beta 1 PRE3; proteasome endopeptidase complex PRE10 (proteasome subunit α type 3); proteasome endopeptidase complex PRE5 (proteasome subunit α type-related); proteasome assembly chaperone 2; proteasome core particle subunit β4; proteasome core particle subunit β6; proteasome regulatory particle base subunit RPN1; proteasome endopeptidase complex PRE6; proteasome endopeptidase complex PUP2; 26S proteasome non-ATPase regulatory subunit 13 (proteasome regulatory particle lid subunit); proteasome endopeptidase complex SCL1 (proteasome subunit α type-6)
Endoplasmic Reticulum	α-1,2-Mannosidase (mannosyl-oligosaccharide 1,2-α-mannosidase) MNS1; endoplasmic reticulum vesicle protein 25 (ERV5); UDP-glucose:glycoprotein glucosyltransferase Kre5; dolichyl-diphosphooligosaccharide-protein glycosyltransferase subunit 1 OST1; dolichyl-diphosphooligosaccharide-protein glycosyltransferase subunit WBP1; dolichyl-diphosphooligosaccharide-protein glycosyltransferase subunit 2; dolichyl-phosphate-mannose-protein mannosyltransferase 2 PMT2; signal peptidase complex subunit SPC2; signal peptidase complex catalytic subunit Sec11.	α-1,2-Mannosidase (mannosyl-oligosaccharide 1,2-α-mannosidase) MNS1; endoplasmic reticulum vesicle protein 25 (ERV5); UDP-Gucose:glycoprotein glucosyltransferase Kre5; dolichyl-diphosphooligosaccharide-protein glycosyltransferase subunit 1 OST1; dolichyl-diphosphooligosaccharide-protein glycosyltransferase subunit WBP1; dolichyl-diphosphooligosaccharide-protein glycosyltransferase subunit 2; dolichyl-phosphate-mannose-protein mannosyltransferase 2 PMT2; signal peptidase complex subunit SPC2; signal peptidase complex catalytic subunit Sec11
Golgi Apparatus	SNAP receptor complex member 1; Golgi phosphoprotein 3 Sauron homolog; α-1,6-mannosyltransferase HOC1 α-1,2-mannosyltransferase MNT1	SNAP receptor complex member 1; Golgi phosphoprotein 3 Sauron homolog; α-1,6-mannosyltransferase HOC1 α-1,2-mannosyltransferase MNT1
Secretion	Protein transport proteins Sec61 subunit β, Sec31; translocation protein Sec63; coatomer subunit γ2 Sec21; Ras-related protein Sec4; signal peptidase complex catalytic subunit Sec11	Protein transport proteins Sec61 subunit β, Sec31; translocation protein Sec63; coatomer subunit γ2 Sec21; Ras-related protein Sec4; signal peptidase complex catalytic subunit Sec11
Vacuole	V-type proton ATPase subunit a VPH1; V-type proton ATPase catalytic subunit A TFP1; V-type proton ATPase subunit C VMA5; H(+)-transporting V0 sector ATPase subunit d	V-type proton ATPase subunit a VPH1; V-type proton ATPase catalytic subunit A TFP1; V-type proton ATPase subunit C VMA5; H(+)-transporting V0 sector ATPase subunit d

Cytoskeleton-related proteins	CDC10 cell-division-related protein; actin-related protein 2/3 complex subunit 3 (ARC18), subunit 4 (ARC19) and subunit 5 (ARC15); MLC1 (a protein related to cytoskeleton with function in the cellular bud neck contractile ring); septin CDC3 and septin CDC12; dynamin-like GTPase VSP1 (vacuolar protein sorting-associated protein 1); SLP2 (stomatin-like protein 2 mitochondrial); tubulin α chain TUB1; tubulin β chain TUB2; fimbrin SAC6; actin ACT1	CDC10 cell division related protein; Actin-related protein 2/3 complex subunit 3 (ARC18), subunit 4 (ARC19) and subunit 5 (ARC15); MLC1 (a protein related to cytoskeleton with function in the cellular bud neck contractile ring); Septin CDC3 and Septin CDC12; Dynamin-like GTPase VSP1 (vacuolar protein sorting-associated protein 1); SLP2 (Stomatin-like protein 2 mitochondrial); Tubulin α chain TUB1; Tubulin β chain TUB2; Fimbrin SAC6; Actin ACT1
Ribosomal and mitochondrial proteins	37S ribosomal proteins S9 and MRP51; 39S ribosomal protein L21; NAD-dependent malic enzyme; Protein FYV4 mitochondrial; 28S ribosomal proteins S7, S18C, S23, S29 and S36; aconitate hydratase; Mitochondrial 54S ribosomal protein YmL7/YmL5; 54S ribosomal protein L3; 39S ribosomal proteins L13, L15, L19, L24, L27, L40, L41, L43, and L49; aspartate transaminase; aspartate aminotransferase AAT22; cytochrome c peroxidase; D-lactate dehydrogenase; enoyl-(acyl-carrier-protein) reductase ETR mitochondrial; enoyl-(acyl-carrier-protein) reductase MRF1 mitochondrial; TOM20; COX assembly protein homolog (UniProtKB ID_Q5ALU3); 37S ribosomal proteins S8, S10, S16, and S17; 39S ribosomal proteins L11, L27, and L43; ATP synthase assembly factor FMC1; succinate dehydrogenase assembly factor 4; HSP60 mitochondrial; arginine-tRNA ligase; succinate dehydrogenase assembly factor 2; HSP10 mitochondrial; aconitate hydratase; 4-aminobutyrate transaminase UGA1; acetolactate synthase regulatory subunit ILV6; TOM40; mitochondrial outer membrane protein OM14; 2-methoxy-6-polyprenyl-1,4-benzoquinol methylase mitochondrial COQ5; H/ACA ribonucleoprotein complex subunit 1 GAR1; cytochrome c CYC1; α-ketoglutarate dehydrogenase KGD1; mitochondrial fission 1 protein FIS1; altered inheritance of mitochondria protein 19.	37S ribosomal proteins S9 and MRP51; 39S ribosomal protein L21; NAD-dependent malic enzyme; protein FYV4 mitochondrial; 28S ribosomal proteins S7, S18C, S23, S29 and S36; aconitate hydratase; mitochondrial 54S ribosomal protein YmL7/YmL5; 54S ribosomal protein L3; 39S ribosomal proteins L13, L15, L19, L24, L27, L40, L41, L43, and L49; aspartate transaminase; aspartate aminotransferase AAT22; Cytochrome c peroxidase; D-lactate dehydrogenase; enoyl-(acyl-carrier-protein) reductase ETR mitochondrial; enoyl-(acyl-carrier-protein) reductase MRF1 mitochondrial; TOM20; COX assembly protein homolog (UniProtKB ID_Q5ALU3); 37S ribosomal proteins S8, S10, S16 and S17; 39S ribosomal proteins L11, L27 and L43; ATP synthase assembly factor FMC1; Succinate dehydrogenase assembly factor 4; HSP60 mitochondrial; Arginine-tRNA ligase; Succinate dehydrogenase assembly factor 2; HSP10 mitochondrial; aconitate hydratase; 4-aminobutyrate transaminase UGA1; acetolactate synthase regulatory subunit ILV6; TOM40; mitochondrial outer membrane protein OM14; 2-methoxy-6-polyprenyl-1,4-benzoquinol methylase mitochondrial COQ5; H/ACA ribonucleoprotein complex subunit 1 GAR1; Cytochrome c CYC1; α-ketoglutarate dehydrogenase KGD1; mitochondrial fission 1 protein FIS1; altered inheritance of mitochondria protein 19

Table 1. Proteins found in several intracellular compartments that were identified in the 2% SDS-extracted cell wall fraction by tandem mass spectrometry according to [29].

304 (25.4%) to organelles. Proteins related to nucleus, proteasome, endoplasmic reticulum, Golgi apparatus, secretion, vacuole, cytoskeleton, and mitochondria are shown in **Table 1**.

The zymolyase-extracted proteins were classified in cellular component as follows: 18 proteins (48.6%) to cell part, one (2.7%) to extracellular region, nine (24.3%) to macromolecular complex, one (2.7%) to membrane, and eight (21.6%) to organelles. The proteins solubilized with chitinase were classified in cell part (23, 43.4%), macromolecular complex (14, 26.4%), membrane proteins (3, 5.7%), and from organelles (13, 24.5%). Three proteins were exclusively found in the fraction of proteins extracted with zymolyase, including the cell surface Cu-only superoxide dismutase 5 (SOD5), the extracellular glycosidase CRH11, and the covalently linked cell wall protein 14 (SSR1 or CCW14).

According to our results and those from other groups, the cell wall of *C. albicans* contains many intracellular proteins that are commonly found in nucleus, cytoplasm, endoplasmic reticulum, Golgi apparatus, and vacuole. Many of these proteins have no signal peptide for entry into the classic secretory pathway involving the endoplasmic reticulum and the Golgi apparatus. For these type of proteins, an unconventional secretory pathway has been reported [19, 55, 78–81]. Some of these proteins are glycolytic enzymes, which perform different functions inside and outside the cell [82], including enolase and phosphoglucose isomerase, which do not have enzymatic activity outside the cell.

The secretory pathway of glycolytic enzymes is yet to be characterized. This pathway seems to be nonconventional since glycolytic enzymes have no secretion signal. One of this nonconventional pathways has been revealed for the acyl coenzyme A-binding protein (AcbA) of *Dictyostelium discoideum*, which has full requirement of the Golgi reassembly stacking protein (GRASP), a protein attached to the cytosolic face of the Golgi apparatus membrane [83]. The Acb1 protein of *S. cerevisiae* also requires the yeast GRASP protein for its secretion, which is also mediated by proteins involved in autophagy and Sso1 protein, a t-SNARE localized in the plasma membrane [84]. The *S. cerevisiae* Eno2 protein does not possess a secretion signal sequence, since Eno2 and Pgi1 proteins were secreted in a *sec23–1* mutant at 37°C. These data provide evidence that the secretion of some glycolytic enzymes is independent of the conventional secretory pathway. Analysis of deletion mutants showed that the knock-in of the SNARE coding *TLG2* gene inhibits the enolase secretion. However, other proteins might be involved, since the inhibition of secretion is not complete [85]. One of the best studied nonconventional secretion pathways is the secretion of the a-factor mating pheromone, which has a full requirement of the Ste6 protein, a transport protein with homology to MDR proteins [86]. Other protein that does not follow the classical secretion pathway is galectin-1, a mammalian lactose-binding lectin that was expressed in yeast [87]. The galectin-1 expression was dependent on the yeast genes *NCE101, NCE102, NCE103,* while the corresponding *C. albicans* are XP_019330961.1(CaNCE1), XP_723171.1 (CaNCE2), and XP_721792.1(CaNCE3, also named carbonate dehydratase), and therefore, the existence of this protein export pathway is possible in *C. albicans.*

Results obtained by mass spectrometry of the cell walls of *C. albicans, C. glabrata, C. krusei,* and *C. parapsilosis* exposed to different concentrations of H_2O_2 have allowed the identification of wall proteins known as multifunctional or *moonlighting*. This class of proteins includes: (i) glycolytic enzymes, (ii) thermal shock, (iii) oxidative stress response proteins, (iv) general metabolic enzymes, and highly conserved proteins, being over- or subexpressed in the presence or absence of reactive species of oxygen (ROS). Some of the identified glycolytic enzymes are the triphosphate isomerase (Tpi1), glyceraldehyde 3-phosphate dehydrogenase (GADPH), fructose bisphosphate aldolase (Fba1), phosphoglycerate kinase (Pgk), phosphoglycerate mutase (Gpm1), pyruvate kinase (Pk), and enolase 1 (Eno1) [88].

Finally, we want to call the attention to the fact that ^{14}C-labeled proteins labeled by TGase activity and extracted with SDS showed a molecular mass lower than 50 kDa [19]. However, the theoretical molecular mass of many of them was in the range between 50 and 220 kDa. Therefore, fragments of these proteins are being transported to the extracellular milieu and can be cross linked each other, either to β-1,3- or β-1,6-glucan, or chitin to strengthen the protection properties of the cell wall. This observation was evidenced by the inhibition of TGase by cystamine that facilitated the action of zymolyase on *C. albicans* yeasts with the consequent reduction of the osmoprotective properties of the cell wall [19]. However, more studies are needed to support this hypothesis.

Acknowledgements

This work was supported in part by the National Council of Science and Technology (Conacyt) of Mexico and the Department of Cell Biology at Cinvestav-IPN (to Dr. Juan Pedro Luna-Arias), and the Spanish Ministry of Innovation and Science Grant AGL 2011-29382 (to Dr. Salvador Mormeneo). We also thank Conacyt for a Ph.D. scholarship award to Dr. Elizabeth Reyna-Beltrán.

Author details

Elizabeth Reyna-Beltrán[1], César Isaac Bazán Méndez[1], María Iranzo[2], Salvador Mormeneo[2] and Juan Pedro Luna-Arias[1]*

*Address all correspondence to: jpluna@cinvestav.mx

1 Departamento de Biología Celular, Centro de Investigación y de Estudios Avanzados del Instituto Politécnico Nacional (Cinvestav-IPN), Mexico City, Mexico

2 Department de Microbiologia i Ecologia, Facultad de Farmacia, Unidad de Microbiología, Universitat de València, Valencia, Spain

References

[1] Gaffi. Global Action Fund For Fungal Infections. Available from: https://www.gaffi.org/

[2] Bongomin F, Gago S, Oladele RO, Denning DW. Global and multi-national prevalence of fungal diseases-estimate precision. Journal of Fungi (Basel, Switzerland). 2017;**3**(4):E57

[3] Poulain D. *Candida albicans*, plasticity and pathogenesis. Critical Reviews in Microbiology. 2015;**41**:208-217

[4] Höfs S, Mogavero S, Hube B. Interaction of *Candida albicans* with host cells: Virulence factors, host defense, escape strategies, and the microbiota. Journal of Microbiology. 2016;**54**:149-169

[5] Ruiz-Herrera J, Elorza MV, Valentín E, Sentandreu R. Molecular organization of the cell wall of *Candida albicans* and its relation to pathogenicity. FEMS Yeast Research. 2006;**6**:14-29

[6] Chaffin WL. *Candida albicans* cell wall proteins. Microbiology and Molecular Biology Reviews. 2008;**72**:495-544

[7] Free SJ. Fungal cell wall organization and biosynthesis. Advances in Genetics. 2013; **81**:33-82

[8] Ecker M, Deutzmann R, Lehle L, Mrsa V, Tanner W. Pir proteins of *Saccharomyces cerevisiae* are attached to β-1,3-glucan by new protein-carbohydrate linkage. The Journal of Biological Chemistry. 2006;**281**:11523-11529

[9] Lesage G, Bussey H. Cell wall assembly in *Saccharomyces cerevisiae*. Microbiology and Molecular Biology Reviews. 2006;**70**:317-343

[10] Popolo L, Degani G, Camilloni C, Fonzi WA. The PHR family: The role of extracellular transglycosylases in shaping *Candida albicans* cells. Journal of Fungi (Basel, Switzerland). 2017;**3**(4):E59

[11] Vai M, Orlandi I, Cavadini P, Alberghina L, Popolo L. *Candida albicans* homologue of *GGP1/GAS1* gene is functional in *Saccharomyces cerevisiae* and contains the determinants for glycosylphosphatidylinositol attachment. Yeast. 1996;**12**:361-368

[12] Griffin M, Casadio R, Bergamini CM. Transglutaminase: Nature's biological glues. Biochemical Journal. 2002;**368**:377-396

[13] Lorand L, Graham RM. Transglutaminases: Crosslinking enzymes with pleiotropic functions. Nature Reviews. Molecular Cell Biology. 2003;**4**:140-156

[14] Akimov SS, Krylov D, Fleischman LF, Belkin AM. Tissue transglutaminase is an integrin-binding adhesion co-receptor for fibronectin. The Journal of Cell Biology. 2000; **148**:825-838

[15] Eckert RL, Kaartinen MT, Nurminskaya M, Belkin AM, Colak G, Johnson GV, et al. Transglutaminase regulation of cell function. Physiological Reviews. 2014;**94**:383-417

[16] Waffenschmidt S, Kusch T, Woessner JP. A transglutaminase immunologically related to tissue transglutaminase catalyzes cross-linking of cell wall proteins in *Chlamydomonas reinhardtii*. Plant Physiology. 1999;**121**:1003-1015

[17] Ruiz-Herrera J, Iranzo M, Elorza MV, Sentandreu R, Mormeneo S. Involvement of transglutaminase in the formation of covalent cross-links in the cell wall of *Candida albicans*. Archives of Microbiology. 1995;**164**:186-193

[18] Iranzo M, Aguado C, Pallotti C, Cañizares JV, Mormeneo S. Transglutaminase activity is involved in *Saccharomyces cerevisiae* wall construction. Microbiology. 2002;**148**: 1329-1334

[19] Reyna-Beltrán E, Iranzo M, Calderón-González KG, Mondragón-Flores R, Labra-Barrios ML, Mormeneo S, et al. The *Candida albicans ENO1* gene encodes a transglutaminase

involved in growth, cell division, morphogenesis, and osmotic protection. The Journal of Biological Chemistry. 2018;**293**:4304-4323

[20] Latgé JP. The cell wall: A carbohydrate armour for the fungal cell. Molecular Microbiology. 2007;**66**:279-690

[21] Kapteyn JC, Hoyer LL, Hecht JE, Müller WH, Andel A, Verkleij AJ, et al. The cell wall architecture of *Candida albicans*. Molecular Microbiology. 2000;**35**:601-611

[22] Taff HT, Nett JE, Zarnowski R, Ross KM, Sanchez H, Cain MT, et al. A *Candida* biofilm-induced pathway for matrix glucan delivery: Implications for drug resistance. PLoS Pathogens. 2012;**8**(8):e1002848

[23] Nett JE, Crawford K, Marchillo K, Andes DR. Role of Fks1p and matrix glucan in *Candida albicans* biofilm resistance to an Echinocandin, pyrimidine, and Polyene. Antimicrobial Agents and Chemotherapy. 2010;**54**:3505-3508

[24] Munro CA, Richard ML. The cell wall: Glycoproteins, remodeling, and regulation. In: Calderone RA, Clancy CJ, editors. *Candida* and Candidiasis. 2nd ed. Washington DC, USA: ASM Press; 2012

[25] Mio T, Adachi-Shimizu M, Tachibana Y, Tabuchi H, Inoue SB, Yabe T, et al. Cloning of the *Candida albicans* homolog of *Saccharomyces cerevisiae* GSC1/FKS1 and its involvement in beta-1,3-glucan synthesis. Journal of Bacteriology. 1997;**179**:4096-4105

[26] Denning DW. Echinocandin antifungal drugs. Lancet. 2003;**362**:1142-1151

[27] Imtiaz T, Lee KK, Munro CA, Maccallum DM, Shankland GS, Johnson EM, et al. Echinocandin resistance due to simultaneous FKS mutation and increased cell wall chitin in a *Candida albicans* bloodstream isolate following brief exposure to caspofungin. Journal of Medical Microbiology. 2012;**61**(Pt 9):1330-1334

[28] Yang F, Zhang L, Wakabayashi H, Myers J, Jiang Y, Cao Y, et al. Tolerance to Caspofungin in *Candida albicans* is associated with at least three distinctive mechanisms that govern expression of FKS genes and cell wall remodeling. Antimicrobial Agents and Chemotherapy. 2017;**61**:e00071-e00017

[29] Suwunnakorn S, Wakabayashi H, Kordalewska M, Perlin DS, Rustchenko E. *FKS2* and *FKS3* genes of opportunistic human pathogen *Candida albicans* influence echinocandin susceptibility. Antimicrobial Agents and Chemotherapy. 2018;**62**:e02299-e02217

[30] Luna-Arias JP, Andaluz E, Ridruejo JC, Olivero I, Larriba G. The major exoglucanase from *Candida albicans*: A non-glycosylated secretory monomer related to its counterpart from *Saccharomyces cerevisiae*. Yeast. 1991;**7**:833-841

[31] Stubbs HJ, Brasch DJ, Emerson GW, Sullivan PA. Hydrolase and transferase activities of the beta-1,3-exoglucanase of *Candida albicans*. European Journal of Biochemistry/FEBS. 1999;**263**:889-895

[32] González MM, Díez-Orejas R, Molero G, Alvarez AM, Pla J, Nombela C, et al. Phenotypic characterization of a *Candida albicans* strain deficient in its major exoglucanase. Microbiology. 1997;**143**:3023-3032

[33] Sarthy AV, McGonigal T, Coen M, Frost DJ, Meulbroek JA, Goldman RC. Phenotype in *Candida albicans* of a disruption of the *BGL2* gene encoding a 1,3-beta-glucosyltransferase. Microbiology. 1997;**143**:367-376

[34] Esteban PF, Ríos I, García R, Dueñas E, Plá J, Sánchez M, et al. Characterization of the *CaENG1* gene encoding an endo-1,3-beta-glucanase involved in cell separation in *Candida albicans*. Current Microbiology. 2005;**51**:385-392

[35] Lussier M, Sdicu AM, Shahinian S, Bussey H. The *Candida albicans KRE9* gene is required for cell wall beta-1, 6-glucan synthesis and is essential for growth on glucose. Proceedings of the National Academy of Sciences of the United States of America. 1998;**95**:9825-9830

[36] Chattaway FW, Holmes MR, Barlow AJ. Cell wall composition of the mycelial and blastospore forms of *Candida albicans*. Journal of General Microbiology. 1968;**51**:367-376

[37] Christodoulidou A, Briza P, Ellinger A, Bouriotis V. Yeast ascospore wall assembly requires two chitin deacetylase isozymes. FEBS Letters. 1999;**460**:275-279

[38] Neiman AM. Sporulation in the budding yeast *Saccharomyces cerevisiae*. Genetics. 2011; **189**:737-765 (7)

[39] Roncero C. The genetic complexity of chitin synthesis in fungi. Current Genetics. 2002; **41**:367-378

[40] Munro CA, Whitton RK, Hughes HB, Rella M, Selvaggini S, Gow NA. *CHS8*-a fourth chitin synthase gene of *Candida albicans* contributes to in vitro chitin synthase activity, but is dispensable for growth. Fungal Genetics and Biology. 2003;**40**:146-158

[41] Preechasuth K, Anderson JC, Peck SC, Brown AJ, Gow NA, Lenardon MD. Cell wall protection by the *Candida albicans* class I chitin synthases. Fungal Genetics and Biology. 2015;**82**:264-276

[42] Lenardon MD, Whitton RK, Munro CA, Marshall D, Gow NA. Individual chitin synthase enzymes synthesize microfibrils of differing structure at specific locations in the *Candida albicans* cell wall. Molecular Microbiology. 2007;**66**:1164-1173

[43] Wiederhold NP. Paradoxical echinocandin activity: A limited in vitro phenomenon? Medical Mycology. 2009;**47**(Suppl 1):S369-S375

[44] Walker LA, Gow NA, Munro CA. Elevated chitin content reduces the susceptibility of *Candida* species to caspofungin. Antimicrobial Agents and Chemotherapy. 2013;**57**: 146-154

[45] Chaffin WL, López-Ribot JL, Casanova M, Gozalbo D, Martínez JP. Cell wall and secreted proteins of *Candida albicans*: Identification, function, and expression. Microbiology and Molecular Biology Reviews. 1998;**62**:130-180

[46] Aebi M. N-linked protein glycosylation in the ER. Biochimica et Biophysica Acta. 2013;**1833**:2430-2437

[47] Conde R, Cueva R, Pablo G, Polaina J, Larriba G. A search for hyperglycosylation signals in yeast glycoproteins. The Journal of Biological Chemistry. 2004;**279**:43789-43798

[48] Hall RA, Gow NA. Mannosylation in *Candida albicans*: Role in cell wall function and immune recognition. Molecular Microbiology. 2013;**90**:1147-1161

[49] Fabre E, Hurtaux T, Fradin C. Mannosylation of fungal glycoconjugates in the Golgi apparatus. Current Opinion in Microbiology. 2014;**20**:103-110

[50] Diaz-Jimenez DF, Mora-Montes HM, Hernandez-Cervantes A, Luna-Arias JP, Gow NA, Flores-Carreon A. Biochemical characterization of recombinant *Candida albicans* mannosyltransferases Mnt1, Mnt2 and Mnt5 reveals new functions in *O*- and *N*-mannan biosynthesis. Biochemical and Biophysical Research Communications. 2012;**419**:77-82

[51] Mora-Montes HM, Bates S, Netea MG, Castillo L, Brand A, Buurman ET, et al. A multifunctional mannosyltransferase family in *Candida albicans* determines cell wall mannan structure and host–fungus interactions. The Journal of Biological Chemistry. 2010; **285**:12087-12095

[52] Neubert P, Strahl S. Protein *O*-mannosylation in the early secretory pathway. Current Opinion in Cell Biology. 2016;**41**:100-108

[53] Skrzypek MS, Binkley J, Binkley G, Miyasato SR, Simison M, Sherlock G. The *Candida* genome database (CGD): Incorporation of assembly 22, systematic identifiers and visualization of high throughput sequencing data. Nucleic Acids Research. 2017;**45**(D1): D592-D596

[54] Pitarch A, Sánchez M, Nombela C, Gil C. Sequential fractionation and two-dimensional gel analysis unravels the complexity of the dimorphic fungus *Candida albicans* cell wall proteome. Molecular & Cellular Proteomics. 2002;**1**:967-982

[55] Urban C, Sohn K, Lottspeich F, Brunner H, Rupp S. Identification of cell surface determinants in *Candida albicans* reveals Tsa1p, a protein differentially localized in the cell. FEBS Letters. 2003;**544**:228-235

[56] de Groot PW, de Boer AD, Cunningham J, Dekker HL, de Jong L, Hellingwerf KJ, et al. Proteomic analysis of *Candida albicans* cell walls reveals covalently bound carbohydrate-active enzymes and adhesins. Eukaryotic Cell. 2004;**3**:955-965

[57] Castillo L, Calvo E, Martínez AI, Ruiz-Herrera J, Valentín E, Lopez JA, et al. A study of the *Candida albicans* cell wall proteome. Proteomics. 2008;**8**:3871-3881

[58] Martínez-Gomariz M, Perumal P, Mekala S, Nombela C, Chaffin WL, Gil C. Proteomic analysis of cytoplasmic and surface proteins from yeast cells, hyphae, and biofilms of *Candida albicans*. Proteomics. 2009;**9**:2230-2252

[59] Heilmann CJ, Sorgo AG, Siliakus AR, Dekker HL, Brul S, de Koster CG, et al. Hyphal induction in the human fungal pathogen *Candida albicans* reveals a characteristic wall protein profile. Microbiology. 2011;**157**:2297-2307

[60] Sorgo AG, Heilmann CJ, Dekker HL, Bekker M, Brul S, de Koster CG, et al. Effects of fluconazole on the secretome, the wall proteome, and wall integrity of the clinical fungus *Candida albicans*. Eukaryotic Cell. 2011;**10**:1071-1081

[61] Sosinska GJ, de Koning LJ, de Groot PW, Manders EM, Dekker HL, Hellingwerf KJ, et al. Mass spectrometric quantification of the adaptations in the wall proteome of *Candida albicans* in response to ambient pH. Microbiology. 2011;**157**:136-146

[62] Ene IV, Heilmann CJ, Sorgo AG, Walker LA, de Koster CG, Munro CA, et al. Carbon source-induced reprogramming of the cell wall proteome and secretome modulates the adherence and drug resistance of the fungal pathogen *Candida albicans*. Proteomics. 2012;**12**:3164-3179

[63] Heilmann CJ, Sorgo AG, Mohammadi S, Sosinska GJ, de Koster CG, Brul S, et al. Surface stress induces a conserved cell wall stress response in the pathogenic fungus *Candida albicans*. Eukaryotic Cell. 2013;**12**:254-264

[64] Navarro-Garcia F, Sanchez M, Pla J, Nombela C. Functional characterization of the *MKC1* gene of *Candida albicans*, which encodes a mitogen-activated protein kinase homolog related to cell integrity. Molecular and Cellular Biology. 1995;**15**:2197-2206

[65] Caminero A, Calvo E, Valentín E, Ruiz-Herrera J, López JA, Sentandreu R. Identification of *Candida albicans* wall mannoproteins covalently linked by disulphide and/or alkali-sensitive bridges. Yeast. 2014;**31**:137-144

[66] Hernández R, Nombela C, Diez-Orejas R, Gil C. Two-dimensional reference map of *Candida albicans* hyphal forms. Proteomics. 2004;**4**:374-382

[67] Ahmed R, Kodgire S, Santhakumari B, Patil R, Kulkarni M, Zore G. Serum responsive proteome reveals correlation between oxidative phosphorylation and morphogenesis in *Candida albicans* ATCC10231. Journal of Proteomics. 2018;**15**(185):25-38

[68] Hernáez ML, Ximénez-Embún P, Martínez-Gomariz M, Gutiérrez-Blázquez MD, Nombela C, Gil C. Identification of *Candida albicans* exposed surface proteins in vivo by a rapid proteomic approach. Journal of Proteomics. 2010;**73**:1404-1409

[69] Vialás V, Perumal P, Gutierrez D, Ximénez-Embún P, Nombela C, Gil C, et al. Cell surface shaving of *Candida albicans* biofilms, hyphae, and yeast form cells. Proteomics. 2012;**12**:2331-2339

[70] Gil-Bona A, Parra-Giraldo CM, Hernáez ML, Reales-Calderon JA, Solis NV, Filler SG, et al. *Candida albicans* cell shaving uncovers new proteins involved in cell wall integrity, yeast to hypha transition, stress response and host–pathogen interaction. Journal of Proteomics. 2015;**127**:340-351

[71] Marín E, Parra-Giraldo CM, Hernández-Haro C, Hernáez ML, Nombela C, Monteoliva L, et al. *Candida albicans* shaving to profile human serum proteins on hyphal surface. Frontiers in Microbiology. 2015;**6**:1343

[72] Medrano-Diaz CL, Vega-Gonzalez A, Ruiz-Baca E, Moreno A, Cuellar-Cruz M. Moonlighting proteins induce protection in a mouse model against *Candida* species. Microbial Pathogenesis. 2018;**124**:21-29

[73] Lopez-Ribot JL, Chaffin WL. Members of the Hsp70 family of proteins in the cell wall of *Saccharomyces cerevisiae*. Journal of Bacteriology. 1996;**178**:4724-4726

[74] Röhm M, Lindemann E, Hiller E, Ermert D, Lemuth K, Trkulja D, et al. A family of secreted pathogenesis-related proteins in *Candida albicans*. Molecular Microbiology. 2013;**87**:132-151

[75] Pardini G, De Groot PW, Coste AT, Karababa M, Klis FM, de Koster CG, et al. The CRH family coding for cell wall glycosylphosphatidylinositol proteins with a predicted trans-glycosidase domain affects cell wall organization and virulence of *Candida albicans*. The Journal of Biological Chemistry. 2006;**281**:40399-40411

[76] Urban C, Xiong X, Sohn K, Schröppel K, Brunner H, Rupp S. The moonlighting protein Tsa1p is implicated in oxidative stress response and in cell wall biogenesis in *Candida albicans*. Molecular Microbiology. 2005;**57**:1318-1341

[77] Shin DH, Jung S, Park SJ, Kim YJ, Ahn JM, Kim W, et al. Characterization of thiol-specific antioxidant 1 (TSA1) of *Candida albicans*. Yeast. 2005;**22**:907-918

[78] Gil-Navarro I, Gil ML, Casanova M, O'Connor JE, Martinez JP, Gozalbo D. The glycolytic enzyme glyceraldehyde-3-phosphate dehydrogenase of *Candida albicans* is a surface antigen. Journal of Bacteriology. 1997;**179**:4992-4999

[79] Angiolella L, Facchin M, Stringaro A, Maras B, Simonetti N, Cassone A. Identification of a glucan-associated enolase as a main cell wall protein of *Candida albicans* and an indirect target of lipopeptide antimycotics. Journal of Infectious Diseases. 1996;**173**:684-690

[80] Nombela C, Gil C, Chaffin WL. Non-conventional protein secretion in yeast. Trends in Microbiology. 2006;**14**:15-21

[81] Sorgo AG, Heilmann CJ, Dekker HL, Brul S, de Koster CG, Klis FM. Mass spectrometric analysis of the secretome of *Candida albicans*. Yeast. 2010;**27**:661-672

[82] Tristan C, Shahani N, Sedlak TW, Sawa A. The diverse functions of GAPDH: Views from different subcellular compartments. Cellular Signalling. 2011;**23**:317-323

[83] Kinseth MA, Anjard C, Fuller D, Guizzunti G, Loomis WF, Malhotra V. The Golgi-associated protein GRASP is required for unconventional protein secretion during development. Cell. 2007;**130**:524-534

[84] Duran JM, Anjard C, Stefan C, Loomis WF, Malhotra V. Unconventional secretion of Acb1 is mediated by autophagosomes. The Journal of Cell Biology. 2010;**188**:527-536

[85] Miura N, Kirino A, Endo S, Morisaka H, Kuroda K, Takagi M, et al. Tracing putative trafficking of the glycolytic enzyme enolase via SNARE-driven unconventional secretion. Eukaryotic Cell. 2012;**11**:1075-1082

[86] Kuchler K, Sterne RE, Thorner J. *Saccharomyces cerevisiae STE6* gene product: A novel pathway for protein export in eukaryotic cells. The EMBO Journal. 1989;**8**:3973-3984

[87] Cleves AE, Cooper DN, Barondes SH, Kelly RB. A new pathway for protein export in *Saccharomyces cerevisiae*. The Journal of Cell Biology. 1996;**133**:1017-1026

[88] Serrano-Fujarte I, Lopez-Romero E, Cuellar-Cruz M. Moonlight-like proteins of the cell wall protect sessile cells of *Candida* from oxidative stress. Microbial Pathogenesis. 2016;**90**:22-33

Section 3

Interaction of Cells Host and Other Microbe

Chapter 5

The Role of Phagocytes in Immunity to *Candida albicans*

Annabelle G. Small, Jovanka R. King,
Deborah A. Rathjen and Antonio Ferrante

Additional information is available at the end of the chapter

http://dx.doi.org/10.5772/intechopen.80683

Abstract

Body clearance of fungi such as *Candida albicans* involves phagocytosis by fixed tissue macrophages as well as infiltrating monocytes and neutrophils. Through phagocytosis, the fungi are confined and killed by the oxidative and non-oxidative anti-microbial systems. These include oxygen derived reactive species, generated from the activation of the NADPH oxidase complex and granule constituents. These same mechanisms are responsible for the damage to hyphal forms of *C. albicans*. Complement promotes phagocytosis, through their interaction with a series of complement receptors including the recently described complement receptor immunoglobulin. However, it is also evident that under other conditions, the killing of yeast and hyphal forms can occur in a complement-independent manner. Phagocytosis and killing of *Candida* is enhanced by the cytokine network, such as tumour necrosis factor and interferon gamma. Patients with primary immunodeficiency diseases who have phagocytic deficiencies, such as those with defects in the NADPH oxidase complex are predisposed to fungal infections, providing evidence for the critical role of phagocytes in anti-fungal immunity. Secondary immunodeficiencies can arise as a result of treatment with anti-cancer or other immunosuppressive drugs. These agents may also predispose patients to fungal infections due to their ability to compromise the anti-microbial activity of phagocytes.

Keywords: *Candida albicans*, macrophages, neutrophils, complement, innate immunity, phagocytosis, fungal killing mechanisms, cytokines, trained immunity, immunodeficiency, immunopharmacology

1. Introduction

C. albicans is considered to be the most common fungus causing both skin and disseminated disease, particularly in immunodeficient and immunocompromised patients. Phagocytes, particularly neutrophils, play an important role in clearing candidal infections. The importance of neutrophils in immunity to *C. albicans* is clearly evident from the increased rate of infection seen in patients with severe neutropenia [1].

In neutrophils, the major response associated with phagocytosis of microbial pathogens is the oxygen-dependent respiratory burst and the generation of reactive oxygen species (ROS). Several decades ago it became evident that neutrophils displayed a unique respiratory burst in the absence of mitochondria, where the generation of ATP comes mainly from glycolysis (reviewed in [2]). It also became apparent that the majority of the oxygen consumed was converted to superoxide ($O\cdot_2^-$) which is then converted to further oxygen intermediates, including singlet oxygen and H_2O_2. The enzyme which catalyses the conversion of O_2 to $O\cdot_2^-$ is assembled in the phagocytic vacuole membrane, facilitating its release into the bacteria or fungus-containing vacuole. In neutrophils, the release of the azurophilic granule content simultaneously into the phagocytic vacuole leads to the generation of HOCl, a highly potent anti-microbial agent, as a result of the action of myeloperoxidase on H_2O_2 in the presence of chloride ions. In addition, ingestion of microbial pathogens and their confinement to the vacuolar environment may restrict the supply of essential nutrients necessary for growth.

The NADPH oxidase complex is responsible for the respiratory burst and consists of a number of different proteins which assemble in the neutrophil vacuole membrane following cell stimulation. This is typically initiated during phagocytosis of bacteria and fungi [3]. The complex consists of the oxidase-specific phox proteins $gp91^{phox}$, $p22^{phox}$, $p40^{phox}$, $p47^{phox}$, $p67^{phox}$ and the small GTPases, Rac1 and Rac2. Cell activation leads to the assembly of these components in the membrane and the initiation of enzymatic activity.

The non-oxidative microbicidal system complements the respiratory burst. Components of the azurophilic granules in neutrophils have been shown to have anti-microbial activity. These include defensins, serprocidins and bactericidal/permeability increasing protein (BPI). Defensins are cationic peptides with broad spectrum antimicrobial activity [2]. The seroprodins, elastase, azurocidin and cathepsin G have antimicrobial activity independent of their enzymatic activity [2].

As with neutrophils, most bacteria and fungi are confined and killed within phagosomes by macrophages [4], involving a variety of agents such as toxic metabolites, peptides and enzymes. These may act either alone or synergistically. In addition, macrophages can produce ROS which have anti-microbial action but unlike monocytes, macrophages lack MPO. Most striking is the marked heterogeneity of macrophages enabling these leukocytes to perform functions relevant to specific tissues in which they are located.

The extrusion of neutrophil extracellular traps (NETs) is also considered to be a defence mechanism against microbial pathogens. NETs are structures composed of DNA as well as anti-microbial substances, elastase, calprotectin and MPO [5]. NETs not only trap the microbial pathogens, but also kill them. Interestingly, it has been reported that the formation of NETs requires the presence of ROS [6].

Effective recognition of microbial pathogens by neutrophils and macrophages requires receptors which bind peptides deposited on bacterial and fungal surfaces which have been generated through the activation of complement, namely C3b and iC3b. Receptors recognising iC3b include CR3 (CD18/CD11b) and CR4 (CD18/CD11c), which are present on both neutrophils and macrophages. Recently, another complement receptor type, complement immunoglobulin receptor (CRIg), expressed only by a subpopulation of macrophages has been described, which binds both iC3b and C3b (reviewed in [7]). It has been shown that this receptor plays an important role in clearance of bacteria from the circulation by liver Kupffer cells [8] and may also be a pattern recognition receptor, facilitating clearance of bacteria in the absence of complement [9].

Antibody bound to microbial pathogens also promotes phagocytosis through the Fcγ receptors, FcγRI (CD64), FcγRIIA (CD32) and FcγRIIIB (CD16), all of which engage the Fc domain of Immunoglobulin G (IgG). The FcαRI which binds the Fc domain of Immunoglobulin A (IgA) also promotes microbial phagocytosis and killing [2].

Apart from the integrins and FcγRs, neutrophils and macrophages express a range of pattern recognition receptors (PPR) which recognise conserved microbial pathogen structures, such as lipoteichoic acid, β-glucans and lipopolysaccharide. Families of PPRs include those found in serum (pentraxins, collectins, complement), those which are membrane bound (classic C-type lectins, non-classic C-type lectin leucine-rich proteins, scavenger receptors) and those which are located intracellularly (NODs, interferon induced proteins).

2. Complement dependent and independent phagocytosis of *C. albicans*

Despite the importance of complement-independent mechanisms for host anti-candidal immunity, it is evident that complement is required for optimal resistance to fungal infection [10–12]. It was also evident in these studies that complement could be activated by *C. albicans* by the alternative pathway. Activation of complement leads to the generation of chemotactic peptides and C5a, which attracts neutrophils to the site of candidal infection [13, 14]. Thong and Ferrante [11], in their studies on the generation of chemotaxis promoting factors by serum treated with *C. albicans*, showed that this activity was totally dependent on heat-labile factors and activation of complement via the alternative pathway. Chemotaxis of neutrophils towards fungus-treated serum was abolished when the serum was either heated at 56°C for 20 min or was C2 deficient (where the alternative but not the classical pathway can be activated). The subsequent step, phagocytosis, was also highly dependent on heat labile opsonins [12]. However, while the chemotactic response was totally dependent on serum complement, the heat labile opsonins only acted to enhance other phagocytosis-promoting mechanisms. Thus, significant phagocytosis was still observed in the presence of heat-inactivated serum. In both of these studies on chemotaxis and phagocytosis-promoting activity of serum, it was shown that these principles applied to a wide-range of clinical isolates of *C. albicans* from patients and both including Serotypes A and B [11, 12].

Zymosan A is a yeast cell wall glucan and, like *C. albicans* derived β(1,3) (1,6)-glucan, is an agonist to TLR2 and Dectin-1 [15]. Using commercially available labelled zymosan A bioparticles which are non-fluorescent outside of the cell and fluoresce once taken into acidic phagosomes, we showed that neutrophils require opsonising conditions to phagocytose particles efficiently (**Figure 1**). This supports the findings of [16], and demonstrates that like monocytes and macrophages, neutrophils require complement for the rapid phagocytosis of yeast particles. Interestingly, the complement dependency of phagocytosis diminished at incubation times of 45–60 min, where complement-independent mechanisms of phagocytosis become more prominent (**Figure 1C**).

The classical complement pathway is likely to be activated by mannan-specific antibodies found in human serum [19] whereas the lectin pathway is activated by the binding of mannose-binding lectin to mannan on the cell wall of the fungus [20]. However, it has also been shown that *C. albicans* can bind the complement regulatory protein, C4b-binding protein (C4BP), thereby inactivating C4b and hence preventing complement activation on the yeast surface. As a result, the microbial pathogen will evade complement activation via the classical and lectin pathways, but the alternative pathway remains operative, generating chemotactic factors and opsonins. Furthermore, *C. albicans* has the ability to regulate the alternative pathway and factors H and FHL-1 [21]. The binding of these regulators is seen with both the cellular and hyphal forms of *C. albicans* [22].

The unique complement receptor CRIg is a member of the transmembrane protein of the type 1 immunoglobulin superfamily, encoded by *VSIG4*. Although two spliced forms of CRIg have been described, a long (L) and short (S) form [8], we have recently identified five forms based on expression of transcripts and western blot analysis [23]. The receptor is expressed selectively by a subpopulation of macrophages, probably of the M2 type, and is abundant in fixed tissue macrophages such as liver Kupffer cells and resident peritoneal macrophages [24, 25]. Unlike CR3 and CR4 which require prior activation, CRIg is naturally active and its activity is controlled by its recycling pattern from the endoplasmic reticulum [8]. Our studies have demonstrated that cytokines alter CRIg expression in human macrophages and this was associated with a corresponding change in ability of neutrophils to phagocytose *C. albicans* in a complement-dependent manner [23, 26, 27].

While CR3 and FcRγ mediate phagocytosis of complement and antibody opsonised *C. albicans*, in the absence of these opsonins, adherence and phagocytosis by neutrophils and macrophages is promoted by C-Type Lectin Receptors (CLRs), in particular Dectin-1 [28–31]. The targets for Dectin-1 are β-1,3 glucan polymers, major components of the fungal cell wall. In *C. albicans* hyphae, this polymer is masked and appears to be different in the yeast form [32].

Cells of the phagocytic system are able to recognise *C. albicans* through multiple classes of receptors [33]. These include pattern recognition receptors (PRRs) such as Toll-like receptor (TLRs) 4 and 2 [34, 35], and CLRs such as Dectin-1 and the mannose receptor [36]. While these receptors are able to induce phagocytosis independently of complement, efficiency of uptake in both macrophages and neutrophils can be significantly increased when *C. albicans* is opsonised [16]. Under these conditions CR3 present on phagocytes is able to recognise iC3b deposited on the fungal cell surface and promote phagocytosis. In macrophages, this process is also able to occur through CRIg [27]. Agents such as dexamethasone that promote the upregulation of CRIg protein expression are also able to induce increased levels of phagocytosis of *C albicans* [23], suggesting that CRIg rather than CR3 plays an important role in the phagocytosis of *C. albicans* in macrophages.

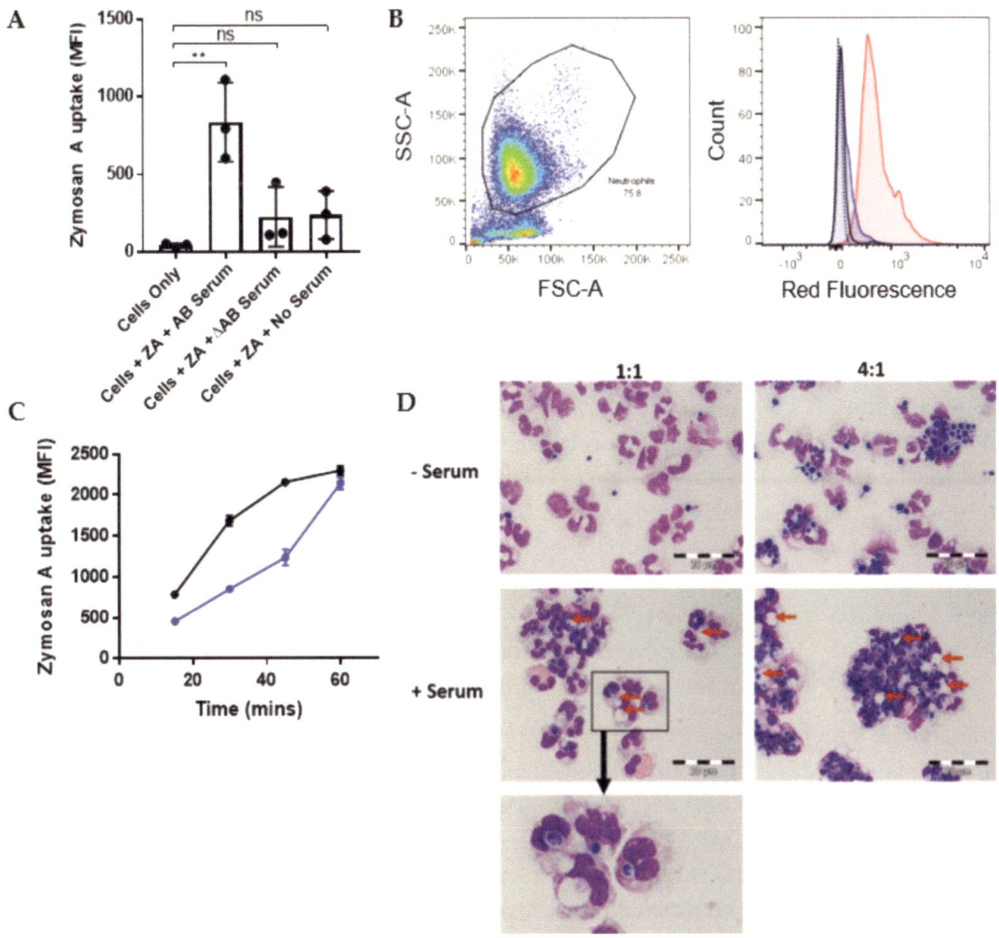

Figure 1. Complement-dependent and -independent phagocytosis of zymosan A bioparticles by human neutrophils. (A) Phagocytosis of *C. albicans* under the different treatment conditions indicated in the x-axis. Results are the mean ± SD of three experiments. (B) Representative histogram/gating strategy for these experimental runs. In these experiments the reaction was terminated at 30 min. Neutrophils only are shown by the dashed line, no serum in black, heat-inactivated serum shown in blue, and native AB serum shown in red. (C) Phagocytosis kinetics over a 60 min incubation period in the presence or absence of serum. Results are presented as mean ± SD of triplicate reactions. Neutrophils were prepared from human peripheral blood from healthy volunteers, using the high density gradient method [17]. Phagocytosis was assayed using pHrodo™Red Zymosan A bioparticles (ThermoFisher, Walter MA, Cat no. P35364) as described previously [18]. Human AB serum was prepared from peripheral blood of healthy volunteers. The serum was shown to have normal levels of complement activity using the CH50 assay. AB serum heated at 56°C for 20 min was confirmed to lack complement activity. The cell samples were analysed using a FACSCanto I flow cytometer (BD). The work was approved by the Human Research Ethics Committee of the Women's and Children's Hospital Network, Adelaide. Statistical analyses were carried out by ANOVA followed by Dunnet's post hoc test. (D) Photomicrographs of the interaction of *C. albicans* with neutrophils in the presence or absence of serum at a 1:1 and 1:4 neutrophil:fungal ratio. Red arrows indicate phagocytic vacuoles following the digestion of the yeast or non-degraded yeast particles (following 30 min of incubation).

Neutrophils recognise *C. albicans* through the PRRs TLR2, TLR4 and Dectin-1, and also under opsonising conditions through FcγR and CR3 [37]. Similar to macrophage phagocytosis of *C. albicans*, uptake of isolated fungal zymosan A is more efficient in opsonising conditions, with phagocytosis after a 15-min incubation time being three times higher in reactions with complement compared to no serum and heat-inactivated serum controls (**Figure 1**).

C. albicans is able to exist in multiple forms, as a single-celled budding yeast or in pseudohyphal or hyphal filamentous forms [38]. While in its unicellular form, the fungus can be tolerated as a commensal organism by the oral or vaginal epithelium. However, when it converts to its hyphal form, the fungus displays pathogenic properties. The host is able to discriminate against the potential danger [37] through MAPK-based recognition in the epithelial cells [39], which leads to mitogen-activated protein kinase phosphatase 1 (MKP1) and c-Fos activation. Neutrophils also play a role in this protection through TLR4-mediated recognition [40].

3. Trained macrophage immunity in anti-fungal immunity

Trained immunity (TI) refers to the ability of innate immune cells to exhibit 'memory' and prevent reinfection of previously encountered invading pathogens [41]. Termed by Netea and colleagues [42], TI induces a state of enhanced antimicrobial action in cells of the innate immune system, particularly monocytes and macrophages, which is distinct from both typical innate immunity and the memory of the adaptive immune system. Alternatively, TI refers to the enhanced response to reinfection against the initial invading microorganisms and cross-protection against different pathogens. Although the concept of TI is relatively new, the phenomenon of protection afforded by previous infection in a manner distinct from adaptive immunity has long been known, particularly in plant and insect systems [43, 44].

TI has been shown to have a role in infection and immunity against *C albicans*. Bistoni et al. [45] demonstrated that not only did injection with a non-pathogenic strain of *C. albicans* induce protection against reinfection, but also cross-protected against the other pathogens *Candida tropicalis* and *Staphylococcus aureus*. This protection was determined to be macrophage-dependent, as transfer of adherent splenic cells from mice administered with the non-pathogenic strain conferred protection to the recipient mice. Two decades later, Quintin et al. [46] expanded on this concept, demonstrating that mice injected with low doses of *C. albicans* showed increased survival rates when administered lethal infection loads, and increased proinflammatory cytokine production upon secondary exposure. This protection was also shown in mice deficient in T and B cells and not in mice lacking *CCR2*, indicating that similar to the results of Bistoni et al. [45], the observed protection was monocyte-dependent. It was also shown that training of monocytes could be induced through purified β-glucan, a polysaccharide that makes up the cell wall of selected bacteria and fungi [47]. The group further investigated the mechanisms behind this protection by analysis of the genome-wide binding pattern of the methylation marks on histone 3 lysine 4 (H3K4me3) and on histone 3 lysine 27 (H3K27me3), and concluded that protection was controlled at the epigenetic level through H3K4me3 in known genes involved in innate immunity. Furthermore, mRNA levels of TNF and IL-6 were higher in trained monocytes compared with non-trained control cells.

While other molecules such as fungal chitin have also been shown to induce TI [48], β-glucan remains the most well-studied molecule with respect to *C. albicans*, which has been shown to induce TI in both human and murine systems [46, 49, 50]. Along with its antimicrobial priming, β-glucan-induced TI has also been investigated in anti-tumour immunity [51].

4. Killing of *C. albicans* by neutrophils and macrophages

Ferrante [52] demonstrated that killing of yeast forms of *C. albicans* and *Candida glabrata* was associated with release of the ROS, superoxide, and constituents of azurophilic granules and specific granules. During this interaction the generation of HOCl occurred, another potent anti-fungal agent. The importance of ROS was demonstrated by the finding that inhibitors of superoxide and H_2O_2 decreased intracellular killing of *C. albicans* [53]. Further proof of the role of ROS generation in the killing of *C. albicans* came from the demonstration that neutrophils and macrophages from patients with chronic granulomatous disease (CGD) (who have defective NADPH oxidase activity), were unable to effectively kill the fungi [54]. However, whether ROS *per se* are responsible for the killing of *C. albicans* remains to be established [55]. The reaction of H_2O_2 with MPO, in the presence of chloride ions, forms a very potent anti-microbial system. We have previously demonstrated that opsonised *C. albicans* induces the release of both H_2O_2 and MPO, thereby establishing an anti-candidal environment [52]. The importance of MPO in the killing of *C. albicans* is supported by the finding that neutrophils and monocytes from MPO deficient patients failed to kill *C. albicans* [56, 57].

In vivo the absence of MPO in macrophages may be overcome by the cells incorporating MPO released by neutrophils at infection sites. Thus, resident peritoneal mouse macrophages in the presence of recombinant MPO caused an increase in intracellular killing of *C. albicans* [58]. However, it is noteworthy that using mouse models of X-linked CGD and MPO deficiency, susceptibility was most evident in the former, suggesting that ROS are the major mediators of candidicidal activity [59]. In comparison, the neutrophil-mediated damage to *C. albicans* pseudohyphae was found to be mediated by the oxidative burst and MPO [60]. Interestingly, this neutrophil-mediated damage occurred in the absence of serum complement.

Two distinct mechanisms for human neutrophil-mediated killing have been documented, depending on the state of fungal opsonisation. Using *in vitro* fungicidal assays, Gazendam et al. [61] showed that killing of un-opsonised *C. albicans* was dependent on CR3 and phosphatidylinositol-3-kinase (PI3K) signalling, but was independent of NADPH oxidase activation. However, the killing of antibody opsonised *C. albicans* by neutrophils was dependent on Fcγ receptors and protein kinase C (PKC) in addition to NADPH.

5. Intracellular signalling required for killing of *C. albicans*

Approximately two decades ago it was demonstrated that human neutrophil-mediated killing of *C. albicans*, in a complement-dependent manner, required the activation of the extracellular signal-regulated protein kinase cascade [62]. More recently it has been reported that PKCδ activation downstream of the receptors Dectin-1 and Mac-1 is important in the neutrophil-mediated resistance to *C. albicans* and fungi-induced ROS generation [63]. In contrast, while PKCδ deficiency in macrophages prevented the stimulation of production of ROS induced by *C. albicans*, this did not affect the killing of the fungus. It has been demonstrated that BTK and

Vav1 are Dectin-1 interacting proteins [64]. These were found to be recruited to phagocytic cups containing yeast or hyphae of *C. albicans*, at the less mature stage of phagosome development. These contribute to the Dectin-1 dependent phagocytosis of *C. albicans*.

In comparison, Gazendam et al. [61] demonstrated that neutrophils display two different mechanisms in the killing of *C. albicans* by evaluating patients with Dectin-1 deficiency, CARD9 deficiency or NADPH deficiency. One of these mechanisms was CR3, PI3K and CARD9 dependent, but independent of ROS generation. The other was selectively dependent on Fcγ, PKC and ROS generation. Both of these candidicidal pathways required Syk tyrosine activation but were independent of Dectin-1.

6. Neutrophil extracellular traps in immunity to *C. albicans*

C. albicans has been shown to induce NET extrusion in phagocytes, particularly in neutrophils. While the formation of this structure is considered as part of cell death or NETosis [65], Byrd et al. [66] reported that the rapid extrusion of NETs in response to *C. albicans* occurs in the absence of cell death. However, others have demonstrated that the yeast forms of *C. albicans* stimulated NETs through autophagy and ROS generation in the early stage of the interaction (first 15 min) [67]. However, with the hyphal forms, NET formation occurred via autophagy and not ROS generation. In the longer term (4 h), only the hyphae stimulated NETs. Interestingly, they found less killing of yeast forms by NETs compared to the high level of damage to the hyphae forms. Other strategic functions of extracellular protrusions of neutrophils have been demonstrated for *Plasmodium falciparum*. Here, the neutrophils were observed to 'throw out' protrusions which penetrated the parasitophorous vacuole containing the intraerythrocytic stage of the parasite and withdrawing the parasite without damaging the erythrocyte [68].

7. Cytokine priming in phagocyte-mediated killing of *C. albicans*

Over three decades ago it became evident that neutrophil responses to microbial pathogens could be significantly increased if the cells were pre-sensitised with products released by activated lymphocytes and macrophages [69], a process dependent on the presence of TNF [70, 71]. The importance of cytokine priming in killing of *C. albicans* by neutrophils was also observed [72]. Thus, neutrophil mediated killing of *C. albicans* and a related fungus, *Candida glabrata* was significantly increased if the phagocytes had been pre-treated with either TNF or GM-CSF [52, 73]. The TNF treatment also increased the candida-induced release of ROS and MPO, consistent with the increased anti-fungal activity induced by the cytokines [52]. The mechanism by which TNF primes neutrophils for increased killing of *C. albicans* has not been studied. However, these mechanisms can be inferred from studies with other microbial pathogens. Kowanko et al. [74] demonstrated that the TNF-induced effects responsible for increased microbial killing could be mediated by both oxygen-dependent and oxygen-independent mechanisms, with respect to killing of opsonised *S. aureus* and *Plasmodium falciparum* infected erythrocytes, respectively. Furthermore, studies with the pathogenic soil amoeba,

N. fowleri have shown that the TNF-enhanced killing requires a functional H_2O_2-MPO-halide system [75]. The priming of neutrophils by TNF is reflected by an increase in expression of CR3 and CR4 on the surface of these cells. The enhanced killing of *S. aureus* was dependent on these receptors, given that this was not seen upon the addition of anti-CD11b and -CD11c monoclonal antibodies [76].

The use of TNF to enhance immunity against various microbial infections has not been considered appropriate because of the highly toxic and tissue damaging effects of TNF. In an effort to harness the anti-infective properties of TNF and exclude some of its tissue damaging properties, we synthesised short peptides representative of the TNF sequence [77]. One of these elevenmer peptides, TNF_{70-80}, was found to activate neutrophils and macrophages to increase microbial killing both *in vitro* and *in vivo* [77–81].

Our studies with *C. albicans* demonstrated that TNF_{70-80} also protected against infections with this fungus (**Tables 1** and **2**). In the first set of experiments, the effect of administering either TNF or TNF_{70-80} to mice infected with *C. albicans* was examined. The recovery of fungi from

Treatment	No. mice/group	Log CFU/g kidney (M ± SD)
PBS	23	7.3 ± 0.6
Amphotericin B	15	$2.7 \pm 2.4^{***}$
TNF (0.1 mg/kg)	29	$5.6 \pm 1.2^{***}$
TNF_{70-90} (4 mg/kg)	9	$5.75 \pm 1.7^{**}$

Eight week old Balb/c mice were challenged with 5×10^5 CFU *C. albicans* intravenously. Treatment of mice commenced 24 h prior to infection, and continued with daily administration until 2 days post-infection. Mice were sacrificed on day 2 and kidney preparations plated on Sabouraud agar. The degree of infection was determined by enumeration of the number of organisms in the kidney at the time of euthanisation ($^{**}p < 0.01$, $^{***}p < 0.001$, 1-way ANOVA, SNK test). The research received approval from the Women's and Children's Hospital Animal Ethics Committee.

Table 1. The effect of TNF and TNF_{70-80} on ***C. albicans*** infection in mice.

Treatment	Route	Dose (mg/kg)	Survivors 10 days post-infection
Vehicle control	IP	—	8
Cyclophosphamide	PO	30	2^{*}
TNF_{70-80} + cyclophosphamide	IP	100	7^{ns}
TNF_{70-80} + cyclophosphamide	IP	10	4^{ns}
TNF_{70-80} + cyclophosphamide	IP	1	4^{ns}
TNF_{70-90} + cyclophosphamide	IP	0.1	2^{*}
Azimexone + cyclophosphamide	IP	100	6^{ns}

Balb/c mice (10/group) were treated with 3 doses of oral (OP) cyclophosphamide (30 mg/kg) and infected with *C. albicans* as described in **Table 1**. Mice were also treated with three doses of TNF_{70-80} at the schedule described in **Table 1**. Azimexone (used as a positive control) was administered intraperitoneally (IP) (n = 10 mice, $^{*}p < 0.05$, ns: not significant, one-sided Fisher's exact test). The research received approval from by Women's and Children's Hospital Animal Ethics Committee.

Table 2. Effect of TNF_{70-80} on ***C. albicans*** infection in immunocompromised mice.

the kidneys of these mice was significantly lower than in non-treated control mice (**Table 1**). In the second experimental set-up, mice treated with cyclophosphamide became highly susceptible to *C. albicans* with the survival of mice dropping from 80 to 20%, 10 days after infection. If the mice had been treated with TNF_{70-80}, survival was increased with 70% survival observed at the highest dose (**Table 2**).

Cytokines also influence the ability of macrophages to phagocytose and kill fungi. Human monocyte-derived macrophages (MDMs) treated with interferon gamma showed increased ability to phagocytose and kill yeast forms of *C. albicans* [82]. The cytokine treated cells showed a corresponding increase in ROS production when challenged with the fungus. This effect of interferon gamma was evident with non-opsonised *C. albicans* and was independent of CR3. These effects of interferon gamma were reproduced with mouse peritoneal macrophages [83]. M-colony stimulating factor has also been shown to increase macrophage phagocytosis and killing of *C. albicans* yeast forms and cause damage to hyphae [84].

From the described studies, it is evident that when considering killing of microbial pathogens including *C. albicans*, this needs to be interpreted in terms of the cytokine milieu generated during the infection. It is evident from other published work that several cytokines regulate phagocyte-mediated microbial killing properties, including interferon gamma, lymphotoxin and interleukin-1 [71, 85].

8. Primary immunodeficiency diseases associated with susceptibility to fungal infection

Primary immunodeficiency diseases (PID) are a heterogeneous group of inborn errors of immunity. Affected individuals develop severe, unusual or recurrent infections, and may also develop features of immune dysregulation with autoimmune manifestations. There are currently over 320 described molecular genetic causes of PID, which can be categorised according to presenting phenotypic features [86]. The International Union of Immunological Sciences (IUIS) classify PID into the following disease categories: immunodeficiencies affecting cellular and humoral immunity, combined immunodeficiencies (CID) with associated or syndromic features, predominantly antibody deficiencies, diseases of immune dysregulation, congenital defects of phagocyte number, function or both, defects in intrinsic and innate immunity, auto-inflammatory disorders, complement deficiencies and phenocopies of PID [86].

Intact immunological processes and pathways are required to mount an effective immune response against fungi, incorporating both innate and adaptive components [87]. Several immune cells and immunological mediators such as cytokines are of critical importance to maintenance of anti-fungal immunity. These include phagocytes, dendritic cells, T cells (particularly T helper 1 (TH1) and T helper 17 (TH17) cells) [87]. The importance of these effectors is evidenced by patients with PID affecting cellular or phagocytic immunity developing severe, invasive or recurrent fungal infections [1].

Primary phagocytic disorders result from mutations in genes encoding key proteins that are essential for normal phagocytic development and function. These disorders may be classified according to whether phagocyte number, function or both are affected, and by the presence or absence of associated syndromic features [86]. These disorders and their underlying, causative genetic abnormality are summarised in **Table 3**.

Congenital defects of phagocytic number, function or both			
Associated with syndromic features		**Not associated with syndromic features**	
Disorder	**Gene(s)**	**Disorder**	**Gene(s)**
Shwachman-Diamond syndrome	*SBDS, DNAJC21*	Elastase deficiency (SCN1)	*ELANE*
G6PC3 deficiency (SCN4)	*G6PC3*	Kostmann disease (HAX1 deficiency; SCN3)	*HAX1*
Glycogen storage disease type 1b	*G6PT1*	GFI1 deficiency (SCN2)	*GFI1*
Cohen syndrome	*COH1*	X-linked neutropaenia/myelodysplasia WAS GOF	*WAS*
Barth syndrome (3-methylglutaconic aciduria type II)	*TAZ*	G-CSF receptor deficiency	*CSF3R*
Clericuzio syndrome (poikiloderma with neutropaenia)	*C16ORF57 (USB1)*	Neutropaenia with combined immune deficiency	*MKL1*
VPS45 deficiency (SCN5)	*VPS45*		
P14/LAMTOR2 deficiency	*LAMTOR2*		
JAGN1 deficiency	*JAGN1*		
3-methylglutaconic aciduria	*CLPB*		
SMARCD2 deficiency	*SMARCD2*		
WDR1 deficiency	*WDR1*		
HYOU1 deficiency	*HYOU1*		
Congenital defects of phagocytic function			
Associated with syndromic features		**Not associated with syndromic features**	
Disorder	**Gene(s)**	**Disorder**	**Gene(s)**
Cystic fibrosis	*CFTR*	Chronic granulomatous disease	*CYBB, NCF1, CYBA, NCF4, NCF2*
Papillon-Lefevre syndrome	*CTSC*	Rac2 deficiency	*RAC2*
Localised juvenile periodontitis	*FPR1*	G6PD deficiency Class 1	*G6PD*
Leukocyte adhesion deficiency (LAD) 1	*ITGB2*	GATA2 deficiency (MonoMac syndrome)	*GATA2*
Leukocyte adhesion deficiency (LAD) 2	*SLC35C1*	Specific granule deficiency	*C/EBPE*
Leukocyte adhesion deficiency (LAD) 3	*FERMT3*	Pulmonary alveolar proteinosis	*CSF2RA, CSF2RB*

Adapted from [86].

SCN = severe congenital neutropaenia, WAS = Wiskott-Aldrich Syndrome, GOF = gain of function.

Table 3. Primary immunodeficiency diseases affecting phagocytic number and/or function.

Of the described primary immunodeficiency diseases of phagocytic number or function, recurrent or invasive candidal disease has been reported in cases of chronic granulomatous disease and myeloperoxidase deficiency [1] and GATA2 deficiency [88]. Candidosis is reported but tends to be less common in leukocyte adhesion deficiency and congenital neutropaenic syndromes [1].

Chronic granulomatous disease (CGD) occurs as a result of defects in components of the NADPH oxidase system, resulting in defective neutrophil oxidative burst and susceptibility to a narrow range of organisms, particularly those which are catalase-producing. As well as the predisposition to infection, patients with CGD develop a hyperinflammatory response and granuloma formation [89]. X-linked CGD occurs due to mutations in the *CYBB* gene which encodes the NADPH oxidase complex component $gp91^{phox}$ [86]. Autosomal recessive forms of CGD are less common, and occur due to mutations in the *NCF1*, *CYBA*, *NCF4* or *NCF2* genes, which encode for other components of the complex, namely $p47^{phox}$, $p22^{phox}$, $p40^{phox}$ and $p67^{phox}$, respectively [86, 89].

Candidosis is well described in CGD patients, with candidal species implicated in episodes of meningitis, fungaemia, suppurative adenitis, pneumonia, subcutaneous abscesses and liver abscess reported in a cohort of 368 patients with CGD [90]. Although the majority of these infections were expected to be due to underlying, impaired phagocytic function, additional factors such as steroid use likely increase the risk of invasive candidiasis. Candidal oesophagitis, keratitis and disseminated infection (particularly affecting young infants) have also been described, however mucocutaneous candidiasis is uncommon in CGD patients [1].

Patients with $gp40^{phox}$ mutations have been noted to have a distinct clinical phenotype as compared with those with other forms of CGD, with a milder clinical course and lower frequency of invasive fungal infection [91]. There is no impairment in the ability of the neutrophils of affected patients to kill candida, suggesting residual NADPH oxidase activity and a potential $gp40^{phox}$-independent process for reactive oxygen species production. Furthermore, monocyte and monocyte-derived macrophage NADPH oxidase generation appears to occur independently of $gp40^{phox}$ [91]. In patients with CGD, a correlation has been shown between residual production of reactive oxygen intermediates (ROI) and improved long-term survival [92]. The specific mutation in NADPH oxidase predicts the amount of residual production of ROI [92].

CGD may be conservatively managed with antibiotic and antifungal prophylaxis, along with adjunctive therapies including subcutaneous interferon therapy. CGD is curable by haematopoietic stem cell transplantation (HSCT), and trials are underway to evaluate the role of gene therapy as an alternative definitive management strategy [93].

MPO deficiency is autosomal recessive with variable penetrance, may be complete or partial, and has an estimated incidence of between 1:2000 and 1:4000 individuals [94]. Most individuals are clinically asymptomatic, although severe infections are reported in around 5% of those affected. MPO-deficient phagocytes have an impaired capacity to kill *C. albicans*, as evidenced by severe infection in MPO-deficient mice.

GATA2 encodes a zinc finger transcription factor which is critical for haematopoetic cell development [95]. Mutations in this gene give rise to a syndrome also known as 'MonoMac', which refers to the monocytopaenia and predisposition to mycobacterial infection which are characteristic of this condition [95, 96]. In addition, affected patients have other haematological anomalies including thrombocytopaenia and neutropaenia, predisposition to haematological malignancy and severe mycobacterial, fungal and human papilloma viral infections [88, 96]. In a recent study of 79 French and Belgian patients with *GATA2* mutations, 16 patients were reported to have had 18 episodes of fungal infection, 5 of which were candidoses [88]. Eight of the 18 infections were associated with chemotherapy or HSCT. The neutrophils from some *GATA2* deficient patients were noted to have reduced granularity [97]. When stimulated with PHA (phytohaemagglutinin), patient PBMCs (peripheral blood mononuclear cells) demonstrated reduced lymphocyte proliferative and cytokine production capacity, which normalised after addition of monocytes [96], highlighting the important role of these cells in eliciting an effective immune response.

In addition to the critical role of phagocytes in anti-fungal immunity, defects in other immune cells and immunologic pathways also give rise to susceptibility to infection with candida and other fungi. A range of single-gene inborn errors of immunity resulting in severe or recurrent superficial or invasive candidiasis have been described [86, 98]. Cell-mediated immunity is essential for anti-fungal immunity. This is evidenced by the predisposition to severe fungal infection in infants with severe combined immunodeficiency (SCID), a life-threatening condition manifested by low, absent or severely dysfunctional T cells [86]. Other forms of combined immunodeficiency, for example, those due to deficiencies in CD25, NEMO/IKBG, DOCK8, TCR-α, ORAI1, MST1/STK4, MHC Class II, along with *IKBA* gain of function mutations and idiopathic CD4+ T cell lymphopaenia are associated with chronic mucocutaneous candidiasis (CMC) [98]. In addition, CMC is a feature of several PID with syndromic features, including STAT3 deficiency (autosomal dominant hyper-immunoglobulin E syndrome), APECED (autoimmune polyendocrinopathy-candidiasis-ectodermal dysplasia), also known as APS-1 (autoimmune polyglandular syndrome type 1) which occurs due to mutations in the *AIRE* gene), and deficiencies of IL12Rβ, IL-12p40 and CARD9 [98, 99]. The importance of the TH17 pathway and IL-17 signalling in anti-candidal immunity has become apparent [100, 101], with severe CMC described in patients with deficiencies of IL-17RA, IL-17F, RORC and *STAT1* gain of function mutations [98, 102]. In particular, AIRE has been demonstrated to have a key role in anti-candidal immunity, as evidenced by its role in fungal synapse formation which is required for initial macrophage recognition of fungal hyphae [103]. AIRE, along with Dectin-1, Dectin-2, Syk and CARD9 are required for formation of the fungal synapse upon stimulation of macrophage-like THP-1 cells after stimulation with *C. albicans* [103].

9. Secondary immunodeficiency diseases associated with disorders of phagocyte number or function

Immunosuppression is a well-described risk factor for infection with candida and other fungal species [98]. Corticosteroids are commonly used in the management of a range of inflammatory and malignant conditions, and use of these agents is a known risk factor for fungal infection [104]. The precise mechanisms by which corticosteroids lead to impaired anti-candidal immunity remain unclear, and this is likely multifactorial [105]. In terms of phagocytic cell function, corticosteroids appear to alter leukocyte differentiation programs. They induce monocytes and macrophages to adopt an anti-inflammatory phenotype. This is modulated by the cytokine environment (including increased IL-10 expression on macrophages), increased apoptotic activity and induction of transcription of anti-inflammatory genes which impact upon chemotaxis, phagocytosis and resistance to oxidative stress [105]. However, despite these observations it has been recently shown that dexamethasone increases the expression of CRIg on human MDMs but not CR3 or CR4, and that this increase was associated with an increase in phagocytosis of complement opsonised *C. albicans* [23, 26, 27].

Cancer patients are at an increased risk of systemic candidiasis, and *C. albicans* is reported to be one of the most common causes of sepsis in this patient group [104]. This predisposition to fungal infection is multifactorial, and may be due to a secondary immunodeficiency caused by the underlying malignancy itself, or due to the effects of chemotherapeutic agents. Chemotherapeutic drugs may induce neutropaenia or affect neutrophil function, thereby impairing anti-candidal immunity. Neutrophil function may be impaired as a result of reduced trafficking, chemotaxis or phagocytic activity. For example, chemotherapeutics targeting microtubule structures likely impair cytoskeletal processes and actin polymerisation, thereby reducing neutrophil chemotaxis and phagocytosis. Chemotherapeutic agents can also interfere in NETosis, which is important for antimicrobial activity. Some drugs may also induce monocytopaenia and impaired monocytic function, further increasing the risk of candidal infection [104].

Patients with liver disease are at an increased risk of fungal infection. Those with cirrhosis have been found to have reduced complement levels and impaired monocyte activation and neutrophil mobilisation [106]. Patients with liver disease are at risk for infectious peritonitis, and *C. albicans* and *C. neoformans* were amongst the main species isolated in these cases. Renal disease is also a risk factor for invasive fungal disease [104]. Neonatal candidal sepsis has been reported in association with jaundice [107]. Interestingly, unconjugated bilirubin in hyperbilirubinemia has also been linked to reduced phagocytic cell function; phagocytosis and killing of fungi [108, 109]. Burns patients are at increased risk of fungal infection owing to a breached skin barrier and use of antimicrobial agents, with candidal infection in particular being associated with increased morbidity and mortality in these patients [106]. In addition to these disease states, other physical factors, alone or in combination, such as the use of intravenous catheters and mechanical ventilation also increase the risk of invasive fungal disease [98, 104].

Finally, it is also evident that anti-fungal drugs *per se* can compromise immunity [109–111]. Several of the imidazoles were found to inhibit neutrophil functions, chemotaxis, phagocytosis and microbial killing of bacteria and candida [110].

Acknowledgements

We are grateful to Christ Stewart for technical assistance with the mouse work. We are also indebted to our colleagues who have contributed to the listed publications. Our research has been supported by grants obtained from the NHMRC of Australia and the Women's and Children's Hospital Network, South Australia.

Conflicts of interest

Authors AGS and JRK declare no conflicts of interest. Authors DAR and AF declare that they are inventors on patent relating to TNF_{70-80} technology.

Author details

Annabelle G. Small[1,2], Jovanka R. King[1,3], Deborah A. Rathjen[1,4] and Antonio Ferrante[1,2]*

*Address all correspondence to: antonio.ferrante@adelaide.edu.au

1 Department of Immunopathology, SA Pathology at Women's and Children's Hospital Campus, The Robinson Research Institute and School of Medicine, University of Adelaide, South Australia, Australia

2 School of Biological Science, University of Adelaide, South Australia, Australia

3 Division of Clinical Immunology, Department of Laboratory Medicine, Karolinska University Hospital Huddinge, Stockholm, Sweden

4 Bionomics, Pty Ltd, Thebarton, South Australia, Australia

References

[1] Antachopoulos C, Walsh TJ, Roilides E. Fungal infections in primary immunodeficiencies. European Journal of Pediatrics. 2007;**166**(11):1099-1117. DOI: 10.1007/s00431-007-0527-7

[2] Ferrante A. In: Kaufmann S, Steward M, editors. Phagocytes Part 2: Neutrophils. 10th ed. London: Hodder Arnold; 2005. pp. 35-54

[3] Quinn MT, Gauss KA. Structure and regulation of the neutrophil respiratory burst oxidase: Comparison with nonphagocyte oxidases. Journal of Leukocyte Biology. 2004;**76**(4):760-781. DOI: 10.1189/jlb.0404216

[4] Brown GD, Gordon S. In: Kaufmann S, Steward M, editors. Phagocytes Part 1: Macrophages. 10th ed. London: Hodder Arnold; 2005. pp. 21-34

[5] Brinkmann V, Reichard U, Goosmann C, Fauler B, Uhlemann Y, Weiss DS, et al. Neutrophil extracellular traps kill bacteria. Science. 2004;**303**(5663):1532-1535. DOI: 10.1126/science.1092385

[6] Fuchs TA, Abed U, Goosmann C, Hurwitz R, Schulze I, Wahn V, et al. Novel cell death program leads to neutrophil extracellular traps. The Journal of Cell Biology. 2007;**176**(2):231-241. DOI: 10.1083/jcb.200606027

[7] Small AG, Al-Baghdadi M, Quach A, Hii C, Ferrante A. Complement receptor immunoglobulin: A control point in infection and immunity, inflammation and cancer. Swiss Medical Weekly. 2016;**146**:w14301. DOI: 10.4414/smw.2016.14301

[8] Helmy KY, Katschke KJ Jr, Gorgani NN, Kljavin NM, Elliott JM, Diehl L, et al. CRIg: A macrophage complement receptor required for phagocytosis of circulating pathogens. Cell. 2006;**124**(5):915-927. DOI: 10.1016/j.cell.2005.12.039

[9] Zeng Z, Surewaard BG, Wong CH, Geoghegan JA, Jenne CN, Kubes P. CRIg functions as a macrophage pattern recognition receptor to directly bind and capture blood-borne gram-positive bacteria. Cell Host & Microbe. 2016;**20**(1):99-106. DOI: 10.1016/j.chom.2016.06.002

[10] Ray TL, Wuepper KD. Activation of the alternative (properdin) pathway of complement by *Candida albicans* and related species. The Journal of Investigative Dermatology. 1976;**67**(6):700-703

[11] Thong YH, Ferrante A. Alternative pathway of complement activation by *Candida albicans*. Australian and New Zealand Journal of Medicine. 1978;**8**(6):620-622

[12] Thong YH, Ness D, Ferrante A. Effect of bilirubin on the fungicidal capacity of human neutrophils. Sabouraudia. 1979;**17**(2):125-129

[13] Hurley DL, Fauci AS. Disseminated candidiasis. I. An experimental model in the Guinea pig. The Journal of Infectious Diseases. 1975;**131**(5):516-527

[14] Maibach HI, Kligman AM. The biology of experimental human cutaneous moniliasis (albicans). Archives of Dermatology. 1962;**85**(2):233-257. DOI: 10.1001/archderm.1962.01590020073009

[15] Dillon S, Agrawal S, Banerjee K, Letterio J, Denning TL, Oswald-Richter K, et al. Yeast zymosan, a stimulus for TLR2 and dectin-1, induces regulatory antigen-presenting cells and immunological tolerance. The Journal of Clinical Investigation. 2006;**116**(4):916-928. DOI: 10.1172/JCI27203

[16] Marodi L, Korchak HM, Johnston RB Jr. Mechanisms of host defense against Candida species. I. Phagocytosis by monocytes and monocyte-derived macrophages. Journal of Immunology. 1991;**146**(8):2783-2789

[17] Quach A, Ferrante A. The application of dextran sedimentation as an initial step in neutrophil purification promotes their stimulation, due to the presence of monocytes. Journal of Immunology Research. 2017;**2017**:10. DOI: 10.1155/2017/1254792

[18] Small A, Lansdown N, Al-Baghdadi M, Quach A, Ferrante A. Facilitating THP-1 macrophage studies by differentiating and investigating cell functions in polystyrene test tubes. Journal of Immunological Methods. 2018;**461**:73-77. DOI: 10.1016/j.jim.2018.06.019

[19] Zhang MX, Kozel TR. Mannan-specific immunoglobulin G antibodies in normal human serum accelerate binding of C3 to *Candida albicans* via the alternative complement pathway. Infection and Immunity. 1998;**66**(10):4845-4850

[20] Kozel TR. Activation of the complement system by pathogenic fungi. Clinical Microbiology Reviews. 1996;**9**(1):34-46

[21] Meri T, Hartmann A, Lenk D, Eck R, Wurzner R, Hellwage J, et al. The yeast *Candida albicans* binds complement regulators factor H and FHL-1. Infection and Immunity. 2002;**70**(9):5185-5192

[22] Meri T, Blom AM, Hartmann A, Lenk D, Meri S, Zipfel PF. The hyphal and yeast forms of *Candida albicans* bind the complement regulator C4b-binding protein. Infection and Immunity. 2004;**72**(11):6633-6641. DOI: 10.1128/iai.72.11.6633-6641.2004

[23] Munawara U, Small AG, Quach A, Gorgani NN, Abbott CA, Ferrante A. Cytokines regulate complement receptor immunoglobulin expression and phagocytosis of *Candida albicans* in human macrophages: A control point in anti-microbial immunity. Scientific Reports. 2017;**7**(1):4050. DOI: 10.1038/s41598-017-04325-0

[24] Gorgani NN, He JQ, Katschke KJ Jr, Helmy KY, Xi H, Steffek M, et al. Complement receptor of the Ig superfamily enhances complement-mediated phagocytosis in a subpopulation of tissue resident macrophages. Journal of Immunology. 2008;**181**(11):7902-7908

[25] Irvine KM, Banh X, Gadd VL, Wojcik KK, Ariffin JK, Jose S, et al. CRIg-expressing peritoneal macrophages are associated with disease severity in patients with cirrhosis and ascites. JCI Insight. 2016;**1**(8):e86914. DOI: 10.1172/jci.insight.86914

[26] Gorgani NN, Thathaisong U, Mukaro VR, Poungpair O, Tirimacco A, Hii CS, et al. Regulation of CRIg expression and phagocytosis in human macrophages by arachidonate, dexamethasone, and cytokines. The American Journal of Pathology. 2011;**179**(3):1310-1318. DOI: 10.1016/j.ajpath.2011.05.021

[27] Ma Y, Usuwanthim K, Munawara U, Quach A, Gorgani NN, Abbott CA, et al. Protein kinase Cα regulates the expression of complement receptor Ig in human monocyte–derived macrophages. The Journal of Immunology. 2015;**194**(6):2855

[28] Herre J, Marshall AS, Caron E, Edwards AD, Williams DL, Schweighoffer E, et al. Dectin-1 uses novel mechanisms for yeast phagocytosis in macrophages. Blood. 2004;**104**(13):4038-4045. DOI: 10.1182/blood-2004-03-1140

[29] Herre J, Willment JA, Gordon S, Brown GD. The role of Dectin-1 in antifungal immunity. Critical Reviews in Immunology. 2004;**24**(3):193-203

[30] Miramon P, Kasper L, Hube B. Thriving within the host: Candida spp. interactions with phagocytic cells. Medical Microbiology and Immunology. 2013;**202**(3):183-195. DOI: 10.1007/s00430-013-0288-z

[31] Naglik JR. Candida immunity. New Journal of Science. 2014;**2014**:27. DOI: 10.1155/2014/390241

[32] Lowman DW, Greene RR, Bearden DW, Kruppa MD, Pottier M, Monteiro MA, et al. Novel structural features in *Candida albicans* hyphal glucan provide a basis for differential innate immune recognition of hyphae versus yeast. The Journal of Biological Chemistry. 2014;**289**(6):3432-3443. DOI: 10.1074/jbc.M113.529131

[33] Cheng SC, Joosten LA, Kullberg BJ, Netea MG. Interplay between *Candida albicans* and the mammalian innate host defense. Infection and Immunity. 2012;**80**(4):1304-1313. DOI: 10.1128/iai.06146-11

[34] Jouault T, Ibata-Ombetta S, Takeuchi O, Trinel PA, Sacchetti P, Lefebvre P, et al. *Candida albicans* phospholipomannan is sensed through toll-like receptors. The Journal of Infectious Diseases. 2003;**188**(1):165-172. DOI: 10.1086/375784

[35] Tada H, Nemoto E, Shimauchi H, Watanabe T, Mikami T, Matsumoto T, et al. Saccharomyces cerevisiae- and *Candida albicans*-derived mannan induced production of tumor necrosis factor alpha by human monocytes in a CD14- and Toll-like receptor 4-dependent manner. Microbiology and Immunology. 2002;**46**(7):503-512

[36] Netea MG, Gow NAR, Munro CA, Bates S, Collins C, Ferwerda G, et al. Immune sensing of *Candida albicans* requires cooperative recognition of mannans and glucans by lectin and Toll-like receptors. Journal of Clinical Investigation. 2006;**116**(6):1642-1650. DOI: 10.1172/JCI27114

[37] Miramón P, Dunker C, Windecker H, Bohovych IM, Brown AJP, Kurzai O, et al. Cellular responses of *Candida albicans* to phagocytosis and the extracellular activities of neutrophils are critical to counteract carbohydrate starvation, oxidative and nitrosative stress. PLoS One. 2012;**7**(12):e52850. DOI: 10.1371/journal.pone.0052850

[38] Sudbery PE. Growth of *Candida albicans* hyphae. Nature Reviews Microbiology. 2011; **9**:737. DOI: 10.1038/nrmicro2636

[39] Moyes DL, Runglall M, Murciano C, Shen C, Nayar D, Thavaraj S, et al. A biphasic innate immune MAPK response discriminates between the yeast and hyphal forms of *Candida albicans* in epithelial cells. Cell Host & Microbe. 2010;**8**(3):225-235. DOI: 10.1016/j.chom.2010.08.002

[40] Weindl G, Naglik JR, Kaesler S, Biedermann T, Hube B, Korting HC, et al. Human epithelial cells establish direct antifungal defense through TLR4-mediated signaling. The Journal of Clinical Investigation. 2007;**117**(12):3664-3672. DOI: 10.1172/jci28115

[41] Cassone A. The case for an expanded concept of trained immunity. MBio. 2018;**9**(3). DOI: 10.1128/mBio.e00570-18

[42] Netea M, Quintin J, van der Meer JW. Trained immunity: A memory for innate host defense. Cell Host & Microbe. 2011;**9**(5):355-361. DOI: 10.1016/j.chom.2011.04.006

[43] Durrant WE, Dong X. Systemic acquired resistance. Annual Review of Phytopathology. 2004;**42**:185-209. DOI: 10.1146/annurev.phyto.42.040803.140421

[44] Pham LN, Dionne MS, Shirasu-Hiza M, Schneider DS. A specific primed immune response in Drosophila is dependent on phagocytes. PLoS Pathogens. 2007;**3**(3):e26. DOI: 10.1371/journal.ppat.0030026

[45] Bistoni F, Vecchiarelli A, Cenci E, Puccetti P, Marconi P, Cassone A. Evidence for macrophage-mediated protection against lethal *Candida albicans* infection. Infection and Immunity. 1986;**51**(2):668-674

[46] Quintin J, Saeed S, Martens JHA, Giamarellos-Bourboulis EJ, Ifrim DC, Logie C, et al. *Candida albicans* infection affords protection against reinfection via functional reprogramming of monocytes. Cell Host & Microbe. 2012;**12**(2):223-232. DOI: 10.1016/j.chom.2012.06.006

[47] Akramiene D, Kondrotas A, Didziapetriene J, Kevelaitis E. Effects of beta-glucans on the immune system. Medicina (Kaunas, Lithuania). 2007;**43**(8):597-606

[48] Rizzetto L, Ifrim DC, Moretti S, Tocci N, Cheng S-C, Quintin J, et al. Fungal chitin induces trained immunity in human monocytes during cross-talk of the host with *Saccharomyces cerevisiae*. The Journal of Biological Chemistry. 2016;**291**(15):7961-7972. DOI: 10.1074/jbc.M115.699645

[49] Garcia-Valtanen P, Guzman-Genuino RM, Williams DL, Hayball JD, Diener KR. Evaluation of trained immunity by beta-1, 3 (d)-glucan on murine monocytes in vitro and duration of response in vivo. Immunology and Cell Biology. 2017;**95**(7):601-610. DOI: 10.1038/icb.2017.13

[50] Ifrim DC, Quintin J, Meerstein-Kessel L, Plantinga TS, Joosten LA, van der Meer JW, et al. Defective trained immunity in patients with STAT-1-dependent chronic mucocutaneaous candidiasis. Clinical and Experimental Immunology. 2015;**181**(3):434-440. DOI: 10.1111/cei.12642

[51] Alexander MP, Fiering SN, Ostroff GR, Cramer RA, Mullins DW. Beta-glucan-induced trained innate immunity mediates antitumor efficacy in the mouse lung. The Journal of Immunology. 2016;**196**(1 Supplement):142.124

[52] Ferrante A. Tumor necrosis factor alpha potentiates neutrophil antimicrobial activity: Increased fungicidal activity against Torulopsis glabrata and *Candida albicans* and associated increases in oxygen radical production and lysosomal enzyme release. Infection and Immunity. 1989b;**57**(7):2115-2122

[53] Thompson HL, Wilton JM. Interaction and intracellular killing of *Candida albicans* blastospores by human polymorphonuclear leucocytes, monocytes and monocyte-derived macrophages in aerobic and anaerobic conditions. Clinical and Experimental Immunology. 1992;**87**(2):316-321

[54] Leijh PCJ, van den Barselaar MT, van Furth R. Kinetics of phagocytosis and intracellular killing of *Candida albicans* by human granulocytes and monocytes. Infection and Immunity. 1977;**17**(2):313-318

[55] Vázquez-Torres A, Balish E. Macrophages in resistance to candidiasis. Microbiology and Molecular Biology Reviews. 1997;**61**(2):170-192

[56] Lehrer RI. The fungicidal mechanisms of human monocytes. I. Evidence for myeloperoxidase-linked and myeloperoxidase-independent candidacidal mechanisms. The Journal of Clinical Investigation. 1975;**55**(2):338-346. DOI: 10.1172/JCI107937

[57] Lehrer RI, Cline MJ. Leukocyte myeloperoxidase deficiency and disseminated candidiasis: The role of myeloperoxidase in resistance to Candida infection. The Journal of Clinical Investigation. 1969;**48**(8):1478-1488. DOI: 10.1172/JCI106114

[58] Lefkowitz SS, Gelderman MP, Lefkowitz DL, Moguilevsky N, Bollen A. Phagocytosis and intracellular killing of *Candida albicans* by macrophages exposed to myeloperoxidase. The Journal of Infectious Diseases. 1996;**173**(5):1202-1207

[59] Aratani Y, Kura F, Watanabe H, Akagawa H, Takano Y, Suzuki K, et al. Relative contributions of myeloperoxidase and NADPH-oxidase to the early host defense against

pulmonary infections with *Candida albicans* and *Aspergillus fumigatus*. Medical Mycology. 2002;**40**(6):557-563

[60] Diamond RD, Krzesicki R, Jao W. Damage to pseudohyphal forms of *Candida albicans* by neutrophils in the absence of serum in vitro. The Journal of Clinical Investigation. 1978;**61**(2):349-359. DOI: 10.1172/jci108945

[61] Gazendam RP, van Hamme JL, Tool ATJ, van Houdt M, Verkuijlen PJJH, Herbst M, et al. Two independent killing mechanisms of *Candida albicans* by human neutrophils: Evidence from innate immunity defects. Blood. 2014;**124**(4):590

[62] Hii CST, Stacey K, Moghaddami N, Murray AW, Ferrante A. Role of the extracellular signal-regulated protein kinase Cascade in human neutrophil killing of *Staphylococcus aureus* and *Candida albicans* and in migration. Infection and Immunity. 1999;**67**(3):1297-1302

[63] Li X, Cullere X, Nishi H, Saggu G, Durand E, Mansour MK, et al. PKC-delta activation in neutrophils promotes fungal clearance. Journal of Leukocyte Biology. 2016;**100**(3):581-588. DOI: 10.1189/jlb.4A0915-405R

[64] Strijbis K, Tafesse FG, Fairn GD, Witte MD, Dougan SK, Watson N, et al. Bruton's tyrosine kinase (BTK) and Vav1 contribute to Dectin1-dependent phagocytosis of *Candida albicans* in macrophages. PLoS Pathogens. 2013;**9**(6):e1003446. DOI: 10.1371/journal.ppat.1003446

[65] Steinberg BE, Grinstein S. Unconventional roles of the NADPH oxidase: Signaling, ion homeostasis, and cell death. Science's STKE. 2007;**2007**(379):pe11. DOI: 10.1126/stke.3792007pe11

[66] Byrd AS, O'Brien XM, Johnson CM, Lavigne LM, Reichner JS. An extracellular matrix-based mechanism of rapid neutrophil extracellular trap formation in response to *Candida albicans*. Journal of Immunology. 2013;**190**(8):4136-4148. DOI: 10.4049/jimmunol.1202671

[67] Kenno S, Perito S, Mosci P, Vecchiarelli A, Monari C. Autophagy and reactive oxygen species are involved in neutrophil extracellular traps release induced by *C. albicans* morphotypes. Frontiers in Microbiology. 2016;**7**:879

[68] Kumaratilake LM, Ferrante A, Kumaratilake JS, Allison AC. Extraction of intraerythrocytic malarial parasites by phagocytic cells. Parasitology Today. 1994;**10**(5):193-196. DOI: 10.1016/0169-4758(94)90029-9

[69] Ferrante A, Mocatta TJ. Human neutrophils require activation by mononuclear leucocyte conditioned medium to kill the pathogenic free-living amoeba, *Naegleria fowleri*. Clinical and Experimental Immunology. 1984;**56**(3):559-566

[70] Ferrante A. Augmentation of the neutrophil response to *Naegleria fowleri* by tumor necrosis factor alpha. Infection and Immunity. 1989a;**57**(10):3110-3115

[71] Ferrante A. In: Coffey RG, editor. Activation of Neutrophils by Interleukins-1 and -2 and Tumour Necrosis Factors. New York, USA: Marcel Dekker Inc; 1992. pp. 417-436

[72] Murphy JW, Wu-Hsieh BA, Singer-Vermes LM, Ferrante A, Moser S, Russo M, et al. Cytokines in the host response to mycotic agents. Journal of Medical and Veterinary Mycology. 1994;**32**(Suppl 1):203-210

[73] Kowanko IC, Ferrante A, Harvey DP, Carman KL. Granulocyte-macrophage colony-stimulating factor augments neutrophil killing of *Torulopsis glabrata* and stimulates neutrophil respiratory burst and degranulation. Clinical and Experimental Immunology. 1991;**83**(2):225-230

[74] Kowanko IC, Ferrante A, Clemente G, Kumaratilake LM. Tumor necrosis factor primes neutrophils to kill *Staphylococcus aureus* by an oxygen-dependent mechanism and *Plasmodium falciparum* by an oxygen-independent mechanism. Infection and Immunity. 1996;**64**(8):3435-3437

[75] Ferrante A, Hill NL, Abell TJ, Pruul H. Role of myeloperoxidase in the killing of *Naegleria fowleri* by lymphokine-altered human neutrophils. Infection and Immunity. 1987;**55**(5):1047-1050

[76] Ferrante A, Martin AJ, Bates EJ, Goh DH, Harvey DP, Parsons D, et al. Killing of *Staphylococcus aureus* by tumor necrosis factor-alpha-activated neutrophils. The role of serum opsonins, integrin receptors, respiratory burst, and degranulation. Journal of Immunology. 1993;**151**(9):4821-4828

[77] Rathjen DA, Ferrante A, Aston R. Differential effects of small tumour necrosis factor-alpha peptides on tumour cell cytotoxicity, neutrophil activation and endothelial cell procoagulant activity. Immunology. 1993;**80**(2):293-299

[78] Kumaratilake LM, Rathjen DA, Mack P, Widmer F, Prasertsiriroj V, Ferrante A. A synthetic tumor necrosis factor-alpha agonist peptide enhances human polymorphonuclear leukocyte-mediated killing of *Plasmodium falciparum* in vitro and suppresses *Plasmodium chabaudi* infection in mice. Journal of Clinical Investigation. 1995;**95**(5):2315-2323

[79] Britton WJ, Meadows N, Rathjen DA, Roach DR, Briscoe H. A tumor necrosis factor mimetic peptide activates a murine macrophage cell line to inhibit mycobacterial growth in a nitric oxide-dependent fashion. Infection and Immunity. 1998;**66**(5):2122-2127

[80] Roach DR, Briscoe H, Baumgart K, Rathjen DA, Britton WJ. Tumor necrosis factor (TNF) and a TNF-mimetic peptide modulate the granulomatous response to *Mycobacterium bovis* BCG infection in vivo. Infection and Immunity. 1999;**67**(10):5473-5476

[81] Mukaro VR et al. Small tumor necrosis factor receptor biologics inhibit the tumor necrosis factor-p38 signalling axis and inflammation. Nature Communications. 2018;**9**(1):1365

[82] Maródi L, Schreiber S, Anderson DC, MacDermott RP, Korchak HM, Johnston RB. Enhancement of macrophage candidacidal activity by interferon-gamma. Increased phagocytosis, killing, and calcium signal mediated by a decreased number of mannose receptors. Journal of Clinical Investigation. 1993;**91**(6):2596-2601

[83] Brummer E, Morrison CJ, Stevens DA. Recombinant and natural gamma-interferon activation of macrophages in vitro: Different dose requirements for induction of killing activity against phagocytizable and nonphagocytizable fungi. Infection and Immunity. 1985;**49**(3):724-730

[84] Roilides E, Lyman CA, Sein T, Gonzalez C, Walsh TJ. Antifungal activity of splenic, liver and pulmonary macrophages against *Candida albicans* and effects of macrophage colony-stimulating factor. Medical Mycology. 2000;**38**(2):161-168

[85] Kumaratilake LM, Ferrante A. IL-4 inhibits macrophage-mediated killing of *Plasmodium falciparum* in vitro. A possible parasite-immune evasion mechanism. Journal of Immunology. 1992;**149**(1):194-199

[86] Bousfiha A, Jeddane L, Picard C, Ailal F, Bobby Gaspar H, Al-Herz W, et al. The 2017 IUIS phenotypic classification for primary immunodeficiencies. Journal of Clinical Immunology. 2018;**38**(1):129-143. DOI: 10.1007/s10875-017-0465-8

[87] Antachopoulos C. Invasive fungal infections in congenital immunodeficiencies. Clinical Microbiology and Infection. 2010;**16**(9):1335-1342. DOI: 10.1111/j.1469-0691.2010.03289.x

[88] Donadieu J, Lamant M, Fieschi C, Sicre de Fontbrune F, Caye A, Ouachee M, et al. Natural history of GATA2 deficiency in a survey of 79 French and Belgian patients. Haematologica. 2018;**103**(8):1278-1287

[89] Rider NL, Jameson MB, Creech CB. Chronic granulomatous disease: Epidemiology, pathophysiology, and genetic basis of disease. Journal of the Pediatric Infectious Diseases Society. 2018;**7**(suppl_1):S2-S5. DOI: 10.1093/jpids/piy008

[90] Winkelstein JA, Marino MC, Johnston RB, Boyle J, Curnutte J, Gallin JI, et al. Chronic granulomatous disease. Report on a national registry of 368 patients. Medicine (Baltimore). 2000;**79**(3):155-169

[91] van de Geer A, Nieto-Patlan A, Kuhns DB, Tool AT, Arias AA, Bouaziz M, et al. Inherited p40phox deficiency differs from classic chronic granulomatous disease. The Journal of Clinical Investigation. 2018;**128**(9):3957-3975. DOI: 10.1172/jci97116

[92] Kuhns DB, Alvord WG, Heller T, Feld JJ, Pike KM, Marciano BE, et al. Residual NADPH oxidase and survival in chronic granulomatous disease. The New England Journal of Medicine. 2010;**363**(27):2600-2610. DOI: 10.1056/NEJMoa1007097

[93] Keller MD, Notarangelo LD, Malech HL. Future of care for patients with chronic granulomatous disease: Gene therapy and targeted molecular medicine. Journal of the Pediatric Infectious Diseases Society. 2018;**7**(suppl_1):S40-S44. DOI: 10.1093/jpids/piy011

[94] Milligan KL, Mann D, Rump A, Anderson VL, Hsu AP, Kuhns DB, et al. Complete myeloperoxidase deficiency: Beware the "false-positive" dihydrorhodamine oxidation. The Journal of Pediatrics. 2016;**176**:204-206. DOI: 10.1016/j.jpeds.2016.05.047

[95] Spinner MA, Sanchez LA, Hsu AP, Shaw PA, Zerbe CS, Calvo KR, et al. GATA2 deficiency: A protean disorder of hematopoiesis, lymphatics, and immunity. Blood. 2014;**123**(6):809-821. DOI: 10.1182/blood-2013-07-515528

[96] Vinh DC, Patel SY, Uzel G, Anderson VL, Freeman AF, Olivier KN, et al. Autosomal dominant and sporadic monocytopenia with susceptibility to mycobacteria, fungi, papillomaviruses, and myelodysplasia. Blood. 2010;**115**(8):1519-1529

[97] Li J, Vinh DC, Casanova J-L, Puel A. Inborn errors of immunity underlying fungal diseases in otherwise healthy individuals. Current Opinion in Microbiology. 2017;**40**:46-57. DOI: 10.1016/j.mib.2017.10.016

[98] Lanternier F, Cypowyj S, Picard C, Bustamante J, Lortholary O, Casanova JL, et al. Primary immunodeficiencies underlying fungal infections. Current Opinion in Pediatrics. 2013;**25**(6):736-747. DOI: 10.1097/mop.0000000000000031

[99] Alves de Medeiros AK, Lodewick E, Bogaert DJ, Haerynck F, Van Daele S, Lambrecht B, et al. Chronic and invasive fungal infections in a family with CARD9 deficiency. Journal of Clinical Immunology. 2016;**36**(3):204-209. DOI: 10.1007/s10875-016-0255-8

[100] Conti HR, Bruno VM, Childs EE, Daugherty S, Hunter JP, Mengesha BG, et al. IL-17 receptor signaling in oral epithelial cells is critical for protection against oropharyngeal candidiasis. Cell Host & Microbe. 2016;**20**(5):606-617. DOI: 10.1016/j.chom.2016.10.001

[101] Patel DD, Kuchroo VK. Th17 cell pathway in human immunity: Lessons from genetics and therapeutic interventions. Immunity. 2015;**43**(6):1040-1051. DOI: 10.1016/j.immuni.2015.12.003

[102] Okada S, Markle JG, Deenick EK, Mele F, Averbuch D, Lagos M, et al. Impairment of immunity to *Candida* and *Mycobacterium* in humans with bi-allelic *RORC* mutations. Science. 2015;**349**(6248):606

[103] Albuquerque JATD, Banerjee PP, Castoldi A, Ma R, Zurro NB, Ynoue LH, et al. The role of AIRE in the immunity against *Candida albicans* in a model of human macrophages. Frontiers in Immunology. 2018;**9**(567). DOI: 10.3389/fimmu.2018.00567

[104] Teoh F, Pavelka N. How chemotherapy increases the risk of systemic candidiasis in cancer patients: Current paradigm and future directions. Pathogens. 2016;**5**(1). DOI: 10.3390/pathogens5010006

[105] Coutinho AE, Chapman KE. The anti-inflammatory and immunosuppressive effects of glucocorticoids, recent developments and mechanistic insights. Molecular and Cellular Endocrinology. 2011;**335**(1):2-13

[106] Pilmis B, Puel A, Lortholary O, Lanternier F. New clinical phenotypes of fungal infections in special hosts. Clinical Microbiology and Infection. 2016;**22**(8):681-687. DOI: 10.1016/j.cmi.2016.05.016

[107] Celebi S, Hacimustafaoglu M, Koksal N, Ozkan H, Cetinkaya M, Ener B. Neonatal candidiasis: Results of an 8 year study. Pediatrics International. 2012;**54**(3):341-349. DOI: 10.1111/j.1442-200X.2012.03574.x

[108] Thong YH, Ferrante A, Ness D. Neutrophil phagocytic and bactericidal dysfunction induced by bilirubin. Australian Paediatric Journal. 1977;**13**(4):287-289

[109] Ferrante A, Thong YH. Amphotericin B-induced immunosuppression in tumor-bearing mice. International Journal of Immunopharmacology. 1979;**1**(4):299-301

[110] Rowan-Kelly B, Ferrante A, Thong YH. Modification of polymorphonuclear leucocyte function by imidazoles. International Journal of Immunopharmacology. 1984;**6**(4):389-393. DOI: 10.1016/0192-0561(84)90059-6

[111] Thong YH, Ferrante A, Secker LK. Suppression of delayed-type hypersensitivity in tumour-bearing mice by treatment with miconazole. Immunology Letters. 1980;**2**(2):119-121. DOI: 10.1016/0165-2478(80)90062-0

Interactions of *Candida albicans* Cells with Aerobic and Anaerobic Bacteria during Formation of Mixed Biofilms in the Oral Cavity

Maria Rapala-Kozik, Marcin Zawrotniak,
Mariusz Gogol, Dominika Bartnicka, Dorota Satala,
Magdalena Smolarz, Justyna Karkowska-Kuleta and
Andrzej Kozik

Additional information is available at the end of the chapter

http://dx.doi.org/10.5772/intechopen.81537

Abstract

Biofilm is a compact coating formed on various artificial and physiologic surfaces by a population of microorganisms which in this habitat establish a close cooperation, exploiting both the physical interactions that stabilize the community and chemical cooperation, engaging numerous agents to modify the environment, i.e., to influence the acidity, nutrient acquisition, or oxygen availability. Microorganisms can also communicate using quorum-sensing molecules carrying specific messages. Some microbes temporarily dominate, while others are constantly replaced by different community members. But these co-operations or competitions have a deep sense—they serve to protect the whole community against the defense system of the host to assure survival. The oral cavity is inhabited by diverse microorganisms, including bacteria, but also yeast-like fungi from the genus *Candida* that stay under a tight control of the host as long as its immune system is not weakened; then these relatively mild commensals convert to dangerous pathogens that start the invasion often in collaboration with other microbes. Elongated hyphal forms of fungal cells favor the biofilm type of growth and communication with other microbes supporting immune resistance of the biofilm. In this chapter, we discuss the mechanisms of interactions between bacteria and *C. albicans* in the oral cavity, their communication, host responses, and possible strategies of anti-biofilm treatment.

Keywords: *Candida albicans*, biofilm, aerobic and anaerobic bacteria, quorum-sensing, host responses, anti-biofilm therapies

1. The oral cavity: a common place for polymicrobial biofilm formation

The oral cavity comprises the most complex niches of the human body colonized by a wide variety of bacteria and fungi species. These commensal or often opportunistic and pathogenic colonizers tend to form biofilms—structured microbial communities, attached to natural or artificial surfaces, which are directly attributable to the virulence of these microorganisms and their ability to cause infections [1].

The variety of microbial species inhabiting the oral cavity results from the presence of two different functional surfaces, the mucosal surface and the teeth, representing various conditions in terms of nutrient and oxygen availability [2]. The microorganisms that early colonize the salivary pellicle on the tooth surface are streptococci such as the oral commensal—*Streptococcus gordonii.* Further biofilm formation involves some bridging microorganisms such as *Fusobacterium nucleatum* [3]. As the biofilm extends below the gum line and becomes subgingival plaque, more pathogenic, Gram-negative anaerobes such as *Porphyromonas gingivalis* or *Tannerella forsythia* are embedded [4].

For polymicrobial growth and survival in the human oral cavity, establishing a well-functioning community, i.e., biofilm, is essential. The formation of biofilm increases a resistance to antimicrobial agents and nutritional changes. However, the transition from planktonic type of growth to biofilm community requires many transcriptional and proteomic changes. Most of them concern co-aggregation/-adhesion processes, sensing diffusible signals, and metabolic interactions. Such a developed biofilm is still exposed to changes of nutrition and oxygen availability, pH fluctuation, antimicrobial properties of saliva, and is also modified by the contact with host tissues [5].

The latest studies of oral microbiome pointed at an opportunistic inhabitant of oral mucosa—the *Candida albicans* yeast-like fungus—as an important biofilm player among microbiota that contacts with mucosal tissues of the host. Under colonization or infection conditions, *C. albicans* adheres to tissues, interacting with a variety of host extracellular matrix molecules that promote adhesion to the host surfaces [6]. The adherence is strictly dependent on *C. albicans* ability to switch morphology between yeast and hyphal forms [7, 8].

Numerous observations supported a hypothesis that fungi have a beneficial or favorable role in maintaining the healthy balance between microbes and the host. On the other hand, *C. albicans* well adapted to constantly changing demands in the human host environment [9] seems to be able to use the different colonizing strategy under situations that emerge in the pathological disparities.

Tracking the yeast oral infections showed that the same initiating and bridging microorganisms composing biofilms were involved in the interaction with hyphal filaments of *C. albicans*, promoting co-colonization of these surfaces by yeast [10]. Moreover, the interactions between yeast and streptococci appear to be synergistic. In addition to providing adhesion sites, streptococci excrete lactate that can act as a carbon source for yeast growth [11]. On the other hand, *C. albicans* may provide bacteria with growth stimulatory factors, resulting from the nutrition metabolism [12] and can reduce the oxygen pressure to the level, preferred by streptococci.

C. albicans also co-aggregates with an obligatory anaerobe, *F. nucleatum* [13], or a facultative anaerobe—*Actinomyces oris* [14].

Current studies [15–17] report that *C. albicans* biofilms protect the obligatory anaerobes, like *P. gingivalis* and *T. forsythia*, under aerobic culture conditions. It is possible because oxygen depletion within the structured fungal biofilm or its fast consumption by fungal cells results in creation of anaerobic micro-niches that help strict anaerobic bacteria to survive and proliferate. The observed depletion of oxygen could depend on the number of *C. albicans* cells or their respiratory rate.

2. Mechanisms and structures involved in formation of *Candida albicans* biofilms

The ability of *C. albicans* to form a biofilm is closely related to the virulence potential of pathogenic form of *C. albicans* and is characterized by a high heterogeneity among different clinical isolates [18, 19]. This form of fungal community depends on the cell surrounding and proceeds via sequential steps. The process starts with the initial adhesion of single yeast-like cells to the artificial or mucosal surface (**Figure 1**) and formation of a basal yeast cells layer [20–22]. During this phase both the surface properties on which microorganism form aggregate, its structure, charge, hydrophobicity, or roughness, as well as the structure of molecules present at the surface of pathogen cells play important roles [23]. In the further cell proliferation step, the fungi develop the filamentous hyphal form of the cells accompanied by production of an extracellular polysaccharide matrix in mature biofilm, which protects them, and strengthens the biofilm structure [24]. These processes lead to a significant increase in the thickness of the biofilm, and its maturation is controlled by at least nine master transcription regulators (Ndt80, Bcr1, Efg1, Rfx2, Flo8, Rob1, Brg1, Gal4, and Tec1) that supervise the network of about 1000 of targeted genes involved in biofilm formation [21, 25]. Then, the dispersion of biofilm-associated yeast-like cells can occur with further fungal cell dissemination, often associated with invasive diseases [24, 26].

Numerous different mechanisms and molecules are involved in the overall complex process of *C. albicans* biofilm formation [27]. Initially, the general physicochemical properties of *C. albicans* cell surface and subsequent activity of cell wall adhesive proteins play extremely important roles, allowing the cells to adhere to the targeted substrates or materials [28]. This essential group of molecules responsible for in vitro and in vivo biofilm development includes several proteins covalently bound to the fungal cell wall and equipped with a signal peptide for classical secretion and glycosylphosphatidylinositol (GPI)-anchor site, i.e., hyphal cell wall protein Hwp1 [29], proteins from agglutinin-like sequence (Als) protein family, such as Als1 and Als3 [28], and hyphally regulated cell wall protein Hyr1 [28]. Their transcription is regulated by the transcription factor Bcr1 [30] and is primarily associated with the morphological transition from yeast cells to filamentous forms, thus implicating their association mainly with the cell wall of hyphae [31–33]. Additionally, other adhesins are required for *C. albicans* adhesion and proper biofilm formation, including Eap1 (enhanced adhesion to polystyrene 1) protein present at the cell surface of both yeast cells and hyphal forms [34, 35]. Adhesion-related proteins are important not only for the binding of fungal cells to the receptors on host

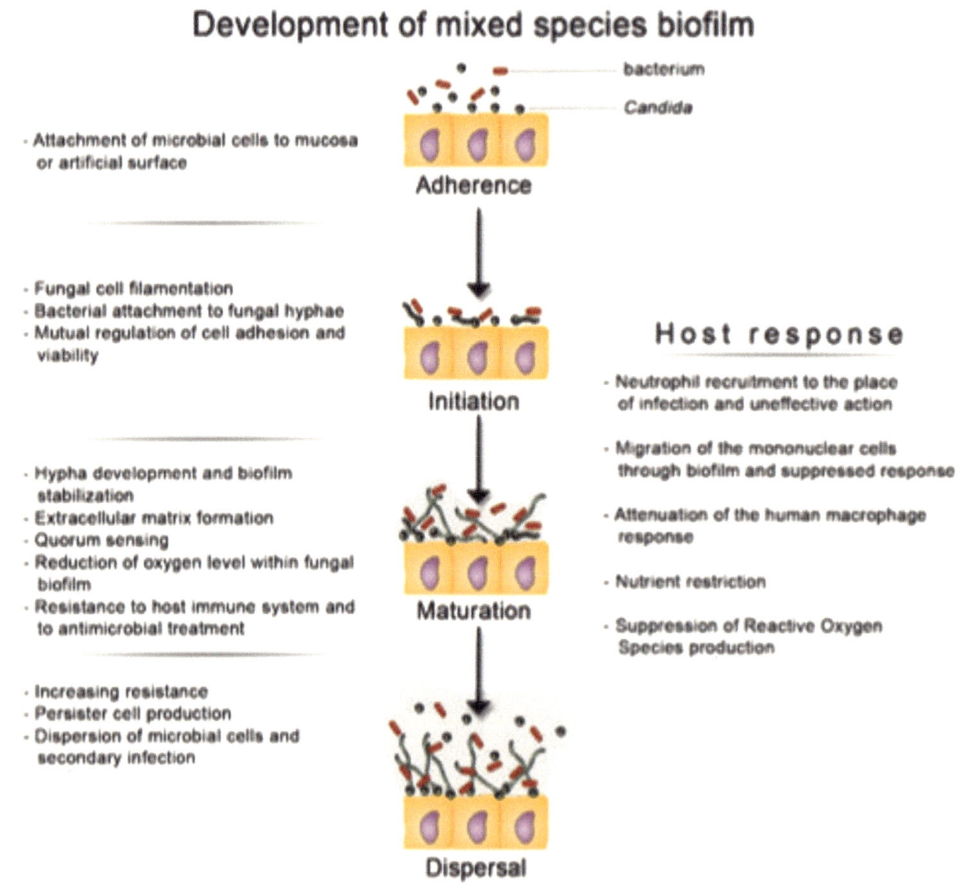

Figure 1. Polymicrobial biofilm: stages of development and host responses.

tissues or to the artificial surfaces, but also for maintaining the cell-cell interactions within the biofilm that allow further stabilization of the structure and avoiding the removal of fungal cells by the action of host defense mechanisms such as the salivary flow. In the process of aggregation and intracellular interactions between fungal cells, fragments of adhesins that consist of amino acid sequences predicted to form amyloid-like β-aggregates and mediating amyloid formation may participate, as described for the protein Rbt1 (repressed by TUP1), a GPI-anchored cell wall protein with a similarity to Hwp1 [36, 37].

After the adhesion step, further proliferation of cells and production of filamentous forms lead to the enhanced development of biofilm [38], processes related not only to the change in the surface properties of fungal cells and the increase of their adhesiveness, but also to the production of further virulence factors and biofilm matrix components [24]. Among the large repertoire of extracellular hydrolytic enzymes produced by *C. albicans* that play a pivotal role during the invasion on host tissues during the infection and are involved in biofilm-related pathogenesis, representatives of families of lipases, phospholipases, and secreted aspartyl proteinases (Saps) can be included [39]. The major biofilm-associated Saps are hypha-specific Sap5 and Sap6, responsible for the acquisition of nutrients, aggregation of fungal cells, intracellular communication, and production of extracellular matrix during biofilm development [40–42].

The highly complicated and heterogeneous extracellular biofilm matrix (ECM) is composed of numerous proteins (55%), carbohydrates (25%), mainly branched α-1,6-mannans, unbranched β-1,6-glucans, and β-1,3-glucans, as well as lipids (15%), including neutral glycerolipids, polar glycerolipids and sphingolipids, and nucleic acids (5%) [43]. The matrix that strengthens the biofilm significantly contributes to the development of biofilm resistance to adverse environmental conditions and penetration of antifungal drugs [43–45]. Several matrix-associated proteins have been identified, both involved in the basic cellular metabolism, as well as the proteins responsible for the rearrangement of the matrix structure and maintenance of its functionality (glucan-modifying enzymes and protein mannosyltransferases, i.e., Xog1, Exg1, Bgl2, Pmt1, Pmt2, Pmt4, Pmt6) [46, 47]. Extracellular DNA (eDNA) detected in *C. albicans* biofilm matrix is probably mainly responsible for the structural integrity [48].

In the process of biofilm dispersion, the molecular chaperone, heat shock protein 90 (Hsp90) is strongly involved [49], affecting the morphogenetic transition from yeast cells to hyphal forms and repressing Ras1/PKA (cAMP-dependent protein kinase) signaling cascade [50]. Furthermore, the conserved histone deacetylase complex, including Set3, Hos2, Snt1, and Sif2 proteins also participates in the dispersal of biofilm, modulating the transcription kinetics of the genes that regulate biofilm maturation [51].

3. *C. albicans* interactions with bacteria during mutual biofilm formation

C. albicans colonizes the oral cavity, presenting the commensal or pathogenic properties that can be modified by direct or indirect interactions with different types of bacteria, depending on the localization of the microbial communities, such as the supragingival plaque, subgingival plaque, and tongue coating. The metabolic activity of microorganisms that colonize the supragingival sites, i.e., nonmutans streptococci and *Actinomyces* enriches the environment in lactic acid, creating a temporarily acidic environment that favors the entrance of the more cariogenic microorganisms, mutans streptococci into the ecosystem. Conversely, at subgingival sites, colonized by *Fusobacteria* and *Prevotella*, a neutral pH and anaerobic environment dominate and facilitate the establishment of a less acid-tolerant but periodontopathogenic bacterium, *P. gingivalis* [52].

The mitis group of streptococci (MGS), including *Streptococcus mitis*, *Streptococcus oralis*, *Streptococcus gordonii*, and *Streptococcus sangunis* [53, 54] belongs to the early colonizers of oral cavity and appears to interact synergistically with *C. albicans* hyphal filaments providing the physical and metabolic interactions by the exposition of specific adhesion sites and excreting lactate that serves as a carbon source for yeast growth [55]. On the other hand, *C. albicans* can provide bacteria with growth stimulatory factors, resulting from the fungal nutrition metabolism [12] and reduces the oxygen pressure to the level, preferred by streptococci. Moreover, a mutual collaboration of *C. albicans* and *S. oralis* increases the inflammatory host responses compared to monospecies infections [56, 57]. The consequence for the host is provided by the precise interactions between the cell surface proteins of *S. oralis*, *S. mitis* and *S. gordonii*, especially SspA and SspB [58, 59] and the candidal cell wall adhesins of agglutinin-like family, mainly Als3, presented on the surface of fungal hyphae [60]. Some parts of these proteins seem to be particularly important for the interactions but a precise mechanism of the interaction requires further research, especially to clarify a significance of cell wall mannosylation [57, 61] in this process.

Moreover, the polysaccharides of both types of interacting microorganisms that compose ECM not only protect the cells, but also create a new platform for mutual interactions between fungal glucans and mannans and bacterial glucosyltransferases within the ECM matrix structure [46, 62, 63].

The best example of microbial cooperation for increased pathogenic properties of mixed biofilm is represented by the interactions of *C. albicans* with *S. aureus*, identified in oral cavity and being a source of systemic infection [64–67]. The bacteria prefer hyphal filaments of *C. albicans* [68, 69] for adhesion, but its localization within the biofilm seems to depend on the surface colonization sequence. When bacteria are the first colonizer, the development of fungal biofilm is slower and bacteria cells are spread in whole three-dimensional biofilm structure [68]. On the other hand, a simultaneous contact of microorganisms with the surface favors the rapid formation of the mixed biofilm with *S. aureus* localization within upper layers of fungal biofilm and with involvement of multiple microbial proteins. However, the role of fungal adhesins from Als family in these interactions has been questioned [70]. The formed biofilm and its ECM with extracellular fungal DNA protect the bacteria against the antibiotic treatment [71–73].

C. albicans co-aggregates with an obligatory anaerobe *F. nucleatum* [13], with engagement of a mannan receptor on the *C. albicans* surface [74]. A facultative anaerobe—*A. oris*—makes its own carbohydrate-containing surface molecules available to the interaction with *C. albicans* [14]. Last studies have also shown that *C. albicans* is able to interact with keystone pathogen of subgingival plaque—*P. gingivalis*—the obligatory anaerobe. However, it is difficult to judge whether the pathogens apply a synergistic or concurrence style of interactions. It was demonstrated that *P. gingivalis* suppressed *Candida* biofilm formation by a reduction of fungal cell viability [75]. Recently, the gingipain activity has been suggested to be the main destructive force influencing fungal cells wall within the mixed bacterial-fungal biofilm (unpublished data). On the other hand, *P. gingivalis* was also shown to induce germ-tube formation by *C. albicans* cells, generating a more invasive phenotype of fungal cells [76]. But such an effect could also result from fungal protection toward contacting bacteria and their virulence potential (unpublished data). The mutual interactions are supported by the involvement of adhesins, especially fungal Als3, Mp65 and surface-located enolase, aand a bacterial internalin, InlJ or gingipains ([77], unpublished data). The important role in the interaction between *C. albicans* and *P. gingivalis* has been also assigned to peptydylarginine transferase of *P. gingivalis* (PPAD), the enzyme capable of modifying Arg residues to citrullines. Its action can directly contribute to the change in the spatial structure of the molecule [78]. A bacterial mutant deprived of PPAD forms a reduced mixed biofilm compared to the wild-type strain. Potential molecules whose citrullination may affect the effectiveness of biofilm formation include arginine-specific cysteine proteinase (RgpA) and adhesive Mfa1 fimbrilin [17, 79].

The interaction of both microbes seems to exert marked consequences to the host. However, it was presented that both microbes appear to have antagonistic effects on one another, as *P. gingivalis* inhibited the adhesion of *C. albicans* to buccal epithelial cells [80]. But the presence of *C. albicans* did not enhance adhesion of *P. gingivalis* to gingival epithelial cells or gingival fibroblasts. On the other hand, a pre-exposure of gingival epithelial cells and fibroblasts to *C. albicans* enhanced the cell invasion by *P. gingivalis* [81].

4. Host responses to the candidal biofilms

Clinical candidal oral biofilm inhabiting mucosal surfaces or artificial devices may trigger more or less similar host responses (**Figure 1**). Regardless of the biofilm origin, it remains under the influence of immune factors produced by contacting epithelial cells [82].

Dongari-Bagtzoglou et al. [83] analyzed the candidal biofilm in a murine model and found that this fungal cell community induced a hyperkeratotic response and epithelial cell desquamation. Moreover, the matrix that surrounded the fungal biofilm was enriched in keratin and desquamated cells.

Also, in a rat model of chronic denture gingival dermatitis, the host proteins were prominent in the extracellular matrix, including amylase, hemoglobin, and antimicrobial peptides [84, 85].

The oral biofilm elicits responses of human immune cells (**Figure 1**). The neutrophil migration and their deeper localization within the biofilm were identified, but these defense cells were not effective in clearing these infections [82]. An analysis of neutrophil responses in the ex vivo models of their contact with *C. albicans* biofilm also showed a diminished activity of neutrophils against this structured fungal community, compared to the responses against the planktonic form of fungal cells [86].

Upon a contact with pathogens, neutrophils can activate many mechanisms of response to suppress the infection. These include degranulation, phagocytosis, and neutrophil extracellular traps (NETs) formation [87]. The latter process aims at entrapment of large objects such as fungal hyphae [88]. Nevertheless, a group of Johnson showed that neutrophils failed to release NETs in contact with fungal biofilm [89]. These results were described as *an immunological silence*, where host immune system ignored contacting biofilm because of its shielding by the matrix components [90]. The biofilm matrix prevents the exposition of so called pathogen-associated molecular patterns (PAMPs) that can be recognized by highly specialized pattern recognition receptors (PRRs) of human immune cells [91, 92].

Another explanation proposed to interpret the evasion of host response by the biofilm was *an immune deviation* that could result from action of yet unidentified fungal compounds. They could act directly or indirectly by triggering host immunomodulatory factors that transform the immune response into ineffective form [93]. Such hypothesis was supported by the observation that *C. albicans* cell wall components were able to induce the expression of Il-10 influencing Th2 response [94].

A further explanation of biofilm survival was represented by a model of *immune resistance* proposed in [95], where GPI-anchored cell wall protein Hyr1 could play an important role

[96]. Moreover, all of proposed mechanisms or their combinations could be involved in local paucity of PMN responses. Katragkou et al. [86, 97] and Xie et al. [98] documented that developed biofilm covered by ECM exposed fungal β-glucans that were involved in hindered neutrophil responses to cytokine priming of PMNs or fungal cell opsonization. Such an argument was also strengthened by the observation that pre-treatment of PMNs with interferon-γ or granulocyte colony-stimulating factor (G-CSF) did not significantly enhance their activity against opsonized or nonopsonized *C. albicans* biofilms. Moreover, neutrophils contacting the mature biofilm did not produce reactive oxygen species, necessary for triggering of phagocytosis, or one of the pathways of NET production [99]. Nevertheless, the precise mechanism of this phenomenon remained to be clarified.

Similar, diminished responses were also observed for a contact of fungal biofilm with mononuclear cells, compared to their co-culture with fungal planktonic form [100]. Although the migration of the mononuclear cells through biofilm was detected with their main compaction in the basal part of biofilm, their phagocytosis properties were suppressed, and the production of pro-inflammatory cytokines in response to biofilm decreased. Surprisingly, the mononuclear cells augmented biofilm proliferation, increasing the biofilm thickness over two-fold [97, 101].

Most of the presented observations were made concerning host response to contact with fungal biofilm, but the host immune system usually has to face an ongoing polymicrobial infection [102] about which the information are rather scarce [103]. An example of the cross-kingdom infection of the human host was represented by *C. albicans* biofilm contacting gingival anaerobic bacteria, *P. gingivalis.* In this case, an attenuation of the human macrophage responses was observed [17]. Moreover, some studies presented that the host responses can vary depending on the pathogen that contacts the fungal biofilm [104]. The pathogen interactions can be synergistic as well as antagonistic. For example, in a rat model, the colonization of the airway by *C. albicans* impaired functions of alveolar macrophages and, in consequence, led to the reduced clearance of *Pseudomonas aeruginosa* [105]. On the other hand, Lopez-Medina et al. [106] showed in a mouse gut model that the co-infection of *P. aeruginosa* with *Candida* cells suppressed the expression of bacterial genes responsible for iron acquisition, and thus suppressed the infection. Nevertheless, our understanding of host responses to mixed biofilm formed between different type of pathogens remains still at its infancy and needs many further studies.

5. Quorum sensing within the mixed biofilm and its significance for the host

An important phenomenon occurring in the process of biofilm formation is also the transmission of signals between microbial cells located within the biofilm, thus stimulating them to further growth and dispersion of the cells or, in contrary, suppressing them. In addition, signaling molecules can also affect microbial cells of other species that inhabit the same niche in the host organism, and thus promote synergistic or antagonistic interactions between different pathogens which can result in clinical outcomes. This phenomenon of the communication between microorganisms through the secretion of low molecular weight compounds, referred

http://dx.doi.org/10.5772/intechopen.81537

to as the quorum sensing (QS) [107], involves specific chemical compounds whose increasing concentration is a signal to change the expression of selected genes in the cells of the entire biofilm population [108].

C. albicans produces autoregulatory substances involved in quorum sensing (quorum-sensing molecules, QSMs) that affect important virulence traits, such as transformation of the morphological forms [109]. One of them is farnesol—an alcohol from the terpene group, secreted by *C. albicans* in the later stages of biofilm formation, with a function of blocking the formation of filamentous forms of this yeast [110]. A function opposite to farnesol has a second fungal QS compound, tyrosol, which stimulates the phase of active growth of the *C. albicans* cell population and the formation of hyphae in the initial phases of biofilm formation, thus increasing the thickness of the biofilm [111–113].

When the concentration of farnesol is higher than that of tyrosol, the conversion of yeast form to hyphae is inhibited and a release of individual cells from the biofilm is stimulated. Such effect indicates possible interactions between these two QS systems in the process of biofilm building [112]. Additionally, *C. albicans* secretes two aromatic alcohols, phenylethyl alcohol and tryptophol, also identified as QSM [113].

The role of QSM seems to be particularly important in mixed biofilms, in which the coexistence of fungi and bacteria is associated with their mutual communications. QSM secreted by the bacteria can exert both stimulatory and inhibitory effects on the cell morphology and biofilm formation by *C. albicans* cells. It is likely that a combination of these contradictory signals orchestrates the balance between the cellular and filamentous form in biofilms, preventing the excessive growth of *C. albicans* within these communities [114].

One example of cross-species communication using QS signals are biofilms formed between *C. albicans* and Gram-negative bacteria *P. aeruginosa*. It has been shown that the presence of farnesol produced by *C. albicans* inhibits functioning of bacteria, and suppresses the production of a bacterial quinolone signaling molecule—PQS, and the piocyjanin—an important bacterial virulence factor [115]. On the other hand, under formation of mixed biofilms, *P. aeruginosa* produces homoserine lactone that may fulfill a role similar to farnesol, reducing the production of fungal hyphae *in vitro* [116].

Another example is the biofilm with participation of *S. mutans*, in which the inhibition of biofilm formation was observed in response to a high concentration of farnesol (>100 μM), while a low level of farnesol (~25 μM) promoted bacterial growth [117–119].

In other studies, *S. mutans* could both, reduce the farnesol production by *C. albicans* [119] and inhibit the formation of filamentous form of *C. albicans* by the competence-peptide CSP, produced on the early stages of biofilm development [120, 121].

The same peptide produced by *S. gordonii* inhibited the formation of *C. albicans* biofilm, but not the hyphal growth [122]. In contrast, other bacterial QSM—the autoinducer-2 (AI-2), as well as H_2O_2 secreted by *S. gordonii* affected the morphogenesis and production of farnesol. The strains with the deletion of the LuxS quorum-sensing system responsible for AI-2 production in *S. gordonii* presented a reduced ability to stimulate the growth of *C. albicans* hyphae and thereby a general reduction of biofilm biomass. The identified responses correlated with an invasion into the host epithelial cells [10, 27].

An interesting interspecies communication is presented by the Gram-negative *Aggregatibacter actinomycetemcomitans*, acting in periodontal disease, which can inhibit the formation of *C. albicans* biofilm by producing AI-2. Although AI-2 has been described as QSM of different bacteria, other species give off other AI derivatives, so that the results obtained for different species do not have to be identical to one another. Interestingly, *A. actinomycetemcomitans* is one of the bacterium having a dual inhibitory system acting toward *C. albicans* biofilm. In addition to QSM, it also includes cytolethal distending toxin (CDT). One of the emerging hypotheses suggests that secreted QSM is a warning signal for *C. albicans* against a competitor that secretes the toxins [120, 123].

QSM also plays an important role in a communication between *C. albicans* and the Gram-positive bacterium—*S. aureus*. Farnesol, secreted by *C. albicans* inhibits the formation of *S. aureus* biofilm and increases its susceptibility to antibiotics [124, 125]. There were also studies, indicating that *S. aureus* stimulated the growth of *C. albicans* biofilm possibly by QSM [69]. It was also proposed that in the presence of farnesol, *S. aureus* acquires a resistant phenotype that induces oxidative stress, resulting in the upregulation of bacterial drug efflux pumps [126].

QS production within the biofilm has also an impact on the efficiency and the functioning of host defense systems. The gingival epithelial cells presented an upregulation of the toll-like receptor TLR2, and a decrease of the expression of TLR4 and TLR6 upon treatment with farnesol, suggesting the resulting activation of antifungal defense. Considering the role of epithelial cells in the secretion of pro-inflammatory cytokines, it was also shown that farnesol increased the secretion of IL-6 and IL-8. Moreover, farnesol modulated the secretion of antimicrobial peptides by epithelial cells, including hBD1 and hBD2. *C. albicans* cells, via production of farnesol, suppressed the epithelial secretion of hBD1, with a simultaneous increase in hBD2 secretion. Since both peptides have a high efficacy in *C. albicans* killing, the results suggest that farnesol may be a key factor in promoting host defense [127]. An additional function performed by farnesol includes its ability to activate neutrophils and monocytes and to reduce the phagocytic activity of mouse macrophages. Farnesol also impairs the differentiation of monocytes into dendritic cells and decreases their ability to activate and expand T cells, which consequently reduce the induction of IL-12 [128].

In summary, mutual QS interactions between fungi and bacteria may play an important role as a virulence mechanism that mediates the communication between the host and the formed biofilm, and could inspire future applications in diagnostics and biofilm treatment.

6. Resistance of oral biofilm

The biofilm formed on mucosal or artificial surfaces in oral cavity is difficult to eliminate since the biofilm structure protects the pathogenic cells against antimicrobial drugs, especially against antifungal agents, and suppresses immune responses [129]. Moreover, cooperating invaders often present increasing virulence resulting from synergistic and complex interactions between microorganisms [130]. It has been demonstrated that *Staphylococcus* adheres to yeast and hyphal forms, and this interaction benefits the growth and antibiotic resistance of *S. epidermidis*. In addition, the components of the biofilm extracellular matrix produced by the wild-type of *S. epidermidis* prevent the effective penetration of antimycotic molecules such as fluconazole into the biofilm and promote the spread of yeast infection [131].

The low susceptibility of biofilms to medical treatment is attributed to multifactorial events, represented by upregulation of efflux pumps, the presence of extracellular matrix and appearance of recalcitrant persister cells [132].

The two classes of fungal efflux pumps (FEP: Cdr1, Cdr2, and Mdr1) are activated in planktonic cells in contact with antifungal drugs but in biofilms FEP are upregulated probably in response to contact with other partners that compose the biofilm. Such an explanation was supported by an observation that FEP efficient function appeared shortly after the cell surface adhesion and remains upregulated during whole process of mixed biofilm formation [21].

Another contributor to mixed biofilm resistance is the extracellular matrix and its components. This three-dimensional complex structure effectively inhibits antibiotic and antimycotic diffusion [133]. Moreover, the biofilm-composing polysaccharides not only mask the biofilm against its recognition by the host receptors, but also can directly bind and inactivate the drugs, as it was presented in a case of antifungal-acting amphotericin B, sequestered by β-1,3-glucan, composing ECM [134].

Also, eDNA is an especially important biofilm component, whose viscosity and negative electric charge influences the structural integrity and stability of biofilms but also contributes to drug resistance via acting as drug chelator. eDNA also binds magnesium ions, whose decreased level serves as a signal, inducing PhoPQ and PmrAB systems, responsible for *P. aeruginosa* resistance to antimicrobial peptides and to aminoglycosides [135].

An important phenomenon that plays a key role in the development of drug resistance by oral microbiome is the horizontal gen transfer (HGT) [131]. The biofilm structure provides a suitable environment for gene exchange, because the microbial cells are in close proximity and the virulence genes are dynamically spread between different species of bacteria composing biofilm. The most popular mobile genetic element in oral microflora is the conjugative transposon *Tn*916, which contains genes encoding ribosomal protection proteins [131]. These proteins inhibit the action of tetracycline, the most popular antibiotic used in periodontal disease treatment, by preventing the binding of this antibiotic to the bacterial ribosome [136]. Another biofilm protective function is carried out by membrane vesicles (MVs), present in ECM [137], which protect bacteria against some antibiotics by the degradative properties of MV enzymes, such as β-lactamase [138].

The important factors that contribute to the biofilm resistance are the persister cells detected in bacterial and fungal biofilm [20, 139]. The persister cells are a minor subset of metabolically dormant cells presented within biofilms that possess extreme resistance to antimicrobial agents and are responsible for the severe chronic infectious disease. However, the mechanism of this resistance of persister cells remains to be discover; they could possibly be a good target for further antimicrobial therapies.

7. The challenges for medical treatment of mixed oral biofilm

As no biofilm-specific drugs exist today, the treatment of infections caused by mixed species community remains a major challenge for contemporary medical biotechnology and the developing of new effective strategies for biofilm eradication becomes critical.

One of the strategies for combating biofilms formed by bacteria and yeasts can be a degradation of ECM. It has been demonstrated that enzymatic degradation of some biofilm-forming components facilitates the penetration of antibiotic and antimycotic molecules and affects the biofilm structural integrity [140, 141]. For example, a study demonstrated that a combined use of deoxyribonuclease and amphotericin B reduced the survival of *C. albicans* cells.

An effective alternative to antibiotic therapy may be a treatment with anti-biofilm peptides. These compounds easily penetrate the structure of multispecies biofilm and inhibit the growth of Gram-positive and Gram-negative bacteria. An example of such an anti-biofilm compound is a short synthetic peptide 1018 (amino acid sequence: VRLIVAVRIWRR), which blocks a stress response through an activation of the stress-signaling nucleotide degradation [142]. Another example of an anti-biofilm compound is D-enantiomeric peptide DJK-5 that has a similar mechanism of action to peptide 1018 [143]. The main advantage of the DJK-5 is its resistance to proteases produced by the host and bacteria. Moreover, DJK-5 possesses a higher biological activity than peptide 1018 and kills most of the oral biofilm-forming bacteria in a few minutes. It has been demonstrated that the use of anti-biofilm peptides in combination with conventional antibiotics both increases the effectiveness of treatment and reduces the required concentration of antibiotics [144].

Several natural products have been also proposed for fungal biofilm treatment. An example of plant metabolites with antifungal activity are terpenoids, such as xanthorrhizol extracted from *Curcuma xanthorrhiza* [145]. It has been demonstrated that this compound effectively inhibits the development of mature biofilms formed by various *Candida* species. Moreover, in contrast to commonly used antifungal drugs, xanthorrhizol is nontoxic to human cells even at very high concentrations.

Also, chemical signal molecules involved in quorum sensing possess a potential for the therapy of oral infections disease. There are two main mechanisms of action of the known QS inhibitors [146]. Some of these cause an enzymatic degradation of signaling molecules. The enzymes—AHL-lactonases and AHL-acylase can be classified to this group. Other inhibitors such as furanones that are produced by red marine algae are structural analogs that prevent bacterial biofilm development via binding to LuxR [147]. In the case of oral *C. albicans* infections, the use of farnesol has been proposed [148]. In vivo studies have shown that the addition of farnesol suppresses the hyphal growth on the mouse tongue at the first step of biofilm formation, and as a result prevents the invasion of mucosal membrane by the yeast and bacteria.

An interesting proposal for the treatment of mixed biofilm can be the photodynamic antimicrobial chemotherapy (PACT) that applies the nontoxic dye (photosensitizer) activated by visible light [149]. Singlet oxygen, which is effectively produced during this process, effectively kills pathogen cells. This novel method has been successfully used against *C. albicans* biofilm and can be a promising antimicrobial therapy that has many advantages such as the high target specificity. What is more, the development of resistance to PACT is unlikely because microorganisms have no resistance mechanism against singlet oxygen [150].

A better understanding of the molecular mechanisms underlying the formation and maintenance of the mixed species biofilm is crucial for the development of their effective treatments in the future.

Acknowledgements

This work was supported in part by the National Science Center of Poland (grant UMO-2015/17/B/NZ6/02078 awarded to M.R.-K).

Author details

Maria Rapala-Kozik[1]*, Marcin Zawrotniak[1], Mariusz Gogol[2], Dominika Bartnicka[1], Dorota Satala[1], Magdalena Smolarz[1], Justyna Karkowska-Kuleta[1] and Andrzej Kozik[2]

*Address all correspondence to: maria.rapala-kozik@uj.edu.pl

1 Department of Comparative Biochemistry and Bioanalytics, Faculty of Biochemistry, Biophysics and Biotechnology, Jagiellonian University, Krakow, Poland

2 Department of Analytical Biochemistry, Faculty of Biochemistry, Biophysics and Biotechnology, Jagiellonian University, Krakow, Poland

References

[1] Huse SM, Ye Y, Zhou Y, Fodor AA. A core human microbiome as viewed through 16S rRNA sequence clusters. PLoS One. 2012;**7**(6):e34242

[2] Simón-Soro A, Belda-Ferre P, Cabrera-Rubio R, Alcaraz LD, Mira A. A tissue-dependent hypothesis of dental caries. Caries Research. 2013;**47**(6):591-600

[3] He X, Hu W, Kaplan CW, Guo L, Shi W, Lux R. Adherence to *Streptococci* facilitates *Fusobacterium nucleatum* integration into an oral microbial community. Microbial Ecology. 2012;**63**(3):532-542

[4] Kuboniwa M, Lamont RJ. Subgingival biofilm formation. Periodontology 2000. 2010; **52**(1):38-52

[5] Jakubovics NS, Kolenbrander PE. The road to ruin: The formation of disease-associated oral biofilms. Oral Diseases. 2010;**16**(8):729-739

[6] Cannon RD, Chaffin WL. Colonization is a crucial factor in oral candidiasis. Journal of Dental Education. 2001;**65**(8):785-787

[7] Mayer FL, Wilson D, Hube B. *Candida albicans* pathogenicity mechanisms. Virulence. 2013;**4**(2):119-128

[8] Polke M, Hube B, Jacobsen ID. *Candida* survival strategies. Advances in Applied Microbiology. 2015;**91**:139-235

[9] Li D, Bernhardt J, Calderone R. Temporal expression of the *Candida albicans* genes CHK1 and CSSK1, adherence, and morphogenesis in a model of reconstituted human esophageal epithelial candidiasis. Infection and Immunity. 2002;**70**(3):1558-1565

[10] Bamford CV, D'Mello A, Nobbs AH, Dutton LC, Vickerman MM, Jenkinson HF. *Streptococcus gordonii* modulates *Candida albicans* biofilm formation through intergeneric communication. Infection and Immunity. 2009;**77**(9):3696-3704

[11] Holmes AR, Gopal PK, Jenkinson HF. Adherence of *Candida albicans* to a cell surface polysaccharide receptor on *Streptococcus gordonii*. Infection and Immunity. 1995; **63**(5):1827-1834

[12] Jenkinson HFDL. *Candida* interactions with bacterial biofilms. In: Brogden KAGJ, editor. Polymicrobial Infections and Disease. Washington, DC: ASM Press; 2002. pp. 357-373

[13] Grimaudo NJ, Nesbitt WE. Coaggregation of *Candida albicans* with oral *Fusobacterium* species. Oral Microbiology and Immunology. 1997;**12**(3):168-173

[14] Grimaudo NJ, Nesbitt WE, Clark WB. Coaggregation of *Candida albicans* with oral Actinomyces species. Oral Microbiology and Immunology. 1996;**11**(1):59-61

[15] Fox EP, Cowley ES, Nobile CJ, Hartooni N, Newman DK, Johnson AD. Anaerobic bacteria grow within *Candida albicans* biofilms and induce biofilm formation in suspension cultures. Current Biology. 2014;**24**(20, 20):2411-2416

[16] Janus MM, Crielaard W, Volgenant CMC, der Veen MH, Brandt BW, Krom BP. *Candida albicans* alters the bacterial microbiome of early in vitro oral biofilms. Journal of Oral Microbiology. 2017;**9**(1):1-10

[17] Karkowska-Kuleta J, Bartnicka D, Zawrotniak M, Zielinska G, Kieronska A, Bochenska O, et al. The activity of bacterial peptidylarginine deiminase is important during formation of dual-species biofilm by periodontal pathogen *Porphyromonas gingivalis* and opportunistic fungus *Candida albicans*. Pathogens and Disease. 2018;**76**(4)

[18] Sherry L, Rajendran R, Lappin DF, Borghi E, Perdoni F, Falleni M, et al. Biofilms formed by *Candida albicans* bloodstream isolates display phenotypic and transcriptional heterogeneity that are associated with resistance and pathogenicity. BMC Microbiology. 2014;**14**(1):182

[19] Rajendran R, Sherry L, Nile CJ, Sherriff A, Johnson EM, Hanson MF, et al. Biofilm formation is a risk factor for mortality in patients with *Candida albicans* bloodstream infection—Scotland, 2012-2013. Clinical Microbiology and Infection. 2016;**22**(1):87-93

[20] Mathé L, Van Dijck P. Recent insights into *Candida albicans* biofilm resistance mechanisms. Current Genetics. 2013;**59**(4):251-264

[21] Nobile CJ, Fox EP, Nett JE, Sorrells TR, Mitrovich QM, Hernday AD, et al. A recently evolved transcriptional network controls biofilm development in *Candida albicans*. Cell. 2012;**148**(12):126-138

[22] Tournu H, Van Dijck P. *Candida* biofilms and the host: models and new concepts for eradication. International Journal of Microbiology. 2012;**2012**:845352

[23] Fox EP, Nobile CJ. A sticky situation. Transcription. 2012;**3**(6):315-322

[24] Tsui C, Kong EF, Jabra-Rizk MA. Pathogenesis of *Candida albicans* biofilm. Mobley H, editor. Pathogens and Disease. 2016;**74**(4):ftw018

[25] Fox EP, Bui CK, Nett JE, Hartooni N, Mui MC, Andes DR, et al. An expanded regulatory network temporally controls *Candida albicans* biofilm formation. Molecular Microbiology. 2015;**96**(6):1226-1239

[26] Uppuluri P, Chaturvedi AK, Srinivasan A, Banerjee M, Ramasubramaniam AK, Köhler JR, et al. Dispersion as an important step in the *Candida albicans* biofilm developmental cycle. Doering TL, editor. PLoS Pathogens. 2010;**6**(3):e1000828

[27] Lohse MB, Gulati M, Johnson AD, Nobile CJ. Development and regulation of single- and multi-species *Candida albicans* biofilms. Nature Reviews. Microbiology. 2018;**16**(1):19-31

[28] Kriznik A, Bouillot M, Coulon J, Gaboriaud F. Morphological specificity of yeast and filamentous *Candida albicans* forms on surface properties. Comptes Rendus Biologies. 2005;**328**(10-11):928-935

[29] Nobile CJ, Nett JE, Andes DR, Mitchell AP. Function of *Candida albicans* adhesin Hwp1 in biofilm formation. Eukaryotic Cell. 2006;**5**(10):1604-1610

[30] Nobile CJ, Schneider HA, Nett JE, Sheppard DC, Filler SG, Andes DR, et al. Complementary adhesin function in *C. albicans* biofilm formation. Current Biology. 2008; **18**(14):1017-1024

[31] Bailey DA, Feldmann PJ, Bovey M, Gow NA, Brown AJ. The *Candida albicans* HYR1 gene, which is activated in response to hyphal development, belongs to a gene family encoding yeast cell wall proteins. Journal of Bacteriology. 1996;**178**(18):5353-5360

[32] Ohkuni K, Hayashi M, Yamashita I. Bicarbonate-mediated social communication stimulates meiosis and sporulation of *Saccharomyces cerevisiae*. Yeast. 1998;**14**(7):623-631

[33] Argimón S, Wishart JA, Leng R, Macaskill S, Mavor A, Alexandris T, et al. Developmental regulation of an adhesin gene during cellular morphogenesis in the fungal pathogen *Candida albicans*. Eukaryotic Cell. 2007;**6**(4):682-692

[34] Li F, Palecek SP. EAP1, a *Candida albicans* gene involved in binding human epithelial cells. Eukaryotic Cell. 2003;**2**(6):1266-1273

[35] Li F, Svarovsky MJ, Karlsson AJ, Wagner JP, Marchillo K, Oshel P, et al. Eap1p, an adhesin that mediates *Candida albicans* biofilm formation in vitro and in vivo. Eukaryotic Cell. 2007;**6**(6):931-939

[36] Monniot C, Boisramé A, Da Costa G, Chauvel M, Sautour M, Bougnoux M-E, et al. Rbt1 Protein domains analysis in *Candida albicans* brings insights into hyphal surface modifications and Rbt1 potential role during adhesion and biofilm Formation. Arkowitz RA, editor. PLoS One. 2013;**8**(12):e82395

[37] Lipke PN, Klotz SA, Dufrene YF, Jackson DN, Garcia-Sherman MC. Amyloid-like β-aggregates as force-sensitive switches in fungal biofilms and infections. Microbiology and Molecular Biology Reviews. 2018;**82**(1):e00035-17

[38] Baillie GS, Douglas LJ. Role of dimorphism in the development of *Candida albicans* biofilms. Journal of Medical Microbiology. 1999;**48**(7):671-679

[39] Nailis H, Kucharíková S, Ricicová M, Van Dijck P, Deforce D, Nelis H, et al. Real-time PCR expression profiling of genes encoding potential virulence factors in *Candida albicans* biofilms: identification of model-dependent and -independent gene expression. BMC Microbiology. 2010;**10**:114

[40] Joo MY, Shin JH, Jang H-C, Song ES, Kee SJ, Shin MG, et al. Expression of SAP5 and SAP9 in *Candida albicans* biofilms: comparison of bloodstream isolates with isolates from other sources. Medical Mycology. 2013;**51**(8):892-896

[41] Kumar R, Saraswat D, Tati S, Edgerton M. Novel aggregation properties of *Candida albicans* secreted aspartyl proteinase sap6 mediate virulence in oral Candidiasis. Infection and Immunity. 2015;**83**(7):2614-2626

[42] Winter MB, Salcedo EC, Lohse MB, Hartooni N, Gulati M, Sanchez H, et al. Global Identification of biofilm-specific proteolysis in *Candida albicans*. MBio. 2016;**7**(5):e01514-16

[43] Zarnowski R, Westler WM, Lacmbouh GA, Marita JM, Bothe JR, Bernhardt J, et al. Novel entries in a fungal biofilm matrix encyclopedia. MBio. 2014;**5**(4):5

[44] Dominguez E, Zarnowski R, Sanchez H, Covelli AS, Westler WM, Azadi P, et al. Conservation and divergence in the *Candida* species biofilm matrix mannan-glucan complex structure, function, and genetic control. MBio. 2018;**9**(2):e00451-18

[45] Al-Fattani MA. Biofilm matrix of *Candida albicans* and *Candida tropicalis*: Chemical composition and role in drug resistance. Journal of Medical Microbiology. 2006;**55**(8):999-1008

[46] Taff HT, Nett JE, Zarnowski R, Ross KM, Sanchez H, Cain MT, et al. A *Candida* biofilm-induced pathway for matrix glucan delivery: Implications for drug resistance. Doering TL, editor. PLoS Pathogens. 2012;**8**(8):e1002848

[47] Peltroche-Llacsahuanga H, Goyard S, D'Enfert C, Prill SK-H, Ernst JF. Protein o-mannosyltransferase isoforms regulate biofilm formation in *Candida albicans*. Antimicrobial Agents and Chemotherapy. 2006;**50**(10):3488-3491

[48] Martins M, Uppuluri P, Thomas DP, Cleary IA, Henriques M, Lopez-Ribot JL, et al. Presence of extracellular DNA in the *Candida albicans* biofilm matrix and its contribution to biofilms. Mycopathologia. 2010;**169**(5):323-331

[49] Robbins N, Uppuluri P, Nett J, Rajendran R, Ramage G, Lopez-Ribot JL, et al. Hsp90 governs dispersion and drug resistance of fungal biofilms. May RC, editor. PLoS Pathogens. 2011;**7**(9):e1002257

[50] Shapiro RS, Uppuluri P, Zaas AK, Collins C, Senn H, Perfect JR, et al. Hsp90 orchestrates temperature-dependent *Candida albicans* morphogenesis via Ras1-PKA signaling. Current Biology. 2009;**19**(8):621-629

[51] Nobile CJ, Fox EP, Hartooni N, Mitchell KF, Hnisz D, Andes DR, et al. A histone deacetylase complex mediates biofilm dispersal and drug resistance in Candida albicans. MBio. 2014;**5**(3):e01201-14

[52] Takahashi N. Microbial ecosystem in the oral cavity: Metabolic diversity in an ecological niche and its relationship with oral diseases. International Congress Series. 2005;**1284**:103-112

[53] Kawamura Y, Hou XG, Sultana F, Miura H, Ezaki T. Determination of 16S rRNA sequences of *Streptococcus mitis* and *Streptococcus gordonii* and phylogenetic relationships among members of the genus *Streptococcus*. International Journal of Systematic Bacteriology. 1995;**45**(2):406-408

[54] Diaz PI, Dupuy AK, Abusleme L, Reese B, Obergfell C, Choquette L, et al. Using high throughput sequencing to explore the biodiversity in oral bacterial communities. Molecular Oral Microbiology. 2012;**27**(3):182-201

[55] Holmes AR, van der Wielen P, Cannon RD, Ruske D, Dawes P. *Candida albicans* binds to saliva proteins selectively adsorbed to silicone. Oral Surgery, Oral Medicine, Oral Pathology, Oral Radiology, and Endodontics. 2006;**102**(4):488-494

[56] Xu H, Sobue T, Thompson A, Xie Z, Poon K, Ricker A, et al. *Streptococcal* co-infection augments *Candida* pathogenicity by amplifying the mucosal inflammatory response. Cellular Microbiology. 2014;**16**(2):214-231

[57] Dutton LC, Nobbs AH, Jepson K, Jepson MA, Vickerman MM, Aqeel Alawfi S, et al. O-mannosylation in *Candida albicans* enables development of interkingdom biofilm communities. MBio. 2014;**5**(2):e00911

[58] Jenkinson HF, Lala HC, Shepherd MG. Coaggregation of *Streptococcus sanguis* and other streptococci with *Candida albicans*. Infection and Immunity. 1990;**58**(5):1429-1436

[59] Holmes ARR, McNab R, Jenkinson HFF. *Candida albicans* binding to the oral bacterium *Streptococcus gordonii* involves multiple adhesin-receptor interactions. Infection and Immunity. 1996;**64**(11):4680-4685

[60] Bamford CV, Nobbs AH, Barbour ME, Lamont RJ, Jenkinson HF. Functional regions of *Candida albicans* hyphal cell wall protein Als3 that determine interaction with the oral bacterium *Streptococcus gordonii*. Microbiology. 2015;**161**(Pt 1):18-29

[61] Hoyer LL, Oh S-H, Jones R, Cota E. A proposed mechanism for the interaction between the *Candida albicans* Als3 adhesin and streptococcal cell wall proteins. Frontiers in Microbiology. 2014;**5**:564

[62] Gregoire S, Xiao J, Silva BB, Gonzalez I, Agidi PS, Klein MI, et al. Role of glucosyltransferase B in interactions of *Candida albicans* with *Streptococcus mutans* and with an experimental pellicle on hydroxyapatite surfaces. Applied and Environmental Microbiology. 2011;**77**(18):6357-6367

[63] Falsetta ML, Klein MI, Colonne PM, Scott-Anne K, Gregoire S, Pai C-H, et al. Symbiotic relationship between *Streptococcus mutans* and *Candida albicans* synergizes virulence of plaque biofilms in vivo. Infection and Immunity. 2014;**82**(5):1968-1981

[64] Merghni A, Ben Nejma M, Hentati H, Mahjoub A, Mastouri M. Adhesive properties and extracellular enzymatic activity of *Staphylococcus aureus* strains isolated from oral cavity. Microbial Pathogenesis. 2014;**73**:7-12

[65] Koukos G, Sakellari D, Arsenakis M, Tsalikis L, Slini T, Konstantinidis A. Prevalence of *Staphylococcus aureus* and methicillin resistant *Staphylococcus aureus* (MRSA) in the oral cavity. Archives of Oral Biology. 2015;**60**(9):1410-1415

[66] Schlecht LM, Peters BM, Krom BP, Freiberg JA, Hänsch GM, Filler SG, et al. Systemic *Staphylococcus aureus* infection mediated by *Candida albicans* hyphal invasion of mucosal tissue. Microbiology. 2015;**161**(Pt 1):168-181

[67] Mathews J, Patel M. Bacterial endotoxins and microorganisms in the oral cavities of patients on cancer therapy. Microbial Pathogenesis. 2018;**123**(July):190-195

[68] Peters BM, Jabra-Rizk MA, Scheper MA, Leid JG, Costerton JW, Shirtliff ME. Microbial interactions and differential protein expression in *Staphylococcus aureus-Candida albicans* dual-species biofilms. FEMS Immunology and Medical Microbiology. 2010;**59**(3):493-503

[69] Lin YJ, Alsad L, Vogel F, Koppar S, Nevarez L, Auguste F, et al. Interactions between *Candida albicans* and *Staphylococcus aureus* within mixed species biofilms. Bios. 2013; **84**(1):30-39

[70] Harriott MM, Noverr MC. Ability of *Candida albicans* mutants to induce *Staphylococcus aureus* vancomycin resistance during polymicrobial biofilm formation. Antimicrobial Agents and Chemotherapy. 2010;**54**(9):3746-3755

[71] Pammi M, Liang R, Hicks JM, Barrish J, Versalovic J. Farnesol decreases biofilms of *staphylococcus epidermidis* and exhibits synergy with nafcillin and vancomycin. Pediatric Research. 2011;**70**(6):578-583

[72] Rajendran R, Sherry L, Lappin DF, Nile CJ, Smith K, Williams C, et al. Extracellular DNA release confers heterogeneity in *Candida albicans* biofilm formation. BMC Microbiology. 2014;**14**:303

[73] Sapaar B, Nur A, Hirota K, Yumoto H, Murakami K, Amoh T, et al. Effects of extracellular DNA from *Candida albicans* and pneumonia-related pathogens on *Candida* biofilm formation and hyphal transformation. Journal of Applied Microbiology. 2014;**116**(6):1531-1542

[74] Jabra-Rizk MA, Falkler WA, Merz WG, Kelley JI, Baqui AAMA, Meiller TF. Coaggregation of *Candida dubliniensis* with *Fusobacterium nucleatum*. Journal of Clinical Microbiology. 1999;**37**(5):1464-1468

[75] Thein ZM, Samaranayake YH, Samaranayake LP. Effect of oral bacteria on growth and survival of *Candida albicans* biofilms. Archives of Oral Biology. 2006;**51**(8):672-680

[76] Nair RG, Anil S, Samaranayake LP. The effect of oral bacteria on *Candida albicans* germ-tube formation. APMIS. 2001;**109**(2):147-154

[77] Sztukowska MN, Dutton LC, Delaney C, Ramsdale M, Ramage G, Jenkinson HF, et al. Community development between *Porphyromonas gingivalis and Candida albicans* mediated by InlJ and Als3. MBio. 2018;**9**(2):e00202-18

[78] Goulas T, Mizgalska D, Garcia-Ferrer I, Kantyka T, Guevara T, Szmigielski B, et al. Structure and mechanism of a bacterial host-protein citrullinating virulence factor, *Porphyromonas gingivalis* peptidylarginine deiminase. Scientific Reports. 2015;**5**(1):11969

[79] Stobernack T, Glasner C, Junker S, Gabarrini G, De Smit M, De Jong A, et al. Extracellular proteome and citrullinome of the oral pathogen *Porphyromonas gingivalis*. Journal of Proteome Research. 2016;**15**(12):4532-4543

[80] Nair RG, Samaranayake LP. The effect of oral commensal bacteria on candidal adhesion to human buccal epithelial cells in vitro. Journal of Medical Microbiology. 1996; **45**(3):179-185

[81] Tamai R, Sugamata M, Kiyoura Y. *Candida albicans* enhances invasion of human gingival epithelial cells and gingival fibroblasts by *Porphyromonas gingivalis*. Microbial Pathogenesis. 2011;**51**(4):250-254

[82] Nett JE. The host's reply to *Candida* biofilm. Pathogens. 2016;**5**(1):33

[83] Dongari-Bagtzoglou A, Kashleva H, Dwivedi P, Diaz P, Vasilakos J. Characterization of mucosal *Candida albicans* biofilms. PLoS One. 2009;**4**(11):e7967

[84] Nett JE, Zarnowski R, Cabezas-Olcoz J, Brooks EG, Bernhardt J, Marchillo K, et al. Host contributions to construction of three device-associated *Candida albicans* biofilms. Infection and Immunity. 2015;**83**(12):4630-4638

[85] Nett JE, Marchillo K, Spiegel CA, Andes DR. Development and validation of an in vivo *Candida albicans* biofilm denture model. Infection and Immunity. 2010;**78**(9):3650-3659

[86] Katragkou A, Kruhlak MJ, Simitsopoulou M, Chatzimoschou A, Taparkou A, Cotten CJ, et al. Interactions between human phagocytes and *Candida albicans* biofilms alone and in combination with antifungal agents. The Journal of Infectious Diseases. 2010;**201**(12):1941-1949

[87] Urban CF, Ermert D, Schmid M, Abu-Abed U, Goosmann C, Nacken W, et al. Neutrophil extracellular traps contain calprotectin, a cytosolic protein complex involved in host defense against *Candida albicans*. PLoS Pathogens. 2009;**5**(10):e1000639

[88] Kernien JF, Snarr BD, Sheppard DC, Nett JE. The interface between fungal biofilms and innate immunity. Frontiers in Immunology. 2017;**8**:1968

[89] Johnson CJ, Cabezas-Olcoz J, Kernien JF, Wang SX, Beebe DJ, Huttenlocher A, et al. The extracellular matrix of *Candida albicans* biofilms impairs formation of neutrophil extracellular traps. PLoS Pathogens. 2016;**12**(9):e1005884

[90] Garcia-Perez JE, Mathé L, Humblet-Baron S, Braem A, Lagrou K, Van Dijck P, et al. A framework for understanding the evasion of host immunity by *Candida* biofilms. Frontiers in Immunology. 2018;**9**:538

[91] Williams DW, Jordan RPC, Wei X-Q, Alves CT, Wise MP, Wilson MJ, et al. Interactions of *Candida albicans* with host epithelial surfaces. Journal of Oral Microbiology. 2013;**5**

[92] Höfs S, Mogavero S, Hube B. Interaction of *Candida albicans* with host cells: Virulence factors, host defense, escape strategies, and the microbiota. Journal of Microbiology. 2016;**54**(3):149-169

[93] Netea MG, Joosten LA, van der Meer JW, Kullberg BJ, van de Veerdonk FL. Immune defence against *Candida* fungal infections. Nature Reviews. Immunology. 2015;**15**(10): 630-642

[94] Chaudhry A, Samstein RM, Treuting P, Liang Y, Pils MC, Heinrich J-M, et al. Interleukin-10 signaling in regulatory T cells is required for suppression of Th17 cell-mediated inflammation. Immunity. 2011;**34**(4):566-578

[95] Mitchell KF, Zarnowski R, Sanchez H, Edward JA, Reinicke EL, Nett JE, et al. Community participation in biofilm matrix assembly and function. Proceedings of the National Academy of Sciences of the United States of America. 2015;**112**(13):4092-4097

[96] Mathews J, Patel M. Bacterial endotoxins and microorganisms in the oral cavities of patients on cancer therapy. Davis D, editor. Microbial Pathogenesis. 2018;**123**(1):190-195

[97] Katragkou A, Simitsopoulou M, Chatzimoschou A, Georgiadou E, Walsh TJ, Roilides E. Effects of interferon-γ and granulocyte colony-stimulating factor on antifungal activity of human polymorphonuclear neutrophils against *Candida albicans* grown as biofilms or planktonic cells. Cytokine. 2011;**55**(3):330-334

[98] Xie Z, Thompson A, Sobue T, Kashleva H, Xu H, Vasilakos J, et al. *Candida albicans* biofilms do not trigger reactive oxygen species and evade neutrophil killing. The Journal of Infectious Diseases. 2012;**206**(12):1936-1945

[99] Zawrotniak M, Bochenska O, Karkowska-Kuleta J, Seweryn-Ozog K, Aoki W, Ueda M, et al. Aspartic proteases and major cell wall components in *Candida albicans* trigger the release of neutrophil extracellular traps. Frontiers in Cellular and Infection Microbiology. 2017;**7**(September):1-21

[100] Chandra J, McCormick TS, Imamura Y, Mukherjee PK, Ghannoum MA. Interaction of *Candida albicans* with adherent human peripheral blood mononuclear cells increases *C. albicans* biofilm formation and results in differential expression of pro- and anti-inflammatory cytokines. Infection and Immunity. 2007;**75**(5):2612-2620

[101] Wheeler RT, Fink GR. A drug-sensitive genetic network masks fungi from the immune system. PLoS Pathogens. 2006;**2**(4):e35

[102] Kean R, McKloud E, Townsend EM, Sherry L, Delaney C, Jones BL, et al. The comparative efficacy of antiseptics against *Candida auris* biofilms. International Journal of Antimicrobial Agents. 2018;**S0924-8579**(18):30138-9. In Press

[103] Nobile CJ, Johnson AD. *Candida albicans* biofilms and human disease. Annual Review of Microbiology. 2015;**69**:71-92

[104] Peleg AY, Hogan DA, Mylonakis E. Medically important bacterial–fungal interactions. Nature Reviews. Microbiology. 2010;**8**(5):340-349

[105] Roux D, Gaudry S, Dreyfuss D, El-Benna J, de Prost N, Denamur E, et al. *Candida albicans* impairs macrophage function and facilitates *Pseudomonas aeruginosa* pneumonia in rat. Critical Care Medicine. 2009;**37**(3):1062-1067

[106] Lopez-Medina E, Fan D, Coughlin LA, Ho EX, Lamont IL, Reimmann C, et al. *Candida albicans* inhibits *Pseudomonas aeruginosa* virulence through suppression of pyochelin and pyoverdine Biosynthesis. Hogan DA, editor. PLoS Pathogens. 2015;**11**(8):e1005129

[107] Wongsuk T, Pumeesat P, Luplertlop N. Fungal quorum sensing molecules: Role in fungal morphogenesis and pathogenicity. Journal of Basic Microbiology. 2016;**56**(5):440-447

[108] Williams P. Quorum sensing, communication and cross-kingdom signalling in the bacterial world. Microbiology. 2007;**153**(12):3923-3938

[109] Hazen KC, Cutler JE. Autoregulation of germ tube formation by *Candida albicans*. Infection and Immunity. 1979;**24**(3):661-666

[110] Hornby JM, Jensen EC, Lisec AD, Tasto JJ, Jahnke B, Shoemaker R, et al. Quorum sensing in the dimorphic fungus *Candida albicans* is mediated by farnesol. Applied and Environmental Microbiology. 2001;**67**(7):2982-2992

[111] Ramage G, Saville SP, Wickes BL, López-Ribot JL. Inhibition of *Candida albicans* biofilm formation by farnesol, a quorum-sensing molecule. Applied and Environmental Microbiology. 2002;**68**(11):5459-5463

[112] Alem MAS, Oteef MDY, Flowers TH, Douglas LJ. Production of tyrosol by *Candida albicans* biofilms and its role in quorum sensing and biofilm development. Eukaryotic Cell. 2006;**5**(10):1770-1779

[113] Chen H, Fujita M, Feng Q, Clardy J, Fink GR. Tyrosol is a quorum-sensing molecule in *Candida albicans*. Proceedings of the National Academy of Sciences. 2004; **101**(14):5048-5052

[114] Förster TM, Mogavero S, Dräger A, Graf K, Polke M, Jacobsen ID, et al. Enemies and brothers in arms: *Candida albicans* and gram-positive bacteria. Cellular Microbiology. 2016;**18**(12):1709-1715

[115] Cugini C, Calfee MW, Farrow JM, Morales DK, Pesci EC, Hogan DA. Farnesol, a common sesquiterpene, inhibits PQS production in *Pseudomonas aeruginosa*. Molecular Microbiology. 2007;**65**(4):896-906

[116] Hogan DA, Vik A, Kolter R. A *Pseudomonas aeruginosa* quorum-sensing molecule influences *Candida albicans* morphology. Molecular Microbiology. 2004;**54**(5):1212-1223

[117] Fernandes RA, Monteiro DR, Arias LS, Fernandes GL, Delbem ACB, Barbosa DB. Biofilm formation by *Candida albicans* and *Streptococcus mutans* in the presence of farnesol: A quantitative evaluation. Biofouling. 2016;**32**(3):329-338

[118] Koo H. Inhibition of *Streptococcus* mutans biofilm accumulation and polysaccharide production by apigenin and tt-farnesol. The Journal of Antimicrobial Chemotherapy. 2003;**52**(5):782-789

[119] Kim D, Sengupta A, Niepa THR, Lee B-H, Weljie A, Freitas-Blanco VS, et al. *Candida albicans* stimulates *Streptococcus mutans* microcolony development via cross-kingdom biofilm-derived metabolites. Scientific Reports. 2017;**7**(1):41332

[120] Jarosz LM, Deng DM, van der Mei HC, Crielaard W, Krom BP. *Streptococcus mutans* competence-stimulating peptide inhibits *Candida albicans* hypha formation. Eukaryotic Cell. 2009;**8**(11):1658-1664

[121] Sztajer H, Szafranski SP, Tomasch J, Reck M, Nimtz M, Rohde M, et al. Cross-feeding and interkingdom communication in dual-species biofilms of *Streptococcus mutans* and *Candida albicans*. The ISME Journal. 2014;**8**(11):2256-2271

[122] Jack AA, Daniels DE, Jepson MA, Vickerman MM, Lamont RJ, Jenkinson HF, et al. *Streptococcus gordonii* comCDE (competence) operon modulates biofilm formation with *Candida albicans*. Microbiology. 2015;**161**(Pt 2):411-421

[123] Bachtiar EW, Bachtiar BM, Jarosz LM, Amir LR, Sunarto H, Ganin H, et al. AI-2 of *Aggregatibacter actinomycetemcomitans* inhibits *Candida albicans* biofilm formation. Frontiers in Cellular and Infection Microbiology. 2014;**4**:94

[124] Akiyama H, Oono T, Huh W-K, Yamasaki O, Ogawa S, Katsuyama M, et al. Actions of farnesol and xylitol against *Staphylococcus aureus*. Chemotherapy. 2002;**48**(3):122-128

[125] Unnanuntana A, Bonsignore L, Shirtliff ME, Greenfield EM. The effects of farnesol on *Staphylococcus aureus* biofilms and osteoblasts. An in vitro study. The Journal of Bone and Joint Surgery. American Volume. 2009;**91**(11):2683-2692

[126] Kong EF, Tsui C, Kucharíková S, Van Dijck P, Jabra-Rizk MA. Modulation of *Staphylococcus aureus* response to antimicrobials by the *Candida albicans* quorum sensing molecule farnesol. Antimicrobial Agents and Chemotherapy. 2017;**61**(12):e01573-17

[127] Décanis N, Savignac K, Rouabhia M. Farnesol promotes epithelial cell defense against *Candida albicans* through Toll-like receptor 2 expression, interleukin-6 and human β-defensin 2 production. Cytokine. 2009;**45**(2):132-140

[128] Leonhardt I, Spielberg S, Weber M, Albrecht-Eckardt D, Bläss M, Claus R, et al. The fungal quorum-sensing molecule farnesol activates innate immune cells but suppresses cellular adaptive immunity. MBio. 2015;**6**(2):e00143-15

[129] Gilbert P, Das J, Foley I. Biofilm susceptibility to antimicrobials. Advances in Dental Research. 1997;**11**(1):160-167

[130] Burmølle M, Webb JS, Rao D, Hansen LH, Sørensen SJ, Kjelleberg S. Enhanced biofilm formation and increased resistance to antimicrobial agents and bacterial invasion are caused by synergistic interactions in multispecies biofilms. Applied and Environmental Microbiology. 2006;**72**(6):3916-3923

[131] Roberts AP, Kreth J. The impact of horizontal gene transfer on the adaptive ability of the human oral microbiome. Frontiers in Cellular and Infection Microbiology. 2014;**4**(September):1-9

[132] Gulati M, Nobile CJ. *Candida albicans* biofilms: development, regulation, and molecular mechanisms. Microbes and Infection. 2016;**18**(5):310-321

[133] Stewart PS. Mechanisms of antibiotic resistance in bacterial biofilms. International Journal of Medical Microbiology. 2002;**292**(2):107-113

[134] Vediyappan G, Rossignol T, D'Enfert C. Interaction of *Candida albicans* biofilms with antifungals: Transcriptional response and binding of antifungals to beta-glucans. Antimicrobial Agents and Chemotherapy. 2010;**54**(5):2096-2111

[135] Mulcahy H, Charron-Mazenod L, Lewenza S. Extracellular DNA chelates cations and induces antibiotic resistance in *Pseudomonas aeruginosa* biofilms. PLoS Pathogens. 2008;**4**(11):e1000213

[136] Connell SR, Tracz DM, Nierhaus KH, Taylor DE. Ribosomal protection proteins and their mechanism of tetracycline resistance. Antimicrobial Agents and Chemotherapy. 2003;**47**(12):3675-3681

[137] Domingues S, Nielsen KM. Membrane vesicles and horizontal gene transfer in prokaryotes. Current Opinion in Microbiology. 2017;**38**:16-21

[138] Ciofu O, Beveridge TJ, Kadurugamuwa J, Walther-Rasmussen J, Høiby N. Chromosomal beta-lactamase is packaged into membrane vesicles and secreted from *Pseudomonas aeruginosa*. The Journal of Antimicrobial Chemotherapy. 2000;**45**(1):9-13

[139] Singh S, Singh SK, Chowdhury I, Singh R. Understanding the mechanism of bacterial biofilms resistance to antimicrobial agents. The Open Microbiology Journal. 2017; **11**(1):53-62

[140] Martins M, Henriques M, Lopez-Ribot JL, Oliveira R. Addition of DNase improves the in vitro activity of antifungal drugs against *Candida albicans* biofilms. Mycoses. 2012;**55**(1):80-85

[141] Høiby N, Bjarnsholt T, Givskov M, Molin S, Ciofu O. Antibiotic resistance of bacterial biofilms. International Journal of Antimicrobial Agents. 2010;**35**(4):322-332

[142] Wang Z, De La Fuente-Núñez C, Shen Y, Haapasalo M, Hancock REW. Treatment of oral multispecies biofilms by an anti-biofilm peptide. PLoS One. 2015;**10**(7):1-16

[143] Zhang T, Wang Z, Hancock REW, De La Fuente-Núñez C, Haapasalo M. Treatment of oral biofilms by a D-enantiomeric peptide. PLoS One. 2016;**11**(11):1-16

[144] De La Fuente-Núñez C, Reffuveille F, Mansour SC, Reckseidler-Zenteno SL, Hernández D, Brackman G, et al. D-Enantiomeric Peptides that eradicate wild-type and multi-drug-resistant biofilms and protect against lethal *Pseudomonas aeruginosa* infections. Chemistry & Biology. 2015;**22**(2):196-205

[145] Shankar Raut J, Mohan Karuppayil S. Phytochemicals as inhibitors of *Candida* biofilm. Current Pharmaceutical Design. 2016;**22**(27):4111-4134

[146] Basavaraju M, Sisnity VS, Palaparthy R, Addanki PK. Quorum quenching: Signal jamming in dental plaque biofilms. Journal of Dental Sciences. 2016;**11**(4):349-352

[147] Lazar V. Quorum sensing in biofilms—How to destroy the bacterial citadels or their cohesion/power? Anaerobe. 2011;**17**(6):280-285

[148] Hisajima T, Maruyama N, Tanabe Y, Ishibashi H, Yamada T, Makimura K, et al. Protective effects of farnesol against oral candidiasis in mice. Microbiology and Immunology. 2008;**52**(7):237-333

[149] Rosseti IB, Chagas LR, Costa MS. Photodynamic antimicrobial chemotherapy (PACT) inhibits biofilm formation by *Candida albicans*, increasing both ROS production and membrane permeability. Lasers in Medical Science. 2014;**29**(3):1059-1064

[150] Costa ACBP, Campos Rasteiro VM, Da Silva Hashimoto ESH, Araújo CF, Pereira CA, Junqueira JC, et al. Effect of erythrosine- and LED-mediated photodynamic therapy on buccal candidiasis infection of immunosuppressed mice and *Candida albicans* adherence to buccal epithelial cells. Oral Surgery, Oral Medicine, Oral Pathology, Oral Radiology. 2012;**114**(1):67-74

Section 4

New Therapeutic and Management

Chapter 7

Nanoparticles as New Therapeutic Agents against *Candida albicans*

Hilda Amelia Piñón Castillo,
Laila Nayzzel Muñoz Castellanos,
Rigoberto Martínez Chamorro,
Reyna Reyes Martínez and
Erasmo Orrantia Borunda

Additional information is available at the end of the chapter

http://dx.doi.org/10.5772/intechopen.80379

Abstract

Candida albicans is an opportunistic dimorphic yeast. This organism is pathogen associated to superficial and systemic infections. Actually, *Candida albicans* represents an emergent pathogen especially in a patient with some immunity compromises. Added to this, the use of antifungal in an indiscriminate form has increased the resistance of the existing drugs. In this aspect, the nanotechnology generates the possibility of creating new therapeutic agents. Nanoparticles are structures of 1–100 nm with special physicochemical characteristics that allow it to function as therapeutic agents or as carriers of these. Palladium, silver, and gold metallic nanoparticles and iron, titanium, zinc, and copper oxides have been used as growth inhibitors. These nanoparticles have been proved alone or in form of nanocomposites. The objective of this chapter is to describe the state of the art of the use of nanoparticles as inhibitors of the growth of *Candida albicans*, as well as the most relevant results regarding the mechanisms involved in this inhibition.

Keywords: nanotechnology, resistance, metallic and oxide nanoparticles

1. Introduction

Candidiasis is one of the most common infections worldwide. These diseases are generated by *Candida albicans* and *Candida* non-*albicans* group. *Candida albicans* is an opportunistic dimorphic yeast. It is a commensal organism in gastrointestinal, respiratory, and genitourinary tracts. Normally, it is a harmless organism that can turn into an opportunistic organism in immunocompromised or immunologically deficient individuals [1–3].

Candida albicans represents the predominant yeast in oral, genital, and cutaneous infections (≥90%) [2, 4, 5]. However, the invasion of blood torrent by *Candida* species is the infection more frequently diagnosed. The frequency of *Candida albicans* isolated in candidemia is of 46.3%, followed by C. *glabrata* (24.4%) [3, 6]. The mortality frequency in patients with invasive candidemia ranges from 40 to >70%, generating a loss of about 1.7 billion dollars every year [1, 7].

For the treatment of invasive infection by *Candida,* few options are available. The polyenes (amphotericin B and nystatin) are the chosen treatment by excellence. However other antifungals are used as azoles [8]. The azoles (fluconazole and itraconazole) and new azoles (ravuconazole, posaconazole, and voriconazole) are selected for prophylaxis, empirical therapy, and diagnostic-driven therapy [9]. The use of echinocandins (caspofungin, anidulafungin, and micafungin) is driven to therapy on invasive candidiasis [1].

However, the adverse effects of antifungal therapy as nephrotoxicity, red blood cell (RBC) toxicity, and arrhythmias for amphotericin B, hepatotoxicity associated with fluconazole, cardiotoxicity and gastrointestinal (GIT) disturbances attributed to itraconazole, voriconazole exhibiting neurological and hepatic toxicities, and posaconazole shown to elevate serum aminotransferase levels and cause mild-to-moderate hepatic toxicity limit its use. These adverse effects are attributed to the action mechanism. Azoles and polyenes have ergosterol as a target from the cell membrane, and the ergosterol has a high similarity between ergosterol and human cholesterol [8]. Azoles inhibit the ergosterol biosynthesis, specifically the lanosterol 14α-demethylase. This generates sterol precursor accumulation, resulting in a cell membrane instability and the loss of viability [3].

The echinocandins are more specific for fungi. This therapeutic antifungal inhibits the enzyme 1,3-b-D-glucan synthase and generates the depletion of polymer synthesis and osmotic stability loss [1, 10].

Due to the discriminated antifungal agent use, *Candida albicans* has developed different resistance mechanisms. Three principal mechanisms have been reported: the efflux pump, which is a membrane transport protein that ejects the azoles out of the cell; the structural changes of the enzyme target, mutations on lanosterol 14 α-demethylase, changing the three-dimensional structure, and, as consequence, the decreased affinity for this enzyme; and the last mechanism is the increase in the production of the target enzyme; this generates the inhibition of only a portion of the enzyme, and the rest are still active [1–3, 8, 11].

On the other hand, the survival characteristics of *Candida albicans* in the host make this organism difficult to kill. The virulence factors include the biofilm formation capacity, the yeast-hyphae morphogenesis, the adhesion to surfaces, and the capacity to secrete enzymes as hydrolases.

Because of this, the research for more specific antifungal therapies has been the focus of attention. In this area, the nanotechnology offers an opportunity to design and create a new pharmaceutical agent that can be specific. As pharmaceutical agent nanoparticles such as copper, silver, and palladium, among others, have been employed in order to inhibit the growth of any microorganism, in particular, *Candida albicans*.

The objective of this chapter is to describe the state of the art of the use of nanoparticles as inhibitors of the growth of *Candida albicans*, as well as the most relevant results regarding the mechanisms involved in this inhibition.

2. *Candida albicans* taxonomy

Fungi form one of the largest eukaryotic kingdoms; they have a very broad function of decomposing biopolymers and different compounds in dead or alive hosts, as well as synthesizing bioactive compounds. The yeast *Candida albicans* originally was described as Blastomycetes class belonging to Monilial order and Cryptococcaceae family. Actually, with gene sequencing, the new taxonomy classifies *Candida albicans* into *Ascomycota*, (dikaryon-producing fungi), *Hemiascomycetes* subclass, and *Saccharomycotina* order [12, 13].

Candida albicans is a common commensal on mucosal surfaces under normal conditions, mainly of the gastrointestinal and vaginal tracts. The commensal stage is frequently described as harmless to the host, and the fungus is controlled by the normal microbiota, epithelial barriers, and the innate immune system [14, 15]. This is the most important opportunistic medical pathogenic fungus, causing candidiasis, which is an acute, subacute, or chronic infection involving any part of the body [15]. The weakening of the defense mechanisms of the host and the adaptation capacity of the microorganism turn this fungus into an aggressive pathogen, starting with superficial infections into a systemic process, mainly in immunosuppressed or HIV patients [16].

3. *Candida albicans* cell wall

Candida albicans cell wall is composed mainly of carbohydrates; three basic main constituents are polysaccharides:

1. Branched polymers of glucose with β-1,3 and β-1,6 bonds, they are called β-glucans.
2. Unbranched polymers of N-acetyl-D-glucosamine (GlcNAc) with β-1,4 bonds; the chitin, which contributes to the fiber insolubility; and theβ-1,3 glucan-chitin complex, which is the major constituent of the inner wall. β-1,6 glucan links the components of the inner and outer walls; thus, chitin and β-1,3 are structural polysaccharides and convey strength and shape to the cell wall [15, 17, 18].
3. Finally mannose polymers, mannans, covalently associated with proteins (glucomannan proteins).

The mannan of the outer cell wall is less structured but allows this fungus breakage of epithelial barriers or dysfunction of the immune system converting *Candida albicans* from a commensal to a pathogen. The cell walls also contain proteins, 6–25%, and minor amounts of lipids, 1–7% [15, 17, 18].

This arrangement is responsible for its cell morphology and provides protection to the cell against physical, chemical, and biological (immune host system) aggression [16].

Likewise, the cell wall participates in both growth and morphological transitions between yeast and hyphae; this conversion from the unicellular yeast to multicellular filamentous fungus is essential for *Candida* virulence [19]. It has cell surface proteins linked to glycosylphosphatidylinositol (GPI) [20], including an N-terminal signal sequence and a C-terminal GPI-anchor addition signal; some of these proteins are secreted to the extracellular environment, but do not remain cell associated. Some of them are hydrolytic enzymes, which hydrolyze complex substrates into small units, to carry out into the cell as a source of nutrition. If these polymers are host targets, all of these facilitate colonization or invasion of *Candida albicans*, and then such enzymes also function as virulence factors [17]. For therapeutic strategies, to establish the functional relationships between *Candida albicans* cell wall's proteins and virulence is a challenge [20].

4. *Candida albicans* virulence factors

Candida spp. invade several anatomical hosts, mainly in patients who are immunocompromised or debilitated. This organism expresses several virulence factors that allow the host-pathogen interaction and that lead to the establishment of infection [21, 22].

Virulence factors include host recognition biomolecules (adhesins) due to cell surface hydrophobicity which enhanced adherence and resistance to phagocytosis; another important factor is its ability to grow either in yeast or in hyphal form (dimorphism, especially in the presence of N-acetyl-D-glucosamine) [23], secreting aspartyl proteases and phospholipases [21].

The phenomenon of the quorum sensing (QS) has been described as a contributor to morphogenesis, playing a role in regulating the yeast to filamentous fungus; one of the conditions is the dependence on cell density associated with the regulation by QS. The QS has an important role in the ability of *Candida albicans* to be virulent and survive within the host [23]. *Candida albicans* uses multiple redundant and/or synergistic signaling pathways to integrate host signals to promote hyphal development, tissue invasion, and virulence during infection, exacerbating the pathogenesis of invasive candidiasis [24].

5. *Candida albicans* biofilm formation

Candida virulence is a result of its capacity to form biofilms on the surface of biomaterials and on catheters and prosthetic devices [25–27]. *Candida albicans* biofilms are composed of hyphae, pseudohyphae, yeast cells, and an extracellular matrix, where the hyphae play an integral role within this complex [26]. Biofilm formation presents an initial adherence of yeast cells (0–2 h), followed by germination and microcolony formation (2–4 h), filamentation (4–6 h), monolayer development (6–8 h), proliferation (8–24 h), and maturation (24–48 h). Serum and saliva preconditioning films and *Candida albicans* preconditioning films increased the initial attachment of the fungus but had minimal effect on subsequent biofilm formation [25, 28]. This kind of biofilm has clinical repercussions because it is the target of both chemical and immunological antifungals. Due of their intrinsic resistance to almost all antifungals in clinical use, increased resistance to this therapy, and the ability of cells within biofilms to withstand host immune defenses, the antifungal resistance of biofilms results most probably from the conjunction of several mechanisms that act in a time-dependent manner [26].

6. *Candida albicans* resistance mechanisms

In order for *Candida albicans* to survive, these microorganisms have developed diverse mechanisms. Biofilm formation, enzyme modification, and the pump efflux are the mechanisms reported to antibiotic resistance [29]. As described previously, biofilms are a complex of cells, DNA, and polymeric matrix, among others. The polymeric matrix has a crucial role in the antifungal resistance since it does not allow antibiotic penetration inside the biofilm. This phenomenon helps the internal cell to survive [25].

Resistance to azole antifungal has been described in three forms: [1] change in the affinity of cytochrome P450 sterol 14α-demethylase (gen Erg11p), [2] increase in the enzyme expression, and [3] changes in the sterol desaturase. Other alterations in an enzyme involved in the cell wall construction have been described [29, 30].

The pump efflux ejects the azoles outside the cell, without any effect inside of it. This efflux is mediated by an ATP-binding cassette (ABC) and the major facilitator superfamily. This is a complex family of genes that involve several CDR genes, including core membrane pore composed by ABC segments in the membrane; two segments are in the cytosolic side, which provide energy to pump the antibiotic outside of the cell [25, 29, 30].

It is an important highlight that this resistance could be a derivative in the resistance to other antifungals.

Candida albicans resistance could be due to one or a combination of all the resistance mechanisms described.

7. Nanotechnology as an antifungal

Nanotechnology is a new technology which manipulates the matter in a nanoscale order (1×10^{-9} m). In this order, the nanoparticles have been used as antimicrobial, as antifungal, or as carried molecules. Nanoparticles have a size from 1–100 nm on any of this dimension. In this scale, nanoparticles have particularity physicochemical characteristics [31]. Within these characteristics, their electron configuration confers an extraordinary quantum effect. These quantum effects enhance their pharmacokinetic effect and generate nanoparticles that can be used as a protein, antibodies, or inclusive specific molecule carriers [32].

As an antifungal, nanoparticles show cell wall damage, an oxidative stress increase, and DNA interaction [33]. Nevertheless, the NPs toxicity mechanisms are dependent on the nanoparticle nature, size, and shape and capping nature, among other characteristics.

This chapter reviewed some NPs characteristics such as nanoparticle nature, synthesis, size, and shape and their influence on the toxic mechanisms in *Candida albicans* and their influence on the virulence factors.

7.1. Silver nanoparticles

Silver has been used for several decades as a disinfectant. Inclusively, silver nitrate was used in burns in order to control infections. However, with the discovery of penicillin, the use of silver was slow. Nowadays, the increase of resistance to antifungal puts attention again on the use of silver as an antibiotic, though the use of silver ions implies complications in their solubility and availability [34].

In this sense, the silver nanoparticles (AgNPs) offer a possibility. AgNPs offer the possibility of using minimal doses and act in microorganisms which present antibiotic resistance or decrease the virulence factors.

The use of different AgNPs synthesis methods caused different effects. The general method is the salt reduction by a reducing agent. However, the addition stabilizing agent has been proved. The stabilizing or capping agent limits the NPs' growth, avoids the agglomeration, and controls the shape, size, and superficial charge [35].

The NP synthesis by plant extract or secondary metabolites from microorganisms offers a new possibility. The phenolic compounds, proteins, anthocyanin, and carboxylic acid, among others, work as reducing agent and stabilizing or capping agent [35, 36].

The green AgNPs synthesis through the use of microorganisms as the use of Actinobacteria *Pilimelia columellifera* subsp. pallida [37], *Bacillus* species [38, 39], filamentous fungi *Monascus purpureus* [7], and Macromycetes as *Pleurotus sajor-caju* [40], as well as extract plant as *Tulsi* leaf extract [41] or extract from *Phoenix dactylifera* [34] has been proven. All these methods present inhibition of *Candida albicans*, with different efficiencies (**Table 1**).

The microorganisms *Bacillus amyloliquefaciens* 1853 and *Bacillus subtilis* 10,833 have been used for the AgNPs synthesis. Culture media were employed as a reducing agent. The scanning electronic microscopy has shown AgNPs produced with capping. *Bacillus* strains have been used in the AgNPS production. *Bacillus amyloliquefaciens* strain 1853 produced AgNPs with a size of 1.071 and other of 73.83 nm. The *Bacillus* subtilis strain 10,833 produced tree population sizes of AgNPs at 1.863, 10.866, and 135.78 nm. AgNPs produced by both microorganisms were employed to evaluate the *Candida albicans* inhibition. No differences were found between AgNPs; inclusive addition of antibiotic with the AgNPs does not improve the inhibition [38].

http://dx.doi.org/10.5772/intechopen.80379

Inhibition test	NPs size (nm)	Synthesis method	Capped	Reference
18 mm (inhibition zone)	15–20	*Cassia roxburghii* + *Ketoconazole*	NR	Moteriya et al. 2017 [42]
64 μg/mL (MIC)	12.7	*Pilimelia columellifera* subsp. pallida SL19	Proteins	Wypij et al. 2017 [37]
16.7 ± 0.25 mm (inhibition zone)	2–8	*Monascus purpureus*	NR	El-Baz et al., 2016 [7]
10 μg/mL (BIC)	40–55	*Dodonaea viscosa* and *Hyptis suaveolens*	Detection of functional groups as phenol, alcohol, proteins, heterocyclic amines	Muthamil et al. 2018 [43]
40 μg/mL (MIC)	5–95	*Bacillus safensis*	Proteins	Lateef et al. 2015 [38]
12 mm (inhibition zone with 30 μL of solution)	18–46	*Gracilaria corticata*	Detection of functional groups as alcohol, carbonyl, amino groups	Kumar et al. 2013 [44]
250 μg/mL (MIC) and 500 μg/mL (MFC)	3.77–33.22	*Pleurotus sajor-caju*	NR	Musa et al. 2018 [40]
60 μg/mL (MIC_{90}) and 120 (MFC)	2–7	Tulsi leaf extract	NR	Khatoon et al., 2015 [45]
20 mm (inhibition zone with 80 μg/20 mL)	21.65–41.05	*Phoenix dactylifera*	Proteins	Oves et al. 2018 [34]
75 ppm reduce 83%/150 ppm (MFC)	70	Red cabbage	Anthocyanin compounds	Ocsoy et al. 2017 [46]
08 μg/mL (MIC)	9–130	*Enterococcus faecalis*	NR	Ashajyothi 2016 [47]
40–60 μg/mL(MIC)	4–39	Citrus lemon aqueous Juice + CTAB	Surrounded by a layer	Rahisuddin et al. 2015 [48]
10.8 ± 0.8 (inhibition zone)	12–85	Clove extract	Detection of functional groups such as methoxy, alcohol, carboxilic acid, aliphatic group	Parlinska-Wojtan et al. 2017 [49]
21–12 cm in combination with antifungal (inhibition zone)	10–90	Flower broth of *Tagetes erecta*	NR	Padalia et al. 2015 [50]
12.14 μg/mL (MIC)	47.0 ± 2.0	*J. curcas*	Protein layer	Kumar et al. 2017 [51]
10.78 μg/mL (MIC)	7.6 ± 0.5	*L. grandis*		

Inhibition test	NPs size (nm)	Synthesis method	Capped	Reference
70% fold increase (combination with antifungal, inhibition zone)	1.836–135.78	*B. subtilis* 10,833	NR	Ghiut et al. 2018 [39]
50% fold increase (combination with antifungal, inhibition zone)	1.012–73.83	*B. amyloliquefaciens*	NR	
14 mm(inhibition zone)	5–10	*Ocimum sanctum* + CTAB	Not visible	Aazam and Zaheer 2016 [35]
30 μg/mL (MIC)	4–36	Actinobacterial strain SF23	NR	Anasane et al. 2016 [52]
40 μg/mL (MIC)	8–60	Actinobacteria C9	NR	

MIC, minimal inhibition concentration.
MFC, minimal fungicide concentration.
NR, not reported.
BIC, biofilm inhibitory concentration.

Table 1. Biogenic synthesized AgNPs against ***Candida albicans.***

Clove extract was used in order to produce AgNPs. The high content of eugenol, β-caryophyllene, humulene, chavicol, methyl salicylate, α-ylangene, and eugenone; flavonoids such as eugenin, rhamnetin, kaempferol, and eugenitin; triterpenoids like oleanolic acid, stigmasterol, and campesterol; and several sesquiterpenes allowed the clove to act as a reducing and capping agent in AgNPs synthesis. The DLS analysis showed AgNPs size of 12 and 85 nm. The AgNPs produced to generate a complete inhibition of *Candida albicans* in 40-fold dilution were applied. The results obtained suggest that the functional groups' incorporation into the AgNPs generates a synergism between clove and NPs [49].

In order to evaluate the AgNPs effect on biofilms, some AgNPs have been evaluated. As described, biofilms are a *Candida albicans* virulence factor. The change from yeast to hyphae is crucial for the biofilm process. In this process, *Candida albicans* produces a polysaccharide matrix, carbohydrates, proteins, and signaling molecules [43]. Changes in the structural biofilm conformation of *Candida albicans* exposed to AgNPs were reported [28, 51, 53, 54].

The use of latex plants in the AgNPs green synthesis was proved. *J. curcas* and *L. grandis* extracts produced AgNPs with a size of 47 ± 2.0 nm and 7.6 ± 0.5 nm, respectively. The Fourier transform infrared (FTIR) spectra suggest the incorporation of phenolic OH, amides, amines, and aldehyde in the AgNPs as a stabilizing agent. The smallest AgNPs showed a better effect in plankton and biofilms of *Candida albicans* than the big ones. The AgNPs synthesized proved to be efficient in the inhibition of biofilm formation, changes in morphology and cell lysis when candida albicans was exposed [51].

Also, silver nitrate chemical reduction by sodium citrate was evaluated in intermediate and mature *Candida albicans* biofilms. The preformatted biofilms were exposed to 54 μg/mL of AgNPs. After 5 h of exposition, biofilms showed a decrease in the number of viable cells, and no differences were found in intermediate or mature biofilm [28].

The results obtained show the importance of the NPs size, being the smallest nanoparticles which show the highest inhibition [51]. Lara et al. (2015) showed that biofilm exposed to AgNPs exhibits a few *Candida albicans* filamentous formations. Electron microscopy analysis showed AgNPs accumulation and internalization [55].

Besides the AgNPs effect in biofilm, the AgNPs have been used alone showing the inhibition, cell wall damage, and incorporation of Ag into the cell. It was used in combination with different antifungals amphotericin b [51, 56] or nystatin and chlorhexidine digluconate, showing a synergic effect [54]. Other molecules as cationic carboxilane have been used, proving a high *Candida albicans* inhibition with small AgNPs concentration [1.8 mg/mL; [57]]. Also, the AgNPs have been capped with L-3,4-dihydroxyphenyl-alanine, showing a minimal fungicide concentration of 31.2–62.5 μg/mL [58].

The direction in the investigation with silver nanoparticles has been driven in the green synthesis due to the low toxicity generated in the human cell and the increase of the toxicity in bacteria and yeast [59]. Their chemical characteristics as a metallic oxidative state and superficial area, among others, allow the NPs to interact with the microorganism and inhibit their growth.

7.2. Zinc oxide nanoparticles

Due to ZnO low toxicity, ZnO has been used in a medical device such as drug carriers, antibacterial agents, and bioimaging probes, among others. For this reason, the ZnO nanoparticles (ZnONPs) have been used in *Candida albicans* control. The ZnONPs application is reflected by the annual production between 550 and 5500. This is more than 10–100 times higher than any other nanoparticle [60]. As the AgNPs, the ZnONPs characteristics depend on the size, synthesis method, and superficial charge, among others [61]. In order to control these characteristics, the green synthesis method has been used to fabricate ZnONPs with a size, shape, and superficial charge with the desirable characteristic [62, 63].

A promissory ZnONPs application area is their use as antimicrobial. Their effect has been evaluated in soil, plants, bacteria, and fungi [60, 64–66] ZnONPs different synthesis methods in ZnONPs have a big influence in the *Candida albicans* inhibition [68].

The use of *Atalantia monophylla* extracts produced ZnONPs 30 nm that have effect inhibition in *Candida albicans* growth. The author proposed the ZnONPs interaction with the cell wall, penetration to the cell, ion liberation, and the oxidative stress production, triggering the protein, lipid, and DNA oxidation [67].

The microwave aqueous extract from *Suaeda aegyptiaca* produced ZnONPs with a size of 80 nm with a capping produced by metabolites from the plant. The effect of the capping was studied. ZnONPs produced by precipitation method were studied and compared with the plant extract produced. ZnONPs *Suaeda aegyptiaca* produced a major inhibitory effect than the ZnONPs without capping agent, suggesting a synergic effect [68].

Extracts for different plants have been proved to synthesize ZnONPs in order to evaluate the toxicity of the resulting nanoparticles in *Candida albicans* inhibition (**Table 2**).

ZnONPs hybrid has been synthesized using chitosan, gelatin, and polystyrene, among others [75–77].

Dhillon et al. 2014 [75] used chitosan as a capping and bridge to carry on citric acid, glycerol, starch, and whey powder using the nanospray drying method. The ZnONPs produced have a size range of 93.2–402.5 nm. The smallest ZnONPs were the chitosan starch NPs. Aggregation effect was observed in the largest ZnONPs; this phenomenon could be explained by the superficial charge. However, these ZnONPs have a small effect in *Candida albicans* inhibition.

Collagen was used as a reducing and capping agent. The ZnONPs produced with this method showed a size between 20 and 50 nm negatively charged. ZnONPs coated by collagen

Inhibition test	NPs size (nm)	Synthesis method	Capped	Reference
24 mm (inhibition zone)	30	*Atalantia monophylla*	NR	Vijayakumar et al. 2018 [67]
80 μM (MIC)/>2590 μM (MFC)	8	*Chelidonium majus*	NR	Dobrucka et al. 2018 [69]
34 ± 1.28 mm (inhibition zone with 100 g/ml)	32–40	*Glycosmis pentaphylla*	Alkane, C=O stretching, other groups	Vijayakumar et al., 2018 [70]
25 μg/mL (MIC)/50 μg/mL (MFC)	18	*Azadirachta indica*	Proteins	Elumalai and Velmurugan 2015 [71]
Strong activity	36	*S. myriocystum*	Protein, carbohydrates, flavonoids, tannins, mannitol	Nagarajan and Kuppusamy 2013 [72]
12.5 mm (inhibition zone with 150 mg/ml)	68.64	*Cassia auriculata*	Alkene, alcohol, ether, and alkane	Padalia et al. 2018 [73]
20.3 ± 0.57 mm (inhibition zone with 200 μg/mL)/25 μg/mL (MIC) 50 μg/mL (MFC)	10–30	*Vitex trifolia*	Proteins, alkanes, aromatic group	Elumalai et al. 2015 [71]
18.01 mm (inhibition zone 100 μg/mL)	80	*Suaeda aegyptiaca*	Alcohol, aldehyde, amine	Rajabi et al. 2018 [68]
Inhibition test	**NPs size (nm)**	**Synthesis method**	**Capped**	**Reference**
24 mm (zone inhibition)	30 nm	*Atalantia monophylla*	NR	Vijayakumar et al. 2018 [67]
80 μM (MIC)/>2590 μM (MFC)	8 nm	*Chelidonium majus*	NR	Dobrucka et al. 2018 [74]

MIC, minimal inhibition concentration.
MFC, minimal fungicide concentration.
NR, not reported.

Table 2. Biogenic synthesized ZnONPs against *Candida albicans*.

showed a MIC of 99.7 ± 0.99 μg/mL, whereas the control with zinc acetate showed an MIC of 297.9 ± 2.0 μg/mL. Biofilms exposed to increase the concentration of ZnONPs showed inhibition as a result of the ZnONPs concentration [78].

Gelatin was used for the ZnONPs as well. ZnONPs produced with gelatin showed a diameter of 20 nm with a negative superficial charge. The ZnONPs produced were employed to biofilm exposition. ZnONPs in a concentration of 50 μg/mL showed a thickness diminution, resulting in a weak adherence biofilm compared with the biofilm treated with fluconazole [72].

Also, egg albumin was used as a ZnONPs template. The synthesis method produced spherical ZnONPs with a range between 10 and 20 nm. The FITR analysis showed the interaction of egg albumin and ZnONPs. In this work, ZnONPs characterization in the culture media showed importance. The use of 45 μg/ml in Sabouraud's dextrose (SD) nutrient media does not significantly affect ZnONPs stability, size, and integrity. The synthesized ZnONPs showed *Candida albicans* inhibition in a dose-dependent manner. Changes in the *Candida albicans* morphology show cavity formation in cells exposed to 15 μg/ml [79].

ZnONPs were evaluated both alone and as a part of nanocomposites with polystyrene. The use of ZnONPs alone showed *Candida albicans* inhibition. However, when ZnONPs are used in nanocomposites, a major concentration of ZnONPs to inhibit *Candida albicans* is necessary [77].

Nanocomposites with ZnONPs have been produced with different materials in order to increase the growth inhibition of microorganisms [45, 46]. However, the results of the antifungal activity were contradictory; ZnONPs capped by surfactant in cotton fiber have proved the inhibition of *Candida albicans* in 94% [68], while the incorporation of ZnONPs in polylactic acid films does not inhibit the *Candida albicans* growth [80].

The principal ZnONPs toxicity mechanisms reported are (1) ion liberation, (2) interaction of ZnONPs with the cell wall, and (3) stress oxidative generation [81]. The ion liberation and the interaction of NPs with the cell trigger the generation of oxidative stress; with that the cell lost their viability [79].

7.3. Copper oxide nanoparticles

Copper exhibits good characteristics such as its antimicrobial activity, chemical stability, and thermal resistance. Due to their toxicity characteristic, copper has been used by Egyptians for water disinfection. The Aztecs used copper to treat sore throats, and the Persian and Indians used it to treat eye infection and venereal ulcers [82].

Copper has an advantage in comparison with other materials, is cheaper than any other, and is easy to oxidize to copper oxide nanoparticles (CuONPs). The CuONPs can easily make nanocomposites with polymers, macromolecules inclusive of other metals [83, 84].

The CuONPs have recovered importance in the investigation due to their low cost and the variety of applications. CuONPs have been used in the pharmaceutical, medicine, and electronic industries, among others.

Diverse synthesis methods have been proved to create CuONPs including chemical [85], electrochemical [86], and green synthesis [87].

The synthesis method influences the characteristics as the size, shape, and agglomeration of the CuONPs [88].

CuONP electrochemical production with different solvents and reaction times produces different NPs sizes from 3 to 200 nm. The CuONPs were obtained in water-acetonitrile, with sodium hydroxide as the electrolyte. The CuONPs obtained in these conditions showed a size around 20 nm. The *Candida albicans* exposition to 25 and 50 μg/ml generates a delay in the growth curves. Added to that, 50 μg/ml inhibited completely *Candida albicans* growth [86].

CuONPs prepared by precipitation method coated with acetate were evaluated together with fluconazole. CuONPs coated with acetate were more toxic than CuONPs without coating. CuONPs in combination with fluconazole showed the complete *Candida albicans* inhibition. The fractional inhibitory concentration index (FICI) suggested the additive to moderate synergism between CuONPs and fluconazole [89].

The use of dispersant as ethylene glycol and Tween 80 proved to be efficient in the production of CuONPs capable to the *Candida albicans* growth inhibition. Inhibition with CuONPs was more efficient than the exposition to itraconazole [90]. Also, the effect of CuONPs was evaluated in *Candida albicans* from oral cavities [91], showing that strain isolated from the patients needs until 1000 ppm to generate an inhibitory effect in *Candida albicans*.

CuONPs capped with pyrimidine derivatives have shown a size of 10 nm. The CuONPs showed interaction with DNA and antioxidant activity. However, the CuONPs showed *Candida albicans* inhibition [92].

Nanocomposites with the incorporation of Fe [93] Cd and Ba CuONPs [94], as well as the incorporation of CuONPs into cotton fiber [95, 96], polyurethane [97], polyester [98], copolymer microgel [99], and open polyurethane foams with starch powder [100], have been proved be efficient in the *Candida albicans* growth inhibition.

As in the other nanoparticles, the green synthesis has captured the investigators' attention. Its methodology does not produce toxic subproduct in the nanoparticle production, and it covers the nanoparticles avoiding the agglomeration [31, 84, 87].

Extracts of garlic and ginger have been used for the oxidation of copper. Garlic produces the smallest CuONPs, with a size around 14.62–22.80 nm. However, ginger shows a big quantity of very small nanoparticles. *Candida albicans* has shown major inhibition when exposed to ginger ethanol extract NPs [101]. Few reports about the use of biological extract for the synthesis of CuONPs that inhibited *Candida albicans* were founded (**Table 3**).

Cu^{2+} ions realized from CuONPs are the principal toxicity mechanisms reported. However, it is not the only one. Some authors reported the CuONP accumulation in the cell wall as well as the oxidative stress increase. However, more studies are necessary in order to understand the NPs toxicity mechanisms.

Inhibition test	NPs size (nm)	Synthesis method	Capped	Reference
15.0 c ± 0.57735 (inhibition zone with 62 μg/mL)/1.93 ± 0.76376 (MIC)	31.7	Alginate	NR	El-Batal et al. 2018 [102]
10.5 (zone inhibition)	26 ± 4	*Acalypha indica*	NR	Sivaraj et al. 2014 [103]
22.5 mm (inhibition zone)	Nanostructures ~8000	*Viburnum opulus*	$Cu_3(PO4)_2$ $3H_2O$	Ildiz et al. 2017 [104]
20 mm (inhibition zone with 50 μg/mL)	NR	*Sargassum polycystum*	NR	Vishnu et al. 2016 [105]
0.15 Log_{10} growth reduction	22 ± 1 nm	Cellulose, chitosan, and keratin nanocomposites	Cellulose, chitosan, and keratin	Tram et al., 2017 [106]

Table 3. Biogenic synthesized CuONPs against *Candida albicans*.

7.4. Other nanoparticles

Other metal nanoparticles as gold [AuNPs, [36]], palladium [PdNPs, [91]], and selenium [SeNPs, [92]], among others, have been proven. The microorganisms' elimination by all these NPs has been proved to be effective.

Chemical reduction of palladium salt has produced PdNPs in an average size of 9 nm (±3.9 nm). The PdNPs showed a significative *Candida albicans* growth reduction at 150 ppm [107].

SeNPs were prepared by the reduction of selenium chloride inside *K. pneumoniae*. This synthesis produced SeNPs with a size range from 90 to 320 nm. The concentration necessary for growth inhibition of *Candida albicans* was 2000 μg/mL [109]. Another strategy for the *Candida albicans* inhibition induced the SeNPs formation inside of *Lactobacillus* and then used them as an antifungal. This strategy worked with *L. plantarum* and *L. johnsonii. Candida albicans* exposed to both microorganisms becomes more efficient than bacterium without NPs [108].

Schiff base ligand 2-((4,6-dihethoxypyr-imidine-2-yl)methyleneenamino)-6-methoxyphenol has been used in order to capped AuNPs and platinum nanoparticles (PtNPs). Nanoparticles synthetized presented a layer. The NPs showed a granular and spherical shape with size of 38.14 ± 4.5 and 58.64 ± 3.0 nm, respectively. Both nanoparticles showed greater inhibition than amphotericin [110].

Biogenic PdNP production with watery extract of *Moringa oleifera* flower was evaluated for the *Candida albicans* inhibition. However, these PdNPs had no effect in *Candida albicans* [111].

Candida albicans exposition to tellurium NPs produces complete inhibition with of ≥2000 μg/mL. Besides, the results showed a squalene accumulation and increase in the expression levels of the ERG1 gene of squalene monooxygenase enzyme [112].

7.5. Nanoparticle characteristics and toxicity and *Candida albicans* inhibition growth

The most important NPs characteristics are the size, shape, composition, superficial charge, and hydrodynamic. These characteristics directly influence their capacity to interact with the molecules or cells. Ocsoy et al. (2017) [46] indicate that the size is strictly related to charge density. The smallest of the NPs has the biggest charge density. This explains the big attraction between small NPs and their agglomeration. The NPs agglomeration is a phenomenon that changes their capacity to work as a caring or their toxicokinetic characteristic [33, 107]. In order to improve the distribution and decrease the agglomeration, a surfactant as Pluronic F® has been added to the synthesis process (**Figure 1**). Other surfactants as CTAB, SDS, cation surfactants [113], or PVP have been proved to be efficient [114].

Another synthesis methodology that improves these characteristics is the use of subproducts from plants, algae, bacteria, fungi, yeast, etc. [115]. The high content of proteins, reducing sugars or anthocyanins, works as reducing agent; additionally, some of these molecules are adhered in the NPs surface [31]. This addition gives different characteristics to the NPs produced by chemical reduction. The NPs obtained with biogenic methods offer homogeneous NPs with antioxidant characteristics that allow the addition of other molecules and control their toxicity characteristics.

The biogenic NPs production has proven to be efficient inhibiting the *Candida albicans* growth (**Tables 1–3**). The differences in the concentration necessary to inhibit their growth are directly related to NPs size, the capping nature, and the microorganisms evaluated [116]. These are evident in the work of Anasane et al. 2016 [52] et al. (2013), where two microorganisms were evaluated in the production of AgNPs, obtaining two NPs sizes. The AgNPs production by the strain SF23 was smaller than that produced by the strain C9. *Candida albicans* presented an MIC of 30 μg/mL with the small AgNPs (size 4–36 nm). Wypij et al. (2017) [37] reported the AgNPs synthesis with *Pilimelia columellifera* subsp. *pallida* SL19, with a capping

Figure 1. Scanning electronic microscopy of PdNPs produced by chemical reduction. (A) Without surfactant. (B) With Pluronic F® (5% W/V).

of proteins. The AgNPs size was on average 12.7 nm with a growth inhibition of 64 μg/mL. Tulsi leaf extract has been used to produce AgNPs with a size of 2–7 nm. The fungicide concentration in the AgNPs produced was of 120 μg/mL [25]. The production of ZnONPs by *Azadirachta indica* [61], *Vitex trifolia* [64], and *Suaeda aegyptiaca* [65], among others (**Table 2**), has been evaluated. The extract produced different ZnONPs sizes and capping (**Table 2**) that influence the growth inhibition.

Due to the NPs size (nanometer scale), diverse toxicity mechanisms have been proposed (**Figure 2**). One is the direct interaction of the nanoparticle with the cell membrane. The interaction of NPs microorganism is allowed by the NPs superficial charge but also the microorganism superficial charge. This interaction generates disruption of the cell wall and the leakage of ions and the intracellular material with the microorganism's death [117–119].

Consequent to NPs dilution, the ion interaction with the cell is another toxicity mechanism. The ions are incorporated into the cell, and they could interact with thiol groups of proteins and enzymes leading to the inhibition of crucial biological activities [82]. Also, the NPs internalization has been reported; this generates the interaction of NPs and ions with molecules of biological importance as DNA or enzymes [33, 120]. Inside the cell, the NPs or the ions provoke Fenton's type reaction. Due to this, the oxidative stress increases, and the proteins, lipids, and DNA release (**Figure 2**).

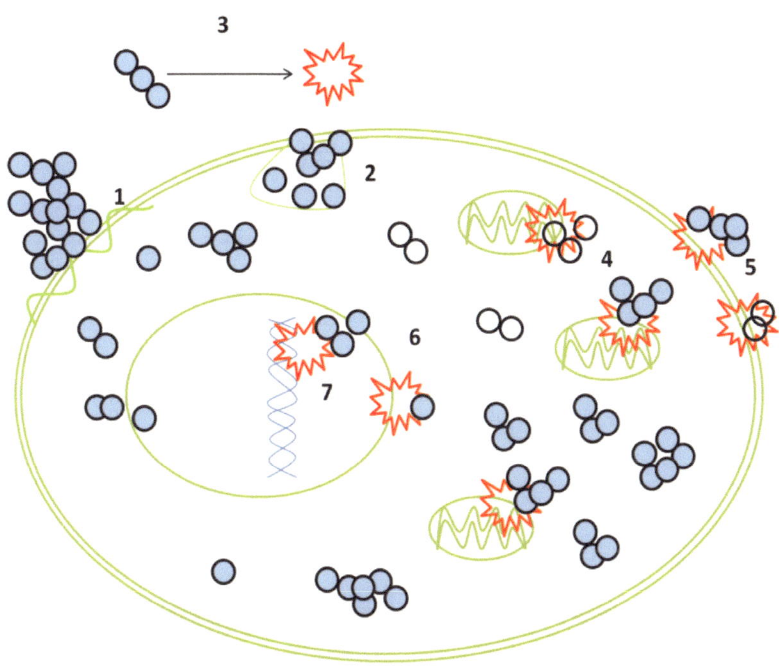

Figure 2. NPs toxicity mechanisms induced in *Candida albicans*. (1) NPs adhesion in the cell membrane causing disruption in the membrane and death. (2) NPs introduction to the cell with their posterior degradation and ion release. (3) NPs ROS generation outside the cell. (4) NPs and ions ROS generation in interaction with mitochondria. (6) Membrane lipid peroxidation with nucleus damage. (7) DNA damage by NPS and ions ROS generation.

Khan et al. (2014) [121] write "when $CoFe_2O_4P$ NCs gained a higher energy than Eg, the electrons (e^-) of $CoFe_2O_4P$ NCs were promoted across the band gap to the conduction band (Ec), which creates a hole (h+) in the valence band (Ec). These e^- in the E_C and holes in the Ev possibly have high reducing and oxidizing powers, respectively." These electron movements generate the NPs reactivity and the oxidative stress generation. The same phenomenon is reported by Setyawati et al. (2014) [122] for TiONPs. These NPs adsorbed UV light, and the adsorbed UV light generates electron excitation, creating a hole in the valence band. The electrons and holes migrate to NPs superficies where they can react with oxygen or water generating oxidative stress. The superficial e^- generates the liberation of hydroxyl increasing the mitochondria membrane depolarization and liberation of cytochrome c. Added to this, the results show DNA damage [120]. However, the oxidative stress is not the only mechanisms involved in the AgNPs toxicity; also the membrane fluidity, ergosterol content, and cellular and ultrastructure morphology are altered [123]. The squalene monooxygenase expression in *Candida albicans* exposed to TeNPs showed an increase [112]. The squalene monooxygenase catalyzes the conversion of squalene to 2,3-oxidosqualene; inhibition in this step generates the squalene increase and the disruption of the *Candida albicans* cell wall [107].

8. Conclusion

The antibiotic resistance increase in many microorganisms has encouraged researchers to focus their efforts on the synthesis and design of multiple effective compounds in order to combat the resistance mechanism of microorganisms. NPs offer a new and effective "antibiotic" that could work against the resistance mechanism in *Candida albicans*.

The biogenic NPs offer a new generation of NPs that work as a carrier. These NPs offer the possibility to drive the toxicity effect to a specific target, in such way the collateral damage could be diminished.

However, the NPs toxicity mechanisms are not completely understood. The oxidative stress is one of the described mechanisms; it is possible to investigate other mechanisms involving similarity to oxidative stress. More studies are necessary to understand the influence that nanoparticles have in the *Candida albicans* resistance mechanisms and if this exposition could not generate new ones.

Acknowledgements

This work was partially financed by F-PROMEP-38/REV-04. SEP-23-005 and ECOS NORD 263456.

Conflict of interest

The authors declare no conflict of interest.

Author details

Hilda Amelia Piñón Castillo[1]*, Laila Nayzzel Muñoz Castellanos[1],
Rigoberto Martínez Chamorro[1], Reyna Reyes Martínez[1] and Erasmo Orrantia Borunda[2]

*Address all correspondence to: hpinon@uach.mx

1 Facultad de Ciencias Químicas, Universidad Autónoma de Chihuahua, Circuito Universitario S/N, Campus II, Chihuahua, Chihuahua, México

2 Centro de Investigación en Materiales Avanzados, Chihuahua, México

References

[1] Simon J, Sun HY, Leong HN, Barez MYC, Huang PY, Talwar D, et al. Echinocandins in invasive candidiasis. Mycoses. 2013;**56**:601-609. DOI: 10.1111/myc.12085

[2] Dadar M, Tiwari R, Karthik K, Chakraborty S, Shahali Y, Dhama K. *Candida albicans*—Biology, molecular characterization, pathogenicity, and advances in diagnosis and control—An update. Microbial Pathogenesis. 2018;**117**:128-138. DOI: 10.1016/j.micpath.2018.02.028

[3] Gonçalves SS, Souza ACR, Chowdhary A, Meis JF, Colombo AL. Epidemiology and molecular mechanisms of antifungal resistance in Candida and Aspergillus. Mycoses. 2016;**59**:198-219. DOI: 10.1111/myc.12469

[4] de Bedout C, Gómez BL. Candida y candidiasis invasora: Un reto continuo para su diagnóstico temprano. Infection. 2010;**14**:159-171. DOI: 10.1016/S0123-9392(10)70133-8

[5] de Sa NP, de Paula LFJ, Lopes LFF, Cruz LIB, Matos TTS, Lino CI, et al. In vivo and in vitro activity of a bis-arylidenecyclo-alkanone against fluconazole-susceptible and resistant isolates of *Candida albicans*. The Journal of Global Antimicrobial Resistance; 2018. DOI: 10.1016/j.jgar.2018.04.012

[6] De la Torre-Saldaña VA, Mayté MV, et al. Factores de riesgo y epidemiología de la candidemia en el Hospital Juárez de México. Medicina Interna de México. 2014;**30**:121-132

[7] El-Baz AF, El-Batal AI, Abomosalam FM, Tayel AA, Shetaia YM, Yang ST. Extracellular biosynthesis of anti-Candida silver nanoparticles using *Monascus purpureus*. Journal of Basic Microbiology. 2016;**56**:531-540. DOI: 10.1002/jobm.201500503

[8] Salci TP, Negri M, Abadio AKR, Svidzinski TIE, Kioshima ÉS. Targeting Candida spp. to develop antifungal agents. Drug Discovery Today. 2018;**23**:802-814. DOI: 10.1016/j.drudis.2018.01.003

[9] Zhou S, Sun Z, Ye Z, Wang Y, Wang L, Xing L, et al. In vitro photodynamic inactivation effects of benzylidene cyclopentanone photosensitizers on clinical fluconazole-resistant *Candida albicans*. Photodiagnosis and Photodynamic Therapy. 2018;**22**:178-186. DOI: 10.1016/j.pdpdt.2018.04.001

[10] Caplan T, Polvi EJ, Xie JL, Buckhalter S, Leach MD, Robbins N, et al. Functional genomic screening reveals core modulators of echinocandin stress responses in *Candida albicans*. Cell Reports. 2018;**23**:2292-2298. DOI: 10.1016/j.celrep.2018.04.084

[11] Mane A, Vidhate P, Kusro C, Waman V, Saxena V, Kulkarni-Kale U, et al. Molecular mechanisms associated with Fluconazole resistance in clinical *Candida albicans* isolates from India. Mycoses. 2016;**59**:93-100. DOI: 10.1111/myc.12439

[12] Arenas R. Micología Médica Ilustrada3era edici. McGraw-Hill; 2014

[13] Choi J, Kim S-H. A genome tree of life for the fungi kingdom. Proceedings of the National Academy of Sciences of the United States of America. 2017;**114**:201711939. DOI: 10.1073/pnas.1711939114

[14] Chaffin WL. *Candida albicans* cell wall proteins. Microbiology and Molecular Biology Reviews. 2008;**72**:495-544. DOI: 10.1128/MMBR.00032-07

[15] Gow NAR, Hube B. Importance of the *Candida albicans* cell wall during commensalism and infection. Current Opinion in Microbiology. 2012;**15**:406-412. DOI: 10.1016/j.mib.2012.04.005

[16] Ruiz-Herrera J, Victoria Elorza M, Valentín E, Sentandreu R. Molecular organization of the cell wall of *Candida albicans* and its relation to pathogenicity. FEMS Yeast Research. 2006;**6**:14-29. DOI: 10.1111/j.1567-1364.2005.00017.x

[17] Chaffin WL, López-ribot JL, Casanova M, Gozalbo D, Martínez JP, Lo L. Cell wall and secreted proteins of *Candida albicans*: Identification, function, and expression. Microbiology and Molecular Biology Reviews. 1998;**62**:130-180

[18] Lipke PN, Ovalle R. Cell wall architecture in yeast: New structure and new challenges. Journal of Bacteriology. 1998;**180**:3735-3740

[19] Ramos AP, Desgarennes CP. Artículo de revisión Factores de virulencia en; 2005. pp. 12-27

[20] Spreghini E, Davis DA, Subaran R, Kim M, Mitchell AP. Roles of *Candida albicans* Dfg5p and Dcw1p cell surface proteins in growth and hypha formation. Eukaryotic Cell. 2003;**2**:746-755. DOI: 10.1128/EC.2.4.746-755.2003

[21] Calderone RA, Fonzi WA. Virulence factors of *Candida Albicans*. Trends in Microbiology. 2001;**9**:327-335

[22] Younes S, Bahnan W, Dimassi HI, Khalaf RA. The *Candida albicans* Hwp2 is necessary for proper adhesion, biofilm formation and oxidative stress tolerance. Microbiological Research. 2011;**166**:430-436. DOI: 10.1016/j.micres.2010.08.004

[23] Kruppa M. Quorum sensing and *Candida albicans*. Mycoses. 2009;**52**:1-10. DOI: 10.1111/j.1439-0507.2008.01626.x

[24] Kadosh D, Lopez-Ribot JL. *Candida albicans*: Adapting to succeed. Cell Host & Microbe. 2013;**14**:483-485. DOI: 10.1016/j.chom.2013.10.016

[25] Ramage G, Vandewalle K, Wickes BL, López-Ribot JL. Characteristics of biofilm formation by *Candida albicans* (xtt). Revista Iberoamericana de Micología. 2001;**18**:163-170 DOI: 200118163 [pii]

[26] Coleman JJ, Okoli I, Tegos GP, Holso EB, Wagner FF, Hamblin MR, et al. Characterization of plant-derived Saponin natural products against *Candida albicans*. ACS Chemical Biology. 2010;**5**:321-332

[27] Monteiro DR, Silva S, Negri M, Gorup LF, de Camargo ER, Oliveira R, et al. Silver colloidal nanoparticles: Effect on matrix composition and structure of *Candida albicans* and *Candida glabrata* biofilms. Journal of Applied Microbiology. 2013;**114**:1175-1183. DOI: 10.1111/jam.12102

[28] Monteiro DR, Takamiya AS, Feresin LP, Gorup LF, de Camargo ER, Delbem ACB, et al. Susceptibility of *Candida albicans* and *Candida glabrata* biofilms to silver nanoparticles in intermediate and mature development phases. Journal of Prosthodontic Research. 2015;**59**:42-48. DOI: 10.1016/j.jpor.2014.07.004

[29] Salari S, Khosravi AR, Mousavi SAA, Nikbakht-Brojeni GH. Mechanisms of resistance to fluconazole in *Candida albicans* clinical isolates from Iranian HIV-infected patients with oropharyngeal candidiasis. The Journal de Mycologie Médicale. 2016;**26**:35-41. DOI: 10.1016/j.mycmed.2015.10.007

[30] Maebashi K, Niimi M, Kudoh M, Fischer FJ, Makimura K, Niimi K, et al. Mechanisms of fluconazole resistance in *Candida albicans* isolates from Japanese AIDS patients. The Journal of Antimicrobial Chemotherapy. 2001;**47**:527-536. DOI: 10.1093/jac/47.5.527

[31] Saratale RG, Saratale GD, Shin HS, Jacob JM, Pugazhendhi A, Bhaisare M, et al. New insights on the green synthesis of metallic nanoparticles using plant and waste biomaterials: Current knowledge, their agricultural and environmental applications. Environmental Science and Pollution Research. 2018;**25**:10164-10183. DOI: 10.1007/s11356-017-9912-6

[32] Carpenter AW, Slomberg DL, Rao KS, Schoenfisch MH. Influence of scaffold size on bactericidal activity of nitric oxide-releasing silica nanoparticles. ASC Nano. 2011;**5**:7235-7244. DOI: 10.1111/j.1743-6109.2008.01122.x.Endothelial

[33] Carlos A-Q, Blanca S-R, Roco I-R, Amelia P-CH, Lorena M-FS, Alberto D-M, et al. Toxicity effects in pathogen microorganisms exposed to silver nanoparticles. Nanoscience and Nanotechnology Letters. 2017;**9**:165-173. DOI: 10.1166/nnl.2017.2275

[34] Oves M, Aslam M, Rauf MA, Qayyum S, Qari HA, Khan MS, et al. Antimicrobial and anticancer activities of silver nanoparticles synthesized from the root hair extract of Phoenix dactylifera. Materials Science and Engineering: C. 2018;**89**:429-443. DOI: 10.1016/j.msec.2018.03.035

[35] Aazam ES, Zaheer Z. Growth of Ag-nanoparticles in an aqueous solution and their antimicrobial activities against Gram positive, Gram negative bacterial strains and Candida fungus. Bioprocess and Biosystems Engineering. 2016;**39**:575-584. DOI: 10.1007/s00449-016-1539-3

[36] Shankar PD, Shobana S, Karuppusamy I, Pugazhendhi A, Ramkumar VS, Arvindnarayan S, et al. A review on the biosynthesis of metallic nanoparticles (gold and silver) using bio-components of microalgae: Formation mechanism and applications. Enzyme and Microbial Technology. 2016;**95**:28-44. DOI: 10.1016/j.enzmictec.2016.10.015

[37] Wypij M, Czarnecka J, Dahm H, Rai M, Golinska P. Silver nanoparticles from Pilimelia columellifera subsp. pallida SL19 strain demonstrated antifungal activity against fungi causing superficial mycoses. Journal of Basic Microbiology. 2017;**57**:793-800. DOI: 10.1002/jobm.201700121

[38] Lateef A, Ojo SA, Oladejo SM. Anti-candida, anti-coagulant and thrombolytic activities of biosynthesized silver nanoparticles using cell-free extract of Bacillus safensis LAU 13. Process Biochemistry. 2016;**51**:1406-1412. DOI: 10.1016/j.procbio.2016.06.027

[39] Ghiuță I, Cristea D, Croitoru C, Kost J, Wenkert R, Vyrides I, et al. Characterization and antimicrobial activity of silver nanoparticles, biosynthesized using Bacillus species. Applied Surface Science. 2018;**438**:66-73. DOI: 10.1016/j.apsusc.2017.09.163

[40] Musa SF, Yeat TS, Kamal LZM, Tabana YM, Ahmed MA, El Ouweini A, et al. Pleurotus sajor-caju can be used to synthesize silver nanoparticles with antifungal activity against *Candida albicans*. Journal of the Science of Food and Agriculture. 2018;**98**:1197-1207. DOI: 10.1002/jsfa.8573

[41] Khatoon N, Mishra A, Alam H, Manzoor N, Sardar M. Biosynthesis, characterization, and antifungal activity of the silver nanoparticles against pathogenic candida species. BioNanoScience. 2015;**5**:65-74. DOI: 10.1007/s12668-015-0163-z

[42] Moteriya P, Padalia H, Chanda S. Characterization, synergistic antibacterial and free radical scavenging efficacy of silver nanoparticles synthesized using *Cassia roxburghii* leaf extract. Journal, Genetic Engineering & Biotechnology. 2017;**15**:505-513. DOI: 10.1016/j.jgeb.2017.06.010

[43] Muthamil S, Devi VA, Balasubramaniam B, Balamurugan K, Pandian SK. Green synthesized silver nanoparticles demonstrating enhanced in vitro and in vivo antibiofilm activity against Candida spp. Journal of Basic Microbiology. 2018;**58**:343-357. DOI: 10.1002/jobm.201700529

[44] Kumar P, Senthamil Selvi S, Govindaraju M. Seaweed-mediated biosynthesis of silver nanoparticles using Gracilaria corticata for its antifungal activity against Candida spp. Applied Nanoscience. 2013;**3**:495-500. DOI: 10.1007/s13204-012-0151-3

[45] Khatoon UT, Nageswara Rao GVS, Mohan KM, Ramanaviciene A, Ramanavicius A. Antibacterial and antifungal activity of silver nanospheres synthesized by tri-sodium citrate assisted chemical approach. Vacuum. 2017;**146**:259-265. DOI: 10.1016/j.vacuum.2017.10.003

[46] Ocsoy I, Temiz M, Celik C, Altinsoy B, Yilmaz V, Duman F. A green approach for formation of silver nanoparticles on magnetic graphene oxide and highly effective

antimicrobial activity and reusability. Journal of Molecular Liquids. 2017;**227**:147-152. DOI: 10.1016/j.molliq.2016.12.015

[47] Ashajyothi C, Prabhurajeshwar C, Handral HK, Kelmani CR. Investigation of antifungal and anti-mycelium activities using biogenic nanoparticles: An eco-friendly approach. Environmental Nanotechnology, Monitoring and Management. 2016;**5**:81-87. DOI: 10.1016/j.enmm.2016.04.002

[48] Rahisuddin Al-Thabaiti SA, Khan Z, Manzoor N. Biosynthesis of silver nanoparticles and its antibacterial and antifungal activities towards Gram-positive, Gram-negative bacterial strains and different species of Candida fungus. Bioprocess and Biosystems Engineering. 2015;**38**:1773-1781. DOI: 10.1007/s00449-015-1418-3

[49] Parlinska-wojtan M, Depciuch J, Fryc B, Kus-liskiewicz M. Green synthesis and antibacterial effects of aqueous colloidal solutions of silver nanoparticles using clove eugenol. Applied Organometallic Chemistry. 2018;**32**:1-9. DOI: 10.1002/aoc.4276

[50] Padalia H, Moteriya P, Chanda S. Green synthesis of silver nanoparticles from marigold flower and its synergistic antimicrobial potential. Arabian Journal of Chemistry. 2015;**8**:732-741. DOI: 10.1016/j.arabjc.2014.11.015

[51] Kumar S, Bhattacharya W, Singh M, Halder D, Mitra A. Plant latex capped colloidal silver nanoparticles: A potent anti-biofilm and fungicidal formulation. Journal of Molecular Liquids. 2017;**230**:705-713. DOI: 10.1016/j.molliq.2017.01.004

[52] Anasane N, Golińska P, Wypij M, Rathod D, Dahm H, Rai M. Acidophilic actinobacteria synthesised silver nanoparticles showed remarkable activity against fungi-causing superficial mycoses in humans. Mycoses. 2016;**59**:157-166. DOI: 10.1111/myc.12445

[53] Hamid S, Zainab S, Faryal R, Ali N. Deterrence in metabolic and biofilms forming activity of Candida species by mycogenic silver nanoparticles. Journal of Applied Biomedicine. 2017;**15**:249-255. DOI: 10.1016/j.jab.2017.02.003

[54] Monteiro DR, Silva S, Negri M, Gorup LF, de Camargo ER, Oliveira R, et al. Antifungal activity of silver nanoparticles in combination with nystatin and chlorhexidine digluconate against *Candida albicans* and *Candida glabrata* biofilms. Mycoses. 2013;**56**:672-680. DOI: 10.1111/myc.12093

[55] Lara HH, Romero-Urbina DG, Pierce C, Lopez-Ribot JL, Arellano-Jiménez MJ, Jose-Yacaman M. Effect of silver nanoparticles on *Candida albicans* biofilms: An ultrastructural study. Journal of Nanobiotechnology. 2015;**13**:1-12. DOI: 10.1186/s12951-015-0147-8

[56] Kumar CG, Poornachandra Y. Biodirected synthesis of Miconazole-conjugated bacterial silver nanoparticles and their application as antifungal agents and drug delivery vehicles. Colloids and Surfaces B. 2015;**125**:110-119. DOI: 10.1016/j.colsurfb.2014.11.025

[57] Peña-González CE, Pedziwiatr-Werbicka E, Martín-Pérez T, Szewczyk EM, Copa-Patiño JL, Soliveri J, et al. Antibacterial and antifungal properties of dendronized silver

and gold nanoparticles with cationic carbosilane dendrons. International Journal of Pharmaceutics. 2017;**528**:55-61. DOI: 10.1016/j.ijpharm.2017.05.067

[58] Selvaraj M, Pandurangan P, Ramasami N, Rajendran SB, Sangilimuthu SN, Perumal P. Highly potential antifungal activity of quantum-sized silver nanoparticles against *Candida albicans*. Applied Biochemistry and Biotechnology. 2014;**173**:55-66. DOI: 10.1007/s12010-014-0782-9

[59] Li D, Liu Z, Yuan Y, Liu Y, Niu F. Green synthesis of gallic acid-coated silver nanoparticles with high antimicrobial activity and low cytotoxicity to normal cells. Process Biochemistry. 2015;**50**:357-366. DOI: 10.1016/j.procbio.2015.01.002

[60] Rajput VD, Minkina TM, Behal A, Sushkova SN, Mandzhieva S, Singh R, et al. Effects of zinc-oxide nanoparticles on soil, plants, animals and soil organisms: A review. Environmental Nanotechnology, Monitoring and Management. 2018;**9**:76-84

[61] Mirzaeia Hamed DM. Zinc oxide nanoparticles: Biological synthesis and biomedical applications. Ceramics International. 2013;**43**:1-11. DOI: 10.1016/j.ijnurstu.2008.11.006

[62] Chinnasamy C, Tamilselvam P, Karthick B, Sidharth B, Senthilnathan M. Green synthesis, characterization and optimization studies of zinc oxide nano particles using costusigneus leaf extract. Materials Today: Proceedings. 2018;**5**:6728-6735. DOI: 10.1016/j.matpr.2017.11.331

[63] Ahmed S, Chaudhry SA, Ikram S. A review on biogenic synthesis of ZnO nanoparticles using plant extracts and microbes: A prospect towards green chemistry. Journal of Photochemistry and Photobiology B: Biology. 2017;**166**:272-284. DOI: 10.1016/j.jphotobiol.2016.12.011

[64] Hou J, Wu Y, Li X, Wei B, Li S, Wang X. Toxic effects of different types of zinc oxide nanoparticles on algae, plants, invertebrates, vertebrates and microorganisms. Chemosphere. 2018;**193**:852-860. DOI: 10.1016/j.chemosphere.2017.11.077

[65] Madhumitha G, Elango G, Roopan SM. Biotechnological aspects of ZnO nanoparticles: Overview on synthesis and its applications. Applied Microbiology and Biotechnology. 2016;**100**:571-581. DOI: 10.1007/s00253-015-7108-x

[66] Ahmed S, Annu SAC, Ikram S. A review on biogenic synthesis of ZnO nanoparticles using plant extracts and microbes: A prospect towards green chemistry. Journal of Photochemistry and Photobiology B: Biology. 2017;**166**:272-284. DOI: 10.1016/j.jphotobiol.2016.12.011

[67] Vijayakumar S, Mahadevan S, Arulmozhi P, Sriram S, Praseetha PK. Green synthesis of zinc oxide nanoparticles using *Atalantia monophylla* leaf extracts: Characterization and antimicrobial analysis. Materials Science in Semiconductor Processing. 2018;**82**:39-45. DOI: 10.1016/j.mssp.2018.03.017

[68] Rajabi HR, Naghiha R, Kheirizadeh M, Sadatfaraji H, Mirzaei A, Alvand ZM. Microwave assisted extraction as an efficient approach for biosynthesis of zinc oxide nanoparticles:

Synthesis, characterization, and biological properties. Materials Science and Engineering: C. 2017;**78**:1109-1118. DOI: 10.1016/j.msec.2017.03.090

[69] Dobrucka R, Dlugaszewska J, Kaczmarek M. Cytotoxic and antimicrobial effects of biosynthesized ZnO nanoparticles using of Chelidonium majus extract. Biomedical Microdevices. 2018;**20**. DOI: 10.1007/s10544-017-0233-9

[70] Vijayakumar S, Krishnakumar C, Arulmozhi P, Mahadevan S, Parameswari N. Biosynthesis, characterization and antimicrobial activities of zinc oxide nanoparticles from leaf extract of *Glycosmis pentaphylla* (Retz.) DC. Microbial Pathogenesis. 2018;**116**:44-48. DOI: 10.1016/j.micpath.2018.01.003

[71] Elumalai K, Velmurugan S. Green synthesis, characterization and antimicrobial activities of zinc oxide nanoparticles from the leaf extract of *Azadirachta indica* (L.). Applied Surface Science. 2015;**345**:329-336. DOI: 10.1016/j.apsusc.2015.03.176

[72] Nagarajan S, Kuppusamy KA. Extracellular synthesis of zinc oxide nanoparticle using seaweeds of gulf of Mannar, India. Journal of Nanobiotechnology. 2013;**11**:39. DOI: 10.1186/1477-3155-11-39

[73] Padalia H, Moteriya P, Chanda S. Synergistic antimicrobial and cytotoxic potential of zinc oxide nanoparticles synthesized using *Cassia auriculata* leaf extract. BioNanoScience. 2018:196-206

[74] Dobrucka R, Długaszewska J. Biosynthesis and antibacterial activity of ZnO nanoparticles using Trifolium pratense flower extract. Saudi Journal of Biological Sciences. **2015**:517-523. DOI: 10.1016/j.sjbs.2015.05.016

[75] Dhillon GS, Kaur S, Brar SK. Facile fabrication and characterization of chitosan-based zinc oxide nanoparticles and evaluation of their antimicrobial and antibiofilm activity. International Nano Letters. 2014;**4**:107. DOI: 10.1007/s40089-014-0107-6

[76] Divya M, Vaseeharan B, Abinaya M, Vijayakumar S, Govindarajan M, Alharbi NS, et al. Biology biopolymer gelatin-coated zinc oxide nanoparticles showed high antibacterial, antibiofilm and anti-angiogenic activity. The Journal of Photochemistry and Photobiology B: Biology. 2018;**178**:211-218. DOI: 10.1016/j.jphotobiol.2017.11.008

[77] Serkhacheva NS, Yashina NV, Prokopov NI, Gaynanova AA, Kuz'micheva GM, Domoroshchina EN, et al. Bactericidal properties of nanoscale zinc(II) and titanium (IV) oxides of different nature and their nanocomposites with polystyrene. Nanotechnologies in Russia. 2016;**11**:99-109. DOI: 10.1134/S1995078016010146

[78] Vijayakumar S, Vaseeharan B. Antibiofilm, anti cancer and ecotoxicity properties of collagen based ZnO nanoparticles. Advanced Powder Technology. 2018;**29**:2331-2345

[79] Shoeb M, Singh BR, Khan JA, Khan W, Singh BN, Singh HB, et al. ROS-dependent anticandidal activity of zinc oxide nanoparticles synthesized by using egg albumen as a biotemplate. Advances in Natural Sciences: Nanoscience and Nanotechnology. 2013;**4**:1-12. DOI: 10.1088/2043-6262/4/3/035015

[80] Mizielińska M, Łopusiewicz Ł, Mężyńska M, Bartkowiak A. The influence of accelerated UV-A and Q-sun irradiation on the antimicrobial properties of coatings containing ZnO nanoparticles. Molecules. 2017;**22**:12-15. DOI: 10.3390/molecules22091556

[81] Agarwal H, Menon S, Venkat Kumar S, Rajeshkumar S. Mechanistic study on antibacterial action of zinc oxide nanoparticles synthesized using green route. Chemico-Biological Interactions. 2018;**286**:60-70. DOI: 10.1016/J.CBI.2018.03.008

[82] Longano D, Ditaranto N, Sabbatini L, Torsi L. In: Nicola Coffi MR, editor. Nano Antimicrobials. 2012. pp. 85-117

[83] Ingle AP, Duran N, Rai M. Bioactivity, mechanism of action, and cytotoxicity of copper-based nanoparticles: A review. Applied Microbiology and Biotechnology. 2014;**98**:1001-1009. DOI: 10.1007/s00253-013-5422-8

[84] Natan M, Banin E. From Nano to Micro: Using nanotechnology to combat microorganisms and their multidrug resistance. FEMS Microbiology Reviews. 2017;**41**:302-322. DOI: 10.1093/femsre/fux003

[85] Bogdanovic U, Lazic V, Vodnik V, Budimir M, Markovic Z, Dimitrijevic S. Copper nanoparticles with high antimicrobial activity. Materials Letters. 2014;**128**:75-78. DOI: 10.1016/j.matlet.2014.04.106

[86] Katwal R, Kaur H, Sharma G, Naushad M, Pathania D. Electrochemical synthesized copper oxide nanoparticles for enhanced photocatalytic and antimicrobial activity. Journal of Industrial and Engineering Chemistry. 2015;**31**:173-184

[87] Din MI, Arshad F, Hussain Z, Mukhtar M. Green adeptness in the synthesis and stabilization of copper nanoparticles: Catalytic, antibacterial, cytotoxicity, and antioxidant activities. Nanoscale Research Letters. 2017;**12**:638. DOI: 10.1186/s11671-017-2399-8

[88] Tamayo L, Azócar M, Kogan M, Riveros A, Páez M. Copper-polymer nanocomposites: An excellent and cost-effective biocide for use on antibacterial surfaces. Materials Science and Engineering: C. 2016;**69**:1391-1409. DOI: 10.1016/j.msec.2016.08.041

[89] Weitz IS, Maoz M, Panitz D, Eichler S, Segal E. Combination of CuO nanoparticles and fluconazole: Preparation, characterization, and antifungal activity against *Candida albicans*. Journal of Nanoparticle Research. 2015;**17**. DOI: 10.1007/s11051-015-3149-4

[90] Ramyadevi J, Jeyasubramanian K, Marikani A, Rajakumar G, Rahuman AA. Synthesis and antimicrobial activity of copper nanoparticles. Materials Letters. 2012;**71**:114-116. DOI: 10.1016/j.matlet.2011.12.055

[91] Amiri M, Etemadifar Z, Daneshkazemi A, Nateghi M. Antimicrobial effect of copper oxide nanoparticles on some oral bacteria and candida species. Journal of Dental Biomaterials. 2017;**4**:347-352. DOI: 28959764s

[92] Adwin Jose P, Dhaveethu Raja J, Sankarganesh M, Rajesh J. Evaluation of antioxidant, DNA targeting, antimicrobial and cytotoxic studies of imine capped copper and nickel nanoparticles. Journal of Photochemistry and Photobiology B: Biology. 2018;**178**:143-151. DOI: 10.1016/j.jphotobiol.2017.11.005

[93] Pugazhendhi A, Kumar SS, Manikandan M, Saravanan M. Photocatalytic properties and antimicrobial efficacy of Fe doped CuO nanoparticles against the pathogenic bacteria and fungi. Microbial Pathogenesis. 2018;**122**:84-89

[94] Arunadevi R, Kavitha B, Rajarajan M, Suganthi A, Jeyamurugan A. Investigation of the drastic improvement of photocatalytic degradation of Congo red by monoclinic Cd, Ba-CuO nanoparticles and its antimicrobial activities. Surfaces and Interfaces. 2018;**10**:32-44. DOI: 10.1016/j.surfin.2017.11.004

[95] Eremenko AM, Petrik IS, Smirnova NP, Rudenko AV, Marikvas YS. Antibacterial and antimycotic activity of cotton fabrics, impregnated with silver and binary silver/copper nanoparticles. Nanoscale Research Letters. 2016;**11**:1-9. DOI: 10.1186/s11671-016-1240-0

[96] El-nahhal IM, Elmanama AA, Amara N, Qodih FS, Selmane M, Chehimi MM. The efficacy of surfactants in stabilizing coating of nano-structured CuO particles onto the surface of cotton fibers and their antimicrobial activity. Materials Chemistry and Physics. 2018;**215**:221-228

[97] Savelyev Y, Gonchar A, Movchan B, Gornostay A, Vozianov S, Rudenko A, et al. Antibacterial polyurethane materials with silver and copper nanoparticles. Materials Today: Proceedings. 2017;**4**:87-94. DOI: 10.1016/j.matpr.2017.01.196

[98] Ballo MKS, Rtimi S, Kiwi J, Pulgarin C, Entenza JM, Bizzini A. Fungicidal activity of copper-sputtered flexible surfaces under dark and actinic light against azole-resistant *Candida albicans* and *Candida glabrata*. Journal of Photochemistry and Photobiology B: Biology. 2017;**174**:229-234. DOI: 10.1016/j.jphotobiol.2017.07.030

[99] Khan J, Siddiq M, Akram B, Ashraf MA. In-situ synthesis of CuO nanoparticles in P(NIPAM-co-AAA) microgel, structural characterization, catalytic and biological applications. Arabian Journal of Chemistry. 2018;**11**:897-909. DOI: 10.1016/j.arabjc.2017.12.018

[100] Reza H, Saeed M, Dorraji S, Fakhrzadeh V, Eslami H. Starch-based polyurethane/CuO nanocomposite foam: Antibacterial effects for infection control. International Journal of Biological Macromolecules. 2018;**111**:1076-1082

[101] El AA, Gehan R, Ayman AG, El Khateeb Y, Hassaan MM. Eco-friendly synthesis of metal nanoparticles using ginger and garlic extracts as biocompatible novel antioxidant and antimicrobial agents. Journal of Nanostructure in Chemistry. 2018;**8**:71-81. DOI: 10.1007/s40097-018-0255-8

[102] El-Batal AI, Al-Hazmi NE, Mosallam FM, El-Sayyad GS. Biogenic synthesis of copper nanoparticles by natural polysaccharides and *Pleurotus ostreatus* fermented fenugreek using gamma rays with antioxidant and antimicrobial potential towards some wound pathogens. Microbial Pathogenesis. 2018;**118**:159-169. DOI: 10.1016/j.micpath.2018.03.013

[103] Sivaraj R, Rahman PKSM, Rajiv P, Narendhran S, Venckatesh R. Biosynthesis and characterization of Acalypha indica mediated copper oxide nanoparticles and evaluation of its antimicrobial and anticancer activity. Spectrochimica Acta Part A. 2014;**129**:25-258. DOI: 10.1016/j.saa.2014.03.027

[104] Nilay Ildiz IO, Baldemir A, Altinkaynak C, Özdemir N, Yilmazc V. Self assembled snowball-like hybrid nanostructures comprising *Viburnum opulus* L. extract and metal ions for antimicrobial and catalytic applications. Enzyme and Microbial Technology. 2017;**102**:60-66

[105] Vishnu SRI, Ramaswamy P, Narendhran S, Sivaraj R. Potentiating effect of ecofriendly synthesis of copper oxide nanoparticles using brown alga: Antimicrobial and anticancer activities. Bulletin of Materials Science. 2016;**39**:361-364. DOI: 10.1007/s12034-016-1173-3

[106] Tran CD, Makuvaza J, Munson E, Bennett B. Biocompatible copper oxide nanoparticle composites from cellulose and chitosan: Facile synthesis, unique structure, and antimicrobial activity. ACS Applied Materials & Interfaces. 2017;**9**:42503-42515. DOI: 10.1021/acsami.7b11969

[107] Athie-García MS, Piñón-Castillo HA, Muñoz-Castellanos LN, Ulloa-Ogaz AL, Martínez-Varela PI, Quintero-Ramos A, et al. Cell wall damage and oxidative stress in *Candida albicans* ATCC10231 and *Aspergillus niger* caused by palladium nanoparticles. Toxicology In Vitro. 2018;**48**:111-120. DOI: 10.1016/j.tiv.2018.01.006

[108] Kazempour ZB, Yazdi MH, Rafii F, Shahverdi AR. Sub-inhibitory concentration of biogenic selenium nanoparticles lacks post antifungal effect for Aspergillus niger and *Candida albicans* and stimulates the growth of *Aspergillus niger*. Iranian Journal of Microbiology. 2013;**5**:81-85

[109] Kheradmand E, Rafii F, Yazdi MH, Sepahi AA. The antimicrobial effects of selenium nanoparticle-enriched probiotics and their fermented broth against *Candida albicans*. Daru. 2014;**22**:1-6. DOI: 10.1186/2008-2231-22-48

[110] Sankarganesh M, Adwin Jose P, Dhaveethu Raja J, Kesavan MP, Vadivel M, Rajesh J, et al. New pyrimidine based ligand capped gold and platinum nano particles: Synthesis, characterization, antimicrobial, antioxidant, DNA interaction and in vitro anticancer activities. Journal of Photochemistry and Photobiology B: Biology. 2017;**176**:44-53. DOI: 10.1016/j.jphotobiol.2017.09.013

[111] Anand K, Tiloke C, Phulukdaree A, Ranjan B, Chuturgoon A, Singh S, et al. Biology biosynthesis of palladium nanoparticles by using Moringa oleifera flower extract and their catalytic and biological properties. Journal of Photochemistry and Photobiology, B: Biology. 2016;**165**:87-95. DOI: 10.1016/j.jphotobiol.2016.09.039

[112] Zare B, Sepehrizadeh Z, Faramarzi MA, Soltany-Rezaee-Rad M, Rezaie S, Shahverdi AR. Antifungal activity of biogenic tellurium nanoparticles against *Candida albicans* and its effects on squalene monooxygenase gene expression. Biotechnology and Applied Biochemistry. 2014;**6**(1):395-400. DOI: 10.1002/bab.1180

[113] Aiad I, Marzouk MI, Shaker SA, Ebrahim NE, Abd-Elaal AA, Tawfik SM. Antipyrine cationic surfactants capping silver nanoparticles as potent antimicrobial agents against pathogenic bacteria and fungi. Journal of Molecular Liquids. 2017;**243**:572-583. DOI: 10.1016/j.molliq.2017.08.072

[114] Javed R, Ahmed M, ul Haq I, Nisa S, Zia M. PVP and PEG doped CuO nanoparticles are more biologically active: Antibacterial, antioxidant, antidiabetic and cytotoxic perspective. Materials Science and Engineering: C. 2017;**79**:108-115. DOI: 10.1016/j.msec.2017.05.006

[115] Durán N, Nakazato G, Seabra AB. Antimicrobial activity of biogenic silver nanoparticles, and silver chloride nanoparticles: An overview and comments. Applied Microbiology and Biotechnology. 2016;**100**:6555-6570. DOI: 10.1007/s00253-016-7657-7

[116] Monteiro DR, Silva S, Negri M, Gorup LF, De Camargo ER, Oliveira R, et al. Silver nanoparticles: Influence of stabilizing agent and diameter on antifungal activity against *Candida albicans* and *Candida glabrata* biofilms. Letters in Applied Microbiology. 2012;**54**:383-391. DOI: 10.1111/j.1472-765X.2012.03219.x

[117] Kim KJ, Sung WS, Suh BK, Moon SK, Choi JS, Kim JG, et al. Antifungal activity and mode of action of silver nano-particles on *Candida albicans*. Biometals. 2009;**22**:235-242. DOI: 10.1007/s10534-008-9159-2

[118] Dos Santos CA, Seckler MM, Ingle AP, Gupta I, Galdiero S, Galdiero M, et al. Silver nanoparticles: Therapeutical uses, toxicity, and safety issues. Journal of Pharmaceutical Sciences. 2014;**103**:1931-1944. DOI: 10.1002/jps.24001

[119] Qasim M, Singh BR, Naqvi AH, Paik P, Das D. Silver nanoparticles embedded mesoporous SiO2 nanosphere: An effective anticandidal agent against *Candida albicans* 077. Nanotechnology. 2015;**26**. DOI: 10.1088/0957-4484/26/28/285102

[120] Hwang IS, Lee J, Hwang JH, Kim KJ, Lee DG. Silver nanoparticles induce apoptotic cell death in *Candida albicans* through the increase of hydroxyl radicals. The FEBS Journal. 2012;**279**:1327-1338. DOI: 10.1111/j.1742-4658.2012.08527.x

[121] Alam J, Qasim M, Raj B, Khan W, Das D, Naqvi AH. Polyaniline/CoFe2O4 nanocomposite inhibits the growth of *Candida albicans* 077 by ROS production. Comptes Rendus Chimie. 2014;**17**:91-102. DOI: 10.1016/j.crci.2013.08.006

[122] Setyawati MI, Fang W, Chia SL, Leong DT. Nanotoxicology of commonmetal oxide based nanomaterials: their ROS-y and non-ROS-y consequences. Asia-Pacific Journal of Chemical Engineering. 2013;**8**:205-217. DOI: 10.1002/apj.1680

[123] Radhakrishnan VS, Dwivedi SP, Siddiqui MH, Prasad T. In vitro studies on oxidative stress-independent, Ag nanoparticles-induced cell toxicity of *Candida albicans*, an opportunistic pathogen. International Journal of Nanomedicine. 2018;**13**:91-96. DOI: 10.2147/IJN.S125010